ADVANCED CALCULUS

THE APPLETON-CENTURY MATHEMATICS SERIES

Edited by Raymond W. Brink

Intermediate Algebra, 2nd ed., by Raymond W. Brink

College Algebra, 2nd ed., by Raymond W. Brink

Algebra—College Course, 2nd ed., by Raymond W. Brink

A First Year of College Mathematics, 2nd ed., by Raymond W. Brink

The Mathematics of Finance, by Franklin C. Smith

Plane Trigonometry, 3rd ed., by Raymond W. Brink

Spherical Trigonometry, by Raymond W. Brink

Essentials of Analytic Geometry, by Raymond W. Brink

Analytic Geometry, by Edwin J. Purcell

Analytic Geometry, rev. ed., by Raymond W. Brink

Analytic Geometry and Calculus, by Lloyd L. Smail

Calculus, by Lloyd L. Smail

College Geometry, by Leslie H. Miller

Solid Analytic Geometry, by John M. H. Olmsted

Intermediate Analysis, by John M. H. Olmsted

Real Variables, by John M. H. Olmsted

Advanced Calculus, by John M. H. Olmsted

Introduction to the Laplace Transform, by Dio L. Holl, Clair G. Maple, and Bernard Vinograde

Introductory Analysis, by V. O. McBrien

ADVANCED CALCULUS

by

John M. H. Olmsted

PROFESSOR OF MATHEMATICS
SOUTHERN ILLINOIS UNIVERSITY

NEW YORK: APPLETON-CENTURY-CROFTS, INC.

Copyright ©, 1961, by
APPLETON-CENTURY-CROFTS, INC.

All rights reserved. This book, or parts thereof, must not be reproduced in any form without permission of the publisher.

631–1

Library of Congress Card Number: 61–6154

Copyright ©, 1956, in part, under
the title of *Intermediate Analysis*, by
Appleton-Century-Crofts, Inc.

Copyright ©, 1959, in part, under
the title of *Real Variables*, by
Appleton-Century-Crofts, Inc.

PRINTED IN THE UNITED STATES OF AMERICA

PREFACE

This book is a basic text in advanced calculus, providing a clear and well-motivated, yet precise and rigorous, treatment of the essential tools of mathematical analysis at a level immediately following that of a first course in calculus. It is designed to satisfy many needs; it fills gaps that almost always, and properly, occur in elementary calculus courses; it contains all of the material in the standard classical advanced calculus course; and it provides a solid foundation in the "deltas and epsilons" of a modern rigorous advanced calculus. It is well suited for courses of considerable diversity, ranging from "foundations of calculus" to "critical reasoning in mathematical analysis." There is even ample material for a course having a standard advanced course as prerequisite.

Throughout the book attention is paid to the average or less-than-average student as well as to the superior student. This is done at every stage of progress by making maximally available whatever concepts and discussion are both relevant and understandable. To illustrate: limit and continuity theorems whose proofs are difficult are discussed and worked with before they are proved, implicit functions are treated before their existence is established, and standard power series techniques are developed before the topic of uniform convergence is studied. Whenever feasible, if both an elementary and a sophisticated proof of a theorem are possible, the elementary proof is given in the text, with the sophisticated proof possibly called for in an exercise, with hints. Generally speaking, the more subtle and advanced portions of the book are marked with stars (*), prerequisite for which is preceding starred material. This contributes to an unusual flexibility of the book as a text.

The author believes that most students can best appreciate the more difficult and advanced aspects of any field of study if they have thoroughly mastered the relatively easy and introductory parts first. In keeping with this philosophy, the book is arranged so that progress moves from the simple to the complex and from the particular to the general. Emphasis is on the concrete, with abstract concepts introduced only as they are relevant, although the general spirit is modern. The Riemann integral, for example, is studied first with emphasis on relatively direct consequences of basic definitions, and then with more difficult results obtained with the aid of step functions. Later some of these ideas are extended to multiple integrals and to the Riemann-Stieltjes integral. Improper integrals are treated at two levels of sophistication; in Chapter 4 the principal ideas are dominance and the "big O" and "little o" concepts, while in Chapter 14 uniform convergence becomes central, with applications to such topics as evaluations and the gamma and beta functions.

Vectors are presented in such a way that a teacher using this book may almost completely avoid the vector parts of advanced calculus if he wishes to emphasize the "real variables" content. This is done by restricting the use of vectors in the main part of the book to the scalar, or dot, product, with applications to such topics as solid analytic geometry, partial differentiation, and Fourier series. The vector, or cross, product and the differential and integral calculus of vectors are fully developed and exploited in the last three chapters on vector analysis, line and surface integrals, and differential geometry. The now-standard Gibbs notation is used. Vectors are designated by means of arrows, rather than bold-face type, to conform with handwriting custom.

Special attention should be called to the abundant sets of problems—there are over 2440 exercises! These include routine drills for practice, intermediate exercises that extend the material of the text while retaining its character, and advanced exercises that go beyond the standard textual subject matter. Whenever guidance seems desirable, generous hints are included. In this manner the student is led to such items of interest as limits superior and inferior, for both sequences and real-valued functions in general, the construction of a continuous nondifferentiable function, the elementary theory of analytic functions of a complex variable, and exterior differential forms. Analytic treatment of the logarithmic, exponential, and trigonometric functions is presented in the exercises, where sufficient hints are given to make these topics available to all. Answers to all problems are given in the back of the book. Illustrative examples abound throughout.

Standard Aristotelian logic is assumed; for example, frequent use is made of the indirect method of proof. An implication of the form *p implies q* is taken to mean that it is impossible for p to be true and q to be false simultaneously; in other words, that the conjunction of the two statements p and *not q* leads to a contradiction. Any statement of equality means simply that the two objects that are on opposite sides of the equal sign are the same thing. Thus such statements as "equals may be added to equals," and "two things equal to the same thing are equal to each other," are true by definition.

A few words regarding notation should be given. The equal sign $=$ is used for equations, both conditional and identical, and the triple bar $\equiv$ is reserved for definitions. For simplicity, if the meaning is clear from the context, neither symbol is restricted to the indicative mood as in "$(a+b)^2 = a^2 + 2ab + b^2$," or "where $f(x) \equiv x^2 + 5$." Examples of subjunctive uses are "let $x = n$," and "let $\epsilon \equiv 1$," which would be read "let x be equal to n," and "let ϵ be defined to be 1," respectively. A similar freedom is granted the inequality symbols. For instance, the symbol $>$ in the following constructions "if $\epsilon > 0$, then $\cdots$," "let $\epsilon > 0$," and "let $\epsilon > 0$ be given," could be translated "is greater than," "be greater than," and "greater than," respectively. A relaxed attitude is also adopted regarding functional notation, and the tradition ($y = f(x)$) established by Dirichlet has been followed. When there can be no reasonable misinterpretation the notation $f(x)$ is used both to indicate the value of the function f corresponding to a particular value x of the independent variable and also to represent the function f itself (and similarly for $f(x, y)$, $f(x, y, z)$, and the like). This permissiveness has two merits. In the first place it indicates in a simple way the number of independent variables and the letters representing them. In the second place it avoids such elaborate constructions as "the function f defined by the equation $f(x) = \sin 2x$ is periodic," by permitting simply, "$\sin 2x$ is periodic." This practice is in the spirit of such statements as "the line $x + y = 2 \cdots$," instead of "the line that is the graph of the equation $x + y = 2 \cdots$," and "this is John Smith," instead of "this is a man whose name is John Smith."

In a few places parentheses are used to indicate alternatives. The principal instances of such uses are heralded by announcements or footnotes in the text. Here again it is hoped that the context will prevent any ambiguity. Such a sentence as "The function $f(x)$ is integrable from a to b $(a < b)$" would mean that "$f(x)$ is integrable from a to b, where it is assumed that $a < b$," whereas a sentence like "A function having a positive (negative) derivative over an interval is strictly increasing (decreasing) there" is a compression of two statements into one, the parentheses indicating an alternative formulation.

Although this text is almost completely self-contained, it is impossible within the compass of a book of this size to pursue every topic to the extent that might be desired by every reader. Numerous references to other books are inserted to aid the intellectually ambitious and curious. Since many of these references are to the author's *Real Variables* (abbreviated here to *RV*), of this same Appleton-Century Mathematics Series, and since the present *Advanced Calculus* (*AC* for short) and *RV* have a very substantial body of common material, the reader or potential user of either book is entitled to at least a short explanation of the differences in their objectives. In brief, *AC* is designed principally for fairly standard advanced calculus courses, of either the "vector analysis" or the "rigorous" type, while *RV* is designed principally for courses in introductory real variables at either the advanced calculus or the post-advanced calculus level. Topics that are in both *AC* and *RV* include all those of the basic "rigorous advanced calculus." Topics that are in *AC* but not in *RV* include solid analytic geometry, vector analysis, complex variables, extensive treatment of line and surface integrals, and differential geometry. Topics that are in *RV* but not in *AC* include a thorough treatment of certain properties of the real numbers, dominated convergence and measure zero as related to the Riemann integral, bounded variation as related to the Riemann-Stieltjes integral and to arc length, space-filling arcs, independence of parametrization for simple arc length, the Moore-Osgood uniform convergence theorem, metric and topological spaces, a rigorous proof of the transformation theorem for multiple integrals, certain theorems on improper integrals, the Gibbs phenomenon, closed and complete orthonormal systems of functions, and the Gram-Schmidt process.

One note of caution is in order. Because of the rich abundance of material available, complete coverage in one year is difficult. Most of the unstarred sections can be completed in a year's sequence, but many teachers will wish to sacrifice some of the later unstarred portions in order to include some of the earlier starred items. Anybody using the book as a text should be advised to give some advance thought to the main emphasis he wishes to give his course and to the selection of material suitable to that emphasis.

The author wishes to express his deep appreciation of the aid and suggestions given by Professor R. W. Brink of the University of Minnesota in the preparation of the manuscript. He is also indebted to many others for their many helpful comments concerning both the material common to *AC* and *RV* and the chapters new to the present volume.

J. M. H. O.

Carbondale, Illinois

CONTENTS

	PAGE
PREFACE	v

Chapter I
THE REAL NUMBER SYSTEM

SECTION

101.	Introduction	1
102.	Axioms of a field	1
103.	Exercises	3
104.	Axioms of an ordered field	5
105.	Exercises	6
106.	Positive integers and mathematical induction	7
107.	Exercises	11
108.	Integers and rational numbers	13
109.	Exercises	14
110.	Geometrical representation and absolute value	14
111.	Exercises	16
112.	Axiom of completeness	17
113.	Consequences of completeness	19
114.	Exercises	20

Chapter 2
FUNCTIONS, SEQUENCES, LIMITS, CONTINUITY

201.	Functions and sequences	22
202.	Limit of a sequence	26
203.	Exercises	28
204.	Limit theorems for sequences	29
205.	Exercises	33
206.	Limits of functions	34
207.	Limit theorems for functions	38
208.	Exercises	39
209.	Continuity	42
210.	Types of discontinuity	44

211. Continuity theorems 46
212. Exercises 46
213. More theorems on continuous functions 47
214. Existence of $\sqrt{2}$ and other roots....................... 48
215. Monotonic functions and their inverses 49
216. Exercises 50
*217. A fundamental theorem on bounded sequences 53
*218. Proofs of some theorems on continuous functions............. 54
*219. The Cauchy criterion for convergence of a sequence........... 56
*220. Exercises 57
*221. Sequential criteria for continuity and existence of limits......... 59
*222. The Cauchy criterion for functions........................... 60
*223. Exercises 61
*224. Uniform continuity 63
*225. Exercises 65

Chapter 3

DIFFERENTIATION

301. Introduction 67
302. The derivative... 67
303. One-sided derivatives.. 70
304. Exercises 73
305. Rolle's theorem and the Law of the Mean 75
306. Consequences of the Law of the Mean 79
307. The Extended Law of the Mean 80
308. Exercises 82
309. Maxima and minima... 85
310. Exercises 88
311. Differentials 90
312. Approximations by differentials 92
313. Exercises 94
314. L'Hospital's Rule.   Introduction 95
315. The indeterminate form 0/0 95
316. The indeterminate form ∞/∞ 97
317. Other indeterminate forms 99
318. Exercises 101
319. Curve tracing 102
320. Exercises 106
*321. Without loss of generality 108
*322. Exercises 108

Chapter 4

INTEGRATION

401. The definite integral 110
402. Exercises 117
*403. More integration theorems................................... 120
*404. Exercises 125
405. The Fundamental Theorem of Integral Calculus 127
406. Integration by substitution................................... 128
407. Exercises 129

408. Sectional continuity and smoothness	131
409. Exercises	133
410. Reduction formulas	133
411. Exercises	135
412. Improper integrals, introduction	136
413. Improper integrals, finite interval	136
414. Improper integrals, infinite interval	139
415. Comparison tests. Dominance	140
416. Exercises	143
*417. The Riemann-Stieltjes integral	146
*418. Exercises	151

Chapter 5

SOME ELEMENTARY FUNCTIONS

*501. The exponential and logarithmic functions	155
*502. Exercises	155
*503. The trigonometric functions	158
*504. Exercises	158
505. Some integration formulas	161
506. Exercises	163
507. Hyperbolic functions	163
508. Inverse hyperbolic functions	165
509. Exercises	167
*510. Classification of numbers and functions	167
*511. The elementary functions	169
*512. Exercises	170

Chapter 6

FUNCTIONS OF SEVERAL VARIABLES

601. Introduction	172
602. Neighborhoods in the Euclidean plane	173
603. Point sets in the Euclidean plane	174
604. Sets in higher-dimensional Euclidean spaces	178
605. Exercises	179
606. Functions and limits	181
607. Iterated limits	184
608. Continuity	186
609. Limit and continuity theorems	187
610. More theorems on continuous functions	188
611. Exercises	189
612. More general functions. Mappings	190
*613. Sequences of points	193
*614. Point sets and sequences	195
*615. Compactness and continuity	196
*616. Proofs of two theorems	198
*617. Uniform continuity	198
618. Exercises	199

Chapter 7

SOLID ANALYTIC GEOMETRY AND VECTORS

701. Introduction 203
702. Vectors and scalars 204
703. Addition and subtraction of vectors. Magnitude.............. 206
704. Linear combinations of vectors 207
705. Exercises 210
706. Direction angles and cosines 211
707. The scalar or inner or dot product........................ 213
708. Vectors orthogonal to two vectors......................... 216
709. Exercises 217
710. Planes 218
711. Lines 220
712. Exercises 222
713. Surfaces. Sections, traces, intercepts 225
714. Spheres 225
715. Cylinders.. 226
716. Surfaces of revolution................................... 227
717. Exercises 229
718. The standard quadric surfaces............................ 231
719. Exercises 235

Chapter 8

ARCS AND CURVES

801. Duhamel's principle for integrals 238
○ *802. A proof with continuity hypotheses 238
803. Arcs and curves 239
804. Arc length.. 241
805. Integral form for arc length 242
○ *806. Remark concerning the trigonometric functions............. 247
807. Exercises 247
808. Cylindrical and spherical coordinates.................... 249
809. Arc length in rectangular, cylindrical, and spherical coordinates. 251
810. Exercises 251
811. Curvature and radius of curvature in two dimensions........... 252
812. Circle of curvature..................................... 254
○ *813. Evolutes and involutes 255
814. Exercises 257

Chapter 9

PARTIAL DIFFERENTIATION

901. Partial derivatives 261
902. Partial derivatives of higher order 262
○ *903. Equality of mixed partial derivatives 263
904. Exercises 264
905. The fundamental increment formula 266
906. Differentials 268
907. Change of variables. The chain rule 269

*908.	Homogeneous functions. Euler's theorem	272
909.	Exercises	272
*910.	Directional derivatives. Tangents and normals	275
*911.	Exercises	278
912.	The Law of the Mean	279
913.	Approximations by differentials	281
914.	Maxima and minima	282
915.	Exercises	286
916.	Differentiation of an implicit function	288
917.	Some notational pitfalls	290
918.	Exercises	292
919.	Envelope of a family of plane curves	293
920.	Exercises	295
921.	Several functions defined implicitly. Jacobians	296
922.	Coordinate transformations. Inverse transformations	300
923.	Functional dependence	304
924.	Exercises	306
925.	Extrema with one constraint. Two variables	308
926.	Extrema with one constraint. More than two variables	312
927.	Extrema with more than one constraint	314
928.	Lagrange multipliers	317
929.	Exercises	320
*930.	Differentiation under the integral sign. Leibnitz's rule	321
*931.	Exercises	324
*932.	The Implicit Function Theorem	325
*933.	Existence theorem for inverse transformations	330
*934.	Sufficiency conditions for functional dependence	331
*935.	Exercises	333

Chapter 10

MULTIPLE INTEGRALS

1001.	Introduction	335
1002.	Double integrals	335
1003.	Area	337
1004.	Second formulation of the double integral	338
*1005.	Inner and outer area. Criterion for area	339
*1006.	Theorems on double integrals	342
*1007.	Proof of the second formulation	346
1008.	Iterated integrals, two variables	347
*1009.	Proof of the Fundamental Theorem	349
1010.	Exercises	350
1011.	Triple integrals. Volume	352
1012.	Exercises	355
1013.	Double integrals in polar coordinates	356
1014.	Volumes with double integrals in polar coordinates	359
1015.	Exercises	360
1016.	Mass of a plane region of variable density	361
1017.	Moments and centroid of a plane region	362
1018.	Exercises	363
1019.	Triple integrals, cylindrical coordinates	364
1020.	Triple integrals, spherical coordinates	365
1021.	Mass, moments, and centroid of a space region	366

1022. Exercises 368
1023. Mass, moments, and centroid of an arc 369
1024. Attraction 370
1025. Exercises 372
1026. Jacobians and transformations of multiple integrals 373
1027. General discussion 375
1028. Exercises 379

Chapter 11

INFINITE SERIES OF CONSTANTS

1101. Basic definitions 381
1102. Three elementary theorems 382
1103. A necessary condition for convergence 382
1104. The geometric series 383
1105. Positive series 383
1106. The integral test 384
1107. Exercises 385
1108. Comparison tests. Dominance 387
1109. The ratio test 390
1110. The root test 392
1111. Exercises 393
*1112. More refined tests 395
*1113. Exercises 397
1114. Series of arbitrary terms 398
1115. Alternating series 398
1116. Absolute and conditional convergence 399
1117. Exercises 401
1118. Groupings and rearrangements 404
1119. Addition, subtraction, and multiplication of series 405
*1120. Some aids to computation 407
1121. Exercises 410

Chapter 12

POWER SERIES

1201. Interval of convergence 412
1202. Exercises 415
1203. Taylor series 416
1204. Taylor's formula with a remainder 418
1205. Expansions of functions 419
1206. Exercises 421
1207. Some Maclaurin series 421
1208. Elementary operations with power series 425
1209. Substitution of power series 427
1210. Integration and differentiation of power series 431
1211. Exercises 433
1212. Indeterminate expressions 435
1213. Computations 435
1214. Exercises 438
1215. Taylor series, several variables 438
1216. Exercises 440

⋆Chapter 13

UNIFORM CONVERGENCE AND LIMITS

⋆1301. Uniform convergence of sequences	441
⋆1302. Uniform convergence of series	444
⋆1303. Dominance and the Weierstrass M-test	445
⋆1304. Exercises	446
⋆1305. Uniform convergence and continuity	448
⋆1306. Uniform convergence and integration	449
⋆1307. Uniform convergence and differentiation	451
⋆1308. Exercises	453
⋆1309. Power series. Abel's theorem	455
⋆1310. Proof of Abel's theorem	456
⋆1311. Exercises	457
⋆1312. Functions defined by power series. Exercises	458
⋆1313. Uniform limits of functions	459
⋆1314. Three theorems on uniform limits	460
⋆1315. Exercises	461

⋆Chapter 14

IMPROPER INTEGRALS

⋆1401. Introduction. Review	463
⋆1402. Alternating integrals. Abel's test	464
⋆1403. Exercises	466
⋆1404. Uniform convergence	467
⋆1405. Dominance and the Weierstrass M-test	469
⋆1406. The Cauchy criterion and Abel's test for uniform convergence	471
⋆1407. Three theorems on uniform convergence	473
⋆1408. Evaluation of improper integrals	475
⋆1409. Exercises	478
⋆1410. The gamma function	480
⋆1411. The beta function	482
⋆1412. Exercises	484
⋆1413. Infinite products	486
⋆1414. Wallis's infinite product for π	488
⋆1415. Euler's constant	489
⋆1416. Stirling's formula	490
⋆1417. Weierstrass's infinite product for $1/\Gamma(\alpha)$	491
⋆1418. Exercises	493
⋆1419. Improper multiple integrals	493
⋆1420. Exercises	495

Chapter 15

COMPLEX VARIABLES

1501. Introduction	496
1502. Complex numbers	496
1503. Embedding of the real numbers	497
1504. The number i	498
1505. Geometrical representation	498

1506. Polar form	499
1507. Conjugates	501
1508. Roots	502
1509. Exercises	503
1510. Limits and continuity	506
1511. Sequences and series	507
1512. Exercises	509
1513. Complex-valued functions of a real variable	509
1514. Exercises	513
*1515. The Fundamental Theorem of Algebra	515

Chapter 16

FOURIER SERIES

1601. Introduction	517
1602. Linear function spaces	518
1603. Periodic functions. The space $R_{2\pi}$	519
1604. Inner product. Orthogonality. Distance	520
1605. Least squares. Fourier coefficients	524
1606. Fourier series	525
1607. Exercises	528
1608. A convergence theorem. The space $S_{2\pi}$	529
1609. Bessel's inequality. Parseval's equation	532
1610. Cosine series. Sine series	534
1611. Other intervals	535
1612. Exercises	536
*1613. Partial sums of Fourier series	538
*1614. Functions with one-sided limits	539
*1615. The Riemann-Lebesgue Theorem	539
*1616. Proof of the convergence theorem	540
*1617. Fejér's summability theorem	540
*1618. Uniform summability	542
*1619. Weierstrass's theorem	542
*1620. Density of trigonometric polynomials	543
*1621. Some consequences of density	545
*1622. Further remarks	546
*1623. Other orthonormal systems	547
*1624. Exercises	547
1625. Applications of Fourier series. The vibrating string	548
1626. A heat conduction problem	550
1627. Exercises	551

Chapter 17

VECTOR ANALYSIS

1701. Introduction	554
1702. The vector or outer or cross product	554
1703. The triple scalar product. Orientation in space	556
1704. The triple vector product	559
1705. Exercises	560
1706. Coordinate transformations	562
1707. Translations	564

1708.	Rotations	564
1709.	Exercises	567
1710.	Scalar and vector fields. Vector functions	568
1711.	Ordinary derivatives of vector functions	569
1712.	The gradient of a scalar field	570
1713.	The divergence and curl of a vector field	572
1714.	Relations among vector operations	573
1715.	Exercises	574
★1716.	Independence of the coordinate system	576
★1717.	Curvilinear coordinates. Orthogonal coordinates	578
★1718.	Vector operations in orthogonal coordinates	580
★1719.	Exercises	582

Chapter 18

LINE AND SURFACE INTEGRALS

1801.	Introduction	584
1802.	Line integrals in the plane	584
1803.	Independence of path and exact differentials	587
1804.	Exercises	591
1805.	Green's Theorem in the plane	593
1806.	Local exactness	596
1807.	Simply- and multiply-connected regions	598
1808.	Equivalences in simply-connected regions	601
1809.	Exercises	602
★1810.	Analytic functions of a complex variable. Exercises	605
1811.	Surface elements	610
1812.	Smooth surfaces	613
1813.	Schwarz's example	616
1814.	Surface area	617
1815.	Exercises	620
1816.	Surface integrals	621
1817.	Orientable smooth surfaces	623
1818.	Surfaces with edges and corners	624
1819.	The divergence theorem	625
1820.	Green's identities	629
1821.	Harmonic functions	630
1822.	Exercises	631
1823.	Orientable sectionally smooth surfaces	632
1824.	Stokes's Theorem	635
1825.	Independence of path. Scalar potential	638
★1826.	Vector potential	640
1827.	Exercises	641
★1828.	Exterior differential forms. Exercises	643

Chapter 19

DIFFERENTIAL GEOMETRY

1901.	Introduction	646
1902.	Curvature. Osculating plane	646
1903.	Applications to kinematics	649
1904.	Torsion. The Frenet formulas	650

1905. Local behavior	653
1906. Exercises	655
1907. Curves on a surface. First fundamental form	657
1908. Intersections of smooth surfaces	659
1909. Plane sections. Meusnier's theorem	660
1910. Normal sections. Mean and total curvature	662
1911. Second fundamental form	666
1912. Exercises	670
ANSWERS TO PROBLEMS	673
INDEX	693

ADVANCED
CALCULUS

1

The Real Number System

101. INTRODUCTION

The massive structure of mathematical analysis, as it has developed and as it exists today, rests on a foundation known as the *real number system*. It is possible to start with a more primitive concept, that of the *natural numbers* 1, 2, 3, $\cdots$, and, by means of an appropriate set of axioms, to *construct* (a step at a time) the entire system of real and complex numbers.† In this book, however, we shall limit ourselves to a formal *description* of the real number system.‡ By this, we shall mean an enumeration of a *fundamental set* of properties from which all *other* properties can be derived. Such a fundamental set will be known as **axioms** of the system. It turns out that a complete description is given by the simple statement, "The real number system is a complete ordered field."§ Let us see what this means.

102. AXIOMS OF A FIELD

The basic operations of the number system are *addition* and *multiplication*. As shown below, *subtraction* and *division* can be defined in terms of these two. The axioms of the basic operations are further subdivided into three subcategories: *addition, multiplication,* and *addition and multiplication*.

The following axioms I, II, and III define what is known as a **field**. That is, any set of objects (herein referred to simply as **numbers**) with two operations satisfying these axioms is *by definition* a field.

---

† The Italian mathematician G. Peano (1858–1932) presented in 1889 a set of five axioms for the natural numbers. For a detailed discussion of the development of both the real and complex numbers from the Peano axioms, see F. Landau, *Foundations of Analysis* (New York, Chelsea Publishing Company, 1951).

‡ Except for Chapter 15 and § 1810, and occasional passing mention, complex numbers are not used in this book. Unless otherwise specified, the word *number* should be interpreted to mean *real number*.

§ For a proof of this and a discussion of the axioms of the real number system, see G. Birkhoff and S. MacLane, *A Survey of Modern Algebra* (New York, the Macmillan Company, 1953), Chapter 4; and R. C. Buck, *Advanced Calculus* (New York, McGraw-Hill Book Company, Inc., 1956), Appendix. Also see the author's *Real Variables* (New York, Appleton-Century-Crofts, Inc., 1959), Ex. 13, §116.

NOTE. Henceforth, in this book, the abbreviation *RV* will be used to represent the author's *Real Variables* referred to above.

I. Addition

(*i*) Any two numbers, x and y, have a unique sum, $x + y$.

(*ii*) The **associative law** holds. That is, if x, y, and z are any numbers,
$$x + (y + z) = (x + y) + z.$$

(*iii*) There exists a number 0 such that for any number x, $x + 0 = x$.

(*iv*) Corresponding to any number x there exists a number $-x$ such that
$$x + (-x) = 0.$$

(*v*) The **commutative law** holds. That is, if x and y are any numbers,
$$x + y = y + x$$

The number 0 of axiom (*iii*) is called **zero**.

The number $-x$ of axiom (*iv*) is called the **negative** of the number x. The **difference** between x and y is defined:
$$x - y \equiv x + (-y).$$
The resulting operation is called **subtraction**.

Some of the properties that can be derived from Axioms I alone are given in Exercises 1–5, § 103.

II. Multiplication

(*i*) Any two numbers, x and y, have a unique product, xy.

(*ii*) The **associative law** holds. That is, if x, y, and z are any numbers,
$$x(yz) = (xy)z.$$

(*iii*) There exists a number $1 \neq 0$ such that for any number x, $x \cdot 1 = x$.

(*iv*) Corresponding to any number $x \neq 0$ there exists a number x^{-1} such that $x \cdot x^{-1} = 1$.

(*v*) The **commutative law** holds. That is, if x and y are any numbers,
$$xy = yx.$$

The number 1 of axiom (*iii*) is called **one** or **unity**.

The number x^{-1} of axiom (*iv*) is called the **reciprocal** of the number x. The **quotient** of x and y ($y \neq 0$) is defined:
$$\frac{x}{y} \equiv x \cdot y^{-1}.$$
The resulting operation is called **division**.

Some of the properties that can be derived from Axioms II alone are given in Exercises 6–9, § 103.

III. Addition and Multiplication

(*i*) The **distributive law** holds. That is, if x, y, and z are any numbers,
$$x(y + z) = xy + xz.$$

The distributive law, together with Axioms I and II, yields further familiar relations, some of which are given in Exercises 10–24, § 103.

We shall now demonstrate how some of the most familiar properties of numbers are direct consequences of the fact that *the real number system is a field*.

Example 1. Prove the uniqueness of zero. That is, prove that there is only one number having the property of the number 0 of Axiom I(*iii*).

Solution. Assume that the numbers 0 and 0' both have the property. Then simultaneously
$$0' + 0 = 0' \quad \text{and} \quad 0 + 0' = 0.$$
By the commutative law, $0' + 0 = 0 + 0'$, and $0' = 0$.

Example 2. Prove the **law of cancellation for addition:**
$$x + y = x + z \quad \text{implies} \quad y = z.$$
Solution. Let $(-x)$ be a number satisfying Axiom I(*iv*). Then
$$(-x) + (x + y) = (-x) + (x + z).$$
By the associative law,
$$[(-x) + x] + y = [(-x) + x] + z.$$
By the commutative law the quantity in brackets is equal to $x + (-x) = 0$, and the preceding equation becomes $0 + y = 0 + z$. By two further applications of the commutative law we conclude that $y = z$, as desired.

Example 3. Prove that for any number x:
$$x \cdot 0 = 0.$$
Solution. By the distributive law, and I(*iii*):
$$x \cdot 0 = x(0 + 0) = x \cdot 0 + x \cdot 0.$$
On the other hand, by I(*iii*):
$$x \cdot 0 = x \cdot 0 + 0.$$
Consequently we can infer the equality
$$x \cdot 0 + x \cdot 0 = x \cdot 0 + 0,$$
and by the law of cancellation (Example 2), the desired conclusion.

Example 4. Prove that the product of two nonzero numbers is nonzero. That is, if $x \neq 0$ and if $y \neq 0$, then $xy \neq 0$. Equivalently, if $xy = 0$, then either $x = 0$ or $y = 0$ (or both).

Solution. Assume $x \neq 0$, $y \neq 0$, and $xy = 0$. Then
$$x^{-1}(xy) = x^{-1} \cdot 0 = 0.$$
On the other hand, by the associative and commutative laws for multiplication,
$$x^{-1}(xy) = (x^{-1}x)y = (xx^{-1})y = 1 \cdot y = y \cdot 1 = y.$$
Consequently $y = 0$, in contradiction to the hypotheses. This means that in the presence of assumptions $x \neq 0$ and $y \neq 0$ the equation $xy = 0$ is impossible.

103. EXERCISES

In Exercises 1–24, prove the given statement or establish the given equation for an arbitrary field.

1. The negative of a number is unique.   *Hint:* If y has the property of $-x$ in Axiom I(iv), § 102, $x + y = x + (-x) = 0$. Use the law of cancellation, given in Example 2, § 102.
2. $-0 = 0$.
3. $-(-x) = x$.
4. $0 - x = -x$.
5. $-(x + y) = -x - y$; $-(x - y) = y - x$.   *Hint:* By the uniqueness of the negative it is sufficient for the first part to prove that
$$(x + y) + [(-x) + (-y)] = 0.$$
Use the commutative and associative laws.
6. There is only one number having the property of the number 1 of Axiom II(iii).
7. The **law of cancellation for multiplication** holds: $xy = xz$ implies $y = z$ if $x \neq 0$.
8. The reciprocal of a number ($\neq 0$) is unique.
9. $1^{-1} = 1$.
10. Zero has no reciprocal.
11. If $x \neq 0$, then $x^{-1} \neq 0$ and $(x^{-1})^{-1} = x$.
12. $\dfrac{x}{y} = 0$ ($y \neq 0$) if and only if $x = 0$.
13. $\dfrac{1}{x} = x^{-1}$ ($x \neq 0$).
14. If $x \neq 0$ and $y \neq 0$, then $(xy)^{-1} = x^{-1}y^{-1}$, or $\dfrac{1}{xy} = \dfrac{1}{x} \cdot \dfrac{1}{y}$.
15. If $b \neq 0$ and $d \neq 0$, then $\dfrac{a}{b} = \dfrac{ad}{bd}$.
Hint: $(ad)(bd)^{-1} = (ad)(d^{-1}b^{-1}) = a[(dd^{-1})b^{-1}] = ab^{-1}$.
16. If $b \neq 0$ and $d \neq 0$, then $\dfrac{a}{b} \cdot \dfrac{c}{d} = \dfrac{ac}{bd}$.
17. If $b \neq 0$ and $d \neq 0$, then $\dfrac{a}{b} + \dfrac{c}{d} = \dfrac{ad + bc}{bd}$.
Hint: $(bd)^{-1}(ad + bc) = (b^{-1}d^{-1})(ad) + (b^{-1}d^{-1})(bc)$.
18. $(-1)(-1) = 1$.   *Hint:* $(-1)(1 + (-1)) = 0$. The distributive law gives $(-1) + (-1)(-1) = 0$. Add 1 to each member.
19. $(-1)x = -x$.
Hint: Multiply each member of the equation $1 + (-1) = 0$ by x.
20. $(-x)(-y) = xy$.
Hint: Write $-x = (-1)x$ and $-y = (-1)y$.
21. $-(xy) = (-x)y = x(-y)$.
22. $-\dfrac{x}{y} = \dfrac{-x}{y} = \dfrac{x}{-y}$ ($y \neq 0$).
23. $x(y - z) = xy - xz$.
24. The general linear equation $ax + b = 0$, $a \neq 0$, has a unique solution $x = -b/a$.
25. Axioms I(i), (ii), (iii), and (iv), § 102, define a **group**. That is, any set of objects with one operation (in this case addition) satisfying these axioms is *by definition* a group. Show that the set of all nonzero members of a field with the operation of multiplication (instead of addition) is a group.

§ 104]   AXIOMS OF AN ORDERED FIELD   5

26. A **commutative group** is a group for which the commutative law holds. Show that a field can be defined as an additive commutative group whose nonzero members form a multiplicative commutative group and for which the distributive law holds. (Cf. Ex. 25.)

★27. Show that the set of five elements 0, 1, 2, 3, 4, with the following addition and multiplication tables, is a field:

	0	1	2	3	4
0	0	1	2	3	4
1	1	2	3	4	0
2	2	3	4	0	1
3	3	4	0	1	2
4	4	0	1	2	3

Addition

	0	1	2	3	4
0	0	0	0	0	0
1	0	1	2	3	4
2	0	2	4	1	3
3	0	3	1	4	2
4	0	4	3	2	1

Multiplication

Hint: The sum (product) of any two elements is the remainder left after division by 5 of the real-number sum (product) of the corresponding two numbers.

104. AXIOMS OF AN ORDERED FIELD

The axioms of § 102 yield only a relatively small portion of the properties of real numbers. In terms of these axioms alone, for example, it is impossible to infer that there are infinitely many numbers (cf. Ex. 27, § 103), and it is meaningless to say that one number is greater than another. The next set of axioms put further restrictions on the number system (cf. Ex. 20, § 105) and permit the introduction of an *order relation*. These axioms are phrased in terms of the primitive concept of *positiveness*. Any field having these properties is called an **ordered field.**

IV. Order
(i) *Some numbers have the property of being positive.*
(ii) *For any number x exactly one of the following three statements is true: $x = 0$; x is positive; $-x$ is positive.*
(iii) *The sum of two positive numbers is positive.*
(iv) *The product of two positive numbers is positive.*

Definition I. *The symbols $<$ and $>$ (read "less than" and "greater than," respectively) are defined by the statements*

$$x < y \text{ if and only if } y - x \text{ is positive;}$$
$$x > y \text{ if and only if } x - y \text{ is positive.}$$

Definition II. *The number x is **negative** if and only if $-x$ is positive.*

Definition III. *The symbols $\leq$ and $\geq$* (read "less than or equal to" and "greater than or equal to," respectively) *are defined by the statements*

$$x \leq y \text{ if and only if either } x < y \text{ or } x = y;$$
$$x \geq y \text{ if and only if either } x > y \text{ or } x = y.$$

NOTE 1. The two statements $x < y$ and $y > x$ are equivalent. The two statements $x \leq y$ and $y \geq x$ are equivalent.

NOTE 2. The *sense* of an inequality of the form $x < y$ or $x \leq y$ is said to be the **reverse** of that of an inequality of the form $x > y$ or $x \geq y$.

NOTE 3. The simultaneous inequalities $x < y$, $y < z$ are usually written $x < y < z$, and the simultaneous inequalities $x > y$, $y > z$ are usually written $x > y > z$. Similar interpretations are given the compound inequalities $x \leq y \leq z$ and $x \geq y \geq z$.

Example 1. $x > 0$ if and only if x is positive, since $x > 0$ if and only if $x - 0 = x + (-0) = x + 0 = x$ is positive (cf. Ex. 2, § 103).

Example 2. Establish the **transitive law** for both $<$ and $>$:

$$x < y, y < z \text{ imply } x < z,$$
$$x > y, y > z \text{ imply } x > z.$$

Solution for $<$. If both $y - x$ and $z - y$ are assumed to be positive, then so is their sum, $(z - y) + (y - x) = z - x$, by IV(*iii*).

Example 3. Establish the **law of trichotomy:** *for any x and y exactly one of the following holds:*

(1) $$x < y, \quad x = y, \quad x > y.$$

Solution. Consider the number $x - y$, and use IV(*ii*). There are exactly three possibilities:

If $x - y = 0$, then $x = y$, and conversely.
If $x - y$ is positive, then $x > y$, and conversely.
If $-(x - y) = y - x$ is positive, then $x < y$, and conversely.

The implications of the form "if ... then" guarantee that at least one of (1) must hold. The converses ensure the uniqueness.

105. EXERCISES

1. Addition of any number to both members of an inequality preserves the order relation: $x < y$ implies $x + z < y + z$. A similar fact holds for subtraction. *Hint:* $(y + z) - (x + z) = y - x$.

2. $x < 0$ if and only if x is negative.

3. The sum of two negative numbers is negative.

4. The product of two negative numbers is positive. *Hint:* $xy = (-x)(-y)$. (Cf. Ex. 20, § 103.)

5. The square of any nonzero real number is positive.

6. $1 > 0$.

7. If $2 \equiv 1 + 1$, then $2 > 1$ and $2 > 0$.

8. The equation $x^2 + 1 = 0$ has no real root. ($x^2 \equiv x \cdot x$.)

9. The product of a positive number and a negative number is negative.

10. The reciprocal of a positive number is positive. The reciprocal of a negative number is negative.

11. $0 < x < y$ imply $0 < \dfrac{1}{y} < \dfrac{1}{x}$.

12. Multiplication or division of both members of an inequality by a positive number preserves the order relation: $x < y$, $z > 0$ imply $xz < yz$ and $x/z < y/z$.

13. Multiplication or division of both members of an inequality by a negative number reverses the order relation: $x < y$, $z < 0$ imply $xz > yz$ and $x/z > y/z$.

14. $a < b$, $c < d$ imply $a + c < b + d$.

15. $0 \leq a < b$, $0 \leq c < d$ imply $ac < bd$.

16. If x and y are nonnegative numbers, then $x < y$ if and only if $x^2 < y^2$ (cf. Ex. 10, § 107).

17. The transitive law holds for $\leq$ (also for $\geq$): $x \leq y$, $y \leq z$ imply $x \leq z$. (Cf. Example 2, § 104.)

18. $x \leq y$, $y \leq x$ imply $x = y$.

19. If x is a fixed number satisfying the inequality $x < \epsilon$ for every positive number ϵ, then $x \leq 0$. *Hint:* If x were positive one could choose $\epsilon = x$.

20. There is no largest number, and therefore the real number system is infinite. (In fact, every ordered field is infinite.) *Hint:* $x + 1 > x$.

21. If $x < a < y$ or if $x > a > y$, then a is said to be **between** x and y. Prove that if x and y are distinct numbers, their arithmetic mean $\frac{1}{2}(x + y)$ is between them. (The number 2 is defined: $2 \equiv 1 + 1$; cf. § 106.)

22. If $x^2 = y$, then x is called a **square root** of y. By Exercise 5, if such a number x exists, y must be nonnegative, and if $y = 0$, $x = 0$ is the only square root of y. Show that if a positive number y has square roots, it has exactly two square roots, one positive and one negative. The unique positive square root is called **the square root** and is written $\sqrt{y}$. It is shown in § 214 that such square roots do exist. *Hint:* Let $x^2 = z^2 = y$. Then

$$x^2 - z^2 = (x - z)(x + z) = 0.$$

Therefore $x = z$ or $x = -z$.

106. POSITIVE INTEGERS AND MATHEMATICAL INDUCTION

Positive integers, as real numbers (or, more generally, as members of any ordered field), can be thought of as being defined successively (cf. Ex. 7, § 105):

(1) $\quad 1, \quad 2 \equiv 1 + 1, \quad 3 \equiv 2 + 1, \quad 4 \equiv 3 + 1, \quad 5 \equiv 4 + 1, \cdots$.

As a consequence of these and other definitions given earlier these numbers form an increasing progression:

(2) $\qquad\qquad 0 < 1 < 2 < 3 < 4 < 5 < \cdots$.

This brief and simple description has a deceptive appearance of completeness. In an effort to indicate the need for more detailed consideration we shall attempt to frame a few provocative questions: "Just what do the dots in (1) and (2) really mean?" "What is the structure of the set of *all* positive integers?" "Are there any positive integers that we cannot reach in a finite number of steps, according to (1), and what does this question mean anyway?" "Is 1 *really* the smallest positive integer, and if so, why?"

A direct and objective solution to the problem of defining positive integers (in an ordered field) is provided by the concept of *inductive set*:

Definition. *An inductive set of numbers is a set A having the two properties:*
 (i) *The number 1 is a member of A.*
 (ii) *Whenever a number x belongs to A, the number $x + 1$ also belongs to A.*

It is not difficult to prove that there is a *smallest* inductive set *PI*, which is a subset of *every* inductive set—that is, every member of *PI* belongs to every inductive set. (Indeed, this is the definition of *PI*: *A number is a member of PI if and only of it is a member of every inductive set.*) This set *PI is by definition the set of positive integers.* In other words, a number n is a positive integer if and only if n belongs to every inductive set. The idea behind this definition is fairly simple: To start with, 1 is a positive integer since, by (i), 1 is a member of every inductive set. Consequently, by (ii), the number $2 \equiv 1 + 1$ belongs to every inductive set and must be a positive integer. In like fashion we see that 3 is a positive integer, and so are 4, 5, and so forth. The restriction of *PI* to being the *smallest* inductive set excludes numbers like $\frac{1}{3}$ or $\frac{5}{2}$, which may belong to *some* inductive sets (like the set of *all* numbers, or the set of *positive* numbers) but will not belong to *all* inductive sets.

We now list five properties of positive integers which (together with (1) and (2), above) follow from the preceding definitions. (Proofs are left as exercises for the interested reader. Hints are given in the *Outline of proof*, below.)

    I. *The "positive integers" are positive.*
    II. *If n is a positive integer, $n \geq 1$.*
    III. *The sum and product of two positive integers are positive integers.*
    IV. *If m and n are positive integers, and $m < n$, then $n - m$ is a positive integer.*
    V. *If n is a positive integer, there is no positive integer m such that $n < m < n + 1$.*

* *Outline of proof.* I: The set of positive numbers is an inductive set. II: The set of all numbers $x \geq 1$ is an inductive set. III: For every positive integer m the set of all numbers x such that $m + x$ is a positive integer is an inductive set, and the set of all numbers x such that mx is a positive integer is an inductive set. IV for $m = 1$: If $n > 1$ and if $n - 1$ is *not* a positive integer, then n can be deleted from the set *PI* to produce an inductive set that is *smaller* than *PI*. V: The set of all positive integers n such that there is *no* positive integer m between n and $n + 1$ is an inductive set. IV: The set of all positive integers m such that for every positive integer n greater than m the number $n - m$ is a positive integer is an inductive set.

An important property of the positive integers, one which validates mathematical induction, is a rewording of the fifth Peano axiom for the natural numbers. For this reason we retain the word *axiom*, in its title Axiom of Induction, although when the real numbers are defined in terms of ordered fields, this property is a *theorem*. Indeed, as we shall see, it is true in *any* ordered field.

Axiom of Induction. *If S is a set of positive integers with the two properties (i) S contains the number 1, and (ii) whenever S contains the positive integer n it also contains the positive integer $n + 1$, then S contains all positive integers.*

Proof. By assumption, S is a subset of *PI*. On the other hand, this fact and properties (*i*) and (*ii*) imply that S is an inductive set. Therefore *PI* is a subset of S. Consequently, S and *PI* are identical: $S = PI$.

An immediate consequence of the Axiom of Induction is the theorem:

Fundamental Theorem of Finite Mathematical Induction. *For every positive integer n let P(n) be a proposition which is either true or false. If (i) P(1) is true and (ii) whenever this proposition is true for the positive integer n it is also true for the positive integer $n + 1$, then P(n) is true for all positive integers n.*

Proof. Let S be the set of positive integers for which $P(n)$ is true, and use the Axiom of Induction.

An important property of the positive integers is the following:

Well-ordering Principle. *Every nonempty set of positive integers (that is, every set of positive integers which contains at least one member) contains a smallest member.*

Proof. Let T be an arbitrary set of positive integers which contains at least one member, and assume that T has no smallest member. We shall obtain a contradiction by letting S be the set of all positive integers n having the property that every member of T is greater than n. Clearly 1 is a member of S, since every member of T is at *least* equal to 1, and if a member of T were *equal* to 1 it would be the smallest member of T (property II above). Now suppose n is a member of S. Then every member of T is greater than n and therefore (property V above) is at least $n + 1$. But by the same argument as that used above, any member of T equal to $n + 1$ would be the smallest member of T. Therefore every member of T is greater than $n + 1$ and $n + 1$ belongs to S. Consequently every positive integer is a member of S, and T must be empty.

We state two general laws, which are familiar to all. Detailed proofs are given in *RV*, § 115.

General Associative Laws. *Any two sums (products) of the n numbers $x_1, x_2, \cdots, x_n$ in the same order are equal regardless of the manner in which the terms (factors) are grouped by parentheses.*

Illustration. Let $a = x_1(x_2(x_3(x_4x_5)))$ and $b = ((x_1x_2)(x_3x_4))x_5$. We shall show that $b = a$ by using the associative law of § 102 to transform b step by step into a. A similar sequence of steps would transform any product of the five numbers into the "standard" product a, and hence justify the theorem for $n = 5$. We start by thinking of the products (x_1x_2) and (x_3x_4) as single numbers and use the associative law to write $b = (x_1x_2)((x_3x_4)x_5)$. Repeating this method we have:

$$b = (x_1x_2)(x_3(x_4x_5)) = x_1(x_2(x_3(x_4x_5))) = a.$$

NOTE. As a consequence of the Axiom of Induction any sum or product of n numbers exists, and by the general associative laws any such sum or product can be written without parentheses, thus: $x_1 + x_2 + \cdots + x_n$ and $x_1x_2 \cdots x_n$.

General Commutative Laws. *Any two sums (products) of the n numbers $x_1, x_2, \cdots, x_n$ are equal regardless of the order of the terms (factors).*

Illustration. Let $a = x_1x_2x_3x_4x_5$ and $b = x_4x_1x_5x_2x_3$. We shall show that $b = a$ by using the commutative law of § 102 to transform b step by step into a. We first bring x_1 to the left-hand end: $b = (x_4x_1)(x_5x_2x_3) = (x_1x_4)(x_5x_2x_3) = x_1x_4x_5x_2x_3$. Next we take care of x_2: $b = x_1x_4(x_5x_2)x_3 = x_1(x_4x_2)x_5x_3 = x_1x_2x_4x_5x_3$. Finally, after x_3 is moved two steps to the left, the form a is reached.

The following examples illustrate the principles and uses of mathematical induction:

Example 1. Establish the formula

(3) $$1^2 + 3^2 + 5^2 + \cdots + (2n-1)^2 = \frac{n(4n^2 - 1)}{3}$$

for every positive integer n.

Solution. Let $P(n)$ be the proposition (3). Direct substitution shows that $P(1)$ is true. We wish to show that whenever $P(n)$ is true for a *particular* positive integer n it is also true for the positive integer $n + 1$. Accordingly, we assume (3) and wish to establish

(4) $$1^2 + 3^2 + 5^2 + \cdots + (2n-1)^2 + (2n+1)^2 = \frac{(n+1)[4(n+1)^2 - 1]}{3}.$$

On the assumption that (3) is correct (for a particular value of n), we can rewrite the left-hand member of (4) by grouping:

$$[1^2 + 3^2 + \cdots + (2n-1)^2] + (2n+1)^2 = \frac{n(4n^2 - 1)}{3} + (2n+1)^2.$$

Thus verification of (4) reduces to verification of

(5) $$\frac{n(4n^2 - 1)}{3} + (2n+1)^2 = \frac{(n+1)(4n^2 + 8n + 3)}{3},$$

which, in turn, is true (divide by 3) by virtue of

(6) $$4n^3 - n + 3(4n^2 + 4n + 1) = (4n^3 + 8n^2 + 3n) + (4n^2 + 8n + 3).$$

By the Fundamental Theorem of Mathematical Induction, (3) is true for all positive integers n.

Example 2. Prove the *general distributive law:*

(7) $$x(y_1 + y_2 + \cdots + y_n) = xy_1 + xy_2 + \cdots + xy_n.$$

Solution. Let $P(n)$ be the proposition (7). $P(1)$ is a triviality, and $P(2)$ is true by the distributive law III(i), § 102. We wish to show now that the truth of (7) for a particular positive integer n implies the truth of $P(n+1)$:

(8) $\quad x(y_1 + y_2 + \cdots + y_n + y_{n+1}) = xy_1 + xy_2 + \cdots + xy_n + xy_{n+1}.$

By using the distributive law of § 102 and the assumption (7), we can rewrite the left-hand member of (8) as follows:

$$x[(y_1 + \cdots + y_n) + y_{n+1}] = x(y_1 + \cdots + y_n) + xy_{n+1}$$
$$= (xy_1 + \cdots + xy_n) + xy_{n+1}.$$

Since this last expression is equal to the right-hand member of (8), the truth of $P(n)$, or (7), is established for all positive integers n by the Fundamental Theorem of Mathematical Induction.

107. EXERCISES

1. Prove that $2 + 2 = 4$. *Hint:* Use the associative law with $4 = 3 + 1 = (2+1)+1$.
2. Prove that $2 \cdot 3 = 6$.
3. Prove that $6 + 8 = 14$ and that $3 \cdot 4 = 12$.
4. Prove that the sum and product of n positive integers are positive integers.
5. Prove that any product of nonzero numbers is nonzero.
6. Prove that if $x_1 \neq 0, x_2 \neq 0, \cdots, x_n \neq 0$, then $(x_1 x_2 \cdots x_n)^{-1} = x_1^{-1} x_2^{-1} \cdots x_n^{-1}$.
7. Prove that if $x_1 < x_2, x_2 < x_3, \cdots, x_{n-1} < x_n$ (usually written $x_1 < x_2 < \cdots < x_n$), then $x_1 < x_n$.
8. Prove that any sum or product of positive numbers is positive.
9. Use mathematical induction to prove that if n is any positive integer, then $x^n - y^n = (x-y)(x^{n-1} + x^{n-2}y + \cdots + xy^{n-2} + y^{n-1})$. *Hint:* $x^{n+1} - y^{n+1} = x^n(x-y) + y(x^n - y^n)$.
10. Prove that if x and y are nonnegative numbers and n is a positive integer, then $x > y$ if and only if $x^n > y^n$. *Hint:* Use Ex. 9. (Cf. Ex. 16, § 105.)
11. Establish the inequality $2^n > n$, where n is a positive integer.
12. Establish the formula $1 + 2 + \cdots + n = \frac{1}{2}n(n+1)$. (Cf. Ex. 28.)
13. Establish the formula (cf. Ex. 29)
$$1^2 + 2^2 + \cdots + n^2 = \tfrac{1}{6}n(n+1)(2n+1).$$
14. Establish the formula (cf. Ex. 30)
$$1^3 + 2^3 + \cdots + n^3 = \tfrac{1}{4}n^2(n+1)^2 = (1 + 2 + \cdots + n)^2.$$
15. Establish the formula (cf. Ex. 31)
$$1^4 + 2^4 + \cdots + n^4 = \tfrac{1}{30}n(n+1)(2n+1)(3n^2 + 3n - 1).$$
16. Prove that every nonempty finite set of real numbers has a least member and a greatest member. That is, show that the set $x_1, x_2, \cdots, x_n$ of real numbers contains members x_i and x_j such that for every member x_k ($k = 1, 2, \cdots, n$) $x_i \leq x_k \leq x_j$. Discuss uniqueness. The two numbers x_i and x_j are written:
$$x_i = \min(x_1, x_2, \cdots, x_n) = \min_{k=1}^{n}(x_k),$$
$$x_j = \max(x_1, x_2, \cdots, x_n) = \max_{k=1}^{n}(x_k).$$

17. Establish the law of exponents: $a^m a^n = a^{m+n}$, where a is any number and m and n are positive integers. *Hint:* Hold m fixed and use induction on n.

18. Establish the law of exponents:
$$\frac{a^m}{a^n} = \begin{cases} a^{m-n} & \text{if } m > n, \\ \dfrac{1}{a^{n-m}} & \text{if } n > m, \end{cases}$$
where a is any nonzero number and m and n are positive integers.

19. Establish the law of exponents: $(a^m)^n = a^{mn}$, where a is any number and m and n are positive integers.

20. Establish the law of exponents: $(ab)^n = a^n b^n$, where a and b are any numbers and n is a positive integer. Generalize to m factors.

21. Establish the law of exponents $\left(\dfrac{a}{b}\right)^n = \dfrac{a^n}{b^n}$, where a and b are any numbers ($b \neq 0$) and n is a positive integer.

22. A positive integer m is a **factor** of a positive integer p if and only if there exists a positive integer n such that $p = mn$. A positive integer p is called **composite** if and only if there exist integers $m > 1$ and $n > 1$ such that $p = mn$. A positive integer p is **prime** if and only if $p > 1$ and p is not composite. Prove that if m_i, $i = 1, 2, \cdots, n$, are integers > 1, then $m_1 m_2 \cdots m_n > n$. Hence prove that any integer > 1 is either a prime or a product of primes.

23. Two positive integers are **relatively prime** if and only if they have no common integral factor greater than 1. A fraction p/q, where p and q are positive integers, is **in lowest terms** if and only if p and q are relatively prime. Prove that any quotient of positive integers is equal to such a fraction in lowest terms. *Hint:* If p and q have a common factor $m > 1$, reduce p/q by dividing by m. Continue this process. Use Ex. 22 to show that this sequence must terminate after a finite number of steps.

24. If n is a positive integer, **n factorial**, written $n!$, is defined: $n! \equiv 1 \cdot 2 \cdot 3 \cdots \cdot n$. **Zero factorial** is defined: $0! \equiv 1$. If r is a positive integer or zero and if $0 \leq r \leq n$, the binomial coefficient $\dbinom{n}{r}$ (also written $_nC_r$) is defined:
$$\binom{n}{r} = \frac{n!}{(n-r)!\, r!}.$$
Prove that $\dbinom{n}{r}$ is a positive integer. *Hint:* Establish the law of *Pascal's Triangle* (cf. any College Algebra text): $\dbinom{n+1}{r} = \dbinom{n}{r-1} + \dbinom{n}{r}$.

25. Prove the **Binomial Theorem** for positive integral exponents (cf. V, § 1207): If x and y are any numbers and n is a positive integer,
$$(x+y)^n = \binom{n}{0} x^n + \binom{n}{1} x^{n-1} y + \cdots + \binom{n}{r} x^{n-r} y^r + \cdots + \binom{n}{n} y^n.$$
(Cf. Ex. 24.)

26. The **sigma summation notation** is defined:
$$\sum_{k=m}^{n} f(k) \equiv f(m) + f(m+1) + \cdots + f(n), \text{ where } n \geq m. \text{ Prove:}$$
(i) k is a **dummy variable**: $\displaystyle\sum_{k=m}^{n} f(k) = \sum_{i=m}^{n} f(i)$.

(ii) $\sum$ is **additive**: $\sum_{k=m}^{n} [f(k) + g(k)] = \sum_{k=m}^{n} f(k) + \sum_{k=m}^{n} g(k)$.

(iii) $\sum$ is **homogeneous**: $\sum_{k=m}^{n} cf(k) = c \sum_{k=m}^{n} f(k)$.

(iv) $\sum_{k=m}^{n} 1 = n - m + 1$.

*27. A useful summation formula is

(1) $$\sum_{k=1}^{n} [f(k) - f(k-1)] = f(n) - f(0).$$

Establish this by mathematical induction.

*28. By means of Exercises 26 and 27 *derive* the formula of Exercise 12. *Hint:* Let $f(n) \equiv n^2$. Then (1) becomes

$$\sum_{k=1}^{n} [k^2 - (k-1)^2] = \sum_{k=1}^{n} (2k - 1) = n^2, \quad \text{or}$$

$$2 \sum_{k=1}^{n} k = n^2 + n.$$

*29. By means of Exercises 26–28, *derive* the formula of Exercise 13. *Hint:* Let $f(n) \equiv n^3$ in (1).

*30. By means of Exercises 26–29, *derive* the formula of Exercise 14.

*31. By means of Exercises 26–30, *derive* the formula of Exercise 15.

*32. If $a_1, a_2, \cdots, a_n$ and $b_1, b_2, \cdots, b_n$ are real numbers, show that

$$\left(\sum_{i=1}^{n} a_i^2 \right) \left(\sum_{i=1}^{n} b_i^2 \right) - \left(\sum_{i=1}^{n} a_i b_i \right)^2 = \sum_{1 \le i < j \le n} (a_i b_j - a_j b_i)^2.$$

Hence establish the **Schwarz** (or **Cauchy**) **inequality**:

$$\left(\sum_{i=1}^{n} a_i b_i \right)^2 \le \sum_{i=1}^{n} a_i^2 \sum_{i=1}^{n} b_i^2.$$

(For an alternative proof see the method suggested in Ex. 20, § 402. Also cf. Ex. 14, § 1117.)

*33. By use of Exercise 32, establish the **Minkowski inequality**:

$$\sqrt{\sum_{i=1}^{n} (a_i + b_i)^2} \le \sqrt{\sum_{i=1}^{n} a_i^2} + \sqrt{\sum_{i=1}^{n} b_i^2}.$$

Hint: Square both members of the Minkowski inequality, expand each term on the left, cancel identical terms that result, divide by 2, and reverse steps. (Cf. Ex. 21, § 402; Ex. 14, § 1117.)

108. INTEGERS AND RATIONAL NUMBERS

Integers and rational numbers are defined in terms of the more primitive concept of *positive integer*.

Definition I. *A number x is a **negative integer** if and only if $-x$ is a positive integer. A number x is an **integer** if and only if it is 0 or a positive integer or a negative integer.*

14   THE REAL NUMBER SYSTEM   [§ 110

Definition II. *A number x is a **rational number** if and only if there exist integers p and q, where $q \neq 0$, such that $x = p/q$. The real numbers that are not rational are called **irrational**.*

109. EXERCISES

1. Prove that the sum of two integers is an integer. *Hint:* Consider all possible cases of signs. For the case $m > -n > 0$, use Property IV, § 106.

2. Prove that the product of two integers is an integer.

3. Prove that the difference between two integers is an integer.

4. Prove that the quotient of two integers need not be an integer. (The integers, therefore, do not constitute a field. They do, however, form an additive group.)

5. Prove that the integers are rational numbers.

6. Prove that the set of all rational numbers is an ordered field.

7. Define integral powers a^n, where $a \neq 0$ and n is any integer, and establish the laws of exponents of Exercises 17–21, § 107, for integral exponents.

8. The square root of a positive number was defined in Exercise 22, § 105. Existence is proved in § 214. Prove that $\sqrt{2}$ is irrational. That is, prove that there is no positive rational number whose square is 2. *Hint:* Assume $\sqrt{2} = p/q$, where p and q are relatively prime positive integers (cf. Ex. 23, § 107). Then $p^2 = 2q^2$, and p is even and of the form $2k$. Repeat the argument to show that q is also even!

110. GEOMETRICAL REPRESENTATION AND ABSOLUTE VALUE

The reader has doubtless made use of the standard representation of real numbers by means of points on a straight line. It is conventional, when considering this line to lie horizontally as in Figure 101, to adopt a uniform scale, with numbers increasing to the right and decreasing to the left. With an appropriate axiomatic system for Euclidean geometry† there is a one-to-one correspondence between real numbers and points on a line. That is, to any

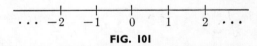

FIG. 101

real number there corresponds precisely one point of the line, and to any point of the line, there corresponds precisely one real number. For this reason it is often immaterial whether one speaks of *numbers* or *points*. In this book we shall frequently use these two words interchangeably, and feel free, for example, to speak of the *point* 3. In this sense, in Figure 101, positive numbers lie to the right of the point 0, and $x < y$ if and only if the point x is to the left of the point y. Again, if $x < z$, then the number y satisfies the simultaneous inequalities $x < y < z$ if and only if the point y is between the points x and z (cf. Ex. 21, § 105). Properties of the real

---

† Cf. D. Hilbert, *The Foundations of Geometry* (La Salle, Ill., The Open Court Publishing Company, 1938).

§ 110] GEOMETRICAL REPRESENTATION 15

numbers, axiomatized and obtained in this chapter, lend strength to our intuitive conviction that a straight line with a number scale furnishes a reliable picture of the real number system.

Definition I. *If a and b are any two real numbers such that $a < b$, the **open interval** from a to b, written (a, b), is the set of all numbers x between a and b, $a < x < b$. The **closed interval** from a to b, written $[a, b]$, includes the points a and b and is the set of all x such that $a \leq x \leq b$. The **half-open intervals** $(a, b]$ and $[a, b)$ are defined by the inequalities $a < x \leq b$ and $a \leq x < b$, respectively. In any of these cases the interval is called a **finite interval** and the points a and b are called **end-points**. **Infinite intervals** are denoted and defined as follows, the point a, where it appears, being the end-point of the interval: $(a, +\infty)$, $x > a$; $[a, +\infty)$, $x \geq a$; $(-\infty, a)$, $x < a$; $(-\infty, a]$, $x \leq a$; $(-\infty, +\infty)$, all x.† Any point of an interval that is not an end-point is called an **interior point** of the interval.*

Definition II. *The **absolute value** of a number x, written $|x|$, is defined:*

$$|x| \equiv \begin{cases} x & \text{if } x \geq 0, \\ -x & \text{if } x < 0. \end{cases}$$

The absolute value of a number can be thought of as its *distance* from the **origin** 0 in Figure 101. Similarly, the absolute value of the difference between two numbers, $|x - y|$, is the distance between the two points x and y. Some of the more useful properties of the absolute value are given below. For hints for some of the proofs, see § 111.

Properties of Absolute Value

I. $|x| \geq 0$; $|x| = 0$ if and only if $x = 0$.
II. $|xy| = |x| \cdot |y|$.
III. $\left|\dfrac{x}{y}\right| = \dfrac{|x|}{|y|}$ $(y \neq 0)$.
IV. If $\epsilon > 0$, then
 (i) the inequality $|x| < \epsilon$ is equivalent to the simultaneous inequalities $-\epsilon < x < \epsilon$;
 (ii) the inequality $|x| \leq \epsilon$ is equivalent to the simultaneous inequalities $-\epsilon \leq x \leq \epsilon$.
V. The **triangle inequality**‡ holds: $|x + y| \leq |x| + |y|$.
VI. $|-x| = |x|$; $|x - y| = |y - x|$.
VII. $|x|^2 = x^2$; $|x| = \sqrt{x^2}$. (Cf. Ex. 22, § 105.)
VIII. $|x - y| \leq |x| + |y|$.
IX. $||x| - |y|| \leq |x - y|$.

† These infinite intervals are also sometimes designated directly by means of inequalities: $a < x < +\infty$, $a \leq x < +\infty$, $-\infty < x < a$, $-\infty < x \leq a$, and $-\infty < x < +\infty$, respectively.
‡ Property V is called the *triangle inequality* because the corresponding inequality for complex numbers states that any side of a triangle is less than or equal to the sum of the other two. (Cf. § 1505; also cf. §§ 703, 1604, Ex. 22, § 605.)

Definition III. *A neighborhood or epsilon-neighborhood of a point a is an open interval of the form* $(a - \epsilon, a + \epsilon)$, *where ϵ is a positive number.*

By Property IV, the neighborhood $(a - \epsilon, a + \epsilon)$ consists of all x satisfying the inequalities $a - \epsilon < x < a + \epsilon$, or $-\epsilon < x - a < \epsilon$, or $|x - a| < \epsilon$. It consists, therefore, of all points whose distance from a is less than ϵ. The point a is the midpoint of each of its neighborhoods.

III. EXERCISES

1. Prove Property I, § 110.
2. Prove Property II, § 110.
3. Prove Property III, § 110. *Hint:* Use Property II with $z = x/y$, so that $x = yz$.
4. Prove Property IV, § 110.
5. Prove Property V, § 110. *Hint:* For the case $x > 0$ and $y < 0$,
$$x + y < x + 0 < x - y = |x| + |y|,$$
$$-(x + y) = -x - y < x - y = |x| + |y|.$$
Use Property IV. Consider all possible cases of sign.
6. Prove Property VI, § 110.
7. Prove Property VII, § 110.
8. Prove Property VIII, § 110.
9. Prove Property IX, § 110. *Hint:* The inequality $|x| - |y| \leq |x - y|$ follows by Property V from $|(x - y) + y| \leq |x - y| + |y|$.
10. Prove that $|x_1 \cdot x_2 \cdots x_n| = |x_1| \cdot |x_2| \cdots |x_n|$.
11. Prove the *general triangle inequality:*
$$|x_1 + x_2 + \cdots + x_n| \leq |x_1| + |x_2| + \cdots + |x_n|.$$
12. Replace by an equivalent single inequality:
$$x > a + b, \quad x > a - b.$$

In Exercises 13–24, find the values of x that satisfy the given inequality or inequalities. Express your answer without absolute values.

13. $|x - 2| < 3$.
14. $|x + 3| \geq 2$.
15. $|x - 5| < |x + 1|$. *Hint:* Square both members. (Cf. Ex. 16, § 105.)
*16. $|x - 4| > x - 2$.
*17. $|x - 4| \leq 2 - x$.
*18. $|x - 2| > x - 4$.
*19. $|x^2 - 2| \leq 1$.
*20. $x^2 - 2x - 15 < 0$. *Hint:* Factor and graph the left-hand member.
*21. $x^2 + 10 < 6x$.
*22. $|x + 5| < 2|x|$.
*23. $x < x^2 - 12 < 4x$.
*24. $|x - 7| < 5 < |5x - 25|$.

In Exercises 25–28, solve for x, and express your answer in a simple form by using absolute value signs.

*25. $\dfrac{x - a}{x + a} > 0$.
*26. $\dfrac{a - x}{a + x} \geq 0$.
*27. $\dfrac{x - 1}{x - 3} > \dfrac{x + 3}{x + 1}$.
*28. $\dfrac{x - a}{x - b} > \dfrac{x + a}{x + b}$.

In Exercises 29–38, sketch the graph.

*29. $y = |x|$.
*30. $y = \dfrac{x}{|x|}$.

★31. $y = x \cdot |x|$.
★32. $y = \sqrt{|x|}$.
★33. $y = ||x| - 1|$.
★34. $|y| = |x|$.
★35. $|y| < |x|$.
★36. $|x| + |y| = 1$.
★37. $|x| + |y| < 1$.
★38. $|x| - |y| \leq 1$.

112. AXIOM OF COMPLETENESS

The remaining axiom of the real number system can be given in any of several forms, all of which involve basically the ordering of the numbers. We present here the one that seems to combine most naturally a fundamental simplicity and an intuitive reasonableness. It is based on the idea of an **upper bound** of a set of numbers; that is, a number that is at least as large as anything in the set. For example, the number 10 is an upper bound of the

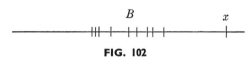

FIG. 102

set consisting of the numbers -15, 0, 3, and 7. It is also an upper bound of the set consisting of the numbers -37, 2, 5, 8, and 10. The number 16 is an upper bound of the open intervals $(-3, 6)$ and $(8, 16)$, and also of the closed intervals $[-31, -2]$ and $[15, 16]$. However, 16 is *not* an upper bound of the set of numbers -6, 13, and 23, nor of the open interval $(8, 35)$, nor of the set of all integers. There is *no* number x that is an upper bound of the set of all positive real numbers, for x is not greater than or equal to the real number $x + 1$. (Cf. Ex. 20, § 105.)

If, for a given set A of numbers, there is some number x that is an upper bound of A, the set A is said to be **bounded above**. This means that the corresponding set B of points on a number scale does not extend indefinitely to the right but, rather, that there is some point x at least as far to the right as any point of the given set (Fig. 102).

Whenever a set is bounded above, it has *many* upper bounds. For example, if x is an upper bound, so is $x + 1$ and, indeed, so is any number greater than x. An appropriate question: "Is there any upper bound *less* than x?" If the answer to this question is "No," then x is the smallest of all possible upper bounds and is called the **least upper bound**† of the set. Since two numbers cannot be such that each is less than the other, there cannot be *more* than one such *least* upper bound. When a number x is a least upper bound of a set, it is therefore called *the* least upper bound of the set. Furthermore, in this case, we say that the set *has* x as its least upper bound, whether x is a member of the set or not. Thus the open interval (a, b) and the closed interval $[a, b]$, where $a < b$, both have b as their least upper bound, whereas b is a member of only the *closed* interval.

† The least upper bound of a set A is also called the **supremum** of A, abbreviated sup(A) (pronounced "supe of A").

The essential question now is whether a given set has a least upper bound. Of course, if a set is not bounded above it has no upper bound and *a fortiori* no *least* upper bound. Suppose a set *is* bounded above. Then does it have a least upper bound? The final axiom gives the answer.

V. Axiom of Completeness. *Any nonempty set of real numbers that is bounded above has a least upper bound.*

This axiom can be thought of geometrically as stating that there are no "gaps" in the number scale. For instance, if a single point x were removed from the number scale (Fig. 103), the remaining numbers would no longer

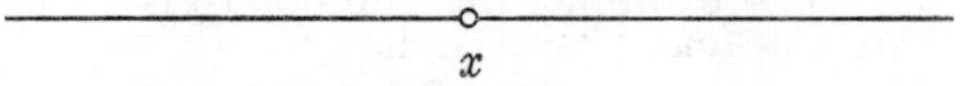

FIG. 103

be complete, since the set A consisting of all numbers less than x would be bounded above (e.g., by $x + 1$), but would have no least upper bound (if y is assumed to be the least upper bound of A and if y is *less* than x, then $y < \frac{1}{2}(x + y) < x$ so that y is not even an upper bound of A; on the other hand, if y is *greater* than x then $y > \frac{1}{2}(x + y) > x$ so that $\frac{1}{2}(x + y)$ is a smaller upper bound than y, and y is not the *least* upper bound of A). (Cf. Ex. 21, § 105.)

It might seem that the axiom of completeness is biased or one-sided. Why should we speak of *upper* bounds and *least* upper bounds, when we might as naturally consider *lower* bounds and *greatest* lower bounds? The answer is that it is immaterial, as far as the axiomatic system is concerned, whether we formulate completeness in terms of upper bounds or lower bounds. This fact is illustrated by the Theorem below, whose statement is actually *equivalent* to the axiom of completeness. **Lower bounds** and **greatest lower bounds** are defined in strict imitation of upper bounds and least upper bounds. A set is **bounded below** if and only if it has a lower bound. A set is **bounded** if and only if it is bounded both below and above; in other words, if and only if it is contained in some finite interval.

Theorem. *Any nonempty set of real numbers that is bounded below has a greatest lower bound.*†

Proof. The set obtained by changing every member of the given set to its negative is bounded above by the negative of any lower bound of the original set. (Look at the number scale with a mirror.) By the axiom of completeness the new set then has a least upper bound, whose negative must be the greatest lower bound of the original set. (Look at the number scale directly again, without the mirror.)

† The greatest lower bound of a set A is called the **infimum** of A, abbreviated inf(A) (pronounced "inf of A").

113. CONSEQUENCES OF COMPLETENESS

The axiom of completeness (taken with the other axioms for real numbers) implies several important basic relations between *certain* real numbers (integers and rational numbers, to be precise) and real numbers *in general*. Some of these fundamental properties will be stated and proved in this section. The first is an algebraic formulation of a basic principle of Euclidean geometry known as the *Archimedean property*.† This principle states that any length (however large) can be exceeded by repeatedly marking off a given length (however small), each successive application starting where the preceding one stopped. (A midget ruler, if used a sufficient number of times, can measure off an arbitrarily large distance.) For real numbers this principle, again called the **Archimedean property**, has the following formal statement and proof.

Theorem I. *If a and b are positive numbers, there is a positive integer n such that $na > b$.*

Proof. If the theorem were false, the inequality $na \leq b$ would hold for all positive integers n. That is, the set $a, 2a, 3a, \cdots$ would be bounded above. Let c be the least upper bound of this set. Then $na \leq c$ for all n, and hence $(n+1)a \leq c$ for all n. Therefore $na + a \leq c$, or $na \leq c - a$, for all n. Thus $c - a$ is an upper bound that is *less* than the *least* upper bound c. This is the desired contradiction.

Corollary I. *If x is any real number there exists a positive integer n such that $n > x$.*

Proof. If $x \leq 0$, let $n = 1$. If $x > 0$, use Theorem I with $a = 1$ and $b = x$.

Corollary II. *If ϵ is any positive number there exists a positive integer n such that*
$$\frac{1}{n} < \epsilon.$$

Proof. Let $n > 1/\epsilon$, by Corollary I.

Corollary III. *If x is any real number there exist integers m and n such that*
$$m < x < n.$$

Proof. According to Corollary I, let n be a positive integer such that $n > x$, let p be a positive integer such that $p > -x$, and let $m = -p$.

Theorem II. *If x is any real number there exists a unique integer n such that*
$$n \leq x < n + 1.$$

Proof. Existence: By Corollary III, Theorem I, there exist integers r and s such that $r < x < s$. Let p be the smallest positive integer such that

† Cf. D. Hilbert, *op. cit.*

$r + p > x$ (such a p exists by the well-ordering principle of § 106). Finally, let $n \equiv r + p - 1$. If $p = 1$, then $n = r < x < r + 1 = n + 1$. If $p \geq 2$, then $n = r + (p - 1) \leq x$ by the minimum property of p. Hence $n \leq x < r + p = n + 1$. *Uniqueness:* If m and n are distinct integers such that $m \leq x < m + 1$ and $n \leq x < n + 1$, assume $n < m$. Then $n < m \leq x < n + 1$, in contradiction to Property V, § 106.

Theorem III. *The rational numbers are **dense** in the system of real numbers. That is, between any two distinct real numbers there is a rational number (in fact, there are infinitely many).*

Proof. Let a and b be two real numbers, where $a < b$, or $b - a > 0$. Let q be a positive integer such that $\dfrac{1}{q} < b - a$, by Corollary II, Theorem I. We now seek an integer p so that $\dfrac{p}{q}$ shall satisfy the relation

$$a < \frac{p}{q} \leq a + \frac{1}{q} < b.$$

This will hold if p is chosen so that $aq < p \leq aq + 1$, or $p \leq aq + 1 < p + 1$, according to Theorem II. With this choice of p and q we have found a rational number $r_1 \equiv p/q$ between a and b. A second rational number r_2 must exist between r_1 and b, a third between r_2 and b, etc.

NOTE. The axioms for a **complete ordered field** (I–III, § 102, IV, § 104, and V, § 112) are *categorical*. This means that any two complete ordered fields are identical except possibly for the symbols used. Each will have a "zero" and a "one," and, more generally, a set of "positive integers," a set of "integers," and a set of "rational numbers," all behaving according to fixed laws. A *one-to-one correspondence* can be set up between the elements of any two complete ordered fields, starting in an obvious way with the "zeros" and the "ones." After this correspondence has been set up for all rational numbers, it is the axiom of completeness that guarantees the extension of the correspondence to all members of the two fields. In more technical terms, any two complete ordered fields are said to be *isomorphic* (they have the same *form* or *structure*). It is in this sense that it is proper to say that the real number system is *completely* described by the statement that it is a complete ordered field. *All* properties of real numbers are consequences of the axioms. (Cf. *RV*, Exs. 12, 13, § 116.)

114. EXERCISES

1.. Prove that if r is a nonzero rational number and x is irrational, then $x \pm r$, $r - x$, xr, x/r, r/x are all irrational. *Hint for $x + r$:* If $x + r = s$, a rational number, then $x = s - r$.

2. Prove that the irrational numbers are dense in the system of real numbers. *Hint:* Let x and y be any two distinct real numbers, assume $x < y$, and find a rational number p/q between $\sqrt{2}x$ and $\sqrt{2}y$. Then divide by $\sqrt{2}$. (Cf. Ex. 8, § 109, and Ex. 1, above.)

3. Prove that the sum of two irrational numbers may be rational. What about their product?

★4. Prove that the **binary numbers**, $p/2^n$, where p is an integer and n is a positive integer, are dense in the system of real numbers. (Cf. Ex. 11, § 107, and Ex. 5, below.)

★5. Prove that the **terminating decimals** $\pm d_{-m} d_{-m+1} \cdots d_{-1} d_0 . d_1 d_2 \cdots d_n$, where d_i is an integral digit ($0 \leq d_i \leq 9$), are dense in the system of real numbers. *Hint:* Each such number is of the form $p/10^q$, where p is an integer and q is a positive integer. (Cf. Ex. 4.)

NOTE. Decimal expansions are treated in Exercises 26–27, § 1111.

6. Prove that if x and y are two fixed numbers and if $y \leq x + \epsilon$ for every positive number ϵ, then $y \leq x$. Prove a corresponding result as a consequence of an inequality of the form $y < x + \epsilon$; of the form $y \geq x - \epsilon$; of the form $y > x - \epsilon$. (Cf. Ex. 19, § 105.)

★7. If x is an irrational number, under what conditions on the rational numbers a, b, c, and d, is $(ax + b)/(cx + d)$ rational?

★8. Prove that the system of integers satisfies the axiom of completeness. (Cf. Ex. 9, below.)

★9. Prove that the system of rational numbers does not satisfy the axiom of completeness. *Hint:* Consider the set S of all rational numbers less than $\sqrt{2}$. Then S has an upper bound in the system of rational numbers (the rational number 2 is one such). Assume that S has a least upper bound r. Use the density of the rational numbers to show that if $r < \sqrt{2}$ then r is not even an upper bound of S, and that if $r > \sqrt{2}$ then r is not the *least* upper bound of S.

★10. Let x be a real number and let S be the set of all rational numbers less than x. Show that x is the least upper bound of S.

★11. Prove that if S is a bounded nonempty set there is a smallest closed interval I containing S. That is, I has the property that if J is any closed interval containing S, then J contains I.

★12. Prove by counterexample that the statement of Exercise 11 is false if the word *closed* is replaced by the word *open*.

★13. Let S be a nonempty set of numbers bounded above, and let x be the least upper bound of S. Prove that x has the two properties corresponding to an arbitrary positive number ϵ: (i) every element s of S satisfies the inequality $s < x + \epsilon$; (ii) at least one element s of S satisfies the inequality $s > x - \epsilon$.

★14. Prove that the two properties of Exercise 13 characterize the least upper bound. That is, prove that a number x subject to these two properties is the least upper bound of S.

★15. Prove **Dedekind's Theorem**: *Let the real numbers be divided into two nonempty sets A and B such that (i) if x is an arbitrary member of A and if y is an arbitrary member of B then $x < y$ and (ii) if x is an arbitrary real number then either x is a member of A or x is a member of B. Then there exists a number c (which may belong to either A or B) such that any number less than c belongs to A and any number greater than c belongs to B.*

★16. Prove that any ordered field in which the Dedekind Theorem (Ex. 15) holds is complete. Conclude that the Dedekind Theorem can be taken as an alternative to the axiom of completeness.

2

Functions, Sequences, Limits, Continuity

••

201. FUNCTIONS AND SEQUENCES

Whenever one says that y is a function of x, one has in mind some mechanism that assigns values to y corresponding to given values of x. The most familiar examples are real-valued functions of a real variable given by formulas, like $y = 3x^2 - 12x$ or $y = \sqrt{x^2 - 4}$. In some cases more than one formula is needed, as for the function defined to be identically zero for negative x and otherwise equal to x^2. In other instances, as shown in examples below, no "formula" in the sense just described is possible. However, there must be *something* that responds to the naming of a value of one variable by producing at least one value of another variable. One of the most satisfactory ways of introducing the concept of *function* without using other similar but undefined terms (like *correspondence*, *rule*, *mechanism*) is by means of ordered pairs, as in the following definition:

Definition I. *Let D and R be two nonempty sets of objects. A **function** with **domain of definition** D and **range of values** R is a set f of ordered pairs*† *(x, y), where x belongs to D and y belongs to R, having the two properties:*

(i) *If x is any member of D there is at least one member y of R such that (x, y) belongs to f.*
(ii) *If y is any member of R there is at least one member x of D such that (x, y) belongs to f.*

*A function f is **single-valued** if and only if whenever the ordered pairs (x, y) and (x, z) belong to f, y and z must be identical: $y = z$.*

† No attempt will be made in this book to define the concept of *ordered pair* in more primitive terms. (Cf. T. M. Apostol, *Mathematical Analysis* (Reading, Mass., Addison-Wesley Publishing Company, 1957), pages 25 ff, for further discussion.) However, it should be noted that *equality* between two ordered pairs (x, y) and (u, v) $((x, y) = (u, v))$ means that $x = u$ and $y = v$, and that if $x \neq y$, then the ordered pairs (x, y) and (y, x) are distinct $((x, y) \neq (y, x))$.

Discussion and Notation. The preceding definition guarantees that if f is a function with domain of definition D and range of values R, then to every "point" x of D there must correspond at least one "point" y of R. If f is single-valued, each x of D determines *exactly one* y of R, and the symbol $f(x)$ denotes the member of R that corresponds to x. If *more than one* y of R corresponds to a particular x of D the symbol $f(x)$ denotes the set of all members of R that correspond to x. Such a functional relationship is also written $y = f(x)$. Conversely, if some formula or rule exists whereby to each point x of D there corresponds at least one point y of R, then the resulting set of all ordered pairs (x, y) such that y corresponds to x constitutes a function as defined above. Whenever there can be no resulting confusion, such an expression as "the function f defined by the equation $y = f(x)$" is compressed to "the function $y = f(x)$" or "the function $f(x)$." In other words, the notation $f(x)$ is used to represent *both* the values of the function f corresponding to the point x *and* the function f itself. This usage is traditional and, when there can be no *reasonable* misunderstanding, convenient.

Definition II. *Let $y = f(x)$ be a function with domain of definition D and range of values R. Then the general member of D and the general member of R are called the **independent** and the **dependent variable**, respectively. In case R consists of just one object, $f(x)$ is called a **constant function**. In case D consists of real numbers, $f(x)$ is called a function of a **real variable**. In case R consists of real numbers, $f(x)$ is called **real-valued**.*

NOTE 1. If a point in the Euclidean plane is *defined* to be an ordered pair of real numbers, a real-valued function of a real variable is simply a nonempty set of points in the plane. Such a function is single-valued if and only if no two distinct points of this set lie on the same vertical line. This set of points, conceived of geometrically, is the "graph" of the given function. Similar geometrical representations are available in higher dimensions. The essential idea of Definition I is to identify the fundamental concepts of *function* and *graph*, without destroying the idea of correspondence associated with a function or that of a geometrical sketch associated with a graph.

NOTE 2. Henceforth, unless explicit statement to the contrary is made (as in Examples 4 and 7, below, Note 1, § 401, Ex. 32, § 611, and Exs. 27, 28, § 1509), it will be implicitly assumed that *all functions considered are single-valued*.

Example 1. The function $y = 3x^2 - 12x$ is defined for all real numbers. If we take D to be the set of all real numbers, R consists of all real numbers ≥ -12, since the function has an absolute minimum (cf. § 309) when $x = 2$. The function is a single-valued real-valued function of a real variable.

Example 2. The function of Example 1 restricted to the domain $D = (1, 5)$ (the open interval from 1 to 5) has range R equal to the half-open interval $[-12, 15)$. This function is not the same as that of Example 1, since it has a different domain. It is also, however, a single-valued real-valued function of a real variable.

Example 3. The function $y = \sqrt{x^2 - 4}$ with domain D consisting of all real numbers x such that $|x| \geq 2$ is a single-valued real-valued function of a real variable, with range R consisting of all nonnegative real numbers. (Cf. § 214.)

Example 4. The function $y = \pm\sqrt{x^2 - 4}$ with the same domain as the function of Example 3 is real-valued, but is single-valued only for $x = \pm 2$. Otherwise it is double-valued. Its range is the set of all real numbers.

Example 5. The **bracket function** or **greatest integer function**, $f(x) \equiv [x]$, is defined to be the largest integer less than or equal to x, with domain all real numbers

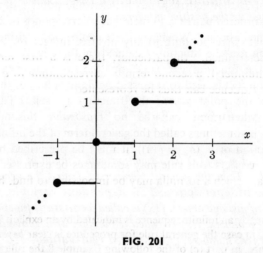

FIG. 201

(Fig. 201). It is a single-valued real-valued function of a real variable. Its range is the set of all integers.

Example 6. Let D be the closed interval $[0, 1]$, and define $f(x)$, for x in D, to be

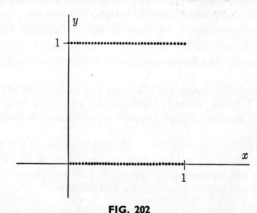

FIG. 202

1 if x is rational and 0 if x is irrational. (See Fig. 202.) Then R consists of the two numbers 0 and 1.

Example 7. Let D be the set of all names of the residents of a certain municipality R and, for a given name x, let y correspond to x if and only if y is a resident of R that possesses the name x. If the municipality is small, this function is likely to be single-valued. However, if the municipality is large, it is very improbable that this function is single-valued *for all x*.

A type of function of particular importance in mathematics is specified in the following definition.

Definition III. *An **infinite sequence** is a (single-valued) function whose domain of definition is the positive integers.*

This means that corresponding to any positive integer there is a unique value or **term** determined. In particular, there is a first term a_1 corresponding to the number 1, a second term a_2 corresponding to the number 2, etc. An infinite sequence can thus be represented:

$$a_1, a_2, \cdots, a_n, \cdots, \quad \text{or} \quad \{a_n\}.$$

The nth term, a_n, is sometimes called the **general term** of the infinite sequence. Since it is a function of n, $(a_n = f(n))$, it must be prescribed by some rule. If the terms are numbers, this rule may sometimes be expressed as a simple algebraic formula. Such a formula may be impossible to find, but a definite rule must exist.

NOTE 3. Frequently an infinite sequence is indicated by an explicit listing of only the first few terms, in case the general rule for procedure is clear *beyond reasonable doubt*. For instance, in part (c) of the following Example 8 the rule that is clearly implied by alternating 1's and 0's for the first six terms is alternating 1's and 0's for all terms, although the ingenious artificer could construct any number of infinite sequences that start with alternating 1's and 0's (the terms could continue by being identically 0, or with alternating 6's and 7's, for example). Such interpretations, we hold, are not only unnatural, but deliberately mischievous.

NOTE 4. For convenience, if the meaning is clear, the single word *sequence* will be used henceforth to mean *infinite sequence*.

Example 8. Give a rule for obtaining the general term for each of the following sequences:

(a) $\frac{1}{2}, -\frac{2}{5}, \frac{3}{8}, -\frac{4}{11}, \cdots$;

(b) $1, \frac{1}{1}, \frac{1}{2}, \frac{1}{3}, \cdots$;

(c) $1, 0, 1, 0, 1, 0, \cdots$;

(d) $1, 2, 3, 1, 2, 3, 1, 2, 3, \cdots$;

(e) $\frac{1}{2}, \frac{1}{3}, \frac{1}{4}, \frac{1}{9}, \frac{1}{8}, \frac{1}{27}, \cdots$.

Solution. (a) The factor $(-1)^n$ or $(-1)^{n+1}$ is a standard device to take care of alternating signs. The general term is $(-1)^{n+1} \dfrac{n}{3n-1}$. (b) If $n = 1$, $a_n = 1$; if $n > 1$, $a_n = \dfrac{1}{n-1}$. (c) First formulation: if n is odd, $a_n = 1$; if n is even, $a_n = 0$. Second formulation: $a_{2n-1} = 1$; $a_{2n} = 0$. Third formulation: $a_n = \frac{1}{2}[(-1)^{n+1} + 1]$. (d) $a_{3n-2} = 1$; $a_{3n-1} = 2$; $a_{3n} = 3$. (e) $a_{2n-1} = \dfrac{1}{2^n}$; $a_{2n} = \dfrac{1}{3^n}$.

202. LIMIT OF A SEQUENCE

A sequence is said to **tend toward**, or **converge to**, a number if and only if the absolute value of the difference between the general term of the sequence and this number is less than any preassigned positive number (however small) whenever the subscript n of the general term is sufficiently large.

Symbolically, this is written

$$\lim_{n \to +\infty} a_n = a \quad \text{or} \quad \lim_{n \to \infty} a_n = a \quad \text{or} \quad a_n \to a,$$

where a_n is the nth term of the sequence and a is the number to which it converges. If $\{a_n\}$ converges to a, a is called the **limit** of the sequence.

A more concise form of the definition given above is the following:

Definition I. *The sequence $\{a_n\}$ has the **limit** a, written $\lim_{n \to +\infty} a_n = a$, if and only if corresponding to an arbitrary positive number ϵ there exists a number $N = N(\epsilon)$ such that $|a_n - a| < \epsilon$ whenever $n > N$.*

NOTE 1. In conformity with the discussion following Definition III, § 110, the statement that $\{a_n\}$ converges to a is equivalent to the statement that every neighborhood of a contains all of the terms of $\{a_n\}$ *from some point on*, and is also equivalent to the statement that every neighborhood of a contains all but a finite number of the terms of $\{a_n\}$ (that is, all of the terms except for a finite number of the subscripts).

If a sequence converges to some number, the sequence is said to be **convergent**; otherwise it is **divergent**.

The concept of an *infinite limit* is important, and will be formulated in precise symbolic form. As an exercise, the student should reformulate the following definition in his own words, without the use of mathematical symbols.

Definition II. *The sequence $\{a_n\}$ has the limit $+\infty$, written*

$$\lim_{n \to +\infty} a_n = +\infty, \quad \text{or} \quad a_n \to +\infty,$$

if and only if corresponding to an arbitrary number B (however large) there exists a number $N = N(B)$ such that $a_n > B$ whenever $n > N$; the sequence $\{a_n\}$ has the limit $-\infty$, written

$$\lim_{n \to +\infty} a_n = -\infty, \quad \text{or} \quad a_n \to -\infty,$$

if and only if corresponding to an arbitrary number B (however large its negative) there exists a number $N = N(B)$ such that $a_n < B$ whenever $n > N$; the sequence has the limit ∞ (unsigned infinity), written

$$\lim_{n \to +\infty} a_n = \infty, \quad \text{or} \quad a_n \to \infty,$$

if and only if $\lim_{n \to +\infty} |a_n| = +\infty$.

NOTE 2. Although the word *limit* is applied to both the finite and infinite cases, the word *converge* is used only for finite limits. Thus, a sequence tending toward $+\infty$ diverges.

NOTE 3. In any extensive treatment of limits there are numerous statements which can be interpreted to apply to both finite and infinite cases, and which are of such a nature that the proofs for the finite and infinite particularizations are in essence identical. In such instances these proofs can be combined into a single proof by appropriate extensions of the word *neighborhood*. We define "neighborhoods of infinity" as follows: (*i*) a **neighborhood of** $+\infty$ is an open interval of the form $(a, +\infty)$; (*ii*) a **neighborhood of** $-\infty$ is an open interval of the form $(-\infty, b)$; (*iii*) a **neighborhood of** ∞ is the set of all x satisfying an inequality of the form $|x| > a$. With these conventions, for example, all cases of Definitions I and II can be included in the following single formulation for $\lim\limits_{n \to +\infty} a_n = a$ (where a may be a number, or $+\infty$, $-\infty$, or ∞): *Corresponding to every neighborhood N_a of a there exists a neighborhood $N_{+\infty}$ of $+\infty$ such that whenever n belongs to $N_{+\infty}$, a_n belongs to N_a.* In the sequel we formulate theorems and proofs separately for the finite and infinite forms, but suggest that the student interested in exploring the simplifying techniques available with general neighborhoods try his hand at combining the separate formulations into unified ones. A word of warning is in order, however: Do not confuse infinite symbols with numbers, and write such nonsense as $|a_n - \infty| < \epsilon$ when dealing with an infinite limit! It is to avoid such possible confusion of ideas that we have adopted the policy of maintaining (in the main) the separation of the finite and infinite.

Definition III. *A **subsequence** of a sequence is a sequence whose terms are terms of the original sequence arranged in the same order. That is, a subsequence of a sequence $\{a_n\}$ has the form $a_{n_1}, a_{n_2}, a_{n_3}, \cdots$, where*

$$n_1 < n_2 < n_3 < \cdots.$$

It is denoted by $\{a_{n_k}\}$.

Example 1. The sequence $\frac{1}{2}, \frac{1}{4}, \frac{1}{8}, \cdots$ is a subsequence of the sequence $\frac{1}{2}, \frac{1}{3}, \frac{1}{4}, \frac{1}{9}, \frac{1}{8}, \cdots$ of Example 8, § 201. The sequence 0, 1, 0, 1, 0, 1, $\cdots$ is a subsequence of the sequence 1, 0, 1, 0, 0, 1, 0, 0, 0, 1, $\cdots$.

Example 2. Show that the following sequences converge to 0:

(a) $1, \frac{1}{2}, \frac{1}{3}, \cdots, \frac{1}{n}, \cdots$; (b) $\frac{1}{2}, \frac{1}{4}, \frac{1}{8}, \cdots, \frac{1}{2^n}, \cdots$; (c) $\frac{1}{2}, -\frac{1}{4}, \frac{1}{8}, \cdots, \frac{(-1)^{n+1}}{2^n}, \cdots$.

Solution. (a) Since $|a_n - a| = \left|\frac{1}{n} - 0\right| = \frac{1}{n}$, and since $\frac{1}{n} < \epsilon$ whenever $n > 1/\epsilon$, we can choose as the function $N(\epsilon)$ of Definition I the expression $1/\epsilon$. (Cf. § 113.) (b) By Ex. 11, § 107, $2^n > n$ for all positive integers, so that we can choose $N(\epsilon) = 1/\epsilon$. (c) This reduces immediately to (b).

Example 3. Find the limit of each sequence: (a) $\frac{1}{2}, \frac{3}{4}, \frac{7}{8}, \cdots, 1 - \frac{1}{2^n}, \cdots$. (b) 3, 3, 3, $\cdots$, 3, $\cdots$; (c) $1, \frac{1}{2}, 1, \frac{3}{4}, 1, \frac{7}{8}, \cdots$.

Solution. (a) The expression $|a_n - 1|$ is equal to $\frac{1}{2^n}$, which is less than any preassigned positive number whenever n is sufficiently large, as shown in Example 2, (b).

Therefore the limit is 1. (b) The absolute value of the difference between the general term and 3 is identically zero, which is less than any preassigned positive number for *any* n, and certainly for n sufficiently large. Therefore the limit is 3. (c) By combining the reasoning in parts (a) and (b) we see that the general term differs numerically from 1 by less than any preassigned positive number if n is sufficiently large. The odd-numbered terms form a subsequence identically 1, while the even-numbered terms form a subsequence which is the sequence of part (a). The limit is 1.

Example 4. Show that each of the following sequences diverges:
(a) $1, 2, 1, 2, 1, 2, \cdots$;   (b) $1, 2, 4, 8, 16, \cdots$;
(c) $1, 2, 1, 3, 1, 4, \cdots$;   (d) $1, -2, 4, -8, 16, \cdots$.

Solution. (a) If $\{a_n\}$ converges to a, *every* neighborhood of a must contain all terms from some point on, and therefore must contain both numbers 1 and 2. On the other hand, no matter what value a may have, a neighborhood of a of length less than 1 cannot contain both of these points! (b) No finite interval about any point can contain all terms of this sequence, from some point on. The limit is $+\infty$. (c) The comment of part (b) applies to this sequence, since there is a subsequence tending toward $+\infty$. This sequence has no limit, finite or infinite. (d) The subsequence of the odd-numbered terms tends toward $+\infty$, and that of the even-numbered terms tends toward $-\infty$. The sequence of absolute values tends toward $+\infty$, so that the sequence itself has the limit ∞.

203. EXERCISES

In Exercises 1–10, draw the graph of the given function, assuming the domain of definition to be as large as possible. Give in each case the domain and the range of values. The bracket function $[x]$ is defined in Example 5, § 201, and square roots are discussed in § 214. (Also cf. Exs. 5–10, § 216.)

1. $y = \sqrt{x^2 - 9}$.
2. $y = \pm\sqrt{25 - x^2}$.
3. $y = \sqrt{-x}$.
4. $y = \pm\sqrt{|x|}$.
5. $y = \sqrt{4x - x^2}$.
6. $y = \sqrt{|x^2 - 16|}$.
7. $y = x - [x]$.
8. $y = (x - [x])^2$.
9. $y = \sqrt{x - [x]}$.
10. $y = [x] + \sqrt{x - [x]}$.

In Exercises 11–18, give a rule for finding the general term of the sequence.

11. $2, \frac{4}{3}, \frac{6}{5}, \frac{8}{7}, \cdots$.
12. $\frac{1}{3}, -\frac{1}{6}, \frac{1}{11}, -\frac{1}{18}, \frac{1}{27}, \cdots$.
13. $1, -1, \frac{1}{2}, -\frac{1}{6}, \frac{1}{24}, -\frac{1}{120}, \cdots$.
14. $1, 2, 24, 720, 40320, \cdots$.
15. $1 \cdot 3, 1 \cdot 3 \cdot 5, 1 \cdot 3 \cdot 5 \cdot 7, 1 \cdot 3 \cdot 5 \cdot 7 \cdot 9, \cdots$.
16. $1, 2, 3, 2, 1, 2, 3, 2, 1, \cdots$.
17. $-1, 1, 1, -2, 2, 2, -3, 3, 3, -4, 4, 4, \cdots$.
18. $1, 2 \cdot 4, 1 \cdot 3 \cdot 5, 2 \cdot 4 \cdot 6 \cdot 8, 1 \cdot 3 \cdot 5 \cdot 7 \cdot 9, \cdots$.

In Exercises 19–24, find the limit of the sequence and justify your contention (cf. Exs. 25–30).

19. $2, 2, 2, 2, 2, \cdots$.
20. $\frac{3}{2}, \frac{5}{4}, \frac{7}{6}, \cdots, \frac{2n+1}{2n}, \cdots$.
21. $\frac{3}{5}, \frac{3}{7}, \frac{5}{9}, \frac{5}{11}, \frac{7}{13}, \frac{7}{15}, \cdots$.
22. $1, 4, 9, 16, \cdots, n^2, \cdots$.
23. $\frac{3}{7}, -\frac{8}{7}, \frac{13}{7}, -\frac{18}{7}, \cdots$.
24. $9, 16, 21, 24, \cdots, 10n - n^2, \cdots$.

§ 204] LIMIT THEOREMS FOR SEQUENCES

In Exercises 25–30, give a simple explicit function $N(\epsilon)$ or $N(B)$, in accord with Definition I or II, for the sequence of the indicated Exercise.
★25. For Ex. 19.   ★26. For Ex. 20.   ★27. For Ex. 21.
★28. For Ex. 22.   ★29. For Ex. 23.   ★30. For Ex. 24.

In Exercises 31–34, prove that the given sequence has no limit, finite or infinite.
31. $1, 5, 1, 5, 1, 5, \cdots$,   32. $1, 2, 3, 1, 2, 3, 1, 2, 3, \cdots$.
33. $1, 2, 1, 4, 1, 8, 1, 16, \cdots$.   34. $2^1, 2^{-2}, 2^3, 2^{-4}, 2^5, 2^{-6}, \cdots$.

204. LIMIT THEOREMS FOR SEQUENCES

Theorem I. *The alteration of a finite number of terms of a sequence has no effect on convergence or divergence or limit. In other words, if $\{a_n\}$ and $\{b_n\}$ are two sequences and if M and N are two positive integers such that $a_{M+n} = b_{N+n}$ for all positive integers n, then the two sequences $\{a_n\}$ and $\{b_n\}$ must either both converge to the same limit or both diverge; in case of divergence either both have the same infinite limit or neither has an infinite limit.*

Proof. If $\{a_n\}$ converges to a, then every neighborhood of a contains all but a finite number of the terms of $\{a_n\}$, and therefore all but a finite number of the terms of $\{b_n\}$. Proof for the case of an infinite limit is similar.

Theorem II. *If a sequence converges, its limit is unique.*

Proof. Assume $a_n \to a$ and $a_n \to a'$, where $a \neq a'$. Take neighborhoods of a and a' so small that they have no points in common. Then each must contain all but a finite number of the terms of $\{a_n\}$. This is clearly impossible.

Theorem III. *If all terms of a sequence, from some point on, are equal to a constant, the sequence converges to this constant.*

Proof. Any neighborhood of the constant contains the constant and therefore all but a finite number of the terms of the sequence.

Theorem IV. *Any subsequence of a convergent sequence converges, and its limit is the limit of the original sequence.* (Cf. Ex. 12, § 205.)

Proof. Assume $a_n \to a$. Since every neighborhood of a contains all but a finite number of terms of $\{a_n\}$ it must contain all but a finite number of terms of any subsequence.

Definition I. *A sequence is **bounded** if and only if all of its terms are contained in some finite interval. Equivalently, the sequence $\{a_n\}$ is bounded if and only if there exists a positive number P such that $|a_n| \leq P$ for all n.*

Theorem V. *Any convergent sequence is bounded.* (Cf. Ex. 2, § 205.)

Proof. Assume $a_n \to a$, and choose a definite neighborhood of a, say the open interval $(a - 1, a + 1)$. Since this neighborhood contains all but a finite number of terms of $\{a_n\}$, a suitable enlargement will contain these missing terms as well.

FUNCTIONS, SEQUENCES, LIMITS, CONTINUITY [§ 204

Definition II. *If $\{a_n\}$ and $\{b_n\}$ are two sequences, the sequences $\{a_n + b_n\}$, $\{a_n - b_n\}$, and $\{a_n b_n\}$ are called their* **sum, difference,** *and* **product,** *respectively. If $\{a_n\}$ and $\{b_n\}$ are two sequences, where b_n is never zero, the sequence $\{a_n/b_n\}$ is called their* **quotient.** *The definitions of sum and product extend to any finite number of sequences.*

Theorem VI. *The sum of two convergent sequences is a convergent sequence, and the limit of the sum is the sum of the limits:*
$$\lim_{n \to +\infty} (a_n + b_n) = \lim_{n \to +\infty} a_n + \lim_{n \to +\infty} b_n.$$
(Cf. Ex. 4, § 205.) *This rule extends to the sum of any finite number of sequences.*

Proof. Assume $a_n \to a$ and $b_n \to b$, and let $\epsilon > 0$ be given. Choose N so large that the following two inequalities hold *simultaneously* for $n > N$:
$$|a_n - a| < \tfrac{1}{2}\epsilon, \ |b_n - b| < \tfrac{1}{2}\epsilon.$$
Then, by the triangle inequality, for $n > N$
$$|(a_n + b_n) - (a + b)| = |(a_n - a) + (b_n - b)|$$
$$\leq |a_n - a| + |b_n - b| < \tfrac{1}{2}\epsilon + \tfrac{1}{2}\epsilon = \epsilon.$$
The extension to the sum of an arbitrary number of sequences is provided by mathematical induction. (Cf. Ex. 3, § 205.)

Theorem VII. *The difference of two convergent sequences is a convergent sequence, and the limit of the difference is the difference of the limits:*
$$\lim_{n \to +\infty} (a_n - b_n) = \lim_{n \to +\infty} a_n - \lim_{n \to +\infty} b_n.$$
Proof. The details are almost identical with those of the preceding proof. (Cf. Ex. 6, § 205.)

Theorem VIII. *The product of two convergent sequences is a convergent sequence and the limit of the product is the product of the limits:*
$$\lim_{n \to +\infty} (a_n b_n) = \lim_{n \to +\infty} a_n \cdot \lim_{n \to +\infty} b_n.$$
(Cf. Ex. 5, § 205.) *This rule extends to the product of any finite number of sequences.*

Proof. Assume $a_n \to a$ and $b_n \to b$. We wish to show that $a_n b_n \to ab$ or, equivalently, that $a_n b_n - ab \to 0$. By addition and subtraction of the quantity ab_n and by appeal to Theorem VI, we can use the relation
$$a_n b_n - ab = (a_n - a)b_n + a(b_n - b)$$
to reduce the problem to that of showing that both sequences $\{(a_n - a)b_n\}$ and $\{a(b_n - b)\}$ converge to zero. The fact that they do is a consequence of the following lemma:

Lemma. *If $\{c_n\}$ converges to 0 and $\{d_n\}$ converges, then $\{c_n d_n\}$ converges to 0.*

Proof of lemma. By Theorem V the sequence $\{d_n\}$ is bounded, and there exists a positive number P such that $|d_n| \leq P$ for all n. If $\epsilon > 0$ is given, choose N so large that $|c_n| < \epsilon/P$ for $n > N$. Then for $n > N$,

$$|c_n d_n| = |c_n| \cdot |d_n| < (\epsilon/P) \cdot P = \epsilon.$$

This inequality completes the proof of the lemma, and hence of the theorem.

The extension to the product of an arbitrary number of sequences is provided by mathematical induction. (Cf. Ex. 3, § 205.)

Theorem IX. *The quotient of two convergent sequences, where the denominators and their limit are nonzero, is a convergent sequence and the limit of the quotient is the quotient of the limits:*

$$\lim_{n \to +\infty} \frac{a_n}{b_n} = \frac{\lim_{n \to +\infty} a_n}{\lim_{n \to +\infty} b_n}.$$

Proof. Assume $a_n \to a$, $b_n \to b$, and that b and b_n are nonzero for all n. Inasmuch as $a_n/b_n = (a_n) \cdot (1/b_n)$, Theorem VIII permits the reduction of this proof to that of showing that $1/b_n \to 1/b$ or, equivalently, that

$$\frac{1}{b_n} - \frac{1}{b} = \frac{b - b_n}{b} \cdot \frac{1}{b_n} \to 0.$$

Let $c_n \equiv (b - b_n)/b$ and $d_n \equiv 1/b_n$ and observe that the conclusion of the Lemma of Theorem VIII is valid (with no change in the proof) when the sequence $\{d_n\}$ is assumed to be merely bounded (instead of convergent). Since the sequence $\{c_n\} = \{(b - b_n) \cdot (1/b)\}$ converges to zero (by this same lemma), we have only to show that the sequence $\{d_n\} = \{1/b_n\}$ is bounded. We proceed now to prove this fact. Since $b \neq 0$, we can choose neighborhoods of 0 and b which have no points in common. Since $b_n \to b$, the neighborhood of b contains all but a finite number of the terms of $\{b_n\}$, so that only a finite number of these terms can lie in the neighborhood of 0. Since b_n is nonzero for all n, there is a (smaller) neighborhood of 0 that excludes *all* terms of the sequence $\{b_n\}$. If this neighborhood is the open interval $(-\epsilon, \epsilon)$, where $\epsilon > 0$, then for all n, $|b_n| \geq \epsilon$, or $|d_n| = |1/b_n| \leq 1/\epsilon$. The sequence $\{d_n\}$ is therefore bounded, and the proof is complete.

Theorem X. *Multiplication of the terms of a sequence by a nonzero constant k does not affect convergence or divergence. If the original sequence converges, the new sequence converges to k times the limit of the original, for any constant k:*

$$\lim_{n \to +\infty} (k a_n) = k \cdot \lim_{n \to +\infty} a_n.$$

Proof. This is a consequence of Theorems III and VIII.

Theorem XI. *If $\{a_n\}$ is a sequence of nonzero numbers, then $a_n \to \infty$ if and only if $1/a_n \to 0$; equivalently, $a_n \to 0$ if and only if $1/a_n \to \infty$.*

Proof. If $|a_n| \to +\infty$ and if $\epsilon > 0$ is given, there exists a number N such that for $n > N$, $|a_n| > 1/\epsilon$, and therefore $|1/a_n| < \epsilon$. Conversely, if $1/a_n \to 0$ and B is any given number, there exists a number N such that for $n > N$, $|1/a_n| < 1/(|B| + 1)$ and therefore $|a_n| > |B| + 1 > |B| \geq B$.

Theorem XII. *If $a > 1$, $\lim\limits_{n \to +\infty} a^n = +\infty$.*

Proof. Let $p \equiv a - 1 > 0$. Then $a = 1 + p$, and by the Binomial Theorem (cf. Ex. 25, § 107), if n is a positive integer,

$$a^n = (1 + p)^n = 1 + np + \frac{n(n-1)}{2} p^2 + \cdots \geq 1 + np.$$

Therefore, if B is a given number and if $n > |B|/p$, then

$$a^n \geq 1 + np > 1 + |B| > B.$$

Theorem XIII. *If $|r| < 1$, $\lim\limits_{n \to +\infty} r^n = 0$.*

Proof. This is a consequence of the two preceding theorems.

Definition III. *A sequence $\{a_n\}$ is **monotonically increasing** (**decreasing**),† written $a_n \uparrow$ ($a_n \downarrow$), if and only if $a_n \leq a_{n+1}$ ($a_n \geq a_{n+1}$) for every n. A sequence is **monotonic** if and only if it is monotonically increasing or monotonically decreasing.*

Theorem XIV. *Any bounded monotonic sequence converges. If $a_n \uparrow$ ($a_n \downarrow$) and if $a_n \leq P$ ($a_n \geq P$) for all n, then $\{a_n\}$ converges; moreover, if $a_n \to a$, then $a_n \leq a \leq P$ ($a_n \geq a \geq P$) for all n.*

Proof. We give the details only for the case $a_n \uparrow$ (cf. Ex. 7, § 205). Since the set A of points consisting of the terms of the sequence $\{a_n\}$ is bounded above, it has a least upper bound a (§ 112), and since P is an upper bound of A, the following inequalities must hold for all n: $a_n \leq a \leq P$. To prove that $a_n \to a$ we let ϵ be a given positive number and observe that there must exist a positive integer N such that $a_N > a - \epsilon$ (cf. Ex. 13, § 114). Therefore, for $n > N$, the following inequalities hold:

$$a - \epsilon < a_N \leq a_n \leq a < a + \epsilon.$$

Consequently $a - \epsilon < a_n < a + \epsilon$, or $|a_n - a| < \epsilon$, and the proof is complete.

NOTE. As a consequence of Theorem XIV we can say in general that *any monotonic sequence has a limit* (finite, $+\infty$, or $-\infty$), and that the limit is finite if and only if the sequence is bounded. (The student should give the details in Ex. 7, § 205.)

---

† Parentheses are used here to indicate an alternative statement. For a discussion of the use of parentheses for alternatives, see the Preface.

Theorem XV. If $a_n \leq b_n$ for all n, and if $\lim\limits_{n \to +\infty} a_n$ and $\lim\limits_{n \to +\infty} b_n$ exist (finite, $+\infty$, or $-\infty$), then $\lim\limits_{n \to +\infty} a_n \leq \lim\limits_{n \to +\infty} b_n$.

Proof. If the two limits are finite we can form the difference
$$c_n \equiv b_n - a_n$$
and, by appealing to Theorem VII, reduce the problem to the special case: if $c_n \geq 0$ for all n and if $C = \lim\limits_{n \to +\infty} c_n$ exists and is finite then $C \geq 0$. By the definition of a limit, for any positive ϵ we can find values of n (arbitrarily large) such that $|c_n - C| < \epsilon$. Now if $C < 0$, let us choose $\epsilon \equiv -C > 0$. We can then find arbitrarily large values of n such that $|c_n - C| = |c_n + \epsilon| < \epsilon$, and hence $c_n + \epsilon < \epsilon$. This contradicts the nonnegativeness of c_n. On the other hand, if in the general theorem it is assumed that $a_n \to +\infty$ and $b_n \to B$ (finite), we may take $\epsilon \equiv 1$ and find first an N_1 such that $n > N_1$ implies $a_n > B + 1$, and then an N_2 such that $n > N_2$ implies $b_n < B + 1$. Again the inequality $a_n \leq b_n$ is contradicted (for n greater than both N_1 and N_2). The student should complete the proof for the cases $a_n \to A$ (finite), $b_n \to -\infty$ and $a_n \to +\infty$, $b_n \to -\infty$.

205. EXERCISES

1. Prove that if two subsequences of a given sequence converge to distinct limits, the sequence diverges.
2. Show by a counterexample that the converse of Theorem V, § 204, is false. That is, a bounded sequence need not converge.
3. Prove the extensions of Theorems VI and VIII, § 204, to an arbitrary finite number of sequences.
4. Prove that if $a_n \to +\infty$ and either $\{b_n\}$ converges or $b_n \to +\infty$, then $a_n + b_n \to +\infty$.
5. Prove that if $a_n \to +\infty$ and either $b_n \to b > 0$ or $b_n \to +\infty$, then $a_n b_n \to +\infty$. Prove that if $a_n \to \infty$ and $b_n \to b \neq 0$, then $a_n b_n \to \infty$.
6. Prove Theorem VII, § 204.
7. Prove Theorem XIV, § 204, for the case $a_n \downarrow$, and the statement of the Note that follows. *Hint:* Let $b_n \equiv -a_n$ and use Theorem XIV, § 204, for the case $b_n \uparrow$. Cf. Ex. 16.
8. Show by counterexamples that the sum (difference, product, quotient) of two divergent sequences need not diverge.
9. Prove that if the sum and the difference of two sequences converge, then both of the sequences converge.
10. Prove that $a_n \to 0$ if and only if $|a_n| \to 0$.
11. Prove that $a_n \to a$ implies $|a_n| \to |a|$. Is the converse true? Prove, or give a counterexample. *Hint:* Use Property IX, § 110.
12. Prove that if a sequence has the limit $+\infty$ ($-\infty$, ∞) then any subsequence has the limit $+\infty$ ($-\infty$, ∞).
13. Prove that if $a_n \leq b$ ($a_n \geq b$) and $a_n \to a$, then $a \leq b$ ($a \geq b$). Show by an example that from the strict inequality $a_n < b$ ($a_n > b$) we cannot infer the strict inequality $a < b$ ($a > b$).

14. Prove that if $0 \leq a_n \leq b_n$ and $b_n \to 0$, then $a_n \to 0$. More generally, prove that if $a_n \leq b_n \leq c_n$ and $\{a_n\}$ and $\{c_n\}$ converge to the same limit, then $\{b_n\}$ also converges to this same limit.

15. Prove that if x is an arbitrary real number, there is a sequence $\{r_n\}$ of rational numbers converging to x. *Hint:* By the density of the rationals (Theorem III, § 113), there is a rational number r_n in the open interval $\left(x - \dfrac{1}{n}, x + \dfrac{1}{n}\right)$.

*16. Prove Theorem XIV, § 204, for the case $a_n \downarrow$, directly (without reference to the case $a_n \uparrow$), using the principle of *greatest lower bound*.

206. LIMITS OF FUNCTIONS

In this section we recall and extend some of the basic limit concepts of elementary calculus. Before formalizing the appropriate definitions for such limits as $\lim\limits_{x \to a} f(x)$ and $\lim\limits_{x \to +\infty} f(x)$, let us agree on one thing: Whenever a limit of a function is concerned, it will be implicitly assumed that *the quantities symbolized exist for at least some values of the independent variable neighboring the limiting value of that variable*. For example, when we write $\lim\limits_{x \to a} f(x)$ we shall assume that every neighborhood of the point a contains at least one point x different from a for which the function $f(x)$ is defined;† and when we write $\lim\limits_{x \to +\infty} f(x)$ we shall assume that for any number N, $f(x)$ exists for some $x > N$.

A function $f(x)$ is said to **tend toward** or **approach** or **have** a limit L as x approaches a number a if and only if the absolute value of the difference between $f(x)$ and L is less than any preassigned positive number (however small) whenever the point x belonging to the domain of definition of $f(x)$ is sufficiently near a but not equal to a. This is expressed symbolically:

$$\lim_{x \to a} f(x) = L.$$

If in this definition the independent variable x is restricted to values greater than a, we say that x approaches a from the **right** or from **above** and write

$$\lim_{x \to a+} f(x) = L.$$

Again, if x is restricted to values less than a, we say that x approaches a from the **left** or from **below** and write

$$\lim_{x \to a-} f(x) = L.$$

The terms *undirected limit* or *two-sided limit* may be used to distinguish the first of these three limits from the other two in case of ambiguity arising from use of the single word *limit*.

---

† In the terminology of Chapter 6, a is a *limit point* of the domain of definition D of $f(x)$. It can be shown that every neighborhood of a contains *infinitely many* points of D.

A more concise formulation for these limits is given in the following definition:

Definition I. *The function $f(x)$ has the limit L as x approaches a, written*

$$\lim_{x \to a} f(x) = L, \quad \text{or} \quad f(x) \to L \quad \text{as} \quad x \to a,$$

if and only if corresponding to an arbitrary positive number ϵ there exists a positive number $\delta = \delta(\epsilon)$ such that $0 < |x - a| < \delta$ implies $|f(x) - L| < \epsilon$, for values of x for which $f(x)$ is defined;† *$f(x)$ has the limit L as x approaches a from the right (left),*‡ *written*

$$\lim_{x \to a+} f(x) = L, \quad \text{or} \quad f(x) \to L \quad \text{as} \quad x \to a+$$

$$(\lim_{x \to a-} f(x) = L, \quad \text{or} \quad f(x) \to L \quad \text{as} \quad x \to a-),$$

if and only if corresponding to an arbitrary positive number ϵ there exists a positive number $\delta = \delta(\epsilon)$ such that $a < x < a + \delta$ ($a - \delta < x < a$) implies $|f(x) - L| < \epsilon$, for values of x for which $f(x)$ is defined. These one-sided limits (if they exist) are also denoted:

$$f(a+) \equiv \lim_{x \to a+} f(x), \quad f(a-) \equiv \lim_{x \to a-} f(x).$$

Since the definition of limit employs only values of x different from a, it is completely immaterial what the value of the function is at $x = a$ or, indeed, whether it is defined there at all. Thus a function can fail to have a limit as x approaches a only by its misbehavior for values of x near a but not equal to a. Since $\lim_{x \to a} f(x)$ exists if and only if $\lim_{x \to a+} f(x)$ and $\lim_{x \to a-} f(x)$ both exist and are equal (cf. Exs. 13–14, § 208), $\lim_{x \to a} f(x)$ may fail to exist either by $\lim_{x \to a+} f(x)$ and $\lim_{x \to a-} f(x)$ being unequal or by either or both of the latter failing to exist in one way or another. These possibilities are illustrated in Example 1 below.

NOTE. An almost immediate consequence of the preceding definition is the fact that whenever the limit of a function is positive (negative) the function itself must be positive (negative) throughout some deleted neighborhood of the limiting value of the independent variable. For a slightly stronger statement, together with a hint for a proof, see Exercise 9, § 208.

Limits as the independent variable becomes infinite have a similar formulation:

---

† An open interval with the midpoint removed is called a **deleted neighborhood** of the missing point. The inequalities $0 < |x - a| < \delta$, then, define a deleted neighborhood of the point a.

‡ Parentheses are used here and in the following two definitions to indicate an alternative statement. For a discussion of the use of parentheses for alternatives, see the Preface.

Definition II. *The function $f(x)$ has the limit L as x becomes positively (negatively) infinite, written*

$$f(+\infty) \equiv \lim_{x \to +\infty} f(x) = L, \quad \text{or} \quad f(x) \to L \text{ as } x \to +\infty$$

$$(f(-\infty) \equiv \lim_{x \to -\infty} f(x) = L, \quad \text{or} \quad f(x) \to L \text{ as } x \to -\infty),$$

if and only if corresponding to an arbitrary positive number ϵ there exists a number $N = N(\epsilon)$ such that $x > N$ ($x < N$) implies $|f(x) - L| < \epsilon$, for values of x for which $f(x)$ is defined.

In an analogous fashion, infinite limits can be defined. Only a sample definition is given here, others being requested in the Exercises of § 208.

Definition III. *The function $f(x)$ has the limit $+\infty$ ($-\infty$) as x approaches a, written*

$$\lim_{x \to a} f(x) = +\infty \, (-\infty), \quad \text{or} \quad f(x) \to +\infty \, (-\infty) \text{ as } x \to a,$$

if and only if corresponding to an arbitrary number B there exists a positive number $\delta = \delta(B)$ such that $0 < |x - a| < \delta$ implies $f(x) > B$ ($f(x) < B$), for values of x for which $f(x)$ is defined.

As with limits of sequences it is often convenient to use an *unsigned infinity*, ∞. When we say that a variable, dependent or independent, tends toward ∞, we shall mean that its absolute value approaches $+\infty$. Thus $\lim_{x \to \infty} f(x) = L$ is defined as in Definition II, with the inequality $x > N$ replaced by $|x| > N$, and $\lim_{x \to a} f(x) = \infty$ is equivalent to $\lim_{x \to a} |f(x)| = +\infty$.

Example 1. Discuss the limits of each of the following functions as x approaches 0, $0+$, and $0-$, and in each case sketch the graph: (a) $f(x) \equiv \dfrac{x^2 + x}{x}$ if $x \neq 0$, undefined for $x = 0$; (b) $f(x) \equiv |x|$ if $x \neq 0$, $f(0) \equiv 3$; (c) the **signum function**, $f(x) \equiv \operatorname{sgn} x \equiv 1$ if $x > 0$, $f(x) \equiv \operatorname{sgn} x \equiv -1$ if $x < 0$, $f(0) \equiv \operatorname{sgn} 0 \equiv 0$; (d)† $f(x) \equiv \sin \dfrac{1}{x}$ if $x \neq 0$, $f(0) \equiv 0$; (e) $f(x) \equiv \dfrac{1}{x}$ if $x \neq 0$, undefined if $x = 0$; (f) $f(x) \equiv \dfrac{1}{x^2}$ if $x \neq 0$, undefined if $x = 0$.

Solution. The graphs are given in Figure 203. In part (a) if $x \neq 0$, $f(x)$ is identically equal to the function $x + 1$, and its graph is therefore the straight line $y = x + 1$ with the single point $(0, 1)$ deleted; $\lim_{x \to 0} f(x) = f(0+) = f(0-) = 1$.

---

† For illustrative examples and exercises the familiar properties of the trigonometric functions will be assumed. An analytic treatment is given in §§ 503–504.

§ 206] LIMITS OF FUNCTIONS 37

In part (b) $\lim_{x \to 0} f(x) = f(0+) = f(0-) = 0$. The fact that $f(0) = 3$ has no bearing on the statement of the preceding sentence. For the signum function (c), $f(0+) = 1, f(0-) = -1$, and $\lim_{x \to 0} f(x)$ does not exist. In part (d) all three limits fail to exist. In part (e) $f(0+) = +\infty$, $f(0-) = -\infty$, and $\lim_{x \to 0} f(x) = \infty$ (unsigned infinity) (cf. Exs. 31–32, § 208). In part (f), $f(0+) = f(0-) = \lim_{x \to 0} f(x) = +\infty$ (cf. Ex. 32, § 208).

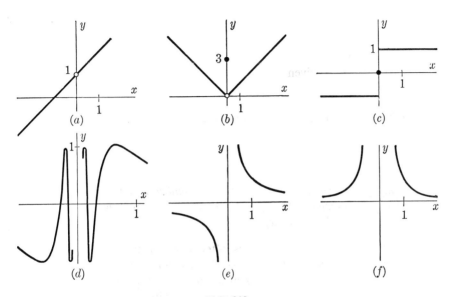

FIG. 203

*Example 2. Show that $\lim_{x \to 2} \dfrac{x^2 - x + 18}{3x - 1} = 4$. Find an explicit function $\delta(\epsilon)$ as demanded by Definition I.

Solution. Form the absolute value of the difference:

(1) $\quad \left| \dfrac{x^2 - x + 18}{3x - 1} - 4 \right| = \left| \dfrac{x^2 - 13x + 22}{3x - 1} \right| = |x - 2| \cdot \left| \dfrac{x - 11}{3x - 1} \right|.$

We wish to show that this expression is small if x is near 2. The first factor, $|x - 2|$, is certainly small if x is near 2; and the second factor, $\left| \dfrac{x - 11}{3x - 1} \right|$, is not dangerously large if x is near 2 and at the same time not too near $\tfrac{1}{3}$. Let us make this precise by first requiring that $\delta \leq 1$. If x is within a distance less than δ of 2, then $1 < x < 3$, and hence also $-10 < x - 11 < -8$ and $2 < 3x - 1 < 8$, so that $|x - 11| < 10$ and $|3x - 1| > 2$. Thus the second factor is less than $\tfrac{10}{2} = 5$. Now let a positive number ϵ be given. Since the expression (1) will be less than ϵ if simultaneously $|x - 2| < \dfrac{\epsilon}{5}$ and $\left| \dfrac{x - 11}{3x - 1} \right| < 5$, we have only to take $\delta = \delta(\epsilon)$ to be the smaller of the two numbers 1 and $\frac{\epsilon}{5}$, $\delta(\epsilon) \equiv \min\left(1, \frac{\epsilon}{5}\right)$. The graph of this function $\delta(\epsilon)$ is shown in Figure 204.

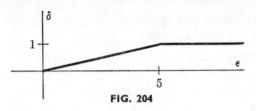

FIG. 204

207. LIMIT THEOREMS FOR FUNCTIONS

Many of the theorems of § 204 are special cases of more general limit theorems which apply to real-valued functions, whether the independent variable tends toward a (finite) number, toward a number from one side only, or toward $+\infty$, $-\infty$ or ∞. Some of these are stated below. Since in each case the statement is essentially the same regardless of the manner in which the independent variable approaches its limit,† this latter behavior is unspecified. Furthermore, since the proofs are mere reformulations of those given in § 204, only one sample is given here. Others are requested in the Exercises of § 208.

Theorem I. *If* $\lim f(x)$ *exists it is unique.*

Theorem II. *If the function $f(x)$ is equal to a constant k, then $\lim f(x)$ exists and is equal to k.*

Theorem III. *If $\lim f(x)$ and $\lim g(x)$ exist and are finite, then $\lim [f(x) + g(x)]$ exists and is finite, and*

$$\lim [f(x) + g(x)] = \lim f(x) + \lim g(x).$$

In short, the limit of the sum is the sum of the limits. This rule extends to the sum of any finite number of functions.

Theorem IV. *Under the hypotheses of Theorem III, the limit of the difference is the difference of the limits:*

$$\lim [f(x) - g(x)] = \lim f(x) - \lim g(x).$$

Theorem V. *Under the hypotheses of Theorem III, the limit of the product is the product of the limits:*

$$\lim [f(x)g(x)] = \lim f(x) \cdot \lim g(x).$$

This rule extends to the product of any finite number of functions.

---

† The word *limit* has been given specific meaning in this chapter only for functions, but it is convenient to extend its use to apply to *both* dependent and independent variables. The reader should recognize, however, that such an isolated statement as "the limit of the variable v is l" is meaningless, and takes on meaning only if v is associated with another variable, thus: $\lim_{x \to a} v(x) = l$, or $\lim_{v \to l} f(v) = k$.

Proof for the case $x \to a$. Assume $f(x) \to L$ and $g(x) \to M$ as $x \to a$. We wish to show that $f(x)g(x) \to LM$ or, equivalently, that
$$f(x)g(x) - LM \to 0, \quad \text{as} \quad x \to a.$$
By addition and subtraction of the quantity $L \cdot g(x)$ and by appeal to Theorem III, we can use the relation
$$f(x)g(x) - LM = [f(x) - L]g(x) + L[g(x) - M]$$
to reduce the problem to that of showing that $[f(x) - L]g(x) \to 0$ and $L[g(x) - M] \to 0$ as $x \to a$. The fact that they do is a consequence of the following lemma:

Lemma. *If* $\phi(x) \to 0$ *and* $\psi(x) \to \mu$ *as* $x \to a$, *then* $\phi(x)\psi(x) \to 0$ *as* $x \to a$.

Proof of lemma. First, by letting $\epsilon = 1$ in the definition of $\lim_{x \to a} \psi(x) = \mu$, we observe that there is a positive number δ_1 such that for x in the deleted neighborhood $0 < |x - a| < \delta_1$ the values of the function $\psi(x)$ lie in the neighborhood $(\mu - 1, \mu + 1)$, and are therefore bounded. Let P be a positive number such that $0 < |x - a| < \delta_1$ implies $|\psi(x)| \leq P$. Now let ϵ be an arbitrary positive number, and let δ be a positive number $\leq \delta_1$ such that $0 < |x - a| < \delta$ implies $|\phi(x)| < \epsilon/P$. Then $0 < |x - a| < \delta$ implies
$$|\phi(x)\psi(x)| < (\epsilon/P) \cdot P = \epsilon,$$
and the proof of the theorem is complete.

Theorem VI. *Under the hypotheses of Theorem III and the additional hypothesis that* $\lim g(x) \neq 0$, *the limit of the quotient is the quotient of the limits:*
$$\lim \frac{f(x)}{g(x)} = \frac{\lim f(x)}{\lim g(x)}.$$

Theorem VII. *If* $f(x) \leq g(x)$ *and if* $\lim f(x)$ *and* $\lim g(x)$ *exist (finite, or* $+\infty$, *or* $-\infty$*), then* $\lim f(x) \leq \lim g(x)$.

208. EXERCISES

In Exercises 1–7, prove the indicated limit theorem for the specified behavior of the independent variable.

1. Theorem I, § 207; $x \to a$, finite or infinite limit.
2. Theorem II, § 207; $x \to \infty$.
3. Theorem III, § 207; $x \to a+$.
4. Theorem III, § 207; $x \to +\infty$.
5. Theorem IV, § 207; $x \to a-$.
6. Theorem V, § 207; $x \to -\infty$.
7. Theorem VII, § 207; $x \to a$.
8. Prove that if $\lim_{x \to a} \phi(x) = 0$, and if $\psi(x)$ is bounded in some deleted neighborhood of a (that is, if there exist positive numbers P and η such that $|\psi(x)| \leq P$ for $0 < |x - a| < \eta$), then $\lim_{x \to a} \phi(x)\psi(x) = 0$. Find $\lim_{x \to 0} x \sin \frac{1}{x}$ (cf. Example 2, § 303).

9. Prove that if $\lim_{x \to a} f(x)$ exists and is positive (negative), then $f(x)$ is positive (negative) in some deleted neighborhood of a. Prove, in fact, that if $\lim_{x \to a} f(x) = m \neq 0$, then for all x within some deleted neighborhood of a, $f(x) > \frac{1}{2}m$ if $m > 0$ and $f(x) < \frac{1}{2}m$ if $m < 0$. Consequently show that the reciprocal of a function is bounded in some deleted neighborhood of any point at which the function has a nonzero limit. Show that this last statement is true even if the nonzero limit is infinite. *Hint:* In case $\lim_{x \to a} f(x) = m > 0$, let δ be a positive number such that $0 < |x - a| < \delta$ implies $|f(x) - m| < \frac{1}{2}m$, so that $m - f(x) < \frac{1}{2}m$.

10. Prove Theorem VI, § 207, for the case $x \to a$. (Cf. Exs. 8 and 9.)

11. Prove Theorem VI, § 207, for the case $x \to +\infty$. (Cf. Exs. 8–10.)

12. Prove that if $f(x) \leq g(x) \leq h(x)$ and if $\lim f(x)$ and $\lim h(x)$ are finite and equal (for the same behavior of the independent variable subject to the restrictions of the first paragraph of § 207), then $\lim g(x)$ exists and is equal to their common value. Extend this result to include infinite limits.

In Exercises 13–20, prove the given statement.

13. $\lim_{x \to a} f(x)$ exists and is finite if and only if $\lim_{x \to a+} f(x)$ and $\lim_{x \to a-} f(x)$ exist and are finite and equal.

14. $\lim_{x \to a} f(x)$ exists (in the finite or infinite sense) if and only if $\lim_{x \to a+} f(x)$ and $\lim_{x \to a-} f(x)$ exist and are equal.

15. Theorems III and V, § 207, hold for any finite number of functions.

16. If k is a constant and $\lim f(x)$ exists and is finite, then $\lim k f(x)$ exists and is equal to $k \lim f(x)$, whatever the behavior of the independent variable x, subject to the restrictions of the first paragraph of § 207.

17. $\lim_{x \to a} x = a$.

18. If n is a positive integer, $\lim_{x \to a} x^n = a^n$. *Hint:* Use Theorem V, § 207, and Ex. 17.

19. If $f(x)$ is a polynomial,
$$f(x) = a_0 x^n + a_1 x^{n-1} + \cdots + a_{n-1} x + a_n,$$
then $\lim_{x \to a} f(x) = f(a)$.

20. If $f(x)$ is a rational function,
$$f(x) = g(x)/h(x),$$
where $g(x)$ and $h(x)$ are polynomials, and if $h(a) \neq 0$, then $\lim_{x \to a} f(x) = f(a)$.

In Exercises 21–26, find the indicated limit.

21. $\lim_{x \to 3} (2x^2 - 5x + 1)$.

22. $\lim_{x \to -2} \dfrac{3x^2 - 5}{2x + 17}$.

23. $\lim_{x \to 2} \dfrac{3x^2 - x - 10}{x^2 + 5x - 14}$. *Hint:* Reduce to lowest terms.

24. $\lim_{x \to -3} \dfrac{x^3 + 27}{x^4 - 81}$. (Cf. Ex. 23.)

25. $\lim_{x \to a} \dfrac{x^3 - a^3}{x - a}$. (Cf. Ex. 23.)

26. $\lim_{x \to a} \dfrac{x^m - a^m}{x - a}$, where m is an integer. (Cf. Ex. 23, above, and Ex. 9, § 107.)

In Exercises 27–30, give a precise definition for the given limit statement.

27. $\lim\limits_{x \to a+} f(x) = -\infty$.

28. $\lim\limits_{x \to -\infty} f(x) = +\infty$.

29. $\lim\limits_{x \to \infty} f(x) = \infty$.

30. $\lim\limits_{x \to a-} f(x) = \infty$.

31. Prove that $\lim\limits_{x \to 0} \dfrac{1}{x} = \infty$ and $\lim\limits_{x \to \infty} \dfrac{1}{x} = 0$. More generally, assuming $f(x)$ to be nonzero except possibly for the limiting value of the independent variable, prove that $\lim f(x) = 0$ if and only if $\lim \dfrac{1}{f(x)} = \infty$. Discuss Theorem VI, § 207, if $\lim f(x) \neq 0$ and $\lim g(x) = 0$. (Cf. Theorem XI, § 204.)

32. Prove that $\lim\limits_{x \to 0+} \dfrac{1}{x} = +\infty$. More generally, assuming $f(x)$ to be positive except possibly for the limiting value of the independent variable, prove that $\lim f(x) = +\infty$ if and only if $\lim \dfrac{1}{f(x)} = 0$.

In Exercises 33 and 34, assuming the standard facts regarding trigonometric functions (cf. §§ 503–504 for analytic definitions of the trigonometric functions), find the specified limits, or establish their nonexistence.

33. (a) $\lim\limits_{x \to \frac{1}{2}\pi+} \tan x$;

(b) $\lim\limits_{x \to \frac{1}{2}\pi-} \tan x$;

(c) $\lim\limits_{x \to \frac{1}{2}\pi} \tan x$;

(d) $\lim\limits_{x \to 0+} \cot x$;

(e) $\lim\limits_{x \to 0-} \cot x$;

(f) $\lim\limits_{x \to 0} \cot x$.

34. (a) $\lim\limits_{x \to +\infty} \sin x$;

(b) $\lim\limits_{x \to -\infty} \cos x^2$;

(c) $\lim\limits_{x \to \infty} \dfrac{\sin x}{x}$;

(d) $\lim\limits_{x \to +\infty} \dfrac{\sec x}{x}$;

(e) $\lim\limits_{x \to -\infty} \dfrac{x - \cos x}{x}$;

(f) $\lim\limits_{x \to \infty} \dfrac{x \sin x}{x^2 - 4}$.

35. Let $f(x)$ be a rational function

$$f(x) = \frac{a_0 x^m + a_1 x^{m-1} + \cdots + a_{m-1} x + a_m}{b_0 x^n + b_1 x^{n-1} + \cdots + b_{n-1} x + b_n},$$

where $a_0 \neq 0$ and $b_0 \neq 0$. Show that $\lim\limits_{x \to \infty} f(x)$ is equal to 0 if $m < n$, to a_0/b_0 if $m = n$, and to ∞ if $m > n$. In particular, show that if $f(x)$ is any nonconstant polynomial, $\lim\limits_{x \to \infty} f(x) = \infty$. *Hint:* Divide every term in both numerator and denominator by the highest power of x present.

36. Discuss the result of Exercise 35 for the case $m > n$ if (i) $x \to +\infty$; (ii) $x \to -\infty$; (iii) $x \to \infty$ and m and n are either both even or both odd. Consider in particular the special case where $f(x)$ is a polynomial.

In Exercises 37–42, find the indicated limit. (Cf. Exs. 35–36.)

37. $\lim\limits_{x \to \infty} \dfrac{5x^2 - 3x + 1}{6x^2 + 5}$.

38. $\lim\limits_{x \to +\infty} (2x^5 - 350x^2 - 10{,}000)$.

39. $\lim\limits_{x \to \infty} (-6x^4 - 9x^3 + x)$.

40. $\lim\limits_{x \to \infty} \dfrac{2x^3 - 5}{3x + 7}$.

41. $\lim_{x \to +\infty} \dfrac{150x + 2000}{x^2 + 3}$.   **42.** $\lim_{x \to -\infty} \dfrac{8x^3 + 13x + 6}{5x^2 + 11}$.

In Exercises 43–50, interpret and prove each relation. For these exercises p designates a positive number, n a negative number, m a nonzero number, and q any number. *Hint for* Ex. 43: This means: If $\lim f(x) = p$ and $\lim g(x) = +\infty$, then $\lim f(x)g(x) = +\infty$. For simplicity let $x \to a$.

43. $p \cdot (+\infty) = +\infty$.   **44.** $n \cdot (+\infty) = -\infty$.
45. $q - (-\infty) = +\infty$.   **46.** $q + (\infty) = \infty$.
47. $(-\infty) - q = -\infty$.   **48.** $(+\infty) + (+\infty) = +\infty$.
49. $\dfrac{0}{\infty} = \dfrac{m}{\infty} = 0$.   **50.** $\dfrac{\infty}{0} = \dfrac{m}{0} = \infty$.

In Exercises 51–56, show by examples that the given expression is indeterminate. (See Hint for Ex. 43.)

51. $\infty + \infty$.   **52.** $(+\infty) - (+\infty)$.

Hint for Ex. 52: Consider the examples (i) $x - x$ and (ii) $x^2 - x$ as $x \to +\infty$.

53. $(+\infty) + (\infty)$.   **54.** $0 \cdot \infty$.
55. $\dfrac{0}{0}$.   **56.** $\dfrac{\infty}{\infty}$.

57. Give an example of a function $f(x)$ satisfying the following three conditions: $\lim_{x \to 0} |f(x)| = 1$, $\lim_{x \to 0-} f(x) = -1$, $\lim_{x \to 0+} f(x)$ does not exist.

**58. Give an example of a function $f(x)$ satisfying the following three conditions: $\lim_{x \to 0+} f(x) = \infty$, $\lim_{x \to 0+} f(x) \neq +\infty$, $\lim_{x \to 0+} f(x) \neq -\infty$. (Cf. Ex. 57.)

In Exercises 59–64, find the required limit, prove that it is the limit by direct use of Definition I, § 206, and obtain explicitly a function $\delta(\epsilon)$ as demanded by that definition. (Cf. Exs. 26–31, § 216.)

**59. $\lim_{x \to 2} 3x$.   **60. $\lim_{x \to -3} x^2$.
**61. $\lim_{x \to 4} (3x^2 - 5x)$.   **62. $\lim_{x \to -5} \dfrac{1}{x}$.
**63. $\lim_{x \to 1} \dfrac{4x^2 - 1}{5x + 2}$.   **64. $\lim_{x \to 2} \dfrac{3x}{4x - 7}$.

In Exercises 65–70, find the required limit, prove that it is the limit by direct use of Definition II or Definition III, § 206, and obtain explicitly a function $N(\epsilon)$ or $\delta(B)$ as demanded by the definition.

**65. $\lim_{x \to +\infty} \dfrac{5}{x}$.   **66. $\lim_{x \to -\infty} \dfrac{1}{x^2}$.
**67. $\lim_{x \to +\infty} \dfrac{3x - 2}{x + 5}$.   **68. $\lim_{x \to +\infty} \dfrac{5x^2 + 1}{3x^2}$.
**69. $\lim_{x \to 0} \dfrac{1}{x^2}$.   **70. $\lim_{x \to 1} \dfrac{2x - 5}{x^3 - 2x^2 + x}$.

209. CONTINUITY

Continuity of a function at a point a can be defined either in terms of limits (Definition I, below), or directly by use of the type of δ-ϵ formulation in which the original limit concepts are framed (Definition II, below). When continuity is couched in terms of limits, we shall make the same implicit assumption that was stated in the first paragraph of § 206, namely, that every neighborhood of the point a contains at least one point x different from a for which the function $f(x)$ is defined. (Cf. Note 2, below.)

Definition I. *A function $f(x)$ is <u>continuous</u> at $x = a$ <u>if and only if the following three conditions are satisfied</u>:*
 (i) *$f(a)$ exists; that is, $f(x)$ is defined at $x = a$;*
 (ii) *$\lim_{x \to a} f(x)$ exists and is finite;*
 (iii) *$\lim_{x \to a} f(x) = f(a)$.*

By inspection of the definition of $\lim_{x \to a} f(x)$, it is possible (cf. Ex. 11, § 212) to establish the equivalence of this definition and the following (in case the implicit assumption of the first paragraph, above, is satisfied).

Definition II. *A function $f(x)$ is <u>continuous</u> at $x = a$ if and only if it is defined at $x = a$ and corresponding to an arbitrary positive number ϵ, there exists a positive number $\delta = \delta(\epsilon)$ such that $|x - a| < \delta$ implies*
$$|f(x) - f(a)| < \epsilon,$$
for values of x for which $f(x)$ is defined.

NOTE 1. Definition I is sometimes called the *limit definition of continuity* and Definition II the *δ-ϵ definition of continuity.*

NOTE 2. The δ-ϵ definition is applicable even when the function is not defined at points neighboring $x = a$ (except at a itself). In this case a is an *isolated point* of the domain of definition, and $f(x)$ is continuous there, although $\lim_{x \to a} f(x)$ has no meaning.

NOTE 3. Each of the following limit statements is a formulation of continuity of $f(x)$ at $x = a$ (Ex. 3, § 212):
(1) $\qquad\qquad f(a + h) - f(a) \to 0 \quad \text{as} \quad h \to 0;$
(2) $\qquad\qquad \text{if} \quad \Delta y = f(a + \Delta x) - f(a), \quad \text{then}$
$$\Delta y \to 0 \quad \text{as} \quad \Delta x \to 0.$$

NOTE 4. An almost immediate consequence of either definition of continuity is the fact that whenever a function is positive (negative) at a point of continuity it must be positive (negative) throughout some neighborhood of that point, to the extent that it is defined there. For a slightly stronger statement, together with a hint for a proof, see Exercise 18, § 212.

A function is said to be **continuous on a set** if and only if it is continuous at every point of that set. In case a function is continuous at every point of its domain of definition it is simply called **continuous**, without further modifying words.

Continuity from the right, or **right-hand continuity**, is defined by replacing, in Definition I, $\lim_{x \to a} f(x)$ by $\lim_{x \to a+} f(x)$. Similarly, **continuity from the left**, or

left-hand continuity, is obtained by replacing $\lim_{x \to a} f(x)$ by $\lim_{x \to a-} f(x)$. Thus $f(x)$ is continuous from the right at $x = a$ if and only if $f(a+) = f(a)$, and $f(x)$ is continuous from the left if and only if $f(a-) = f(a)$, it being assumed that the expressions written down exist.

A useful relation between continuity and limits is stated in the following theorem:

Theorem. *If $f(x)$ is continuous at $x = a$ and if $\phi(t)$ has the limit a, as t approaches some limit (finite or infinite, one-sided or not), and if $f(\phi(t))$ is defined, then*

$$f(\phi(t)) \to f(a).$$

In short, the limit of the function is the function of the limit:

$$\lim f(\phi(t)) = f(\lim \phi(t)).$$

Proof. We shall prove the theorem for the single case $t \to +\infty$. (Cf. Exs. 12–13, § 212.) Accordingly, let ϵ be an arbitrary positive number, and let δ be a positive number such that $|x - a| < \delta$ implies

$$|f(x) - f(a)| < \epsilon.$$

Then choose N so large that $t > N$ implies $|\phi(t) - a| < \delta$. Combining these two implications by setting $x = \phi(t)$ we have the result: $t > N$ implies $|f(\phi(t)) - f(a)| < \epsilon$. This final implication is the one sought.

Example 1. A function whose domain of definition is a closed interval $[a, b]$ is continuous there if and only if it is continuous at each interior point, continuous from the right at $x = a$, and continuous from the left at $x = b$. (See Fig. 205.)

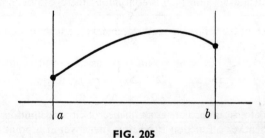

FIG. 205

Example 2. The function $[x]$ (Example 5, § 201) is continuous except when x is an integer. It is everywhere continuous from the right. (See Fig. 206.)

210. TYPES OF DISCONTINUITY

The principal types of discontinuity are the following four:

(i) *The limit of the function exists, but the function either is not defined at the point or has a value different from the limit there.* (Figure 207, (a) and (b); cf. Example 1, § 206.) Such a discontinuity is called a **removable discontinuity**

§ 210] TYPES OF DISCONTINUITY 45

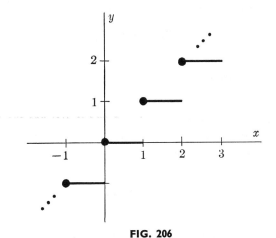

FIG. 206

because if the function is redefined at $x = a$ to have the value $f(a) \equiv \lim_{x \to a} f(x)$, it becomes continuous there.

(ii) *The two one-sided limits exist and are finite, but are not equal.* An example is the signum function (Fig. 207, (c); cf. Example 1, § 206). Such a discontinuity is called a **jump discontinuity.**

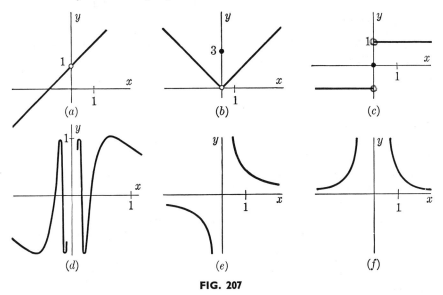

FIG. 207

(iii) *At least one one-sided limit fails to exist.* An example is $\sin \frac{1}{x}$ (Fig. 207, (d); cf. Example 1, § 206).

(iv) *At least one one-sided limit is infinite* (Fig. 207, (e) and (f); cf. Example 1, § 206).

211. CONTINUITY THEOREMS

The limit theorems II–VI of § 207 have (as immediate corollaries) counterparts in terms of continuity:

Theorem I. *Any constant function is continuous.*

Theorem II. *If $f(x)$ and $g(x)$ are continuous at $x = a$, then their sum $f(x) + g(x)$ is also continuous at $x = a$. In short, the sum of two continuous functions is a continuous function. This rule applies to any finite number of functions.*

Theorem III. *Under the hypotheses of Theorem II, $f(x) - g(x)$ is continuous at $x = a$: the difference of two continuous functions is a continuous function.*

Theorem IV. *Under the hypotheses of Theorem II, $f(x)g(x)$ is continuous at $x = a$: the product of two continuous functions is a continuous function. This rule applies to any finite number of functions.*

Theorem V. *Under the hypotheses of Theorem II and the additional hypothesis that $g(a) \neq 0$, $f(x)/g(x)$ is continuous at $x = a$: the quotient of two continuous functions is continuous where the denominator does not vanish.*

A direct consequence of the Theorem of § 209 is the following:

Theorem VI. *A continuous function of a continuous function is a continuous function. More precisely, if $f(x)$ is continuous at $x = a$, if $g(y)$ is continuous at $y = b = f(a)$, and if $h(x) \equiv g(f(x))$, then $h(x)$ is a continuous function of x at $x = a$.*

212. EXERCISES

1. Prove that any polynomial is continuous for all values of the independent variable. (Cf. Ex. 19, § 208.)
2. Prove that any rational function is continuous except where the denominator vanishes. (Cf. Ex. 20, § 208.)
3. Prove the statements in Note 3, § 209.
4. Redefine the signum function (Example 1, § 206) at $x = 0$ so that it becomes everywhere continuous from the right; so that it becomes everywhere continuous from the left.
5. Establish continuity at $x = 0$ for each of the two functions:

(a) $f(x) \equiv x \sin \frac{1}{x}$ if $x \neq 0, f(0) \equiv 0$;   (b) $g(x) \equiv x^2 \sin \frac{1}{x}$ if $x \neq 0$, $g(0) \equiv 0$.

(Cf. Ex. 8, § 208. Also see Examples 2 and 3, § 303, for graphs and further discussion.)

6. Define the function $x \cos \frac{1}{x}$ at $x = 0$ so that it becomes everywhere continuous.

In Exercises 7–10, state the type of discontinuity at $x = 0$.

7. $f(x) = x^2 - 8x$ if $x \neq 0$, $f(0) = 6$.

8. $f(x) = x^3 \cos \dfrac{1}{x^2}$ if $x \neq 0$, undefined if $x = 0$.

9. $f(x) = \dfrac{1}{x}$ if $x > 0$, $f(x) = 0$ if $x \leq 0$.

10. $f(x) = x + 1$ if $x > 0$, $f(x) = -x - 1$ if $x < 0$, $f(0) = 0$.

11. Prove the equivalence of the two definitions of continuity, § 209, under the assumptions of the first paragraph of that section.

12. Prove the Theorem of § 209 (a) for the case $t \to c$; (b) for the case $t \to c-$.

13. Give an example to show that the Theorem of § 209 is false if the continuity assumption is omitted. Can you construct an example where the limit of the function and the function of the limit both exist, but are unequal?

14. Prove that the **negation of continuity** of a function at a point of its domain can be expressed: $f(x)$ is discontinuous at $x = a$ if and only if there is a positive number ϵ having the property that corresponding to an arbitrary positive number δ (however small), there exists a number x such that $|x - a| < \delta$ and $|f(x) - f(a)| \geq \epsilon$.

15. Show that the function defined for all real numbers, $f(x) = 1$ if x is rational, $f(x) = 0$ if x is irrational, is everywhere discontinuous. (Cf. Example 6, § 201.)

16. Prove that $|x|$ is everywhere continuous.

★17. Prove that $|f(x)|$ is continuous wherever $f(x)$ is. Give an example of a function defined for all real numbers which is never continuous but whose absolute value is always continuous. *Hint:* Consider a function like that of Ex. 15, with values ± 1.

18. Prove that if $f(x)$ is continuous and positive at $x = a$, then there is a neighborhood of a in which $f(x)$ is positive. Prove that, in fact, there exist a positive number ϵ and a neighborhood of a such that in this neighborhood $f(x) > \epsilon$. State and prove corresponding facts if $f(a)$ is negative. *Hint:* Cf. Ex. 9, § 208.

★19. Prove that if $f(x)$ is continuous at $x = a$, and if $\epsilon > 0$ is given, then there exists a neighborhood of a such that for any two points in this neighborhood the values of $f(x)$ differ by less than ϵ.

★20. Show that the function with domain the closed interval $[0, 1]$, $f(x) = x$ if x is rational, $f(x) = 1 - x$ if x is irrational, is continuous only for $x = \frac{1}{2}$.

★21. Show that the following example is a counterexample to the following False Theorem, which strives to generalize the Theorem of § 209, and Theorem VI, § 211: If $\lim \phi(t) = a$, where a is finite and t approaches some limit, and if $\lim\limits_{x \to a} f(x) = b$, where b is finite, then $\lim f(\phi(t)) = b$, as t approaches its limit. *Counter example:* $\phi(t) \equiv 0$ for all t, $f(x) = 0$ if $x \neq 0$, $f(0) = 1$; $t \to 0$, $x \to 0$. Prove that this False Theorem becomes a True Theorem with the additional hypothesis that $\phi(t)$ is never equal to a, except possibly for the limiting value of t.

213. MORE THEOREMS ON CONTINUOUS FUNCTIONS

We list below a few important theorems on continuous functions, whose proofs depend on certain rather sophisticated ideas discussed in §§ 217–223, which are starred for possible omission or postponement. The proofs of these particular theorems are given in § 218. The fact expressed in Theorem IV

48   FUNCTIONS, SEQUENCES, LIMITS, CONTINUITY [§ 214

is sometimes referred to as the *intermediate-value* property of continuous functions.

Theorem I. *A function continuous on a closed interval is bounded there. That is, if $f(x)$ is continuous on $[a, b]$, there exists a number B such that $a \leq x \leq b$ implies $|f(x)| \leq B$.*

Theorem II. *A function continuous on a closed interval has a maximum value and a minimum value there. That is, if $f(x)$ is continuous on $[a, b]$, there exist points x_1 and x_2 in $[a, b]$ such that $a \leq x \leq b$ implies $f(x_1) \leq f(x) \leq f(x_2)$.*

Theorem III. *If $f(x)$ is continuous on the closed interval $[a, b]$ and if $f(a)$ and $f(b)$ have opposite signs, there is a point x_0 between a and b for which $f(x_0) = 0$.*

Theorem IV. *A function continuous on an interval assumes (as a value) every number between any two of the values that it assumes on that interval.* (Cf. Ex. 40, § 308.)   *intermediate-value property*

Note. For other properties of a function continuous on a closed interval, see §§ 224 and 401.

214. EXISTENCE OF $\sqrt{2}$ AND OTHER ROOTS

In Chapter 1 (Ex. 8, § 109) $\sqrt{2}$ was mentioned as an example of an irrational number, but proof of its existence was deferred. We are able now, with the aid of the last theorem of the preceding section, to give a simple proof that there exists a positive number whose square is 2. The idea is to consider the function $f(x) \equiv x^2$, which is continuous everywhere (Ex. 1, § 212) and, in particular, on the closed interval from $x = 0$ to $x = 2$. Since the values of the function at the end-points of this interval are $0^2 = 0$ and $2^2 = 4$, and since the number 2 is between these two extreme values, there must be a (positive) number between these end-points for which the value of the function is 2. That is, there is a positive number whose square is 2. The following theorem generalizes this result, and establishes the uniqueness of positive nth roots:

Theorem. *If p is a positive number and n is a positive integer, there exists a unique positive number x such that $x^n = p$. This number is called **the nth root of p** and is written $x = \sqrt[n]{p}$.*

Proof. We establish uniqueness first. If $x^n = y^n = p$, where x and y are positive, then (Ex. 9, § 107) $x^n - y^n = (x - y)(x^{n-1} + x^{n-2}y + \cdots + y^{n-1}) = 0$. Since the second factor is positive (Ex. 8, § 107) the first factor must vanish (Example 4, § 102): $x - y = 0$, or $x = y$.

For the proof of existence, we note first that since $x^n \geq x$ for $x \geq 1$, $\lim_{x \to +\infty} x^n = +\infty$. Therefore there exists a number b such that $b^n > p$.

The number p is thus between the extreme values assumed by x^n on the closed interval $[0, b]$, and therefore, since x^n is continuous, this function must assume the value p at some point x between 0 and b: $x^n = p$.

Corollary. *If y is a real number and n is an odd positive integer, there exists a unique number x such that $x^n = y$. This number is called* **the nth root of y** *and is written* $x = \sqrt[n]{y}$.

215. MONOTONIC FUNCTIONS AND THEIR INVERSES

Definition. *A function $f(x)$ is* **monotonically increasing (decreasing)**, *written $f(x)\uparrow(\downarrow)$, on a set A if and only if whenever a and b are elements of A and $a < b$, then $f(a) \leq f(b)$ ($f(a) \geq f(b)$). In either case it is called* **monotonic**. *Whenever $a < b$ implies $f(a) < f(b)$ ($f(a) > f(b)$), $f(x)$ is called* **strictly increasing (decreasing)**, *and in either case* **strictly monotonic**. (Cf. Fig. 208.)

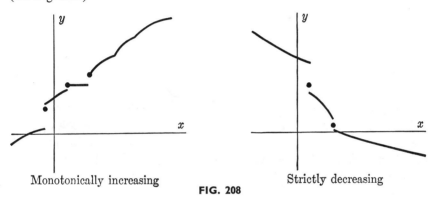

Monotonically increasing   Strictly decreasing

FIG. 208

Theorem XIV, § 204, states facts about monotonic sequences. If we permit the inclusion of infinite limits we can drop the assumption of boundedness and state that any monotonic sequence has a limit (cf. the Note, § 204). In Exercise 23, § 216, the student is asked to generalize this fact and prove that a monotonic function $f(x)$ always has one-sided limits. Consequently the only type of discontinuity that a monotonic function can have (at a point where it is defined) is a finite jump (Ex. 24, § 216).

Consider now the case of a function that is continuous and strictly monotonic on a closed interval $[a, b]$. For definiteness assume that $f(x)$ is strictly increasing there, and let $c = f(a)$ and $d = f(b)$. Then $c < d$, and (Theorem IV, § 213) $f(x)$ assumes in the interval $[a, b]$ every value between c and d. Furthermore, it cannot assume the same value twice, for if $\alpha < \beta$, then $f(\alpha) < f(\beta)$. (Cf. Fig. 209.) The function $f(x)$, therefore, establishes a one-to-one correspondence between the points of the two closed intervals $[a, b]$ and $[c, d]$. Thus to each point y of the closed interval $[c, d]$ corresponds a unique point x of $[a, b]$ such that $y = f(x)$. Since y determines x uniquely,

x can be considered as a single-valued function of y, $x = \phi(y)$, and is called the **inverse function** of $y = f(x)$, the latter being referred to, then, as the **direct function**. The nth roots obtained in the preceding section are inverse functions: if $y = f(x) = x^n$, then $x = \phi(y) = \sqrt[n]{y}$. We now state an important fact regarding the continuity of inverse functions. The proof is given in § 218.

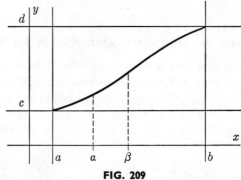

FIG. 209

Theorem. *If $y = f(x)$ is continuous and strictly monotonic on a closed interval $[a, b]$, its inverse function $x = \phi(y)$ is continuous and strictly monotonic on the corresponding closed interval $[c, d]$.*

Corollary. *The function $y = \sqrt[n]{x}$, where n is a positive integer, is a continuous and strictly increasing function of x for $x \geq 0$. If n is an odd positive integer, $y = \sqrt[n]{x}$ is a continuous and strictly increasing function of x for $-\infty < x < +\infty$. (Cf. Ex. 18, § 216.)*

Proof of continuity: Since $\sqrt[n]{x}$ is continuous over every closed interval where it is defined it is continuous wherever it is defined.

216. EXERCISES

1. Show that the equation $x^4 + 2x - 11 = 0$ has a real root. Prove that if $f(x)$ is a real polynomial (real coefficients) of odd degree, the equation $f(x) = 0$ has a real root.

2. Prove that the maximum (minimum) value of a function on a closed interval is unique, but show by examples that the point ξ at which $f(x)$ is a maximum (minimum) may or may not be unique.

3. Prove that $\sqrt{f(x)}$ is continuous wherever $f(x)$ is continuous and positive, and that $\sqrt[3]{f(x)}$ is continuous wherever $f(x)$ is continuous.

4. Assuming continuity and other standard properties of $\sin x$ and $\cos x$ (cf. §§ 503–504), discuss continuity of each function.

(a) $\sqrt{1 + \sin x}$;

(b) $\dfrac{1}{\sqrt{1 - \sin x}}$;

(c) $\sqrt[3]{\cos x^2}$;

(d) $\dfrac{1}{\sqrt[3]{1 + \cos^2 x - \sin^2 x}}$.

In Exercises 5–10, discuss the discontinuities of each function, and draw a graph in each case. The bracket function $[x]$ is defined in Example 5, § 201. (Also cf. Exs. 7–10, § 203.)

5. $[-x]$.
6. $[x] + [-x]$.
7. $[\sqrt{x}]$.
8. $[x^2]$.
9. $[x] + \sqrt{x - [x]}$.
10. $[x] + (x - [x])^2$.

11. Prove the two laws of radicals:

(i) $\sqrt[n]{ab} = \sqrt[n]{a}\sqrt[n]{b}$,

(ii) $\sqrt[n]{\dfrac{a}{b}} = \dfrac{\sqrt[n]{a}}{\sqrt[n]{b}}$, $b \neq 0$,

where n is a positive integer, and in case n is even a and b are nonnegative.

12. Prove that $\lim\limits_{x \to +\infty} \sqrt[n]{x} = +\infty$, where n is a positive integer.

★13. Prove that $\lim\limits_{x \to +\infty} \sqrt{x}(\sqrt{x + a} - \sqrt{x}) = \tfrac{1}{2}a$.

Hint: Multiply by $(\sqrt{x + a} + \sqrt{x})/(\sqrt{x + a} + \sqrt{x})$.

★14. Let $f(x)$ be a polynomial of degree m and leading coefficient $a_0 > 0$, and let $g(x)$ be a polynomial of degree n and leading coefficient $b_0 > 0$, and let k be a positive integer. Prove that the limit

$$\lim_{x \to +\infty} \sqrt[k]{f(x)}/\sqrt[k]{g(x)}$$

is equal to 0 if $m < n$, to $\sqrt[k]{a_0/b_0}$ if $m = n$, and to $+\infty$ if $m > n$. (Cf. Ex. 35, § 208.)

★15. Give an example of a function which is continuous and bounded on the open interval $(0, 1)$, but which has neither maximum nor minimum there.

★16. Give an example of a function which is defined and single-valued on the closed interval $[0, 1]$, but which is not bounded there.

★17. Give an example of a function which is defined, single-valued, and bounded on the closed interval $[0, 1]$, but which has neither maximum nor minimum there.

★18. Prove that $\sqrt[n]{x}$ is an increasing function of n for any fixed x between 0 and 1, and a decreasing function of n for any fixed $x > 1$. *Hint:* Show that the desired order relation between $\sqrt[n]{x}$ and $\sqrt[n+1]{x}$ follows from Ex. 10, § 107, by taking $n(n + 1)$th powers.

★19. Prove that if $x > 0$ then $\lim\limits_{n \to +\infty} \sqrt[n]{x} = 1$. *Hint:* Assume for definiteness that $0 < x < 1$. Then (Ex. 18) $\sqrt[n]{x} \uparrow$ as $n \uparrow$. Since $\sqrt[n]{x} < 1$ for all n, $\lim\limits_{n \to +\infty} \sqrt[n]{x} = L$ exists (Theorem XIV, § 204) and $L \leq 1$. If $L < 1$, then $\sqrt[n]{x} \leq L$ for all n, and $x \leq L^n$ for all n. Use Theorem XIII, § 204.

★20. Define $^{2n-1}\sqrt{x}$ for all x, and show that $\lim\limits_{n \to +\infty} {}^{2n-1}\sqrt{x} = \operatorname{sgn} x$ (cf. Example 1 (c), § 206).

★21. If r is a positive rational number, represented as the quotient of two positive integers, $r = p/q$, prove that the two definitions of x^r, where $x \geq 0$, are equivalent: $x^r = \sqrt[q]{x^p}$; $x^r = (\sqrt[q]{x})^p$. Also show that either definition is independent of the representation of r as a quotient of positive integers. Discuss the function $f(x) = x^r$ as a strictly increasing continuous function of x for $x \geq 0$. Extend to the range $-\infty < r < +\infty$, for $x > 0$.

★22. If $f(x)$ is continuous on $[a, b]$, if $a < c < d < b$, and $K = f(c) + f(d)$, prove that there exists a number ξ between a and b such that $K = 2f(\xi)$. More generally, if m and n are positive numbers, show that $mf(c) + nf(d) = (m + n)f(\xi)$, for some ξ between a and b. Finally, extend this result to the formula
$$m_1 f(c_1) + \cdots + m_k f(c_k) = (m_1 + \cdots + m_k) f(\xi).$$
Hint: Show that $\dfrac{m}{m+n} f(c) + \dfrac{n}{m+n} f(d)$ is between $f(c)$ and $f(d)$, and use Theorem IV, § 213.

★23. Prove that a monotonic function always has one-sided limits (finite or infinite). *Hint:* Use suprema or infima (cf. Th. XIV, § 204; Ex. 16, § 205).

★24. Prove that the only discontinuities that a monotonic function can have at points where it is defined are finite jumps. (Cf. Ex. 23.)

★25. Prove the following converse of the statement of § 215 that a continuous strictly monotonic function has a single-valued inverse: *If $f(x)$ is continuous on a closed interval $[a, b]$ and if $f(x)$ does not assume there any value twice (that is, at distinct points of $[a, b]$ $f(x)$ has distinct values), then $f(x)$ is strictly monotonic there.*

In Exercises 26–31, find a specific function $\delta = \delta(\epsilon)$ as specified by Definition II, § 209, for the given function at the prescribed point. (Cf. Exs. 59–64, § 208.)

★26. $\sqrt{x}$, $a = 3$. *Hint:* If $x > 0$,
$$|\sqrt{x} - \sqrt{3}| = \left| \frac{\sqrt{x} - \sqrt{3}}{1} \cdot \frac{\sqrt{x} + \sqrt{3}}{\sqrt{x} + \sqrt{3}} \right| = \frac{|x - 3|}{\sqrt{x} + \sqrt{3}} < |x - 3|.$$

★27. $\sqrt{x^2 - 4}$, $a = 2$.   **★28.** $\sqrt{3x^2 - x}$, $a = -1$.

★29. $\sqrt[3]{x}$, $a = 5$. *Hint:* If $x > 0$,
$$|\sqrt[3]{x} - \sqrt[3]{5}| = \left| \frac{\sqrt[3]{x} - \sqrt[3]{5}}{1} \cdot \frac{\sqrt[3]{x^2} + \sqrt[3]{5x} + \sqrt[3]{25}}{\sqrt[3]{x^2} + \sqrt[3]{5x} + \sqrt[3]{25}} \right| < |x - 5|.$$

★30. $\dfrac{1}{\sqrt{x}}$, $a = 6$.   **★31.** $\dfrac{1}{\sqrt[3]{x}}$, $a = 1$.

★32. If a is a fixed positive rational number prove that the set of all numbers the form $r + s\sqrt{a}$, where r and s are arbitrary rational numbers, form an ordered field.

★33. Prove that $\lim\limits_{n \to +\infty} \left(1 + \dfrac{1}{n}\right)^n$ exists and is between 2 and 3. (This is one definition of e. For another treatment see §§ 501–502.) *Hint:* First express the binomial expansion of $a_n \equiv \left(1 + \dfrac{1}{n}\right)^n$ in the form
$$1 + 1 + \frac{1}{2!}\left(1 - \frac{1}{n}\right) + \frac{1}{3!}\left(1 - \frac{1}{n}\right)\left(1 - \frac{2}{n}\right) + \cdots,$$
and thereby show that $a_n \uparrow$ and, furthermore, that
$$a_n < 1 + 1 + \frac{1}{2} + \frac{1}{2^2} + \frac{1}{2^3} + \cdots + \frac{1}{2^{n-1}}.$$

*217. A FUNDAMENTAL THEOREM ON BOUNDED SEQUENCES†

One of the properties of the real number system that was assumed as an axiom in Chapter 1 is that of *completeness*. A useful consequence (which is actually one of several alternative formulations of the concept of completeness—cf. R. C. Buck, *Advanced Calculus* (New York, McGraw-Hill Book Company, Inc., 1956), Appendix) is stated in the following theorem:

Theorem. Fundamental Theorem on Bounded Sequences. *Every bounded sequence (of real numbers) contains a convergent subsequence.*

In order to prove this theorem we observe first that we already know (Theorem XIV, § 204) that every bounded *monotonic* sequence converges. Therefore the proof will be complete as soon as we establish the following lemma:

Lemma I. *Every sequence (of real numbers) contains a monotonic subsequence.*

Before proceeding with the details of the proof of this lemma we discuss briefly the idea of *largest term* of a sequence, and give a few illustrative examples.

By a **largest term** of a sequence $a_1, a_2, a_3, \cdots$ we mean any term a_k with the property $a_k \geq a_n$ for every positive integer n. The examples below show that a sequence may not have a largest term, and that if it does have a largest term it may have several. In case a sequence has a largest term, there must be a *first* largest term (first in the order of the terms according to the subscripts). This particular largest term will be called for convenience **the largest term.**

Example 1. The sequence $0, 1, 0, 1, \cdots$ contains the two monotonic constant subsequences $0, 0, 0, \cdots$ and $1, 1, 1, \cdots$. It has a largest term $a_k = 1$, where k may be any even positive integer. *The* largest term of the sequence is the second.

Example 2. The sequence $1, -1, 2, -2, 3, -3, \cdots$ contains the monotonic subsequences $1, 2, 3, \cdots$ and $-1, -2, -3, \cdots$. It has no largest term, but the subsequence $-1, -2, -3, \cdots$ has the largest term -1.

Example 3. The sequence $1, -1, \frac{1}{2}, -\frac{1}{2}, \frac{1}{3}, -\frac{1}{3}, \cdots$ contains the monotonic subsequences $1, \frac{1}{2}, \frac{1}{3}, \cdots$ and $-1, -\frac{1}{2}, -\frac{1}{3}, \cdots$, and has the largest term 1. The subsequence $-1, -\frac{1}{2}, -\frac{1}{3}, \cdots$ has no largest term.

Example 4. The sequence $1, \frac{1}{2}, \frac{2}{3}, \frac{3}{4}, \frac{4}{5}, \cdots$ contains the monotonic subsequence $\frac{1}{2}, \frac{2}{3}, \frac{3}{4}, \cdots$. The sequence has the largest term 1, but no monotonic subsequence has a largest term.

---

† Unless otherwise qualified the word *sequence* should be interpreted in the rest of this chapter to mean *infinite sequence of real numbers*.

As an aid in the proof of Lemma I we establish a simple auxiliary lemma:

Lemma II. *If a sequence S has a largest term equal to x, and if a subsequence T of S has a largest term equal to y, then $x \geqq y$.*

Proof of Lemma II. Let $S = \{a_n\}$. Since T is a subsequence of S, y is equal to some term a_n of S. If a_k is the largest term of S, then $a_k \geqq a_n$, or $x \geqq y$.

We are now ready to prove Lemma I:

Proof of Lemma I. Let $S = a_1, a_2, a_3, \cdots$ be a given sequence of real numbers, and let certain subsequences of S be denoted as follows: $S_0 \equiv S$, $S_1 \equiv a_2, a_3, a_4, \cdots$, $S_2 \equiv a_3, a_4, a_5, \cdots$, etc. There are two cases to consider:

Case I. Every sequence $S_0, S_1, S_2, \cdots$ contains a largest term. Denote by a_{n_1} the largest term of S_0, by a_{n_2} the largest term of S_{n_1}, by a_{n_3} the largest term of S_{n_2}, etc. By construction, $n_1 < n_2 < n_3 \cdots$, so that $a_{n_1}, a_{n_2}, a_{n_3}, \cdots$ is a subsequence of S, and by Lemma II it is a monotonically decreasing subsequence.

Case II. There exists a sequence S_N containing no largest term. Then every term of S_N is followed ultimately by some larger term (for otherwise there would be some term of S_N which could be exceeded only by its predecessors, of which there are at most a finite number, so that S_N would have a largest term). We can therefore obtain inductively an increasing subsequence of S by letting $a_{n_1} \equiv a_{N+1}$, $a_{n_2} \equiv$ the first term following a_{n_1} that is greater than a_{n_1}, $a_{n_3} \equiv$ the first term following a_{n_2} that is greater than a_{n_2}, etc. This completes the proof of Lemma I, and therefore that of the Fundamental Theorem.

*218. PROOFS OF SOME THEOREMS ON CONTINUOUS FUNCTIONS

The Fundamental Theorem on bounded sequences provides us with the means of proving the theorems on continuous functions given in §§ 213 and 215. Some important generalizations of these theorems are obtained in Chapter 6.

Proof of Theorem I, § 213. In order to prove that a function continuous on a closed interval is bounded there, we assume the contrary and seek a contradiction. Let $f(x)$ be continuous on $[a, b]$, and assume it is unbounded there. Under these conditions, for any positive integer n there is a point x_n of the interval $[a, b]$ such that $|f(x_n)| > n$. By the Fundamental Theorem on bounded sequences (§ 217), since the sequence $\{x_n\}$ is bounded, it contains a convergent subsequence $\{x_{n_k}\}$, converging to some point x_0, $x_{n_k} \to x_0$. By Exercise 13, § 205, $a \leqq x_{n_k} \leqq b$ implies $a \leqq x_0 \leqq b$, so that x_0 also belongs to the closed interval $[a, b]$. Since $f(x)$ is continuous at x_0, $f(x_{n_k}) \to f(x_0)$

(Theorem, § 209). But this means that the sequence $\{f(x_{n_k})\}$ is bounded, a statement inconsistent with the inequality
$$|f(x_n)| > n_k.$$
This contradiction establishes Theorem I, § 213.

Proof of Theorem II, § 213. Let $f(x)$ be continuous on the closed interval $[a, b]$. We shall show that it has a maximum value there (the proof for a minimum value is similar). By the preceding theorem, $f(x)$ is bounded on $[a, b]$. Let M be the least upper bound of its values there. We wish to show that there is some number x_0 such that $f(x_0) = M$. By the definition of M, there are values of $f(x)$ arbitrarily close to M. For each positive integer n, let x_n be a point in the interval $[a, b]$ such that $|f(x_n) - M| < \frac{1}{n}$. Then $\lim_{n \to +\infty} f(x_n) = M$. Furthermore, since the sequence $\{x_n\}$ is bounded, it contains a convergent subsequence $\{x_{n_k}\}$ (by the Fundamental Theorem of § 217) converging to some point x_0 of the closed interval (cf. proof of Theorem 1, § 213), $x_{n_k} \to x_0$. Since $f(x)$ is continuous at x_0, $f(x_{n_k}) \to f(x_0)$ (Theorem, §209). Since $\{f(x_{n_k})\}$ is a subsequence of the sequence $\{f(x_n)\}$ and $f(x_n) \to M$, $f(x_{n_k}) \to M$ (Theorem IV, § 204). Finally, by the uniqueness of the limit of the convergent sequence $\{f(x_{n_k})\}$ (Theorem II, § 204), $f(x_0) = M$, and the proof is complete.

Proof of Theorem III, §213. Assume for definiteness that $f(a) < 0$ and $f(b) > 0$. We wish to find a number x_0 between a and b such that $f(x_0) = 0$. Let A be the set of points x of the interval $[a, b]$ where $f(x) < 0$ (this set contains at least the point a and is bounded above by b) and *define* x_0 to be the least upper bound of the set A: $x_0 \equiv \sup(A)$. We shall prove that $f(x_0) = 0$ by showing that *each* of the inequalities $f(x_0) < 0$ and $f(x_0) > 0$ is impossible. First assume that $f(x_0) < 0$. By Note 4, § 209, any function negative and continuous at a point is negative (where defined) in some neighborhood of that point. If $f(x_0) < 0$, therefore, there must exist some neighborhood of x_0 throughout which $f(x) < 0$, or since $x_0 < b$, this means that $f(x) < 0$ for some values of x greater than x_0, in contradiction to the fact that x_0 is an upper bound of A. Finally, by the same argument, if $f(x_0) > 0$ then $f(x) > 0$ for values of x throughout some neighborhood of x_0, so that some number *less* than x_0 is an upper bound of A. This fact is inconsistent with the definition of x_0 as the *least* upper bound of A. The only alternative left is the one sought: $f(x_0) = 0$.

Proof of Theorem IV, § 213. Assume that c lies between the two numbers $f(a)$ and $f(b)$, and consider the function $g(x) \equiv f(x) - c$. Since $g(x)$ is continuous throughout the closed interval $[a, b]$ and has opposite signs at a and b, there must exist, by the theorem just established, a number x_0 between a and b at which $g(x)$ vanishes, so that $g(x_0) = f(x_0) - c = 0$, or $f(x_0) = c$.

Proof of the Theorem of § 215. The only part of the proof that presents any difficulty is the continuity of the inverse function. We shall seek a contradiction to the assumption that $x = \phi(y)$ is discontinuous at some point y_0 of the interval $[c, d]$. Accordingly, by Exercise 14, § 212, there must be a positive number ϵ such that however small the positive number δ may be, there exists a number y of the interval $[c, d]$ such that $|y - y_0| < \delta$ and $|\phi(y) - \phi(y_0)| \geq \epsilon$. We construct the sequence $\{y_n\}$ such that if $x_n \equiv \phi(y_n)$ and $x_0 \equiv \phi(y_0)$, then $|y_n - y_0| < 1/n$ and $|x_n - x_0| \geq \epsilon$. The former inequality guarantees the convergence of $\{y_n\}$ to y_0. The boundedness of the sequence $\{x_n\}$ (each x_n belongs to the interval $[a, b]$), by the Fundamental Theorem of § 217, ensures the convergence of some subsequence $\{x_{n_k}\}$ to some number x_0' of the interval $[a, b]$ (cf. proof of Theorem I, § 213). The inequality $|x_n - x_0| \geq \epsilon$ implies $|x_{n_k} - x_0| \geq \epsilon$, for all k, so that the subsequence $\{x_{n_k}\}$ cannot converge to x_0. In other words, $x_0' \neq x_0$. On the other hand, the direct function $y = f(x)$ is assumed to be continuous at x_0', so that $x_{n_k} \to x_0'$ implies $y_{n_k} = f(x_{n_k}) \to f(x_0')$. But $y_n \to y_0 = f(x_0)$ implies $y_{n_k} \to f(x_0)$ (Theorem IV, § 204). Finally, the uniqueness of limits (Theorem II, § 204) means that $f(x_0) = f(x_0')$, in contradiction to the assumption that $f(x)$ is strictly monotonic.

*219. THE CAUCHY CRITERION FOR CONVERGENCE OF A SEQUENCE

Convergence of a sequence means that there is some number (the limit of the sequence) that has a particular property, formulated in terms of ϵ and N. In order to test the convergence of a given sequence, then, it might seem that one is forced to obtain its limit first. That this is not *always* the case was seen in Theorem XIV, § 204, where the convergence of a *monotonic* sequence is guaranteed by the simple condition of *boundedness*. A natural question to ask now is whether there is some way of testing an *arbitrary* sequence for convergence without knowing in advance what its limit is. It is the purpose of this section to answer this question by means of the celebrated Cauchy Criterion (due to the French mathematician A. L. Cauchy (1789–1857)) which, to put it crudely, says that the terms of a sequence get arbitrarily close to something fixed, for sufficiently large subscripts, if and only if they get arbitrarily close to each other, for sufficiently large subscripts.

Definition. *If $\{a_n\}$ is a sequence (of real numbers), the notation* $\lim_{m,n \to +\infty} (a_m - a_n) = 0$ *means that corresponding to an arbitrary positive number ϵ there exists a number N such that $m > N$ and $n > N$ together imply $|a_m - a_n| < \epsilon$. A sequence $\{a_n\}$ satisfying the condition* $\lim_{m,n \to +\infty} (a_m - a_n) = 0$ *is called a **Cauchy sequence**.*

Theorem. Cauchy Criterion. *A sequence (of real numbers) converges if and only if it is a Cauchy sequence.*

Proof. The "only if" part of the proof is simple. Assume that the sequence $\{a_n\}$ converges, and let its limit be a: $a_n \to a$. We wish to show that $\lim_{m,n \to +\infty} (a_m - a_n) = 0$. Corresponding to a preassigned $\epsilon > 0$, let N be a number such that $n > N$ implies $|a_n - a| < \frac{1}{2}\epsilon$. Then if m and n are both greater than N, we have simultaneously

$$|a_m - a| < \tfrac{1}{2}\epsilon \quad \text{and} \quad |a_n - a| < \tfrac{1}{2}\epsilon,$$

so that by the triangle inequality (§ 110)

$$|a_m - a_n| = |(a_m - a) - (a_n - a)| \leq |a_m - a| + |a_n - a| < \epsilon.$$

For the "if" half of the proof we assume that a sequence $\{a_n\}$ satisfies the Cauchy condition, and prove that it converges. The first step is to show that it is bounded. To establish boundedness we choose N so that $n > N$ implies $|a_n - a_{N+1}| < 1$. (This is the Cauchy Criterion with $\epsilon = 1$ and $m = N+1$.) Then, by the triangle inequality, for $n > N$,

$$|a_n| = |(a_n - a_{N+1}) + a_{N+1}| \leq |a_n - a_{N+1}| + |a_{N+1}| < |a_{N+1}| + 1.$$

Since the terms of the sequence are bounded for all $n > N$, and since there are only a finite number of terms with subscripts less than N, the entire sequence is bounded. According to the Fundamental Theorem (§ 217), then, there must be a convergent subsequence $\{a_{n_k}\}$. Let the limit of this subsequence be a. We shall show that $a_n \to a$. Let ϵ be an arbitrary positive number. Then, by the Cauchy condition, there exists a number N such that for $m > N$ and $n > N$, $|a_m - a_n| < \frac{1}{2}\epsilon$. Let n be an arbitrary positive integer greater than N. We shall show that $|a_n - a| < \epsilon$. By the triangle inequality,

$$|a_n - a| \leq |a_n - a_{n_k}| + |a_{n_k} - a|.$$

This inequality holds for all positive integers k and, in particular, for values of k so large that (i) $n_k > N$ and (ii) $|a_{n_k} - a| < \frac{1}{2}\epsilon$. If we choose any such value for k, the inequality above implies $|a_n - a| < \frac{1}{2}\epsilon + \frac{1}{2}\epsilon = \epsilon$, and the proof is complete.

NOTE. Without the Axiom of Completeness, the Cauchy condition may be satisfied even for some divergent sequences. For example, in the system of rational numbers, where the Axiom of Completeness fails (Ex. 9, § 114), a sequence of rational numbers converging to the irrational number $\sqrt{2}$ (Ex. 15, § 205) satisfies the Cauchy condition, but does not converge *in the system of rational numbers*.

*220. EXERCISES

★1. Prove that a sequence $\{a_n\}$ (of real numbers) converges if and only if corresponding to an arbitrary positive number ϵ there exists a positive integer N such that for all positive integers p and q, $|a_{N+p} - a_{N+q}| < \epsilon$.

*2. Prove that a sequence $\{a_n\}$ (of real numbers) converges if and only if corresponding to an arbitrary positive number ϵ there exists a number N such that $n > N$ implies $|a_n - a_N| < \epsilon$.

*3. Prove that a sequence $\{a_n\}$ (of real numbers) converges if and only if corresponding to an arbitrary positive number ϵ there exists a positive integer N such that for all positive integers p, $|a_{N+p} - a_N| < \epsilon$.

*4. Prove that the condition $\lim\limits_{n \to +\infty} (a_{n+p} - a_n) = 0$ for every positive integer p is necessary but not sufficient for the convergence of the sequence $\{a_n\}$. *Hint:* Consider a sequence suggested by the terms $1, 2, 2\frac{1}{2}, 3, 3\frac{1}{3}, 3\frac{2}{3}, 4, 4\frac{1}{4}, 4\frac{1}{2}, 4\frac{3}{4}, 5, 5\frac{1}{5}, \cdots$.

*5. Find the error in the "theorem" and "proof": *If $a_n \to a$, then $a_n = a$ for sufficiently large n. Proof.* Any convergent sequence is a Cauchy sequence. Therefore, if $\epsilon > 0$ there exists a positive integer N such that for all positive integers m and n, with $m > N$, $|a_m - a_{N+n}| < \epsilon$. But this means that $\lim\limits_{n \to +\infty} a_{N+n} = a_m$, which in turn implies that $\lim\limits_{n \to +\infty} a_n = a = a_m$ for all $m > N$.

*6. If $\{b_n\}$ is a convergent sequence and if $\{a_n\}$ is a sequence such that $|a_m - a_n| \leq |b_m - b_n|$ for all positive integers m and n, prove that $\{a_n\}$ converges.

*7. A number x is called a **limit point** of a sequence $\{a_n\}$ if and only if there exists some subsequence of $\{a_n\}$ converging to x. Prove that x is a limit point of $\{a_n\}$ if and only if corresponding to $\epsilon > 0$ the inequality $|a_n - x| < \epsilon$ holds for infinitely many values of the subscript n. Show by an example that this does not mean that $|a_n - x| < \epsilon$ must hold for infinitely many *distinct* values of a_n.

*8. Prove that a bounded sequence converges if and only if it has exactly one limit point. (Cf. Ex. 7.)

*9. Explain what you would mean by saying that $+\infty$ $(-\infty)$ is a limit point of a sequence $\{a_n\}$. Prove that a sequence is unbounded above (below) if and only if $+\infty$ $(-\infty)$ is a limit point of the sequence. Show how the word "bounded" can be omitted from Exercise 8. (Cf. Ex. 7.)

*10. Let $\{a_n\}$ be a sequence of real numbers, and let A_n be the least upper bound of the set $\{a_n, a_{n+1}, a_{n+2}, \cdots\}$ ($A_n \equiv +\infty$ if this set is not bounded above). Prove that either $A_n = +\infty$ for every n, or A_n is a monotonically decreasing sequence of real numbers, and that therefore $\lim\limits_{n \to +\infty} A_n$ exists ($+\infty$, finite, or $-\infty$). Prove a similar result for the sequence $\{\alpha_n\}$, where α_n is the greatest lower bound of the set $\{a_n, a_{n+1}, a_{n+2}, \cdots\}$.

*11. The **limit superior** and the **limit inferior** of a sequence $\{a_n\}$, denoted $\overline{\lim\limits_{n \to +\infty}} a_n$, or $\limsup\limits_{n \to +\infty} a_n$, and $\underline{\lim\limits_{n \to +\infty}} a_n$, or $\liminf\limits_{n \to +\infty} a_n$, respectively, are defined as the limits of the sequences $\{A_n\}$ and $\{\alpha_n\}$, respectively, of Exercise 10. Justify the following formulations:

$$\overline{\lim\limits_{n \to +\infty}} a_n \equiv \lim\limits_{n \to +\infty} A_n = \inf\limits_{n=1}^{+\infty} \left[\sup\limits_{m=n}^{+\infty} (a_m)\right],$$

$$\underline{\lim\limits_{n \to +\infty}} a_n \equiv \lim\limits_{n \to +\infty} \alpha_n = \sup\limits_{n=1}^{+\infty} \left[\inf\limits_{m=n}^{+\infty} (a_m)\right].$$

(Cf. § 112, Exs. 9–22, § 223.)

*12. Prove that a number L is the limit superior of a sequence $\{a_n\}$ if and only if it has the following two properties, where ϵ is an arbitrary preassigned positive number:
   (i) The inequality $a_n < L + \epsilon$ holds for all but a finite number of terms.
   (ii) The inequality $a_n > L - \epsilon$ holds for infinitely many terms.
State and prove a similar result for the limit inferior.
Prove that the limit superior and limit inferior of a bounded sequence are limit points of that sequence (cf. Exs. 7–9), and that any other limit point of the sequence is between these two. Extend this result to unbounded sequences. (Cf. Exs. 10–11, and Ex. 25, § 618.)

*13. Prove that for any sequence $\{a_n\}$, bounded or not, $\varliminf\limits_{n\to+\infty} a_n \leq \varlimsup\limits_{n\to+\infty} a_n$. Prove that a sequence converges if and only if its limit superior and limit inferior are finite and equal, and that in the case of convergence, $\lim\limits_{n\to+\infty} a_n = \varlimsup\limits_{n\to+\infty} a_n = \varliminf\limits_{n\to+\infty} a_n$. Extend these results to include infinite cases. (Cf. Exs. 10–12.)

In Exercises 14–17, find the limit superior and the limit inferior. (Cf. Exs. 10–13.)

*14. $0, 1, 0, 1, 0, 1, \cdots$.
*15. $1, -2, 3, -4, \cdots, (-1)^{n+1}n, \cdots$.
*16. $\frac{2}{3}, \frac{1}{3}, \frac{3}{4}, \frac{1}{4}, \frac{4}{5}, \frac{1}{5}, \frac{5}{6}, \frac{1}{6}, \cdots$.
*17. $\frac{3}{2}, -\frac{1}{2}, \frac{4}{3}, -\frac{1}{3}, \frac{5}{4}, -\frac{1}{4}, \frac{6}{5}, -\frac{1}{5}, \cdots$.

In Exercises 18–20, establish the inequalities, including all cases except $(-\infty) + (+\infty)$ and $0 \cdot (+\infty)$ (cf. Exs. 43–56, § 208). (The notation $n \to +\infty$ is omitted for conciseness.)

*18. $\varliminf(-a_n) = -\varlimsup a_n$; if $a_n > 0$, $\varliminf(1/a_n) = 1/\varlimsup a_n$.
*19. $\varliminf a_n + \varliminf b_n \leq \varliminf(a_n + b_n) \leq \varlimsup(a_n + b_n) \leq \varlimsup a_n + \varlimsup b_n$.
*20. If $a_n, b_n \geq 0$, $\varliminf a_n \varliminf b_n \leq \varliminf a_n b_n \leq \varlimsup a_n b_n \leq \varlimsup a_n \varlimsup b_n$.

*221. SEQUENTIAL CRITERIA FOR CONTINUITY AND EXISTENCE OF LIMITS

For future purposes it will be convenient to have a necessary and sufficient condition for continuity of a function at a point, in a form that involves only sequences of numbers, and a similar condition for existence of limits. (Cf. the Theorem, § 209.)

Theorem I. *A necessary and sufficient condition for a function $f(x)$ to be continuous at a point a of its domain of definition is that whenever $\{x_n\}$ is a sequence of numbers which converges to a (and for which $f(x)$ is defined), then $\{f(x_n)\}$ is a sequence of numbers converging to $f(a)$; in short, that $x_n \to a$ implies $f(x_n) \to f(a)$.*

Proof. We first establish necessity. Assume that $f(x)$ is continuous at $x = a$, and let $x_n \to a$. We wish to show that $f(x_n) \to f(a)$. Let $\epsilon > 0$ be given. Then there exists a positive number δ such that $|x - a| < \delta$ implies $|f(x) - f(a)| < \epsilon$ (for values of x for which $f(x)$ is defined). Since $x_n \to a$ there exists a number N such that, for $n > N$, $|x_n - a| < \delta$. Therefore, for $n > N$, $|f(x_n) - f(a)| < \epsilon$. This establishes the desired convergence.

60   FUNCTIONS, SEQUENCES, LIMITS, CONTINUITY   [§ 222

Next we prove sufficiency, by assuming that $f(x)$ is discontinuous at $x = a$, and showing that we can then obtain a sequence $\{x_n\}$ converging to a such that the sequence $\{f(x_n)\}$ does not converge to $f(a)$. By Exercise 14, § 212, the discontinuity of $f(x)$ at $x = a$ means that there exists a positive number ϵ such that however small the positive number δ may be, there exists a number x such that $|x - a| < \delta$ and $|f(x) - f(a)| \geq \epsilon$. We construct the sequence $\{x_n\}$ by requiring that x_n satisfy the two inequalities $|x_n - a| < 1/n$ and $|f(x_n) - f(a)| \geq \epsilon$. The former guarantees the convergence of $\{x_n\}$ to a, while the latter precludes the convergence of $\{f(x_n)\}$ to $f(a)$. This completes the proof of Theorem I.

The formulation and the proof of a sequential criterion for the existence of a limit of a function are similar to those for continuity. The statement in the following theorem does not specify the manner in which the independent variable approaches its limit, since the result is independent of the behavior of the independent variable, subject to the restrictions of the first paragraph of § 207. The details of the proof, with hints, are left to the student in Exercises 1 and 2, § 223.

Theorem II. *The limit, $\lim f(x)$, of the function $f(x)$ exists (finite or infinite) if and only if, for every sequence $\{x_n\}$ of numbers having the same limit as x but never equal to this limit (and for which $f(x)$ is defined), the sequence $\{f(x_n)\}$ has a limit (finite or infinite); in short, if and only if $x_n \to \lim x$, $x_n \neq \lim x$ imply $f(x_n) \to$ limit.*

As might be expected, the Cauchy Criterion for convergence of a sequence has its application to the question of the existence of a limit of a function. With the same understanding regarding the behavior of the independent variable as was assumed for Theorem II, we have as an immediate corollary of that theorem the following:

Theorem III. *The limit, $\lim f(x)$, of the function $f(x)$ exists and is finite if and only if, for every sequence $\{x_n\}$ of numbers approaching the same limit as x but never equal to this limit (and for which $f(x)$ is defined), the sequence $\{f(x_n)\}$ is a Cauchy sequence; in short, if and only if $x_n \to \lim x$, $x_n \neq \lim x$ imply $\{f(x_n)\}$ is a Cauchy sequence.*

*222. THE CAUCHY CRITERION FOR FUNCTIONS

The Cauchy Criterion for sequences gives a test for the convergence of a sequence that does not involve explicit evaluation of the limit of the sequence. Similar tests can be formulated for the existence of finite limits for more general functions, where explicit evaluation of the limits is not a part of the test. Such criteria are of great theoretical importance and practical utility whenever direct evaluation of a limit is difficult. Owing to the latitude in the behavior granted the independent variable, we have selected for specific formulation in this section only two particular cases, and the proof of one. The remaining proof and other special cases are treated in the exercises of § 223. The assumptions of the first paragraph of § 206 are still in effect.

Theorem I. *The limit* $\lim_{x \to a} f(x)$ *exists and is finite if and only if corresponding to an arbitrary positive number ϵ there exists a positive number δ such that $0 < |x' - a| < \delta$ and $0 < |x'' - a| < \delta$ imply $|f(x') - f(x'')| < \epsilon$, for values of x' and x'' for which $f(x)$ is defined.*

Proof. "Only if": Assume $\lim_{x \to a} f(x) = L$ and let $\epsilon > 0$. Then there exists $\delta > 0$ such that $0 < |x - a| < \delta$ implies $|f(x) - L| < \frac{1}{2}\epsilon$. If x' and x'' are any two numbers such that $0 < |x' - a| < \delta$ and $0 < |x'' - a| < \delta$, the triangle inequality gives

$$|f(x') - f(x'')| = |(f(x') - L) - (f(x'') - L)|$$
$$\leq |(f(x') - L| + |f(x'') - L| < \tfrac{1}{2}\epsilon + \tfrac{1}{2}\epsilon = \epsilon.$$

"If": Let $\{x_n\}$ be an arbitrary sequence of numbers (for which $f(x)$ is defined) such that $x_n \to a$ and $x_n \neq a$. By Theorem III, § 221, we need only show that the sequence $\{f(x_n)\}$ is a Cauchy sequence:

$$\lim_{m, n \to +\infty} [f(x_m) - f(x_n)] = 0.$$

If ϵ is a preassigned positive number, let δ be a positive number having the assumed property that $0 < |x' - a| < \delta$ and $0 < |x'' - a| < \delta$ imply $|f(x') - f(x'')| < \epsilon$. Since $x_n \to a$ and $x_n \neq a$, there exists a number N such that $n > N$ implies $0 < |x_n - a| < \delta$. Accordingly, if $m > N$ and $n > N$, we have simultaneously $0 < |x_m - a| < \delta$ and $0 < |x_n - a| < \delta$, so that $|f(x_m) - f(x_n)| < \epsilon$. Thus the sequence $\{f(x_n)\}$ is a Cauchy sequence, and the proof is complete.

Theorem II. *The limit* $\lim_{x \to +\infty} f(x)$ *exists and is finite if and only if corresponding to an arbitrary positive number ϵ there exists a number N such that $x' > N$ and $x'' > N$ imply $|f(x') - f(x'')| < \epsilon$ (for values of x' and x'' for which $f(x)$ is defined).*

*223. EXERCISES

*1. Prove Theorem II, § 221, for the case $x \to a$. *Hint:* First show that if $x_n \to a$, $x_n \neq a$ imply that the sequence $\{f(x_n)\}$ has a limit then this limit is unique, by considering, for any two sequences $\{x_n\}$ and $\{x_n'\}$ which converge to a, the compound sequence $x_1, x_1', x_2, x_2', x_3, x_3', \cdots$. For the case $f(x_n) \to +\infty$, assume that $\lim_{x \to a} f(x) \neq +\infty$, and show that there must exist a constant B and a sequence $\{x_n\}$ such that $0 < |x_n - a| < 1/n$ and $f(x_n) \leq B$. For other cases of $\lim_{x \to a} f(x_n)$, proceed similarly.

*2. Prove Theorem II, § 221, for the case $x \to a+$; $x \to a-$; $x \to +\infty$; $x \to -\infty$; $x \to \infty$. (Cf. Ex. 1.)

*3. Assuming only Theorem I, § 221, and the limit theorems for sequences (§ 204), prove the continuity theorems of § 211. *Hint for Theorem IV:* Let $\{x_n\}$ be an arbitrary sequence converging to a. Then $f(x_n) \to f(a)$ and $g(x_n) \to g(a)$, and therefore $f(x_n)g(x_n) \to f(a)g(a)$.

*4. Assuming only Theorem II, § 221, and the limit theorems for sequences (§ 204), prove the limit theorems of § 207. (Cf. Ex. 3.)

*5. Reformulate and prove Theorem I, § 222, for the case $x \to a+$; for the case $x \to a-$.

*6. Prove Theorem II, § 222.

*7. Reformulate and prove Theorem II, § 222, for the case $x \to -\infty$; for the case $x \to \infty$.

*8. If $f(x)$ and $g(x)$ are functions defined for $x > 0$, if $\lim_{x \to 0+} g(x)$ exists and is finite, and if $|f(b) - f(a)| \leq |g(b) - g(a)|$ for all positive numbers a and b, prove that $\lim_{x \to 0+} f(x)$ exists and is finite.

*9. Let $f(x)$ be a real-valued function, defined for at least some values of x neighboring the point $x = a$ (except possibly at a itself). If δ is an arbitrary positive number, let $\phi(\delta)$ and $\psi(\delta)$ be the least upper bound and the greatest lower bound, respectively, of the values of $f(x)$ for all x such that $0 < |x - a| < \delta$ (and for which $f(x)$ is defined). Prove that, in a sense that includes infinite values (cf. Ex. 10, § 220), $\phi(\delta)$ and $\psi(\delta)$ are monotonic functions and therefore have limits as $\delta \to 0+$. (Cf. § 215.)

*10. The limit superior and the limit inferior of a function $f(x)$ at a point $x = a$, denoted $\overline{\lim}_{x \to a} f(x)$, or $\limsup_{x \to a} f(x)$, and $\underline{\lim}_{x \to a} f(x)$, or $\liminf_{x \to a} f(x)$, respectively, are defined as the limits of the functions $\phi(\delta)$ and $\psi(\delta)$, respectively, of Exercise 9. Justify the following formulations (cf. Exs. 11–13, § 220, Exs. 26–28, § 618):

$$\overline{\lim_{x \to a}} f(x) \equiv \lim_{\delta \to 0+} \phi(\delta) = \inf_{\delta > 0}\left[\sup_{0 < |x-a| < \delta} f(x)\right],$$

$$\underline{\lim_{x \to a}} f(x) \equiv \lim_{\delta \to 0+} \psi(\delta) = \sup_{\delta > 0}\left[\inf_{0 < |x-a| < \delta} f(x)\right].$$

*11. Prove that a number L is the limit superior of a function $f(x)$ at $x = a$ if and only if it has the following two properties, where ϵ is an arbitrary preassigned positive number:

(i) The inequality $f(x) < L + \epsilon$ holds for all x in some deleted neighborhood of a: $0 < |x - a| < \delta$.

(ii) The inequality $f(x) > L - \epsilon$ holds for some x in every deleted neighborhood of a: $0 < |x - a| < \delta$.

State and prove a similar result for the limit inferior. (Cf. Exs. 9–10.)

*12. Prove that for any function $f(x)$, $\underline{\lim}_{x \to a} f(x) \leq \overline{\lim}_{x \to a} f(x)$ (cf. Ex. 13, § 220). Prove that $\lim_{x \to a} f(x)$ exists if and only if $\underline{\lim}_{x \to a} f(x) = \overline{\lim}_{x \to a} f(x)$, and in case of equality, $\lim_{x \to a} f(x) = \underline{\lim}_{x \to a} f(x) = \overline{\lim}_{x \to a} f(x)$. (Cf. Exs. 9–11.)

*13. Formulate definitions and state and prove results corresponding to those of Exercises 9–12 for the case $x \to a+$; $x \to a-$.

*14. Formulate definitions and state and prove results corresponding to those of Exercises 9–12 for the case $x \to +\infty$; $x \to -\infty$; $x \to \infty$.

In Exercises 15–20, find the limit superior and the limit inferior. Draw a graph. (Cf. Exs. 9–14.)

*15. $x \sin x$, as $x \to +\infty$.

*16. $\cos \dfrac{1}{x}$, as $x \to 0$.

*17. $\dfrac{x+1}{x} \sin x$, as $x \to +\infty$.

*18. $\dfrac{x+1}{x} \sin x$, as $x \to -\infty$.

*19. $\dfrac{x-1}{x} \cos x$, as $x \to +\infty$.

*20. $(x^2 + 1) \sin \dfrac{1}{x}$, as $x \to 0+$.

*21. State and prove relations for functions corresponding to those of Exercises 18–20, § 220.

*22. A function $f(x)$ is **upper semicontinuous** at a point $x = a$ if and only if $\varlimsup\limits_{x \to a} f(x) \leq f(a)$, and is **lower semicontinuous** there if and only if $\varliminf\limits_{x \to a} f(x) \geq f(a)$. Prove that a function is continuous at a point if and only if it is both upper and lower semicontinuous at the point.

*224. UNIFORM CONTINUITY

The function $\dfrac{1}{x}$ is continuous in the open interval $(0, 1)$ (Ex. 2, § 212). Let us consider this statement from the point of view of the δ-ϵ definition of continuity. Let x_0 be a point in $(0, 1)$, and let ϵ be a preassigned positive number. We wish to find δ as a function of ϵ so that $|x - x_0| < \delta$ implies $\left| \dfrac{1}{x} - \dfrac{1}{x_0} \right| < \epsilon$. In other words, by making the numerator of the fraction $\dfrac{|x - x_0|}{|x| \cdot x_0}$ small, we wish to ensure the smallness of the fraction itself. It is clear, however, that mere smallness of the numerator alone is not going to be enough. If the denominator is also small, the fraction is not restricted. The trick (cf. Exs. 59–64, § 208) is to pin down the denominator first by not permitting x to come too close to 0, and then to tackle the numerator. Without going through any of the details, however, we can see by inspecting Figure 210 that δ is going to depend quite essentially not only on ϵ but on the point x_0 as well. If x_0 is near 1, and an ϵ is given, we can be fairly generous in the size of δ. However, if x_0 is near 0, and the same ϵ is given, the δ required must be considerably smaller.

In some cases it is possible, for a given interval or other set, and for a given ϵ, to choose δ so that the inequalities concerned will hold for all points of the set *without varying the number δ*. When this state exists we have a sort of *uniformity* to the amount of squeezing on $|x - x_0|$ that has to be imposed. This leads to the definition:

Definition. *A function $f(x)$, defined on a set A, is* **uniformly continuous** *on the set A if and only if corresponding to an arbitrary positive number ϵ*

there exists a positive number $\delta = \delta(\epsilon)$ such that for any x' and x'' belonging to A, $|x' - x''| < \delta$ implies $|f(x') - f(x'')| < \epsilon$.

Let us contrast the definitions of continuity and uniform continuity. The most obvious distinction is that continuity is defined *at a point*, whereas uniform continuity is defined *on a set*. These concepts are also distinguished by the order in which things happen. In the case of continuity we have (1) the point x_0, (2) the positive number ϵ, and (3) the positive number δ, which

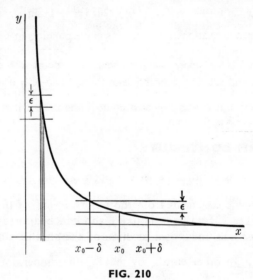

FIG. 210

depends on both x_0 and ϵ. In the case of uniform continuity we have (1) the positive number ϵ, (2) the positive number δ, which depends only on ϵ, and (3) the points x' and x''.

In spite of these contrasts, there are certain relations between the two ideas. The simplest one is stated in the following theorem, whose proof is requested in Exercise 1, § 225.

Theorem I. *A function uniformly continuous on a set is continuous at each point of that set. In brief, uniform continuity implies continuity.*

The converse, as we have seen by an example, is false. The function $\dfrac{1}{x}$ is continuous on the open interval $(0, 1)$, but it is not uniformly continuous there. (Cf. Ex. 9, § 225.) However, it is an important fact that for *closed intervals* there is a valid converse, as given in Theorem II below. (For a more general theorem see § 617.) As an aid in the proof of Theorem II, we first formulate the *negation* of uniform continuity. The student is asked in Exercise 2, § 225, to validate this formulation. (Cf. Ex. 14, § 212.)

Negation of Uniform Continuity. *A function $f(x)$, defined on a set A, fails to be uniformly continuous on A if and only if there exists a positive number ϵ having the property that for any positive number δ there exist points x' and x'' of A such that $|x' - x''| < \delta$ and $|f(x') - f(x'')| \geq \epsilon$.*

Theorem II. *A function continuous on a closed interval is uniformly continuous there.*

Proof. Assume that $f(x)$ is continuous on the closed interval I but fails to be uniformly continuous there. Then (by the Negation above) for any positive integer n there exist $\epsilon > 0$ and points x_n' and x_n'' of I such that $|x_n' - x_n''| < \frac{1}{n}$ and $|f(x_n') - f(x_n'')| \geq \epsilon$. The bounded sequence $\{x_n'\}$ contains a convergent subsequence $\{x_{n_k}'\}$, converging to some point x_0 of I (cf. proof of Theorem I, § 213, given in § 218). We show that the corresponding sequence $\{x_{n_k}''\}$ also converges to x_0 by use of the triangle inequality:

$$|x_{n_k}'' - x_0| = |(x_{n_k}'' - x_{n_k}') + (x_{n_k}' - x_0)|$$
$$\leq |x_{n_k}'' - x_{n_k}'| + |x_{n_k}' - x_0| < 1/n_k + |x_{n_k}' - x_0|.$$

As k becomes infinite each of the last two terms approaches zero, and therefore so does the quantity $|x_{n_k}'' - x_0|$. By continuity of $f(x)$ at x_0, $\lim_{k \to +\infty} f(x_{n_k}') = \lim_{k \to +\infty} f(x_{n_k}'') = f(x_0)$, and therefore $\lim_{k \to +\infty} [f(x_{n_k}') - f(x_{n_k}'')] = 0$. This last statement is inconsistent with the inequality $|f(x_{n_k}') - f(x_{n_k}'')| \geq \epsilon$, which must hold for all values of k and the proof is complete.

*225. EXERCISES

*1. Prove Theorem I, § 224.
*2. Establish the Negation of Uniform Continuity, § 224.

In Exercises 3–8, find an explicit function $\delta = \delta(\epsilon)$ in conformity with the definition of uniform continuity.

*3. $y = x^2$, for $0 \leq x \leq 1$. *Hint:*
$$|x''^2 - x'^2| = |x'' - x'| \cdot |x'' + x'| \leq 2|x'' - x'|.$$

*4. $y = x^2$, for $0 \leq x \leq 2$.

*5. $y = \sqrt{x}$, for $1 \leq x \leq 2$. *Hint:* $\sqrt{x''} - \sqrt{x'} = \dfrac{x'' - x'}{\sqrt{x''} + \sqrt{x'}}$.

*6. $y = \sqrt{x}$, for $0 \leq x \leq 1$. *Hint:*
$$\sqrt{x''} + \sqrt{x'} \geq \sqrt{|x'' - x'|} \quad \text{(cf. Ex. 5)}.$$

*7. $y = \dfrac{1}{x}$, for $x \geq 1$.

*8. $y = \sqrt{1 - x^2}$, for $|x| \leq 1$. *Hint:* For $0 \leq x' < x'' \leq 1$,
$$\sqrt{1 - x'^2} - \sqrt{1 - x''^2} = \frac{(x'' + x')(x'' - x')}{\sqrt{1 + x'}\sqrt{1 - x'} + \sqrt{1 + x''}\sqrt{1 - x''}}$$
$$\leq \frac{2(x'' - x')}{\sqrt{1 - x'} + \sqrt{1 - x''}} \leq \frac{2(x'' - x')}{\sqrt{(1 - x') - (1 - x'')}} \quad \text{(cf. Ex. 6)}.$$

In Exercises 9–12, use the Negation of § 224 to show that the given function is not uniformly continuous on the given interval. (Cf. Exs. 13–14.)

★9. $\dfrac{1}{x}, 0 < x < 1$. ★10. $x^2, x \geq 1$.

★11. $\sin \dfrac{1}{x}, 0 < x < 1$. ★12. $x \sin x, x > 0$.

In Exercises 13–14, find an explicit function $\delta = \delta(\epsilon, x_0)$, in conformity with the δ-ϵ definition of continuity, for the given function at a given point x_0 of the specified interval. Observe that δ depends essentially on x_0. (Cf. Exs. 9–10; also Exs. 59–64, § 208.)

★13. $\dfrac{1}{x}, 0 < x < 1$. ★14. $x^2, x \geq 1$.

★15. If $f(x)$ is uniformly continuous on an open interval (a, b), prove that the two limits $f(a+)$ and $f(b-)$ exist and are finite. *Hint:* Use the Cauchy Criterion for functions.

★16. Prove the following converse to Exercise 15: *If $f(x)$ is continuous on an open interval (a, b) and if the two limits $f(a+)$ and $f(b-)$ exist and are finite, then $f(x)$ is uniformly continuous on (a, b).* *Hint:* Extend $f(x)$ to a function continuous on the closed interval $[a, b]$.

★17. Is the following analogue of Exercise 16 true? (Prove or disprove.) *If $f(x)$ is continuous on $(-\infty, +\infty)$ and if the two limits $f(-\infty)$ and $f(+\infty)$ exist and are finite, then $f(x)$ is uniformly continuous on $(-\infty, +\infty)$.*

★18. Prove that the sum of two functions each of which is uniformly continuous on a given set is uniformly continuous on that set. Prove that the product of two functions each of which is uniformly continuous and bounded on a given set is uniformly continuous on that set. (Cf. Ex. 46, § 308.)

3

Differentiation

301. INTRODUCTION

This and the following chapter contain a review and an amplification of certain topics from a first course in calculus. Some of the theorems that are usually stated without proof in a first introduction are established here. Other results are extended beyond the scope of a first course. On the other hand, many definitions and theorems with which the student can be assumed to be familiar are repeated here for the sake of completeness, without the full discussion and motivation which they deserve when first encountered. For illustrative material we have felt free to use calculus formulas which either are assumed to be well known or are established in later sections. For example, the trigonometric, exponential, and logarithmic functions provide useful examples and exercises for these chapters, but their analytic treatment is deferred to Chapter 5. The only inverse trigonometric functions used in this book are the inverse sine, tangent, cosine, and cotangent, with values restricted to the principal value ranges: $-\frac{1}{2}\pi \leq \text{Arcsin } x \leq \frac{1}{2}\pi$, $-\frac{1}{2}\pi < \text{Arctan } x < \frac{1}{2}\pi$, $0 \leq \text{Arccos } x \leq \pi$ and $0 < \text{Arccot } x < \pi$, respectively. (The upper case *A* indicates principal values.)

302. THE DERIVATIVE

Throughout this chapter we shall consider only single-valued real-valued functions of a real variable defined in a neighborhood of the particular value of the independent variable concerned (or possibly just for values of the independent variable neighboring the particular value on one side).

Definition. *A function $y = f(x)$ is said to have a derivative or be **differentiable** at a point x if and only if the following limit exists and is finite; the function $f'(x)$ defined by the limit is called its **derivative**:*

(1) $$\frac{dy}{dx} \equiv f'(x) \equiv \lim_{h \to 0} \frac{f(x+h) - f(x)}{h}.$$

*If $f'(x)$ is continuous, $f(x)$ is **continuously differentiable**.*

NOTE. On occasion it is convenient to speak of an infinite derivative in the sense that the limit in the definition above is either $+\infty$ or $-\infty$. For simplicity we shall adopt the convention that the word *derivative* refers to a finite quantity unless it is preceded by the word *infinite*.

Let us observe first that any differentiable function is continuous. More precisely, if $f(x)$ has a (finite) derivative at $x = x_0$, it is continuous there. We show this by taking limits of both members of the equation

$$f(x_0 + h) - f(x_0) = h \cdot \frac{f(x_0 + h) - f(x_0)}{h},$$

as $h \to 0$, the limit of the right-hand member being $0 \cdot f'(x_0) = 0$ (cf. Theorem V, § 207, and Note 3, § 209). The finiteness of the derivative is essential.

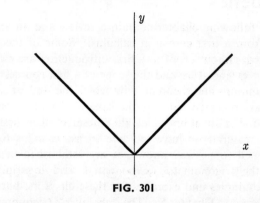

FIG. 301

For example, the signum function (Example 1, § 206) has an infinite derivative at $x = 0$, but is not continuous there.

The converse of the statement of the preceding paragraph is false. A continuous function need not have a derivative at every point. For example, the function $|x|$ (Fig. 301) is everywhere continuous, but has no derivative at $x = 0$. Example 2, § 303, contains a more pathological function. Even more startling is the renowned example of K. W. T. Weierstrass (German, 1815–1897) of a function which is everywhere continuous and nowhere differentiable. Although we shall not present this particular example in this book, every student at the level of advanced calculus should know of its existence. A discussion is given in E. W. Hobson, *The Theory of Functions of a Real Variable* (Washington, Harren Press, 1950). Another example of a continuous nondifferentiable function is presented in Exercise 30, § 1308.

With the notation $\Delta y \equiv f(x + \Delta x) - f(x)$, the definition of a derivative takes the form

(2) $$\frac{dy}{dx} \equiv \lim_{\Delta x \to 0} \frac{\Delta y}{\Delta x}.$$

This fact can be written $\lim_{\Delta x \to 0} \left(\frac{\Delta y}{\Delta x} - \frac{dy}{dx} \right) = 0$. In other words, the expression

(3) $$\epsilon \equiv \frac{\Delta y}{\Delta x} - \frac{dy}{dx},$$

as a function of Δx, is an infinitesimal (that is, its limit as $\Delta x \to 0$ is 0).† With the aid of this infinitesimal, equation (3) can be rewritten in either of the following two ways:

(3) $$\frac{\Delta y}{\Delta x} = \frac{dy}{dx} + \epsilon,$$

(4) $$\Delta y = \frac{dy}{dx} \Delta x + \epsilon \Delta x.$$

Under the assumption that the difference Δx between the values of the independent variable is numerically small, equation (3) states that the difference quotient $\Delta y / \Delta x$ differs but slightly from the derivative, and equation (4) states that the difference Δy can be approximated closely by an expression involving the derivative. Use of this fact is made in § 312.

The derivations of the formulas for the derivative of a constant function and of the sum, product, and quotient of functions will be omitted since they are available in any calculus text. Two standard formulas, however, are often not completely proved in a first course, and we supply their proofs now. The first is the *chain rule* for differentiation of a *composite function* (a function of a function).

Theorem I. Chain Rule. *If y is a differentiable function of u and if u is a differentiable function of x, then y, as a function of x, is differentiable and*

(5) $$\frac{dy}{dx} = \frac{dy}{du} \cdot \frac{du}{dx}.$$

Proof. Let $u \equiv f(x)$ be differentiable at $x = x_0$, $y \equiv g(u)$ be differentiable at $u = u_0 \equiv f(x_0)$, and let $h(x) \equiv g(f(x))$. With the customary notation,

$$\Delta u = f(x_0 + \Delta x) - f(x_0) = f(x_0 + \Delta x) - u_0$$
$$\Delta y = g(u_0 + \Delta u) - g(u_0)$$
$$= g(f(x_0 + \Delta x)) - g(f(x_0)) = h(x_0 + \Delta x) - h(x_0).$$

The usual simple device to make formula (5) seem plausible is the following: Write

(6) $$\frac{\Delta y}{\Delta x} = \frac{\Delta y}{\Delta u} \cdot \frac{\Delta u}{\Delta x}$$

† In general, an **infinitesimal** is any function having a zero limit, as the independent variable approaches its limit.

and take limits as $\Delta x \to 0$. Since u is a continuous function of x at $x = x_0$, $\Delta x \to 0$ implies $\Delta u \to 0$ (Note 3, § 209), and the desired formula is obtained:

$$\lim_{\Delta x \to 0} \frac{\Delta y}{\Delta x} = \lim_{\Delta u \to 0} \frac{\Delta y}{\Delta u} \cdot \lim_{\Delta x \to 0} \frac{\Delta u}{\Delta x}$$

—unless in this process Δu vanishes and makes equation (6) meaningless!

To avoid this difficulty we define a new function of the independent variable Δu:

(7) $$\epsilon(\Delta u) \equiv \begin{cases} \dfrac{\Delta y}{\Delta u} - \dfrac{dy}{du}, & \text{if } \Delta u \neq 0, \\ 0, & \text{if } \Delta u = 0. \end{cases}$$

Then $\lim\limits_{\Delta x \to 0} \epsilon(\Delta u) = \lim\limits_{\Delta u \to 0} \epsilon(\Delta u) = \dfrac{dy}{du} - \dfrac{dy}{du} = 0$.

From the formulation (7), if $\Delta u \neq 0$,

(8) $$\Delta y = \frac{dy}{du} \cdot \Delta u + \epsilon(\Delta u) \cdot \Delta u.$$

In fact, equation (8) holds whether Δu is zero or not! All that remains is to divide by Δx (which is *nonzero*) and take limits:

$$\lim_{\Delta x \to 0} \frac{\Delta y}{\Delta x} = \frac{dy}{du} \cdot \lim_{\Delta x \to 0} \frac{\Delta u}{\Delta x} + \lim_{\Delta x \to 0} \epsilon(\Delta u) \cdot \lim_{\Delta x \to 0} \frac{\Delta u}{\Delta x},$$

or $$\frac{dy}{dx} = \frac{dy}{du} \cdot \frac{du}{dx} + 0 \cdot \frac{du}{dx}.$$

Theorem II. *If $y = f(x)$ is strictly monotonic (§ 215) and differentiable in an interval, and if $f'(x) \neq 0$ in this interval, then the inverse function $x = \phi(y)$ is strictly monotonic and differentiable in the corresponding interval, and*

$$\frac{dx}{dy} = \frac{1}{\dfrac{dy}{dx}}.$$

Proof. By the Theorem of § 215, $\phi(y)$ is a continuous function, and therefore $\Delta x \to 0$ if and only if $\Delta y \to 0$ (Note 3, § 209). By Theorem VI, § 207, therefore,

$$\lim_{\Delta y \to 0} \frac{\Delta x}{\Delta y} = \lim_{\Delta x \to 0} \frac{1}{\dfrac{\Delta y}{\Delta x}} = \frac{1}{\lim\limits_{\Delta x \to 0} \dfrac{\Delta y}{\Delta x}},$$

and the proof is complete.

It will be assumed that the reader is familiar with the definitions and notations for derivatives of higher order than the first.

303. ONE-SIDED DERIVATIVES

It is frequently important to consider one-sided limits in relation to derivatives. There are three principal ways in which this can be done.

We give the three definitions here and call for examples and properties in later exercises. It happens that in many applications the most useful of these three definitions is the second, and accordingly we reserve the term *right-hand* or *left-hand derivative* for that concept rather than for the first, for which it might at first seem more natural. For Definitions I and III we create names for distinguishing purposes.

Definition I. *The derivative from the right (left) of a function $f(x)$ at the point $x = a$ is the one-sided limit*

$$\lim_{h \to 0+} \frac{f(a+h) - f(a)}{h} \quad \left(\lim_{h \to 0-} \frac{f(a+h) - f(a)}{h} \right).$$

Definition II. *The right-hand (left-hand) derivative of a function $f(x)$ at the point $x = a$ is the one-sided limit*

$$\lim_{h \to 0+} \frac{f(a+h) - f(a+)}{h} \quad \left(\lim_{h \to 0-} \frac{f(a+h) - f(a-)}{h} \right),$$

where $f(a+) \equiv \lim_{x \to a+} f(x)$ $(f(a-) \equiv \lim_{x \to a-} f(x))$.

Definition III. *The right-hand (left-hand) limit of the derivative of a function $f(x)$ at the point $x = a$ is the one-sided limit*

$$f'(a+) \equiv \lim_{x \to a+} f'(x) \quad (f'(a-) \equiv \lim_{x \to a-} f'(x)).$$

A function has a derivative at a point if and only if it has equal derivatives from the right and from the left at the point (cf. Ex. 14, § 208). The function $|x|$ (Fig. 301) is an example of a function that has unequal derivatives from the right and from the left. Definitions I, II, and III are related as follows: In case of one-sided continuity, Definitions I and II coincide, and hence if the limit in Definition I exists and is finite, the limit in Definition II exists and is equal to it. If the limit in Definition III exists and is finite, it can be proved (Ex. 45, § 308) that the limit in Definition II exists and is equal to it. Therefore, in case of differentiability, Definitions I and II are consistent with each other and with the Definition of § 302, and in case of continuity of the derivative all four definitions (Definitions I, II, and III of this section and the Definition of § 302) are consistent.

Example 1. The signum function (Example 1, § 206) has infinite derivatives from the right and from the left at $x = 0$. (Fig. 302.) Its right-hand and left-hand derivatives and the right-hand and left-hand limits of the derivative are all zero.

Example 2. The function defined by $y = x \sin \dfrac{1}{x}$ if $x \neq 0$ and $y = 0$ if $x = 0$ is everywhere continuous, even at the point $x = 0$ (cf. Ex. 8, § 208), but it has no derivative of any kind at $x = 0$. (Fig. 303.) Relative to the point $a = 0$, the

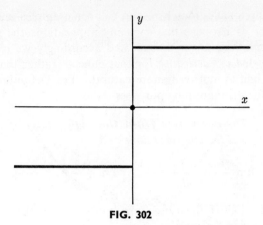

FIG. 302

fraction $\dfrac{\Delta y}{\Delta x}$ oscillates infinitely many times between $+1$ and -1 as $\Delta x \to 0$. For the manner in which Definition III applies to this function, cf. Example 3.

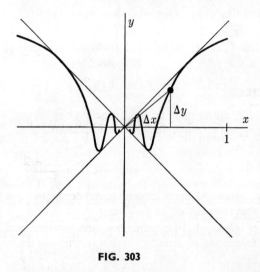

FIG. 303

Example 3. The function defined by $y = x^2 \sin \dfrac{1}{x}$ if $x \neq 0$ and $y = 0$ if $x = 0$ has a derivative for every value of x. When $x = 0$ this derivative has the value

$$f'(0) = \lim_{\Delta x \to 0} \frac{\Delta y}{\Delta x} = \lim_{\Delta x \to 0} \frac{\Delta x^2 \sin\left(\dfrac{1}{\Delta x}\right)}{\Delta x} = \lim_{\Delta x \to 0} \Delta x \sin\left(\dfrac{1}{\Delta x}\right) = 0 \text{ (cf. Ex. 8, § 208).}$$

However, the derivative $f'(x)$ *is not continuous at* $x = 0$ and, in fact, its one-sided limits as $x \to 0$ (Definition III) both fail to exist! To show this, we differentiate $y = x^2 \sin \dfrac{1}{x}$ according to formula: $\dfrac{dy}{dx} = 2x \sin \dfrac{1}{x} - \cos \dfrac{1}{x}$, when $x \neq 0$. The first term of this expression tends to zero as $x \to 0$, but the second term approaches no limit. Therefore $\dfrac{dy}{dx}$ can approach no limit. (Why?) Cf. Exs. 40–41, 45, § 308. (See Fig. 304.)

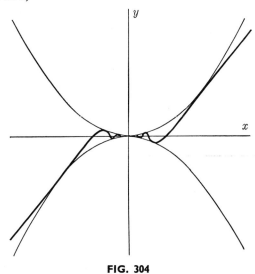

FIG. 304

304. EXERCISES

In Exercises 1–6, differentiate the given function by direct use of the Definition of § 302 (without appeal to any differentiation formulas).

1. $y = x^2 - 4x + 7$.
2. $y = x^3$.
3. $y = \dfrac{1}{x^2}$.
4. $y = \dfrac{3x + 2}{5x - 4}$.
5. $y = \sqrt{x}$. *Hint:* Multiply numerator and denominator of $(\sqrt{x + h} - \sqrt{x})/h$ by a quantity that rationalizes the numerator.
6. $y = \sqrt[3]{x}$. *Hint:* Multiply numerator and denominator of $(\sqrt[3]{x + h} - \sqrt[3]{x})/h$ by a quantity (consisting of three terms) that rationalizes the numerator.
7. Prove that if $f(x)$ is defined in a neighborhood of $x = \xi$, and if $f'(\xi)$ exists and is positive (negative), then within some neighborhood of ξ, $f(x) < f(\xi)$ for $x < \xi$ and $f(x) > f(\xi)$ for $x > \xi$ ($f(x) > f(\xi)$ for $x < \xi$ and $f(x) < f(\xi)$ for $x > \xi$). *Hint:* If $\lim_{\Delta x \to 0} \dfrac{f(\xi + \Delta x) - f(\xi)}{\Delta x} > 0$, then by the Note, § 206, there exists a deleted neighborhood of 0 such that for any Δx within this deleted neighborhood $\dfrac{f(\xi + \Delta x) - f(\xi)}{\Delta x} > 0$.
8. Derive the formula $\dfrac{d}{dx}(x^n) = nx^{n-1}$, if n is a positive integer.
*9. Derive the formula $\dfrac{d}{dx}(x^n) = nx^{n-1}$, if n is a positive rational number and

x is positive (without assuming in the process that x^n is differentiable). *Hint:* With standard notation, $y \equiv x^{p/q}$, $y^q = x^p$, and $(y + \Delta y)^q = (x + \Delta x)^p$, so that $y^q + qy^{q-1}\Delta y + \cdots = x^p + px^{p-1}\Delta x + \cdots$. Cancel first terms, divide by Δx, and let $\Delta x \to 0$. Since $\Delta y \to 0$ (Ex. 21, § 216), the formula follows by algebraic simplification.

*10. Derive the formula $\dfrac{d}{dx}(x^n) = nx^{n-1}$, if n is any rational number and x is positive. *Hint:* Use Ex. 9 to show that x^n is differentiable. If n is negative, let $m \equiv -n$ and differentiate the quotient $1/x^m$.

NOTE 1. The formula $\dfrac{d}{dx}(x^n) = nx^{n-1}$ is shown in Exercise 7, § 502, to be valid for all real values of n for $x > 0$.

11. Prove that any rational function (cf. Ex. 20, § 208) of a single variable is differentiable wherever it is defined.

12. Give an example of a function for which $\Delta x \to 0$ does not imply $\Delta y \to 0$. Give an example of a continuous function for which $\Delta y \to 0$ does not imply $\Delta x \to 0$.

13. By mathematical induction extend the chain rule for differentiating a composite function to the case of n functions: $y = f_1(f_2(f_3 \cdots (f_n(x)) \cdots))$.

14. Prove that the derivative of a monotonically increasing (decreasing) differentiable function $f(x)$ satisfies the inequality $f'(x) \geq 0 \, (\leq 0)$. Does strict increase (decrease) imply a strict inequality?

In Exercises 15–20, discuss differentiability of the given function $f(x)$ (where $f(0) \equiv 0$). Is the derivative continuous wherever it is defined?

15. $x^2 \operatorname{sgn} x$ (cf. Example 1, § 206).

16. $x \cos \dfrac{1}{x}$.

17. $x^2 \cos \dfrac{1}{x}$.

18. $x^2 \sin \dfrac{1}{x^2}$.

*19. $x^{\frac{4}{3}} \sin \dfrac{1}{x}$.

*20. $x^{\frac{5}{3}} \cos \dfrac{1}{\sqrt[3]{x}}$.

*21. Discuss differentiability of the function $f(x) \equiv 0$ if $x \leq 0$, $f(x) \equiv x^n$ if $x > 0$. For what values of n does $f'(x)$ exist for all values of x? For what values of n is $f'(x)$ continuous for all values of x?

*22. If $f(x) \equiv x^n \sin \dfrac{1}{x}$ for $x > 0$ and $f(0) \equiv 0$, find $f'(x)$. For what values of n does $f'(x)$ exist for all nonnegative values of x? For what values of n is $f'(x)$ continuous for all nonnegative values of x?

*23. By mathematical induction establish **Leibnitz's Rule:** *If u and v are functions of x, each of which possesses derivatives of order n, then the product also does and*

$$\frac{d^n}{dx^n}(uv) = \frac{d^n u}{dx^n} \cdot v + \binom{n}{1}\frac{d^{n-1}u}{dx^{n-1}} \cdot \frac{dv}{dx} + \binom{n}{2}\frac{d^{n-2}u}{dx^{n-2}} \cdot \frac{d^2v}{dx^2} + \cdots + u \cdot \frac{d^n v}{dx^n},$$

where the coefficients are the binomial coefficients (Ex. 25, § 107).

*24. If y is a function of u, and u is a function of x, each possessing derivatives of as high an order as desired, establish the following formulas for higher order derivatives of y with respect to x:

$$\frac{d^2y}{dx^2} = \frac{d^2y}{du^2}\left(\frac{du}{dx}\right)^2 + \frac{dy}{du}\frac{d^2u}{dx^2},$$

$$\frac{d^3y}{dx^3} = \frac{d^3y}{du^3}\left(\frac{du}{dx}\right)^3 + 3\frac{d^2y}{du^2}\frac{du}{dx}\frac{d^2u}{dx^2} + \frac{dy}{du}\frac{d^3u}{dx^3},$$

$$\frac{d^4y}{dx^4} = \frac{d^4y}{du^4}\left(\frac{du}{dx}\right)^4 + 6\frac{d^3y}{du^3}\left(\frac{du}{dx}\right)^2\frac{d^2u}{dx^2} + 3\frac{d^2y}{du^2}\left(\frac{d^2u}{dx^2}\right)^2$$
$$+ 4\frac{d^2y}{du^2}\frac{du}{dx}\frac{d^3u}{dx^3} + \frac{dy}{du}\frac{d^4u}{dx^4}.$$

305. ROLLE'S THEOREM AND THE LAW OF THE MEAN

From the fact that a function continuous on a closed interval has a maximum value there (Theorem II, § 213) stem many of the most important propositions of pure and applied mathematics. In this section we initiate a sequence of these theorems.

Theorem I. *If $f(x)$ is continuous on the closed interval $[a, b]$ and differentiable in the open interval (a, b), and if $f(x)$ assumes either its maximum or minimum value for the closed interval $[a, b]$ at an interior point ξ of the interval, then $f'(\xi) = 0$.*

Proof. Assume the hypotheses of the theorem and, for definiteness, let $f(\xi)$ be the *maximum* value of $f(x)$ for the interval $[a, b]$, where $a < \xi < b$. (The student should supply the details for the case where $f(\xi)$ is the *minimum* value of $f(x)$.) Consider the difference quotient

(1) $$\frac{\Delta y}{\Delta x} \equiv \frac{f(\xi + \Delta x) - f(\xi)}{\Delta x}$$

for values of Δx so small numerically that $\xi + \Delta x$ is also in the open interval (a, b). Since $f(\xi)$ is the maximum value of $f(x)$, $f(\xi) \geq f(\xi + \Delta x)$, and $\Delta y \leq 0$. Therefore $\Delta y / \Delta x \geq 0$ for $\Delta x < 0$ and $\Delta y / \Delta x \leq 0$ for $\Delta x > 0$. Hence (in the limit) the derivative from the left at ξ is nonnegative and the derivative from the right at ξ is nonpositive. By hypothesis these one-sided derivatives are equal, and must therefore both equal zero.

Theorem II. Rolle's Theorem. *If $f(x)$ is continuous on the closed interval $[a, b]$, if $f(a) = f(b) \neq 0$, and if $f(x)$ is differentiable in the open interval (a, b), then there is some point ξ of the open interval (a, b) such that $f'(\xi) = 0$.*

Proof. If $f(x)$ vanishes identically the conclusion is trivial. If $f(x)$ is somewhere positive it attains its maximum value at some interior point, and if it is somewhere negative it attains its minimum value at some interior point. In either case, the conclusion follows from Theorem I.

The geometric interpretation of Rolle's Theorem (Fig. 305) is that the graph of the function $f(x)$ has a horizontal tangent for at least one intermediate point.

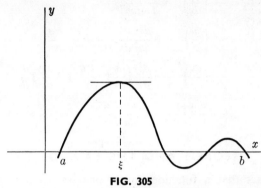

FIG. 305

Theorem III. Law of the Mean (Mean Value Theorem for Derivatives). *If $f(x)$ is continuous on the closed interval $[a, b]$, and if $f(x)$ is differentiable in the open interval (a, b), then there is some point ξ of the open interval (a, b) such that*

(2) $$f'(\xi) = \frac{f(b) - f(a)}{b - a}.$$

Proof. The geometric interpretation is suggested in Figure 306: the tangent line to the graph of the function $f(x)$, at some appropriate point

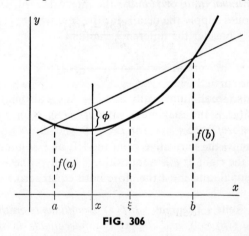

FIG. 306

between a and b, is parallel to the secant line between the points with abscissas a and b. The figure also suggests a proof. Let $K \equiv \dfrac{f(b) - f(a)}{b - a}$ be the slope of the secant line, so that

(3) $$f(b) = f(a) + K(b - a).$$

The equation of this secant line (the straight line through the point $(b, f(b))$ with slope K) can be written in the form

(4) $$y = f(b) - K(b - x).$$

For an arbitrary x on the closed interval $[a, b]$, since the curve has the equation $y = f(x)$, the vertical distance from the curve to the secant (ϕ in Fig. 306) is given by the expression

(5) $$\phi(x) \equiv f(b) - f(x) - K(b - x).$$

It is a simple matter to verify that the function $\phi(x)$ defined by (5) satisfies the conditions of Rolle's Theorem (check the details), so that the conclusion is valid. That is, there exists a number ξ of the open interval (a, b) where the derivative $\phi'(x) = -f'(x) + K$ vanishes, and we have the conclusion sought:

$$f'(\xi) = K = \frac{f(b) - f(a)}{b - a}.$$

A more general form of the Law of the Mean which is useful in evaluating indeterminate forms (§§ 314–317) is given in the following theorem. The proof is requested in Exercise 14, § 308, where hints are given.

Theorem IV. Generalized Law of the Mean (Generalized Mean Value Theorem for Derivatives). *If $f(x)$ and $g(x)$ are continuous on the closed interval $[a, b]$, if $f(x)$ and $g(x)$ are differentiable in the open interval (a, b), and if $g'(x)$ does not vanish in the open interval (a, b), then there is some point ξ of the open interval (a, b) such that*

(6) $$\frac{f'(\xi)}{g'(\xi)} = \frac{f(b) - f(a)}{g(b) - g(a)}.$$

The geometric interpretation is similar to that for the preceding Law of the Mean. The curve this time is given parametrically, where for convenience we relabel the independent variable with the letter t. The coordinates of a point (x, y) on the curve are given in terms of the parameter t by the functions $x = g(t)$ and $y = f(t)$, where $a \leq t \leq b$ (Fig. 307). In this case, by a formula from elementary calculus (cf. Ex. 39, § 308), the slope of the tangent at the point where $t = \xi$ is the left-hand member of (6), while the slope of the secant joining the points corresponding to $t = a$ and $t = b$ is the right-hand member of (6).

We conclude this section with some Notes, whose proofs are requested in the Exercises of § 308.

NOTE 1. Rolle's Theorem and the Law of the Mean remain valid if $f'(x)$ is permitted to be either positively or negatively infinite in the open interval (a, b). (Ex. 11, § 308.)

NOTE 2. Formulas (2) and (6) are valid, under corresponding assumptions, in case $a > b$. (Ex. 12, § 308.)

NOTE 3. If $f(x)$ is continuous at every point of an interval I (open, closed, or half-open) that contains the point a, and if $f(x)$ is differentiable at every interior point of I, then for any x belonging to I and distinct from a there exists a point ξ between a and x such that

(7) $$f(x) = f(a) + f'(\xi)(x - a).$$

If $f'(a)$ exists, then (7) holds for the case $x = a = \xi$ as well. (Ex. 13, § 308.)

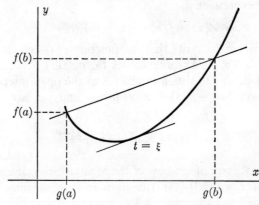

FIG. 307

NOTE 4. If $f(x)$ is continuous at every point of an interval I (open, closed, or half-open) that contains the point a, and if $f(x)$ is differentiable at every interior point of I, then for any nonzero h such that $a + h$ belongs to I there exists a number θ such that $0 < \theta < 1$ and

(8) $$f(a + h) = f(a) + f'(a + \theta h)h.$$

If $f'(a)$ exists, then (8) holds for the case $h = 0$ as well. (Ex. 13, § 308.)

Example 1. Use the Law of the Mean to establish the inequality $\sin x < x$, for $x > 0$.

Solution. If $a = 0$, $h = x$, and $f(x) = \sin x$, the equation $f(a + h) = f(a) + f'(a + \theta h) \cdot h$ becomes $\sin x = \cos(\theta x) \cdot x$. If $0 < x \leq 1$, then $0 < \theta x < 1$ and $\cos(\theta x) < 1$, so that $\sin x < x$. If $x > 1$, $\sin x \leq 1 < x$.

Example 2. Use the Law of the Mean to establish the inequalities

(9) $$\frac{h}{1 + h} < \ln(1 + h) < h,$$

if $h > -1$ and $h \neq 0$.

Solution. If $a = 1$, and $f(x) = \ln x$, the equation $f(a + h) = f(a) + f'(a + \theta h) \cdot h$ becomes $\ln(1 + h) = \dfrac{h}{1 + \theta h}$. If $h > 0$, the inequalities $0 < \theta < 1$ imply $1 < 1 + \theta h < 1 + h$, and hence $\dfrac{1}{1 + h} < \dfrac{1}{1 + \theta h} < 1$, whence (9) follows. If $-1 < h < 0$, the inequalities $0 < \theta < 1$ imply $1 > 1 + \theta h > 1 + h > 0$, and hence $\dfrac{1}{1 + h} > \dfrac{1}{1 + \theta h} > 1$, whence (9) follows.

306. CONSEQUENCES OF THE LAW OF THE MEAN

It is a trivial fact that a constant function has a derivative that is identically zero. The converse, though less trivial, is also true.

Theorem I. *A function with an identically vanishing derivative throughout an interval must be constant in that interval.*†

Proof. If $f(x)$ is differentiable and nonconstant in an interval, there are two points, a and b, of that interval where $f(a) \neq f(b)$. By the Law of the Mean there must be a point ξ between a and b such that

$$f'(\xi) = [f(b) - f(a)]/(b - a) \neq 0,$$

in contradiction to the basic assumption.

Theorem II. *Two differentiable functions whose derivatives are equal throughout an interval must differ by a constant in that interval.*

Proof. This is an immediate consequence of Theorem I, since the difference of the two functions has an identically vanishing derivative and must therefore be a constant.

A direct consequence of the definition of a derivative is that monotonic differentiable functions have derivatives of an appropriate sign (Ex. 14, § 304). A converse is stated in the following theorem:

Theorem III. *If $f(x)$ is continuous over an interval and differentiable in the interior, and if $f'(x) \geq 0 \ (\leq 0)$ there, then $f(x)$ is monotonically increasing (decreasing) on the interval. If furthermore $f'(x) > 0 \ (< 0)$, then $f(x)$ is strictly increasing (decreasing).*

Proof. Let x_1 and x_2 be points of the interval such that $x_1 < x_2$. Then, by the Law of the Mean, there is a number x_3 between x_1 and x_2 such that $f(x_2) - f(x_1) = f'(x_3)(x_2 - x_1)$. The resulting inequalities give the desired conclusions.

An important relation between monotonicity of a function and the non-vanishing of its derivative is contained in the following theorem:

Theorem IV. *A function continuous over an interval and having a nonzero derivative throughout at least the interior of that interval is strictly monotonic (over the interval), and its derivative is of constant sign (wherever it is defined in the interval).*

Proof. By Theorem III, it is sufficient to show that the derivative is of constant sign. Accordingly, we seek a contradiction to the assumption that there exist two points, a and b $(a < b)$, of an interval throughout which the function $f(x)$ has a nonzero derivative, and such that $f'(a)$ and $f'(b)$ have

† Throughout this section the interval under consideration is arbitrary—finite or infinite, closed or open or neither.

opposite signs. On the closed interval $[a, b]$ the continuous function $f(x)$ has a maximum value at some point ξ of $[a, b]$ and a minimum value at some point η of $[a, b]$, where $\xi \neq \eta$ (why must ξ and η be distinct?). By Theorem I, § 305, neither ξ nor η can be an interior point of $[a, b]$. Therefore either $\xi = a$ and $\eta = b$ or $\xi = b$ and $\eta = a$. However, if $\xi = a$ and $\eta = b$, $f'(a) \leq 0$ and $f'(b) \leq 0$ (why?), whereas if $\xi = b$ and $\eta = a$, $f'(a) \geq 0$ and $f'(b) \geq 0$. Either conclusion is a contradiction to the assumption that $f'(a)$ and $f'(b)$ have opposite signs.

Corollary. *If a function has a nonzero derivative over an interval, the inverse function exists and is differentiable, and consequently the formula* $\dfrac{dx}{dy} = 1 \Big/ \dfrac{dy}{dx}$ *is valid whenever its right-hand member exists over an interval.*

307. THE EXTENDED LAW OF THE MEAN

In order to motivate an extension of the Law of the Mean to include higher order derivatives, we consider heuristically the problem of trying to approximate a given function $f(x)$ in a neighborhood of a point $x = a$ by means of a polynomial $p(x)$. The higher the degree of $p(x)$, the better we should expect to be able to approximate $f(x)$ (assuming, as we shall for this introductory discussion, that $f(x)$ not only is continuous but has derivatives of as high an order as we wish to consider, for x in the neighborhood of a). If $p(x)$ is a constant (degree at most zero—the polynomial identically equal to zero is sometimes said to be of degree $-\infty$), a reasonable value for this constant, if it is to approximate $f(x)$ for x near a, is clearly $f(a)$. If $p(x)$ is a polynomial of (at most) the first degree, it should certainly approximate $f(x)$ if its graph is the line tangent to the graph of $y = f(x)$ at the point $(a, f(a))$, that is, if $p(a) = f(a)$ and $p'(a) = f'(a)$. In this case, $p(x)$ has the form

(1) $$p(x) = f(a) + f'(a)(x - a).$$

More generally, let us approximate $f(x)$ in the neighborhood of $x = a$ by a polynomial $p(x)$ of degree $\leq n$ with the property that $p(a) = f(a)$, $p'(a) = f'(a), \cdots, p^n(a) = f^n(a)$. We first express $p(x)$ by means of the substitution of $a + (x - a)$ for x, and subsequent expansion, in the form

(2) $$p(x) = p_0 + p_1(x - a) + \cdots + p_n(x - a)^n.$$

Successive differentiation and substitution in the equations $p(a) = f(a)$, $p'(a) = f'(a), \cdots, p^{(n)}(a) = f^{(n)}(a)$ lead to an evaluation of the coefficients in (2) in terms of the function $f(x)$, and hence to the following expression for $p(x)$ (check the details):

(3) $$p(x) = f(a) + f'(a)(x - a) + \frac{f''(a)}{2!}(x - a)^2 + \cdots + \frac{f^{(n)}(a)}{n!}(x - a)^n.$$

The Law of the Mean, in the form

(4) $$f(x) = f(a) + f'(\xi)(x - a),$$

states that $f(x)$ can be represented by an expression which resembles the approximating polynomial (1) of (at most) the first degree, differing from it only by the substitution of ξ for a in the coefficient of the last term. It is not altogether unreasonable to expect that $f(x)$ can be represented more generally by an expression which resembles the approximating polynomial (3) of degree $\leq n$, differing from it only by the substitution of ξ for a in the coefficient of the last term. Our objective in this section is to show that this is indeed the case.

Theorem. Extended Law of the Mean (Mean Value Theorem). *If $f(x), f'(x), \cdots, f^{n-1}(x)$ are continuous on the closed interval $[a, b]$, and if $f^{(n)}(x)$ exists in the open interval (a, b), then there is some point ξ of the open interval (a, b) such that*

(5) $f(b) = f(a) + f'(a)(b - a) + \dfrac{f''(a)}{2!}(b - a)^2 + \cdots$
$$+ \dfrac{f^{(n-1)}(a)}{(n-1)!}(b-a)^{n-1} + \dfrac{f^{(n)}(\xi)}{n!}(b-a)^n.$$

Proof. The methods used in establishing the Law of the Mean in § 305 can be extended to the present theorem. Let the constant K be defined by the equation

(6) $f(b) = f(a) + f'(a)(b - a) + \dfrac{f''(a)}{2!}(b - a)^2 + \cdots$
$$+ \dfrac{f^{(n-1)}(a)}{(n-1)!}(b-a)^{n-1} + \dfrac{K}{n!}(b-a)^n,$$

and define a function $\phi(x)$ by replacing a by x in (6), and rearranging terms:

(7) $\phi(x) \equiv f(b) - f(x) - f'(x)(b - x) - \dfrac{f''(x)}{2!}(b - x)^2 - \cdots$
$$- \dfrac{f^{(n-1)}(x)}{(n-1)!}(b-x)^{n-1} - \dfrac{K}{n!}(b-x)^n.$$

This function $\phi(x)$, for the interval $[a, b]$, satisfies the conditions of Rolle's Theorem (check this), and therefore its derivative must vanish for some point ξ of the open interval (a, b):

(8) $$\phi'(\xi) = 0$$

Routine differentiation of (7) gives

(9) $\phi'(x) = -f'(x) + f'(x) - f''(x)(b - x) + f''(x)(b - x) - \cdots$
$$- \dfrac{f^{(n)}(x)}{(n-1)!}(b-x)^{n-1} + \dfrac{K}{(n-1)!}(b-x)^{n-1};$$

in a form where all of the terms except the last two cancel in pairs. Equation (8) becomes, therefore, on simplification:

(10) $$K = f^{(n)}(\xi),$$

and the proof is complete.

The Extended Law of the Mean (5) is also called **Taylor's Formula with a Remainder** (cf. § 1204).

Notes similar to those of § 405 apply to this section:

NOTE 1. The Extended Law of the Mean remains valid if $f^{(n)}(x)$ is permitted to be either positively or negatively infinite in the open interval (a, b). (Ex. 15, § 408.)

NOTE 2. Formula (5) is valid, under corresponding assumptions, in case $a > b$. (Ex. 16, § 408.)

NOTE 3. If $f^{(n-1)}(x)$ is continuous at every point of an interval I (open, closed, or half-open) that contains the point a, and if $f^{(n)}(x)$ exists at every interior point of I, then for any x belonging to I and distinct from a there exists a point ξ between a and x such that

(11) $f(x) = f(a) + f'(a)(x - a) + \dfrac{f''(a)}{2!}(x - a)^2 + \cdots$
$+ \dfrac{f^{(n-1)}(a)}{(n-1)!}(x - a)^{n-1} + \dfrac{f^{(n)}(\xi)}{n!}(x - a)^n.$

If $f^{(n)}(a)$ exists, then (11) holds for the case $x = a = \xi$ as well. (Ex. 17, § 308.)

NOTE 4. If $f^{(n-1)}(x)$ is continuous at every point of an interval I (open, closed, or half-open) that contains the point a, and if $f^{(n)}(x)$ exists at every interior point of I, then for any nonzero h such that $a + h$ belongs to I there exists a number θ such that $0 < \theta < 1$ and

(12) $f(a + h) = f(a) + f'(a)h + \dfrac{f''(a)}{2!}h^2 + \cdots$
$+ \dfrac{f^{(n-1)}(a)}{(n-1)!}h^{n-1} + \dfrac{f^{(n)}(a + \theta h)}{n!}h^n.$

If $f^{(n)}(a)$ exists, then (12) holds for the case $h = 0$ as well. (Ex. 17, § 308.)

308. EXERCISES

In Exercises 1–2, find a value for ξ as prescribed by Rolle's Theorem. Draw a figure.

1. $f(x) \equiv \cos x$, for $\frac{1}{2}\pi \leq x \leq \frac{7}{2}\pi$.
2. $f(x) \equiv x^3 - 6x^2 + 6x - 1$, for $\frac{1}{2}(5 - \sqrt{21}) \leq x \leq 1$.

In Exercises 3–4, find a value for ξ as prescribed by the Law of the Mean, (2), § 305. Draw a figure.

3. $f(x) \equiv \ln x$, for $1 \leq x \leq e$.
4. $f(x) \equiv px^2 + qx + r$, for $a \leq x \leq b$.

In Exercises 5–6, find a value for θ as prescribed by the Law of the Mean, (8), § 305. Draw a figure.

5. $f(x) \equiv \ln x$, for $a = e$, $h = 1 - e$.
6. $f(x) \equiv px^2 + qx + r$, a and h arbitrary.

In Exercises 7–8, find a value for ξ as prescribed by the Generalized Law of the Mean, (6), § 305. Draw a figure.

7. $f(x) \equiv 2x + 5$, $g(x) \equiv x^2$, for $0 < b \leq x \leq a$.
8. $f(x) \equiv x^3$, $g(x) \equiv x^2$, for $1 \leq x \leq 3$.

In Exercises 9–10, find a value for ξ as prescribed by the Extended Law of the Mean, (5), § 307.

9. $f(x) \equiv \dfrac{1}{1-x}$, $a = 0$, arbitrary n, and $b < 1$.

10. $f(x) \equiv \ln x$, $a = 1$, $b = 3$, $n = 3$.

11. Prove Note 1, § 305. Explain why the function $x^{\frac{2}{3}}$ on the closed interval $[-1, 1]$ is excluded.

12. Prove Note 2, § 305.

13. Prove Notes 3 and 4, § 305.

14. Prove Theorem IV, § 305. *Hint:* First show that $g(b) \neq g(a)$, by using Rolle's Theorem with the function $h(x) \equiv g(x) - g(a)$. Then let
$$K = (f(b) - f(a))/(g(b) - g(a))$$
and define the function $\phi(x) \equiv f(x) - f(a) - K[g(x) - g(a)]$. Proceed as with the proof of the Law of the Mean.

15. Prove Note 1, § 307.

16. Prove Note 2, § 307.

17. Prove Notes 3 and 4, § 307.

18. The functions $f(x) \equiv \dfrac{1}{x}$ and $g(x) \equiv \dfrac{1}{x} + \operatorname{sgn} x$ (Example 1, § 206) have identical derivatives, but do not differ by a constant. Explain how this is possible in the presence of Theorem II, § 306.

In Exercises 19–30, use the Law of the Mean to establish the given inequalities. (Assume the standard properties of the transcendental functions.)

19. $\tan x > x$ for $0 < x < \frac{1}{2}\pi$.

20. $|\sin a - \sin b| \leq |a - b|$.

21. $\dfrac{b-a}{b} < \ln \dfrac{b}{a} < \dfrac{b-a}{a}$, for $0 < a < b$.

22. $\sqrt{1+h} < 1 + \frac{1}{2}h$, for $-1 < h < 0$ or $h > 0$. More generally, for these values of h and $0 < p < 1$, $(1+h)^p < 1 + ph$.

23. $(1+h)^p > 1 + ph$, for $-1 < h < 0$ or $h > 0$, and $p > 1$ or $p < 0$.

24. $\dfrac{h}{1+h^2} < \operatorname{Arctan} h < h$, for $h > 0$.

25. $x < \operatorname{Arcsin} x < \dfrac{x}{\sqrt{1-x^2}}$, for $0 < x < 1$.

26. $\operatorname{Arctan} (1+h) \leq \dfrac{\pi}{4} + \dfrac{h}{2}$, for $h > -1$.

27. $\left|\dfrac{\cos ax - \cos bx}{x}\right| \leq |a - b|$, for $x \neq 0$.

28. $\dfrac{\sin px}{x} < p$, for $p > 0$ and $x > 0$.

29. $e^a(b-a) < e^b - e^a < e^b(b-a)$, for $a < b$.

30. $ae^{-ax} < \dfrac{1 - e^{-ax}}{x} < a$, for $a > 0$ and $x > 0$.

In Exercises 31–34, use the Extended Law of the Mean to establish the given inequalities.

31. $\cos x \geq 1 - \dfrac{x^2}{2}$.

32. $\cos x > 1 - \dfrac{x^2}{2}$, for $x \neq 0$.

33. $x - \dfrac{x^3}{6} < \sin x < x$, for $x > 0$.

34. $1 + x + \dfrac{x^2}{2} < e^x < 1 + x + \dfrac{x^2}{2} e^x$, for $x > 0$.

*35. Prove that $\dfrac{2}{\pi} < \dfrac{\sin x}{x} < 1$, for $0 < x < \dfrac{\pi}{2}$. *Hint:* For the first inequality, show that $\sin x / x$ is a decreasing function.

*36. Prove that $\dfrac{4}{\pi} > \dfrac{\tan x}{x} > 1$, for $0 < x < \dfrac{\pi}{4}$. (Cf. Ex. 35.)

*37. Let $f(x)$ be differentiable, with $f'(x) \geq 0$ ($f'(x) \leq 0$), on an interval, and assume that on no subinterval does $f'(x)$ vanish identically. Prove that $f(x)$ is strictly increasing (decreasing) on the interval. Formulate and prove a converse.

*38. Prove that a function differentiable at every point of an interval is monotonic there if and only if its derivative does not change sign there.

39. Let $x = g(t)$ and $y = f(t)$ be continuous over a closed interval $a \leq t \leq b$ and differentiable in the interior, and assume that $g'(t)$ does not vanish there. Prove that y, as a function of x, is continuous over a corresponding interval and differentiable in the interior, and that for any interior point

$$\frac{dy}{dx} = \frac{f'(t)}{g'(t)}.$$

Interpret the results geometrically.

*40. It was shown in Example 3, § 303, that although a function may be differentiable for all values of the independent variable, its derivative may not be continuous. In spite of this fact, a derivative shares with continuous functions the intermediate-value property of Theorem IV, § 213. Prove the theorem: *If $f(x)$ is the derivative of some function $g(x)$, on an interval, then $f(x)$ assumes (as a value) every number between any two of its values.* *Hint:* Let c and d be any two distinct values of $f(x)$, let r be any number between c and d, and apply Theorem IV, § 306, to the function $\phi(x) \equiv g(x) - rx$.

*41. Show by the example $f(x) \equiv x^2 \sin \dfrac{1}{x^2}$ ($f(0) \equiv 0$) that derivatives do not always share with continuous functions the property of being bounded on closed intervals (Theorem I, § 213). That is, exhibit a closed interval at every point of which $f(x)$ is differentiable but on which $f'(x)$ is unbounded.

*42. Show by the example $f(x) \equiv x + 2x^2 \sin(1/x)$ ($f(0) \equiv 0$) that the hypotheses (i) $f'(x)$ exists in a neighborhood of the point $x = a$, and (ii) $f'(a) > 0$, do not imply that there exists some neighborhood of $x = a$ throughout which $f'(x)$ is positive, nor do they imply that there exists some neighborhood of $x = a$ throughout which $f(x)$ is increasing.

*43. Prove that if a function has a bounded derivative on an interval it is uniformly continuous there.

*44. Prove that if $f(x)$ has a bounded derivative on an open interval (a, b), then $f(a+)$ and $f(b-)$ exist and are finite. *Hint:* Cf. Ex. 43, and Ex. 15, § 225.

*45. Prove that if the right-hand (left-hand) limit of the derivative of a function (Definition III, § 303) exists and is finite, then the right-hand (left-hand) derivative (Definition II, § 303) exists and is equal to it. Conclude that among the discontinuities discussed in § 210 for functions in general, *derivatives existing throughout an interval* can have discontinuities only of type (*iii*), where *not both* one-sided limits exist. *Hint:* For the case $x \to a+$, show that there exists an interval of the form $(a, a + h)$, where $h > 0$, in which $f'(x)$ is bounded (cf. Ex. 9, § 208), and use Exercise 44 to infer that $f(a+) \equiv \lim_{x \to a+} f(x)$ exists. Redefine $f(a) \equiv f(a+)$ and use the Law of the Mean in the form $[f(a + h) - f(a+)]/h = f'(a + \theta h)$.

*46. By means of the example $x \sin x$ on the interval $(0, +\infty)$, show that the product of two functions each of which is uniformly continuous on a given set and only one of which is bounded there may not be uniformly continuous on that set. (Cf. Exs. 12, 18, § 225, Ex. 43 above.)

309. MAXIMA AND MINIMA

We shall assume that the reader is familiar with the standard routine of finding maximum and minimum values of a function $y = f(x)$: (*i*) differentiate; (*ii*) set the derivative equal to zero; (*iii*) solve the equation $f'(x) = 0$ for x; (*iv*) test the values of x thus obtained, by using either the first or second derivative of the function; and (*v*) substitute in $f(x)$ the appropriate values of x to find the maximum and minimum values of $y = f(x)$.

Inasmuch as this routine gives only a partial answer to the story of maxima and minima, we present in this section a more complete summary of the pertinent facts and tests.

Let $f(x)$ be a function defined over a set A, and let ξ be a point of A. If the inequality $f(\xi) \geq f(x)$ ($f(\xi) \leq f(x)$) holds for every x in A, we say that $f(x)$ has a **maximum (minimum)** value on A equal to $f(\xi)$. According to Theorem II, § 213, such maximum and minimum values exist if A is a closed interval and $f(x)$ is continuous on A. If a function has a maximum (minimum) value on its domain of definition, this value is called the **absolute maximum (minimum)** value of the function. If ξ is a point where $f(x)$ is defined, and if within some neighborhood of ξ the inequality $f(\xi) \geq f(x)$ ($f(\xi) \leq f(x)$) holds whenever $f(x)$ is defined, we say that $f(\xi)$ is a **relative maximum (minimum)** value of $f(x)$. By a **critical value** of x for a function $f(x)$ we mean any point c of the domain of definition of $f(x)$ such that either (*i*) $f'(c)$ does not exist (as a finite quantity) or (*ii*) $f'(c) = 0$.

Theorem I. *If a function has a maximum or minimum value on an interval at a point ξ of the interval, then ξ is either an end-point of the interval or a critical value for the function.*

Proof. This theorem extends Theorem I, § 305, to an arbitrary (not necessarily closed) interval and drops continuity and differentiability assumptions. The details of the proof, however, are identical.

Some different kinds of maxima for a function continuous on a closed interval are illustrated in Figure 308.

Theorem II. First Derivative Test. *If $f(x)$ is continuous at a point $x = \xi$ and differentiable in a deleted neighborhood of ξ, and if in this deleted neighborhood $f'(x) > 0$ for $x < \xi$ and $f'(x) < 0$ for $x > \xi$ ($f'(x) < 0$ for $x < \xi$ and $f'(x) > 0$ for $x > \xi$), then $f(x)$ has a relative maximum (minimum) value at $x = \xi$. If, on the other hand, $f'(x)$ is of constant sign throughout the deleted neighborhood, $f(x)$ has neither a relative maximum nor a relative minimum value at $x = \xi$.*

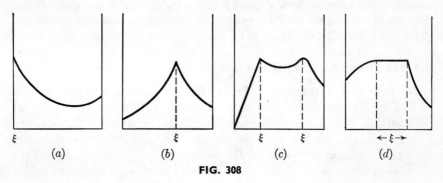

FIG. 308

Proof. Let x be an arbitrary point in the deleted neighborhood of ξ. By the Law of the Mean, § 305, there exists a point ξ' between ξ and x such that $f(x) - f(\xi) = f'(\xi')(x - \xi)$. Examination of each individual case leads to an appropriate inequality of the form $f(x) > f(\xi)$ or $f(x) < f(\xi)$. (Check the details.)

NOTE 1. The conditions assumed in Theorem II are sufficient but not necessary, even if $f'(x)$ is continuous and $f'(\xi) = 0$ (cf. Exs. 22–23, § 310).

Theorem III. Second Derivative Test. *If $f(x)$ is differentiable in a neighborhood of a critical value ξ, and if $f''(\xi)$ exists and is negative (positive), then $f(x)$ has a relative maximum (minimum) value at $x = \xi$.*

Proof. Assume $f''(\xi) < 0$. Then, by Exercise 7, § 304, with in some neighborhood of ξ, $f'(x) > f'(\xi) = 0$ for $x < \xi$ and $f'(x) < f'(\xi) = 0$ for $x > \xi$. By the First Derivative Test, $f(x)$ has a relative maximum value at $x = \xi$. (Supply the corresponding details for the case $f''(\xi) > 0$.)

NOTE 2. No conclusion regarding maximum or minimum of a function can be drawn from the vanishing of the second derivative at a critical value—the function may have a maximum value or a minimum value or neither at such a point (cf. Ex. 9, § 310).

A useful extension of the Second Derivative Test is the following:

Theorem IV. *Let $f(x)$ be a function which, in some neighborhood of the point $x = \xi$, is defined and has derivatives $f'(x), f''(x), \cdots, f^{(n-1)}(x)$, of order $\leq n - 1$, where $n > 1$. If $f'(\xi) = f''(\xi) = \cdots = f^{(n-1)}(\xi) = 0$, and if $f^{(n)}(\xi)$*

exists and is different from zero, then (i) if n is even, $f(x)$ has a relative maximum value or a relative minimum value at $x = \xi$ according as $f^{(n)}(\xi)$ is negative or positive, and (ii) if n is odd, $f(x)$ has neither a relative maximum value nor a relative minimum value at $x = \xi$.

Proof. Owing to the vanishing of the derivatives of order $< n - 1$ at the point ξ, the Extended Law of the Mean provides the formula

$$(1) \qquad f(x) - f(\xi) = \frac{f^{(n-1)}(\xi')}{(n-1)!}(x - \xi)^{n-1},$$

where x is in a suitably restricted deleted neighborhood of ξ, and ξ' is between ξ and x. The proof resolves itself into determining what happens to the sign of the right-hand member of (1), as x changes from $x < \xi$ to $x > \xi$, according to the various possibilities for the sign of $f^{(n)}(\xi)$ and the parity of n. We give the details for the case $f^{(n)}(\xi) < 0$, and suggest that the student furnish the corresponding details for the case $f^n(\xi) > 0$. By Exercise 7, § 304 (applied to the function $f^{(n-1)}(x)$), as x changes from $x < \xi$ to $x > \xi$, $f^{(n-1)}(x)$ changes from $+$ to $-$. Therefore if n is even, the right-hand member of (1) is negative whether $x < \xi$ or $x > \xi$, whence $f(x) < f(\xi)$ for x in the deleted neighborhood of ξ, and $f(\xi)$ is a relative maximum value of $f(x)$. If n is odd, the right-hand member of (1) is positive for $x < \xi$ and negative for $x > \xi$, so that $f(\xi)$ is neither a relative maximum value nor a relative minimum value of $f(x)$ at $x = \xi$.

Example. Examine the function $f(x) \equiv \dfrac{x^{\frac{2}{3}}}{x^2 + 8}$ for critical values of x and relative and absolute maxima and minima. Find its maximum and minimum values (when they exist) for the intervals $[1, 3]$, $(-1, 2)$, and $[1, +\infty)$.

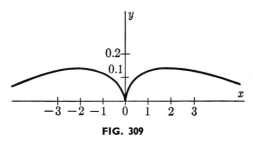

FIG. 309

Solution. The graph of $f(x)$ (Fig. 309) is symmetrical with respect to the y-axis, $\lim\limits_{x \to \infty} f(x) = 0$, and from the definition of one-sided derivatives, the derivative from the right at $x = 0$ is $+\infty$ and the derivative from the left at $x = 0$ is $-\infty$. For $x \neq 0$,

$$f'(x) = \frac{4(4 - x^2)}{3\sqrt[3]{x}(x^2 + 8)^2}.$$

The critical values of x are $x = 0, 2,$ and -2. At $x = 0$ the function has both a relative and an absolute minimum value of 0. At $x = \pm 2$ the function has a relative and an absolute maximum value of $\frac{1}{12}\sqrt[3]{4} = 0.1326$ (approximately, to four decimal places). On the interval $[1, 3]$ the function has a maximum of $f(2)$ and a minimum of $f(1) = \frac{1}{9} = 0.1111$ ($f(3) = 0.1224$). On the interval $(-1, 2)$ $f(x)$ has a minimum of $f(0) = 0$, but no maximum. On the interval $[1, +\infty)$ $f(x)$ has a maximum of $f(2)$, but no minimum.

310. EXERCISES

In Exercises 1–4, find the relative and absolute maximum and minimum values of the function, and the intervals within which the function is increasing, or decreasing. Draw a figure.

1. $x^3 - 6x^2 + 9x + 5$.
2. $\dfrac{2x}{1 + x^2}$.
3. $x^{\frac{2}{3}}(1 - x)$.
4. $\sqrt{x(x-1)^2}$.

In Exercises 5–8, find the maximum and minimum values of the given function (whenever they exist) for the designated interval. Draw a figure.

5. $x + 2x^2 - 4x^3$, $[-\frac{1}{3}, 1]$.
6. $\dfrac{x^2 + 100}{x^2 - 25}$, $[-1, 3]$.
7. $\cos x + \cos 2x$, $(-\infty, +\infty)$.
8. xe^x, $(-\infty, 0)$.

9. Show that the three functions x^4, $-x^4$, and x^3 all have the property of possessing a continuous second derivative which vanishes at $x = 0$, whereas at this point the functions have an absolute minimum, an absolute maximum, and neither a maximum nor a minimum, respectively. Thus justify Note 2, § 309.

10. Show that the function $x^m(x - 1)^n$, where m and n are positive integers, has a relative minimum regardless of the values of m and n, and that it has an absolute minimum if and only if m and n have the same parity ($m + n$ is even). Draw figures.

★11. Show that the function $|x|^p \cdot |x - 1|^q$, where p and q are positive, has a relative maximum value of $p^p q^q/(p + q)^{p+q}$.

★12. Show that the function $\dfrac{ax + b}{cx + d}$ has neither a relative maximum nor a relative minimum unless it is a constant.

★13. If $a_1, a_2, \cdots, a_n$ are given constants, show that the expression $\sum_{i=1}^{n}(x - a_i)^2$ is minimized if and only if x is their arithmetic mean, $x = \dfrac{1}{n}\sum_{i=1}^{n} a_i$.

★14. Show that the function $x^3 + 3px + q$ has either a relative maximum and a relative minimum or neither. Draw figures corresponding to the various cases.

15. A publisher is planning on putting out a new magazine, and an efficiency expert has estimated that if the price per copy is x cents, the profit is given by a formula of the type
$$\text{Profit} = K\left[\frac{x - a}{x^2 + 25} - b\right],$$
where K, a, and b are positive constants. Determine what price should be set for maximum profit, in each of the following cases:

(a) x is a multiple of 5, $a = 10$, $b = .02$;
(b) x is a multiple of 5, $a = 11$, $b = .02$;
(c) x is an integer, $a = 10$, $b = .02$;
(d) x is an integer, $a = 11$, $b = .03$.

16. A truck has a top speed of 60 miles per hour and, when traveling at the rate of x miles per hour, consumes gasoline at the rate of $\dfrac{1}{200}\left(\dfrac{400}{x} + x\right)$ gallons per mile. This truck is to be taken on a 200 mile trip by a driver who is to be paid at the rate of b dollars per hour plus a commission of c dollars. Since the time required for this trip, at x miles per hour, is $200/x$, the total cost, if gasoline costs a dollars per gallon, is

$$\left(\dfrac{400}{x} + x\right)a + \dfrac{200}{x}b + c.$$

Find the most economical possible speed under each of the following sets of conditions:

(a) $b = 0$;
(b) $a = .25$, $b = 1.25$, $c = 5$;
(c) $a = .20$, $b = 4$ (a crew).

17. A problem in maxima and minima can frequently be simplified by such devices as eliminating constant terms and factors, squaring, and taking reciprocals. Suppose, for example, we wish to find the values of x on the interval $1 \le x \le 2$ that maximize and minimize the function $17x/50\sqrt{x^4 + 2x^2 + 2}$. The value of x that maximizes (minimizes) this function is the same as that which maximizes (minimizes) the function $x/\sqrt{x^4 + 2x^2 + 2}$ or $x^2/(x^4 + 2x^2 + 2)$, and is the same as that which minimizes (maximizes) the reciprocal $(x^4 + 2x^2 + 2)/x^2$ or, equivalently, $x^2 + 2x^{-2}$. Let us now substitute $t = x^2$, and seek the value of t ($1 \le t \le 4$) that minimizes (maximizes) the function $g(t) = t + 2t^{-1}$. Since $g'(t) = 1 - 2t^{-2}$, $g(t)$ is minimized on the interval by $t = \sqrt{2}$, and is maximized by $t = 4$. Therefore the original function is maximized on $1 \le x \le 2$ by $x = \sqrt[4]{2}$ and minimized by $x = 2$. Discuss this technique in general for a function $y = f(x)$ defined on a set A, a strictly monotonic function $v = \phi(y)$ over the range of $f(x)$, and a suitably restricted substitution function $t = t(x)$. For the function $\phi(y)$, treat in particular $y + k$, py, y^n, $\sqrt[n]{y}$, and $1/y$, where k, p, and n are constants. (Be careful to explain under what circumstances squaring is legitimate.)

*__18.__ Prove that the function
$$a|x| + b|x - 1|, \text{ where } a \ne b,$$
has a relative minimum value for the interval $(-\infty, +\infty)$ if and only if $a + b \ge 0$ and that if $a + b \ge 0$ its minimum value is the smaller of the two numbers a and b.

*__19.__ Find the value of x that minimizes the function $t\sqrt{x^2 + a^2} + s|b - x|$ on the interval $(-\infty, +\infty)$, where a, b, s, and t are positive constants. Thus solve the following problem: The shore of a lake extends for a considerable distance along the x-axis of a coordinate system. The lake lies in the first two quadrants and has an island on the y-axis at the point A: $(0, a)$. A man on the island wishes to go as quickly as possible to the point B: $(b, 0)$ on the positive half of the x-axis by rowing straight to the point P: $(x, 0)$ at a rate of s miles an hour, and walking from P to B at a rate of t miles an hour. What course should he steer?

*__20.__ Find the minimum value of the function $|x^2 - ax| + rx^2$ on the interval $[0, +\infty)$, where a is a positive number and r is a real number.

*__21.__ Assume that $x_1 \le x_2 \le x_3 \le \cdots \le x_n$. Prove that the function
$$|x - x_1| + |x - x_2| + \cdots + |x - x_n|$$

is minimized by the following values of x: (i) if n is odd and equal to $2m - 1$, $x = x_m$; (ii) if n is even and equal to $2m$, $x_m \leqq x \leqq x_{m+1}$.

★**22.** Show that the function $f(x)$ defined to be $x^4 \left(2 + \sin \dfrac{1}{x} \right)$ when $x \neq 0$ and 0 when $x = 0$ has an absolute strict minimum at $x = 0$ ($f(x) > f(0)$ if $x \neq 0$) and a continuous derivative everywhere, but that this derivative does not change sign from $-$ to $+$ as x changes from $x < 0$ to $x > 0$. Thus show that the conditions of the First Derivative Test (§ 309) are only sufficient and not necessary for a maximum (minimum). (Cf. Ex. 23 for a converse of the First Derivative Test.) Sketch a graph of the function given above. (It oscillates infinitely many times between the two curves $y = x^4$ and $y = 3x^4$.)

★**23.** Prove the following converse of the First Derivative Test: *Let $f(x)$ be differentiable in a deleted neighborhood of a point $x = \xi$ at which $f(x)$ is continuous and has a relative maximum (minimum) value, and assume that in this deleted neighborhood $f'(x)$ vanishes at only a finite number of points. Then there exists a neighborhood of $x = \xi$ within which $f'(x) > 0$ for $x < \xi$ and $f'(x) < 0$ for $x > \xi$ ($f'(x) < 0$ for $x < \xi$ and $f'(x) > 0$ for $x > \xi$).* (Cf. Ex. 22.)

311. DIFFERENTIALS

The student of calculus becomes familiar with the differential notation, and learns to appreciate its convenience in the treatment of composite, inverse, and implicit functions and functions defined parametrically, and in the technique of integration by substitution. Differentials also lend themselves simply and naturally to such procedures as the solving of differential equations. Such techniques and their legitimacy will not be discussed here. In the present section we restrict ourselves to basic definitions and theoretical facts.

If $y = f(x)$ is a differentiable function of x, we introduce two symbols, dx and dy, devised for the purpose of permitting the derivative symbol to be regarded and manipulated as a fraction.

To this end we let dx denote an arbitrary real number, and define $dy = d(f(x))$ to be a function of the *two* independent variables x and dx, prescribed by the equation

(1) $$dy \equiv f'(x)\, dx.$$

The differentials dx and dy are interpreted geometrically in Figure 310.

Although calculus was the common invention of Sir Isaac Newton (1642–1727) and Gottfried Wilhelm von Leibnitz (1646–1716), the differential notation is due to Leibnitz. Its principal importance lies ultimately in the fact that formula (1), initially true under the hypothesis that y is a function of the independent variable x, remains true under any possible reinterpretation of the dependence or independence of the variables x and y. This fact is given explicit formulation in the two theorems that follow. (For illustrative examples, see Exs. 3–10, § 313.) Before proceeding to these theorems we point out a further justification of the differential notation which is of so elementary a character that it is frequently overlooked, but which follows directly from the definition (1): If two variables are related by the identity relation, $y \equiv x$, their differentials are also related by the identity relation, $dy \equiv dx$ (Ex. 1, § 313).

Theorem I. *If $x = \phi(y)$ is a differentiable function of the independent variable y in a certain interval, if $\phi'(y) \neq 0$ in this interval, and if dy and dx denote the differentials of the independent variable y and the dependent variable x, respectively, related by definition by the equation $dx \equiv \phi'(y)\,dy$, then if*

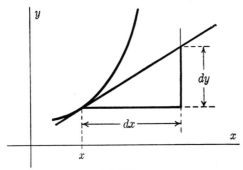

FIG. 310

$y = f(x)$ denotes the inverse function of $x = \phi(y)$, these differentials are also related by equation (1): $dy = f'(x)\,dx$.

Proof. By the Corollary to Theorem IV, § 306, the derivatives dx/dy and dy/dx are reciprocals, so that $\phi'(y) = 1/f'(x)$. Therefore

$$dx = (1/f'(x))\,dy,$$

and $dy = f'(x)\,dx$.

Theorem II. *If $y = f(x)$ is a differentiable function of the variable x, and if x is a differentiable function of the variable t, then if x and y are both regarded as dependent variables, depending on the independent variable t, their differentials are related by equation (1): $dy = f'(x)\,dx$.*

Proof. Let $x = \phi(t)$ and $y = \psi(t) \equiv f(\phi(t))$ denote the dependence of x and y on the independent variable t. Then by definition, $dx \equiv \phi'(t)\,dt$ and $dy \equiv \psi'(t)\,dt$. By the Chain Rule (Theorem I, § 302),

$$dy/dt = (dy/dx)(dx/dt),$$

or $\psi'(t) = f'(x)\phi'(t) = f'(\phi(t))\phi'(t)$, so that

$$dy = \psi'(t)\,dt = f'(\phi(t))(\phi'(t)\,dt) = f'(x)\,dx.$$

This completes the proof and shows that Theorem II is, in essence, simply a reformulation of the Chain Rule.

*312. APPROXIMATIONS BY DIFFERENTIALS

In order to compare the differentials dx and dy, on the one hand, and the increments Δx and Δy, on the other, we observe that whereas the differentials can be associated with the *tangent* to the curve $y = f(x)$ (Fig. 310), the increments are associated with the *curve* itself. It is often convenient to regard the real number dx as an increment of the variable x; that is, $dx = \Delta x$. In this case the quantities just discussed find their geometric interpretation in Figure 311. Throughout this section we shall identify dx and Δx. It should be noted, however, that under this assumption dy and Δy are *not* in general the same.

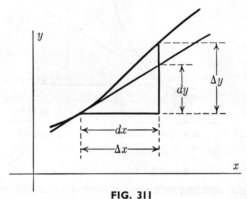

FIG. 311

Although in theory the increment Δy is a simpler concept than the differential dy, in practice the differential is usually easier to compute than the increment, and dy is often a useful approximation to Δy. The statement that for a numerically small increment $dx = \Delta x$, the quantity dy is a good approximation to the quantity Δy, has a precise formulation. The relation (under the assumption that the given function is differentiable at the point in question) is given by equation (4), § 302, rewritten in the form

(1) $$\Delta y = dy + \epsilon \Delta x,$$

where, *for a fixed x*, $\epsilon \equiv \epsilon(\Delta x)$ is an infinitesimal function of Δx ($\epsilon \to 0$ as $\Delta x \to 0$).

Using equation (1) we can formulate the statement that (for small $|dx| = |\Delta x|$) dy is a good approximation to Δy by means of the equation

(2) $$\lim_{\Delta x \to 0} \frac{\Delta y - dy}{\Delta x} = 0,$$

or, in case dy is not identically zero, by the equation

(3) $$\lim_{\Delta x \to 0} \frac{\Delta y}{dy} = 1 \quad \text{(Ex. 2, § 313).}$$

Example 1. If $y = f(x) \equiv x^3$, then $\Delta y = f(x + \Delta x) - f(x) = 3x^2 \Delta x + 3x \Delta x^2 + \Delta x^3$, and $dy = 3x^2\, dx$. Therefore, in equation (1), the function $\epsilon(\Delta x)$ is $3x\,\Delta x + \Delta x^2$, equation (2) takes the form $\lim_{\Delta x \to 0} (3x\,\Delta x + \Delta x^2) = 0$, and equation (3) becomes

$$\lim_{\Delta x \to 0} \frac{3x^2 + 3x\,\Delta x + \Delta x^2}{3x^2} = 1.$$

Example 2. Prove that for numerically small h the quantity $\sqrt{1 + h}$ is closely approximated by $1 + \tfrac{1}{2}h$.

Solution. Consider the function $y = f(x) \equiv \sqrt{x}$. Take $x = 1$ and $x + \Delta x = 1 + h$, so that $y = \sqrt{1} = 1$ and $y + \Delta y = \sqrt{1 + h}$. Then $dx = \Delta x = h$, and $dy = \dfrac{dx}{2\sqrt{x}} = \dfrac{h}{2}$. Approximately, then,

$$\sqrt{1 + h} = y + dy = 1 + \tfrac{1}{2}h.$$

The Extended Law of the Mean provides a measure of the accuracy of approximation of dy for Δy. Under the assumption that $f''(x)$ exists in the neighborhood of $x = a$, the Extended Law of the Mean ((12), § 307, for $n = 2$) ensures the existence of a number θ between 0 and 1 such that

(4) $\quad f(a + h) = f(a) + f'(a)h + \tfrac{1}{2}f''(a + \theta h)h^2,$

for numerically small h. Expressed in terms of differentials, with $\Delta x = h$, $\Delta y = f(a + h) - f(a)$ and $dy = f'(a)h$, (4) becomes

(5) $\quad \Delta y - dy = \tfrac{1}{2}f''(a + \theta \Delta x)\Delta x^2.$

If B is a bound for the absolute value of $f''(x)$ (that is, $|f''(x)| \leq B$) for x in a certain neighborhood of $x = a$, and if Δx is so restricted that $a + \Delta x$ (and therefore $a + \theta \Delta x$) is in this neighborhood, then (5) gives the inequality

(6) $\quad |\Delta y - dy| \leq \tfrac{1}{2} B\, \Delta x^2$

Example 3. Find an estimate for the accuracy of the approximation established in Example 2, if $|h| \leq 0.1$.

Solution. If $f(x) = \sqrt{x}$, then $f'(x) = \dfrac{1}{2\sqrt{x}}$, $f''(x) = -\dfrac{1}{4x\sqrt{x}}$, and formula (5) becomes

(7) $\quad \Delta y - dy = (\sqrt{1 + h} - 1) - (\tfrac{1}{2}h) = -\dfrac{h^2}{8(1 + \theta h)^{3/2}}.$

If h is positive (whether $h \leq 0.1$ or $h > 0.1$), $1 + \theta h > 1$ and therefore the third member of (7) is between $-h^2/8$ and 0. We thus have the inequalities

(8) $\quad 1 + \tfrac{1}{2}h - \tfrac{1}{8}h^2 < \sqrt{1 + h} < 1 + \tfrac{1}{2}h, \quad h > 0.$

(Illustration: $1.058 < \sqrt{1.12} < 1.06$.)

If $-0.1 \leq h < 0$, $1 + \theta h > 0.9$, and the third member of (7) is between $-h^2/6$ and 0. Therefore

(9) $\quad 1 + \tfrac{1}{2}h - \tfrac{1}{6}h^2 < \sqrt{1 + h} < 1 + \tfrac{1}{2}h, \quad -0.1 \leq h < 0.$

(Illustration: $0.9694 < \sqrt{0.94} < 0.97$.)

NOTE. Formula (7) permits sharper results than those just obtained. For example, if h is between -0.1 and 0, the third member of (7) is between $-h^2/6$ and

$-h^2/8$, so that $0.9694 < \sqrt{0.94} < 0.9696$. For computations where a high degree of accuracy is desired, however, the methods of this section are inadequate, and the reader is referred to Chapter 12.

313. EXERCISES

1. Prove that if two variables are related by the identity relation $y \equiv x$, their differentials are also related by the identity relation $dy \equiv dx$.

2. If y is a function of x which is differentiable for a particular value of x, prove the limit statements (2) and (3), § 312: As $\Delta x \to 0$, $\dfrac{\Delta y - dy}{\Delta x} \to 0$, and if furthermore $dy \neq 0$, $\dfrac{\Delta y}{dy} \to 1$.

In Exercises 3–6, express the given relation $x = \phi(y)$ in the form $y = f(x)$, write down both equations $dx = \phi'(y)\,dy$ and $dy = f'(x)\,dx$, and thus verify Theorem I, § 311. Draw figures.

3. $x = y^2,\ y > 0$.   **4.** $x = y^2 - 4y + 5,\ y < 2$.
5. $x = \ln(y^2 + 1),\ y > 0$.   **6.** $x = \cos y,\ 0 < y < \pi$.

In Exercises 7–10, use the given relations $x = \phi(t)$ and $y = \psi(t)$ to express y as a function of x, $y = f(x)$, write down the three equations $dx = \phi'(t)\,dt$, $dy = \psi'(t)\,dt$, and $dy = f'(x)\,dx$, and thus verify Theorem II, § 311. Draw figures.

7. $x = t,\ y = 5t^2 - 7t - 6$.   **8.** $x = t^2,\ y = t^3 - t,\ t < 0$.
9. $x = \ln t,\ y = e^t,\ t > 0$.   **10.** $x = \cos t,\ y = \sin t,\ 0 < t < \tfrac{1}{2}\pi$.

In Exercises 11–13, find Δy, dy, and $\epsilon(\Delta x)$, and verify the limit statements (2) and (3), § 312.

11. $y = x^4 - 5x^2 + 7$.   **12.** $y = \dfrac{1}{x}$.

13. $y = \sqrt{x}$. *Hint:* Rationalize a numerator (cf. Ex. 5, § 304).

14. Prove that if $y = f(x)$ has a continuous second derivative in a neighborhood of the point $x = a$, then
$$f''(a) = \lim_{\Delta x \to 0} 2 \cdot \frac{\Delta y - dy}{\Delta x^2} = \lim_{h \to 0} \frac{f(a+h) + f(a-h) - 2f(a)}{h^2}.$$
Use these relations to obtain the second derivative of each of the functions of Exercises 11–13. (Cf. Ex. 10, § 322.)

In Exercises 15–26, use differentials to obtain an approximation to the given number, or the given function for values of the variable near the specified value.

15. $\sqrt{110}$ (use $10.5^2 = 110.25$).   **16.** $\ln(1+h)$, h near 0.
17. $\ln(0.94)$ (cf. Ex. 16).   **18.** $\sin x$, x near 0.

19. $\tan x$, x near 0.   **20.** $\cos x$, x near $\dfrac{\pi}{3}$.

21. $\operatorname{Arctan} x$, x near 0.   **22.** $\sqrt[n]{1+h}$, h near 0, n an integer > 1.

23. $\ln \cos x$, x near 0.   **24.** $\dfrac{\ln(1+h)}{1+h}$, h near 0.

25. $e^{\sin x}$, x near 0.   **26.** $f(x) \equiv x^2 \sin \dfrac{1}{x}$ ($f(0) \equiv 0$), x near 0.

In Exercises 27–38, verify the given inequalities, which give estimates of the errors in the approximations of Exercises 15–26.

★27. $10.48808 < \sqrt{110} < 10.48810$.
★28. $h - \frac{1}{2}h^2 < \ln(1+h) < h, h > 0$;
$h - \frac{2}{3}h^2 < \ln(1+h) < h, -0.1 \leq h < 0$.
★29. $-0.0622 < \ln 0.94 < -0.06$.
★30. $x - \frac{x^3}{6} < \sin x < x, x > 0$. *Hint:* Show that
$$\Delta y - dy = -\frac{1}{6}\cos(\theta x)x^3, \quad 0 < \theta < 1.$$
★31. $x + \frac{1}{3}x^3 < \tan x < x + \frac{1}{2}x^3, 0 < x < 0.1$. *Hint:* Show that
$$\Delta y - dy = \frac{1}{3}(1 + 3t^2)(1 + t^2)x^3, \text{ where } t = \tan(\theta x), \quad 0 < \theta < 1.$$
★32. $\frac{1}{2} - \frac{1}{2}\sqrt{3}(x - \frac{1}{3}\pi) - \frac{1}{2}(x - \frac{1}{3}\pi)^2 \leq \cos x \leq \frac{1}{2} - \frac{1}{2}\sqrt{3}(x - \frac{1}{3}\pi)$,
$$0 \leq x \leq \frac{1}{2}\pi.$$
★33. $x - \frac{1}{3}x^3 < \text{Arctan } x < x, 0 < x < \frac{1}{2}\pi$.
★34. $1 + \frac{h}{n} - \frac{(n-1)h^2}{2n^2} < \sqrt[n]{1+h} < 1 + \frac{h}{n}, h > 0$;
$1 + \frac{h}{n} - \frac{(n-1)h^2}{n^2} < \sqrt[n]{1+h} < 1 + \frac{h}{n}, -\frac{1}{4} < h < 0$.
★35. $-x^2 < \ln \cos x < -\frac{1}{2}x^2, 0 < x < \frac{1}{4}\pi$.
★36. $h - \frac{9}{4}h^2 < \frac{\ln(1+h)}{1+h} < h - h^2, 0 < |h| < 0.1$.
★37. $1 + x + 0.4x^2 < e^{\sin x} < 1 + x + 0.61x^2, 0 < |x| < 0.1$.
★38. $x^2 \sin \frac{1}{x} \leq x^2$.

314. L'HOSPITAL'S RULE. INTRODUCTION

In the following three sections we consider the most important types of indeterminate forms, with some proofs given in the text, and others deferred to the following Exercises. The principles established in these sections are called upon in many of the problems in curve tracing that follow. Evaluation of indeterminate expressions by means of infinite series is discussed in § 1212.

315. THE INDETERMINATE FORM 0/0

The statement that $0/0$ is an indeterminate form means that the fact that two functions have the limit 0, as the independent variable approaches some limit, does not in itself imply anything about the limit of their quotient. The four examples x/x, x^2/x, x/x^2, and $\left(x \sin \frac{1}{x}\right)/x$, show that as $x \to 0+$, the quotient of functions, each tending toward zero, may have a limit 1, or 0, or $+\infty$, or it may have no limit at all, finite or infinite. That the involvement of infinity or the apparent division by zero does not in itself constitute an indeterminacy was seen in Exercises 31, 32, 50, § 208. Furthermore, the fact that $0/0$ is an indeterminate form, in the sense explained above, certainly does not mean that a quotient of functions, each tending toward zero, cannot have a limit. Indeed, the simple examples above show this, as does any evaluation of a derivative as the limit of the quotient of two increments.

Frequently limits of quotients of functions, each tending toward zero, can be determined by the device known as l'Hospital's Rule. This is stated first in general form. The behavior of the independent variable is then specified in the separate cases. The letter a represents a real number.

Theorem. L'Hospital's Rule. *If $f(x)$ and $g(x)$ are differentiable functions and $g'(x) \neq 0$ for x in a deleted neighborhood of its limit, if $\lim f(x) = \lim g(x) = 0$, and if*

$$\lim \frac{f'(x)}{g'(x)} = L \text{ (finite, } +\infty, -\infty, \text{ or } \infty\text{)},$$

then

$$\lim \frac{f(x)}{g(x)} = L.$$

Case 1. $x \to a+$. (Proof below.)
Case 2. $x \to a-$. (Ex. 31, § 318.)
Case 3. $x \to a$. (Ex. 32, § 318.)
Case 4. $x \to +\infty$. (Proof below.)
Case 5. $x \to -\infty$. (Ex. 33, § 318.)
Case 6. $x \to \infty$. (Ex. 34, § 318.)

Proof of Case 1. Let $f(a)$ and $g(a)$ be defined (or redefined if necessary) to be zero. Then $f(x)$ and $g(x)$ are both continuous on some closed interval $[a, a + \epsilon]$, where $\epsilon > 0$. The number ϵ can be chosen so small that $g'(x)$ does not vanish in the open interval $(a, a + \epsilon)$ and the conditions of the Generalized Law of the Mean (Theorem IV, § 305) are fulfilled for any x such that $a < x \leq a + \epsilon$. Thus, for any such x there exists a number ξ such that $a < \xi < x$ and

$$\frac{f(x)}{g(x)} = \frac{f'(\xi)}{g'(\xi)}.$$

As $x \to a+$, $\xi \to a+$ and the limit of the right-hand member of this equation exists (finite or infinite) by hypothesis. Therefore the limit of the left-hand member of the equation also exists (finite or infinite) and is equal to it.

Proof of Case 4. Use reciprocals:

$$L = \lim_{x \to +\infty} \frac{f'(x)}{g'(x)} = \lim_{t \to 0+} \frac{f'(1/t)}{g'(1/t)} = \lim_{t \to 0+} \frac{-f'(1/t)t^{-2}}{-g'(1/t)t^{-2}}$$

(multiplying and dividing by $-t^{-2}$)

$$= \lim_{t \to 0+} \frac{\dfrac{d}{dt} f(1/t)}{\dfrac{d}{dt} g(1/t)} = \lim_{t \to 0+} \frac{f(1/t)}{g(1/t)} = \lim_{x \to +\infty} \frac{f(x)}{g(x)}.$$

The next-to-the-last equality is true by Case 1. The sequence of equalities implies that the limit under consideration exists and is equal to L.

Example 1. $\lim\limits_{x \to 0} \dfrac{\sin x}{x} = \lim\limits_{x \to 0} \dfrac{\cos x}{1} = 1$. (Cf. Ex. 16, § 504.)

Example 2.
$$\lim_{x \to 0} \frac{\sin x - x}{x^3} = \lim_{x \to 0} \frac{\cos x - 1}{3x^2} = \lim_{x \to 0} \frac{-\sin x}{6x} = \lim_{x \to 0} \frac{-\cos x}{6} = -\frac{1}{6}.$$
In this case l'Hospital's Rule is iterated. The existence of each limit implies that of the preceding and their equality.

Example 3. $\lim\limits_{x \to +\infty} \dfrac{e^{-x}}{\dfrac{1}{x}} = \lim\limits_{x \to +\infty} \dfrac{-e^{-x}}{-\dfrac{1}{x^2}} = \lim\limits_{x \to +\infty} \dfrac{e^{-x}}{\dfrac{2}{x^3}}$.

Things are getting worse! See Example 1, § 316.

It is important before applying l'Hospital's Rule to check on the indeterminacy of the expression being treated. The following example illustrates this.

Example 4. A routine and thoughtless attempt to apply l'Hospital's Rule may yield an incorrect result as follows:
$$\lim_{x \to 1} \frac{2x^2 - x - 1}{x^2 - x} = \lim_{x \to 1} \frac{4x - 1}{2x - 1} = \lim_{x \to 1} \frac{4}{2} = 2.$$
The first equality is correct, and the answer is obtained by direct substitution of 1 for x in the continuous function $(4x - 1)/(2x - 1)$, to give the correct value of 3.

316. THE INDETERMINATE FORM ∞/∞

The symbol ∞/∞ indicates that a limit is being sought for the quotient of two functions, each of which is becoming infinite (the absolute value approaches $+\infty$) as the independent variable approaches some limit. L'Hospital's Rule is again applicable, but the proof is more difficult.

Theorem. L'Hospital's Rule. *If $f(x)$ and $g(x)$ are differentiable functions and $g'(x) \neq 0$ for x in a deleted neighborhood of its limit, if $\lim f(x) = \lim g(x) = \infty$, and if*
$$\lim \frac{f'(x)}{g'(x)} = L \text{ (finite, } +\infty, -\infty, \text{ or } \infty\text{),}$$
then
$$\lim \frac{f(x)}{g(x)} = L.$$

Case 1. $x \to a+$. (Ex. 35, § 318.)
Case 2. $x \to a-$. (Ex. 36, § 318.)
Case 3. $x \to a$. (Ex. 37, § 318.)
Case 4. $x \to +\infty$. (Proof below.)
Case 5. $x \to -\infty$. (Ex. 38, § 318.)
Case 6. $x \to \infty$. (Ex. 39, § 318.)

Proof of Case 4. Observe that whenever x is sufficiently large to prevent the vanishing of $f(x)$ and $g(x)$, and N_1 is sufficiently large to prevent the vanishing of $g'(\xi)$ for $\xi > N_1$, the generalized law of the mean guarantees the relation

$$(1) \quad \frac{f(x)}{g(x)} \cdot \frac{1 - f(N_1)/f(x)}{1 - g(N_1)/g(x)} = \frac{f(x) - f(N_1)}{g(x) - g(N_1)} = \frac{f'(\xi)}{g'(\xi)},$$

and therefore,

$$(2) \quad \frac{f(x)}{g(x)} = \frac{f'(\xi)}{g'(\xi)} \frac{1 - g(N_1)/g(x)}{1 - f(N_1)/f(x)},$$

for $x > N_1$ and a suitable ξ between x and N_1. First choose N_1 so large that if $\xi > N_1$, then $f'(\xi)/g'(\xi)$ is within a specified degree of approximation of L. (If L is infinite, the term *approximation* should be interpreted liberally, in accordance with the definitions of infinite limits, § 206.) Second, using the hypotheses that $\lim_{x \to +\infty} |f(x)| = \lim_{x \to +\infty} |g(x)| = +\infty$, let N_2 be so large that if $x > N_2$, then the fraction $[1 - g(N_1)/g(x)]/[1 - f(N_1)/f(x)]$ is within a specified degree of approximation of the number 1. In combination, by equation (2), these two approximations guarantee that $f(x)/g(x)$ approximates L. This completes the outline of the proof, but for more complete rigor, we present the "epsilon" details for the case where L is finite. (Cf. Ex. 40, § 318.)

Let L be an arbitrary real number and ϵ an arbitrary positive number. We shall show first that there exists a positive number δ such that $|y - L| < \frac{1}{2}\epsilon$ and $|z - 1| < \delta$ imply $|yz - L| < \epsilon$. To do this we use the triangle inequality to write

$$|yz - L| \leq |yz - y| + |y - L| = |y| \cdot |z - 1| + |y - L|.$$

If $|y - L| < \frac{1}{2}\epsilon$, $|y| < |L| + \frac{1}{2}\epsilon$, so that if $\delta \equiv \frac{1}{2}\epsilon(|L| + \frac{1}{2}\epsilon)^{-1}$ the assumed inequalities imply

$$|yz - L| < (|L| + \tfrac{1}{2}\epsilon)\frac{\epsilon}{2(|L| + \tfrac{1}{2}\epsilon)} + \frac{\epsilon}{2} = \epsilon,$$

which is the desired result. The rest is simple. First choose N_1 such that

$$\xi > N_1 \quad \text{implies} \quad \left|\frac{f'(\xi)}{g'(\xi)} - L\right| < \tfrac{1}{2}\epsilon,$$

and second choose $N_2 > N_1$ such that

$$x > N_2 \quad \text{implies} \quad \left|\frac{1 - g(N_1)/g(x)}{1 - f(N_1)/f(x)} - 1\right| < \delta.$$

Then, by (2),

$$x > N_2 \quad \text{implies} \quad \left|\frac{f(x)}{g(x)} - L\right| < \epsilon.$$

Example 1. Show that $\lim_{x \to +\infty} \dfrac{x^a}{e^x} = 0$ for any real a.

Solution. If $a \leq 0$ the expression is not indeterminate. Assume $a > 0$. Then $\lim_{x \to +\infty} \dfrac{x^a}{e^x} = \lim_{x \to +\infty} \dfrac{ax^{a-1}}{e^x}$, and if this process is continued, an exponent for x is ultimately found that is zero or negative. This example shows that e^x increases, as $x \to +\infty$, faster than any power of x, and therefore faster than any polynomial.

Example 2. Show that $\lim\limits_{x \to +\infty} \dfrac{\ln x}{x^a} = 0$ for any $a > 0$.

Solution. $\lim\limits_{x \to +\infty} \dfrac{\ln x}{x^a} = \lim\limits_{x \to +\infty} \dfrac{\frac{1}{x}}{ax^{a-1}} = \lim\limits_{x \to +\infty} \dfrac{1}{ax^a} = 0$. (Cf. Ex. 7, § 502.) In other words, $\ln x$ increases, as $x \to +\infty$, more slowly than any positive power of x.

Example 3. Show that $\lim\limits_{n \to +\infty} \dfrac{e^n}{n!} = 0$.

Solution. This is an indeterminate form to which l'Hospital's Rule does not apply, since $n!$ (unless the gamma function (§ 1410) is used to define $n!$ for all positive real numbers) cannot be differentiated. We can establish the limit as follows: Let n be greater than 3. Then

$$\frac{e^n}{n!} = \left(\frac{e}{1} \cdot \frac{e}{2} \cdot \frac{e}{3}\right)\left(\frac{e}{4} \cdots \frac{e}{n}\right) < \frac{e^3}{6} \cdot \left(\frac{e}{4}\right)^{n-3}.$$

As $n \to +\infty$, the last factor approaches zero.

Example 4. Criticize: $\lim\limits_{x \to +\infty} \dfrac{\sin x}{x} = \lim\limits_{x \to +\infty} \dfrac{\cos x}{1}$, and therefore does not exist!

Solution. The given expression is not indeterminate, and the hypotheses of l'Hospital's Rule are invalid. Since $|\sin x/x| \leq 1/x$, $x > 0$, the limit is 0.

317. OTHER INDETERMINATE FORMS

In the sense discussed in § 315, the forms

$$0 \cdot \infty, \quad \infty - \infty, \quad 0^0, \quad \infty^0, \quad \text{and} \quad 1^\infty$$

are indeterminate (Ex. 42, § 318). The first type can often be evaluated by writing the product $f(x) \cdot g(x)$ as a quotient and then using l'Hospital's Rule (Example 1). The second type sometimes lends itself to rearrangement, use of identities, or judicious multiplication by unity (Examples 2–3). The remaining three forms are handled by first taking a logarithm: if $y = f(x)^{g(x)}$, then $\ln y = g(x) \ln (f(x))$, and an indeterminacy of the first type above results. Then, by continuity of the exponential function (cf. § 502), $\lim y = \lim e^{(\ln y)} = e^{\lim (\ln y)}$. (Examples 4–6). Finally, other devices, including separation of determinate from indeterminate expressions and substitution of a reciprocal variable, are possible (Examples 7–9).

Example 1. Find $\lim\limits_{x \to 0+} x^a \ln x$.

Solution. If $a \leq 0$, the expression is not indeterminate (cf. Ex. 7, § 502), and the limit is $-\infty$. If $a > 0$, the limit can be written

$$\lim\limits_{x \to 0+} \frac{\ln x}{x^{-a}} = \lim\limits_{x \to 0+} \frac{1/x}{-ax^{-a-1}} = \lim\limits_{x \to 0+} \frac{x^a}{-a} = 0.$$

In other words, whenever the expression $x^a \ln x$ is indeterminate, the algebraic factor "dominates" the logarithmic factor (cf. Example 2, § 316).

Example 2. $\lim\limits_{x \to \frac{\pi}{2}} (\sec x - \tan x) = \lim\limits_{x \to \frac{\pi}{2}} \dfrac{1 - \sin x}{\cos x} = \lim\limits_{x \to \frac{\pi}{2}} \dfrac{-\cos x}{-\sin x} = 0.$

Alternatively, without the use of l'Hospital's Rule,

$$\lim_{x \to \frac{\pi}{2}} (\sec x - \tan x) = \lim_{x \to \frac{\pi}{2}} \frac{\sec^2 x - \tan^2 x}{\sec x + \tan x} = \lim_{x \to \frac{\pi}{2}} \frac{1}{\sec x + \tan x} = 0.$$

Example 3. $\lim\limits_{x \to \infty} [\sqrt{x^2 - a^2} - |x|] = \lim\limits_{x \to \infty} \dfrac{\sqrt{x^2 - a^2} - |x|}{1} \cdot \dfrac{\sqrt{x^2 - a^2} + |x|}{\sqrt{x^2 - a^2} + |x|}$

$$= \lim_{x \to \infty} \frac{-a^2}{\sqrt{x^2 - a^2} + |x|} = 0.$$

Example 4. Find $\lim\limits_{x \to 0+} x^x$.

Solution. Let $y = x^x$. Then $\ln y = x \ln x$ and, by Example 1, $\ln y \to 0$. Therefore, by continuity of the function e^x, $y = e^{\ln y} \to e^0 = 1$.

Example 5. Find $\lim\limits_{x \to +\infty} (1 + ax)^{\frac{1}{x}}$, $a > 0$.

Solution. If $y = (1 + ax)^{\frac{1}{x}}$, $\ln y = \dfrac{\ln (1 + ax)}{x} \to 0$.

Therefore $y = e^{\ln y} \to e^0 = 1$.

Example 6. Show that $\lim\limits_{x \to 0} (1 + ax)^{\frac{1}{x}} = e^a$.

Solution. If $a \neq 0$ and if $y = (1 + ax)^{\frac{1}{x}}$,

$$\lim_{x \to 0} \ln y = \lim_{x \to 0} \frac{\ln (1 + ax)}{x} = \lim_{x \to 0} \frac{a}{1 + ax} = a,$$

and $y = e^{\ln y} \to e^a$. (Cf. Ex. 10, § 502.)

Example 7. Find $\lim\limits_{x \to 0+} xe^{\frac{1}{x}}$.

Solution. If this is written $\lim\limits_{x \to 0+} \dfrac{e^{\frac{1}{x}}}{\frac{1}{x}}$, differentiation leads to the answer.

However, the limit can be written $\lim\limits_{t \to +\infty} \dfrac{e^t}{t} = +\infty$.

Example 8. $\lim\limits_{x \to 0} (\csc^2 x - x \csc^3 x) = \lim\limits_{x \to 0} \dfrac{\sin x - x}{x^3} \cdot \dfrac{x^3}{\sin^3 x}$

$$= \left(-\frac{1}{6}\right) \cdot \left(\lim_{x \to 0} \frac{x}{\sin x}\right)^3 = -\frac{1}{6},$$

by Example 2, § 315, and the continuity of the function x^3 (the limit of the cube is the cube of the limit).

Example 9. $\lim\limits_{x \to 1} \dfrac{12 \sin \dfrac{\pi}{2x} \ln x}{(x^3 + 5)(x - 1)}$

$= \lim\limits_{x \to 1} \dfrac{12 \sin \dfrac{\pi}{2x}}{x^3 + 5} \cdot \lim\limits_{x \to 1} \dfrac{\ln x}{x - 1} = \dfrac{12 \cdot 1}{6} \cdot \lim\limits_{x \to 1} \dfrac{\dfrac{1}{x}}{1} = 2.$

318. EXERCISES

In Exercises 1–30, evaluate the limit.

1. $\lim\limits_{x \to 2} \dfrac{3x^2 + x - 14}{x^2 - x - 2}.$

2. $\lim\limits_{x \to -3} \dfrac{x^3 + x + 30}{4x^3 + 11x^2 + 9}.$

3. $\lim\limits_{x \to 1} \dfrac{\ln x}{x^2 + x - 2}.$

4. $\lim\limits_{x \to 1} \dfrac{\cos \tfrac{1}{2}\pi x}{x - 1}.$

5. $\lim\limits_{x \to 0} \dfrac{\cos x - 1 + \tfrac{1}{2}x^2}{x^4}.$

6. $\lim\limits_{x \to 0} \dfrac{\ln(1 + x) - x}{\cos x - 1}.$

7. $\lim\limits_{x \to \pi} \dfrac{\sin^2 x}{\tan^2 4x}.$

8. $\lim\limits_{x \to \infty} \dfrac{\sin(1/x)}{\operatorname{Arc\,tan}(1/x)}.$

9. $\lim\limits_{x \to 0} \dfrac{a^x - 1}{b^x - 1}.$

10. $\lim\limits_{x \to 0} \dfrac{\tan x - x}{\operatorname{Arc\,sin} x - x}.$

11. $\lim\limits_{x \to \infty} \dfrac{8x^5 - 5x^2 + 1}{3x^5 + x}.$

12. $\lim\limits_{x \to \tfrac{1}{2}\pi} \dfrac{\tan x - 6}{\sec x + 5}.$

13. $\lim\limits_{x \to \tfrac{1}{2}\pi^-} \dfrac{\ln \sin 2x}{\ln \cos x}.$

14. $\lim\limits_{x \to \pi} \dfrac{\ln \sin x}{\ln \sin 2x}.$

15. $\lim\limits_{x \to +\infty} \dfrac{\cosh x}{e^x}$ (cf. § 507).

16. $\lim\limits_{x \to \tfrac{1}{2}^-} \dfrac{\ln(1 - 2x)}{\tan \pi x}.$

17. $\lim\limits_{x \to +\infty} \dfrac{(\ln x)^n}{x}, \ n > 0.$

18. $\lim\limits_{x \to +\infty} \dfrac{a^x}{x^b}, \ a > 1, b > 0.$

19. $\lim\limits_{x \to 0+} x(\ln x)^n, \ n > 0.$

20. $\lim\limits_{x \to a} (x - a) \tan \dfrac{\pi x}{2a}.$

21. $\lim\limits_{x \to \tfrac{1}{2}\pi} \left[x \tan x - \dfrac{\pi}{2} \sec x\right].$

22. $\lim\limits_{x \to 1+} \left[\dfrac{x}{x - 1} - \dfrac{1}{\ln x}\right].$

23. $\lim\limits_{x \to +\infty} (1 + x^2)^{\tfrac{1}{x}}.$

24. $\lim\limits_{x \to 0} (1 + 2 \sin x)^{\cot x}.$

25. $\lim\limits_{x \to 0+} x^{\tfrac{1}{\ln x}}.$

26. $\lim\limits_{x \to 0} (x + e^{2x})^{\tfrac{1}{x}}.$

27. $\lim\limits_{x \to 0+} x^{x^x} \ (x^{x^x} = x^{(x^x)}).$

28. $\lim\limits_{x \to 0+} [\ln(1 + x)]^x.$

29. $\lim\limits_{x \to 0} (\cos 2x)^{\tfrac{1}{x^2}}.$

30. $\lim\limits_{x \to 0+} \left[\dfrac{\ln x}{(1 + x)^2} - \ln \dfrac{x}{1 + x}\right].$

31. Prove Case 2 of l'Hospital's Rule, § 315.
32. Prove Case 3 of l'Hospital's Rule, § 315. *Hint:* Make sensible use of Cases 1 and 2.
33. Prove Case 5 of l'Hospital's Rule, § 315.
34. Prove Case 6 of l'Hospital's Rule, § 315.

35. Prove Case 1 of l'Hospital's Rule, § 316. *Hint:* Apply Case 4 (cf. proof of Case 4, § 315).

★36. Prove Case 2 of l'Hospital's Rule, § 316.

★37. Prove Case 3 of l'Hospital's Rule, § 316.

★38. Prove Case 5 of l'Hospital's Rule, § 316.

★39. Prove Case 6 of l'Hospital's Rule, § 316.

★40. For Case 4 of l'Hospital's Rule, § 316, supply the "epsilon" details for the case $L = +\infty$.

41. Prove that the forms $(+0)^{+\infty}$, $(+\infty)^{+\infty}$, and $a^{+\infty}$ (where $a > 0$ and $a \neq 1$) are determinate. What can you say about $(+0)^{-\infty}$? $(+\infty)^{-\infty}$? $(+0)^{\infty}$? $(+\infty)^{\infty}$?

42. Show by examples that all of the forms of § 317 are indeterminate, as stated.

43. Criticize the following alleged "proof" of l'Hospital's Rule for the form $0/0$ as $x \to a+$: By the Law of the Mean, for any $x > a$, there exist ξ_1 and ξ_2 between a and x such that $\dfrac{f(x) - f(a+)}{x - a} = f'(\xi_1)$ and $\dfrac{g(x) - g(a+)}{x - a} = g'(\xi_2)$. Therefore

$$\frac{f(x)}{g(x)} = \frac{f(x) - f(a+)}{g(x) - g(a+)} = \frac{f'(\xi_1)}{g'(\xi_2)} \to \frac{f'(a+)}{g'(a+)} = \lim_{x \to a+} \frac{f'(x)}{g'(x)}.$$

319. CURVE TRACING

It is not our purpose in this section to give an extensive treatment of curve tracing. Rather, we wish to give the reader an opportunity to review in

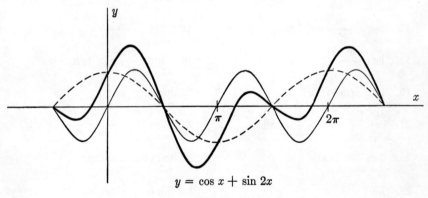

$y = \cos x + \sin 2x$

FIG. 312

practice such topics from differential calculus as increasing and decreasing functions, maximum and minimum points, symmetry, concavity, and points of inflection. Certain basic principles we do wish to recall explicitly, however.

(i) Composition of ordinates. The graph of a function represented as the sum of terms can often be obtained most simply by graphing the separate terms, and adding the ordinates visually, as indicated in Figure 312.

(ii) *Dominant terms.* If different terms "dominate" an expression for different values of the independent variable, the general shape of the curve can often be inferred. For example, for positive values of x, the function $x + x^{-1}$ is dominated by the second term if x is small and by the first term if x is large (Fig. 313).

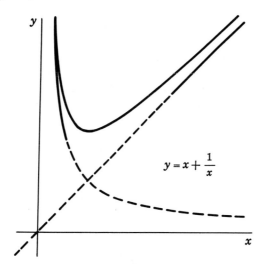

FIG. 313

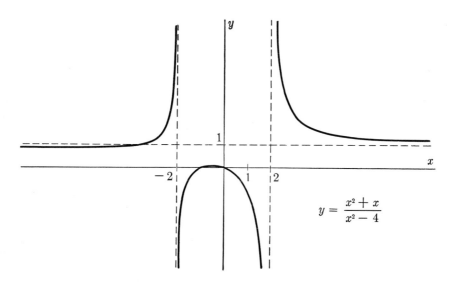

FIG. 314

(*iii*) *Vertical and horizontal asymptotes.* A function represented as a quotient $f(x)/g(x)$ of continuous functions has a vertical asymptote at a point a where $g(a) = 0$ and $f(a) \neq 0$. If $\lim f(x)/g(x)$, as x becomes infinite ($+\infty$, $-\infty$, or ∞), exists and is a finite number b, then $y = b$ is a horizontal asymptote. (Fig. 314.)

(*iv*) *Other asymptotes.* If $f(x) - mx - b$ approaches zero as x becomes infinite ($+\infty$, $-\infty$, or ∞), then the line $y = mx + b$ is an asymptote for the graph of $f(x)$. For the function $x + e^x$, for example, the line $y = x$ is an asymptote as $x \to -\infty$. (Fig. 315.)

(*v*) *Two factors.* Certain principles used for functions represented as sums have their applications to functions represented as products. The functions $e^{-ax} \sin bx$ and $e^{-ax} \cos bx$, useful in electrical theory, are good examples. (Fig. 316.)

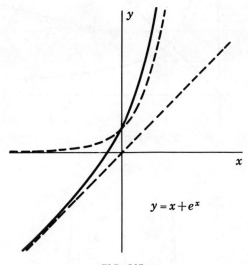

$y = x + e^x$

FIG. 315

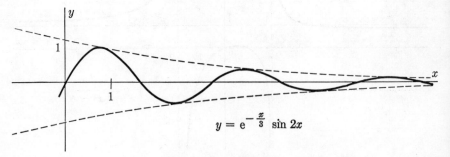

$y = e^{-\frac{x}{3}} \sin 2x$

FIG. 316

Vanishing factors often determine the general shape of a curve in neighborhoods of points where they vanish. For example, the graph of $y^2 = x^2(2 - x)$ is approximated by that of $y^2 = 2x^2$ for x near 0 and by that of $y^2 = 4(2 - x)$ for x near 2. (Fig. 317.)

A further aid in graphing an equation like that of Figure 317, of the form $y^2 = f(x)$, is graphing the function $f(x)$ itself to determine the values of x for which $f(x)$ is positive, zero, or negative, and hence for which y is double-valued, zero, or imaginary. (Fig. 318.)

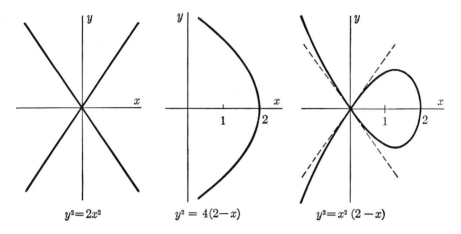

FIG. 317

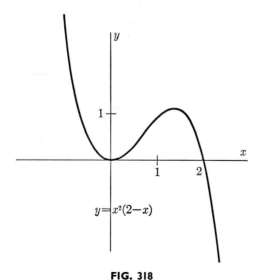

FIG. 318

(vi) *Parametric equations.* If $x = f(t)$ and $y = g(t)$, we recall two formulas:

(1) $$y' = \frac{dy}{dx} = \frac{g'(t)}{f'(t)};$$

(2) $$y'' = \frac{d^2y}{dx^2} = \frac{\dfrac{dy'}{dt}}{f'(t)}.$$

The folium of Descartes,

(3) $$x = \frac{3at}{t^3 + 1}, \quad y = \frac{3at^2}{t^3 + 1},$$

is illustrated in Figure 319. Since

(4) $$\frac{dy}{dx} = \frac{t(t^3 - 2)}{2t^3 - 1},$$

horizontal tangents correspond to the values of t for which the numerator of (4) vanishes: $t = 0$ (the point $(0, 0)$) and $t = \sqrt[3]{2}$ (the point $(a\sqrt[3]{2}, a\sqrt[3]{4})$).

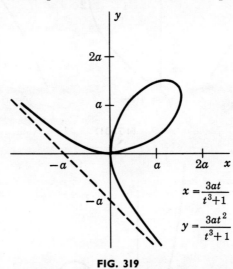

FIG. 319

Vertical tangents correspond to the values of t for which (4) becomes infinite: $t = \infty$ (the point $(0, 0)$) and $t = \sqrt[3]{\frac{1}{2}}$ (the point $(a\sqrt[3]{4}, a\sqrt[3]{2})$). Since $x + y = 3at/(t^2 - t + 1)$, the line $x + y + a = 0$ is an asymptote (let $t \to -1$).

320. EXERCISES

In Exercises 1–50, graph the equation, showing essential shape and asymptotic behavior.

1. $y = x^2(x^2 - 9)$.

2. $y = \dfrac{2}{x^2 + 1}$.

3. $y = \dfrac{x^2 - 4}{x^2 - 9}$.

4. $y = \dfrac{x^2 - 9}{x^2 - 4}$.

5. $y = x + \dfrac{4}{x}$.

6. $y = x^2 + \dfrac{2}{x}$.

7. $y = x + \dfrac{4}{x^2}$.

8. $y^2 = x^2(x^2 - 9)$.

9. $y^2 = \dfrac{x^2 - 4}{x^2 - 9}$.

10. $y^2 = \dfrac{x^2 - 9}{x^2 - 4}$.

11. $y^2 = (x - 1)(x - 2)^2(x - 3)^3$.
12. $y^2 = (x - 1)^3(x - 2)(x - 3)^2$.
13. $x^2 + xy + y^2 = 4$.
14. $x^2 + 4xy + y^2 = 4$.
15. $y = e^{-x^2}$.
16. $y = e^{1/x}$.
17. $y = xe^x$.
18. $y = x^2 e^{-x}$.
19. $y = xe^{-x^2}$.
20. $y = e^x - x$.
21. $y = \ln |x|$.
22. $y = x \ln x$.
23. $y = \dfrac{\ln x}{x}$.
24. $y = \dfrac{x}{\ln x}$.
25. $y = \dfrac{1}{x \ln x}$.
26. $y = x^2 \ln x$.
27. $y = x + \sin x$.
28. $y = \ln x \cdot e^{-x}$.
29. $y = \ln(1 + x^2)$.
30. $y = \ln \sin x$.
31. $y = e^{-x} \cos 2x$.
32. $y = \dfrac{g}{k}(1 - e^{-kx})$.

33. The Witch of Agnesi,
 $x^2 y = 4a^2(2a - y)$.
34. The Cissoid of Diocles,
 $y^2(2a - x) = x^3$.
35. The Catenary,
 $y = \tfrac{1}{2}a(e^{x/a} + e^{-x/a})$.
36. The Folium of Descartes,
 $x^3 + y^3 = 3axy$.
37. The Parabola,
 $\pm x^{\frac{1}{2}} \pm y^{\frac{1}{2}} = a^{\frac{1}{2}}$.
38. The Hypocycloid,
 $x^{\frac{2}{3}} + y^{\frac{2}{3}} = a^{\frac{2}{3}}$.
39. The Lemniscate of Bernoulli,
 $(x^2 + y^2)^2 = a^2(x^2 - y^2)$.
40. The Ovals of Cassini,
 $(x^2 + y^2 + a^2)^2 - 4a^2 x^2 = c^4$.
41. $x = \dfrac{t^2 - 1}{t^2 + 1}$, $y = \dfrac{2t}{t^2 + 1}$.
42. $x = t + \ln t$, $y = t + e^t$.
43. The Ellipse,
 $x = a \cos t$, $y = b \sin t$.
44. The Cycloid,
 $x = a(t - \sin t)$,
 $y = a(1 - \cos t)$.
45. The Hypocycloid,
 $x = a \cos^3 t$, $y = a \sin^3 t$.
46. The Cardioid,
 $x = a(2 \cos t - \cos 2t)$,
 $y = a(2 \sin t - \sin 2t)$.
47. The Serpentine,
 $x = a \cot t$, $y = b \sin t \cos t$.
48. The Witch of Agnesi,
 $x = a \cot t$, $y = a \sin^2 t$.
49. The Hypocycloid of Three Cusps,
 $x = 2a \cos t + a \cos 2t$
 $y = 2a \sin t - a \sin 2t$.
50. The Hyperbolic Spiral,
 $x = \dfrac{a}{t} \cos t$, $y = \dfrac{a}{t} \sin t$.

51. Graph $f(x) \equiv \dfrac{1}{1 + e^{-1/x}}$, and show that $f(0+) = 1$ and $f(0-) = f'(0+) = f'(0-) = 0$.

52. Graph $f(x) \equiv e^{-1/x^2}$, $x \neq 0$, $f(0) \equiv 0$. Prove that $\lim_{x \to 0} x^{-n} f(x) = 0$ for every positive integer n, and hence show that $f(x)$ has everywhere continuous derivatives of all orders, all of which vanish at $x = 0$. (Cf. Example 2, § 1205, for a use of this function as a counterexample.)

★53. Show by a graph that a function $f(x)$ exists having the following properties: (i) $f(x)$ is strictly increasing, (ii) $f'(x)$ exists for every real x, (iii) $f(x)$ is bounded above, (iv) the statement $\lim_{x \to +\infty} f'(x) = 0$ is false. (construct a function)

★54. Show that the function $f(x) \equiv x/(1 + e^{1/x})$, $x \neq 0$, $f(0) \equiv 0$, is everywhere continuous, and has unequal right-hand and left-hand derivatives at $x = 0$.

★55. Graph $(y - x^2)^2 = x^5$, with particular attention to a neighborhood of the origin.

★321. WITHOUT LOSS OF GENERALITY

One of the standard techniques of an analytic proof is to reduce a general proposition to a special case "without loss of generality." This means that it is possible to construct a proof of the general theorem on the assumption that the special form of that theorem is true. Following the establishment of this inference, proof of the special case implies proof of the general proposition. This device was used in two proofs given in § 204: in the proof that the limit of the product of two sequences is the product of their limits (Theorem VIII) we saw that it could be assumed without loss of generality that one of the limits is zero; in the proof that the limit of the quotient of two sequences is the quotient of their limits (Theorem IX) we assumed without loss of generality that the numerators were identically equal to unity. Another instance of the same principle is the proof of the Law of the Mean (§ 305) by reducing it to the special case called Rolle's Theorem.

In the following Exercises are a few problems in showing that the special implies the general.

★322. EXERCISES

In Exercises 1–9, prove the stated proposition on the basis of the assumption given in the braces { }.

★1. If $I_1, I_2, \cdots, I_n$ are open intervals and if I is the set of all x such that x is a member of every I_k, $k = 1, 2, \cdots, n$, then I is either empty or an open interval. $\{n = 2.\}$

★2. The Schwarz inequality (Ex. 32, § 107). {All of the numbers $a_1, \cdots, a_n$, $b_1, \cdots, b_n$ are positive.}

★3. $\dfrac{\sin px}{x} < p$, for $p > 0$ and $x > 0$. $\{p = 1.\}$

★4. $\lim_{x \to 0+} x^p \ln x = 0, p > 0.$ $\{p = 1.\}$

★5. Any bounded sequence $\{a_n\}$ that does not converge to a contains a subsequence $\{a_{n_k}\}$ converging to some $b \neq a$. $\{|a_n - a| \geq \epsilon > 0$ for all $n.\}$

★6. The trigonometric functions are continuous where they are defined. {$\sin x$ and $\cos x$ are continuous at $x = 0$.}

★7. If $f(x)$ and $g(x)$ are continuous in a neighborhood of a and if $f(a) > g(a)$ then $f(x) > g(x)$ in some neighborhood of a. $\{g(x) = 0$ identically.$\}$

★8. If $f'(x) \leq g'(x)$ for $a \leq x \leq b$, then $f(x) - f(a) \leq g(x) - g(a)$. $\{f(x) = 0$ for $a \leq x \leq b.\}$

★9. Theorem III, § 213. $\{a = 0, b = 1.\}$

★10. Show that in proving that if $f(x)$ and $f'(x)$ are defined in a neighborhood of $x = a$ and if $f''(a)$ exists, then

(1) $$f''(a) = \lim_{h \to 0} \frac{f(a + h) + f(a - h) - 2f(a)}{h^2},$$

it may be assumed without loss of generality, that $a = 0$, $f(0) = 0$, $f'(0) = 0$, and $f''(0) = 0$. Hence prove (1). *Hints:* To show that one may assume $f(0) = 0$, define a new function $g(x) \equiv f(x) - f(0)$. To show that one may assume $f'(0) = 0$, define a new function $g(x) \equiv f(x) - f'(0) \cdot x$. (Cf. Ex. 14, § 313.)

4
Integration

401. THE DEFINITE INTEGRAL

It will be assumed that the reader is already familiar with some of the properties and many of the applications of the definite integral. It is our purpose in this section to give a precise definition and a few of the simpler properties of the integral, with analytical proofs that do not depend on the persuasion of a geometrical picture.

We shall be dealing in the main with a fixed closed interval $[a, b]$. On such an interval a finite set of points $a = a_0 < a_1 < a_2 < \cdots < a_n = b$ is called

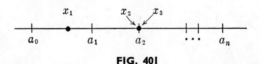

FIG. 401

a **net**, and denoted $\mathfrak{N}$. (Cf. Fig. 401.) The closed intervals $[a_{i-1}, a_i]$, $i = 1, 2, \cdots, n$, are called the **subintervals** of $[a, b]$ for the net $\mathfrak{N}$, and their lengths are denoted $\Delta x_i \equiv a_i - a_{i-1}$, $i = 1, 2, \cdots, n$. The maximum length of the subintervals is called the **norm** of the net $\mathfrak{N}$ and denoted $|\mathfrak{N}|$: $|\mathfrak{N}| \equiv \max \Delta x_i$, $i = 1, 2, \cdots, n$.

Let $f(x)$ be defined (and single-valued) over the closed interval $[a, b]$, and, for a net $\mathfrak{N}$, let x_i be an arbitrary point of the ith subinterval $(a_{i-1} \leq x_i \leq a_i)$, $i = 1, 2, \cdots, n$ (Fig. 401). Consider the sum

$$(1) \qquad \sum_{i=1}^{n} f(x_i)\, \Delta x_i = f(x_i)\, \Delta x_1 + \cdots + f(x_n)\, \Delta x_n.$$

The definite integral of the function $f(x)$ is defined as the limit of sums of the form (1). We shall first explain what is meant by such a limit.

Definition I. *The limit statement*

$$(2) \qquad \lim_{|\mathfrak{N}| \to 0} \sum_{i=1}^{n} f(x_i)\, \Delta x_i = I,$$

where I is a (finite) number, means that corresponding to $\epsilon > 0$ there exists

§ 401]   THE DEFINITE INTEGRAL   111

$\delta > 0$ such that for any net $\mathfrak{N}$ of norm less than δ and any choice of points x_i such that $a_{i-1} \leq x_i \leq a_i$, $i = 1, 2, \cdots, n$, the inequality

(3) $$\left| \sum_{i=1}^{n} f(x_i) \Delta x_i - I \right| < \epsilon$$

holds.

NOTE 1. Closely though Definition I may resemble the definition of the limit of a single-valued function of a real variable as the independent variable approaches 0, the type of limit just introduced should be recognized as a new concept. Although, for a given net $\mathfrak{N}$ and points $x_1, x_2, \cdots, x_n$, the sum (1) is uniquely determined, the limit (2) is formed with respect to the norm alone as the independent variable. For a given positive number $p < b - a$ there are infinitely many nets $\mathfrak{N}$ of norm $|\mathfrak{N}|$ equal to p, and for each such net there are infinitely many choices of the points $x_1, x_2, \cdots, x_n$. In other words, as a function of the independent variable $|\mathfrak{N}|$, the sum (1) is an infinitely many valued function,† and it is the limit of such a function that is prescribed in Definition I. It should not be forgotten, however, that each sum (1) appearing in the inequality (3) is simply a number obtained by adding together a finite number of terms.

Definition II. *A function $f(x)$, defined on $[a, b]$, is* **integrable**‡ *there if and only if the limit (2) exists (and is finite). In case the limit exists it is called the* **definite integral**‡ *of the function, and denoted*

(4) $$\int_a^b f(x)\, dx \equiv \lim_{|\mathfrak{N}| \to 0} \sum_{i=1}^{n} f(x_i) \Delta x_i.$$

Definition III. *If $b < a$,*

(5) $$\int_a^b f(x)\, dx \equiv -\int_b^a f(x)\, dx,$$

in case the latter integral exists. Furthermore,

(6) $$\int_a^a f(x)\, dx \equiv 0.$$

Certain questions naturally come to mind. Does the limit (2) always exist? If it does not always exist, under what circumstances does it exist? When it does exist is it unique?

Let us remark first that the limit (2) *never* exists for unbounded functions. In other words, <u>every integrable function is bounded</u>. To see this, let $\mathfrak{N}$ be an arbitrary net and let $f(x)$ be unbounded in the kth subinterval $[a_{k-1}, a_k]$. Then whatever may be the choice of points x_i for $i \neq k$, the point x_k can be

---

† The expression "infinitely many valued" should be interpreted here to mean "possibly infinitely many valued" since for any constant function $f(x)$, the sums (1) can have only one value (cf. Theorem VII). It can be shown that for any function $f(x)$ that is not constant on $[a, b]$, and for any positive number δ less than half the length of the interval $[a, b]$, the sum (1), as a function of $|\mathfrak{N}|$, is strictly infinitely many valued for the particular value $|\mathfrak{N}| = \delta$.

‡ The terms *Riemann integrable* and *Riemann integral* are also used, especially if it is important to distinguish the integral defined in this section from some other type, such as the Riemann-Stieltjes integral (§ 417) or the Lebesgue integral (not discussed in this book).

picked so that the sum (1) is numerically larger than any preassigned quantity.

On the other hand, not all bounded functions are integrable. For example, the function of Example 6, § 201, which is 1 on the rational numbers from 0 to 1 and 0 on the irrational numbers from 0 to 1, is not integrable on [0, 1]. For, no matter how small the norm of a net $\mathfrak{N}$ may be, every subinterval contains both rational and irrational points (Theorem III, § 113, and Ex. 2, § 114) and the sums (1) can be made to have either the extreme value 1 (if every point x_i is chosen to be rational) or the extreme value 0 (if every point x_i is chosen to be irrational). The limit (2), then, cannot exist for this function (let $\epsilon \equiv \frac{1}{2}$).

One answer to the question of integrability lies in the concept of *continuity*. The function just considered, which is not integrable, is *nowhere* continuous. At the opposite extreme is a function that is *everywhere* continuous (on a closed interval). We shall see that such a function is *always* integrable (Theorem VIII). Between these two extremes are bounded functions that are somewhere but not everywhere continuous. It will be seen that such functions are certainly integrable if they have only *finitely* many discontinuities (Theorem IX) and that they *may* be integrable even with *infinitely* many discontinuities (Example 2, § 403). A *criterion* for integrability lies in the intriguing concept of *continuity almost everywhere* (cf. RV, Ex. 54, § 503).

We proceed now to the establishment of some of the simpler properties of the definite integral.

Theorem I. *If* $\lim_{|\mathfrak{N}| \to 0} \sum_{i=1}^{n} f(x_i) \Delta x_i$, *exists, the limit is unique.*

Proof. Assume that

$$\lim_{|\mathfrak{N}| \to 0} \sum_{i=1}^{n} f(x_i) \Delta x_i = I \quad \text{and} \quad \lim_{|\mathfrak{N}| \to 0} \sum_{i=1}^{n} f(x_i) \Delta x_i = J,$$

where $I > J$, and let $\epsilon \equiv \frac{1}{2}(I - J)$. Then there exists a positive number δ so small that for any net $\mathfrak{N}$ of norm less than δ, and for any choice of points $x_1, x_2, \cdots, x_n$, the following inequalities hold simultaneously:

$$I - \epsilon < \sum_{i=1}^{n} f(x_i) \Delta x_i < J + \epsilon.$$

But this implies $I - \epsilon < J + \epsilon$, or $\epsilon > \frac{1}{2}(I - J)$. (Contradiction.)

Theorem II. *If* $f(x)$ *and* $g(x)$ *are integrable on* $[a, b]$, *and if* $f(x) \leq g(x)$ *there, then*

$$\int_a^b f(x)\, dx \leq \int_a^b g(x)\, dx.$$

Proof. Let $I \equiv \int_a^b f(x)\, dx$ and $J \equiv \int_a^b g(x)\, dx$, assume $I > J$, and let

$\epsilon \equiv \frac{1}{2}(I - J)$. Then there exists a positive number δ so small that for any net $\mathfrak{N}$ of norm less than δ and any choice of points $x_1, x_2, \cdots, x_n$, the following inequalities hold simultaneously:

$$\sum_{i=1}^{n} g(x_i) \Delta x_i < J + \epsilon = I - \epsilon < \sum_{i=1}^{n} f(x_i) \Delta x_i.$$

But this implies an inequality inconsistent with the assumed inequality $f(x) \leq g(x)$.

Theorem III. *If $f(x)$ is integrable on $[a, b]$ and if k is a constant, then $kf(x)$ is integrable on $[a, b]$ and*

(7) $$\int_a^b kf(x)\, dx = k \int_a^b f(x)\, dx.$$

Proof. If $k = 0$, the proof is trivial. If $k \neq 0$, and if $\epsilon > 0$ is given, let $\delta > 0$ be such that $|\mathfrak{N}| < \delta$ implies $\left| \sum_{i=1}^{n} f(x_i) \Delta x_i - I \right| < \epsilon/|k|$, and therefore

$$\left| \sum_{i=1}^{n} kf(x_i) \Delta x_i - kI \right| < |k|\epsilon/|k| = \epsilon.$$

Theorem IV. *If $f(x)$ and $g(x)$ are integrable on $[a, b]$, then so are their sum and difference, and*

(8) $$\int_a^b [f(x) \pm g(x)]\, dx = \int_a^b f(x)\, dx \pm \int_a^b g(x)\, dx.$$

Proof. Let $I \equiv \int_a^b f(x)\, dx$ and $J \equiv \int_a^b g(x)\, dx$, and if $\epsilon > 0$ is given, let $\delta > 0$ be such that $|\mathfrak{N}| < \delta$ implies simultaneously the inequalities

$$\left| \sum_{i=1}^{n} f(x_i) \Delta x_i - I \right| < \tfrac{1}{2}\epsilon, \quad \left| \sum_{i=1}^{n} g(x_i) \Delta x_i - J \right| < \tfrac{1}{2}\epsilon,$$

and hence, by the triangle inequality,

$$\left| \sum_{i=1}^{n} [f(x_i) \pm g(x_i)] \Delta x_i - (I \pm J) \right|$$

$$\leq \left| \sum_{i=1}^{n} f(x_i) \Delta x_i - I \right| + \left| \sum_{i=1}^{n} g(x_i) \Delta x_i - J \right| < \epsilon.$$

Theorem V. *If $a < b < c$, and if $f(x)$ is integrable on the intervals $[a, b]$ and $[b, c]$, then it is integrable on the interval $[a, c]$ and*

(9) $$\int_a^c f(x)\, dx = \int_a^b f(x)\, dx + \int_b^c f(x)\, dx.$$

Proof. Let $|f(x)| < K$ for $a \leq x \leq c$, let

$$I \equiv \int_a^b f(x)\, dx \quad \text{and} \quad J \equiv \int_b^c f(x)\, dx,$$

and let $\epsilon > 0$ be given. Let δ be a positive number less than $\epsilon/4K$ and so small that for any net on $[a, b]$ or $[b, c]$ with norm less than δ any sum of the form (1) differs from I or J, respectively, by less than $\frac{1}{4}\epsilon$. We shall show that for $[a, c]$, $|\mathfrak{N}| < \delta$ implies $\left|\sum_{i=1}^{n} f(x_i) \Delta x_i - (I + J)\right| < \epsilon$. Accordingly, for such a net $\mathfrak{N}$ let $a_{k-1} < b \leq a_k$ (the kth subinterval is the first containing the point b), and write the sum (1) in the form

$$S \equiv \sum_{i=1}^{k-1} f(x_i) \Delta x_i + f(x_k) \Delta x_k + \sum_{i=k+1}^{n} f(x_i) \Delta x_i.$$

The following sum,

$$S' \equiv \left[\sum_{i=1}^{k-1} f(x_i) \Delta x_i + f(b)(b - a_{k-1})\right] + \left[f(b)(a_k - b) + \sum_{i=k+1}^{n} f(x_i) \Delta x_i\right],$$

can be considered as made up of two parts, which approximate I and J, each by less than $\frac{1}{4}\epsilon$. Thus $|S' - (I + J)| < \frac{1}{2}\epsilon$. On the other hand, $|S - S'| = |f(x_k) - f(b)| \Delta x_i < 2K(\epsilon/4K) = \frac{1}{2}\epsilon$. Therefore,

$$|S - (I + J)| \leq |S - S'| + |S' - (I + J)| < \tfrac{1}{2}\epsilon + \tfrac{1}{2}\epsilon = \epsilon.$$

NOTE 2. By virtue of Definition III, the relation (9) is universally true whenever the three integrals exist, whatever may be the order relation between the numbers a, b, and c. For example, if $c < a < b$, then

$$\int_c^b f(x)\, dx = \int_c^a f(x)\, dx + \int_a^b f(x)\, dx.$$

Hence

$$\int_a^c f(x)\, dx = -\int_c^a f(x)\, dx = \int_a^b f(x)\, dx - \int_c^b f(x)\, dx = \int_a^b f(x)\, dx + \int_b^c f(x)\, dx.$$

By mathematical induction, the relation (9) can be extended to an arbitrary number of terms:

(10) $$\int_{a_0}^{a_n} f(x)\, dx = \sum_{i=1}^{n} \int_{a_{i-1}}^{a_i} f(x)\, dx,$$

where $a_0, a_1, \cdots, a_n$ are any $n + 1$ real numbers, and where every integral of (10) is assumed to exist. (The student should satisfy himself regarding (9), by considering other order relations between a, b, and c, including possible equalities of some of these letters, and he should give the details of the proof of (10). (Cf. Ex. 1, § 402.)

Theorem VI. *If the values of a function defined on a closed interval are changed at a finite number of points of the interval, neither the integrability nor the value of the integral is affected.*

Proof. Thanks to mathematical induction, the proof can (and will) be reduced to showing the following: If $f(x)$ is integrable on $[a, b]$, with integral I, and if $g(x)$ is defined on $[a, b]$ and equal to $f(x)$ at every point of $[a, b]$ except for one point c, then $g(x)$ is integrable on $[a, b]$ with integral I. For any net $\mathfrak{N}$, the terms of the sum $\sum_{i=1}^{n} g(x_i) \Delta x_i$ must be identical with the terms of the sum $\sum_{i=1}^{n} f(x_i) \Delta x_i$ with the exception of at most two terms (in case $x_{i-1} = x_i = c$ for some i). Therefore

$$\left| \sum_{i=1}^{n} g(x_i) \Delta x_i - \sum_{i=1}^{n} f(x_i) \Delta x_i \right| \leq 2(|f(c)| + |g(c)|) \cdot |\mathfrak{N}|.$$

Thus, for a given $\epsilon > 0$, let δ be a positive number less than

$$\epsilon/4(|f(c)| + |g(c)|)$$

and so small that $|\mathfrak{N}| < \delta$ implies $\left| \sum_{i=1}^{n} f(x_i) \Delta x_i - I \right| < \tfrac{1}{2}\epsilon.$ Then $|\mathfrak{N}| < \delta$ implies

$$\left| \sum_{i=1}^{n} g(x_i) \Delta x_i - I \right| \leq \left| \sum_{i=1}^{n} g(x_i) \Delta x_i - \sum_{i=1}^{n} f(x_i) \Delta x_i \right|$$
$$+ \left| \sum_{i=1}^{n} f(x_i) \Delta x_i - I \right| < \tfrac{1}{2}\epsilon + \tfrac{1}{2}\epsilon = \epsilon.$$

NOTE 3. Theorem VI makes it possible to define integrability and integral for a function that is defined on a closed interval except for a finite number of points. This is done by assigning values to the function at the exceptional points in any manner whatsoever. Theorem VI assures us that the result of applying Definition II is independent of the values assigned. Since the assignment of values does not affect the value of the integral, where it exists, we shall assume that the definition is extended to include such functions even though they remain undefined at the exceptional points.

Theorem VII. *If $f(x)$ is constant, $f(x) \equiv k$, on the interval $[a, b]$, then $f(x)$ is integrable there and*

$$\int_a^b f(x)\, dx = k(b - a).$$

Proof. For any net $\mathfrak{N}$, $\sum_{i=1}^{n} f(x_i) \Delta x_i = k \sum_{i=1}^{n} \Delta x_i = k(b - a)$.

For the sake of convenience and accessibility we state now the three best-known sufficient conditions for integrability (the first being a special case of the second). The proofs are given in § 403 and Exercise 1, § 404.

Theorem VIII. *A function continuous on a closed interval is integrable there.*

Theorem IX. *A function defined and bounded on a closed interval and continuous there except for a finite number of points is integrable there.*

Theorem X. *A function defined and monotonic on a closed interval is integrable there.*

Example. Prove that $\int_0^b x\, dx = \tfrac{1}{2}b^2$ if $b > 0$, and, more generally, that

$$\int_a^b x\, dx = \tfrac{1}{2}(b^2 - a^2),$$

where a and b are any real numbers.

Solution. Let us first observe that since the function $f(x) \equiv x$ is everywhere continuous the integrals exist. The problem is one of *evaluation*. To evaluate $\int_0^b x \, dx$, where $b > 0$, we form a particular simple net by means of the points

$$a_0 \equiv 0, \ a_1 \equiv b/n, \cdots, a_i \equiv ib/n, \cdots, a_n \equiv nb/n = b,$$

and choose $x_i \equiv a_i$, $i = 1, 2, \cdots, n$. The sum $\sum_{i=1}^{n} f(x_i) \Delta x_i$ becomes

$$\left[\frac{b}{n} + \frac{2b}{n} + \cdots + \frac{nb}{n}\right] \cdot \frac{b}{n} = \frac{b^2}{n^2}[1 + 2 + \cdots + n] = \frac{b^2}{n^2} \cdot \frac{n(n+1)}{2},$$

the last equality having been obtained in Exercise 12, § 107. Therefore $\int_0^b x \, dx = \lim_{n \to +\infty} \tfrac{1}{2}b^2 \frac{n+1}{n} = \tfrac{1}{2}b^2$, as stated. It is left to the student to show, first, that $\int_0^b x \, dx = \tfrac{1}{2}b^2$ if $b \leq 0$ (cf. Exs. 7–8, § 402) and then, by using formula (9) of Theorem V and formula (5) of Definition III, that $\int_a^b x \, dx = \tfrac{1}{2}(b^2 - a^2)$.

Other evaluations of this type are given in Exercises 12–15, § 402.

In conclusion, we state a mean value theorem for integrals, three theorems on integrability, and a theorem on a special kind of limit of a sum that turns out to be an integral. For hints on proofs, see Exercise 5, § 402, and Exercises 2–5, § 404. For other mean value theorems for integrals, see Exercise 6, § 402, Exercise 14, § 407, and Exercises 22, 24–26, § 418. Also see Exercise 9, § 407. For applications of Bliss's Theorem, see Exercise 7, § 404, and Theorem I, § 417.

Theorem XI. First Mean Value Theorem for Integrals. *If $f(x)$ is continuous on $[a, b]$ there exists a point ξ such that $a < \xi < b$ and*

$$(11) \qquad \int_a^b f(x) \, dx = f(\xi) \cdot (b - a).$$

Theorem XII. *A function integrable on a closed interval is integrable on any closed subinterval.*

Theorem XIII. *If a function $f(x)$ is integrable on an interval $[a, b]$, then so is its absolute value $|f(x)|$.*

Theorem XIV. *If two functions $f(x)$ and $g(x)$ are integrable on an interval $[a, b]$, then so is their product $f(x)g(x)$.*

Theorem XV. Bliss's Theorem:[†] *If $f(x)$ and $g(x)$ are integrable on $[a, b]$, then the limit $\lim_{|\mathfrak{N}| \to 0} \sum_{i=1}^{n} f(x_i) g(x_i') \Delta x_i$, where $a_{i-1} \leq x_i \leq a_i$ and $a_{i-1} \leq x_i' \leq a_i$, $i = 1, 2, \cdots, n$, the limit being interpreted in the sense of Definition I, exists, and is equal to $\int_a^b f(x)g(x) \, dx$.*

† Cf. G. A. Bliss, "A Substitute for Duhamel's Theorem," *Annals of Mathematics*, Vol. 16 (1914–15), pp. 45–49.

402. EXERCISES

1. Extend Theorems IV and V, § 401, to an arbitrary finite number of terms, and Theorem V to an arbitrary order relation between a, b, and c. (Cf. Note 2, § 401.)

2. Assuming that $f(x)$ is integrable on $[a, b]$ and that $|f(x)| \leq K$ there, prove that
$$\left| \int_a^b f(x)\,dx \right| \leq \int_a^b |f(x)|\,dx \leq K(b - a).$$

Hint: Use Theorems II and XIII, § 401, and the inequalities
$$-|f(x)| \leq f(x) \leq |f(x)|.$$

3. Prove that if $f(x)$ is continuous on $[a, b]$ and $f(x) \geq 0$ but not identically zero there, then $\int_a^b f(x)\,dx > 0$. *Hint:* Use Exercise 18, § 212.

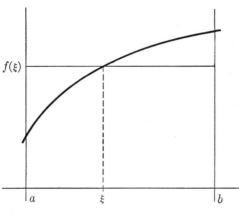

FIG. 402

4. Prove that if $f(x)$ and $g(x)$ are continuous on $[a, b]$ and $f(x) \leq g(x)$ but $f(x)$ and $g(x)$ are not identical there, then $\int_a^b f(x)\,dx < \int_a^b g(x)\,dx$. (Cf. Ex. 3.)

5. Prove and give a geometric interpretation to the First Mean Value Theorem for Integrals (Theorem XI, § 401). Also consider the case $a > b$. (See Fig. 402.) *Hint:* If m and M are the minimum and maximum values, respectively, of $f(x)$ on $[a, b]$ and if $f(x)$ is not a constant, use Exercise 4 to show that
$$m < \left[\int_a^b f(x)\,dx \right] \bigg/ (b - a) < M.$$
Conclude by applying Theorem IV, § 213.

6. Prove the following generalized form of the First Mean Value Theorem for Integrals (Theorem XI, § 401): *If $f(x)$ and $g(x)$ are continuous on $[a, b]$ and if $g(x)$ never changes sign there, then there exists a point ξ such that $a < \xi < b$ and*
$$\int_a^b f(x)g(x)\,dx = f(\xi) \int_a^b g(x)\,dx.$$
Also consider the case $a > b$. (Cf. Ex. 5.)

7. A function $f(x)$ is said to be **even** if and only if the equality $f(-x) = f(x)$ holds for all x in the domain of definition of the function. (Examples: constants, x^{2n}, $\cos x$, $|x|$.) Prove that if $f(x)$ is even on $[-a, a]$, and integrable on $[0, a]$, where $a > 0$, then $f(x)$ is integrable on $[-a, a]$ and $\int_{-a}^{a} f(x)\, dx = 2\int_{0}^{a} f(x)\, dx$. Prove that if $f(x)$ is even on $(-\infty, +\infty)$ and a and b are any real numbers, then $\int_{-b}^{-a} f(x)\, dx = \int_{a}^{b} f(x)\, dx$, whenever the integrals exist.

8. A function $f(x)$ is said to be **odd** if and only if the equality $f(-x) = -f(x)$ holds for all x in the domain of definition of the function. (Examples: $10x$, x^{2n+1}, $\sin x$, the signum function of Example 1, § 206.) Prove that if $f(x)$ is odd on $[-a, a]$, and integrable on $[0, a]$, where $a > 0$, then $f(x)$ is integrable on $[-a, a]$ and $\int_{-a}^{a} f(x)\, dx = 0$. Prove that if $f(x)$ is odd on $(-\infty, +\infty)$ and a and b are any real numbers, then $\int_{-a}^{-b} f(x)\, dx = \int_{a}^{b} f(x)\, dx$, whenever the integrals exist.

9. Prove that the only function that is both even and odd is identically zero. (Cf. Exs. 7–8.)

10. Prove that any function whose domain contains the negative of every one of its members is uniquely representable as the sum of an even function and an odd function. (Cf. Exs. 7–9.) *Hint:*
$$f(x) = \tfrac{1}{2}[f(x) + f(-x)] + \tfrac{1}{2}[f(x) - f(-x)].$$

11. Prove that if $f(x)$ is even (odd) and differentiable, then $f'(x)$ is odd (even). (Cf. Exs. 7–10.)

12. Prove that if $m = 2, 3,$ or 4, then
$$\int_{a}^{b} x^m\, dx = \frac{1}{m+1}(b^{m+1} - a^{m+1}).$$
(Cf. the Example, § 401, and Exs. 13–15, § 107.)

*13. Prove that $\int_{a}^{b} \sin x\, dx = \cos a - \cos b$. *Hint:* As in the Example, § 401, let $b > 0$ and $a = 0$, and write
$$\int_{0}^{b} \sin x\, dx = \lim_{n \to +\infty}\left(\sin\frac{b}{n} + \cdots + \sin\frac{nb}{n}\right)\cdot\frac{b}{n}.$$
Multiply each term by $2\sin\dfrac{b}{2n}$, and use the identity
$$2\sin A \sin B = \cos(A - B) - \cos(A + B).$$

*14. Prove that $\int_{a}^{b} \cos x\, dx = \sin b - \sin a$. (Cf. Ex. 13.)

*15. Prove that $\int_{a}^{b} e^x\, dx = e^b - e^a$.

*16. Prove the **Trapezoidal Rule** for approximating a definite integral: If $f(x)$ is integrable on $[a, b]$, if $[a, b]$ is subdivided into n equal subintervals of length Δx, and if the values of $f(x)$ at the $n + 1$ points of subdivision, $x_0, x_1, x_2, \cdots, x_n$,

are $y_0, y_1, y_2, \cdots, y_n$, respectively, then
$$\int_a^b f(x)\,dx = \lim_{n \to +\infty} (\tfrac{1}{2}y_0 + y_1 + y_2 + \cdots + y_{n-1} + \tfrac{1}{2}y_n)\,\Delta x.$$
Also prove the following estimate for the error in the trapezoidal formula, assuming existence of $f''(x)$ on $[a, b]$:
$$\int_a^b f(x)\,dx - [\tfrac{1}{2}y_0 + y_1 + \cdots + \tfrac{1}{2}y_n]\,\Delta x = -\frac{b-a}{12} f''(\xi)\,\Delta x^2,$$
where $a < \xi < b$.

Hints: For the first part, write the expression in brackets in the form $(y_1 + \cdots + y_n) + (\tfrac{1}{2}y_0 - \tfrac{1}{2}y_n)$. For the second part, reduce the problem to that of approximating $\int_a^b f(x)\,dx$ by a *single* trapezoid, and assume without loss of generality that the interval $[a, b]$ is $[-h, h]$. (Cf. Ex. 22, § 216, Ex. 40, § 308.) The problem reduces to that of evaluating K, defined by the equation
$$\int_a^b f(x)\,dx = \frac{b-a}{2}[f(b) + f(a)] + K(b-a)^3,$$
or
$$\int_{-h}^h f(t)\,dt = h[f(h) + f(-h)] + 8Kh^3.$$
Define the function
$$\phi(x) \equiv \int_{-x}^x f(t)\,dt - x[f(x) + f(-x)] - 8Kx^3.$$
Show that $\phi(h) = \phi(0) = 0$, and hence there must exist, by Rolle's Theorem (§ 305), a number x_1 between 0 and h such that $\phi'(x_1) = 0$. (Cf. Ex. 2, § 407.)

*17. Prove **Simpson's Rule** for approximating a definite integral: *Under the assumptions and notation of Exercise 16, where n is even,*
$$\int_a^b f(x)\,dx = \lim_{n \to +\infty} \tfrac{1}{3}(y_0 + 4y_1 + 2y_2 + 4y_3 + 2y_4 + \cdots + 4y_{n-1} + y_n)\,\Delta x.$$
Also prove the following estimate for the error in Simpson's Rule, assuming existence of $f'''(x)$ on $[a, b]$:
$$\int_a^b f(x)\,dx - \tfrac{1}{3}[y_0 + 4y_1 + \cdots + y_n]\,\Delta x = -\frac{b-a}{180} f'''(\xi)\,\Delta x^4,$$
where $a < \xi < b$. (Cf. Example 5, § 1213.)

Hint: Proceed as in Exercise 16, making use of the auxiliary function
$$\phi(x) \equiv \int_{-x}^x f(t)\,dt - \frac{x}{3}[f(-x) + 4f(0) + f(x)] - Kx^5.$$
Evaluate $\phi''(x_1)$. (Cf. Ex. 2, § 407.)

*18. Evaluate $\lim\limits_{n \to +\infty} \left(\dfrac{1}{n+1} + \dfrac{1}{n+2} + \cdots + \dfrac{1}{2n} \right)$. *Hint:* The sum, when rewritten $\left(\dfrac{1}{1+\frac{1}{n}} + \dfrac{1}{1+\frac{2}{n}} + \cdots + \dfrac{1}{1+\frac{n}{n}} \right) \cdot \dfrac{1}{n}$, can be interpreted as one of the approximating sums for $\int_0^1 \dfrac{dx}{1+x}$. (For the specific evaluation of this as $\ln 2$, see § 405.)

*19. Evaluate $\lim\limits_{n \to +\infty} n \cdot \left(\dfrac{1}{n^2 + 1^2} + \dfrac{1}{n^2 + 2^2} + \cdots + \dfrac{1}{2n^2} \right)$. (Cf. Ex. 18.)

*20. Assuming that $f(x)$ and $g(x)$ are integrable on $[a, b]$, establish the **Schwarz** (or **Cauchy**) **inequality** (cf. Ex. 32, § 107, Ex. 14, § 1117):

$$\left[\int_a^b f(x)g(x)\,dx \right]^2 \leq \int_a^b [f(x)]^2\,dx \cdot \int_a^b [g(x)]^2\,dx.$$

Hint: First show that $[f(x) + tg(x)]^2$ is integrable for all real t. With the notation $A \equiv \int_a^b [f(x)]^2\,dx$, $B \equiv \int_a^b f(x)g(x)\,dx$, and $C \equiv \int_a^b [g(x)]^2\,dx$, show that $A + 2Bt + Ct^2 \geq 0$ for all real t. If $C = 0$, then $B = 0$, and if $C \neq 0$ the discriminant of $A + 2Bt + Ct^2$ must be nonpositive.

*21. Use the Schwarz inequality to establish the **Minkowski inequality** (cf. Ex. 33, § 107, Ex. 14, § 1117), assuming that $f(x)$ and $g(x)$ are integrable on $[a, b]$:

$$\left\{ \int_a^b [f(x) + g(x)]^2\,dx \right\}^{\frac{1}{2}} \leq \left\{ \int_a^b [f(x)]^2\,dx \right\}^{\frac{1}{2}} + \left\{ \int_a^b [g(x)]^2\,dx \right\}^{\frac{1}{2}}.$$

Hint: Use the hint of Exercise 33, § 107.

*22. Show that the value of the constant c that minimizes

(1) $$\int_a^b [f(x) - c]^2\,dx,$$

where $f(x)$ is integrable on the interval $[a, b]$, is the **mean value** of $f(x)$ over $[a, b]$:
$c = \left[\int_a^b f(x)\,dx \right] / (b - a)$. As a corollary show that

(2) $$\left[\int_a^b f(x)\,dx \right]^2 \leq (b - a) \int_a^b [f(x)]^2\,dx,$$

equality holding if and only if the integral (1) vanishes for some constant c.

*403. MORE INTEGRATION THEOREMS

Definition. A *step-function* is a function, defined on a closed interval $[a, b]$, that is constant in the interior of each subinterval of some net on $[a, b]$. (Fig. 403.)

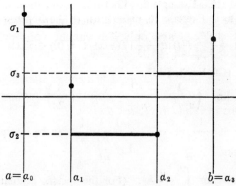

FIG. 403

Notation

Step-function: $\sigma(x)$ [sigma] or $\tau(x)$ [tau].
Net: $\mathfrak{N} : a = \alpha_0 < \alpha_1 < \cdots < \alpha_m = b$.
Constant values: σ_i or τ_i, $i = 1, \cdots, m$.

Theorem I. *Any step-function is integrable and, with the preceding notation,*
$$\int_a^b \sigma(x)\, dx = \sum_{i=1}^m \sigma_i(\alpha_i - \alpha_{i-1}).$$

Proof. We need prove this only for the case $m = 1$, since Theorem V, § 401, and mathematical induction will extend this special case to the general result. Accordingly, let $\sigma(x)$ be identically equal to a constant k for $a < x < b$, and redefine $\sigma(x)$, if necessary, at the end-points of the interval so that

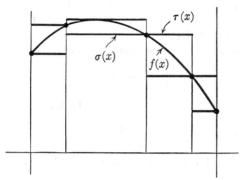

FIG. 404

$\sigma(x) \equiv k$ for $a \leq x \leq b$. Since the new function is integrable with integral $k(b - a)$ (Theorem VII, § 401), the original step-function is also integrable with the same integral (Theorem VI, § 401), and the proof is complete.

The purpose of introducing step-functions in our discussion of the definite integral is to make use of the fact established in the following theorem that a function is integrable if and only if it can be appropriately "squeezed" between two step-functions. (Cf. Fig. 404.)

Theorem II. *A function $f(x)$, defined on a closed interval $[a, b]$, is integrable there if and only if, corresponding to an arbitrary positive number ϵ, there exist step-functions $\sigma(x)$ and $\tau(x)$ such that*

(1) $$\sigma(x) \leq f(x) \leq \tau(x),$$

for $a \leq x \leq b$, and

(2) $$\int_a^b [\tau(x) - \sigma(x)]\, dx < \epsilon.$$

Proof. We first establish the "if" part by assuming, for $\epsilon > 0$, the existence of step-functions $\sigma(x)$ and $\tau(x)$ satisfying (1) and (2). Let us observe initially that for *any* step-functions satisfying (1),

$$\int_a^b \sigma(x)\,dx \le \int_a^b \tau(x)\,dx$$

(Theorem II, § 401), so that the *least upper bound* I of the integrals $\int_a^b \sigma(x)\,dx$, for all $\sigma(x) \le f(x)$, and the *greatest lower bound* J of the integrals $\int_a^b \tau(x)\,dx$, for all $\tau(x) \ge f(x)$, both exist and are finite, and $I \le J$ (supply the details). Because of (2) and the arbitrariness of $\epsilon > 0$, integrals of the form $\int_a^b \sigma(x)\,dx$ and $\int_a^b \tau(x)\,dx$ can be found arbitrarily close to each other, so that I cannot be *less* than J, and therefore $I = J$. Now let $\epsilon > 0$ be given, and choose step-functions $\sigma(x)$ and $\tau(x)$ satisfying (1) and (2) and such that

$$\int_a^b \sigma(x)\,dx > I - \tfrac{1}{2}\epsilon \quad \text{and} \quad \int_a^b \tau(x)\,dx < J + \tfrac{1}{2}\epsilon.$$

Then choose $\delta > 0$ so that $|\mathfrak{N}| < \delta$ implies

$$\sum_{i=1}^n \sigma(x_i)\,\Delta x_i > \int_a^b \sigma(x)\,dx - \tfrac{1}{2}\epsilon,$$

$$\sum_{i=1}^n \tau(x_i)\,\Delta x_i < \int_a^b \tau(x)\,dx + \tfrac{1}{2}\epsilon.$$

For any such net $\mathfrak{N}$,

$$I - \epsilon < \sum_{i=1}^n \sigma(x_i)\,\Delta x_i \le \sum_{i=1}^n f(x_i)\,\Delta x_i \le \sum_{i=1}^n \tau(x_i)\,\Delta x_i < I + \epsilon.$$

Therefore (Definition II, § 401), $f(x)$ is integrable on $[a, b]$, and

$$\int_a^b f(x)\,dx = I = J.$$

We now prove the "only if" part by assuming that $f(x)$ is integrable on $[a, b]$, with integral $I \equiv \int_a^b f(x)\,dx$, and letting $\epsilon > 0$ be given. Choose a net $\mathfrak{N}$, to be held fixed, of such a small norm that

$$I - \tfrac{1}{3}\epsilon < \sum_{i=1}^n f(x_i)\,\Delta x_i < I + \tfrac{1}{3}\epsilon,$$

for all x_i in $[a_{i-1}, a_i]$, $i = 1, 2, \cdots, n$. For each $i = 1, 2, \cdots, n$, let σ_i be the greatest lower bound of $f(x)$ for $a_{i-1} \le x \le a_i$ ($f(x)$ is bounded since it is integrable) and let τ_i be the least upper bound of $f(x)$ for $a_{i-1} \le x \le a_i$. Then (give the details in Ex. 6, § 404)

$$I - \tfrac{1}{3}\epsilon \le \sum_{i=1}^n \sigma_i\,\Delta x_i \le \sum_{i=1}^n f(x_i)\,\Delta x_i \le \sum_{i=1}^n \tau_i\,\Delta x_i \le I + \tfrac{1}{3}\epsilon,$$

and if the step-functions $\sigma(x)$ and $\tau(x)$ are defined to have the values σ_i and

τ_i, respectively, for $a_{i-1} < x < a_i$, $i = 1, 2, \cdots, n$, and the values $f(a_i)$ for $x = a_i$, $i = 0, 1, \cdots, n$, then $\sigma(x) \leq f(x) \leq \tau(x)$ and

$$\int_a^b [\tau(x) - \sigma(x)] \, dx = \sum_{i=1}^n (\tau_i - \sigma_i) \Delta x_i \leq \tfrac{2}{3}\epsilon < \epsilon.$$

NOTE. The definite integral $\int_a^b f(x) \, dx$ is defined as the limit of sums of the form $\sum_{i=1}^n f(x_i) \Delta x_i$. If, for a given net $\mathfrak{N}$ and points x_i from the subintervals $[a_{i-1}, a_i]$, $i = 1, 2, \cdots, n$, a step-function $\sigma(x)$ is defined to have the values $f(x_i)$ for $a_{i-1} < x < a_i$, $i = 1, 2, \cdots, n$, and arbitrary values at the points a_i, $i = 0, 1, \cdots, n$ (cf. Fig. 404), then $\sum_{i=1}^n f(x_i) \Delta x_i = \int_a^b \sigma(x) \, dx$, and the definite integral of $f(x)$ can be thought of as the limit of the definite integrals of such "approximating" step-functions:

$$\int_a^b f(x) \, dx = \lim_{|\mathfrak{N}| \to 0} \int_a^b \sigma(x) \, dx.$$

With the aid of the theorem just established it is easy to prove that *any function continuous on a closed interval is integrable there*:

Proof of Theorem VIII, § 401. Let $f(x)$ be continuous on the closed interval $[a, b]$. Then $f(x)$ is uniformly continuous there (Theorem II, § 224), and therefore, if ϵ is any positive number there exists a positive number δ such that $|x' - x''| < \delta$ implies $|f(x') - f(x'')| < \epsilon/(b - a)$. Let $\mathfrak{N}$ be any net of norm less than δ, and let σ_i and τ_i, for $i = 1, 2, \cdots, n$, be defined as the minimum and maximum values, respectively, of $f(x)$ on the subinterval $[a_{i-1}, a_i]$. If x_i' and x_i'' are points of $[a_{i-1}, a_i]$ such that $f(x_i') = \sigma_i$ and $f(x_i'') = \tau_i$, $i = 1, 2, \cdots, n$, and if the step-functions $\sigma(x)$ and $\tau(x)$ are defined to have the values σ_i and τ_i, respectively, for $a_{i-1} < x < a_i$, $i = 1, 2, \cdots, n$, and the values $f(a_i)$ for $x = a_i$, $i = 0, 1, \cdots, n$ (see Fig. 404), then $\sigma(x) \leq f(x) \leq \tau(x)$ and

$$\int_a^b [\tau(x) - \sigma(x)] \, dx = \sum_{i=1}^n (\tau_i - \sigma_i) \Delta x_i$$
$$= \sum_{i=1}^n |f(x_i') - f(x_i'')| \Delta x_i < \frac{\epsilon}{b-a} \sum_{i=1}^n \Delta x_i = \epsilon.$$

We generalize the theorem just proved, to permit a finite number of discontinuities:

Proof of Theorem IX, § 401. Thanks to Theorem V, § 401, and mathematical induction we need prove this only for the case of a function $f(x)$ that is defined and bounded on a closed interval and continuous in the interior. Because of the theorem just established, this special case is a simple consequence of the following theorem.

Theorem III. *If $f(x)$ is bounded on $[a, b]$ and integrable on every $[c, d]$, where $a < c < d < b$, then $f(x)$ is integrable on $[a, b]$.*

Proof. If $|f(x)| < K$ for $a \leq x \leq b$, and if $\epsilon > 0$, let $0 < \eta < \epsilon/8K$. Construct step-functions $\sigma(x)$ and $\tau(x)$ such that $\sigma(x) = -K$ and $\tau(x) = K$ on the intervals $[a, a + \eta)$ and $(b - \eta, b]$ (see Fig. 405), and such that for the closed interval $[a + \eta, b - \eta]$, $\sigma(x) \leq f(x) \leq \tau(x)$ and

$$\int_{a+\eta}^{b-\eta} [\tau(x) - \sigma(x)] \, dx < \tfrac{1}{2}\epsilon$$

(see Theorem II). Then for the interval $[a, b]$, $\sigma(x) \leq f(x) \leq \tau(x)$ and

$$\int_a^b [\tau(x) - \sigma(x)] \, dx < 2K\eta + \tfrac{1}{2}\epsilon + 2K\eta < \epsilon.$$

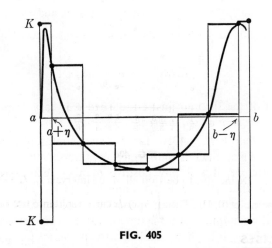

FIG. 405

Example 1. The function $\sin \dfrac{1}{x}$ (whether or however it is defined at $x = 0$) is integrable on the interval $[0, 1]$.

Example 2. Let $f(x)$ be a function defined on $[0, 1]$ as follows: If x is irrational let $f(x) \equiv 0$; if x is a positive rational number equal to p/q, where p and q are relatively prime positive integers (cf. Ex. 23, § 107), let $f(x) \equiv 1/q$; let $f(0) \equiv 1$. (See Fig. 406.) Prove that $f(x)$ is continuous at every irrational point of $[0, 1]$ and discontinuous with a removable discontinuity at every rational point of $[0, 1]$. Prove that in spite of having infinitely many discontinuities $f(x)$ is integrable on $[0, 1]$. Show, in fact, that $\int_0^1 f(x) \, dx = 0$.

Solution. If $x_0 = p/q$ we have only to take $\epsilon = 1/q$ to show that $f(x)$ is discontinuous at $x = x_0$, since there are points arbitrarily near x_0 where $f(x) = 0$. Now let x_0 be an irrational number and let $\epsilon > 0$. Since there are only a finite number of positive integers $\leq 1/\epsilon$, there are only a finite number of rational numbers p/q of the interval $[0, 1]$ for which $f(p/q) \geq \epsilon$. If $\delta > 0$ is chosen so small that the interval $(x_0 - \delta, x_0 + \delta)$ excludes all of these finitely many rational points, then $|x - x_0| < \delta$ implies $f(x) < \epsilon$, whether x is rational or irrational, and continuity at x_0 is established. If x_0 is rational and if $f(x_0)$ is redefined to be 0, continuity at x_0 is established by the same argument. Finally, if $\epsilon > 0$, since $f(x) \geq 0$ all that remains to be shown is the existence of a step-function $\tau(x) \geq f(x)$ such that $\int_0^1 \tau(x)\,dx < \epsilon$. Let $x_1, x_2, \cdots, x_n$ be the rational points of $[0, 1]$ such that $f(x_i) \geq \frac{1}{2}\epsilon$, $i = 1, 2, \cdots, n$, and let I_i be a neighborhood of x_i of length $< \epsilon/2n$,

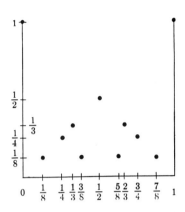

FIG. 406

$i = 1, 2, \cdots, n$. Define $\tau(x)$ to be 1 on all of the intervals $I_1, I_2, \cdots, I_n$ and $\frac{1}{2}\epsilon$ on the remaining points of $[0, 1]$. Then $\int_0^1 \tau(x)\,dx$ can be split into two parts, each $< \frac{1}{2}\epsilon$.

*404. EXERCISES

*1. Prove Theorem X, § 401. *Hint:* Assume for definiteness that $f(x)$ is monotonically increasing on $[a, b]$, and for a given net $\mathfrak{N}$ define the step-functions $\sigma(x) \equiv f(a_{i-1})$ for $a_{i-1} \leq x < a_i$ ($\sigma(b) \equiv f(b)$) and $\tau(x) \equiv f(a_i)$ for $a_{i-1} < x \leq a_i$ ($\tau(a) \equiv f(a)$), $i = 1, 2, \cdots, n$. Then $\sigma(x) \leq f(x) \leq \tau(x)$ and

$$\int_a^b [\tau(x) - \sigma(x)]\,dx = \sum_{i=1}^n [f(a_i) - f(a_{i-1})]\Delta x_i$$

$$\leq |\mathfrak{N}| \cdot \sum_{i=1}^n [f(a_i) - f(a_{i-1})] = |\mathfrak{N}| \cdot (f(b) - f(a)).$$

*2. Prove Theorem XII, § 401. *Hint:* Use Theorem II, § 403.

*3. Prove Theorem XIII, § 401. *Hint:* If $c \leq y \leq d$, then

$$\max(0, c, -d) \leq |y| \leq \max(|c|, |d|),$$
$$\max(|c|, |d|) - \max(0, c, -d) \leq d - c.$$

*4. Prove Theorem XIV, § 401. *Hint:* Show that it may be assumed without loss of generality that $f(x)$ and $g(x)$ are nonnegative. If $0 \leq f(x) \leq K$ and $0 \leq g(x) \leq K$, let σ, τ, ϕ, and ψ be step-functions such that

$$0 \leq \sigma(x) \leq f(x) \leq \tau(x) \leq K, \quad 0 \leq \phi(x) \leq g(x) \leq \psi(x) \leq K.$$

Write $\tau\psi - \sigma\phi = \tau(\psi - \phi) + \phi(\tau - \sigma)$.

★5. Prove Theorem XV, §401. *Hint:* Use the identity
$$\sum f(x_i)g(x_i') \Delta x_i = \sum f(x_i)g(x_i) \Delta x_i + \sum f(x_i)[g(x_i') - g(x_i)] \Delta x_i,$$
and the inequality
$$\left|\sum f(x_i)[g(x_i') - g(x_i)] \Delta x_i\right| \leq K \cdot \sum |g(x_i') - g(x_i)| \Delta x_i.$$
Finally, let $\xi_i \equiv x_i$ or x_i' so that $g(\xi_i) = \max(g(x_i), g(x_i'))$, and let $\eta_i \equiv x_i$ or x_i' so that $g(\eta_i) = \min(g(x_i), g(x_i'))$, $i = 1, 2, \cdots, n$. Then
$$\sum |g(x_i') - g(x_i)| \Delta x_i = \sum g(\xi_i) \Delta x_i - \sum g(\eta_i) \Delta x_i.$$

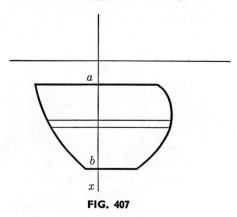

FIG. 407

★6. Supply the details requested in the second part of the proof of Theorem II, §403. *Hint:* If $\sum \tau_i \Delta x_i > I + \frac{1}{3}\epsilon$, define $\eta \equiv \frac{1}{n}[\sum \tau_i \Delta x_i - (I + \frac{1}{3}\epsilon)] > 0$. For each $i = 1, \cdots, n$, choose x_i such that $f(x_i) > \tau_i - (\eta/\Delta x_i)$, and obtain a contradiction.

★7. Bliss's Theorem (Theorem XV, §401) is used in such applied problems as work performed by a variable force and total fluid force on a submerged plate. In the latter case, for example, assume that a vertical plane area is submerged in a liquid of constant density ρ, and that the width at depth x is $w(x)$ (cf. Fig. 407). Then (with appropriate continuity assumptions) the element of area, between x_i and $x_i + \Delta x_i$ is neither $w(x_i) \Delta x_i$ nor $w(x_i + \Delta x_i) \Delta x_i$, but $w(\eta_i) \Delta x_i$ for some appropriate intermediate η_i (cf. the Mean Value Theorem for Integrals, and Ex. 5, §402). Furthermore, the total force on this element of area is greater than the area times the pressure at the top of the strip, and less than the area times the pressure at the bottom of the strip. It is therefore equal to the product of the area and an intermediate pressure, $\rho \xi_i$. Therefore the total force on the plate is the sum of the individual elements of force: $F = \sum_{i=1}^{n} \rho \xi_i w(\eta_i) \Delta x_i$. According to Bliss's Theorem, the limit of this sum can be expressed as a definite integral, and we have the standard formula, $F = \rho \int_a^b xw(x)\, dx$. Discuss the application of Bliss's Theorem to the work done in pumping the fluid contents of a tank to a certain height above the tank.

405. THE FUNDAMENTAL THEOREM OF INTEGRAL CALCULUS

Evaluation of a definite integral by actually taking the limit of a sum, as in the Example of § 401 and Exercises 12–15, § 402, is usually extremely arduous. Historically, many of the basic concepts of the definite integral as the limit of a sum were appreciated by the ancient Greeks long before the invention of the Differential Calculus by Newton and Leibnitz. Before the invention of derivatives, an area or a volume could be computed only by such a limiting process as is involved in the definition of the definite integral. The introduction of the notion of a derivative provided a spectacular impetus to the development of mathematics, not only by its immediate application as a rate of change, but also by permitting the evaluation of a definite integral by the process of reversing differentiation, known as *antidifferentiation* or *integration*. It is the purpose of this section to study certain basic relations between the two fundamental concepts of calculus, the derivative and the definite integral.

Theorem I. *Let $f(x)$ be defined and continuous on a closed interval $[a, b]$ or $[b, a]$, and define the function $F(x)$ on this interval:*

(1) $$F(x) \equiv \int_a^x f(t) \, dt.†$$

Then $F(x)$ is differentiable there with derivative $f(x)$:

(2) $$F'(x) = f(x).$$

(Cf. Exs. 21–22, § 407.)

Proof. Let x_0 and $x_0 + \Delta x$, where $\Delta x \neq 0$, belong to the given interval. Then, by Theorem V and Note 2, § 401,

$$F(x_0 + \Delta x) - F(x_0) = \int_a^{x_0 + \Delta x} f(t) \, dt - \int_a^{x_0} f(t) \, dt = \int_{x_0}^{x_0 + \Delta x} f(t) \, dt.$$

By the First Mean Value Theorem for Integrals (Theorem XI, § 401),

$$\frac{F(x_0 + \Delta x) - F(x_0)}{\Delta x} = \frac{1}{\Delta x} \int_{x_0}^{x_0 + \Delta x} f(t) \, dt = f(\xi)$$

---

† Since the letter x designates the upper limit of integration, it might be confusing to use this same symbol simultaneously in a second role as the variable of integration, and write $\int_a^x f(x) \, dx$. Note that the integral $\int_a^b f(x) \, dx$ is *not* a function of x. Rather, the letter x in this case serves only as a *dummy variable* (cf. Ex. 26, § 107), and any other letter could be substituted, thus: $\int_a^b f(x) \, dx = \int_a^b f(t) \, dt = \int_a^b f(u) \, du$. Therefore, instead of $\int_a^x f(x) \, dx$ we write $\int_a^x f(t) \, dt$, $\int_a^x f(u) \, du$, etc.

for some number ξ between x_0 and $x_0 + \Delta x$. Therefore if $\Delta x \to 0$, $\xi \to x_0$ and

$$F'(x_0) = \lim_{\Delta x \to 0} \frac{F(x_0 + \Delta x) - F(x_0)}{\Delta x} = f(x_0),$$

as stated in the theorem.

If a **primitive** or **antiderivative** or **indefinite integral** of a given function is defined to be any function whose derivative is the given function, Theorem I asserts that *any continuous function has a primitive.* The next theorem gives a method for evaluating the definite integral of any continuous function in terms of a given primitive.

Theorem II. Fundamental Theorem of Integral Calculus. *If $f(x)$ is continuous on the closed interval $[a, b]$ or $[b, a]$ and if $F(x)$ is any primitive of $f(x)$ on this interval, then*

(3) $$\int_a^b f(x)\, dx = F(b) - F(a).$$

(Cf. Exs. 19–20, § 407.)

Proof. By Theorem I and the hypotheses of Theorem II, the two functions $\int_a^x f(t)\, dt$ and $F(x)$ have the same derivative on the given interval. Therefore, by Theorem II, § 306, they differ by a constant:

(4) $$\int_a^x f(t)\, dt - F(x) = C.$$

Substitution of $x = a$ gives the value of this constant: $C = -F(a)$. Upon substitution for C in (4) we have $\int_a^x f(t)\, dt = F(x) - F(a)$ which, for the particular value $x = b$, is the desired result.

By virtue of the Fundamental Theorem of Integral Calculus, the integral symbol $\int$, suggested by the letter S (the definite integral is the limit of a *sum*), is appropriate for the *indefinite* integral as well as the *definite* integral. If $F(x)$ is an arbitrary indefinite integral of a function $f(x)$, we write the equation

(5) $$\int f(x)\, dx = F(x) + C,$$

where C is an arbitrary **constant of integration**, and call the symbol $\int f(x)\, dx$ **the indefinite integral** of $f(x)$. The Fundamental Theorem can thus be considered as an expression of the relation between the two kinds of integrals, definite and indefinite.

406. INTEGRATION BY SUBSTITUTION

The question before us is essentially one of appropriateness of notation: In the expression $\int f(x)\, dx$, does the "dx," which looks like a differential, really behave like one? Happily, the answer is in the affirmative:

Theorem. *If $f(v)$ is a continuous function of v and if $v(x)$ is a continuously differentiable function of x, then*

(1) $$\int f(v(x))v'(x)\,dx = \int f(v(x))\,d(v(x)) = \int f(v)\,dv.$$

If $v(a) = c$ and $v(b) = d$, then

(2) $$\int_a^b f(v(x))v'(x)\,dx = \int_c^d f(v)\,dv.$$

Proof. Equation (1) is a statement that a function $F(v)$ whose derivative with respect to v is $f(v)$ has a derivative with respect to x equal to $f(v)\dfrac{dv}{dx}$, and this statement is a reformulation for the rule for differentiating a function of a function. Equation (2) follows from (1) by the Fundamental Theorem of Integral Calculus: $F(v(b)) - F(v(a)) = F(d) - F(c)$.

Example. $\displaystyle\int_2^3 xe^{x^2}\,dx = \frac{1}{2}\int_2^3 e^{x^2}\,d(x^2) = \frac{1}{2}\int_4^9 e^v\,dv = \frac{e^9 - e^4}{2}.$

407. EXERCISES

1. Let the constant a and the variable x belong to an interval I throughout which the function $f(x)$ is continuous. Prove that

(1) $$\frac{d}{dx}\int_a^x f(t)\,dt = f(x), \quad \frac{d}{dx}\int_x^a f(t)\,dt = -f(x).$$

Hint: Use Theorem V and Note 2, § 401.

2. Let $u(x)$ and $v(x)$ be differentiable functions whose values lie in an interval I throughout which $f(t)$ is continuous. Prove that

(2) $$\frac{d}{dx}\int_{u(x)}^{v(x)} f(t)\,dt = f(v(x))v'(x) - f(u(x))u'(x).$$

Hint: Write $\displaystyle\int_{u(x)}^{v(x)} f(t)\,dt = \int_a^{v(x)} f(t)\,dt - \int_a^{u(x)} f(t)\,dt$, and define $F(v) \equiv \int_a^v f(t)\,dt$, with $v = v(x)$.

In Exercises 3–8, find the derivative with respect to x. The letters a and b represent constants. (Cf. Exs. 1–2.)

3. $\displaystyle\int_a^b \sin x^2\,dx.$ **4.** $\displaystyle\int_a^x \sin t^2\,dt.$

5. $\displaystyle\int_x^b \sin t^2\,dt.$ **6.** $\displaystyle\int_a^{x^3} \sin t^2\,dt.$

7. $\displaystyle\int_{x^3}^{x^4} \sin t^2\,dt.$ **8.** $\displaystyle\int_{\sin^2 x}^{\sin x^2} \sin t^2\,dt.$

9. Assume necessary conditions of continuity on $f(x)$ and $f'(x)$ in an interval $[a, b]$ and use the Fundamental Theorem of Integral Calculus to show that the

Mean Value Theorems for Integrals and Derivatives are merely alternative expressions of the same fact. *Hint:* Note that $f(x)$ is a primitive for $f'(x)$.

10. Show that the **integration by parts** formula,

$$\text{(3)} \qquad \int u\,dv = uv - \int v\,du,$$

is equivalent to the formula for the differential of the product of two functions: $d(uv) = u\,dv + v\,du$. On occasion integrations by parts can be more readily evaluated by judicious differentiation of products than by use of (3). (Cf. Exs. 11–12.)

11. Obtain the integrals

$$\text{(4)} \qquad \int e^{ax} \sin bx\,dx = \frac{e^{ax}}{a^2 + b^2}(a \sin bx - b \cos bx) + C,$$

$$\text{(5)} \qquad \int e^{ax} \cos bx\,dx = \frac{e^{ax}}{a^2 + b^2}(b \sin bx + a \cos bx) + C,$$

as follows: (*i*) Differentiate the products $e^{ax} \sin bx$ and $e^{ax} \cos bx$; (*ii*) multiply both members of each of the resulting equations by constants so that addition or subtraction eliminates either the terms involving $\cos bx$ or the terms involving $\sin bx$. (Cf. Ex. 10.)

12. Find a primitive of $x^4 e^x$ by the methods suggested in Exercise 10–11. Devise additional examples to illustrate both methods of integrating by parts.

13. Establish the following integration by parts formula where $u(x)$, $v(x)$, $u'(x)$, and $v'(x)$ are continuous on $[a, b]$:

$$\text{(6)} \qquad \int_a^b u(x)v'(x)\,dx = u(b)v(b) - u(a)v(a) - \int_a^b v(x)u'(x)\,dx.$$

14. Prove the **Second Mean Value Theorem for Integrals.** If $f(x)$, $\phi(x)$, and $\phi'(x)$ are continuous on the closed interval $[a, b]$, and if $\phi(x)$ is monotonic there (equivalently, by Exercise 38, § 308, $\phi'(x)$ does not change sign there), *then there exists a number ξ such that $a < \xi < b$ and*

$$\text{(7)} \qquad \int_a^b f(x)\phi(x)\,dx = \phi(a)\int_a^\xi f(x)\,dx + \phi(b)\int_\xi^b f(x)\,dx.$$

(For a generalization that eliminates assumptions on $\phi'(x)$ see Ex. 25, § 418; also cf. Ex. 26, § 418.) *Hint:* Use integration by parts, letting $F(x) \equiv \int_a^x f(t)\,dt$:

$$\int_a^b f(x)\phi(x)\,dx = \phi(x)F(x)\Big]_a^b - \int_a^b F(x)\phi'(x)\,dx,$$

and apply the generalized form of the First Mean Value Theorem (Ex. 6, § 402) to this last integral.

15. Show that the function $f(x)$ defined on $[0, 1]$ to be 1 if x is rational and 0 if x is irrational (Example 6, § 201) has neither integral nor primitive there. *Hint:* Cf. Exercise 40, § 308.

16. Show that although $\int_a^x f(t)\,dt$ is always a primitive of a given continuous function $f(x)$, not every primitive of $f(x)$ can be written in the form $\int_a^x f(t)\,dt$. *Hint:* Consider $f(x) \equiv \cos x$.

17. Show that the function $f(x) \equiv 0$ for $0 \leq x \leq 1$ and $f(x) \equiv 1$ for $1 < x \leq 2$ is integrable on the interval $[0, 2]$, although it has no primitive there. (Cf. Ex. 40, § 308.)

18. Show that the function $F(x) \equiv x^2 \sin \dfrac{1}{x^2}$ ($F(0) \equiv 0$) has an unbounded derivative $f(x)$ on $[0, 1]$, so that a function may have a primitive without being integrable.

19. Prove the following general form for the Fundamental Theorem of Integral Calculus: *If $f(x)$ is integrable on the interval $[a, b]$, and if it has a primitive $F(x)$ there, then $\int_a^b f(x)\, dx = F(b) - F(a)$.* Hint: Let $a_0 < a_1 < \cdots < a_n$ be an arbitrary net $\mathfrak{N}$ on $[a, b]$, and write

$$F(b) - F(a) = \sum_{i=1}^{n} \{F(a_i) - F(a_{i-1})\} = \sum_{i=1}^{n} f(x_i)\, \Delta x_i$$

(with the aid of the Law of the Mean, § 305).

*20. Generalize Exercise 19 as follows: *If $F(x)$ is continuous at every point of $[a, b]$, if $F'(x)$ exists at all but a finite number of points of $[a, b]$, and if $f(x) \equiv F'(x)$ is integrable on $[a, b]$, then $\int_a^b f(x)\, dx = F(b) - F(a)$.* (Cf. Ex. 5, § 409.) Hint: Use the method suggested in Exercise 19 but require the points of $\mathfrak{N}$ to include the points where $F'(x)$ fails to exist.

*21. Prove that if $f(x)$ is integrable on $[a, b]$, then $F(x) \equiv \int_a^x f(t)\, dt$ is continuous on $[a, b]$. Hint: $|F(x) - F(x_0)| \leq \left| \int_{x_0}^{x} |f(t)|\, dt \right|$.

*22. Prove that if $f(x)$ is integrable on $[a, b]$, then $F(x) \equiv \int_a^x f(t)\, dt$ is differentiable at every point x at which $f(x)$ is continuous, and at such points of continuity of $f(x)$, $F'(x) = f(x)$. Hint:

$$\left| \frac{F(x) - F(x_0)}{x - x_0} - f(x_0) \right| \leq \left| \frac{1}{x - x_0} \int_{x_0}^{x} |f(t) - f(x_0)|\, dt \right|.$$

408. SECTIONAL CONTINUITY AND SMOOTHNESS

In many applications certain discontinuous functions play an important role. In such applications it is necessary, however, that the discontinuities be within reason, both as to their nature and as to their number. An extremely useful class of such functions, frequently employed in the study of Fourier Series, is given in the following definition.

Definition I. *A function $f(x)$ is **sectionally continuous** on the closed interval $[a, b]$ if and only if it is continuous there except for at most a finite number of removable or jump discontinuities. That is, the one-sided limits $f(a+)$ and $f(b-)$ exist and are finite and $f(x+)$ and $f(x-)$ exist and are finite for all x between a and b, and $f(x+) = f(x-) = f(x)$ for all x between*

a and b with at most a finite number of exceptions. For sectional continuity the function $f(x)$ may or may not be defined at a or b at any of the exceptional points of discontinuity. (See Fig. 408.)

An example of a sectionally continuous function is the signum function (§ 206) on the interval $[-1, 1]$. Another example is the derivative of $|x|$ (on the same interval), which is identical with the signum function, except

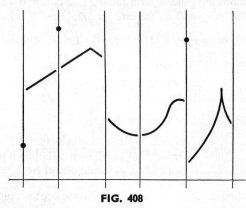

FIG. 408

that it is not defined at $x = 0$. Neither $1/x$ nor $\sin(1/x)$ is sectionally continuous on $[0, 1]$, or $[-1, 1]$, since there is no finite limit for either, as $x \to 0+$.

One of the most important properties of a sectionally continuous function is its integrability. This statement is made more precise and complete in the following theorem, whose proof is requested in Exercise 4, § 409, with hints.

Theorem. *If $f(x)$ is sectionally continuous on the closed interval $[a, b]$, and is continuous there except possibly at the points*

$$a = a_0 < a_1 < \cdots < a_n = b,$$

then the integrals $\int_a^b f(x)\, dx$ *and* $\int_{a_{i-1}}^{a_i} f(x)\, dx$, $i = 1, 2, \cdots, n$, *exist and*

(1) $$\int_a^b f(x)\, dx = \sum_{i=1}^n \int_{a_{i-1}}^{a_i} f(x)\, dx.$$

The values of the integrals in formula (1) are not affected by any values that might be assigned or reassigned to $f(x)$ at the exceptional points.

Closely related to sectional continuity is the more restrictive property of sectional smoothness:

Definition II. *A function $f(x)$ is **sectionally smooth** on the closed interval $[a, b]$ if and only if both $f(x)$ and $f'(x)$ are sectionally continuous there.* (Cf. Ex. 7, § 409.)

REDUCTION FORMULAS

Examples. On the interval $[-1, 1]$ the following functions are sectionally smooth: (a) the signum function (§ 206); (b) $|x|$; (c) $f(x) \equiv x - 1$ if $x > 0$, $f(x) \equiv x + 1$ if $x < 0$. On $[-1, 1]$ the following functions are sectionally continuous but not sectionally smooth: (d) $\sqrt{1 - x^2}$; (e) $\sqrt[3]{x^2}$; (f) $x^2 \sin(1/x)$ (Example 3, § 303).

409. EXERCISES

1. Show that the function $[x]$ (Example 5, § 201) is sectionally smooth on any finite closed interval. Draw the graph of its derivative.

2. Show that $|\sin x|$ is sectionally smooth on any finite interval. Draw the graphs of the function and its derivative.

3. Prove that a function that is sectionally continuous on a closed interval is bounded there. Must it have a maximum value and a minimum value there? Must it take on all values between any two of its values?

4. Prove the Theorem of § 408. *Hint:* For the interval $[a_{i-1}, a_i]$, if $f(x)$ is defined or redefined: $f(a_{i-1}) \equiv f(a_{i-1}+)$ and $f(a_i) \equiv f(a_i-)$, then $f(x)$ becomes continuous there. Apply Theorem V, § 401, and induction.

5. Let $F(x)$ be sectionally smooth on $[a, b]$, with exceptional points a_i, $i = 1, 2, \cdots, n - 1$, and possibly $a_0 = a$ and $a_n = b$, and denote by J_i the jump of $F(x)$ at a_i: $J_i \equiv F(a_i+) - F(a_i-)$, $i = 1, 2, \cdots, n - 1$. Prove that

$$(1) \qquad \int_a^b F'(x)\, dx = F(b-) - F(a+) - \sum_{i=1}^{n-1} J_i.$$

(Cf. Ex. 20, § 407.)

6. Prove that if $f(x)$ and $g(x)$ are sectionally continuous (smooth) then so are $f(x) + g(x)$ and $f(x)g(x)$. Extend by induction to n terms and n factors.

★7. Prove that if $f'(x)$ is sectionally continuous on $[a, b]$, then $f(x)$ is sectionally smooth there. Furthermore, prove that $f(x)$ has a right-hand and a left-hand derivative at each point between a and b, and a right-hand derivative at a and a left-hand derivative at b. (Cf. Exs. 44–45, § 308.)

★8. If $f(x)$ is sectionally continuous on the closed interval $[a, b]$, then $F(x) \equiv \int_a^x f(t)\, dt$ is continuous there and $F'(x)$ exists and is equal to $f(x)$ there with at most a finite number of exceptions (Exs. 21–22, § 407). Prove that any function $\Phi(x)$ that is continuous on $[a, b]$ such that $\Phi'(x)$ exists and is equal to $f(x)$ there with at most a finite number of exceptions must differ from $F(x)$ by at most a constant. If $\Phi(x)$ is not everywhere continuous, what can you say?

410. REDUCTION FORMULAS

It often happens that the routine evaluation of an integral involves repeated applications of integration by parts, all such integrations by parts being of the same tedious type. For example, in evaluating $I = \int \sin^{10} x\, dx$, we might proceed:

$$I = \int \sin^9 x\, d(-\cos x) = -\sin^9 x \cos x + 9 \int \sin^8 x \cos^2 x\, dx$$
$$= -\sin^9 x \cos x + 9 \int \sin^8 x\, dx - 9I,$$

so that

(1) $$I = -\tfrac{1}{10} \sin^9 x \cos x + \tfrac{9}{10} \int \sin^8 x \, dx.$$

We have succeeded in reducing the exponent from 10 to 8. We could repeat this labor to reduce the new exponent from 8 to 6; then from 6 to 4; etc.

A more satisfactory method is to establish a *single formula* to handle all integrals of a single type. We present a few derivations of such **reduction formulas** in Examples, and ask for more in the Exercises of the following section. Since differentiation is basically a simpler process than integration, we perform our integrations by parts by means of differentiating certain products.

Example 1. Express $\int \sin^m x \cos^n x \, dx$, where $m + n \neq 0$, in terms of an integral with reduced exponent on $\sin x$ (cf. Exs. 1–2, §411).

Solution. The derivative of the product $\sin^p x \cos^q x$ is
$$p \sin^{p-1} x \cos^{q+1} x - q \sin^{p+1} x \cos^{q-1} x$$
$$= p \sin^{p-1} x \cos^{q-1} x (1 - \sin^2 x) - q \sin^{p+1} x \cos^{q-1} x$$
$$= p \sin^{p-1} x \cos^{q-1} x - (p + q) \sin^{p+1} x \cos^{q-1} x.$$

In other words,

(2) $$p \int \sin^{p-1} x \cos^{q-1} x \, dx - (p+q) \int \sin^{p+1} x \cos^{q-1} x \, dx = \sin^p x \cos^q x + C.$$

Letting $p = m - 1$ and $q = n + 1$, and absorbing the constant of integration with the $\int$, we have the formula sought:

(3) $$\int \sin^m x \cos^n x \, dx = -\frac{\sin^{m-1} x \cos^{n+1} x}{m+n} + \frac{m-1}{m+n} \int \sin^{m-2} x \cos^n x \, dx.$$

Equation (1) is a special case of (3), with $m = 10$, $n = 0$.

Often it is desirable to increase a negative exponent.

Example 2. Establish the reduction formula ($m \neq 1$) (cf. Exs. 1–2, §411):

(4) $$\int \frac{\cos^n x \, dx}{\sin^m x} = -\frac{\cos^{n+1} x}{(m-1)\sin^{m-1} x} - \frac{n-m+2}{m-1} \int \frac{\cos^n x \, dx}{\sin^{m-2} x}.$$

Solution. In (2), let $p = -m + 1$ and $q = n + 1$.

Example 3. Establish the reduction formula ($n \neq 1$):

(5) $$\int \frac{dx}{(a^2 + x^2)^n} = \frac{x}{2(n-1)a^2(a^2+x^2)^{n-1}} + \frac{2n-3}{2(n-1)a^2} \int \frac{dx}{(a^2+x^2)^{n-1}}.$$

Solution. The derivative of the product $x(a^2 + x^2)^m$ is
$$(a^2 + x^2)^m + 2mx^2(a^2 + x^2)^{m-1} = (a^2 + x^2)^m + 2m[(a^2 + x^2) - a^2](a^2 + x^2)^{m-1}$$
$$= (1 + 2m)(a^2 + x^2)^m - 2ma^2(a^2 + x^2)^{m-1}.$$

In other words,

(6) $$(1 + 2m) \int (a^2 + x^2)^m \, dx - 2ma^2 \int (a^2 + x^2)^{m-1} \, dx = x(a^2 + x^2)^m + C.$$

Now let $m = -n + 1$, from which we obtain (5).

Useful reduction formulas are given in nearly every Table of Integrals.

411. EXERCISES

In Exercises 1–12, establish the reduction formula.

1. $\int \sin^m x \cos^n x \, dx = \dfrac{\sin^{m+1} x \cos^{n-1} x}{m+n}$
$+ \dfrac{n-1}{m+n} \int \sin^m x \cos^{n-2} x \, dx \qquad (m+n \neq 0).$

2. $\int \dfrac{\sin^m x \, dx}{\cos^n x} = \dfrac{\sin^{m+1} x}{(n-1)\cos^{n-1} x} - \dfrac{m-n+2}{n-1} \int \dfrac{\sin^m x \, dx}{\cos^{n-2} x} \qquad (n \neq 1).$

3. $\int \tan^n x \, dx = \dfrac{\tan^{n-1} x}{n-1} - \int \tan^{n-2} x \, dx \qquad (n \neq 1).$

4. $\int \cot^n x \, dx = -\dfrac{\cot^{n-1} x}{n-1} - \int \cot^{n-2} x \, dx \qquad (n \neq 1).$

5. $\int \sec^n x \, dx = \dfrac{\sec^{n-2} x \tan x}{n-1} + \dfrac{n-2}{n-1} \int \sec^{n-2} x \, dx \qquad (n \neq 1).$

6. $\int \csc^n x \, dx = -\dfrac{\csc^{n-2} x \cot x}{n-1} + \dfrac{n-2}{n-1} \int \csc^{n-2} x \, dx \qquad (n \neq 1).$

7. $\int x^n e^x \, dx = x^n e^x - n \int x^{n-1} e^x \, dx.$

8. $\int x^n \sin x \, dx = -x^n \cos x + n x^{n-1} \sin x - n(n-1) \int x^{n-2} \sin x \, dx.$

9. $\int x^n \cos x \, dx = x^n \sin x + n x^{n-1} \cos x - n(n-1) \int x^{n-2} \cos x \, dx.$

10. $\int x^m (ax+b)^n \, dx = \dfrac{x^m (ax+b)^{n+1}}{a(m+n+1)}$
$- \dfrac{mb}{a(m+n+1)} \int x^{m-1}(ax+b)^n \, dx \qquad (m+n+1 \neq 0).$

11. $\int x^m (ax+b)^n \, dx = \dfrac{x^{m+1}(ax+b)^n}{m+n+1}$
$+ \dfrac{nb}{m+n+1} \int x^m (ax+b)^{n-1} \, dx \qquad (m+n+1 \neq 0).$

12. $\int \dfrac{x^n \, dx}{\sqrt{ax^2+bx+c}} = \dfrac{x^{n-1}}{an} \sqrt{ax^2+bx+c}$
$- \dfrac{b(2n-1)}{2an} \int \dfrac{x^{n-1} \, dx}{\sqrt{ax^2+bx+c}} - \dfrac{c(n-1)}{an} \int \dfrac{x^{n-2} \, dx}{\sqrt{ax^2+bx+c}} \qquad (n \neq 0).$

13. $\int x^n \sqrt{2ax - x^2} \, dx = -\dfrac{x^{n-1}(2ax-x^2)^{\frac{3}{2}}}{n+2}$
$+ \dfrac{a(2n+1)}{n+2} \int x^{n-1} \sqrt{2ax-x^2} \, dx \qquad (n \neq -2).$

In Exercises 14–21, perform the integration, using reduction formulas above.

14. $\int \sin^6 x \, dx.$

15. $\int \cos^5 x \, dx.$

16. $\int \cot^5 \frac{x}{5} \, dx$.

17. $\int \sec^7 x \, dx$.

18. $\int x^4 \sin 2x \, dx$.

19. $\int x^{\frac{2}{3}}(x+4)^{\frac{1}{2}} \, dx$.

20. $\int \frac{x^3 \, dx}{\sqrt{x^2+x+1}}$.

21. $\int x^3 \sqrt{6x-x^2} \, dx$.

*22. Use mathematical induction to verify the formula:
$$\int_a^b (x-a)^m (b-x)^n \, dx = (b-a)^{m+n+1} \frac{m! \, n!}{(m+n+1)!} \text{ (m and n positive integers)}.$$

*23. Establish the reduction formula, for $\alpha, \beta, n \geq 1$:
$$\int_0^1 x^n x^{\alpha-1} (1-x)^{\beta-1} \, dx = \frac{n+\alpha-1}{n+\alpha+\beta-1} \int_0^1 x^{n-1} x^{\alpha-1} (1-x)^{\beta-1} \, dx.$$

412. IMPROPER INTEGRALS, INTRODUCTION

The definite integral $\int_a^b f(x) \, dx$, as defined in § 401, has meaning only if a and b are finite and $f(x)$ is bounded on the interval $[a, b]$. In the following two sections we shall extend the definition so that under certain circumstances the symbol $\int_a^b f(x) \, dx$ shall be meaningful even when the interval of integration is infinite or the function $f(x)$ is unbounded.

For the sake of conciseness, parentheses will be used in some of the definitions to indicate alternative statements. For a discussion of the use of parentheses for alternatives, see the Preface.

A more extensive treatment of improper integrals is given in Chapter 14.

413. IMPROPER INTEGRALS, FINITE INTERVAL

Definition I. *Let $f(x)$ be (Riemann-) integrable on the interval $[a, b-\epsilon]$ ($[a+\epsilon, b]$) for every number ϵ such that $0 < \epsilon < b-a$, but not integrable on the interval $[a, b]$, and assume that*

$$\lim_{\epsilon \to 0+} \int_a^{b-\epsilon} f(x) \, dx \; \left(\lim_{\epsilon \to 0+} \int_{a+\epsilon}^b f(x) \, dx \right)$$

*exists. Under these conditions the **improper integral** $\int_a^b f(x) \, dx$ is defined to be this limit:*

(1) $$\int_a^b f(x) \, dx \equiv \lim_{\epsilon \to 0+} \int_a^{b-\epsilon} f(x) \, dx \; \left(\lim_{\epsilon \to 0+} \int_{a+\epsilon}^b f(x) \, dx \right).$$

*If the limit in (1) is finite the integral $\int_a^b f(x) \, dx$ is **convergent** to this limit and the function $f(x)$ is said to be **improperly integrable** on the half-open interval $[a, b)$ ($(a, b]$); if the limit in (1) is infinite or does not exist, the integral is **divergent**.*

§ 413]   IMPROPER INTEGRALS, FINITE INTERVAL

Note 1. A function improperly integrable on a half-open interval is necessarily unbounded there. In fact, it is unbounded in every neighborhood of the end-point of the interval that is not included. (Cf. Theorem III, § 403.)

Definition II. *Let $[a, b]$ be a given finite interval, let $a < c < b$, and let both integrals $\int_a^c f(x)\, dx$ and $\int_c^b f(x)\, dx$ be convergent improper integrals in the sense of Definition I. Then the **improper integral** $\int_a^b f(x)\, dx$ is convergent and defined to be:*

(2) $$\int_a^b f(x)\, dx \equiv \int_a^c f(x)\, dx + \int_c^b f(x)\, dx.$$

If either integral on the right-hand side of (2) diverges, so does $\int_a^b f(x)\, dx$.

Note 2. There are four ways in which the integrals (2) may be improper, corresponding to the points in the neighborhoods of which $f(x)$ is unbounded: (i) $c-$ and $c+$, (ii) $a+$ and $c+$, (iii) $c-$ and $b-$, and (iv) $a+$ and $b-$. In this last case (iv), the definition (2) is meaningful only if the value of $\int_a^b f(x)\, dx$ is independent of the interior point c. This independence is indeed a fact (cf. Ex. 11, § 416). As illustration of this independence of the point c, see Example 4, below.

Note 3. In case the function $f(x)$ is (Riemann-) integrable on the interval $[a, b]$ it is often convenient to refer to the integral $\int_a^b f(x)\, dx$ as a **proper** integral, in distinction to the improper integrals defined above, and to call the integral $\int_a^b f(x)\, dx$ **convergent,** even though it is not improper. (Cf. Example 3, below.)

Example 1. Evaluate $\int_4^5 \dfrac{dx}{\sqrt{x-4}}$.

Solution. The integrand becomes infinite as $x \to 4+$. The given improper integral has the value

$$\lim_{\epsilon \to 0+} \int_{4+\epsilon}^5 \frac{dx}{\sqrt{x-4}} = \lim_{\epsilon \to 0+} \left[2\sqrt{x-4}\right]_{4+\epsilon}^5 = \lim_{\epsilon \to 0+} [2 - 2\sqrt{\epsilon}] = 2.$$

Example 2. Evaluate $\int_{-1}^1 \dfrac{dx}{x^2}$.

Solution. A thoughtless, brash, and incorrect "evaluation" would yield the ridiculous negative result:

$$\int_{-1}^1 \frac{dx}{x^2} = \left[-\frac{1}{x}\right]_{-1}^1 = [-1] - [1] = -2.$$

Since the integrand becomes infinite as $x \to 0$, a correct evaluation is

$$\int_{-1}^1 \frac{dx}{x^2} = \lim_{\epsilon \to 0+} \int_{-1}^{-\epsilon} \frac{dx}{x^2} + \lim_{\epsilon \to 0+} \int_\epsilon^1 \frac{dx}{x^2}$$

$$= \lim_{\epsilon \to 0+} \left(\frac{1}{\epsilon} - 1\right) + \lim_{\epsilon \to 0+} \left(-1 + \frac{1}{\epsilon}\right) = +\infty,$$

and the integral diverges.

Example 3. For what values of p does $\int_0^1 \dfrac{dx}{x^p}$ converge?

Solution. If $p \leq 0$, the integral is proper, and therefore converges (Note 3). If $0 < p < 1$,
$$\int_0^1 x^{-p}\, dx = \lim_{\epsilon \to 0+} \left[\frac{x^{1-p}}{1-p}\right]_\epsilon^1 = \frac{1}{1-p},$$
and the integral converges. If $p = 1$, $\lim_{\epsilon \to 0+}\left[\ln x\right]_\epsilon^1 = \lim_{\epsilon \to 0+}(-\ln \epsilon) = +\infty$, and if $p > 1$, $\lim_{\epsilon \to 0+}\left[\dfrac{x^{1-p}}{1-p}\right]_\epsilon^1 = +\infty$. Therefore the given integral converges if and only if $p < 1$.

Example 4. Evaluate $\int_{-2}^{2} \dfrac{dx}{\sqrt{4 - x^2}}$.

Solution. The integrand becomes infinite at both end-points of the interval $[-2, 2]$, and we therefore evaluate according to Definition II, choosing some number c in the interior of this interval. The simplest value of c is 0. We find, then, that
$$\int_{-2}^0 \frac{dx}{\sqrt{4-x^2}} = \lim_{\epsilon \to 0+}\left[\text{Arcsin}\frac{x}{2}\right]_{-2+\epsilon}^0 = -\text{Arcsin}(-1) = \frac{\pi}{2},$$
and
$$\int_0^2 \frac{dx}{\sqrt{4-x^2}} = \lim_{\epsilon \to 0+}\left[\text{Arcsin}\frac{x}{2}\right]_0^{2-\epsilon} = \text{Arcsin}(1) = \frac{\pi}{2},$$
and the value of the given integral is π. Notice that any other value of c between -2 and 2 could have been used (Note 2, above):
$$\int_{-2}^c \frac{dx}{\sqrt{4-x^2}} + \int_c^2 \frac{dx}{\sqrt{4-x^2}} = \left(\text{Arcsin}\frac{c}{2} + \frac{\pi}{2}\right) + \left(\frac{\pi}{2} - \text{Arcsin}\frac{c}{2}\right) = \pi.$$

NOTE 4. Under certain circumstances a student may evaluate an improper integral on a finite interval, with correct result, without forming a limit, or even without recognizing that the given integral is improper. This would be the case with Example 1, above—but not with Example 2. The following theorem justifies such a method and simplifies many evaluations:

Theorem. *If $f(x)$ is continuous in the open interval (a, b), and if there exists a function $F(x)$ which is continuous over the closed interval $[a, b]$ and such that $F'(x) = f(x)$ in the open interval (a, b), then the integral $\int_a^b f(x)\, dx$, whether proper or improper, converges and*

(Cf. Ex. 15, § 416.) $\qquad \int_a^b f(x)\, dx = F(b) - F(a).$

Proof. By the Fundamental Theorem of Integral Calculus, for any c between a and b, and sufficiently small positive ϵ and η,
$$\int_{a+\epsilon}^c f(x)\, dx + \int_c^{b-\eta} f(x)\, dx = F(b - \eta) - F(a + \epsilon),$$
and the result follows from the continuity of $F(x)$ at a and b.

Example 5. In Example 1, above, let $F(x) \equiv 2\sqrt{x-4}$. Then
$$\int_4^5 \frac{dx}{\sqrt{x-4}} = \left[2\sqrt{x-4}\right]_4^5 = 2.$$

414. IMPROPER INTEGRALS, INFINITE INTERVAL

Definition I. *Let $f(x)$ be (Riemann-) integrable on the interval $[a, u]$ for every number $u > a$, and assume that $\lim_{u \to +\infty} \int_a^u f(x)\, dx$ exists. Under these conditions the* **improper integral** $\int_a^{+\infty} f(x)\, dx$ *is defined to be this limit:*

(1) $$\int_a^{+\infty} f(x)\, dx \equiv \lim_{u \to +\infty} \int_a^u f(x)\, dx.$$

If the limit in (1) is finite the improper integral is **convergent** *to this limit and the function $f(x)$ is said to be* **improperly integrable** *on the interval $[a, +\infty)$; if the limit in (1) is infinite or does not exist, the integral is* **divergent**.

A similar definition holds for the improper integral $\int_{-\infty}^a f(x)\, dx$.

Definition II. *Let $f(x)$ be (Riemann-) integrable on every finite closed interval, and assume that both improper integrals $\int_0^{+\infty} f(x)\, dx$ and $\int_{-\infty}^0 f(x)\, dx$ converge. Then the* **improper integral** $\int_{-\infty}^{+\infty} f(x)\, dx$ *is convergent and defined to be:*

(2) $$\int_{-\infty}^{+\infty} f(x)\, dx \equiv \int_{-\infty}^0 f(x)\, dx + \int_0^{+\infty} f(x)\, dx.$$

If either integral on the right-hand side of (2) diverges, so does $\int_{-\infty}^{+\infty} f(x)\, dx$.

NOTE 1. The improper integral (2) could have been defined unambiguously:
$$\int_{-\infty}^{+\infty} f(x)\, dx = \int_{-\infty}^c f(x)\, dx + \int_c^{+\infty} f(x)\, dx,$$ where c is an arbitrary number. (Cf. Ex. 12, § 416.)

Improper integrals on finite and infinite intervals are often combined:

Definition III. *Let $f(x)$ be improperly integrable on the interval $(a, c]$ and on the interval $[c, +\infty)$, where c is any constant greater than a. Then the* **improper integral** $\int_a^{+\infty} f(x)\, dx$ *is convergent and defined to be:*

(3) $$\int_a^{+\infty} f(x)\, dx \equiv \int_a^c f(x)\, dx + \int_c^{+\infty} f(x)\, dx.$$

If either integral on the right-hand side of (3) diverges, so does $\int_a^{+\infty} f(x)\, dx$. Similar statements hold for a similarly improper integral $\int_{-\infty}^a f(x)\, dx$.

NOTE 2. The improper integral (3) is independent of c. (Cf. Ex. 12, § 416.)

Example 1. Evaluate $\int_0^{+\infty} e^{-ax}\,dx$, $a > 0$.

Solution. $\int_0^{+\infty} e^{-ax}\,dx = \lim_{u \to +\infty} \left[\dfrac{e^{-ax}}{-a}\right]_0^u = \lim_{u \to +\infty} \left[\dfrac{1}{a} - \dfrac{e^{-au}}{a}\right] = \dfrac{1}{a}.$

Example 2. Evaluate $\int_{-\infty}^{+\infty} \dfrac{dx}{a^2 + x^2}$, $a > 0$.

Solution.

$$\int_{-\infty}^{+\infty} \frac{dx}{a^2 + x^2} = \lim_{u \to -\infty} \left[\frac{1}{a} \operatorname{Arctan} \frac{x}{a}\right]_u^0 + \lim_{v \to +\infty} \left[\frac{1}{a} \operatorname{Arctan} \frac{x}{a}\right]_0^v$$

$$= \lim_{u \to -\infty} \left[-\frac{1}{a} \operatorname{Arctan} \frac{u}{a}\right] + \lim_{v \to +\infty} \left[\frac{1}{a} \operatorname{Arctan} \frac{v}{a}\right] = \frac{1}{a} \cdot \frac{\pi}{2} + \frac{1}{a} \cdot \frac{\pi}{2} = \frac{\pi}{a}.$$

Example 3. For what values of p does $\int_1^{+\infty} \dfrac{dx}{x^p}$ converge?

Solution. If $p \neq 1$, $\int_1^u x^{-p}\,dx = \left[\dfrac{x^{1-p}}{1-p}\right]_1^u$, and the integral converges if $p > 1$ and diverges if $p < 1$. Similarly, $\int_1^{+\infty} \dfrac{dx}{x} = \lim_{u \to +\infty} \ln u = +\infty$. Therefore the given integral converges if and only if $p > 1$, and its value, for such p, is $1/(p-1)$.

Example 4. For what values of p does $\int_0^{+\infty} \dfrac{dx}{x^p}$ converge?

Solution. For this improper integral to converge, both $\int_0^1$ and $\int_1^{+\infty}$ must converge. But they never converge for the same value of p. (Examples 3, §§ 413, 414.) Answer: none.

415. COMPARISON TESTS. DOMINANCE

For a nonnegative function, convergence of an improper integral means (in a sense determined by the definition of the improper integral concerned) that *the function is not too big*: on a finite interval the function does not become infinite too fast, and on an infinite interval the function does not approach 0 too slowly. This means that whenever a nonnegative function $g(x)$ has a convergent improper integral, any well-behaved nonnegative function $f(x)$ less than or equal to $g(x)$ also has a convergent improper integral. We make these ideas precise:

Definition I. *The statement that a function $g(x)$ **dominates** a function $f(x)$ on a set A means that both functions are defined for every member x of A and that for every such x, $|f(x)| \leq g(x)$.*

NOTE 1. Any dominating function is automatically nonnegative, although a dominated function may have negative values.

§ 415] COMPARISON TESTS. DOMINANCE

Theorem I. Comparison Test. *If a function $g(x)$ dominates a nonnegative function $f(x)$ on the interval $[a, b)$ ($[a, +\infty)$), if both functions are integrable on the interval $[a, c]$ for every c such that $a < c < b$ ($a < c$), and if the improper integral $\int_a^b g(x)\,dx$ $\left(\int_a^{+\infty} g(x)\,dx\right)$ converges, then so does $\int_a^b f(x)\,dx$ $\left(\int_a^{+\infty} f(x)\,dx\right)$.*

Similar statements apply to other types of improper integrals defined in §§ 413–414.

Proof. Since the two functions are nonnegative, the two integrals $\int_a^c f(x)\,dx$ and $\int_a^c g(x)\,dx$ are monotonically increasing functions of c, and both have limits as $c \to b-$ ($c \to +\infty$). (Cf. § 215.) The inequalities $0 \leq \int_a^c f(x)\,dx \leq \int_a^c g(x)\,dx$ imply, thanks to Theorem VII, § 207, the inequalities $0 \leq \int_a^b f(x)\,dx \leq \int_a^b g(x)\,dx$ $\left(0 \leq \int_a^{+\infty} f(x)\,dx \leq \int_a^{+\infty} g(x)\,dx\right)$, and the proof is complete.

Example 1. Since, for $x \geq 1$, $\dfrac{1}{x^2 + 5x + 17} < \dfrac{1}{x^2}$, the convergence of $\int_1^{+\infty} \dfrac{dx}{x^2}$ (Example 3, § 414) implies that of $\int_1^{+\infty} \dfrac{dx}{x^2 + 5x + 17}$.

NOTE 2. An interchange of the roles of $f(x)$ and $g(x)$ in Theorem I furnishes a comparison test for divergence.

Example 2. Since, for $0 < x \leq 1$, $\dfrac{1}{x^2 + 5x} \geq \dfrac{1}{6x}$, the divergence of $\int_0^1 \dfrac{dx}{x}$ (Example 3, § 413) implies that of $\int_0^1 \dfrac{dx}{x^2 + 5x}$.

A convenient method for establishing dominance, and hence the convergence of an improper integral, can be formulated in terms of the "big O" notation (this is an upper case letter O derived from the expression "order of magnitude"):

Definition II. *The notation $f = O(g)$ (read "f is big O of g"), or equivalently $f(x) = O(g(x))$, as x approaches some limit (finite or infinite), means that within some deleted neighborhood of that limit, $f(x)$ is dominated by some positive constant multiple of $g(x)$: $|f(x)| \leq K \cdot g(x)$. If simultaneously $f = O(g)$ and $g = O(f)$, the two functions $f(x)$ and $g(x)$ are said to be of the **same order of magnitude** as x approaches its limit.†*

Example 3. As $x \to +\infty$, $\sin x = O(1)$ and $\ln x = O(x)$. (Cf. Ex. 26, § 416.)

---

† Similar to the "big O" notation is the "little o" notation: $f = o(g)$ means that within some deleted neighborhood of the limiting value of x, an inequality of the form $|f(x)| \leq K(x)g(x)$ holds, where $K(x) \geq 0$ and $K(x) \to 0$. Obviously $f = o(g)$ implies $f = O(g)$.

Big O and order of magnitude relationships are usually established by taking limits:

Theorem II. *Let $f(x)$ and $g(x)$ be positive functions and assume that $\lim \dfrac{f(x)}{g(x)} = L$ exists, in a finite or infinite sense. Then:*
  (i) $0 \leq L < +\infty$ implies $f = O(g)$,
  (ii) $0 < L \leq +\infty$ implies $g = O(f)$,
  (iii) $0 < L < +\infty$ implies that $f(x)$ and $g(x)$ are of the same order of magnitude.

(Give the details of the proof in Ex. 25, § 416.)

Example 4. As $x \to +\infty$, $\dfrac{1}{x+1} - \dfrac{1}{x} = O\left(\dfrac{1}{x^2}\right)$, since $\left[\dfrac{1}{x+1} - \dfrac{1}{x}\right] / \left[\dfrac{1}{x^2}\right] = \dfrac{-x^2}{x^2+x}$, and $\lim\limits_{x \to +\infty} \left|\dfrac{-x^2}{x^2+x}\right| = 1 < +\infty$.

NOTE 3. The "big O" notation is also used in the following sense: $f(x) = g(x) + O(h(x))$ means that $f(x) - g(x) = O(h(x))$. Thus, as $x \to +\infty$, $\dfrac{1}{x+1} = \dfrac{1}{x} + O\left(\dfrac{1}{x^2}\right)$, by Example 4.

Theorem III. *If $f(x)$ and $g(x)$ are nonnegative on $[a, b]$ and integrable on $[a, c]$ for every c such that $a < c < b$, and if $f = O(g)$ as $x \to b-$, then the convergence of $\displaystyle\int_a^b g(x)\,dx$ implies that of $\displaystyle\int_a^b f(x)\,dx$.*

Similar statements apply to other types of improper integrals defined in §§ 413–414.

Proof. The convergence of $\displaystyle\int_c^b K g(x)\,dx$ implies that of $\displaystyle\int_c^b f(x)\,dx$ by Theorem I.

NOTE 4. An interchange of the roles of $f(x)$ and $g(x)$ in Theorem III furnishes a test for divergence. If $f(x)$ and $g(x)$ are of the same order of magnitude, then their two integrals either both converge or both diverge.

Example 5. Test for convergence or divergence: $\displaystyle\int_0^1 \dfrac{dx}{1-x^3}$.

Solution. The integrand, $f(x)$, becomes infinite at $x = 1$. Write
$$f(x) = \dfrac{1}{(1-x)(1+x+x^2)},$$
and compare it with $g(x) \equiv \dfrac{1}{1-x}$. Since
$$\lim_{x \to 1-} \dfrac{f(x)}{g(x)} = \lim_{x \to 1-} \dfrac{1}{1+x+x^2} = \tfrac{1}{3} > 0,$$
$g = O(f)$ as $x \to 1-$, and since $\displaystyle\int_0^1 g(x)\,dx$ diverges, $\displaystyle\int_0^1 f(x)\,dx$ diverges.

Example 6. Test for convergence or divergence $\int_0^5 \dfrac{dx}{\sqrt[3]{7x + 2x^4}}$.

Solution. The integrand, $f(x)$, becomes infinite at $x = 0$. Write
$$f(x) = \dfrac{1}{\sqrt[3]{x}\sqrt[3]{7 + 2x^3}}$$
and compare it with $g(x) \equiv \dfrac{1}{\sqrt[3]{x}}$. Since
$$\lim_{x \to 0+} \dfrac{f(x)}{g(x)} = \lim_{x \to 0+} \dfrac{1}{\sqrt[3]{7 + 2x^3}} = \dfrac{1}{\sqrt[3]{7}} < +\infty,$$
$f = O(g)$ as $x \to 0+$, and since $\int_0^5 g(x)\, dx$ converges, $\int_0^5 f(x)\, dx$ converges.

Example 7. Test for convergence or divergence: $\int_1^{+\infty} \dfrac{dx}{\sqrt{x^3 + 2x + 2}}$.

Solution. The integrand, as $x \to +\infty$, is of the same order of magnitude as $x^{\frac{3}{2}}$. Therefore the given integral converges.

Example 8. (The *Beta Function.*) Determine the values of p and q for which
$$B(p, q) \equiv \int_0^1 x^{p-1}(1 - x)^{q-1}\, dx$$
converges. (Many of the important properties of the beta function are considered in § 1411.)

Solution. The integrand $f(x)$ has possible discontinuities at $x = 0$ and $x = 1$. As $x \to 0+$, $f(x)$ is of the same order of magnitude as x^{p-1}, and as $x \to 1-$, $f(x)$ is of the same order of magnitude as $(1 - x)^{q-1}$. Therefore convergence of both $\int_0^c f(x)\, dx$ and $\int_c^1 f(x)\, dx$, for $0 < c < 1$, is equivalent to the two inequalities $p - 1 > -1$ and $q - 1 > -1$. Therefore $\int_0^1 f(x)\, dx$ converges if and only if both p and q are positive.

Example 9. (The *Gamma Function.*) Determine the values of α for which
$$\Gamma(\alpha) \equiv \int_0^{+\infty} x^{\alpha-1} e^{-x}\, dx$$
converges. (Many of the important properties of the gamma function are considered in § 1410.)

Solution. If $f(x)$ is the integrand, $f(x)$ is of the same order of magnitude as $x^{\alpha-1}$ as $x \to 0+$, and $f(x) = O\left(\dfrac{1}{x^2}\right)$ as $x \to +\infty$ (this is true by l'Hospital's Rule in case $\alpha > -1$). Since $\int_0^1 x^{\alpha-1}\, dx$ converges if and only if $\alpha > 0$, and $\int_1^{+\infty} \dfrac{dx}{x^2}$ converges, the given integral converges if and only if α is positive.

416. EXERCISES

In Exercises 1–10, evaluate every convergent improper integral and specify those that diverge.

1. $\int_0^3 \dfrac{dx}{\sqrt{9-x^2}}.$

2. $\int_{-1}^1 \dfrac{dx}{\sqrt[3]{x}}.$

3. $\int_{-2}^2 \dfrac{dx}{x^3}.$

4. $\int_0^{\pi/2} \sqrt{\sin x \tan x}\, dx.$

5. $\int_0^{+\infty} \dfrac{dx}{\sqrt{e^x}}.$

6. $\int_0^{+\infty} \sin x\, dx.$

7. $\int_1^{+\infty} \dfrac{dx}{x\sqrt{x^2-1}}.$

8. $\int_2^{+\infty} \dfrac{dx}{x(\ln x)^k}.$

9. $\int_0^{+\infty} \dfrac{e^{-\sqrt{x}}}{\sqrt{x}}\, dx.$

10. $\int_{-\infty}^{+\infty} \dfrac{dx}{1+4x^2}.$

11. Prove that the existence and the value of the improper integral (2), § 413, does not depend on the value of c in case (iv) of Note 2, § 413. *Hint:* Let $a + \epsilon < c < d < b - \eta$. Then
$$\int_{a+\epsilon}^c + \int_c^{b-\eta} = \int_{a+\epsilon}^c + \int_c^d + \int_d^{b-\eta} = \int_{a+\epsilon}^d + \int_d^{b-\eta}.$$

12. Prove the independence of c of the improper integrals referred to in the two Notes of § 414. (Cf. Ex. 11.) Illustrate each by an example.

13. Let $f(v)$ be continuous for $c < v < d$, let $v(x)$ have values between c and d, and a continuous derivative $v'(x)$, for $a < x < b$, and assume that $v(x) \to c$ as $x \to a+$ and $v(x) \to d$ as $x \to b-$. Prove that if the integral $\int_c^d f(v)\, dv$ converges (whether proper or improper) then the left-hand member in the following integration by substitution formula also converges and

(1) $$\int_a^b f(v(x))v'(x)\, dx = \int_c^d f(v)\, dv.$$

Evaluate the integral of Example 4, § 413, by use of this formula and the substitution $x = 2 \sin v$. Illustrate with other examples. *Hint:* If $F(v)$ is a primitive of $f(v)$, $c < v < d$,
$$\int_{a+\epsilon}^{b-\eta} f(v(x))v'(x)\, dx = F(v(b-\eta)) - F(v(a+\epsilon)).$$

14. Discuss the integration by substitution formula (1) for cases involving infinite intervals. Illustrate with examples.

In Exercises 15–24, merely establish convergence or divergence. Do not try to evaluate. (Cf. §§ 1408, 1419.)

15. $\int_0^{+\infty} \dfrac{dx}{\sqrt{1+x^4}}.$

16. $\int_2^{+\infty} \dfrac{x\, dx}{\sqrt{x^4-1}}.$

17. $\int_{-\infty}^{+\infty} e^{-x^2}\, dx.$

18. $\int_0^1 \dfrac{\ln x\, dx}{\sqrt{x}}.$

19. $\int_0^1 \dfrac{dx}{\sqrt{x}\ln x}.$

20. $\int_{-\infty}^0 e^x \ln |x|\, dx.$

21. $\int_{-\infty}^{+\infty} \dfrac{2x\, dx}{e^x - e^{-x}}.$

22. $\int_0^1 \dfrac{\ln x\, dx}{1-x}.$

23. $\displaystyle\int_{-1}^{1} e^{1/x}\,dx.$

24. $\displaystyle\int_{0}^{1} \sqrt[3]{x \ln(1/x)}\,dx.$

25. Prove Theorem II, §415.

26. Prove that if p is an arbitrary positive number, then

(2) $\qquad \ln x = o(x^p), \quad x = o(e^{px}), \quad \text{as} \quad x \to +\infty;$

(3) $\qquad \ln x = o(x^{-p}), \quad \dfrac{1}{x} = o(e^{p/x}), \quad \text{as} \quad x \to 0+.$

27. Show that the functions in any "big O" relationship can be multiplied or divided by any function whose values are positive: If $f(x) = O(g(x))$ and $h(x) > 0$, then $f(x)h(x) = O(g(x)h(x))$; conversely, if $f(x)h(x) = O(g(x)h(x))$, then $f(x) = O(g(x))$. Is the same thing true for the "little o" relation?

28. Prove that $\displaystyle\int_{2}^{+\infty} \dfrac{dx}{x(\ln x)^p}$ converges if and only if $p > 1$.

29. Prove that $\displaystyle\int_{3}^{+\infty} \dfrac{dx}{x(\ln x)(\ln \ln x)^p}$ converges if and only if $p > 1$. More generally, prove that $\displaystyle\int_{a}^{+\infty} \dfrac{dx}{x(\ln x)(\ln \ln x) \cdots (\ln \ln \cdots \ln x)^p}$, where a is sufficiently large, converges if and only if $p > 1$.

★30. The **Cauchy principal value** of the improper integral $\displaystyle\int_{-\infty}^{+\infty} f(x)\,dx$ is denoted and defined

$$(P)\int_{-\infty}^{+\infty} f(x)\,dx \equiv \lim_{u \to +\infty} \int_{-u}^{u} f(x)\,dx,$$

provided the integral $\displaystyle\int_{-u}^{u} f(x)\,dx$ and its limit exist. Prove that whenever the improper integral $\displaystyle\int_{-\infty}^{+\infty} f(x)\,dx$ converges, its Cauchy principal value exists and is equal to it. Give examples to show that the converse is false. (Cf. Ex. 31.)

★31. For each of the cases (i) and (ii), below, define a Cauchy principal value of the improper integral $\displaystyle\int_{-a}^{a} f(x)\,dx$ assuming $f(x)$ is integrable on $[u, v]$ for every u and v such that

(i) $-a < u < v < a;$ (ii) $-a \leq u < v < 0$ and $0 < u < v \leq a.$

Give a sufficient condition for the existence of the integral in each case. (Cf. Ex. 30.)

In Exercises 32–35, establish the given relation.

★32. $\displaystyle\int_{0}^{1} \dfrac{x^{p-1}}{x+1}\,dx = \int_{1}^{+\infty} \dfrac{x^{-p}\,dx}{x+1}, \; p > 0.$

★33. $\displaystyle\int_{0}^{+\infty} \dfrac{x^{p-1}\,dx}{x+1} = \int_{0}^{+\infty} \dfrac{x^{-p}\,dx}{x+1}, \; 0 < p < 1.$

★34. $\displaystyle\int_{0}^{1} x^{p-1}(1-x)^q\,dx = \dfrac{q}{p}\int_{0}^{1} x^p(1-x)^{q-1}\,dx, \; p > 0, q > 0.$

★35. $\displaystyle\int_{0}^{+\infty} \dfrac{dx}{(x+p)\sqrt{x}} = \dfrac{\pi}{\sqrt{p}}, \; p > 0.$

146   INTEGRATION   [§ 417

In Exercises 36 and 37, use integration by parts and mathematical induction to verify the formulas.

★36. Wallis's Formulas.

$$\int_0^1 \frac{x^n \, dx}{\sqrt{1-x^2}} = \int_0^{\frac{\pi}{2}} \sin^n x \, dx = \int_0^{\frac{\pi}{2}} \cos^n x \, dx$$

$$= \begin{cases} \dfrac{2 \cdot 4 \cdot 6 \cdots (n-1)}{3 \cdot 5 \cdot 7 \cdots n}, & \text{if } n \text{ is an odd integer} > 1; \\[2mm] \dfrac{1 \cdot 3 \cdot 5 \cdots (n-1)}{2 \cdot 4 \cdot 6 \cdots n} \cdot \dfrac{\pi}{2}, & \text{if } n \text{ is an even integer} > 0. \end{cases}$$

★37. $\int_0^{+\infty} x^n e^{-ax} \, dx = \dfrac{n!}{a^{n+1}}$ (n a positive integer, $a > 0$).

In Exercises 38–41, state and prove the analogue for improper integrals of the specified theorem of § 401.

★38. Theorem II.   **★39.** Theorem III.
★40. Theorem IV.   **★41.** Theorem V.

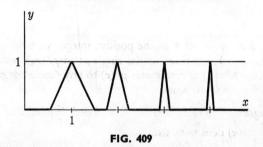

FIG. 409

★42. Show by an example that a continuous function improperly integrable on $[0, +\infty)$ need not have a zero limit at $+\infty$, in contrast to the fact that the general term of a convergent infinite series must tend toward zero (§ 1103). (For another example see Ex. 43.) *Hint:* See Figure 409.

★43. Prove that $\int_0^{+\infty} \cos x^2 \, dx$ converges, by integrating $\int_u^v \cos x^2 \, dx$ by parts to obtain $\dfrac{1}{2v} \sin v^2 - \dfrac{1}{2u} \sin u^2 + \dfrac{1}{2} \int_u^v \dfrac{\sin x^2}{x^2} \, dx$. Also consider $\int_0^{+\infty} x \cos x^4 \, dx$ (where the integrand is unbounded). (Cf. Ex. 42.)

★417. THE RIEMANN-STIELTJES INTEGRAL

A generalization of the definite integral, with useful applications in many applied fields including Physics and Statistics, is given in the following definition:

Definition I. *Let $f(x)$ and $g(x)$ be defined and bounded on a closed interval $[a, b]$. Then $f(x)$ is **Riemann-Stieltjes integrable** with respect to $g(x)$ on*

[a, b], with **Riemann-Stieltjes integral I**, if and only if corresponding to $\epsilon > 0$ there exists $\delta > 0$ such that for any net
$$\mathfrak{N} = \{a = a_0 < a_1 < \cdots < a_n = b\}$$
of norm $|\mathfrak{N}| < \delta$ and any choice of points x_i, $a_{i-1} \leq x_i \leq a_i$, $i = 1, 2, \cdots, n$:

(1) $$\left| \sum_{i=1}^{n} f(x_i)[g(a_i) - g(a_{i-1})] - I \right| < \epsilon.$$

Letting $\Delta g_i \equiv g(a_i) - g(a_{i-1})$ and $\int_a^b f(x) \, dg(x) \equiv I$, we write (1) in limit notation:

(2) $$\int_a^b f(x) \, dg(x) = \lim_{|\mathfrak{N}| \to 0} \sum_{i=1}^{n} f(x_i) \Delta g_i.$$

NOTE 1. If $g(x) \equiv x$, the Riemann-Stieltjes integral (2) reduces to the Riemann or definite integral:

(3) $$\int_a^b f(x) \, dg(x) = \int_a^b f(x) \, dx, \quad \text{if} \quad g(x) \equiv x.$$

Example 1. If $f(x)$ is defined and bounded on $[a, b]$ and continuous at $x = c$ ($a < c < b$), and if $g(x) \equiv 0$ for $a \leq x < c$ and $g(x) \equiv p$ for $c \leq x \leq b$, show that $f(x)$ is Riemann-Stieltjes integrable with respect to $g(x)$ on $[a, b]$, and
$$\int_a^b f(x) \, dg(x) = pf(c).$$

Solution. For any net $\mathfrak{N}$, let k be the positive integer such that $a_{k-1} < c \leq a_k$. Then the sum $\Sigma f(x_i) \Delta g_i$ reduces to the single term $pf(x_k)$, which (because of continuity of $f(x)$ at $x = c$) approximates $pf(c)$ to any desired degree of accuracy if the norm of $\mathfrak{N}$ is sufficiently small.

Example 2. If $f(x)$ and $g(x)$ are both $\equiv 0$ for $0 \leq x < \frac{1}{2}$ and $\equiv 1$ for $\frac{1}{2} \leq x \leq 1$, show that $\int_0^1 f(x) \, dg(x)$ does not exist.

Solution. For any net $\mathfrak{N}$, let k be the positive integer such that $a_{k-1} < \frac{1}{2} \leq a_k$. Then the sum $\Sigma f(x_i) \Delta g_i$ reduces to the single term $f(x_k)$, which has the value 0 or 1 according as $x_k < \frac{1}{2}$ or $x_k \geq \frac{1}{2}$. Thus, for $\epsilon = \frac{1}{2}$ there is no $\delta > 0$ guaranteeing the inequality (1), whatever the number I may be!

NOTE 2. Example 1, above, illustrates the effect of a jump discontinuity in the function $g(x)$ at a point of continuity of the function $f(x)$ (Ex. 9, §418), while Example 2 illustrates the general principle that whenever $f(x)$ and $g(x)$ have a common point of discontinuity the integral $\int_a^b f(x) \, dg(x)$ fails to exist (Ex. 10, §418).

The following theorem establishes the differential nature of the symbol $dg(x)$, and shows how certain Riemann-Stieltjes integrals can be written as standard Riemann integrals.

Theorem I. *If $f(x)$ is defined and $g(x)$ is differentiable at every point of a closed interval $[a, b]$, and if $f(x)$ and $g'(x)$ are integrable there, then $f(x)$ is Riemann-Stieltjes integrable with respect to $g(x)$ there, and*
$$\int_a^b f(x) \, dg(x) = \int_a^b f(x) g'(x) \, dx.$$

Proof. If $I \equiv \int_a^b f(x)g'(x)\,dx$ (cf. Theorem XIV, § 401), and $\epsilon > 0$, we wish to establish an inequality of the form (1), which can be rewritten with the aid of the Law of the Mean in the form

(4) $$\left|\sum_{i=1}^n f(x_i)g'(\xi_i)\Delta x_i - I\right| < \epsilon.$$

To this end we appeal to Bliss's Theorem (Theorem XV, § 401), and produce the required $\delta > 0$.

The following theorem shows the reciprocal relation between $f(x)$ and $g(x)$ in the definition of Riemann-Stieltjes integrability (cf. Ex. 11, § 418, for hints on a proof):

Theorem II. Integration by Parts. *If $f(x)$ is integrable with respect to $g(x)$ on $[a, b]$, then $g(x)$ is integrable with respect to $f(x)$ there and*

(5) $$\int_a^b f(x)\,dg(x) + \int_a^b g(x)\,df(x) = f(b)g(b) - f(a)g(a).$$

Example 3. Evaluate the Riemann-Stieltjes integral $\int_0^{\frac{\pi}{2}} x\,d\sin x$.

Solution. By Theorem I, this integral is equal to $\int_0^{\frac{\pi}{2}} x\cos x\,dx$. However, it is simpler to use the integration by parts formula (Theorem II) directly:

$$\int_0^{\frac{\pi}{2}} x\,d\sin x = \frac{\pi}{2}\sin\frac{\pi}{2} - 0\sin 0 - \int_0^{\frac{\pi}{2}} \sin x\,dx = \frac{\pi}{2} - 1.$$

Example 4. Evaluate the Riemann-Stieltjes integral $\int_0^3 e^{2x}\,d[x]$, where $[x]$ is the bracket function of § 201.

Solution. The integration by parts formula gives

$$\int_0^3 e^{2x}\,d[x] = 3e^6 - 0e^0 - \int_0^3 [x]\,de^{2x},$$

which, by Theorem I, is equal to

$$3e^6 - 2\int_1^2 e^{2x}\,dx - 4\int_2^3 e^{2x}\,dx = e^2 + e^4 + e^6.$$

The result could be obtained directly by using the fact that the function $[x]$ in the original form of the integral makes contributions only at its jumps, of amounts determined by the values of e^{2x} and the size of the jumps (cf. Example 1, and Exs. 8–9, § 418.)

The question of existence of the integral of a function $f(x)$ with respect to a function $g(x)$ in case the function $g(x)$ is not differentiable, or possibly not even continuous, naturally arises. The simplest useful sufficient condition corresponds to the condition of continuity of $f(x)$ for the existence of the definite integral of $f(x)$:

Theorem III. *If, on the interval $[a, b]$, one of the functions $f(x)$ and $g(x)$ is continuous and the other monotonic, the integral $\int_a^b f(x)\,dg(x)$ exists.*

Proof. Thanks to Theorem II and the similarity in behavior of monotonically increasing and decreasing functions, we shall assume without loss of generality that $f(x)$ is continuous and $g(x)$ monotonically increasing (with $g(b) > g(a)$ since the case $g(b) = g(a)$ is trivial).

Our object is to find a number I which is approximated by all possible sums $\Sigma f(x_i)(g(a_i) - g(a_{i-1})) = \Sigma f(x_i)\,\Delta g_i$ associated with nets of sufficiently small norm. For a given net $\mathfrak{N}$, let m_i and M_i be the minimum and maximum values, respectively, of the continuous function $f(x)$ for $a_{i-1} \leq x \leq a_i$, $i = 1, 2, \cdots, n$. For any choice of x_i such that $a_{i-1} \leq x_i \leq a_i$, $i = 1, 2, \cdots, n$, $m_i \leq f(x_i) \leq M_i$, and since $g(x)$ is assumed to be monotonically increasing, $\Delta g_i \geq 0$, $i = 1, 2, \cdots, n$. Therefore

$$(6) \quad \sum_{i=1}^n m_i\,\Delta g_i \leq \sum_{i=1}^n f(x_i)\,\Delta g_i \leq \sum_{i=1}^n M_i\,\Delta g_i.$$

The extreme left-hand and right-hand terms of (6) are called the **lower** and **upper sums** for the net $\mathfrak{N}$, and written $L(\mathfrak{N})$ and $U(\mathfrak{N})$, respectively.

In our quest for the desired number I, we shall define two numbers, I and J, and then prove that they are equal:

$$(7) \quad \begin{cases} I \equiv \sup \sum_{i=1}^n m_i\,\Delta g_i = \sup \text{ (all lower sums)}, \\ \\ J \equiv \inf \sum_{i=1}^n M_i\,\Delta g_i = \inf \text{ (all upper sums)}. \end{cases}$$

In order to establish the equality of the numbers I and J, we first affirm that $I \leq J$ and that this inequality is a consequence of the fact that *every lower sum is less than or equal to every upper sum:* If $\mathfrak{M}$ and $\mathfrak{N}$ are any two nets on $[a, b]$,

$$(8) \quad L(\mathfrak{M}) \leq U(\mathfrak{N}).$$

Reasonable though (8) may appear, it needs proof, and detailed hints are given in Exercise 12, § 418. In Exercise 13, § 418, the student is asked to prove that (8) implies $I \leq J$.

Finally, with $I \leq J$ established, we wish to show that $I = J$ and that if $\epsilon > 0$ there exists $\delta > 0$ such that

$$(9) \quad \left| \sum_{i=1}^n f(x_i)\,\Delta g_i - I \right| < \epsilon$$

whenever $|\mathfrak{N}| < \delta$. By use of the inequalities (6) and

$$(10) \quad L(\mathfrak{N}) \leq I \leq J \leq U(\mathfrak{N}),$$

the desired equality $I = J$ and inequality (9) follow from the fact that if δ is chosen so small that (thanks to the uniform continuity of $f(x)$)

$$|x' - x''| < \delta \quad \text{implies} \quad |f(x') - f(x'')| < \frac{\epsilon}{g(b) - g(a)},$$

then $|\mathfrak{N}| < \delta$ implies

$$U(\mathfrak{N}) - L(\mathfrak{N}) = \sum_{i=1}^{n}(M_i - m_i)\Delta g_i < \frac{\epsilon}{g(b) - g(a)} \sum_{i=1}^{n} \Delta g_i = \epsilon.$$

(Details are requested in Ex. 14, § 418.)

A few facts about the result of adding or subtracting functions or of multiplying by constants are herewith assembled (proofs are requested in Ex. 15, § 418):

Theorem IV. *If each of the functions $f(x), f_1(x)$, and $f_2(x)$ is (Riemann-Stieltjes) integrable with respect to each of the functions $g(x), g_1(x)$, and $g_2(x)$, and if c is any constant, then $f_1(x) + f_2(x)$ is integrable with respect to $g(x)$, $f(x)$ is integrable with respect to $g_1(x) + g_2(x)$, $cf(x)$ is integrable with respect to $g(x), f(x)$ is integrable with respect to $cg(x)$, and (with a simplified notation suggested by $\int f \, dg \equiv \int_a^b f(x) \, dg(x)$):*

(11) $$\int (f_1 + f_2) \, dg = \int f_1 \, dg + \int f_2 \, dg,$$

(12) $$\int f \, d(g_1 + g_2) = \int f \, dg_1 + \int f \, dg_2,$$

(13) $$\int cf \, dg = \int f \, d(cg) = c \int f \, dg.$$

Definition II. *If $b < a$,*

(14) $$\int_a^b f(x) \, dg(x) \equiv -\int_b^a f(x) \, dg(x),$$

in case the latter integral exists. Furthermore,

(15) $$\int_a^a f(x) \, dg(x) \equiv 0.$$

For integrals over adjoining intervals we have:

Theorem V. *For any three numbers $a, b,$ and c,*

(16) $$\int_a^c f(x) \, dg(x) = \int_a^b f(x) \, dg(x) + \int_b^c f(x) \, dg(x),$$

provided the three integrals exist.

Proof. We shall assume $a < b < c$ (cf. Exs. 16–17, § 418). For $\epsilon > 0$, let $\delta > 0$ be such that $|\mathfrak{N}| < \delta$ implies that any sum corresponding to any of the three integrals approximates that integral within $\frac{1}{3}\epsilon$. Then the sum of two such approximating sums for the two integrals on the right-hand side of (16) must approximate the left-hand side within $\frac{2}{3}\epsilon$. Therefore the two sides of (16) are constants differing by less than ϵ, and must consequently be equal.

NOTE 3. At a more advanced level of mathematical analysis than that customarily reached in a course in advanced calculus, the concept of *bounded variation* becomes important. *A function $f(x)$, defined on a closed interval $[a, b]$, is **of bounded variation** there if and only if, for all possible nets $\mathfrak{N}$, consisting of points $a = a_0 < a_1 < a_2 < \cdots < a_n = b$, the sums $\sum_{i=1}^{n} |f(a_i) - f(a_{i-1})|$ are bounded.* Although no attempt will be made to investigate this definition here, it should be of value to record the fact that *a function is of bounded variation on a closed interval if and only if it can be represented there as the difference between two monotonically increasing functions.* Therefore, by Theorem IV, the word *monotonic* in the statement of Theorem III can be replaced by the words *of bounded variation*. A step-function is a simple example of a function of bounded variation. For a discussion of bounded variation and its relation to the Riemann-Stieltjes integral, see *RV*, §§ 516–518, 1526.

*418. EXERCISES

The notation $[x]$ indicates the bracket function of § 201.
In Exercises 1–6, evaluate the Riemann-Stieltjes integral.

★1. $\displaystyle\int_0^1 x\, dx^2$.

★2. $\displaystyle\int_0^2 x\, d[x]$.

★3. $\displaystyle\int_0^{\pi/2} \cos x\, d\sin x$.

★4. $\displaystyle\int_0^3 e^x\, d\{x - [x]\}$.

★5. $\displaystyle\int_{-1}^1 e^x\, d|x|$.

★6. $\displaystyle\int_0^\pi x\, d|\cos x|$.

★7. Prove that if $g(x)$ is defined on $[a, b]$, then $\displaystyle\int_a^b dg(x)$ exists and is equal to $g(b) - g(a)$.

★8. Prove that if $f(x)$ is continuous on $[0, n]$, then $\displaystyle\int_0^n f(x)\, d[x] = f(1) + f(2) + \cdots + f(n)$.

★9. For a net $\mathfrak{N}$: $a = a_0 < a_1 < \cdots < a_n = b$, let $g(x)$ be a step-function constant for $a_{i-1} < x < a_i$, $i = 1, 2, \cdots, n$, and having jumps $J_0 \equiv g(a+) - g(a)$, $J_i \equiv g(a_i+) - g(a_i-)$, $i = 1, 2, \cdots, n-1$, and $J_n \equiv g(b) - g(b-)$. If $f(x)$ is continuous on $[a, b]$, prove that

(1) $$\int_a^b f(x)\, dg(x) = \sum_{i=0}^n J_i f(a_i).$$

Hint: Since the integral exists (Note 3, § 417) choose a net of arbitrarily small norm such that each a_i, $i = 1, 2, \cdots, n-1$, is interior to some subinterval. Then let each a_i, $i = 0, 1, \cdots, n$, be a point chosen for evaluating $f(x)$.

★10. Prove that if $f(x)$ and $g(x)$ are both discontinuous at a point c, where $a \leq c \leq b$, then $\displaystyle\int_a^b f(x)\, dg(x)$ does not exist. *Hint:* For $a < c < b$, and c nonremovable for $g(x)$, let $\mathfrak{N}$ be a net not containing c and let $a_{k-1} < c < a_k$. Then $|\Delta g_k|$ can always be made $\geq \eta$, a fixed positive number. Also, for two suitable numbers, x_k and x_k', in $[a_{k-1}, a_k]$, $|f(x_k) - f(x_k')| \geq \xi$, a fixed positive number. Hence, for any $\delta > 0$, there is a net $\mathfrak{N}$ of norm $< \delta$, and two sums $\Sigma f(x_i)\,\Delta g_i$ which differ numerically by at least $\eta \xi$ (for $i \neq k$, let $x_i = x_i'$).

*11. Prove Theorem II, § 417. *Hint:* Expansion of $\sum_{i=1}^{n} g(x_i)[f(a_i) - f(a_{i-1})]$ and rearrangement of terms leads to the following identical formulation, where $x_0 \equiv a$ and $x_{n+1} \equiv b$:
$$f(b)g(b) - f(a)g(a) - \sum_{i=0}^{n} f(a_i)(g(x_{i+1}) - g(x_i)).$$

*12. Prove (8), § 417. *Hint:* If $\mathfrak{M}$ and $\mathfrak{N}$ are arbitrary nets on $[a, b]$, let $\mathfrak{P}$ be the net consisting of all points appearing in either $\mathfrak{M}$ or $\mathfrak{N}$ (or both). Then, since $\mathfrak{P}$ contains both $\mathfrak{M}$ and $\mathfrak{N}$,
$$L(\mathfrak{M}) \leq L(\mathfrak{P}) \leq U(\mathfrak{P}) \leq U(\mathfrak{N}).$$

*13. Prove that (8), § 417, implies $I \leq J$.

*14. Supply the details requested at the end of the proof of Theorem III, § 417.

*15. Prove Theorem IV, § 417.

*16. Show that formula (16), Theorem V, § 417, holds whatever the order relation between a, b, and c may be.

*17. Show by the example
$$f(x) \equiv 0 \text{ for } 0 \leq x < 1, \quad f(x) \equiv 1 \text{ for } 1 \leq x \leq 2,$$
$$g(x) \equiv 0 \text{ for } 0 \leq x \leq 1, \quad g(x) \equiv 1 \text{ for } 1 < x \leq 2$$
that the existence of the integrals on the right-hand side of (16), Theorem V, § 417, do not imply the existence of the integral on the left-hand side.

*18. Prove the *integration by substitution* formula for Riemann-Stieltjes integrals: If $\phi(t)$ is continuous and strictly monotonic on $[a, b]$, and if $\phi(a) = c$ and $\phi(b) = d$, then the equality
$$(2) \qquad \int_c^d f(x)\, dg(x) = \int_a^b f(\phi(t))\, dg(\phi(t))$$
holds, the existence of either integral implying that of the other.

*19. Prove that if $f(x) \geq 0$ and $g(x)$ is monotonically increasing on $[a, b]$, then the integral $\int_a^b f(x)\, dg(x)$, if it exists, is nonnegative. State and prove a generalization of this for which $f_1(x) \leq f_2(x)$ implies $\int_a^b f_1(x)\, dg(x) \leq \int_a^b f_2(x)\, dg(x)$.

*20. Prove that if $f(x)$ is continuous and nonnegative on $[a, b]$ but not identically 0 there, and if $g(x)$ is strictly increasing on $[a, b]$, then
$$\int_a^b f(x)\, dg(x) > 0.$$
State and prove a generalization of this for which $f_1(x) \leq f_2(x)$ implies
$$\int_a^b f_1(x)\, dg(x) < \int_a^b f_2(x)\, dg(x)$$
(Cf. Ex. 19.)

*21. Prove the analogue of Exercise 20 for the case $f(x)$ continuous and positive on $[a, b]$, and $g(x)$ monotonically increasing but nonconstant there. Also generalize to $f_1(x)$ and $f_2(x)$.

*22. Prove the **First Mean Value Theorem** for Riemann-Stieltjes integrals (cf. Exs. 5–6, § 402): *If $f(x)$ is continuous and $g(x)$ is monotonic (strictly monotonic) on $[a, b]$, then there exists a point ξ such that $a \leq \xi \leq b$ $(a < \xi < b)$, and such that*
$$(3) \qquad \int_a^b f(x)\, dg(x) = f(\xi)[g(b) - g(a)].$$

Hint: Assume $g(x)$ increasing, and let $M \equiv \sup f(x)$ and $m \equiv \inf f(x)$ for $a \leq x \leq b$. Then
$$m[g(b) - g(a)] \leq \int_a^b f(x)\, dg(x) \leq M[g(b) - g(a)].$$
(See Ex. 20 for the strict inequalities.)

*23. Show by an example that equation (3) may not hold for $a < \xi < b$ if $g(x)$ is not *strictly* monotonic. *Hint:* Let $g(x) \equiv [x]$ on $[0, 1]$.

*24. Prove the **Second Mean Value Theorem** for Riemann-Stieltjes integrals (cf. Ex. 14, § 407): *If $f(x)$ is monotonic (strictly monotonic) and $g(x)$ is continuous on $[a, b]$, then there exists a point ξ such that $a \leq \xi \leq b$ ($a < \xi < b$) and such that*

(4) $$\int_a^b f(x)\, dg(x) = f(a)[g(\xi) - g(a)] + f(b)[g(b) - g(\xi)].$$

Hint: Use (3), Exercise 22, and the integration by parts formula (Theorem II, § 417).

*25. Prove the following form of the Second Mean Value Theorem for Riemann integrals (cf. Ex. 14, § 407): *If $f(x)$ is monotonic (strictly monotonic) and if $h(x)$ is continuous on $[a, b]$, then there exists a point ξ such that $a \leq \xi \leq b$ ($a < \xi < b$) such that*

(5) $$\int_a^b f(x)h(x)\, dx = f(a) \int_a^\xi h(x)\, dx + f(b) \int_\xi^a h(x)\, dx.$$

Hint: Let $g(x)$ be a primitive of $h(x)$, and use (4). (Cf. Ex. 28.)

*26. Prove the following Bonnet form of the Second Mean Value Theorem for Riemann integrals: *If $f(x) \geq 0$ and is monotonically decreasing (or, alternatively, $f(x) \leq 0$ and is monotonically increasing) and if $h(x)$ is continuous on $[a, b]$, then there exists a point ξ on $[a, b]$ such that*

(6) $$\int_a^b f(x)h(x)\, dx = f(a) \int_a^\xi h(x)\, dx.$$

If $f(x) \geq 0$ and is monotonically increasing (or, alternatively, $f(x) \leq 0$ and is monotonically decreasing) and if $h(x)$ is continuous on $[a, b]$, then there exists a point ξ of $[a, b]$ such that

(7) $$\int_a^b f(x)h(x)\, dx = f(b) \int_\xi^b h(x)\, dx.$$

Hint: Use (5), redefining $f(x)$ to be 0 at either b or a. (Cf. Ex. 28.)

*27. Prove that if $f(x)$ and $h(x)$ are Riemann integrable on a closed interval $[a, b]$, and if the function $g(x)$ is defined on $[a, b]$ by the formula:
$$g(x) \equiv \int_a^x h(t)\, dt,$$
then $f(x)$ is Riemann-Stieltjes integrable with respect to $g(x)$, and
$$\int_a^b f(x)\, dg(x) = \int_a^b f(x)h(x)\, dx.$$

Hint: Show first that it may be assumed without loss of generality that $f(x) \geq 0$, with $0 \leq f(x) < K$, and let $\epsilon > 0$. Next find step-functions σ and τ such that $\sigma(x) \leq h(x) \leq \tau(x)$ and $\int (\tau - \sigma)\, dx < \epsilon/2K$. Then let $\delta > 0$ be such that $|\mathfrak{N}| < \delta$

implies $|\Sigma f(x_i)\sigma(x_i') \Delta x_i - \int f\sigma \, dx| < \frac{1}{2}\epsilon$ and $|\Sigma f(x_i)\tau(x_i'') \Delta x_i - \int f\tau \, dx| < \frac{1}{2}\epsilon$. Then (in part)

$$\Sigma f(x_i)\Delta g_i = \Sigma f(x_i)\int_{a_{i-1}}^{a_i} h(x)dx \leq \Sigma f(x_i)\tau(x_i'') \Delta x_i < \int f\tau \, dx + \tfrac{1}{2}\epsilon < \int fh \, dx + \epsilon.$$

*28. Use Exercise 27 to show that in the Second Mean Value Theorems of Exercises 25 and 26 the assumption that $h(x)$ is continuous may be replaced by the weaker assumption that $h(x)$ is merely integrable.

5

Some Elementary Functions

*501. THE EXPONENTIAL AND LOGARITHMIC FUNCTIONS

The function x^n has been defined for integral exponents and nonzero x in Chapter 1 (Ex. 7, § 109), and for positive rational exponents and positive x in Chapter 2 (Ex. 21, § 216). In this section and the Exercises of the next, we study the expression a^b ($a > 0$, b real) and the logarithmic function.

One method of procedure is to define the exponential function a^x, first for integral exponents, then for all rational exponents, and finally by a limiting process, for all real exponents. Having defined the exponential function one can then define and study its inverse, the logarithmic function. For simplicity, we have chosen to reverse the order and define the logarithmic function first, by means of an integral, and then the exponential function as its inverse. We finally show that the exponential expression thus obtained agrees with the special cases previously studied. All of this is presented in the following section, by means of exercises arranged in logical order. (For another method cf. § 1312.)

*502. EXERCISES

*1. Prove that the logarithmic function *defined* by the equation

(1) $$\ln x \equiv \int_1^x \frac{dt}{t}, \quad x > 0,$$

has the following five properties (cf. Fig. 501):
 (i) $\ln 1 = 0$;
 (ii) $\ln x$ is strictly increasing;
 (iii) $\ln x$ is continuous and, in fact, differentiable, with derivative $\frac{1}{x}$;
 (iv) $\lim\limits_{x \to +\infty} \ln x = +\infty$;
 (v) $\lim\limits_{x \to 0+} \ln x = -\infty$.

Hints for (iv) *and* (v):

$$\int_1^{2^n} \frac{dt}{t} = \int_1^2 \frac{dt}{t} + \int_2^4 \frac{dt}{t} + \int_4^8 \frac{dt}{t} + \cdots + \int_{2^{n-1}}^{2^n} \frac{dt}{t}$$

$$> \int_1^2 \frac{dt}{2} + \int_2^4 \frac{dt}{4} + \cdots + \int_{2^{n-1}}^{2^n} \frac{dt}{2^n} = \frac{n}{2}.$$

$$\int_{1/n}^1 \frac{dt}{t} = \int_n^1 \frac{d(1/u)}{1/u} = \int_1^n \frac{du}{u}.$$

★2. Prove the following laws for $\ln x$:
  (*i*) $\ln (xy) = \ln x + \ln y$, $x > 0, y > 0$;
  (*ii*) $\ln \left(\dfrac{x}{y}\right) = \ln x - \ln y$, $x > 0, y > 0$;
  (*iii*) $\ln (x^n) = n \ln x$, $x > 0$, n an integer;
  (*iv*) $\ln (\sqrt[n]{x}) = \dfrac{1}{n} \ln x$, $x > 0$, n a positive integer;
  (*v*) $\ln (x^r) = r \ln x$, $x > 0$, r rational.

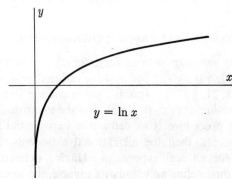

FIG. 501

(Cf. § 214 and Ex. 21, § 216.) It follows from Exercise 6, below, that (*v*) holds for any real *r*. *Hints*:
  (*i*): $\displaystyle\int_1^{xy} \dfrac{dt}{t} = \int_1^x \dfrac{dt}{t} + \int_x^{xy} \dfrac{dt}{t} = \int_1^x \dfrac{dt}{t} + \int_x^{xy} \dfrac{d(t/x)}{t/x}$;
  (*ii*): use (*i*) with $x = y \cdot \dfrac{x}{y}$;
  (*iv*): use (*iii*) with $x = (\sqrt[n]{x})^n$.

★3. Let the function $e^x = \exp(x)$ be *defined* as the inverse of the logarithmic function:
$$y = e^x = \exp(x) \quad \text{if and only if} \quad x = \ln y.$$

Prove that e^x has the following three properties (cf. Fig. 502):
  (*i*) e^x is defined, positive, and strictly increasing for all real x;
  (*ii*) $e^0 = 1$, $\displaystyle\lim_{x \to +\infty} e^x = +\infty$, $\displaystyle\lim_{x \to -\infty} e^x = 0$;
  (*iii*) e^x is continuous and, in fact, differentiable, with derivative e^x.

★4. Prove the following laws for e^x:
  (*i*) $e^x e^y = e^{x+y}$,
  (*ii*) $\dfrac{e^x}{e^y} = e^{x-y}$,
  (*iii*) if r is rational, $(e^x)^r = e^{rx}$.

Hints: (*i*): let $a \equiv e^x$, $b \equiv e^y$, $c \equiv e^{x+y}$; then $x = \ln a$, $y = \ln b$, $x + y = \ln c = \ln (ab)$; (*ii*): use (*i*) with e^{x-y} and e^y.

★5. Define the number e as the value of e^x, for $x = 1$, $e \equiv \exp(1)$, or, equivalently, as the number whose (natural) logarithm is 1. Prove that if x is rational, then the exponential function e^x, defined in Exercise 3, is identical with the function e^x previously defined in Exercise 21, § 216.

*6. For an arbitrary positive base a, define the exponential function:
$$a^x \equiv e^{x \ln a}.$$
Prove the following ten properties of a^x:
  (i) a^x is defined and positive for all real x;
  (ii) if $a = 1$, $a^x = 1$ for all x;
  (iii) if $a = e$, $a^x = e^x$ (Ex. 3), for all x;
  (iv) if x is rational, a^x, as defined here, is identical with the function a^x previously defined in Exercise 21, § 216;

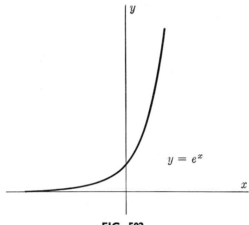

FIG. 502

  (v) if $a > 1$ ($a < 1$), a^x is strictly increasing (decreasing);
  (vi) $a^x a^y = a^{x+y}$;
  (vii) $\dfrac{a^x}{a^y} = a^{x-y}$;
  (viii) $(a^x)^y = a^{xy}$;
  (ix) $a^x b^x = (ab)^x$, $a > 0$, $b > 0$;
  (x) a^x is continuous and, in fact, differentiable, with derivative $\ln a \cdot a^x$.

*7. The function x^a, for $x > 0$ and arbitrary real a, is defined as in Exercise 6:
$$x^a \equiv e^{a \ln x}.$$
Prove the following properties of x^a:
  (i) If $a > 0$ ($a < 0$), x^a is strictly increasing (decreasing);
  (ii) if $a > 0$ ($a < 0$), $\lim\limits_{x \to +\infty} x^a = +\infty$ (0) and $\lim\limits_{x \to 0+} x^a = 0$ $(+\infty)$;
  (iii) x^a is continuous and, in fact, differentiable, with derivative ax^{a-1}.

*8. If $a > 0$, prove that the function x^a, defined as in Exercise 7 for $x > 0$, and defined to be 0 when $x = 0$, has the three properties stated in Exercise 7, except that if $0 < a < 1$, the derivative of x^a at $x = 0$ is $+\infty$.

*9. Prove that $e = \lim\limits_{x \to 0} (1 + x)^{\frac{1}{x}}$. *Hint:* At $x = 1$,
$$\frac{d}{dx} \ln x = \lim_{h \to 0} \frac{\ln(1+h) - \ln(1)}{h} = \lim_{h \to 0} \ln(1+h)^{\frac{1}{h}} = 1.$$

⋆10. Prove that $\lim_{x \to 0} (1 + ax)^{\frac{1}{x}} = e^a$.

⋆11. Define $\log_a x$, where $a > 0$ and $a \neq 1$, $x > 0$, as the inverse of the function a^x. That is, $\log_a x$ is that unique number y such that $a^y = x$. Prove the standard laws of logarithms, and the change of base formulas:
$$\log_a x = \log_a b \, \log_b x = \frac{\log_b x}{\log_b a}.$$
Prove that $a^{\log_a x} = \log_a (a^x) = x$.

⋆503. THE TRIGONOMETRIC FUNCTIONS

In a first course in trigonometry the six basic trigonometric functions are defined. Their definitions there and their subsequent treatment in calculus, however, are usually based on geometric arguments and intuitive appeal unfortified by a rigorous analytic background. It is the purpose of this section and the following section of exercises to present purely analytic definitions and discussion of the trigonometric functions. That these definitions correspond to those of the reader's previous experience is easily shown after the concept of arc length is available. (Cf. § 806.)

The development here is restricted to the sine and cosine functions, since the remaining four trigonometric functions are readily defined in terms of those two. Furthermore, the calculus properties of the other four are immediately obtainable, once they are established for $\sin x$ and $\cos x$. These properties, as well as those of the inverse trigonometric functions, will be assumed without specific formulation here. (For another method cf. § 1312.)

⋆504. EXERCISES

⋆1. Prove that the function defined by the equation
$$\text{Arcsin } x \equiv \int_0^x \frac{dt}{\sqrt{1 - t^2}}, \quad -1 < x < 1,$$
is strictly increasing, continuous, and, in fact, differentiable with derivative $(1 - x^2)^{-\frac{1}{2}}$. (Fig. 503.)

⋆2. Let the function $\sin x$ be *defined* as the inverse of the function prescribed in Exercise 1:
$$y = \sin x \quad \text{if and only if} \quad x = \text{Arcsin } y, \quad \text{for} \quad -\text{Arcsin } \tfrac{3}{4} < x < \text{Arcsin } \tfrac{3}{4}.$$
If the number π is defined: $\pi \equiv 4 \text{ Arcsin } \tfrac{1}{2}\sqrt{2}$, and if the number b is defined: $b \equiv \text{Arcsin } \tfrac{3}{4}$, prove that over the interval $(-b, b)$, containing $\tfrac{1}{4}\pi$, $\sin x$ is strictly increasing, continuous, and, in fact, differentiable with derivative $(1 - \sin^2 x)^{\frac{1}{2}} \equiv [1 - (\sin x)^2]^{\frac{1}{2}}$. (Fig. 504.) Also, $\sin 0 = 0$, $\sin \tfrac{1}{4}\pi = \tfrac{1}{2}\sqrt{2}$.

⋆3. Let the function $\cos x$ be *defined*, on the interval $(-b, b)$ of Exercise 2, as the positive square root of $1 - \sin^2 x$. Prove that $\cos x$ is differentiable and that the derivatives of $\sin x$ and $\cos x$ are $\cos x$ and $-\sin x$, respectively. Also, $\cos 0 = 1$ and $\cos \tfrac{1}{4}\pi = \tfrac{1}{2}\sqrt{2}$. *Hint:* If $y = \cos x \neq 0$, $y^2 = 1 - \sin^2 x$, $2 \cos x \, dy/dx = -2 \sin x \cos x$.

*4. Let $s(x)$ and $c(x)$ be any two functions defined and differentiable on an open interval $(-a, a)$, $a > 0$, and possessing there the following four properties:

(1) $\qquad (s(x))^2 + (c(x))^2 = 1,$

(2) $\qquad \dfrac{d}{dx}(s(x)) = c(x),$

(3) $\qquad \dfrac{d}{dx}(c(x)) = -s(x),$

(4) $\qquad c(0) = 1.$

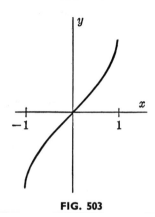

FIG. 503

Prove that if α and β are any two numbers which, together with their sum $\alpha + \beta$, belong to the interval $(-a, a)$, then the following two identities hold:

(5) $\qquad s(\alpha + \beta) = s(\alpha)c(\beta) + c(\alpha)s(\beta),$
(6) $\qquad c(\alpha + \beta) = c(\alpha)c(\beta) - s(\alpha)s(\beta).$

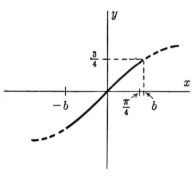

FIG. 504

Hence prove that if α and 2α belong to $(-a, a)$, then
(7) $\qquad s(2\alpha) = 2s(\alpha)c(\alpha),$
(8) $\qquad c(2\alpha) = (c(\alpha))^2 - (s(\alpha))^2.$

Hint: For (5), let $\gamma \equiv \alpha + \beta$, and prove that the function $s(x)c(\gamma - x) + c(x)s(\gamma - x)$ has an identically vanishing derivative and is therefore a constant. Then let $x = \alpha$ and $x = 0$, in turn.

*5. Let $s(x)$ and $c(x)$ be any two functions satisfying (1)–(4) of Exercise 4, on the open interval $(-a, a)$, $a > 0$. Define two new functions in the open interval $(-2a, 2a)$ by means of the formulas

(9) $\quad\quad\quad\quad\quad\quad S(x) \equiv 2s(\tfrac{1}{2}x)c(\tfrac{1}{2}x),$

(10) $\quad\quad\quad\quad\quad\quad C(x) \equiv (c(\tfrac{1}{2}x))^2 - (s(\tfrac{1}{2}x))^2.$

Prove that $S(x) = s(x)$ and $C(x) = c(x)$ on $(-a, a)$, and that properties (1)–(4), and therefore also (5)–(8), hold for $S(x)$ and $C(x)$ on $(-2a, 2a)$ (where s and c are replaced by S and C, respectively).

*6. Prove that a repeated application of the extension definitions of Exercise 5 provide two functions, $\sin x$ and $\cos x$, defined and differentiable for all real numbers, possessing properties (1)–(8) (where $s(x)$ and $c(x)$ are replaced by $\sin x$ and $\cos x$, respectively), and agreeing with the originally defined $\sin x$ and $\cos x$ on the interval $(-b, b)$ of Exercise 2.

*7. Prove that $\sin x$ and $\cos x$, as defined for all real numbers x, are odd and even functions, respectively. (Cf. Exs. 7–8, § 402.)

*8. Prove that $\sin \tfrac{1}{2}\pi = 1$, and $\cos \tfrac{1}{2}\pi = 0$, and that $\tfrac{1}{2}\pi$ is the smallest positive number whose sine is 1 and also the smallest positive number whose cosine is 0. Hence show that in the closed interval $[0, \tfrac{1}{2}\pi]$, $\sin x$ and $\cos x$ are strictly increasing and decreasing, respectively. *Hint:* If $2k \leq \tfrac{1}{2}\pi$, and if $\sin 2k = 1$, then $\cos 2k = \cos^2 k - \sin^2 k = 0$, and $\cos k = \sin k = \tfrac{1}{2}\sqrt{2}$, and $k = \tfrac{1}{4}\pi$.

*9. Prove that $\sin \pi = 0$ and $\cos \pi = -1$, and that π is the smallest positive number whose sine is 0 and also the smallest positive number whose cosine is -1.

*10. Prove that $\sin 2\pi = 0$ and $\cos 2\pi = 1$, and that 2π is the smallest positive number whose cosine is 1.

*11. Prove that $\sin x$ and $\cos x$ are periodic with period 2π. That is, $\sin(x + 2\pi) = \sin x$ and $\cos(x + 2\pi) = \cos x$ for all x, and if either $\sin(x + k) = \sin x$ or $\cos(x + k) = \cos x$ for all x, then $k = 2n\pi$, for some integer n.

*12. Prove that $\sin x$ and $\cos x$ have the familiar signs in the appropriate "quadrants"; namely, $+, +, -, -,$ and $+, -, -, +$, respectively, for x in the ranges $(0, \tfrac{1}{2}\pi), (\tfrac{1}{2}\pi, \pi), (\pi, \tfrac{3}{2}\pi), (\tfrac{3}{2}\pi, 2\pi)$.

*13. Prove that if λ and μ are any two real numbers such that $\lambda^2 + \mu^2 = 1$, then there is a unique value of x in the half open interval $[0, 2\pi)$ such that $\sin x = \lambda$ and $\cos x = \mu$. *Hint:* Let x_0 be the unique number in the closed interval $[0, \tfrac{1}{2}\pi]$ whose sine is $|\lambda|$, and choose for x the appropriate number according to Exercise 12: $x_0, \pi - x_0, \pi + x_0, 2\pi - x_0$.

*14. Prove that $-1 \leq \sin x \leq 1$ and $-1 \leq \cos x \leq 1$ for all x.

*15. Prove that $\sin x$ and $\cos x$ have, everywhere, continuous derivatives of all orders.

*16. Prove that $\lim\limits_{x \to 0} \dfrac{\sin x}{x} = 1$.

Hint: Let $f(x) \equiv \sin x$. Then $f'(0) = \lim\limits_{h \to 0} \dfrac{f(0 + h) - f(0)}{h}$.

*17. Prove that the function $f(x) \equiv \dfrac{\sin x}{x}$, $x \neq 0$, $f(0) \equiv 1$, (*i*) is everywhere continuous; (*ii*) is everywhere differentiable; (*iii*) has everywhere a continuous nth derivative for all positive integers n.

505. SOME INTEGRATION FORMULAS

We start by asking two questions. (i) If $\int \frac{dx}{x} = \ln x + C$, and if $\ln x$ is defined only for $x > 0$, how does one evaluate the simple integral $\int_{-3}^{-2} \frac{dx}{x}$?

(ii) If $\int \frac{dx}{a^2 - x^2} = \frac{1}{2a} \ln \frac{a+x}{a-x} + C$ and $\int \frac{dx}{x^2 - a^2} = \frac{1}{2a} \ln \frac{x-a}{x+a} + C'$,

where $a > 0$, why cannot each of these integration formulas be obtained from the other by a mere change in sign? In other words, why does their sum give formally the result

$$\frac{1}{2a} \ln \left(\frac{a+x}{a-x} \cdot \frac{x-a}{x+a} \right) + C + C' \quad \text{or} \quad \frac{1}{2a} \ln(-1) + C + C'$$

which does not even exist!

The answers to these questions lie most simply in the use of absolute values. We know that if $x > 0$, then $\frac{d}{dx}(\ln(x)) = \frac{1}{x}$ and that if $x < 0$, then $\frac{d}{dx}(\ln(-x)) = \frac{1}{-x}(-1) = \frac{1}{x}$. That is, $\ln x$ and $\ln(-x)$ both have the same derivative, formally, but have completely distinct domains of definition, $\ln x$ being defined for $x > 0$ and $\ln(-x)$ for $x < 0$. The function $\ln |x|$ encompasses both and is defined for any nonzero x. Thus the single integration formula

(1) $$\int \frac{dx}{x} = \ln |x| + C$$

is applicable under all possible circumstances. The integration of question (i) is therefore simple: $\int_{-3}^{-2} \frac{dx}{x} = \left[\ln |x| \right]_{-3}^{-2} = \ln |-2| - \ln |-3| = \ln \frac{2}{3}$.

For similar reasons the formulas of question (ii) are only apparently incompatible, since they apply to different domains of definition. The first is applicable if $x^2 < a^2$ and the second if $x^2 > a^2$. (Ex. 11, § 506.) However, again a single integration formula is available which is universally applicable, and can be written alternatively in the two forms:

(2) $$\int \frac{dx}{a^2 - x^2} = \frac{1}{2a} \ln \left| \frac{a+x}{a-x} \right| + C;$$

(2') $$\int \frac{dx}{x^2 - a^2} = \frac{1}{2a} \ln \left| \frac{x-a}{x+a} \right| + C.$$

The seeming paradox is resolved by the fact that $|-1| = 1$, whose logarithm certainly exists (although it vanishes).

More generally, any of the standard integration formulas that involve logarithms become universally applicable when absolute values are inserted.

The student should establish this fact in detail for the following formulas (Ex. 12, § 506). (In formulas (7)–(14), a represents a positive constant.)

(3) $\int \tan x \, dx = -\ln |\cos x| + C = \ln |\sec x| + C;$

(4) $\int \cot x \, dx = \ln |\sin x| + C = -\ln |\csc x| + C;$

(5) $\int \sec x \, dx = \ln |\sec x + \tan x| + C;$

(6) $\int \csc x \, dx = \ln |\csc x - \cot x| + C;$

(7) $\int \dfrac{dx}{\sqrt{x^2 \pm a^2}} = \ln |x + \sqrt{x^2 \pm a^2}| + C, |x| > a$ for the $-$ case;

(8) $\int \sqrt{x^2 \pm a^2} \, dx = \tfrac{1}{2}[x\sqrt{x^2 \pm a^2} \pm a^2 \ln |x + \sqrt{x^2 \pm a^2}|] + C, |x| > a$ for the $-$ case;

(9) $\int \dfrac{dx}{x\sqrt{a^2 \pm x^2}} = \dfrac{1}{a} \ln \left| \dfrac{\sqrt{a^2 \pm x^2} - a}{x} \right| + C, |x| < a$ for the $-$ case.

Since the derivative of $\sec x \tan x$ is

$\sec^3 x + \sec x \tan^2 x = \sec^3 x + \sec x(\sec^2 x - 1) = 2 \sec^3 x - \sec x,$

and since by (5), above, the derivative of $\ln |\sec x + \tan x|$ is $\sec x$, we have, by addition,

(10) $\int \sec^3 x \, dx = \tfrac{1}{2} \sec x \tan x + \tfrac{1}{2} \ln |\sec x + \tan x| + C.$

Finally, we give four more integration formulas, which involve inverse trigonometric functions. These are valid for the principal value ranges specified. The student should draw the graphs of the functions involved and verify the statements just made, as well as show that the formulas that follow are not all valid if the range of the inverse function is unrestricted (Ex. 13, § 506).

(11) $\int \dfrac{dx}{\sqrt{a^2 - x^2}} = \text{Arcsin} \dfrac{x}{a} + C, -a < x < a, -\dfrac{\pi}{2} \leq \text{Arcsin} \dfrac{x}{a} \leq \dfrac{\pi}{2};$

(12) $\int \dfrac{dx}{a^2 + x^2} = \dfrac{1}{a} \text{Arctan} \dfrac{x}{a} + C, -\dfrac{\pi}{2} < \text{Arctan} \dfrac{x}{a} < \dfrac{\pi}{2};$

(13) $\int \sqrt{a^2 - x^2} \, dx = \tfrac{1}{2}x\sqrt{a^2 - x^2} + \tfrac{1}{2}a^2 \text{Arcsin} \dfrac{x}{a} + C, |x| < a;$

(14) $\int \dfrac{dx}{x\sqrt{x^2 - a^2}} = \begin{cases} -\dfrac{1}{a} \text{Arcsin} \left(\dfrac{a}{x}\right) + C, x > a, \\ \dfrac{1}{a} \text{Arcsin} \left(\dfrac{a}{x}\right) + C, x < -a. \end{cases}$

506. EXERCISES

In Exercises 1–10, perform the integration, and specify any limitations on the variable x.

1. $\displaystyle\int \tan 5x \, dx.$

2. $\displaystyle\int \sec 4x \, dx.$

3. $\displaystyle\int \frac{dx}{\sqrt{2-x^2}}.$

4. $\displaystyle\int \sqrt{x^2 + 4x} \, dx.$

5. $\displaystyle\int \frac{dx}{\sqrt{4x^2 - 4x + 5}}.$

6. $\displaystyle\int \sqrt{x - 3x^2} \, dx.$

7. $\displaystyle\int \frac{dx}{x\sqrt{5 - 2x^2}}.$

8. $\displaystyle\int \frac{\sec^3 \sqrt{x}}{\sqrt{x}} \, dx.$

9. $\displaystyle\int \frac{dx}{3x^2 + 5x - 7}.$

10. $\displaystyle\int \frac{dx}{3x^2 + 5x + 7}.$

11. Show that in formula (2), § 505, the absolute value signs can be removed if $|x| < a$, and similarly for formula (2′), § 505, if $|x| > a$. *Hints:* The quotient of $a + x$ and $a - x$ is positive if and only if their product is positive.

12. Establish formulas (3)–(9), § 505, by direct evaluation rather than mere differentiation of the right-hand members. *Hints:* For (3)–(6) express the functions in terms of sines and cosines. For (7)–(9) make trigonometric substitutions and, if necessary, use (10), § 505.

13. Establish formulas (11)–(14), § 505, by direct evaluation. (Cf. Ex. 12.)

507. HYPERBOLIC FUNCTIONS

The hyperbolic functions, called *hyperbolic sine, hyperbolic cosine,* etc., are defined:

(1)
$$\begin{cases} \sinh x \equiv \dfrac{e^x - e^{-x}}{2}, & \coth x \equiv \dfrac{1}{\tanh x}, \\[4pt] \cosh x \equiv \dfrac{e^x + e^{-x}}{2}, & \operatorname{sech} x \equiv \dfrac{1}{\cosh x}, \\[4pt] \tanh x \equiv \dfrac{\sinh x}{\cosh x}, & \operatorname{csch} x \equiv \dfrac{1}{\sinh x}. \end{cases}$$

These six functions bear a close resemblance to the trigonometric functions. For example, $\sinh x$, $\tanh x$, $\coth x$, and $\operatorname{csch} x$ are odd functions and $\cosh x$ and $\operatorname{sech} x$ are even functions. (Ex. 21, § 509.) Furthermore, the hyperbolic functions satisfy identities that are similar to the basic trigonometric identities (verify the details in Ex. 22, § 509):

(2) $\cosh^2 x - \sinh^2 x = 1$;

(3) $1 - \tanh^2 x = \operatorname{sech}^2 x$;

(4) $\coth^2 x - 1 = \operatorname{csch}^2 x$;

(5) $\sinh(x \pm y) = \sinh x \cosh y \pm \cosh x \sinh y$;

(6) $\cosh(x \pm y) = \cosh x \cosh y \pm \sinh x \sinh y$;

(7) $\tanh(x \pm y) = \dfrac{\tanh x \pm \tanh y}{1 \pm \tanh x \tanh y}$;

(8) $\sinh 2x = 2 \sinh x \cosh x$;

(9) $\cosh 2x = \cosh^2 x + \sinh^2 x = 2 \cosh^2 x - 1 = 2 \sinh^2 x + 1$.

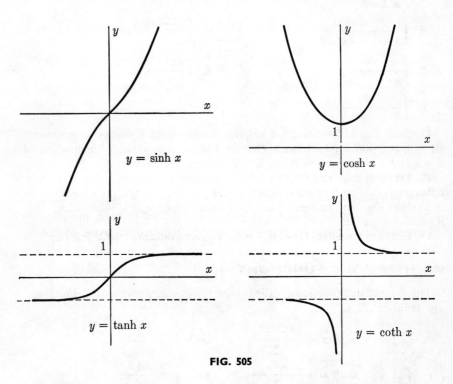

FIG. 505

The differentiation formulas (and therefore the corresponding integration formulas, which are omitted here) also have a familiar appearance (Ex. 23, § 509):

(10) $d(\sinh x)/dx = \cosh x$;

(11) $d(\cosh x)/dx = \sinh x$;

(12) $d(\tanh x)/dx = \operatorname{sech}^2 x$;

(13) $d(\coth x)/dx = -\operatorname{csch}^2 x$;

(14) $d(\operatorname{sech} x)/dx = -\operatorname{sech} x \tanh x$;

(15) $d(\operatorname{csch} x)/dx = -\operatorname{csch} x \coth x$.

The graphs of the first four hyperbolic functions are given in Figure 505.

A set of integration formulas (omitting those that are mere reformulations of the differentiation formulas (10)–(15)) follows (Ex. 24, § 509):

(16) $$\int \tanh x \, dx = \ln \cosh x + C;$$

(17) $$\int \coth x \, dx = \ln |\sinh x| + C;$$

(18) $$\int \operatorname{sech} x \, dx = \operatorname{Arctan}(\sinh x) + C;$$

(19) $$\int \operatorname{csch} x \, dx = \ln \left| \tanh \frac{x}{2} \right| + C.$$

NOTE. The trigonometric functions are sometimes called the **circular functions** because of their relation to a circle. For example, the parametric equations of the circle $x^2 + y^2 = a^2$ can be written $x = a \cos \theta$, $y = a \sin \theta$. In analogy with this, the hyperbolic functions are related to a hyperbola. For example, the parametric equations of the rectangular hyperbola $x^2 - y^2 = a^2$ can be written $x = a \cosh \theta$, $y = a \sinh \theta$. In this latter case, however, the parameter θ does not represent the polar coordinate angle for the point (x, y).

508. INVERSE HYPERBOLIC FUNCTIONS

The integrals of certain algebraic functions are expressed in terms of inverse trigonometric functions (§ 505). In a similar fashion, the integrals of certain other algebraic functions can be expressed in terms of inverse

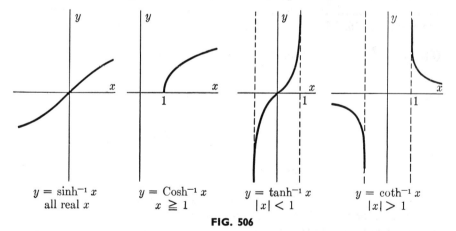

$y = \sinh^{-1} x$ $y = \operatorname{Cosh}^{-1} x$ $y = \tanh^{-1} x$ $y = \coth^{-1} x$
all real x $x \geq 1$ $|x| < 1$ $|x| > 1$

FIG. 506

hyperbolic functions. The four hyperbolic functions whose inverses are the most useful in this connection are the first four, $\sinh x$, $\cosh x$, $\tanh x$, and $\coth x$. The graphs and notation are given in Figure 506. For simplicity, by analogy with the principal value ranges for the inverse trigonometric functions, we choose only the nonnegative values of $\cosh^{-1} x$, and write for these principal values $\operatorname{Cosh}^{-1} x$.

The inverse hyperbolic functions can be expressed in terms of functions discussed previously:

(1) $\sinh^{-1} x = \ln(x + \sqrt{x^2 + 1})$, all real x;

(2) $\operatorname{Cosh}^{-1} x = \ln(x + \sqrt{x^2 - 1})$, $x \geq 1$;

(3) $\tanh^{-1} x = \frac{1}{2} \ln \frac{1+x}{1-x}$; $|x| < 1$;

(4) $\coth^{-1} x = \frac{1}{2} \ln \frac{x+1}{x-1}$; $|x| > 1$.

We prove (2) and leave the rest for the student (Ex. 25, § 509). If $x = \frac{e^y + e^{-y}}{2}$, let the positive quantity e^y be denoted by p. Then $2x = p + \frac{1}{p}$, or $p^2 - 2xp + 1 = 0$. Solving this equation we get $p = x \pm \sqrt{x^2 - 1}$. Since y is to be chosen nonnegative, $p = e^y \geq 1$. If $x > 1$, $x + \sqrt{x^2 - 1} > 1$, and since the product of this and $x - \sqrt{x^2 - 1}$ is equal to 1, $x - \sqrt{x^2 - 1} < 1$. We therefore reject the minus sign and have $p = x + \sqrt{x^2 - 1}$, or equation (2). Notice that the expressions in (3) and (4) exist for the specified values of x. For example, in (3), $(1 + x)$ and $(1 - x)$ have a positive quotient if and only if they have a positive product, and $1 - x^2 > 0$ if and only if $|x| < 1$.

The derivatives of the four inverse hyperbolic functions considered here can either be obtained from the derivatives of the corresponding hyperbolic functions (Ex. 26, § 509) or from formulas (1)–(4) (Ex. 27, § 509). They are:

(5) $\frac{d}{dx} \sinh^{-1} x = \frac{1}{\sqrt{1 + x^2}}$, all real x;

(6) $\frac{d}{dx} \operatorname{Cosh}^{-1} x = \frac{1}{\sqrt{x^2 - 1}}$, $x > 1$;

(7) $\frac{d}{dx} \tanh^{-1} x = \frac{1}{1 - x^2}$, $|x| < 1$;

(8) $\frac{d}{dx} \coth^{-1} x = \frac{1}{1 - x^2}$, $|x| > 1$.

The corresponding integration formulas are mere reformulations of formulas (2) and (7) of § 505, with appropriate ranges specified. They can be established independently by differentiations of the right-hand members (Ex. 28, § 509). The letter a represents a positive number throughout.

(9) $\int \frac{dx}{\sqrt{a^2 + x^2}} = \sinh^{-1} \frac{x}{a} + C$, all real x;

(10) $\int \frac{dx}{\sqrt{x^2 - a^2}} = \operatorname{Cosh}^{-1} \frac{x}{a} + C$, $x > a$;

(11) $$\int \frac{dx}{a^2 - x^2} = \frac{1}{a} \tanh^{-1} \frac{x}{a} + C, |x| < a;$$

(12) $$\int \frac{dx}{x^2 - a^2} = -\frac{1}{a} \coth^{-1} \frac{x}{a} + C, |x| > a.$$

509. EXERCISES

In Exercises 1–10, differentiate the given function.

1. $\cosh 3x$.
2. $\sinh^2 x$.
3. $\tanh (2 - x)$.
4. $x \coth x^2$.
5. $\ln \sinh 2x$.
6. $e^{ax} \cosh bx$.
7. $\sinh^{-1} 4x$.
8. $\cosh^{-1} e^x$.
9. $\tanh^{-1} x^2$.
10. $\coth^{-1} (\sec x)$.

In Exercises 11–20, perform the indicated integration, expressing your answer in terms of hyperbolic functions or their inverses.

11. $\int \tanh 6x \, dx$.
12. $\int e^x \coth e^x \, dx$.
13. $\int \cosh^3 x \, dx$.
14. $\int \tanh^2 10x \, dx$.
15. $\int \sinh^2 x \, dx$.
16. $\int x^3 \tanh^{-1} x \, dx$.
17. $\int \frac{dx}{\sqrt{x^2 - 2}}$.
18. $\int \frac{dx}{\sqrt{4x^2 - 4x + 5}}$.
19. $\int \frac{dx}{3x^2 + 5x - 7}$, $(3x^2 + 5x - 7 > 0)$.
20. $\int \frac{dx}{7 - 5x - 3x^2}$, $(7 - 5x - 3x^2 > 0)$.

21. Prove that $\cosh x$ and $\operatorname{sech} x$ are even functions and that the other four hyperbolic functions are odd (cf. Exs. 7–8, § 402).
22. Establish the identities (2)–(9), § 507.
23. Establish the differentiation formulas (10)–(15), § 507.
24. Establish the integration formulas (16)–(19), § 507. *Hint for* (19):
$$\int \frac{dx}{\sinh x} = \int \frac{du}{u^2 - 1}, \text{ where } u = \cosh x, \text{ and } \frac{\cosh x - 1}{\cosh x + 1} = \frac{2 \sinh^2 \frac{1}{2} x}{2 \cosh^2 \frac{1}{2} x}.$$
25. Establish formulas (1), (3), (4), § 508.
26. Establish formulas (5)–(8), § 508, by use of (10)–(13), § 507. *Hint for* (5): Let $y = \sinh^{-1} x$. Then $x = \sinh y$, $dx/dy = \cosh y$, and $dy/dx = 1/\sqrt{1 + \sinh^2 y}$.
27. Establish formulas (5)–(8), § 508, by use of (1)–(4), § 508.
28. Establish formulas (9)–(12), § 508.

*29. Show that $\int \operatorname{csch} x \, dx = -\coth^{-1} (\cosh x) + C$.

*30. Establish formulas (7)–(9), § 505, by means of hyperbolic substitutions.

*510. CLASSIFICATION OF NUMBERS AND FUNCTIONS

Certain classes of numbers (integers, rational numbers, and irrational numbers) were defined in Chapter 1. Another important class of numbers is the algebraic numbers, an **algebraic number** being defined to be a root of a **polynomial equation**

(1) $$a_0 x^n + a_1 x^{n-1} + \cdots + a_n = 0,$$

where the coefficients $a_0, a_1, \cdots, a_n$ are integers. Examples of algebraic numbers are $\tfrac{3}{4}$ (a root of the equation $4x - 3 = 0$) (in fact, *every rational number is algebraic*), $-\sqrt[3]{5}$ (a root of the equation $x^3 + 5 = 0$), and $\sqrt[3]{5} + \sqrt{3}$ (a root of the equation $x^6 - 9x^4 - 10x^3 + 27x^2 - 90x - 2 = 0$). It can be shown† that any number that is the sum, difference, product, or quotient of algebraic numbers is algebraic, and that any (real) root of an algebraic number is an algebraic number. Therefore any number (like $\sqrt{3 - \sqrt[6]{8/9}} \big/ 4\sqrt[7]{61}$) that can be obtained from the integers by a finite sequence of sums, differences, products, quotients, powers, and roots is algebraic. However, it should be appreciated that not every algebraic number is of the type just described. In fact, it is shown in Galois Theory (cf. the Birkhoff and MacLane book just referred to) that the general equation of degree 5 or higher cannot be solved in terms of radicals. In particular, the real root of the equation $x^5 - x - 1 = 0$ (this number is algebraic by definition) cannot be expressed in the finite form described above.

A **transcendental number** is any number that is not algebraic. The most familiar transcendental numbers are e and π. The transcendental character of e was established by Hermite in 1873, and that of π by Lindemann in 1882.‡ For a discussion of the transcendence of these two numbers see Felix Klein, *Elementary Mathematics from an Advanced Standpoint* (New York, Dover Publications, 1945). The first number to be proved transcendental was neither e nor π, but a number constructed artificially for the purpose by Liouville in 1844. (Cf. page 413 of the Birkhoff and MacLane reference.) The existence of a vast infinite supply of transcendental numbers was provided by Cantor in 1874 in his theory of transfinite cardinal numbers.

Functions are classified in a manner similar to that just outlined for numbers. The role played by integers above is now played by polynomials. A **rational function** is any function that can be expressed as the quotient of two polynomials, an **algebraic function** is any function $f(x)$ that satisfies (identically in x) a polynomial equation

(2) $$a_0(x)(f(x))^n + a_1(x)(f(x))^{n-1} + \cdots + a_n(x) = 0,$$

where $a_0(x), a_1(x), \cdots, a_n(x)$ are polynomials, and a **transcendental function** is any function that is not algebraic.

---

† Cf. Birkhoff and MacLane, *A Survey of Modern Algebra* (New York, The Macmillan Company, 1944).

‡ It is easy to show that e is irrational (Ex. 34, § 1211). A proof of the irrationality of π was given in 1761 by Lambert. For an elementary proof, and further discussion of both the irrationality and transcendence of π, see Ivan Niven, *Irrational Numbers* (Carus Mathematical Monographs, 1956).

Examples of polynomials are $2x - \sqrt{3}$ and $\pi x^5 + \sqrt{e}$. Examples of rational functions that are not polynomials are $1/x$ and $(x + \sqrt{3})/(5x - 6)$. Examples of algebraic functions that are not rational are $\sqrt{x}$ and $1/\sqrt{x^2 + \pi}$. Examples of transcendental functions are e^x, $\ln x$, $\sin x$, and $\cos x$. (Cf. Exs. 10–12, § 512.)

*511. THE ELEMENTARY FUNCTIONS

Most of the familiar functions which one encounters at the level of elementary calculus, like $\sin 2x$, e^{-x^2}, and $\frac{1}{2}$ Arctan $\frac{1}{2}x$, are examples of what are called *elementary functions*. In order to define this concept, we describe first the elementary operations on functions. The **elementary operations** on functions $f(x)$ and $g(x)$ are those that yield any of the following: $f(x) \pm g(x)$, $f(x)g(x)$, $f(x)/g(x)$, $\{f(x)\}^a$, $a^{f(x)}$, $\log_a f(x)$, and $T(f(x))$, where T is any trigonometric or inverse trigonometric function. The **elementary functions** are those generated by constants and the independent variable by means of a finite sequence of elementary operations. Thus

$$(x^2 + 17\pi)^e \operatorname{Arcsin}[(\log_3 \cos x)(e^{\sqrt{x}})]$$

is an elementary function. Its derivative is also an elementary function. In fact (cf. Ex. 9, § 512), the derivative of any elementary function is an elementary function. Reasonable questions to be asked at this point are, "Is there ever any occasion to study functions that are *not* elementary? What are some examples?" The answer is that nonelementary functions arise in connection with infinite series, differential equations, integral equations, and equations defining functions implicitly. For example, the differential equation $\dfrac{dy}{dx} = \dfrac{\sin x}{x}$ ($= 1$ if $x = 0$) has a nonelementary solution

$$(1) \qquad \int_0^x \frac{\sin t}{t}\, dt$$

whose power series expansion (cf. Chapter 12) is

$$x - \frac{x^3}{3 \cdot 3!} + \frac{x^5}{5 \cdot 5!} - \frac{x^7}{7 \cdot 7!} + \cdots.$$

Other examples of nonelementary functions which are integrals of elementary functions are the *probability integral*,

$$(2) \qquad \int_0^x e^{-t^2}\, dt,$$

the *Fresnel sine integral* and *cosine integral*,

$$(3) \qquad \int_0^x \sin t^2\, dt \quad \text{and} \quad \int_0^x \cos t^2\, dt,$$

and the *elliptic integrals of the first and second kind* (cf. Ex. 24, § 807),

$$(4) \qquad \int_0^x \frac{dt}{\sqrt{1 - k^2 \sin^2 t}} \quad \text{and} \quad \int_0^x \sqrt{1 - k^2 \sin^2 t}\, dt.$$

Extensive tables for these and other nonelementary functions exist.† Proofs that these functions are not elementary are difficult and will not be given in this book.‡ For some evaluations, see Example 3, § 1013, §§ 1408, 1409.

The inverse of an elementary function need not be elementary. For example, the inverse of the function $x - a \sin x$ ($a \neq 0$), the function defined implicitly by *Kepler's equation*

$$(5) \qquad y - a \sin y - x = 0,$$

is not elementary.‡ Again, the proof is omitted.

An important lesson to be extracted from the above considerations is that inability to integrate an elementary function in terms of elementary functions does not mean that the integral fails to exist. In fact, many important nonelementary functions owe their existence to the (Riemann) integration process. This is somewhat similar to the position of the logarithmic, exponential, and trigonometric functions as developed in this chapter, which were defined, ultimately, in terms of Riemann integrals of algebraic functions.

One more final remark. The reader might ask, "Apart from the importance of the functions defined above, if all we seek is an example of a nonelementary function, isn't the bracket function $[x]$ of § 201 one such?" Perhaps the most satisfactory reply is that whereas $[x]$ is elementary in intervals, the functions presented above might be described as "nowhere elementary." A behind-the-scenes principle here is that of analytic continuation, important in the theory of analytic functions of a complex variable. (Cf. § 1810.)

*512. EXERCISES

*1. Prove that $\sqrt{5} - \sqrt{2}$ is algebraic by finding a polynomial equation with integral coefficients of which it is a root.

*2. Prove that $1/x$ is not a polynomial. *Hint:* It is not sufficient to remark that it does not *look* like one. Show that $1/x$ has some property that no polynomial can have. (Cf. Ex. 4.)

*3. Prove that $\dfrac{x^2}{x+1}$ is not a polynomial (cf. Ex. 2).

*4. Prove that $\sqrt{x^2+1}$ is not a polynomial. *Hint:* Differentiate the function a large number of times.

*5. Prove that $\sqrt{x^2+1}$ is not rational. (Cf. Ex. 4.)

*6. Prove that the hyperbolic functions and their inverses are elementary.

*7. Prove that x^x is elementary.

*8. Prove that $\displaystyle\int_0^x t^n \sin t \, dt$, where n is a positive integer, is elementary.

*9. Prove that the derivative of any elementary function is elementary.

---

† Cf. Eugen Jahnke and Fritz Emde, *Tables of Functions* (New York, Dover Publications, 1943).
‡ Cf. J. F. Ritt, *Integration in Finite Terms* (New York, Columbia University Press, 1948).

*10. Prove that e^x is transcendental. *Hint:* Let
$$a_0(x)e^{nx} + a_1(x)e^{(n-1)x} + \cdots + a_{n-1}(x)e^x + a_n(x) = 0,$$
where $a_0(x), \cdots, a_n(x)$ are polynomials and $a_n(x)$ is not identically 0. Form the limit as $x \to -\infty$. Then, surprisingly, $\lim\limits_{x \to -\infty} a_n(x) = 0$.

*11. Prove that $\ln x$ is transcendental. *Hint:* Let $y \equiv \ln x$, and assume that y satisfies identically a polynomial equation, arranged in the form
$$b_0(y)x^n + b_1(y)x^{n-1} + \cdots + b_{n-1}(y)x + b_n(y) = 0,$$
where $b_0(y)$ is not identically 0. Divide every term by x^n, and let $x \to +\infty$.

*12. Prove that $\sin x$ and $\cos x$ are transcendental. *Hint for $\sin x$:* Let
$$a_0(x)\sin^n x + a_1(x)\sin^{n-1} x + \cdots + a_{n-1}(x)\sin x + a_n(x) = 0,$$
where $a_0(x), \cdots, a_n(x)$ are polynomials and $a_n(x)$ is not identically 0. Let $x = k\pi$, where k is so large that $a_n(x) \neq 0$.

*13. Prove that the system of algebraic numbers does not satisfy the axiom of completeness (§ 112). (Cf. Ex. 9, § 114.)

6
Functions of Several Variables

601. INTRODUCTION

In the first five chapters of this book attention has been focused on functions of *one real variable*. We wish now to extend our consideration to functions of *several real variables*—where the word *several* means at *least one*. It will be seen, particularly in this chapter, that many concepts, theorems, and proofs generalize from one to several variables almost without change. Such is the case, for instance, with Theorems I and II, § 213, concerning the boundedness and the attainment of extreme values for a real-valued function continuous on a closed interval. On the other hand, such a topic as integration (Chapter 10), especially when it comes to a change of variables, requires radical reorganization.

We start with some definitions. These include concepts that, although of interest in the space E_1 of real numbers, are of vital importance in spaces of more than one dimension. Most of the examples and detailed formulations are given for the Euclidean plane E_2. The student can readily supply examples of his own invention for Euclidean spaces of dimension one or greater than two. It should be noted that the two concepts of *compact set* and *region* are important and natural generalizations of *closed interval* and *open interval*, respectively. For some special applications to the space of real numbers of the ideas about to be introduced, see Exercises 23–30, § 605.

In the remainder of the book, and particularly in Chapter 10, it will occasionally be convenient to use a small amount of the notation of *point set algebra* (for a more complete presentation, cf. RV, § 1022). **Set membership** is denoted as follows: *x is a member of the set A* is written

$$x \in A.$$

The statement that *a set A is a* **subset** *of a set B* (that is, every member of A is a member of B) is written

$$A \subset B \quad \text{or} \quad B \supset A.$$

(In particular, if the **empty set** is denoted $\emptyset$, then $\emptyset \subset A$ and $A \subset A$ for every set A.) The **union** of the sets A and B, denoted

$$A \cup B,$$

is the set of all points p such that $p \in A$ or $p \in B$ (or both). The **intersection** of the sets A and B, denoted
$$A \cap B,$$
is the set of all points p such that $p \in A$ and $p \in B$. The **difference** between the sets A and B, denoted
$$A - B,$$
is the set of all points of A that do *not* belong to B. For an arbitrary finite collection of sets $A_1, A_2, \cdots, A_n$, the **union**, denoted
$$A_1 \cup A_2 \cup \cdots \cup A_n,$$
is the set of all points p having the property that p belongs to *at least one* set of the collection. The **intersection** of $A_1, A_2, \cdots, A_n$, denoted
$$A_1 \cap A_2 \cap \cdots \cap A_n,$$
is the set of all points p having the property that p belongs to *every* set of the collection. These definitions apply in like manner to arbitrary (finite or infinite) collections of sets. Two sets, A and B, are called **disjoint** in case they have no points in common; that is, $A \cap B = \emptyset$.

602. NEIGHBORHOODS IN THE EUCLIDEAN PLANE

In the Euclidean space E_1 of one dimension—that is, the space of real numbers—a *neighborhood* of a point a is defined (Definition III, § 110) as an open interval of the form $(a - \epsilon, a + \epsilon)$. Equally well, we can describe this neighborhood as the set of all points x whose distance from a is less than ϵ:

(1) $$|x - a| < \epsilon.$$

In the Euclidean plane E_2 we shall speak of *two* kinds of neighborhoods, circular and square. A **circular neighborhood** of a point (a, b) is the set of all points (x, y) whose distance from (a, b) is less than some fixed positive number ϵ:

(2) $$\sqrt{(x - a)^2 + (y - b)^2} < \epsilon.$$

It consists of all points inside a circle of radius ϵ and center (a, b). A **square neighborhood** of a point (a, b) is the set of all points (x, y) whose coordinates satisfy two inequalities of the form

(3) $$|x - a| < \epsilon, \quad |y - b| < \epsilon.$$

It consists of all points in a square whose sides are parallel to the coordinate axes and of length 2ϵ, and whose center is (a, b).

As indicated in Figure 601, any circular neighborhood of a point contains a square neighborhood of the point (for a smaller ϵ, of course), and any square neighborhood of a point contains a circular neighborhood of the point. This fact will be referred to as the **equivalence** of the two kinds of neighborhoods. (Cf. the Note, below.)

By a **neighborhood** of a point we shall mean either a circular neighborhood or a square neighborhood of the point. By a **deleted neighborhood** of a point we shall mean any neighborhood of that point with the point itself removed.

NOTE. In nearly all the following sections, only the single word *neighborhood* is used. The reader should observe that if either word *circular* or *square* is introduced,

FIG. 601

the meaning is unaffected. (Cf. Definitions II and III, § 603.) This is the real significance of the term *equivalence*, as applied to the two types of neighborhoods.

603. POINT SETS IN THE EUCLIDEAN PLANE

Definition I. *If A is a set, the **complement** of A, written A', is the set of all points in the Euclidean plane that are not members of A.*

Example 1. Let A be the set of points (x, y) belonging to a circular neighborhood of some point (a, b): $(x - a)^2 + (y - b)^2 < r^2$, where $r > 0$. Such a set is called an **open disk** (Fig. 602). Then A' consists of all points (x, y) such that

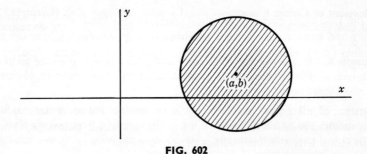

FIG. 602

$(x - a)^2 + (y - b)^2 \geq r^2$, and is made up of the circumference of A together with the points completely outside the circle. The open disk with center $(0, 0)$ and radius 1 is called the **open unit disk**.

Example 2. If A is a **closed disk** defined by an inequality of the form $(x - a)^2 + (y - b)^2 \leq r^2$, where $r > 0$, then A' consists of the points completely outside the circle. All points of the circumference belong to the set A. The closed disk with center $(0, 0)$ and radius 1 is called the **closed unit disk**.

Definition II. *A set A is* **open** *if and only if every member of A has some neighborhood contained entirely in A.*

Example 3. Let A be the set of points (x, y) such that both x and y are positive. This set is called the **open first quadrant** (Fig. 603). Then A is an open set. To prove this, we choose an arbitrary member (x, y) of A. Then $\epsilon \equiv \min(x, y)$, the smaller of the two numbers x and y, is positive. Furthermore, the corresponding neighborhood, whether circular ((2), § 602) or square ((3), § 602), lies entirely within the set A.

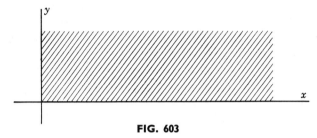

FIG. 603

Discussion of Earlier Examples. Any open disk (Example 1) is open (give the details of the proof). A closed disk (Example 2) is *not* open. The point $(1, 0)$, for instance, belongs to the closed unit disk A, but every neighborhood of this point extends partly outside the circle A. In other words, *no* neighborhood of $(1, 0)$ lies entirely in A. Thus, not *every* member of A has a neighborhood contained entirely in A.

Definition III. *A point p is a* **limit point** *of a set A if and only if every deleted neighborhood of p contains at least one point of A.*

Discussion of Earlier Examples. If A is either an open disk (Example 1) or a closed disk (Example 2), the set B of all limit points of A is the corresponding closed disk.

Example 4. If A is the open first quadrant of Example 3, then the set B of all limit points of A is the **closed first quadrant** consisting of all points (x, y) such that both x and y are nonnegative.

NOTE 1. If p is a limit point of a set A we say that the set A has p as a limit point, whether p is a member of A or not. For example, both the open unit disk and the closed unit disk (Examples 1 and 2) have the point $(1, 0)$ as a limit point, but only the latter contains $(1, 0)$ as a member.

NOTE 2. If p is a limit point of a set A, every neighborhood of p contains infinitely many points of A (Ex. 8, § 605).

Definition IV. *A set is* **closed** *if and only if it contains all of its limit points.*

Discussion of Earlier Examples. Any closed disk (Example 2) and the closed first quadrant of Example 4 are closed sets. Neither any open disk (Example 1) nor the open first quadrant of Example 3 is closed (the point $(1, 0)$ is a limit point of the open unit disk and the open first quadrant, but is a member of neither).

Example 5. Let A be the set of all **integral points** of the plane, that is, the set of all points (x, y) such that *both* x and y are integers. This set has *no* limit points. (One way to see this is to observe that *no* neighborhood of *any* point can contain infinitely many points of A.) It follows that the set A *contains* all of its limit points, since otherwise there must exist a limit point of the set A—in fact, a limit point of the set that is not a member of the set.† Thus A is closed.

Example 6. The reasoning used in Example 5 establishes the fact that *any finite set is closed*.

Example 7. The circumference of a disk, defined by the equation $(x - a)^2 + (y - b)^2 = r^2$, is closed. (Why?) It is not open. (Why?)

Example 8. The set made up of all points of the x-axis together with all points of the y-axis is closed. It is not open (as a set in E_2). (Why?)

Example 9. The positive half of the x-axis, that is, the points (x, y) such that $x > 0$ and $y = 0$, is neither open nor closed. (Why?)

Example 10. Let A be the set of all **rational points** of the plane, that is, the set of all points (x, y) such that *both* x and y are rational. This set is neither open nor closed. (Why?)

NOTE 3. A basic relation between open and closed sets is the following (cf. Ex. 9, § 605): *A set is open if and only if its complement is closed. Equivalently, a set is closed if and only if its complement is open.* It is of interest that in the Euclidean plane the only sets that are *both* open and closed are the empty set ∅ and the entire plane. (For a proof of this remarkable fact, see Ex. 19, § 618.)

Definition V. *A set is **bounded** if and only if it is contained in some disk, or, equivalently, in some square. A set is **compact** if and only if it is closed and bounded.*‡

Discussion of Earlier Examples. The open and closed unit disks (Examples 1 and 2) are both bounded, and the latter is compact. Neither the open nor the closed first quadrant (Examples 3 and 4) is bounded, and therefore neither is compact. The set of integral points of the plane (Example 5) is unbounded. Since any finite set (Example 6) is bounded and closed, it is compact. The circumference of the unit circle (Example 7) is compact. Not one of Examples 8–10 is bounded, and therefore not one is compact.

Definition VI. *A point p, belonging to a set A, is an **interior point** of A if and only if it has a neighborhood lying entirely in A. The set of all interior points of a set A is called its **interior** and denoted $I(A)$. A point p, whether it belongs to a set A or not, is a **frontier point** of A if and only if every neighborhood of p contains points of A and of its complement A'. The set of all frontier*

---

† It may be held by some that the fact that a set with no limit points contains all of them is completely obvious. It is well, however, to be ready with the machinery to convince the skeptic who claims that the alleged fact is *not* obvious to *him*. The persuasive machinery in this case is to follow to a logical consequence the *denial* of the original assertion and to obtain a contradiction that does not rest on any individual state of mind.

‡ This formulation of compactness is suitable for finite dimensional Euclidean spaces, but not for general abstract spaces. For a brief introduction to metric and topological spaces, see *RV*, §§ 1026–1028.

§ 603]   POINT SETS IN THE EUCLIDEAN PLANE   177

*points of a set A is called its **frontier** and denoted F(A). The frontier of a set A is therefore made up of points that are either members of A and limit points of A' or members of A' and limit points of A.*†

Example 11. Let A be a set made up of an open disk (Example 1) together with *some* but *not all* of the points of the circumference. This set lies *between* the open disk and the corresponding closed disk (Example 2). It is neither open nor closed. (Why?) Its interior (as also is the interior of either the open or the closed disk) is the open disk. (Why?) Its frontier (as also is the frontier of either the open or the closed disk) is the circumference of the disk. (Why?)

Discussion of Earlier Examples. The interior of either the open or the closed first quadrant (Examples 3 and 4) is the open first quadrant, and its frontier is the nonnegative x-axis together with the nonnegative y-axis. The interior of the set A of all integral points (Example 5) is empty (that is, there are no interior points), and its frontier is the set A itself. Similar statements hold for any finite set (Example 6). The interior of the circumference A of any disk (Example 7) is empty, and the frontier of A is the set A itself. The interior of the set A of all rational points (Example 10) is empty, and its frontier is the entire plane. The detailed verification of the preceding statements is left as an exercise for the reader (Ex. 10, § 605).

NOTE 4. An interesting relation between the concepts of *open set* and *interior of a set* is that the interior of any set is open. (Give a proof in Ex. 11, § 605.)

Definition VII. *The **closure** $\bar{A}$ of a set A consists of all points of A together with all limit points of A.*

Discussion of Earlier Examples. If A is either an open or closed disk of Examples 1 or 2, or a disk of Example 11, its closure $\bar{A}$ is the corresponding closed disk. The closure of either the open or closed first quadrant (Examples 3 and 4) is the closed first quadrant. The closure of the set of all integral points (Example 5) is the set itself. A similar statement holds for any finite set (Example 6). The closure of the set of all rational points (Example 10) is the entire plane. The reader should supply the details of the verification of the preceding statements (Ex. 12, § 605).

NOTE 5. The term *closure* is appropriate because of the fact that for *any* set A, its closure $\bar{A}$ is always a closed set. (Cf. Ex. 13, § 605.) In a definite sense, $\bar{A}$ is the *smallest* closed set containing A (Ex. 15, § 605).

Definition VIII. *A **region** is a nonempty open set R any two of whose points can be connected by a broken line segment lying entirely in R.* (Cf. Fig. 604.) *A **closed region** is the closure of a region. A **compact region** is a bounded closed region.*

NOTE 6. The single word *region* will be reserved for the meaning defined above, and will always refer to an open set. A closed region, therefore, is a region only if it is the entire plane (cf. Note 3), and a compact region is never a region.

† The frontier of a set is sometimes also called its *boundary*. However, since the word *boundary* has other meanings in combinatorial topology and in the theory of curves and surfaces (cf. Chapter 18), we shall avoid using this latter term here.

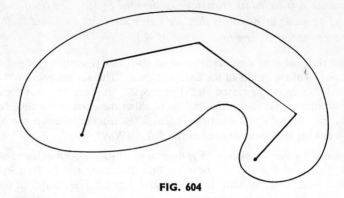

FIG. 604

604. SETS IN HIGHER DIMENSIONAL EUCLIDEAN SPACES

In three-dimensional Euclidean space E_3 the distance between two points $p: (x_1, y_1, z_1)$ and $q: (x_2, y_2, z_2)$ is given by the formula

(1) $$d(p, q) = \sqrt{(x_2 - x_1)^2 + (y_2 - y_1)^2 + (z_2 - z_1)^2}.\dagger$$

More generally, in n-dimensional Euclidean space E_n, where a point is *defined* to be an ordered n-tuple of real numbers, the distance between two points $p: (x_1, x_2, \cdots, x_n)$ and $q: (y_1, y_2, \cdots, y_n)$ is *defined* by means of a formula corresponding to (1):

(2) $$d(p, q) \equiv \sqrt{(y_1 - x_1)^2 + (y_2 - x_2)^2 + \cdots + (y_n - x_n)^2}.$$

In a manner analogous to the treatment of neighborhoods in the Euclidean plane, we speak of two kinds of neighborhoods in E_n. A **spherical neighborhood** of a point $p_0: (a_1, a_2, \cdots, a_n)$ is the set of all points $p: (x_1, x_2, \cdots, x_n)$ whose distance from p_0 is less that some fixed number ϵ:

(3) $$d(p, p_0) < \epsilon.$$

A **cubical neighborhood** of a point $p_0: (a_1, a_2, \cdots, a_n)$ is the set of all points $p: (x_1, x_2, \cdots, x_n)$ whose coordinates satisfy the n inequalities

(4) $$|x_1 - a_1| < \epsilon, \quad |x_2 - a_2| < \epsilon, \quad \cdots, \quad |x_n - a_n| < \epsilon.$$

The Definitions and Notes of § 603 are immediately adaptable to the space E_n, virtually the only change being the replacement of the expression *Euclidean plane* by the expression *n-dimensional Euclidean space*. These details, as well as the construction of examples corresponding to those of § 603, are left to the reader. Perhaps the greatest difficulty in the transition from two- to n-dimensional space lies in the concept of *broken line segment*, in Definition VIII, § 603, of a region. This depends in turn on the question of *segment between two points*. For a brief discussion of this, see Exercise 21, § 605.

† For a short treatment of the analytic geometry of three dimensions, see Chapter 7.

605. EXERCISES

In Exercises 1–6, state whether the given set is (a) open, (b) closed, (c) bounded, (d) compact, (e) a region, (f) a closed region, (g) a compact region. Give its (h) set of limit points, (i) closure, (j) interior, and (k) frontier. The space under consideration in each case is the Euclidean plane.

1. All (x, y) such that $a \leq x < b$, $c \leq y < d$, a, b, c, and d fixed constants such that $a < b$ and $c < d$.
2. All (x, y) such that $xy \neq 0$.
3. All (ρ, θ) such that $\rho^2 \leq \cos 2\theta$, ρ and θ being polar coordinates.
4. All (x, y) such that $y = \sin \dfrac{1}{x}$, $x \neq 0$.
*5. All (x, y) such that $y < [x]$ (cf. Example 5, §201).
*6. All (x, y) such that x and y have the form $x = m + \dfrac{1}{p}$ and $y = n + \dfrac{1}{q}$, m and n integers, p and q integers > 1.

In Exercises 7–20, assume that the space under consideration is the Euclidean plane. More generally prove the corresponding result for the space E_n.

7. Prove that a point p is a limit point of a set A if and only if every open set containing p contains at least one point of A different from p.
8. Prove the statement of Note 2, §603.
9. Prove the statement of Note 3, §603, that a set is open if and only if its complement is closed.
10. Verify the details called for in the *Discussion of Earlier Examples*, following Example 11, §603.
11. Prove the statement of Note 4, §603, that if A is an arbitrary set, its interior is always an open set. *Hint:* Let B denote the interior of A, and let p be an arbitrary point of B. Since p is an interior point of A, it has a neighborhood N_p lying entirely in A. Show that N_p lies entirely in B.
12. Verify the details called for in the *Discussion of Earlier Examples*, following Definition VII, §603.
13. Prove the statement of Note 5, §603, that if A is an arbitrary set, its closure $\bar{A}$ is always a closed set. *Hint:* Assume that $\bar{A}$ is *not* closed, and let p be a limit point of $\bar{A}$ that is not a member of $\bar{A}$. Since p is neither a member of A nor a limit point of A, it has a neighborhood N_p having no points in common with A. Show that this fact is inconsistent with the assumption that p is a limit point of $\bar{A}$, since N_p must contain some point q belonging to $\bar{A}$.
14. Prove that a set is closed if and only if it is equal to its closure.
*15. Prove that $\bar{A}$ is the smallest closed set containing A in the sense that (i) $\bar{A}$ is a closed set containing A, and (ii) if B is any closed set containing A, then B contains $\bar{A}$. (Cf. Ex. 14.)
*16. Prove that the set of all limit points of any set is always closed.
*17. Prove that the frontier of any set is always closed.
*18. Prove that the interior and the frontier of a set never have any points in common, and together constitute the closure of the set.
*19. Prove that the interior of the complement of any set is equal to the complement of the closure of the set.

★20. Prove that any open set is contained in the interior of its closure. Give an example of a region that is not equal to the interior of its closure. (Cf. Theorem I, § 1027.)

★21. If p_0: (a, b) and p_1: (c, d) are two distinct points in the Euclidean plane, and if p: (x, y) is an arbitrary point on the line $p_0 p_1$ (cf. Fig. 605), show that
(1) $\qquad (x - a):(c - a) = (y - b):(d - b),$
with a suitable interpretation in case any quantity vanishes. Letting t represent the common value of the ratios in (1), show that the line $p_0 p_1$ can be represented parametrically:
(2) $\qquad x = a + t(c - a), \quad y = b + t(d - b),$
with consideration of the special case of a vertical or horizontal line. Observe that the segment $p_0 p_1$ corresponds to the values $0 \leq t \leq 1$. Generalize these con-

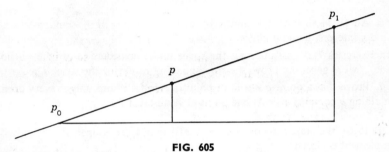

FIG. 605

siderations to the space E_n, both for a line through two points p_0: $(a_1, a_2, \cdots, a_n)$ and p_1: $(b_1, b_2, \cdots, b_n)$ and the segment $p_0 p_1$:
(3) $\qquad (a_1 + t(b_1 - a_1), \; a_2 + t(b_2 - a_2), \; \cdots, \; a_n + t(b_n - a_n)).$

★22. Establish the **triangle inequality** in E_n: if $p, q,$ and r are any three points in E_n, then
(4) $\qquad d(p, r) \leq d(p, q) + d(q, r).$

Hint: Use the Minkowski inequality of Exercise 33, § 107, with $a_i \equiv y_i - x_i$ and $b_i \equiv z_i - y_i$.

In Exercises 23–30, the sets under consideration all belong to the space E_1 but, as subsets of the x-axis, may be thought of as sets lying in the plane E_2.

★23. Let A be any open interval of real numbers, finite or infinite. Prove that A is an open set in E_1, but not an open set in E_2.

★24. Prove that any open set in the space E_1 of real numbers contains both rational and irrational numbers and, in fact, infinitely many of each.

★25. Prove that any finite closed interval and the infinite intervals $[a, +\infty)$ and $(-\infty, b]$, as subsets of E_1, are closed (cf. Ex. 26).

★26. Let A be a set of real numbers. Prove that A as a subset of E_1 is closed if and only if A as a subset of E_2 is closed.

★27. Same as Exercise 26 for compact sets.

★28. Let A be an arbitrary set of real numbers. Prove that the closure $\bar{A}$ of A is the same whether A is considered as a subset of E_1 or as a subset of E_2.

★29. If p is the least upper bound (or the greatest lower bound) of a nonempty set A of real numbers and if p is not a member of A, prove that p is a limit point of A.

(*30) Prove that every nonempty compact set of real numbers has a greatest member and a least member. (Cf. Ex. 29.)

606. FUNCTIONS AND LIMITS

The definition of *function* given in § 201 is sufficiently general to include functions of several independent variables. A function $f(x, y)$ of the two real variables x and y, for example, can be regarded as a function $f(p)$ of the variable point $p : (x, y)$ whose coordinates are x and y. The domain of definition, in this instance, would be a portion of the Euclidean plane; and the range, for a real-valued function, would be a portion of the real number system. In a similar way, a function of n real variables can be conveniently regarded as a function of a point in n-dimensional Euclidean space.

In this section and in following sections, unless explicit statement to the contrary is made, the word *function* will refer to a single-valued real-valued function of a variable point in n-dimensional Euclidean space, defined in a neighborhood of the particular point concerned (or possibly just for points in a set having the particular point as a limit point). For convenience, most definitions and theorems will be formulated for functions of two real variables. The generalizations to n variables are mostly obvious, and will usually be omitted.

We start with the basic definition of a finite limit as the independent variables approach finite limits, phrased in the convenient language of neighborhoods, followed by explicit formulations in terms of deltas and epsilons for both circular and square neighborhoods.

Definition I. *The function $f(x, y)$ has the limit L as the point (x, y) approaches the point (a, b), written*

$$\lim_{(x,y)\to(a,b)} f(x, y) = L, \lim_{\substack{x\to a \\ y\to b}} f(x, y) = L, \text{ or } f(x, y) \to L \text{ as } (x, y) \to (a, b),$$

if and only if corresponding to an arbitrary neighborhood N_L of L, in the space E_1 of real numbers, there exists a deleted neighborhood $N_{(a,b)}$ of the point (a, b) in the Euclidean plane E_2, such that for every point (x, y) of $N_{(a,b)}$, for which $f(x, y)$ is defined, $f(x, y)$ belongs to N_L.

NOTE 1. In terms of circular neighborhoods, Definition I assumes the form: $\lim_{(x,y)\to(a,b)} f(x, y) = L$ if and only if corresponding to an arbitrary positive number ϵ there exists a positive number $\delta = \delta(\epsilon)$ such that $0 < \sqrt{(x - a)^2 + (y - b)^2} < \delta$ implies $|f(x, y) - L| < \epsilon$, for points (x, y) for which $f(x, y)$ is defined. For square neighborhoods the last implication is replaced by $|x - a| < \delta, |y - b| < \delta$, and $0 < |x - a| + |y - b|$ imply $|f(x, y) - L| < \epsilon$, for points (x, y) for which $f(x, y)$ is defined.

NOTE 2. As with the limit of a function of a single variable, it is immaterial what the value of the function $f(x, y)$ is at the point (a, b), or whether it is defined there at all. This fact appears in the preceding definition in the use of *deleted neighborhoods*.

Limits where either the independent variables or the dependent variable may become infinite can be formulated in a manner similar to definitions given in Chapter 2. A few samples are given here, others being requested in Exercises 12–13, § 611. Notice that the Euclidean plane, like the space of real numbers, can be considered to have associated with it different kinds of "points at infinity."

Definition II.
$$\lim_{x,y \to +\infty} f(x, y) = \lim_{\substack{x \to +\infty \\ y \to +\infty}} f(x, y) = L,$$

where L is a real number, if and only if corresponding to $\epsilon > 0$ there exists a number $N = N(\epsilon)$ such that $x > N$ and $y > N$ imply $|f(x, y) - L| < \epsilon$, for points (x, y) for which $f(x, y)$ is defined.†

Definition III.
$$\lim_{(x,y) \to \infty} f(x, y) = L,$$

where L is a real number, if and only if corresponding to $\epsilon > 0$ there exists a number $N = N(\epsilon)$ such that $\sqrt{x^2 + y^2} > N$ implies $|f(x, y) - L| < \epsilon$, for points (x, y) for which $f(x, y)$ is defined.

Definition IV.
$$\lim_{(x,y) \to (a,b)} f(x, y) = -\infty$$

if and only if corresponding to an arbitrary number B there exists a positive number $\delta = \delta(B)$ such that $0 < \sqrt{(x-a)^2 + (y-b)^2} < \delta$ implies $f(x, y) < B$, for points (x, y) for which $f(x, y)$ is defined.

It will be seen that limits for functions of several variables are more varied and interesting than limits for functions of a single real variable. One reason for this is that the variable point $p : (x_1, x_2, \cdots, x_n)$ may be considered as approaching a limit point in a much richer variety of ways. Another reason is that whenever the number of dimensions exceeds one, there is the related question of *iterated limits*, considered in the next section, where first one, and then the other, independent variable approaches its limit.

Postponing to later sections (§ 613, Ex. 28, § 618) the study of *criteria* for the existence of a limit, we give an important *necessary condition* for such existence, with a consequent useful method for establishing *nonexistence* of a limit, depending on the idea of *path of approach*:

Definition V. *Let $\phi(t)$ and $\psi(t)$ be functions that approach finite limits a and b, respectively, but are never simultaneously equal to these limits, as t*

† Compare the definition of Cauchy sequence, § 219.

approaches some limit. Then the point (x, y) is said to **approach the point (a, b) as a limit**, along the path $x = \phi(t)$, $y = \psi(t)$. If a function $f(x, y)$ is defined along the path $x = \phi(t)$, $y = \psi(t)$, and if $f(\phi(t), \psi(t))$ approaches a limit L, finite or infinite, as t approaches its limit, the function $f(x, y)$ is said to **approach the limit L as (x, y) approaches (a, b) along the path $x = \phi(t)$, $y = \psi(t)$**.

A similar definition holds for the case where the point (x, y) approaches an infinite limit. (Cf. Ex. 14, § 611.)

Theorem I. *If a function $f(x, y)$ has a limit L, finite or infinite, as (x, y) approaches a point (a, b), then the function $f(x, y)$ approaches the limit L as (x, y) approaches the point (a, b) along any path.*

A similar theorem holds for the case where the point (x, y) approaches an infinite limit.

Proof. For definiteness, assume that L is finite (cf. Ex. 15, § 611), let (x, y) approach (a, b) along the path $x = \phi(t)$, $y = \psi(t)$, and let N_L be an arbitrary neighborhood of L (in E_1). By assumption, there exists a deleted neighborhood $N_{(a,b)}$ of the point (a, b) (in E_2) such that for any point (x, y) in $N_{(a,b)}$ for which $f(x, y)$ is defined, $f(x, y)$ is in N_L. Since (x, y) is assumed to approach (a, b) along the path $x = \phi(t)$, $y = \psi(t)$, it follows that if t is required to be close enough to *its* limit, the point $(\phi(t), \psi(t))$ must be in $N_{(a,b)}$, and hence the number $f(\phi(t), \psi(t))$ must be in N_L. But by definition, this is what we mean by saying that $f(x, y)$ approaches L as (x, y) approaches (a, b) along the given path.

As a corollary (give the proof in Ex. 16, § 611) we have

Theorem II. *If a function $f(x, y)$ has distinct limits as (x, y) approaches a point (a, b) along two distinct paths, the limit $\lim_{(x,y)\to(a,b)} f(x, y)$ does not exist.*

A similar theorem holds for the case where the point (x, y) approaches an infinite limit.

We consider two examples.

Example 1. Show that $\lim_{(x,y)\to(0,0)} \dfrac{xy}{x^2 + y^2}$ does not exist.

Solution. Let $f(x, y) \equiv xy/(x^2 + y^2)$ for $x^2 + y^2 \neq 0$, $f(0, 0) \equiv 0$. Then $f(x, y)$ is identically zero on each coordinate axis, and therefore its limit, as (x, y) approaches the origin along either axis is 0. On the other hand, let (x, y) approach the origin along the straight line $y = mx$, where m is a nonzero real number. Then

$$f(x, mx) = \frac{mx^2}{x^2 + m^2x^2} = \frac{m}{1 + m^2} \to \frac{m}{1 + m^2}.$$

Since this limit depends on the path of approach, the limit given initially cannot exist. The surface $z = f(x, y)$ can be visualized by considering it as generated by a line through and perpendicular to the z-axis.

Example 2. Consider the limit of the function
$$f(x, y) \equiv \frac{x^2 y}{x^4 + y^2}, \quad \text{when } x^4 + y^2 \neq 0, \quad f(0, 0) \equiv 0,$$
by letting (x, y) approach $(0, 0)$ along different paths.

Solution. As in Example 1, $f(x, y)$ approaches the limit 0 along each coordinate axis. Along the straight line $y = mx$ ($m \neq 0$), the function has the limit
$$\lim_{x \to 0} f(x, mx) = \lim_{x \to 0} \frac{mx^3}{x^4 + m^2 x^2} = \lim_{x \to 0} \frac{mx}{x^2 + m^2} = 0.$$
Is it correct to assume that, since $f(x, y)$ has the limit 0 as (x, y) approaches the origin along an arbitrary straight-line path, the limit $\lim_{(x,y) \to (0,0)} f(x, y)$ exists and is equal to 0? The possibly unexpected answer of "No" is obtained by considering the parabolic path $y = x^2$:
$$\lim_{x \to 0} f(x, x^2) = \lim_{x \to 0} \frac{x^4}{x^4 + x^4} = \frac{1}{2} \neq 0.$$
The surface $z = f(x, y)$, with its parabolic escarpment, is an interesting challenge to visualize.

607. ITERATED LIMITS

As was mentioned in § 606, an iterated limit associated with a limit of a function $f(x, y)$, as the point (x, y) approaches a limit, finite or infinite, is obtained by first allowing one independent variable, and then the other, to approach its limit. Thus, corresponding to the limit $\lim_{(x,y) \to (a,b)} f(x, y)$ are the two iterated limits $\lim_{x \to a} \lim_{y \to b} f(x, y)$ and $\lim_{y \to b} \lim_{x \to a} f(x, y)$. Similarly, associated with a limit of a function $f(x, y, z)$ of three variables are six iterated limits and, more generally, for a function of n independent variables there are $n!$ iterated limits.

The principal relation between limits and iterated limits is stated in the Theorem, below. Other facts are revealed in the Examples that follow below, and are summarized in the Notes. (Also cf. Ex. 32, § 611, and the Note, § 1408.) A more difficult and subtle relation, known as the *Moore-Osgood Theorem*, is discussed in *RV*, § 1014.

Theorem. *Consider the three limits*

(1) $\qquad\qquad\qquad \lim_{(x,y) \to (a,b)} f(x, y),$

(2) $\qquad\qquad\qquad \lim_{x \to a} \lim_{y \to b} f(x, y),$

and

(3) $\qquad\qquad\qquad \lim_{y \to b} \lim_{x \to a} f(x, y).$

The existence, finite or infinite, of (1) *and of either of the two iterated limits* (2) *and* (3) *implies the equality of* (1) *and that iterated limit. Consequently, if all three limits exist, they are all equal.*

Proof. Assume for definiteness that the limit (1) exists and is finite (cf. Ex. 17, § 611), and denote this limit by L. We shall show that if the limit (2) exists, it must also be equal to L. Accordingly, let ϵ be an arbitrary positive number and choose a deleted square neighborhood $N_{(a,b)}$ of (a,b) such that for all points (x, y) in $N_{(a,b)}$ for which $f(x, y)$ is defined, $|f(x, y) - L| < \epsilon$. If δ denotes one-half the length of each side of $N_{(a,b)}$, and if x is a number such that $0 < |x - a| < \delta$ and for which the limit function $g(x) \equiv \lim_{y \to b} f(x, y)$ exists, then, since for y sufficiently near b the point (x, y) must lie in $N_{(a,b)}$, it follows that $|g(x) - L| \leq \epsilon$. (Why *does* it follow?) But this is precisely what we mean by the statement $\lim_{x \to a} g(x) = L$.

NOTE 1. The preceding proof has established more than was requested in the statement of the theorem. We know, in fact, that if the limit (1) exists the existence of either *partial limit* $\lim_{y \to b} f(x, y)$ or $\lim_{x \to a} f(x, y)$ implies the existence of the corresponding limit (2) or (3), and hence equality with (1). (Cf. Ex. 20, § 611.)

Example 1. As shown in Example 1, § 606, if $f(x, y) \equiv xy/(x^2 + y^2)$ if $x^2 + y^2 \neq 0$, then $\lim_{(x,y) \to (0,0)} f(x, y)$ does not exist. However, both iterated limits exist and are equal to 0. For instance, $\lim_{x \to 0} \lim_{y \to 0} xy/(x^2 + y^2) = \lim_{x \to 0} 0 = 0$. (Similar statements apply to the function $x^2y/(x^4 + y^2)$ of Example 2, § 606.)

Example 2. Let $f(x, y) \equiv x + y \sin \dfrac{1}{x}$ if $x \neq 0$, and let $f(0, y)$ be undefined. Then, since $|f(x, y)| \leq |x| + |y|$, where defined, $\lim_{(x,y) \to (0,0)} f(x, y) = 0$ (for, if $\epsilon > 0$ and $\delta = \delta(\epsilon) \equiv \frac{1}{2}\epsilon$, then $|f(x, y)| < \epsilon$ wherever $f(x, y)$ is defined in the square neighborhood $|x| < \delta, |y| < \delta$). Furthermore, $\lim_{x \to 0} \lim_{y \to 0} \left(x + y \sin \dfrac{1}{x} \right) = \lim_{x \to 0} x = 0$. On the other hand, $\lim_{y \to 0} \lim_{x \to 0} \left(x + y \sin \dfrac{1}{x} \right)$ does not exist, since for $y \neq 0$, $\lim_{x \to 0} \left(x + y \sin \dfrac{1}{x} \right)$ does not exist.

NOTE 2. The preceding Examples demonstrate that for no two of the three limits (1), (2), (3) do existence and equality imply the existence of the third.

Example 3. Let $f(x, y) \equiv x \sin \dfrac{1}{y} + y \sin \dfrac{1}{x}$ if $xy \neq 0$, $f(x, y) \equiv 0$ if $xy = 0$. As in Example 2, since $|f(x, y)| \leq |x| + |y|$, $\lim_{(x,y) \to (0,0)} f(x, y) = 0$. Neither iterated limit exists.

Example 4. Let $f(x, y) \equiv \dfrac{xy}{x^2 + y^2} + x \sin \dfrac{1}{y}$, if $y \neq 0$, $f(x, 0)$ undefined. Then $\lim_{(x,y) \to (0,0)} f(x, y)$ and $\lim_{x \to 0} \lim_{y \to 0} f(x, y)$ both fail to exist, and $\lim_{y \to 0} \lim_{x \to 0} f(x, y) = 0$.

NOTE 3. Examples 3 and 4 demonstrate that any one of the three limits (1), (2), (3) may exist while the other two fail to exist.

Example 5. Let $f(m, n) \equiv \dfrac{m - n}{m + n}$, where m and n are positive integers, and let $m, n \to +\infty$. The two iterated limits both exist, but are distinct:

$$\lim_{m \to +\infty} \lim_{n \to +\infty} \frac{m - n}{m + n} = \lim_{m \to +\infty} (-1) = -1,$$

$$\lim_{n \to +\infty} \lim_{m \to +\infty} \frac{m - n}{m + n} = \lim_{n \to +\infty} (1) = 1.$$

By the Theorem, the limit $\lim_{\substack{m \to +\infty \\ n \to +\infty}} \dfrac{m - n}{m + n}$ cannot exist.

NOTE 4. Example 5 illustrates that the two iterated limits (2) and (3) may exist and be distinct. In such a case, however, the limit (1) cannot exist.

608. CONTINUITY

Continuity for a function of several variables is in essence the same concept as continuity for a function of a single variable. As in § 209, we give two equivalent formulations, the *limit-definition* and the *neighborhood-definition*, assuming for the former that the particular point at which continuity is defined is a limit point of the domain of definition of the function. For simplicity, the definitions are formulated for a function of two real variables.

Definition I. *A function $f(x, y)$ is* **continuous** *at (a, b) if and only if the following three conditions are satisfied:*

 (i) $f(a, b)$ exists; that is, $f(x, y)$ is defined at (a, b);

 (ii) $\lim_{(x,y) \to (a,b)} f(x, y)$ exists and is finite;

 (iii) $\lim_{(x,y) \to (a,b)} f(x, y) = f(a, b)$.

Definition II. *A function $f(x, y)$ is* **continuous** *at (a, b) if and only if it is defined there and corresponding to an arbitrary neighborhood $N_{f(a,b)}$ of $f(a, b)$, in E_1, there exists a neighborhood $N_{(a,b)}$ of (a, b), in E_2, such that for every point (x, y) of $N_{(a,b)}$ for which $f(x, y)$ is defined, $f(x, y)$ belongs to $N_{f(a,b)}$.*

NOTE 1. In terms of circular neighborhoods, Definition II assumes the form: $f(x, y)$ is continuous at a point (a, b) of the domain of definition if and only if corresponding to an arbitrary positive number ϵ there exists a positive number $\delta = \delta(\epsilon)$ such that $\sqrt{(x - a)^2 + (y - b)^2} < \delta$ implies $|f(x, y) - f(a, b)| < \epsilon$, for points (x, y) for which $f(x, y)$ is defined. For square neighborhoods the last implication is replaced by $|x - a| < \delta$ and $|y - b| < \delta$ imply $|f(x, y) - f(a, b)| < \epsilon$, for points (x, y) for which $f(x, y)$ is defined.

NOTE 2. Any formulation of continuity employs *full neighborhoods* in distinction to the *deleted neighborhoods* used for limits.

NOTE 3. Definition II is applicable even when the point (a, b) is not a limit point of the domain of definition of $f(x, y)$. In this case (a, b) is called an *isolated point* of the domain of definition, and $f(x, y)$ is automatically continuous there, even though $\lim_{(x,y)\to(a,b)} f(x, y)$ has no meaning.

The Theorem of § 209, which states that for a continuous function *the limit of the function is the function of the limit*, generalizes to continuous functions of any number of variables (cf. Exs. 22, 25–29, § 611):

Theorem. *If $f(x, y)$ is continuous at (a, b) and if $\phi(t)$ and $\psi(t)$ are functions such that $\phi(t) \to a$ and $\psi(t) \to b$, as t approaches some limit, and such that $f(\phi(t), \psi(t))$ is defined, then $f(\phi(t), \psi(t)) \to f(a, b)$:*

$$\lim f(\phi(t), \psi(t)) = f(\lim \phi(t), \lim \psi(t)).$$

In particular, if the point (x, y) approaches (a, b) along a path $x = \phi(t)$, $y = \psi(t)$, then $f(x, y)$ approaches $f(a, b)$ as (x, y) approaches (a, b) along the path.

The proof is left as an exercise (Ex. 21, § 611).

The definitions of *continuity on a set* and *continuity*, given in § 209, remain unchanged, as does the definition of a *removable discontinuity*, in § 210.

609. LIMIT AND CONTINUITY THEOREMS

The limit theorems of § 207 and the continuity theorems of § 211, and their proofs, are altered only in minor details when applied to functions of several variables. We shall feel free, therefore, to make use of them without specific delineation here. (Cf. Exs. 22, 25–29, § 611.)

There is, however, an essentially new problem before us now, and that is the relation between a function $f(x, y)$ of two variables (or more generally, n variables) and this same function $f(x, y)$, considered as a function of just one of the independent variables, the other being held fixed. The two most basic theorems in this connection follow, formulated for simplicity for a function of two variables.

Theorem I. *If $f(x, y)$ is continuous at (a, b), then $f(x, b)$ is a continuous function of x at $x = a$, and $f(a, y)$ is a continuous function of y at $y = b$. That is, if a function of two variables is continuous in both variables together, it is continuous in each variable separately.*

Proof. Let $\epsilon > 0$ be given, and let $\delta > 0$ be such that

$$\sqrt{(x - a)^2 + (y - b)^2} < \delta$$

implies $|f(x, y) - f(a, b)| < \epsilon$. Then the inequality $|x - a| < \delta$, since it is equivalent to $\sqrt{(x - a)^2 + (b - b)^2} < \delta$, implies $|f(x, b) - f(a, b)| < \epsilon$. This proves continuity of $f(x, b)$. The other half of the proof is similar.

It will be shown by the Examples, below, that the converse of Theorem I is false: *It is possible for a function of two variables to be continuous in each separately without being continuous in both together*. A partial converse, however, is contained in the following theorem.

Theorem II. *Let $g(x)$ be defined at $x = a$, and let $f(x, y) \equiv g(x)$ for all values of x for which $g(x)$ is defined, and for all values of y. If b is an arbitrary real number, $f(x, y)$ is continuous at (a, b) if and only if $g(x)$ is continuous at a.*

Proof. Half (the "only if" implication) is true by Theorem I. For the other half, assume $g(x)$ is continuous at a, and determine, for a given $\epsilon > 0$, the number $\delta > 0$ so that $|x - a| < \delta$ implies $|g(x) - g(a)| < \epsilon$. Then $\sqrt{(x - a)^2 + (y - b)^2} < \delta$ implies $|x - a| < \delta$, and therefore also $|g(x) - g(a)| = |f(x, y) - f(a, b)| < \epsilon$.

As a consequence of the theorems mentioned at the beginning of this section, and of Theorem II, functions of several variables that are made up, in certain simple ways, of continuous functions of single variables are continuous. In particular, any polynomial (a polynomial in x and y is a sum of terms of the form $ax^m y^n$, where a is a constant and m and n are nonnegative integers) is continuous, and any rational function (a quotient of polynomials) is continuous wherever defined. As further examples, $e^x \cos y$ and $\cos(nx - y \sin x)$ are continuous functions of x and y together. (Cf. Ex. 22, § 611.)

Examples. The two examples of § 606 are continuous everywhere, except at the origin, where they are discontinuous. However, each is a continuous function of each variable for every value of the other! (Verify this.) Example 3, § 607, is continuous in each open quadrant, and at the origin, but at each point of either axis, different from the origin, the function is discontinuous. At such points continuity is hopeless, since the limit fails to exist. That is, there are no removable discontinuities.

610. MORE THEOREMS ON CONTINUOUS FUNCTIONS

The theorems of § 213 have their generalizations to n-dimensional space. In these theorems *intervals* are replaced by higher dimensional sets, of which open and closed intervals are special one-dimensional cases. The key notion for the first two theorems is *compactness*, and that for the last two is *region*. Proofs of the theorems are given in §§ 615–616. The functions are assumed to be real-valued.

Theorem I. *A function continuous on a nonempty compact set is bounded there. That is, if $f(p)$ is continuous on a nonempty compact set A, there exists a number B such that $|f(p)| \leq B$ for all points p belonging to the set A.*

Theorem II. *A function continuous on a nonempty compact set has a maximum value and a minimum value there. That is, if $f(p)$ is continuous on a nonempty compact set A, there exist points p_1 and p_2 in A such that $f(p_1) \leq f(p) \leq f(p_2)$ for all points p belonging to A.*

Theorem III. *If $f(p)$ is continuous on a region or closed region R, and if p_1 and p_2 are two points of R such that $f(p_1)$ and $f(p_2)$ have opposite signs, then there is a point p_3 in the interior of R for which $f(p_3) = 0$.*

Theorem IV. *A function continuous on a region or closed region assumes in its interior all values between any distinct two of its values.*

611. EXERCISES

In Exercises 1–10, determine whether the indicated limit exists, and if it does, find it. Give reasons.

1. $\lim\limits_{(x,y)\to(0,0)} \dfrac{xy}{x^4 + y^4}$.

2. $\lim\limits_{(x,y)\to(0,0)} \dfrac{x + ye^{-x^2}}{1 + y^2}$.

3. $\lim\limits_{(x,y)\to(0,0)} \dfrac{x^2y^2}{x^2 + y^2}$.

4. $\lim\limits_{(x,y)\to(0,0)} \dfrac{x^3y^2}{x^6 + y^4}$.

5. $\lim\limits_{\substack{m\to+\infty \\ n\to+\infty}} \dfrac{m}{m+n}$, m and n positive integers.

6. $\lim\limits_{\substack{m\to+\infty \\ n\to+\infty}} \dfrac{mn}{m^2 + n^2}$, m and n positive integers.

7. $\lim\limits_{\substack{m\to+\infty \\ n\to+\infty}} \dfrac{m+n}{m^2 + n^2}$, m and n positive integers.

8. $\lim\limits_{\substack{m\to+\infty \\ n\to+\infty}} \dfrac{n}{m^2 e^{n/m}}$, m and n positive integers.

9. $\lim\limits_{(x,y)\to(0,0)} \dfrac{xy^2}{x^4 + y^2}$.

10. $\lim\limits_{(x,y)\to(0,0)} xy \ln(x^2 + y^2)$.

11. Generalize to a function of n independent real variables the formulations of limit given in Definition I and Note 1, § 606.

12. Give an explicit definition for $\lim\limits_{\substack{x\to+\infty \\ y\to+\infty}} f(x, y) = +\infty$.

13. Give an explicit definition for $\lim\limits_{(x,y)\to\infty} f(x, y) = \infty$.

14. Reformulate Definition V, § 606, for the cases (a) $x \to +\infty$, $y \to +\infty$; (b) $(x, y) \to \infty$.

15. Prove Theorem I, § 606, for the cases (a) $(x, y) \to (a, b)$, $L = \infty$; (b) $(x, y) \to \infty$, $L = +\infty$.

16. Prove Theorem II, § 606.

17. Prove the Theorem, § 607, for the case $L = -\infty$.

18. Construct an example illustrating the same principle as that of Example 3, § 607, but where $x \to +\infty$, $y \to +\infty$.

19. Construct an example illustrating the same principle as that of Example 5, § 607, but where the independent variables approach finite limits.

20. Prove that if $\lim\limits_{(x,y)\to(a,b)} f(x, y)$ and $\lim\limits_{y\to b} f(x, y)$ both exist, for all x and y near a and b, respectively, then the iterated limit $\lim\limits_{x\to a} \lim\limits_{y\to b} f(x, y)$ exists and is equal

to $\lim_{(x,y)\to(a,b)} f(x, y)$. Show that this result is still valid if x and y are restricted by the added condition $x \neq a$, $y \neq b$. (Cf. Note 1, § 607.)

21. Prove the Theorem of § 608.

22. Prove the following form of extension of Theorem VI, § 211 ("a continuous function of a continuous function is continuous"): *If $f(x, y)$ is continuous at $x = a$, $y = b$, and if $g(r)$ is continuous at $r = f(a, b)$, then $h(x, y) \equiv g(f(x, y))$ is continuous at (a, b).* State and prove the corresponding extension of the Theorem of § 209 ("the limit of the function is the function of the limit").

23. Show that $\ln(1 + x^2 + y^2) + e^x \sin y$ is everywhere continuous. (Cf. Ex. 22.)

24. Discuss continuity of the function $[x + y]$, where the brackets indicate the bracket or greatest integer function (Example 5, § 201).

25. Prove the following form of extension of Theorem VI, § 211 (cf. Ex. 22): *If $\phi(t)$ and $\psi(t)$ are continuous at $t = t_0$, and if $f(x, y)$ is continuous at $x = a = \phi(t_0)$, $y = b = \psi(t_0)$, then $h(t) \equiv f(\phi(t), \psi(t))$ is continuous at $t = t_0$.* (Cf. the Theorem, § 608.)

In Exercises 26–29, state and prove extensions of the Theorem of § 209 and Theorem VI, § 211, for a composite function of the specified form (cf. Exs. 22, 25).

26. $f(\phi(s), \psi(t))$.   **27.** $f(\phi(s, t), \psi(s, t))$.
28. $f(\phi(s, t), \psi(u, v))$.   **29.** $f(\phi(x), \psi(y, z))$.

30. Formulate and prove the Negation of Continuity of a function of n variables (cf. Ex. 14, § 212).

*31. Let $f(x, y)$ be defined on a closed rectangle R: $a \leq x \leq b$, $c \leq y \leq d$, and assume that for each x, $f(x, y)$ is a continuous function of y, and that for all y, $f(x, y)$ is a *uniformly* continuous function of x in the sense that corresponding to $\epsilon > 0$ there exists $\delta > 0$ such that $|x_1 - x_2| < \delta$ implies $|f(x_1, y) - f(x_2, y)| < \epsilon$ for all x_1 and x_2 of $[a, b]$ and all y of $[c, d]$. Prove that $f(x, y)$ is continuous on R.

*32. Define the multiple-valued function $\overline{\lim}_{y \to b} f(x, y)$ to be the function $g(x)$ having all values satisfying the inequalities $\underline{\lim}_{y \to b} f(x, y) \leq g(x) \leq \overline{\lim}_{y \to b} f(x, y)$, and define $\overline{\lim}_{x \to a} f(x, y)$ similarly. Discuss the **generalized iterated limits** $\lim_{x \to a} \overline{\lim}_{y \to b} f(x, y)$ and $\lim_{y \to b} \overline{\lim}_{x \to a} f(x, y)$. Prove that if the limit $\lim_{(x,y) \to (a,b)} f(x, y)$ exists, then both generalized iterated limits exist and are equal to it.

612. MORE GENERAL FUNCTIONS. MAPPINGS

It was remarked in § 606 that the original definition of function (§ 201) permits consideration of real-valued functions of several real variables. This was accomplished by letting the *domain of definition* of the function belong to the n-dimensional space E_n. In similar fashion it is possible to allow the *range of values* of the function more latitude by letting it form part of a Euclidean space E_m. For distinguishing purposes we shall use the term **mapping** or **transformation** to mean a single-valued function whose values are not assumed to be real numbers, preserving the word *function*

§ 612]   MORE GENERAL FUNCTIONS. MAPPINGS   191

(unless otherwise qualified) to mean "real-valued function." One reason for the use of a distinguishing term is that mappings into higher-dimensional spaces are usually not given by single formulas, but rather by *systems* of real-valued functions:

(1)
$$\begin{cases} y_1 = f_1(x_1, x_2, \cdots, x_n), \\ y_2 = f_2(x_1, x_2, \cdots, x_n), \\ \cdots \cdots \cdots \cdots \cdots \\ y_m = f_m(x_1, x_2, \cdots, x_n). \end{cases}$$

It is convenient for many purposes to use functional notation for mappings and write, for instance, $y = f(x)$, $y = T(x)$, $Y = f(X)$, or $q = T(p)$, where x, y, X, Y, p, and q represent points in appropriate spaces. A slight modification of this notation that is particularly suitable for mappings of a

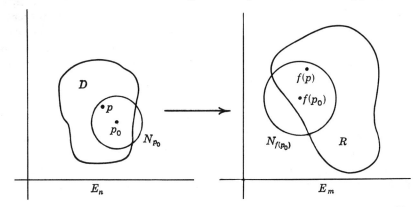

FIG. 606

space into itself is suggested by the example $f((x, y)) = (2x, 3y)$, which transforms an arbitrary point of the Euclidean plane into the point with twice the abscissa and three times the ordinate. Still another method uses arrows, and would indicate the mapping just considered by the symbols $(x, y) \to (2x, 3y)$.

Definitions of limit and continuity, together with many of their simpler properties, are produced by properly interpreting the word *neighborhood* for the appropriate Euclidean space. We shall feel free to use any of these concepts or properties without boring the reader with details that are merely tedious repetitions of work already done. One typical definition should suffice (cf. Fig. 606):

Definition. *A mapping $f(p)$, whose domain of definition D and range of values R belong to E_n and E_m, respectively, is* **continuous** *at the point p_0 in D if and only if corresponding to an arbitrary neighborhood $N_{f(p_0)}$ of $f(p_0)$, in E_m, there exists a neighborhood N_{p_0} of p_0, in E_n, such that for every point p of N_{p_0} that belongs to D, $f(p)$ belongs to $N_{f(p_0)}$.*

How can one determine whether a mapping given by a system of functions is continuous or not? The answer is simple:

Theorem I. *The mapping* (1) *is continuous at a given point if and only if every* f_k, $k = 1, 2, \cdots, m$, *is continuous at that point.*

Proof. Let the given point be $(a_1, a_2, \cdots, a_n)$ and define
$$b_k \equiv f_k(a_1, a_2, \cdots, a_n), \quad k = 1, 2, \cdots, m.$$
The result follows from the inequalities, true for each $k = 1, 2, \cdots, m$:
$$|y_k - b_k| \leq \sqrt{(y_1 - b_1)^2 + \cdots + (y_m - b_m)^2}$$
$$\leq |y_1 - b_1| + \cdots + |y_m - b_m|.$$
(Give the details.)

A mapping f with domain D and range R is said to **map D onto R**. If a set A contains R (A possibly being identical with R), f is said to **map D into A**. The range R is called the **image** of the domain D, and D is called the **preimage** or **inverse image** of R, under the mapping. If f is continuous on D, R is called a **continuous image** of D. A convenient symbolism for "onto" and "into" mappings is the following:

$$D \xrightarrow[\text{onto}]{f} R \quad \text{and} \quad D \xrightarrow[\text{into}]{f} A.$$

If a mapping $y = f(x)$ of D onto R is *one-to-one*, the **inverse mapping** $x = f^{-1}(y)$ of R onto D exists, and the question of its continuity arises (cf. § 215 on monotonic functions and their inverses). The simplest satisfactory statement on this matter follows (cf. § 615 for a proof):

Theorem II. *Any continuous one-to-one mapping whose domain is compact has a continuous inverse.*

NOTE. A mapping that is continuous, and whose inverse is also continuous, is called **bicontinuous** or **topological**. A topological mapping is also called a **homeomorphism**. Two sets are called **topologically equivalent**, or **homeomorphic**, and either set is called a **homeomorph** of the other, in case there exists a homeomorphism having one of the sets as domain and the other as range. (Cf. RV, §§ 1024–1028, for a brief discussion of other topological ideas.)

The question of *existence* of inverse mappings is considered in Chapter 9.

Example 1. The pair of functions
$$(2) \quad \begin{cases} u = x + y, \\ v = x - y, \end{cases}$$
defines a mapping of the space E_2 whose points are designated (x, y) into the space E_2 whose points are designated (u, v). It is possible to solve explicitly for x and y in terms of u and v:
$$(3) \quad \begin{cases} x = \tfrac{1}{2}(u + v), \\ y = \tfrac{1}{2}(u - v). \end{cases}$$

From this fact, and the form of the functions involved, it is clear that the given mapping maps E_2 onto E_2, is one-to-one, and has a continuous inverse. Rewriting (2) in the form

(4)
$$\begin{cases} u = \sqrt{2}\left(x \cos \frac{\pi}{4} + y \sin \frac{\pi}{4}\right), \\ v = -\sqrt{2}\left(-x \sin \frac{\pi}{4} + y \cos \frac{\pi}{4}\right), \end{cases}$$

shows that the mapping is the result of a rotation through 45° together with a change of scale of magnitude $\sqrt{2}$, and a reflection.

Example 2. The mapping

(5)
$$\begin{cases} u = x^2 - y^2, \\ v = 2xy, \end{cases}$$

is continuous everywhere in E_2. Since the points (x, y) and $(-x, -y)$ are both mapped into the same point, it is clear that the mapping is one-to-one only at the origin. To study this mapping more thoroughly we write the coordinates in polar form, with $x = r \cos \theta$, $y = r \sin \theta$, $u = \rho \cos \phi$, $v = \rho \sin \phi$, and find:

(6)
$$\begin{cases} \rho = r^2, \\ \phi = 2\theta \end{cases}$$

(neglecting multiples of 2π). A point in the xy-plane, therefore, is mapped into a point in the uv-plane having its distance from the origin squared, and its polar angle doubled. Therefore, if the domain of definition of the mapping is sufficiently restricted the mapping becomes one-to-one. For example, the open first quadrant is mapped onto the open upper half-plane in a one-to-one manner. With such a restriction, the inverse mapping is always continuous at any point q other than the origin for the following reason: choose a neighborhood N_p of the corresponding point p in the xy-plane such that N_p subtends an angle at the origin of less than $\pi/4$ (say). Then the image, A, of the closure of N_p contains q in its interior. By Theorem II, the mapping with domain restricted to N_p has its inverse continuous on A, and therefore at q.

Example 3. The mapping

(7)
$$(x, y) \to (x, 0),$$

which transforms the point (x, y) into the point on the x-axis having the same abscissa as that of the original point, is everywhere continuous in E_2, but it has no inverse unless the domain is restricted to such an extent that no vertical line meets it in more than one point. It is called a **projection**.

*613. SEQUENCES OF POINTS

In order to treat adequately the unproved theorems of the preceding sections, we turn our attention to the propositions of §§ 217–225, many of which extend to real-valued functions of n variables and, more generally, to mappings with domains in a Euclidean space E_n and ranges in a Euclidean space E_m.

We start by defining convergence of a sequence of points in the space E_m. Let us remark in passing that this is merely an explicit formulation of a limit of a mapping whose domain is the positive integers (a set in the space E_1) and whose range is in the space E_m.

Definition I. *In the Euclidean space E_m, a sequence of points $\{p_n\}$* **converges** *to a point p, written*

$$\lim_{n \to +\infty} p_n = p \quad \text{or} \quad \lim_{n \to \infty} p_n = p \quad \text{or} \quad p_n \to p,$$

if and only if corresponding to an arbitrary positive number ϵ there exists a number $N = N(\epsilon)$ such that $n > N$ implies $d(p_n, p) < \epsilon$.

NOTE. The last part of this definition can be phrased: $\cdots$ *if and only if every neighborhood of p contains all but a finite number of terms of the sequence.*

Our first task with sequences will be to establish a simple relation between convergence of sequences of points in E_m and convergence of sequences of real numbers. For notational convenience we formulate the statement and proof for the special case $m = 2$. (Cf. Ex. 9, § 618.)

Theorem I. *A sequence of points $\{p_n\} = \{(x_n, y_n)\}$ in E_2 converges if and only if both sequences of real numbers $\{x_n\}$ and $\{y_n\}$ converge. More precisely, (i) if $p_n \to p = (a, b)$, then $x_n \to a$ and $y_n \to b$; and (ii) if $x_n \to a$ and $y_n \to b$, then $(x_n, y_n) \to (a, b)$.*

Proof. The theorem is a consequence of the inequalities

$$\left. \begin{array}{c} |x_n - a| \\ |y_n - b| \end{array} \right\} \leq \sqrt{(x_n - a)^2 + (y_n - b)^2} \leq |x_n - a| + |y_n - b|.$$

(Cf. Theorem I, § 612.)

The fundamental theorem on bounded sequences (§ 217) applies to Euclidean spaces in general (a **bounded sequence** being one all of whose terms are contained in some neighborhood):

Theorem II. Fundamental Theorem on Bounded Sequences. *Every bounded sequence in E_m contains a convergent subsequence.*

Proof for Two Dimensions. If $\{x_n, y_n\}$ is bounded, so are the sequences $\{x_n\}$ and $\{y_n\}$. By Theorem I, § 217, the sequence $\{x_n\}$ contains a convergent subsequence. The *corresponding* subsequence of $\{y_n\}$ is again bounded, and contains a convergent subsubsequence, $y_{n_1}, y_{n_2}, y_{n_3}, \cdots$. Then (Theorem IV, § 204) $x_{n_1}, x_{n_2}, x_{n_3}, \cdots$ converges. Therefore the subsequence $\{x_{n_k}, y_{n_k}\}$ of the original sequence converges, by Theorem I.

The Cauchy criterion for convergence of a sequence of points in E_m has formulation and proof almost identical with those of § 219, the only change being the use of the distance $d(p, q)$ between two points (in place of the absolute value of their difference, in the one-dimensional case). We repeat the statements, but omit the proof (Ex. 11, § 618).

Definition II. *A **Cauchy sequence** of points in E_k† is a sequence $\{p_n\}$ such that*
$$\lim_{m,n\to+\infty} d(p_m, p_n) = 0.$$

Theorem III. Cauchy Criterion. *A sequence of points in E_k converges if and only if it is a Cauchy sequence.*

As a final step in setting the stage for two of the most important fundamental theorems in the theory of mappings, and consequent proofs of earlier theorems, we have the *sequential criterion for continuity of mappings* (cf. § 221; give a proof in Ex. 11, § 618):

Theorem IV. *A necessary and sufficient condition for a mapping $f(p)$ to be continuous at a point $p = p_0$ of its domain of definition is that whenever $\{p_n\}$ is a sequence of points that converges to p_0 (and for which $f(p)$ is defined), then $\{f(p_n)\}$ is a sequence of points converging to $f(p_0)$; in short, that $p_n \to p_0$ implies $f(p_n) \to f(p_0)$.*

*614. POINT SETS AND SEQUENCES

In order to make available for higher-dimensional spaces the sequential techniques of §§ 217–224, we establish a few results relating sequences to certain set properties. In particular, we shall see that the key concept that permits the useful extension of the Fundamental Theorem for sequences to more general sets than intervals is compactness. This is shown in Theorem IV of this section. All sets under consideration are assumed to lie in a Euclidean space E_m.

Definition. *A sequence $\{a_n\}$ of points is called a **sequence of distinct points** if and only if no two terms are equal; that is, if and only if $m \neq n$ implies $a_m \neq a_n$.*

Theorem I. *If a sequence $\{a_n\}$ converges to a point a, then either all but a finite number of the terms are equal to a or there exists a subsequence of distinct terms converging to a.*

Proof. Assume that $a_n \to a$ and that for every N there exists an $n > N$ such that $a_n \neq a$. The subsequence sought can be obtained inductively. Let a_{n_1} be the first term different from a. Take $\epsilon = d(a_{n_1}, a) > 0$ and let a_{n_2} be the first term a_n satisfying the inequalities $0 < d(a_n, a) < \epsilon = d(a_{n_1}, a)$, let a_{n_3} be the first term a_n satisfying the inequalities
$$0 < d(a_n, a) < d(a_{n_2}, a), \text{ etc.}$$
By construction, $n_1 < n_2 < n_3 < \cdots$, so that $\{a_{n_k}\}$ is a subsequence, and no two terms are equal.

† The subscript k is used, in place of m, to permit the customary use of the letter m in this formulation.

Theorem II. *A point p is a limit point of a set A if and only if there exists a sequence $\{a_n\}$ of distinct points of A converging to p.*

Proof. If $\{a_n\}$ is a sequence of distinct points of A converging to p, then every neighborhood of p contains all points of the sequence from some index on, and therefore infinitely many points of A, so that p is a limit point of A. On the other hand, if p is a limit point of A, we can find, for each positive integer n, a point p_n of A such that $0 < d(p_n, p) < 1/n$. The sequence $\{p_n\}$ therefore converges to p, and since none of the terms are equal to p it contains (by Theorem I) a subsequence $\{p_{n_k}\}$ no two terms of which are equal, such that $p_{n_k} \to p$. Let $a_k \equiv p_{n_k}$.

Theorem III. *A set A is closed if and only if the limit of every convergent sequence $\{a_n\}$ of points of A is a point of A.*

Proof. "*If*": Assume that the limit of every convergent sequence of points of A is a point of A and let p be a limit point of A. We wish to show that p is a point of A. Since p is a limit point of A, Theorem II guarantees the existence of a sequence of points of A converging to p, so that p must belong to A.

"*Only if*": Let A be a closed set, and let $\{a_n\}$ be a sequence of points of A converging to the point a. We wish to show that a belongs to A. According to Theorem I there are two possibilities. Either $a_n = a$ for all but a finite number of n (in which case a must belong to A) or the sequence $\{a_n\}$ contains a subsequence of distinct terms (in which case, by Theorem II, a must be a limit point of A). In either case, since A contains all of its limit points, a must belong to A.

Theorem IV. *A set A is compact if and only if every sequence $\{a_n\}$ of points of A contains a subsequence converging to a point of A.*

Proof. "*If*": Assume that every sequence $\{a_n\}$ of points of A contains a convergent subsequence whose limit belongs to A. We wish to show that A is bounded and closed. If A were unbounded, then for any fixed point a there would exist a sequence $\{a_n\}$ of points of A such that $d(a_n, a) > n$, so that no subsequence could converge to *any* point. If A were not closed, there would exist (by Theorem III) a sequence $\{a_n\}$ of points of A converging to a point p not a member of A. Since every subsequence would also converge to p, no subsequence could converge to a point of A. These contradictions show that A must be both bounded and closed, and therefore compact.

"*Only if*": Assume that A is compact and let $\{a_n\}$ be an arbitrary sequence of points of A. Since A is bounded, $\{a_n\}$ contains a convergent subsequence $\{a_{n_k}\}$ (Theorem II, § 613), and since A is closed, by Theorem III the limit of this subsequence must belong to A.

*615. COMPACTNESS AND CONTINUITY

Theorems I and II, § 213, on the boundedness and extrema of a real-valued function continuous on a closed interval generalize to mappings in general

Euclidean spaces in a most satisfactory manner. To show this we make use of the results of §§ 613–614.

Theorem I. *Any continuous image of a compact set is compact. That is, if $q = f(p)$ is a continuous mapping whose domain D is a compact subset of a Euclidean space E and whose range R is a subset of a Euclidean space F, then R is compact.*

Proof. We wish to show that an arbitrary sequence $\{q_n\}$ of points of R contains a subsequence converging to a point of R. By definition of the range of a mapping, each point q_n is the image of at least one point p_n of D: $q_n = f(p_n)$. Since D is compact there exists a subsequence $\{p_{n_k}\}$ converging to a point p_0 of D. Continuity of the mapping f at the point p_0 guarantees that $\lim_{k \to +\infty} q_{n_k} = \lim_{k \to +\infty} f(p_{n_k}) = f(p_0)$, by Theorem IV, § 613. In other words $\{q_{n_k}\}$ is the subsequence desired, since it converges to the point $f(p_0)$ which belongs to R, and the proof is complete.

If the range R is a subset of the Euclidean space E_1 of real numbers we can conclude immediately that any real-valued function continuous on a compact set is bounded there. This is Theorem I, § 610. Theorem II, § 610 follows directly from Exercise 30, § 605.

The techniques just applied to the general mapping theorem for compact sets can also be used to prove Theorem II, § 612, on the continuity of an inverse mapping in case the original mapping $q = f(p)$ is one-to-one, with a compact domain D lying in a Euclidean space E and a range R lying in a Euclidean space F:

Proof of Theorem II, § 612. Let $p = \phi(q)$ denote the inverse mapping (with the notation given directly above) and assume that ϕ fails to be continuous at a point q_0 of R. By Theorem IV, § 613, there must exist a sequence $\{q_n\}$ of points of R converging to q_0 such that the sequence $\{p_n\}$, where $p_n \equiv \phi(q_n)$ does *not* converge to the point $p_0 \equiv \phi(q_0)$. The fact that $\{p_n\}$ *fails* to converge to p_0 means that there must be a neighborhood N_{p_0} of p_0 such that infinitely many terms of the sequence $\{p_n\}$ lie *outside* N_{p_0} (cf. the Note, § 613). In other words, there must exist a positive number ϵ and a subsequence of $\{p_n\}$ such that for every term p_n of this subsequence the inequality $d(p_n, p_0) \geq \epsilon$ holds. Since D is compact, the subsequence of $\{p_n\}$ under discussion must itself contain a convergent subsequence, converging to a point $\tilde{p}$ of D. Denoting this subsubsequence by $\{p_{n_k}\}$, we have $p_{n_k} \to \tilde{p}$. We can now infer from the inequality $d(p_{n_k}, p_0) \geq \epsilon$ that $d(\tilde{p}, p_0) \geq \epsilon$. (A contradiction is easily obtained from the triangle inequality of Ex. 22, § 605, $d(p_{n_k}, p_0) \leq d(p_{n_k}, \tilde{p}) + d(\tilde{p}, p_0)$, if the contrary inequality $d(\tilde{p}, p_0) < \epsilon$ is assumed.) The fact that $\tilde{p}$ and p_0 are distinct points gives us the desired contradiction: Since $f(p)$ is continuous at $\tilde{p}$, $p_{n_k} \to \tilde{p}$ implies $q_{n_k} \to \tilde{q} \equiv f(\tilde{p})$. On the other hand, since $q_n \to q_0$, $q_{n_k} \to q_0 = f(p_0)$. Therefore $\tilde{q}$ and q_0, being limits of the same sequence of points, must be identical, in contradiction to the assumption that the mapping $q = f(p)$ is one-to-one.

*616. PROOFS OF TWO THEOREMS

The general mapping theorem of § 615 has a companion theorem, where the property of *compactness* is replaced by that of *connectedness*. (For a discussion of connected sets, and their relation to continuous mappings, see *RV*, §§ 309, 1019, 1020.) Although this general concept will not be treated in this book, a kind of connectedness known as *arc-wise connectedness* is involved in the definition of a region (Definition VIII, § 603). We make use of this now in establishing Theorems III and IV, § 610:

Proof of Theorem III, § 610. Assume the hypotheses of the theorem as stated. If R is a *region*, denote it also by the letter S. If R is a *closed region*, let it be the closure of a region S. In either case, since $f(p)$ is continuous at p_1 and since p_1 is a limit point of points of S, there must exist a point p_1' of S such that $f(p_1')$ and $f(p_1)$ have the same sign. Similarly, let p_2' be a point of S such that $f(p_2')$ and $f(p_2)$ have the same sign. Then $f(p_1')$ and $f(p_2')$ have opposite signs, by assumption. Now let p_1' and p_2' be joined by a broken line segment L lying in S. If $f(p)$ vanishes at one of the vertices of L, let p_3 be this vertex, and we are through. Otherwise, there must exist two consecutive vertices, a and b, at which $f(p)$ has opposite signs. The next step is to construct from $f(p)$, which is a real-valued function continuous on the segment $\overline{ab}$ and having opposite signs at the endpoints, a real-valued function of a real variable having similar properties. For *simplicity of notation* (and not because of any intrinsic significance of dimension) assume that the region S is located in the plane E_2, and let $a = (x_1, y_1)$ and $b = (x_2, y_2)$. Then either $x_1 \neq x_2$ or $y_1 \neq y_2$. Assuming for definiteness that $x_1 < x_2$, define the function $\phi(x) \equiv f(p)$, where x is the x-coordinate of p and $x_1 \leq x \leq x_2$. Then $\phi(x)$ is a continuous function of x on the interval $[x_1, x_2]$ (this is true since $f(p)$ is a continuous function of x and y, the coordinates of p, and y is a continuous linear function of x—cf. Ex. 25, § 611) with opposite signs at the endpoints. Therefore, by Theorem III, § 213, there is a point x_3 between x_1 and x_2 at which $\phi(x)$ vanishes. Let p_3 be the corresponding point of $\overline{ab}$. Finally, since p_3 belongs to S it must be an interior point of R. (Why? Cf. Ex. 20, § 605.)

Proof of Theorem IV, § 610. If c is a real number between $f(p_1)$ and $f(p_2)$, where $f(p)$ is continuous on a region or closed region R and p_1 and p_2 are two points of R, consider the function $g(p) \equiv f(p) - c$. By Theorem III, § 610, $g(p)$ must vanish at some interior point p_3 of R: $f(p_3) - c = 0$.

*617. UNIFORM CONTINUITY

The concept of uniform continuity for mappings in general is virtually the same as that for functions of a single variable:

Definition. *A mapping $f(p)$, defined on a set A, is **uniformly continuous** on A if and only if corresponding to an arbitrary positive number ϵ there exists* a positive number $\delta = \delta(\epsilon)$ such that any two points, p_1 and p_2, of A such that $d(p_1, p_2) < \delta$ map into points $f(p_1)$ and $f(p_2)$ such that $d(f(p_1), f(p_2)) < \epsilon$.

The fact that a real-valued function continuous on a closed interval is uniformly continuous there generalizes to the following (cf. Ex. 12, § 618, for hints on a proof):

Theorem. *Any mapping continuous on a compact set is uniformly continuous there.*

618. EXERCISES

In the following problems all points are assumed to lie in Euclidean spaces.

1. Show that the mappings
$$f((x, y)) = (ax, by), \quad ab \neq 0,$$
$$g((x, y)) = (x + y, y),$$
continuously map E_2 onto E_2, and have continuous inverses. Show that the first is produced by changes of scale and possible reflections, and that the second can be thought of as a kind of *shearing* transformation.

2. Show that the mappings
$$f((x, y)) = (e^x, e^y),$$
$$g((x, y)) = (e^x, \text{Arc tan } y),$$
$$h((x, y)) = (\text{Arc tan } x, \text{Arc tan } y),$$
continuously map E_2 onto the open first quadrant, an open half-infinite strip, and an open square, respectively, and have continuous inverses.

3. Show that $(x, y) \to \left(\dfrac{x}{1 + \sqrt{x^2 + y^2}}, \dfrac{y}{1 + \sqrt{x^2 + y^2}} \right)$ continuously maps E_2 onto the open unit circle, and has a continuous inverse.

4. Show that $(x, y) \to \left(\dfrac{x}{x^2 + y^2}, \dfrac{y}{x^2 + y^2} \right)$, $x^2 + y^2 \neq 0$, is continuous and equal to its own inverse.

5. Show that the mapping
$$(x, y) \to (e^x \cos y, e^x \sin y)$$
is continuous on E_2, but not one-to-one there. By restricting the domain obtain a single-valued continuous inverse.

6. Show that the mapping
$$f(t) = (\cos t, \sin t),$$
with domain $0 \leq t < 2\pi$ in E_1, and range the circumference of the unit circle in E_2, is continuous and one-to-one, but that the inverse mapping is not continuous.

7. Formulate the *Negation of Continuity* for mappings, corresponding to the Definition of § 612.

*8. Prove that if a sequence of points in a Euclidean space converges, the sequence is bounded and its limit is unique.

*9. Formulate the statement and proof of Theorem I, § 613, for the space E_m. *Hint:* Use superscripts to denote terms of a sequence: $p_n = (x_1{}^n, x_2{}^n, \cdots, x_m{}^n)$.

*10. Prove Theorem II, § 613. (Cf. Ex. 9.)

*11. Prove Theorems III and IV, § 613. (Cf. Exs. 9–10.)

★12. Prove the Theorem of § 617. *Hint:* Follow the method of the proof of Theorem II, § 224, replacing absolute values of differences by distances between points, and making double use of the triangle inequality (Ex. 22, § 605).

★13. Formulate and prove the analogue of the Cauchy Criterion (Theorem I, § 222) for mappings.

In Exercises 14–15, find $\delta(\epsilon)$ explicitly, according to the Definition, § 617, of uniform continuity.

★14. $x^2 + y^2$, for $x^2 + y^2 \leq 1$.

★15. $e^x \sin y$, for $x \leq a$.

★16. The **distance between a point p and a nonempty set** A, written $\delta(p, A)$, is defined to be the greatest lower bound of the set of distances $d(p, a)$ between p and arbitrary points a of A. Prove that the distance between p and A is 0 if and only if p belongs to the closure of A. Prove that if p is a point that is not a member of a nonempty closed set A, then there is a point a of A such that $\delta(p, A) = d(p, a) > 0$. *Hint:* Let $\{a_n\}$ be a sequence of points of A such that $d(p, a_n) < \delta(p, A) + \dfrac{1}{n}$ for every positive integer n, and let $a_{n_k} \to a$.

★17. The **distance between two nonempty sets** A and B, written $\delta(A, B)$, is defined to be the greatest lower bound of the set of distances $d(a, b)$ between arbitrary points a of A and b of B. Prove that $\delta(A, B) \geq 0$, and is always zero unless A and B are disjoint. Show that $\delta(A, B)$ may be zero if A and B are disjoint, and even if A and B are disjoint closed sets. Prove that if A and B are nonempty disjoint closed sets at least one of which is bounded (compact), then there exist points a of A and b of B such that $\delta(A, B) = d(a, b)$. Hence prove that $\delta(A, B) > 0$. *Hint:* Assume that A is compact, and choose points a_n of A and b_n of B such that $d(a_n, b_n) < \delta(A, B) + \dfrac{1}{n}$. Let $a_{n_k} \to a$, and choose a convergent subsequence of $\{b_{n_k}\}$.

★18. The **diameter** of a nonempty bounded set A, denoted $\delta(A)$, is defined to be the supremum of all distances $d(p, q)$ for p and q in A: $\delta(A) \equiv \sup_{p,q \in A} d(p, q)$. Prove that the diameter of a nonempty compact set is actually attained as a maximum distance between two of its points. *Hint:* For every $n = 1, 2, \cdots$, let p_n and q_n be points of A such that $d(p_n, q_n) > \delta(A) - \dfrac{1}{n}$. Let $p_{n_k} \to p$, and choose a convergent subsequence of $\{q_{n_k}\}$. (Cf. Exs. 16–17.)

★19. Prove the statement of Note 3, § 603, that in the Euclidean plane the only sets that are both open and closed are the empty set $\varnothing$ and the entire plane. Generalize to n dimensions. *Hint:* Let p belong to a set A that is both open and closed, and let q be a point of the complement A' such that $d(p, q) = \delta(p, A')$. (Cf. Ex. 16.)

★20. By means of the example $y = \sin x$ in E_1, show that a continuous image of an open interval may be open, or closed, or neither. By means of the example $y = e^x$ in E_1, show that a continuous image of a closed set may be closed, or open, or neither.

★21. Let $f(x)$ be real-valued and continuous for all real numbers x, and let c be a constant. Prove that the following sets are open: (*i*) all x such that $f(x) > c$; (*ii*) all x such that $f(x) < c$. Prove that the following sets are closed: (*iii*) all x such that $f(x) \geq c$; (*iv*) all x such that $f(x) \leq c$; (*v*) all x such that $f(x) = c$. If

$f(x)$ is bounded, must any of these sets be bounded? (Prove or give a counterexample.)

⋆**22.** Give an example of a real-valued function, defined for all real numbers x, such that the set of all points of continuity is (*i*) open but not closed; (*ii*) closed but not open; (*iii*) neither open nor closed.

⋆**23.** Define the **graph** G of a real-valued function of a real variable, $y = f(x)$, to be the set of all points (x, y) in E_2 such that x belongs to the domain D of $f(x)$ and $y = f(x)$. Prove that the mapping that carries the point x of D (in E_1) into the point $(x, f(x))$ of G (in E_2) is continuous on D if and only if $f(x)$ is continuous on D. Generalize.

⋆**24.** Prove that the graph of a real-valued function continuous on a compact set is compact (cf. Ex. 23).

⋆**25.** A *limit point* of a sequence $\{a_n\}$ of real numbers was defined in Exercise 7, § 220, to be a number x to which some subsequence converges. Prove that the set of all limit points of a given sequence of real numbers is closed, and that the set of all limit points of a bounded sequence of real numbers is compact. Hence show that any bounded sequence of real numbers has a largest limit point and a smallest limit point (cf. Ex. 30, § 605). Prove that these are the limit superior and limit inferior, respectively, of the sequence. (Cf. Exs. 11–12, § 220.) Extend these results to unbounded sequences.

⋆**26.** Define the limit superior and limit inferior of a real-valued function $f(p)$ at a point p_0, where the domain is in E_n:

$$\overline{\lim_{p \to p_0}} f(p) = \limsup_{p \to p_0} f(p) \equiv \inf_{\delta > 0} \left[\sup_{0 < d(p, p_0) < \delta} f(p) \right],$$

$$\underline{\lim_{p \to p_0}} f(p) = \liminf_{p \to p_0} f(p) \equiv \sup_{\delta > 0} \left[\inf_{0 < d(p, p_0) < \delta} f(p) \right],$$

and prove that they always exist in a finite or infinite sense. (Cf. Exs. 9–10, § 223.)

⋆**27.** State and prove results similar to those of Exercise 11, § 223, for functions of n variables. (Cf. Ex. 26.)

⋆**28.** Prove that for any real-valued function $f(p)$, where p is a point in E_n, $\underline{\lim}_{p \to p_0} f(p) \leq \overline{\lim}_{p \to p_0} f(p)$. Prove that $\lim_{p \to p_0} f(p)$ exists if and only if $\underline{\lim}_{p \to p_0} f(p) = \overline{\lim}_{p \to p_0} f(p)$, and in case of equality,

$$\lim_{p \to p_0} f(p) = \underline{\lim_{p \to p_0}} f(p) = \overline{\lim_{p \to p_0}} f(p).$$

(Cf. Ex. 12, § 223.)

⋆**29.** Prove the **Bolzano-Weierstrass Theorem** for E_n: *Any infinite bounded set has a limit point.* *Hint:* If A is an infinite bounded set, let $\{a_n\}$ be a sequence of distinct points of A, and let $\{a_{n_k}\}$ be a convergent subsequence of $\{a_n\}$ converging to a point p. Show that p is a limit point of A.

⋆**30.** A sequence $\{A_n\}$ of sets is called **monotonically decreasing** if and only if every set A_n of the sequence contains its successor A_{n+1}. This property is symbolized $A_n \downarrow$. (A **constant sequence**, where $A_n \equiv A$, for all n, is an extreme example.) Prove the theorem for E_m: *If $\{A_n\}$ is a monotonically decreasing sequence of nonempty compact sets there exists a point p common to every set of the sequence.* The **nested intervals theorem** is the special case of this theorem where every compact set

A_n is a (finite) closed interval in E_1. Show that this theorem is false if the assumption of compactness is replaced by either boundedness or closedness alone. *Hint:* For every positive integer n let a_n be a point of A_n, and let $a_{n_k} \to p$. For any N, a_{n_k} belongs to A_N for sufficiently large k, so that p also belongs to A_N.

★31. A collection of open sets is said to **cover** a set A if and only if *every* point of A is a member of *some* open set of the collection. Such a collection of open sets is called an **open covering** of A. An open covering of a set A is said to be **reducible to a finite covering** if and only if there exists some finite subcollection of the open sets of the covering which also covers A. Prove the **Heine-Borel Theorem:** *Any open covering of a compact set is reducible to a finite covering.* *Hint for E_1:* First prove the theorem for the special case where the compact set is a (finite) closed interval I, as follows: Assume that F is a collection of open sets (each *member* of F is an open set) which covers I and which is not reducible to a finite covering of I. Consider the two closed intervals into which I is divided by its midpoint. At least one of these two subintervals cannot be covered by any finite collection of sets of F. Call this closed subinterval I_2, and relabel $I = I_1$. Let I_3 be a closed half of I_2 that is not covered by any finite collection of sets of F, and repeat the process to obtain a decreasing sequence $\{I_n\}$ of closed intervals whose lengths tend toward zero. If x is a point common to every interval I_n (Ex. 30), let B be an open set of the family F that contains x. Then B contains a neighborhood $(x - \epsilon, x + \epsilon)$ of x which, in turn, must contain one of the intervals I_n. But this means that I_n is covered by a finite collection of sets of F, namely, the single set B. This contradiction establishes the special case. Now let A be an arbitrary compact set and let F be an arbitrary open covering of A. Let I be a closed interval containing A, and adjoin to the family F the open set A' (A' is the complement of A). This larger collection is an open covering of I, and accordingly is reducible to a finite covering of I, by the first part of the proof. Those sets of F that belong to this finite covering of I cover A. *Hint for E_2:* Make repeated subdivisions of closed rectangles into quarters.

★32. Prove, for E_m, the **Converse of the Heine-Borel Theorem** of Exercise 31: *If every open covering of a set is reducible to a finite covering, the set is compact.* *Hint:* Assume that the set A contains a sequence $\{p_n\}$ of points containing no subsequence converging to a point of A. Let C_n be the closure of the set consisting of the points $p_{n+1}, p_{n+2}, \cdots$, and let U_n be the complement of C_n. Then the family of all U_n, $n = 1, 2, \cdots$, is an open covering of A that is not reducible to a finite covering.

★33. A set A in a Euclidean space is **convex** if and only if for every pair of points p and q of A the straight line segment joining p and q is a subset of A. (Cf. Ex. 21, § 605.) Prove that every open or closed spherical neighborhood, $d(p, p_0) < \epsilon$ and $d(p, p_0) \leq \epsilon$, in E_n, is convex. Prove that the closure of any convex set is a convex set. (Cf. *RV*, Ex. 39, § 1025.)

7

Solid Analytic Geometry and Vectors

701. INTRODUCTION

In this section will be presented some of the elementary portions of the geometry of three dimensions, and a brief introduction to the algebra and geometry of vectors in E_3. It will be assumed that the reader has had at least some brief encounter with the manner in which points in E_3 are determined by a rectangular coordinate system.† We merely recall, therefore,

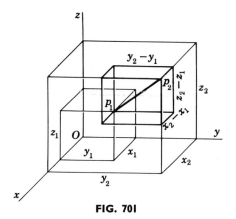

FIG. 701

the one-to-one correspondence existing between points and ordered triads of real numbers, provided by a system of three mutually perpendicular directed **coordinate axes** intersecting in a point O called the **origin**, the formula for distance between two points p: (x_1, y_1, z_1) and q: (x_2, y_2, z_2) (cf. Fig. 701):

(1) $\qquad d(p, q) = \sqrt{(x_2 - x_1)^2 + (y_2 - y_1)^2 + (z_2 - z_1)^2}$,

† For a review of the elements of solid analytic geometry see any good text on Analytic Geometry or Analytic Geometry and Calculus, such as any of the books bearing these titles, by R. W. Brink, E. J. Purcell, and L. L. Smail, in the Appleton-Century Mathematics Series listed in the front of the present volume. For a more complete treatment see the author's *Solid Analytic Geometry*, also in the Appleton-Century Mathematics Series.

and the fact that the **first octant** is defined as the portion of space every point of which has only positive coordinates.

For the sake of applications, a *right-handed* system of axes will be used consistently, according to the following definition:

Definition. *A rectangular coordinate system is* **right-handed** *if and only if the axes are arranged so that if the xy-plane is viewed from a point on the positive half of the z-axis, a counterclockwise ninety-degree rotation in this plane about the origin carries points on the positive half of the x-axis into points on the positive half of the y-axis. The system is* **left-handed** *if and only if it is not right-handed.*†

NOTE 1. Using the concept of a right-handed screw (that is, one that is threaded in such a way that when it is turned about its axis in a clockwise sense it moves away from the observer), we can describe a right-handed rectangular coordinate system as one for which a right-handed screw whose axis coincides with the z-axis advances in the positive z direction when turned through the right angle xOy.

NOTE 2. The relationship between the axes expressed in this definition is a cyclic one. The x-, y-, and z-axes could be replaced (in the definition) by the y-, z-, and x-axes, respectively, or by the z-, x-, and y-axes, respectively, without changing the effect of the definition.

702. VECTORS AND SCALARS

A **vector** in E_3 is an ordered triad of real numbers, called **components** or **coordinates** of the vector, with the notation:

(1) $$\vec{v} = (v_1, v_2, v_3).$$

The vector (1) is said to be **represented by** a directed line segment pq where $p: (x_1, y_1, z_1)$ and $q: (x_2, y_2, z_2)$ are two points of E_3, if and only if the orthogonal projections of pq on the coordinate axes are equal to the corresponding vector components (cf. Fig. 701):

(2) $$x_2 - x_1 = v_1, \quad y_2 - y_1 = v_2, \quad z_2 - z_1 = v_3.$$

Thus, any two parallel and similarly directed line segments of the same length represent the same vector. The statement that two vectors $\vec{u} = (u_1, u_2, u_3)$ and $\vec{v} = (v_1, v_2, v_3)$ are **equal** means that they are the same vector (although they may be represented by different line segments):

$$\vec{u} = \vec{v} \quad \text{if and only if} \quad u_1 = v_1, \quad u_2 = v_2, \quad \text{and} \quad u_3 = v_3.$$

A single vector equation therefore is equivalent to three real-number equations.

It is often convenient to compress a complete but cumbersome expression into an abridged form if the meaning is clear. For example, the phrase "the vector $\vec{pq}$" is merely a short way of saying "the vector represented by the

† For a brief discussion of the relative and intrinsic aspects of right-handedness and left-handedness, see Note 4, § 1703.

directed line segment pq." It is in this spirit that we denote by $\vec{Op}$ the **radius vector** of the point p: (x, y, z):

(3) $$\vec{Op} \equiv (x, y, z).$$

The vector $\vec{pq}$ is said to have **initial point** p and **terminal point** q.

The **magnitude** or **absolute value** of a vector $\vec{v} = (v_1, v_2, v_3)$ is the nonnegative real number defined and denoted:†

(4) $$|\vec{v}| \equiv \sqrt{v_1^2 + v_2^2 + v_3^2}.$$

The **zero vector** $\vec{0} \equiv (0, 0, 0)$ is therefore the unique vector with zero magnitude. A **unit vector** is a vector with magnitude equal to 1. A unit vector is also called a **direction**.

The term **scalar** is a synonym for *real number* that is often used when it is important to distinguish between a number and a vector. The product of a scalar x and a vector $\vec{v} = (v_1, v_2, v_3)$, called a **scalar multiple** of $\vec{v}$, is defined:

(5) $$x\vec{v} \equiv \vec{v}x \equiv (xv_1, xv_2, xv_3).$$

The **negative** of a vector $\vec{v} = (v_1, v_2, v_3)$ is:

(6) $$-\vec{v} \equiv (-1)\vec{v} \equiv (-v_1, -v_2, -v_3).$$

Thus all scalar multiples of a nonzero vector represented by a directed line segment Op can be represented by directed line segments all of which lie on the same line as Op. Any such vectors that can be represented by directed line segments of a single line are called **collinear**. If the scalar factor x is positive, $x\vec{v}$ and $\vec{v}$ are **similarly directed**, and if x is negative, $x\vec{v}$ and $\vec{v}$ are **oppositely directed**. Any vector and its negative are thus oppositely directed vectors of the same magnitude.

Division of a vector by a nonzero scalar is defined:

$$\frac{\vec{v}}{x} \equiv \frac{1}{x}\vec{v}.$$

The result of dividing any nonzero vector by its magnitude:

(7) $$\frac{\vec{v}}{|\vec{v}|}$$

is a unit vector collinear with it, and similarly directed. The vector (7) is the result of **normalizing** $\vec{v}$.

Example 1. If $\vec{v} = (6, -1, 4)$,
$$5\vec{v} = (30, -5, 20),$$
$$-\vec{v} = (-6, 1, -4).$$

Example 2. $x\vec{0} = \vec{0}$, for all scalars x,
$0\vec{v} = \vec{0}$, for all vectors $\vec{v}$.

† The magnitude of a vector is thus the length of any segment representing it.

703. ADDITION AND SUBTRACTION OF VECTORS. MAGNITUDE

Vector addition is defined as follows: If $\vec{u} = (u_1, u_2, u_3)$ and $\vec{v} = (v_1, v_2, v_3)$ are any two vectors, then $\vec{w} = \vec{u} + \vec{v}$ is defined as the vector (w_1, w_2, w_3), where:

(1) $$w_1 \equiv u_1 + v_1, \quad w_2 \equiv u_2 + v_2, \quad w_3 \equiv u_3 + v_3.$$

The vector $\vec{u} + \vec{v}$ is called the **sum** or **resultant** of $\vec{u}$ and $\vec{v}$.

Vector subtraction is defined:

(2) $$\vec{u} - \vec{v} \equiv \vec{u} + (-\vec{v}).$$

The vector $\vec{u} - \vec{v}$ is called the **difference** between $\vec{u}$ and $\vec{v}$.

The sum of two vectors is customarily represented either by the completion of a parallelogram (in case the vectors have a common initial point) or by

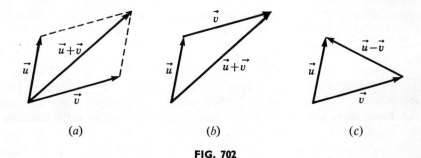

FIG. 702

placing one vector so that its initial point is at the terminal point of the other (Fig. 702). The difference, $\vec{u} - \vec{v}$, between two vectors $\vec{u}$ and $\vec{v}$ having a common initial point is similarly represented by the directed line segment drawn from the terminal point of $\vec{v}$ to that of $\vec{u}$. In each case the vectors are all represented by directed line segments lying in a plane. Such vectors and, in general, any vectors that can be represented by coplanar line segments, are called **coplanar**. (Justify the preceding statements in Ex. 1, § 705.)

The magnitude of a vector has some of the properties of the absolute value of a real number (cf. § 110; also cf. § 1505 for corresponding properties for complex numbers). (Give proofs in Ex. 2, § 705.)

Properties of Magnitude

I. $|\vec{v}| \geq 0$; $|\vec{v}| = 0$ if and only if $\vec{v} = \vec{0}$.

II. $|x\vec{v}| = |x||\vec{v}|$.

III. $\left|\dfrac{\vec{v}}{x}\right| = \dfrac{|\vec{v}|}{|x|}$.

§ 704] LINEAR COMBINATIONS OF VECTORS 207

    IV. *The **triangle inequality**† holds:* $|\vec{u} + \vec{v}| \leq |\vec{u}| + |\vec{v}|$.

    V. $|-\vec{v}| = |\vec{v}|$; $|\vec{u} - \vec{v}| = |\vec{v} - \vec{u}|$.‡

    VI. $|\vec{u} - \vec{v}| \leq |\vec{u}| + |\vec{v}|$.

    VII. $||\vec{u}| - |\vec{v}|| \leq |\vec{u} - \vec{v}|$.

NOTE. The set V of all vectors of E_3 satisfy the following four laws (give a proof in Ex. 3, § 705):

    (i) V is a *commutative group with respect to addition* (cf. Exs. 25, 26, § 103);
    (ii) *the following two **distributive laws** hold for all scalars x and y and all vectors $\vec{u}$ and $\vec{v}$:*

$$x(\vec{u} + \vec{v}) = x\vec{u} + x\vec{v},$$
$$(x + y)\vec{u} = x\vec{u} + y\vec{u};$$

    (iii) *the following **associative law** holds for all scalars x and y and vectors $\vec{v}$:*

$$x(y\vec{v}) = (xy)\vec{v};$$

    (iv) $1\vec{v} = \vec{v}$, *for all vectors $\vec{v}$.*

Any set S of objects that satisfies the preceding four properties, for a given field of scalars for which all scalar multiples of members of S are defined and are members of S, is called a **linear space** or **vector space** over that field. Thus, the vectors of E_3 form a linear space over the field of real numbers. They also form a linear space over the field of rational numbers. (Cf. Ex. 3, § 705.)

704. LINEAR COMBINATIONS OF VECTORS

A **linear combination** of a set of vectors $\vec{v}_1, \vec{v}_2, \cdots, \vec{v}_n$ is a vector of the form

(1) $$a_1\vec{v}_1 + a_2\vec{v}_2 + \cdots + a_n\vec{v}_n,$$

where $a_1, a_2, \cdots, a_n$ are scalars. The vectors $\vec{v}_1, \vec{v}_2, \cdots, \vec{v}_n$ are **linearly dependent** if and only if there exist scalars $a_1, a_2, \cdots, a_n$ not all zero such that

(2) $$a_1\vec{v}_1 + a_2\vec{v}_2 + \cdots + a_n\vec{v}_n = \vec{0}.$$

If no such constants exist, that is, if (2) holds only for constants all of which are zero, then the vectors $\vec{v}_1, \vec{v}_2, \cdots, \vec{v}_n$ are **linearly independent**.

We now prove two fundamental theorems on linear combinations of vectors:

Theorem I. *If $\vec{u}$ and $\vec{v}$ are any two noncollinear vectors, then any vector $\vec{w}$ coplanar with $\vec{u}$ and $\vec{v}$ is a unique linear combination of $\vec{u}$ and $\vec{v}$; that is, there exist unique scalars a and b for which*

(3) $$\vec{w} = a\vec{u} + b\vec{v}.$$

$\vec{u}, \vec{v} \neq 0$

Proof. Represent $\vec{u}, \vec{v}$, and $\vec{w}$ by directed line segments having a common initial point (cf. Fig. 703a). Then lines drawn through the terminal point of

---

† In geometrical terms, no side of a triangle can exceed the sum of the other two (cf. Fig. 702b; also cf. §§ 110, 1505, 1604).
‡ The absolute value of the difference between two vectors having a common initial point is the distance between their terminal points (cf. Fig. 702c).

$\vec{w}$ parallel to $\vec{u}$ and $\vec{v}$ produce a parallelogram with sides $a\vec{u}$ and $b\vec{v}$ and diagonal $\vec{w}$. To establish uniqueness, assume $\vec{w} = a_1\vec{u} + b_1\vec{v} = a_2\vec{u} + b_2\vec{v}$. Then $(a_1 - a_2)\vec{u} = (b_2 - b_1)\vec{v}$, and unless both $(a_1 - a_2)$ and $(b_2 - b_1)$ are zero, the vectors $\vec{u}$ and $\vec{v}$ are collinear. Hence $a_1 = a_2$ and $b_1 = b_2$.

Theorem II. *If $\vec{u}$, $\vec{v}$, and $\vec{w}$ are any three noncoplanar vectors, then any vector $\vec{g}$ is a unique linear combination of $\vec{u}$, $\vec{v}$, and $\vec{w}$; that is, there exist unique scalars a, b, and c for which*

(4) $$\vec{g} = a\vec{u} + b\vec{v} + c\vec{w}.$$

Proof. Represent $\vec{u}$, $\vec{v}$, $\vec{w}$, and $\vec{g}$ by directed line segments having a common initial point (cf. Fig. 703b). Let planes be drawn through the terminal point of $\vec{g}$ parallel to the planes of $\vec{u}$ and $\vec{v}$, $\vec{u}$ and $\vec{w}$, and $\vec{v}$ and $\vec{w}$, completing a

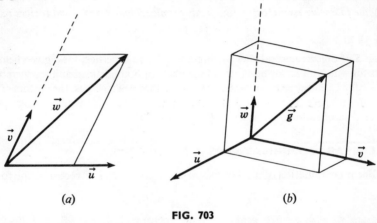

(a)   (b)

FIG. 703

parallelepiped with edges $a\vec{u}$, $b\vec{v}$, and $c\vec{w}$ and diagonal $\vec{g}$. To establish uniqueness, assume $\vec{g} = a_1\vec{u} + b_1\vec{v} + c_1\vec{w} = a_2\vec{u} + b_2\vec{v} + c_2\vec{w}$. Then $(a_1 - a_2)\vec{u} + (b_1 - b_2)\vec{v} + (c_1 - c_2)\vec{w} = \vec{0}$, and unless all three coefficients vanish, at least one of $\vec{u}$, $\vec{v}$, and $\vec{w}$ can be obtained as a linear combination of the other two. But this means that $\vec{u}$, $\vec{v}$, and $\vec{w}$ are coplanar! Hence, $a_1 = a_2$, $b_1 = b_2$, and $c_1 = c_2$.

Owing to the property expressed in Theorem II, any set of three noncoplanar vectors are called a **basis** or **base set of vectors**.

The **basic triad** of unit vectors are defined and denoted:

(5) $$\vec{i} \equiv (1, 0, 0), \quad \vec{j} \equiv (0, 1, 0), \quad \vec{k} \equiv (0, 0, 1).$$

For this basis, the representation of an arbitrary vector $\vec{v} = (v_1, v_2, v_3)$ is particularly simple:

(6) $$\vec{v} = v_1\vec{i} + v_2\vec{j} + v_3\vec{k}.$$

We conclude with some basic facts regarding linear dependence and independence for vectors in E_3 (some of which hold in general vector spaces). Proofs are requested in the indicated Exercises of § 705:

NOTE. A necessary and sufficient condition for a set of vectors to be linearly dependent is that at least one of them is a linear combination of the others (Ex. 4, § 705). A necessary and sufficient condition for a set of vectors to be linearly independent is that any particular linear combination of them has a unique representation as a linear combination (Ex. 5, § 705). A single vector is linearly dependent if and only if it is the zero vector (Ex. 6, § 705). Two vectors are linearly dependent if and only if they are collinear (Ex. 7, § 705). Three vectors are linearly dependent if and only if they are coplanar (Ex. 8, § 705). Four or more vectors are always linearly dependent (Ex. 9, § 705). Any subset of a set of linearly independent vectors are linearly independent (Ex. 10, § 705).

Example. Prove that the diagonals of a parallelogram bisect each other.

First Solution. Let $ABCD$ (Fig. 704) be the given parallelogram, and let M be the midpoint of the diagonal from A to C. By assumption, then, $\overrightarrow{AM} = \overrightarrow{MC}$.

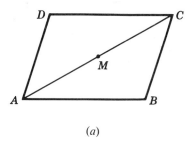
(a)

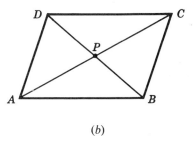
(b)

FIG. 704

We wish to prove that M is also the midpoint of the diagonal from D to B or, in vector notation, that $\overrightarrow{DM} = \overrightarrow{MB}$. Using the equality $\overrightarrow{CB} = \overrightarrow{DA}$, we have

$$\overrightarrow{DM} = \overrightarrow{DA} + \overrightarrow{AM},$$
$$\overrightarrow{MB} = \overrightarrow{MC} + \overrightarrow{CB} = \overrightarrow{AM} + \overrightarrow{DA}.$$

By the commutative law (Note (i), § 703) the desired equality is obtained.

Second Solution. Since

$$\overrightarrow{AC} = \overrightarrow{AB} + \overrightarrow{BC}, \quad \overrightarrow{DB} = \overrightarrow{AB} - \overrightarrow{BC},$$

and $\overrightarrow{AB}$ and $\overrightarrow{BC}$ are linearly independent, so are $\overrightarrow{AC}$ and $\overrightarrow{DB}$ ($x\overrightarrow{AC} + y\overrightarrow{DB} = (x+y)\overrightarrow{AB} + (x-y)\overrightarrow{BC} = \vec{0}$ implies $x = y = 0$). Therefore the lines $\overrightarrow{AC}$ and $\overrightarrow{DB}$ intersect in some point P (Fig. 704). If $\overrightarrow{AP} = s\overrightarrow{AC}$ and $\overrightarrow{DP} = t\overrightarrow{DB}$, then

(7)
$$\begin{cases} \overrightarrow{AB} = \overrightarrow{AP} + \overrightarrow{PB} = s\overrightarrow{AC} + (1-t)\overrightarrow{DB}, \\ \overrightarrow{DC} = \overrightarrow{PC} + \overrightarrow{DP} = (1-s)\overrightarrow{AC} + t\overrightarrow{DB}. \end{cases}$$

Since $\vec{AB} = \vec{DC}$, the coefficients of AC and those of DB must be respectively equal: $s = 1 - s$, $1 - t = t$, and $s = t = \frac{1}{2}$.

Third Solution. $\vec{AC} + \vec{DB} = \vec{AB} + \vec{BC} + \vec{DC} + \vec{CB}$
$$= \vec{AB} + \vec{DC} = 2\vec{AB}.$$

Therefore $\frac{1}{2}\vec{AC} + \frac{1}{2}\vec{DB} = \vec{AB}$, or:
$$\frac{1}{2}\vec{AC} = \vec{AB} + \frac{1}{2}\vec{BD}.$$

This equation can be interpreted as stating that the point reached by starting at A and going halfway to C is the same as the point reached by starting at A, going to B, and then proceeding halfway to D.

705. EXERCISES

1. Justify the representations of vector addition and subtraction as indicated in Figure 702 and requested in § 703.

2. Prove the Properties of Magnitude, § 703. *Hint for* IV: Express both sides of the inequality in terms of components, square, cancel terms, and repeat. Ultimately, justify reversing of all steps (cf. Ex. 33, § 107).

3. Prove that the vectors of E_3 form a linear space over the field of real numbers, and also a linear space over the field of rational numbers, as stated in the Note, § 703.

In Exercises 4–10, prove the statements of the Note, § 704, as requested in the Note.

In Exercises 11–14, express the first vector as a linear combination of the remaining vectors:

11. $(1, -1, 0), (0, 1, -1), (1, 0, -1)$.
12. $(1, 1, -2), (2, -1, 1), (1, 4, -7)$.
13. $(2, 3, 3), (1, 2, 4), (1, 1, 1), (2, 1, -2)$.
14. $(1, 1, 5), (2, -1, -2), (1, -1, -3), (1, 0, 1)$.

In Exercises 15–18, determine whether the vectors are (*i*) collinear, (*ii*) coplanar but not collinear, or (*iii*) noncoplanar.

15. $(1, 1, -2), (3, -1, 4), (3, 1, -1)$.
16. $(10, 5, -15), (2, 1, -3), (-6, -3, 9)$.
17. $(3, 10, -4), (1, -1, 0), (4, -7, -1)$.
18. $(5, -8, 7), (3, -5, 4), (1, -1, 2)$.

19. Use vector methods to prove that the segment joining the midpoints of two sides of any triangle is parallel to the third side and half its length.

20. Use vector methods to prove that the medians of a triangle (lines joining the vertices to the midpoints of opposite sides) intersect in a point of trisection of each.

21. Use vector methods to prove that the midpoints of the sides of any quadrilateral (whether planar or not) are the vertices of a parallelogram.

22. Use vector methods to prove that the lines from any vertex of a parallelogram to the midpoints of the opposite sides trisect the diagonal they intersect.

23. Prove that if M is the midpoint of a line segment AB, and Z is any point, then
$$\vec{ZM} = \frac{\vec{ZA} + \vec{ZB}}{2}.$$

In particular, show that the radius vector of the midpoint p of the segment joining the two points $p_1 : (x_1, y_1, z_1)$ and $p_2 : (x_2, y_2, z_2)$ is

$$\overrightarrow{Op} = \left(\frac{x_1 + x_2}{2}, \frac{y_1 + y_2}{2}, \frac{z_1 + z_2}{2}\right).$$

24. Generalize Exercise 23 to an arbitrary point of division of the segment AB. Discuss in particular the points of trisection of AB.

25. Prove that the sum of any number of vectors is zero if and only if they can be represented by the sides of a closed polygon (possibly self-crossing) directed in cyclic sequence.

26. Prove that if $|\overrightarrow{AB}| = |\overrightarrow{AC}|$, then a point P lies on the bisector of the angle BAC if and only if for any point Q in space there exists a scalar x such that

$$\overrightarrow{QP} = \overrightarrow{QA} + x(\overrightarrow{AB} + \overrightarrow{AC}).$$

27. A **centroid** of a set of points $p_1, p_2, \cdots, p_n$ is any point p such that

$$\overrightarrow{pp_1} + \overrightarrow{pp_2} + \cdots + \overrightarrow{pp_n} = \vec{0}.$$

Show that p exists, is unique, and has radius vector

$$\overrightarrow{Op} = \frac{\overrightarrow{Op_1} + \overrightarrow{Op_2} + \cdots + \overrightarrow{Op_n}}{n}.$$

Show that for three noncollinear points the centroid is the point of intersection of the medians of the associated triangle.

28. Prove that if

$$\vec{v}_1, \vec{v}_2, \cdots, \vec{v}_n, \vec{v}_{n+1}$$

where $n = 1, 2,$ or 3, are vectors such that

$$\vec{v}_1, \vec{v}_2, \cdots, \vec{v}_n$$

are linearly independent and

(1) $$\vec{v}_1 + \vec{v}_2 + \cdots + \vec{v}_{n+1} = \vec{0},$$

then the equation

$$a_1 \vec{v}_1 + a_2 \vec{v}_2 + \cdots + a_{n+1} \vec{v}_{n+1} = \vec{0}.$$

implies $a_1 = a_2 = \cdots = a_{n+1}$.

Formulate and prove a converse.

Conclude that whenever equation (1) holds for $n + 1$ vectors of which a *particular set of n* are linearly independent, then *every set of n* are linearly independent.

706. DIRECTION ANGLES AND COSINES

The angle θ between two directed lines L and M is defined by considering the directed lines L' and M' parallel, respectively, to L and M and passing through the origin. If L' and M' are identical and similarly directed, $\theta = 0$; if L' and M' are coincident but oppositely directed, $\theta = \pi$. Otherwise, consider the plane Π determined by L' and M', and the rotations about an axis through the origin and perpendicular to Π that carry either L' or M' into identical coincidence with the other. Then θ is defined to be the (unique) angle between 0 and π of such a rotation. In general, then,

(1) $$0 \leq \theta \leq \pi.$$

212   SOLID ANALYTIC GEOMETRY AND VECTORS   [§ 706

A similar definition applies to the angle between any two nonzero vectors $\vec{u}$ and $\vec{v}$. If either vector is zero, the angle θ is indeterminate. It follows directly from the definition that the concept of "angle between" is symmetric; that is, the angle between $\vec{u}$ and $\vec{v}$ is the same as the angle between $\vec{v}$ and $\vec{u}$. This angle is also referred to as the angle that either vector makes with the other.

The angles α, β, and γ that a nonzero vector $\vec{v}$ makes with the three directed coordinate axes are called the **direction angles** of $\vec{v}$. Their cosines:

(2) $$\lambda \equiv \cos \alpha, \quad \mu \equiv \cos \beta, \quad \nu \equiv \cos \gamma,$$

are called the **direction cosines** of $\vec{v}$. Any three numbers, not all zero, proportional to the direction cosines:

(3) $$l = k\lambda, \quad m = k\mu, \quad n = k\nu, \quad k \neq 0,$$

are called a set of **direction numbers** of $\vec{v}$.†

If $\vec{v} = (v_1, v_2, v_3)$ is represented by the directed line segment from $p: (x_1, y_1, z_1)$ to $q: (x_2, y_2, z_2)$, then (cf. Fig. 701):

(4) $$\lambda = \frac{x_2 - x_1}{d(p,q)}, \quad \mu = \frac{y_2 - y_1}{d(p,q)}, \quad \nu = \frac{z_2 - z_1}{d(p,q)},$$

(5) $$\lambda = \frac{v_1}{|\vec{v}|}, \quad \mu = \frac{v_2}{|\vec{v}|}, \quad \nu = \frac{v_3}{|\vec{v}|}.$$

From these formulas we see that the direction cosines of any nonzero vector $\vec{v}$ form the components of the unit normalized vector that is collinear with $\vec{v}$ and similarly directed. In particular, the following identity always holds:

(6) $$\lambda^2 + \mu^2 + \nu^2 = 1.$$

Conversely, any three numbers satisfying (6) are a set of direction cosines (of the unit vector having components λ, μ, and ν). Furthermore, a set of direction numbers is a set of direction cosines if and only if the sum of the squares of these numbers is 1. The three numbers

(7) $$x_2 - x_1, \quad y_2 - y_1, \quad z_2 - z_1$$

are a set of direction numbers of the vector $\overrightarrow{pq}$, and (equivalently) the components of any nonzero vector are a set of direction numbers of that vector. Finally, by (5), any set of direction numbers is converted to a set of direction cosines if each is divided by the square root of the sum of their squares.

Example 1. Find the direction cosines of the vector from $(13, -5, 0)$ to $(11, 3, 16)$.

Solution. By (4), $\lambda = \dfrac{11 - 13}{18} = -\dfrac{1}{9}$, $\mu = \dfrac{3 + 5}{18} = \dfrac{4}{9}$, $\nu = \dfrac{16 - 0}{18} = \dfrac{8}{9}$.

Example 2. Find direction cosines of a line with direction numbers $6, -2, 3$.

† Direction angles, cosines, and numbers of a *directed line* are similarly defined. It should be noted that direction *numbers*, in contrast to direction *cosines*, do *not* determine direction on a line.

§ 707]   THE SCALAR OR INNER OR DOT PRODUCT   213

Solution. $\pm\sqrt{36+4+9} = \pm 7$. Therefore, by (5), the two sets of direction cosines are $\frac{6}{7}, -\frac{2}{7}, \frac{3}{7}$, and $-\frac{6}{7}, \frac{2}{7}, -\frac{3}{7}$. The first set applies if the line is directed upward, making an acute angle with the z-axis, and the second set applies if the line is directed downward.

707. THE SCALAR OR INNER OR DOT PRODUCT

The **scalar product** or **inner product** or **dot product** of two vectors $\vec{u} = (u_1, u_2, u_3)$ and $\vec{v} = (v_1, v_2, v_3)$ is defined and denoted:

(1) $$\vec{u} \cdot \vec{v} \equiv u_1 v_1 + u_2 v_2 + u_3 v_3.$$

In this section we shall discuss the geometric significance of this concept.†

If either $\vec{u}$ or $\vec{v}$ is the zero vector, $\vec{u} \cdot \vec{v} = 0$. Assuming for present purposes that neither $\vec{u}$ nor $\vec{v}$ is the zero vector, let θ be the angle between $\vec{u}$ and $\vec{v}$, and let $(\lambda_1, \mu_1, \nu_1)$ and $(\lambda_2, \mu_2, \nu_2)$ be the unit vectors whose components are their

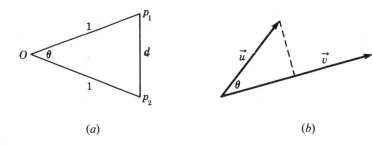

(a)   (b)

FIG. 705

direction cosines (§ 706). Applying the law of cosines to the triangle whose vertices are the origin O and the two points $p_1: (\lambda_1, \mu_1, \nu_1)$ and $p_2: (\lambda_2, \mu_2, \nu_2)$, and denoting by d the distance between p_1 and p_2 (cf. Fig. 705a), we have:

$$\cos \theta = \frac{2 - d^2}{2} = \frac{2 - [(\lambda_1 - \lambda_2)^2 + (\mu_1 - \mu_2)^2 + (\nu_1 - \nu_2)^2]}{2}$$

$$= \frac{2 - [(\lambda_1^2 + \mu_1^2 + \nu_1^2) + (\lambda_2^2 + \mu_2^2 + \nu_2^2) - 2(\lambda_1 \lambda_2 + \mu_1 \mu_2 + \nu_1 \nu_2)]}{2},$$

or, after simplification:

(2) $$\cos \theta = \lambda_1 \lambda_2 + \mu_1 \mu_2 + \nu_1 \nu_2.$$

Upon substitution into (1) of the expressions $u_1 = \lambda_1 |\vec{u}|$, $u_2 = \mu_1 |\vec{u}|$, $u_3 = \nu_1 |\vec{u}|$, $v_1 = \lambda_2 |\vec{v}|$, $v_2 = \mu_2 |\vec{v}|$, and $v_3 = \nu_2 |\vec{v}|$ (cf. (5), § 702), we obtain:

(3) $$\boxed{\vec{u} \cdot \vec{v} = |\vec{u}| \cdot |\vec{v}| \cdot \cos \theta.}$$

This product is called a *scalar product* because the result is always a real number. In Chapter 17, another kind of product of two vectors, called the *vector* or *outer* or *cross product*, is introduced and used. In this latter case the result is always a vector.

If, for the case where either $\vec{u}$ or $\vec{v}$ is the zero vector we permit the indeterminate angle θ to be arbitrarily assigned, we have the following general theorem:

Theorem. *The scalar product of two vectors $\vec{u}$ and $\vec{v}$ is equal to the product of their magnitudes and the cosine of the angle between them.*

NOTE 1. The dot product of a vector and itself is sometimes denoted by means of an exponent 2. In this special case ($\vec{u} = \vec{v}$), formula (3) gives:

(4) $$\vec{u}^2 \equiv \vec{u} \cdot \vec{u} = |\vec{u}|^2.$$

If θ is the angle between two nonzero vectors $\vec{u}$ and $\vec{v}$, the **projection** (more precisely, the **orthogonal projection**) of $\vec{u}$ on $\vec{v}$ is a vector collinear with $\vec{v}$ with magnitude $||\vec{u}|\cos\theta|$, defined and denoted:

(5) $$\operatorname{proj}_v \vec{u} \equiv |\vec{u}|\cos\theta \, \frac{\vec{v}}{|\vec{v}|}.$$

The **component** of $\vec{u}$ in the direction of $\vec{v}$ is the scalar:

(6) $$\operatorname{comp}_v \vec{u} \equiv |\vec{u}|\cos\theta.$$

(Cf. Fig. 705b.) From formulas (3), (5), (6) we have:

(7) $$\vec{u}\cdot\vec{v} = (\operatorname{proj}_v \vec{u})\cdot\vec{v} = (\operatorname{proj}_u \vec{v})\cdot\vec{u}$$
$$= (\operatorname{comp}_v \vec{u})|\vec{v}| = (\operatorname{comp}_u \vec{v})|\vec{u}|.$$

In particular, the scalar product of any vector $\vec{u}$ and a *unit vector* $\vec{v}$ is equal to the component of $\vec{u}$ in the direction of $\vec{v}$.

NOTE 2. Familiar laws of the real number system find their counterparts in the algebra of scalar products (give a proof, and a discussion of the geometric significance, in Ex. 1, § 709):

(i) *The following **commutative law** holds for all vectors $\vec{u}$ and $\vec{v}$:*
$$\vec{u}\cdot\vec{v} = \vec{v}\cdot\vec{u}.$$

(ii) *The following **homogeneous law** holds for all scalar x and y and vectors $\mathbf{u}$ and $\vec{v}$:* $(x\vec{u})\cdot(y\vec{v}) = (xy)(\vec{u}\cdot\vec{v}).$

(iii) *The following **distributive laws** hold for all vectors $\vec{u}$, $\vec{v}$, and $\vec{w}$:*
$$\vec{u}\cdot(\vec{v}+\vec{w}) = \vec{u}\cdot\vec{v} + \vec{u}\cdot\vec{w},$$
$$(\vec{u}+\vec{v})\cdot\vec{w} = \vec{u}\cdot\vec{w} + \vec{v}\cdot\vec{w}.$$

NOTE 3. The zero vector is by definition orthogonal (or perpendicular) to all vectors (including itself). Therefore it can be stated in complete generality that *two vectors $\vec{u}$ and $\vec{v}$ are orthogonal if and only if their scalar product vanishes:*
$$\vec{u}\cdot\vec{v} = u_1v_1 + u_2v_2 + u_3v_3 = 0.$$

NOTE 4. The associative law for scalar products is meaningless since the scalar product of two vectors is not a vector.

NOTE 5. The following multiplication table holds for scalar products of the basic unit vectors $\vec{i}, \vec{j}$, and $\vec{k}$:

(8)

	$\vec{i}$	$\vec{j}$	$\vec{k}$
$\vec{i}$	1	0	0
$\vec{j}$	0	1	0
$\vec{k}$	0	0	1

§ 707]   THE SCALAR OR INNER OR DOT PRODUCT   215

NOTE 6. A vector, as defined in § 702, is an ordered triad of real numbers. This triad is necessarily associated with a particular rectangular coordinate system. A change of coordinate system may produce a change in this triad, that is, in the components of a given vector. For example, if the axes of a given right-handed coordinate system are rotated about the z-axis through an angle of 90° in such a way that the positive x-axis is transformed into the positive y-axis, then the vector $\vec{j} = (0, 1, 0)$ in the original system is represented by the triad $(1, 0, 0)$ in the new system. (Such coordinate transformations are studied more extensively in §§ 1706–1708.) However, since the magnitude of any vector and the angle between any two vectors are unchanged by any rectangular coordinate transformation, the theorem of this section establishes the fact that *the scalar product of any two vectors is independent of the rectangular coordinate system*. (Cf. Example 1, § 1708.)

Example 1. If $\vec{u} = (1, 1, 4)$ and $\vec{v} = (0, 1, -1)$.

$$|\vec{u}| = \sqrt{18} = 3\sqrt{2}, \quad |\vec{v}| = \sqrt{2}.$$

$$\vec{u} \cdot \vec{v} = 0 + 1 - 4 = -3,$$

$$\cos \theta = \frac{-3}{3\sqrt{2} \cdot \sqrt{2}} = -\frac{1}{2}, \quad \theta = \frac{2}{3}\pi,$$

$$\text{proj}_v \vec{u} = -\frac{3}{2}\sqrt{2}\frac{(0, 1, -1)}{\sqrt{2}} = \left(0, -\frac{3}{2}, \frac{3}{2}\right),$$

$$\text{proj}_u \vec{v} = -\frac{\sqrt{2}}{2}\frac{(1, 1, 4)}{3\sqrt{2}} = \left(-\frac{1}{6}, -\frac{1}{6}, -\frac{2}{3}\right).$$

Example 2. Discuss the cancellation law for scalar products:
(9) $\vec{u} \cdot \vec{v} = \vec{u} \cdot \vec{w}$ implies $\vec{v} = \vec{w}$.

Solution. The example $\vec{u} = \vec{i}$, $\vec{v} = \vec{j}$, $\vec{w} = \vec{k}$ shows that (9) is false as it stands. However, if the equation $\vec{u} \cdot \vec{v} = \vec{u} \cdot \vec{w}$ holds for *all* $\vec{u}$, then $\vec{v}$ and $\vec{w}$ must be equal. This follows from the fact that if $\vec{u} \cdot \vec{v} = \vec{u} \cdot \vec{w}$ for all $\vec{u}$, then $\vec{u} \cdot (\vec{v} - \vec{w}) = 0$, for all $\vec{u}$. In particular, this last equation holds when $\vec{u} = \vec{v} - \vec{w}$, and therefore, by (4):

$$(\vec{v} - \vec{w}) \cdot (\vec{v} - \vec{w}) = |\vec{v} - \vec{w}|^2 = 0,$$

$\vec{v} - \vec{w} = \vec{0}$, and $\vec{v} = \vec{w}$.

Example 3. If $\vec{u}_1$, $\vec{u}_2$, and $\vec{u}_3$ are three linearly independent vectors, and if $\vec{v}$ and $\vec{w}$ are any two vectors, prove that the equality
(10) $\vec{v} = \vec{w}$
holds if and only if
(11) $\vec{u}_m \cdot \vec{v} = \vec{u}_m \cdot \vec{w}, \quad m = 1, 2, 3$.

Solution. It is obvious that (10) implies (11). On the other hand, if (11) holds, then the equation $\vec{u} \cdot \vec{v} = \vec{u} \cdot \vec{w}$ holds for every vector $\vec{u}$ that can be written as a linear combination of $\vec{u}_1$, $\vec{u}_2$, and $\vec{u}_3$. But by Theorem II, § 704, every vector $\vec{u}$ can be so represented. Therefore, by Example 2, $\vec{v} = \vec{w}$.

Example 4. Prove that the sum of the squares of the diagonals of any parallelogram is the sum of the squares of the four sides.

Solution. If the vectors $\vec{u}$ and $\vec{v}$ represent two adjacent sides of a parallelogram, we are to show that

$$|\vec{u} + \vec{v}|^2 + |\vec{u} - \vec{v}|^2 = 2|\vec{u}|^2 + 2|\vec{v}|^2.$$

This follows from (4) and the laws of Note 2:
$$(\vec{u} + \vec{v}) \cdot (\vec{u} + \vec{v}) + (\vec{u} - \vec{v}) \cdot (\vec{u} - \vec{v}) = \vec{u}^2 + 2\vec{u} \cdot \vec{v} + \vec{v}^2$$
$$+ \vec{u}^2 - 2\vec{u} \cdot \vec{v} + \vec{v}^2 = 2\vec{u}^2 + 2\vec{v}^2.$$

708. VECTORS ORTHOGONAL TO TWO VECTORS

Let $\overrightarrow{Op}$ and $\overrightarrow{Oq}$ be two noncollinear unit vectors (the angle θ between them is neither 0 nor π), where O is the origin and p and q are the points $(\lambda_1, \mu_1, \nu_1)$ and $(\lambda_2, \mu_2, \nu_2)$, respectively. The problem of finding a vector orthogonal to both $\overrightarrow{Op}$ and $\overrightarrow{Oq}$ reduces to that of finding a point (x, y, z) other than O on the line Λ that passes through O and is perpendicular to the plane of O, p, and q. The assumption that the radius vector of (x, y, z) is orthogonal to both $\overrightarrow{Op}$ and $\overrightarrow{Oq}$ takes the form (by Note 3, § 707) of the simultaneous system in the three unknowns x, y, and z:

(1) $$\begin{cases} \lambda_1 x + \mu_1 y + \nu_1 z = 0, \\ \lambda_2 x + \mu_2 y + \nu_2 z = 0. \end{cases}$$

The vanishing of two determinants:

(2) $$\begin{vmatrix} \lambda_1 & \mu_1 & \nu_1 \\ \lambda_1 & \mu_1 & \nu_1 \\ \lambda_2 & \mu_2 & \nu_2 \end{vmatrix} = \begin{vmatrix} \lambda_2 & \mu_2 & \nu_2 \\ \lambda_1 & \mu_1 & \nu_1 \\ \lambda_2 & \mu_2 & \nu_2 \end{vmatrix} = 0,$$

expressed specifically by expansions in terms of the first row of each, states that a particular solution of (1) is given by the cofactors in these expansions:

(3) $$x = \begin{vmatrix} \mu_1 & \nu_1 \\ \mu_2 & \nu_2 \end{vmatrix}, \quad y = \begin{vmatrix} \nu_1 & \lambda_1 \\ \nu_2 & \lambda_2 \end{vmatrix}, \quad z = \begin{vmatrix} \lambda_1 & \mu_1 \\ \lambda_2 & \mu_2 \end{vmatrix}.$$

It remains, now, only to show that the three second order determinants of (3) cannot all vanish simultaneously. We show this as a simple corollary of the following theorem:

Theorem I. *If θ is the angle between the unit vectors $(\lambda_1, \mu_1, \nu_1)$ and $(\lambda_2, \mu_2, \nu_2)$, then*

(4) $$\sin^2 \theta = \begin{vmatrix} \mu_1 & \nu_1 \\ \mu_2 & \nu_2 \end{vmatrix}^2 + \begin{vmatrix} \nu_1 & \lambda_1 \\ \nu_2 & \lambda_2 \end{vmatrix}^2 + \begin{vmatrix} \lambda_1 & \mu_1 \\ \lambda_2 & \mu_2 \end{vmatrix}^2.$$

Proof. The right-hand member of (4) is identically equal to
$$(\lambda_1^2 + \mu_1^2 + \nu_1^2)(\lambda_2^2 + \mu_2^2 + \nu_2^2) - (\lambda_1 \lambda_2 + \mu_1 \mu_2 + \nu_1 \nu_2)^2,$$
and this latter expression is equal to $1 - \cos^2 \theta$.

As a further corollary we have (cf. Ex. 2, § 709):

Theorem II. *If $\vec{u} = (u_1, u_2, u_3)$ and $\vec{v} = (v_1, v_2, v_3)$ are any two vectors, the vector*
$$\vec{w} \equiv \left(\begin{vmatrix} u_2 & u_3 \\ v_2 & v_3 \end{vmatrix}, \begin{vmatrix} u_3 & u_1 \\ v_3 & v_1 \end{vmatrix}, \begin{vmatrix} u_1 & u_2 \\ v_1 & v_2 \end{vmatrix} \right)$$
is orthogonal to both $\vec{u}$ and $\vec{v}$. If $\vec{u}$ and $\vec{v}$ are noncollinear (and hence nonzero), then any vector orthogonal to both $\vec{u}$ and $\vec{v}$ must be a scalar multiple of $\vec{w}$.

Example 1. Find the sine of the angle between the unit vectors $(\tfrac{2}{3}, \tfrac{1}{3}, \tfrac{2}{3})$ and $(-\tfrac{4}{9}, \tfrac{7}{9}, -\tfrac{4}{9})$.

First Solution. The scalar product of these two vectors is
$$\cos \theta = -\tfrac{8}{27} + \tfrac{7}{27} - \tfrac{8}{27} = -\tfrac{9}{27} = -\tfrac{1}{3}.$$
Therefore $\sin^2 \theta = 1 - \tfrac{1}{9} = \tfrac{8}{9}$, and $\sin \theta = \tfrac{2}{3}\sqrt{2}$.

Second Solution. By formula (4)
$$\sin^2 \theta = (\tfrac{2}{3})^2 + (0)^2 + (\tfrac{2}{3})^2 = \tfrac{8}{9}, \quad \sin \theta = \tfrac{2}{3}\sqrt{2}.$$

Example 2. Find all vectors orthogonal to both $(4, -1, -1)$ and $(3, -7, 3)$.

Solution. One vector orthogonal to both is:
$$\left(\begin{vmatrix} -1 & -1 \\ -7 & 3 \end{vmatrix}, \begin{vmatrix} -1 & 4 \\ 3 & 3 \end{vmatrix}, \begin{vmatrix} 4 & -1 \\ 3 & -7 \end{vmatrix} \right) = -5(2, 3, 5).$$
Therefore *every* vector orthogonal to both must have the form
$$t(2, 3, 5) = (2t, 3t, 5t).$$

709. EXERCISES

1. Prove the laws given in Note 2, § 707, and give a discussion of their geometric significance.
2. Prove Theorem II, § 708.
3. Of the following ordered triads of angles, choose those that are sets of direction angles:
 (a) 30°, 45°, 60°; (b) 30°, 90°, 120°; (c) 90°, 0°, 270°;
 (d) 90°, 135°, 45°; (e) 30°, 150°, θ; (f) 120°, 135°, 60°.
4. If two direction angles of a directed line are 60° and 60°, find the third direction angle.

In Exercises 5–9, find the direction cosines of the given directed line.
5. The x-axis. 6. The y-axis. 7. The z-axis.
8. The line $x = y$ in the xy-plane, directed from the origin into the third quadrant.
9. The line that makes equal angles with the coordinate axes, directed from the origin into the first octant.

In Exercises 10–13, find the direction cosines of the given line segment, directed from the first point to the second.
10. $(3, 8, 1), (1, 2, 10)$.   11. $(0, -5, 13), (6, 4, -5)$.
12. $(10, 11, 12), (6, 3, 31)$.   13. $(6, -1, -2), (4, 9, 9)$.
14. Find the point in the first octant 28 units from the origin, whose radius vector has direction cosines satisfying the relations $\lambda = 2\mu = 3\nu$.

In Exercises 15–18, find the cosine of the acute angle between two lines having the given direction numbers.

15. $3, -2, 0;\ 4, 3, 1.$
16. $0, 1, 2;\ 5, -2, -4.$
17. $5, 1, -4;\ 2, -1, 4.$
18. $2, 3, 5;\ -3, 5, 2.$

In Exercises 19–22, determine whether the two given sets of direction numbers are associated with perpendicular lines.

19. $2, 8, 3;\ 5, 1, -6.$
20. $3, -1, 7;\ 5, 1, -2.$
21. $3, 2, -5;\ -2, 1, -1.$
22. $-7, 5, 11;\ -3, -13, 4.$

In Exercises 23–24, find the dot product, $\cos\theta$, and the projection of each vector on the other:

23. $\vec{u} = (6, 3, -3),\ \vec{v} = (-6, 2, 9).$
24. $\vec{u} = (-2, 3, -1),\ \vec{v} = (2, 2, -1).$

In Exercises 25–30, find a set of direction numbers for a line perpendicular to two lines having the given direction numbers.

25. $1, 2, -3;\ 3, -7, 4.$
26. $0, 5, 2;\ 0, 3, -7.$
27. $3, -2, 8;\ -6, 4, 5.$
28. $1, 7, 5;\ 1, 1, 1.$
29. $3, -2, 1;\ -4, 1, 7.$
30. $8, 6, -3;\ -6, 1, 5.$

31. Prove that any set of nonzero vectors every two of which are orthogonal are linearly independent.

710. PLANES

If (a, b, c) is a nonzero vector, and $p_0\colon (x_0, y_0, z_0)$ an arbitrary point, then the point $p\colon (x, y, z)$ lies in the plane Π through p_0 perpendicular to (a, b, c) if and only if the vectors $\overrightarrow{p_0 p}$ and (a, b, c) are perpendicular; that is, if and only if

$$(1) \qquad a(x - x_0) + b(y - y_0) + c(z - z_0) = 0.$$

Equation (1), then, is the equation of the plane† through the point (x_0, y_0, z_0) perpendicular to all lines having direction numbers a, b, and c. If equation (1) is multiplied out, with collecting of terms, we have established the first half of the following theorem (for the second half, cf. Ex. 1, § 712):

Theorem I. *Any plane is the graph of some linear equation*

$$(2) \qquad ax + by + cz + d = 0,$$

where the coefficients a, b, and c are direction numbers of the perpendiculars or normals to the plane. Conversely, the graph of any linear equation (2), where not all of the coefficients a, b, and c vanish, is a plane.

The results of § 708 give important facts concerning pairs of planes. Two of these are stated here, with proofs requested in Exercise 2, § 712:

---

† The use of the expression "the equation of the plane" (as if it were unique) is for convenience. A similarly helpful expediency is the contraction "the plane $a(x - x_0) + \cdots$" in place of "the plane whose equation is $a(x - x_0) + \cdots$."

Theorem II. *A set of direction numbers of the line of intersection of the two planes*

(3) $\quad a_1x + b_1y + c_1z + d_1 = 0, \qquad a_2x + b_2y + c_2z + d_2 = 0$

is

(4) $\quad \begin{vmatrix} b_1 & c_1 \\ b_2 & c_2 \end{vmatrix}, \quad \begin{vmatrix} c_1 & a_1 \\ c_2 & a_2 \end{vmatrix}, \quad \begin{vmatrix} a_1 & b_1 \\ a_2 & b_2 \end{vmatrix}.$

Theorem III. *The equation of the plane perpendicular to the two planes of (3) and passing through the point (x_0, y_0, z_0) is*

(5) $\quad \begin{vmatrix} b_1 & c_1 \\ b_2 & c_2 \end{vmatrix}(x - x_0) + \begin{vmatrix} c_1 & a_1 \\ c_2 & a_2 \end{vmatrix}(y - y_0) + \begin{vmatrix} a_1 & b_1 \\ a_2 & b_2 \end{vmatrix}(z - z_0) = 0,$

or

(6) $\quad \begin{vmatrix} x - x_0 & y - y_0 & z - z_0 \\ a_1 & b_1 & c_1 \\ a_2 & b_2 & c_2 \end{vmatrix} = \begin{vmatrix} x & y & z & 1 \\ x_0 & y_0 & z_0 & 1 \\ a_1 & b_1 & c_1 & 0 \\ a_2 & b_2 & c_2 & 0 \end{vmatrix} = 0.$

We conclude by mentioning the **intercept form** for the equation of any plane that crosses the three coordinate axes:

(7) $\quad \dfrac{x}{r} + \dfrac{y}{s} + \dfrac{z}{t} = 1,$

where r, s, and t are the x-, y-, and z-intercepts, respectively. (Cf. § 713.)

Example 1. Find the equation of the plane parallel to $3x - 2y + 4z + 3 = 0$ passing through the point $(1, 1, -2)$.

Solution. By (1), the equation is $3(x - 1) - 2(y - 1) + 4(z + 2) = 0$, or $3x - 2y + 4z + 7 = 0$.

Example 2. Find the equation of the plane through the three points p_1: $(1, 12, 1)$, p_2: $(2, 7, -1)$, and p_3: $(-4, 5, 3)$.

First Solution. The coefficients a, b, c, in the equation of the plane are direction numbers of a line perpendicular to the plane and therefore to the vectors p_1p_2 and p_1p_3, which have direction numbers $1, -5, -2$ and $5, 7, -2$, respectively. Therefore (by Theorem II, § 708) $a : b : c = 24 : -8 : 32 = 3 : -1 : 4$. The equation thus has the form $3x - y + 4z + d = 0$. Substitution gives the value of d, and the equation is $3x - y + 4z + 5 = 0$.

Second Solution. Let the equation of the plane be $ax + by + cz + d = 0$. Substitution of the coordinates of the three points gives the following system of three equations in the four unknowns, a, b, c, d:

$$a + 12b + c + d = 0$$
$$2a + 7b - c + d = 0$$
$$-4a + 5b + 3c + d = 0.$$

Subtraction of the left members of the first equation and the second, and of the first equation and the third, gives the two equations

$$a - 5b - 2c = 0$$
$$5a + 7b - 2c = 0.$$

As in the first solution, $a:b:c = 3:-1:4$, and the work is completed in the same way. Notice that the algebraic work is practically the same by the two methods.

For a third solution, see Exercise 12, § 712.

711. LINES

The line Λ through the point $p_0: (x_0, y_0, z_0)$ and with direction numbers a, b, and c is the set of all points $p: (x, y, z)$ such that the vector $\overrightarrow{p_0 p}$ is a scalar multiple t of the vector (a, b, c):

(1) $\qquad (x - x_0, y - y_0, z - z_0) = t(a, b, c) = (at, bt, ct).$

This gives immediately the equations of Λ in **parametric form** (t being the parameter):

(2) $\qquad x = x_0 + at, \quad y = y_0 + bt, \quad z = z_0 + ct.$

If Λ is determined by two points $p_0: (x_0, y_0, z_0)$ and $p_1: (x_1, y_1, z_1)$, equations (2) take the form:

(3) $\qquad x = x_0 + (x_1 - x_0)t, \quad y = y_0 + (y_1 - y_0)t, \quad z = z_0 + (z_1 - z_0)t.$

If the collinearity of the vectors $\overrightarrow{p_0 p}$ and (a, b, c) is expressed as a proportion, equations (2) take the following **symmetric form**:

(4) $\qquad \dfrac{x - x_0}{a} = \dfrac{y - y_0}{b} = \dfrac{z - z_0}{c},$

where the vanishing of any denominator is to be interpreted as implying the vanishing of the corresponding numerator, rather than as division by 0.

Equations (2) and (4) can be interpreted as equations of the line through (x_0, y_0, z_0) perpendicular to the plane $ax + by + cz + d = 0$.

One further formula should be included: The equations of the line of intersection of the two planes $a_1 x + b_1 y + c_1 z + d_1 = 0$ and $a_2 x + b_2 y + c_2 z + d_2 = 0$ are:

(5) $\qquad \dfrac{x - x_0}{\begin{vmatrix} b_1 & c_1 \\ b_2 & c_2 \end{vmatrix}} = \dfrac{y - y_0}{\begin{vmatrix} c_1 & a_1 \\ c_2 & a_2 \end{vmatrix}} = \dfrac{z - z_0}{\begin{vmatrix} a_1 & b_1 \\ a_2 & b_2 \end{vmatrix}},$

where (x_0, y_0, z_0) is an arbitrary point of that line.

Example 1. Find equations of the line through each pair of points: (a) $(3, 1, -4)$, $(5, -6, 1)$; (b) $(4, 5, -3)$, $(6, -1, -3)$; (c) $(6, 1, 7)$, $(1, 1, 7)$.

Solution. Direct substitution in equations (1) gives for the first pair
$$\frac{x-3}{2} = \frac{y-1}{-7} = \frac{z+4}{5},$$
for the second pair
$$\frac{x-4}{2} = \frac{y-5}{-6} = \frac{z+3}{0},$$
or
$$\frac{x-4}{1} = \frac{y-5}{-3} \quad \text{and} \quad z = -3,$$
and for the third pair
$$\frac{x-6}{-5} = \frac{y-1}{0} = \frac{z-7}{0},$$
or
$$y = 1 \quad \text{and} \quad z = 7.$$

Example 2. Find equations in symmetric form for the line
$$x - 4y + 2z + 7 = 0, \quad 3x + 3y - z - 2 = 0.$$

First Solution. First, by Theorem II, § 710, we find $\lambda:\mu:\nu = -2:7:15$. To find a point on the line we can eliminate z to obtain $7x + 2y + 3 = 0$. A solution of this equation is easily found by giving either x or y a value. A convenient solution is $x = -1, y = 2$. The corresponding value of z is 1, and the equations are
$$\frac{x+1}{-2} = \frac{y-2}{7} = \frac{z-1}{15}.$$

Second Solution. As in the first solution, eliminate z. Then solve for y in terms of x. In a similar way, eliminate x and solve for y in terms of z. The two resulting equations can be combined as follows:
$$\frac{-7x - 3}{2} = y = \frac{7z + 23}{15},$$
or
$$\frac{x + \frac{3}{7}}{-2} = \frac{y}{7} = \frac{z + \frac{23}{7}}{15}.$$

This last set of equations is in symmetric form and shows that the given line has the direction numbers obtained before and passes through the point $(-\frac{3}{7}, 0, -\frac{23}{7})$. The choice of the variable y was arbitrary.

Third Solution. Using the method of the first solution, we find any two points on the line, say $(-1, 2, 1)$ and $(1, -5, -14)$, and from their coordinates, a set of direction numbers of the line. The equations can then be written down as before.

Example 3. Write equations of the line through the point $(1, 5, -2)$ perpendicular to the plane $7x - y + 8z + 11 = 0$.

Solution. The coefficients in the equation of the plane are direction numbers of the line. Therefore its equations are
$$\frac{x-1}{7} = \frac{y-5}{-1} = \frac{z+2}{8}.$$

Example 4. Write an equation of the plane through $(1, 5, -2)$ perpendicular to the line

$$\frac{x+9}{7} = \frac{y-3}{-1} = \frac{z}{8}.$$

Solution. The denominators can be used as coefficients in the equation of the plane. Determining the constant term by substitution, we have the equation: $7x - y + 8z + 14 = 0$. Alternatively, we could substitute in equation (1), § 710.

Example 5. Show that the plane $x + 4y - z - 1 = 0$ contains the line

$$\frac{x-3}{2} = \frac{y+1}{1} = \frac{z+2}{6}.$$

Solution. Since $2 \cdot 1 + 1 \cdot 4 - 6 \cdot 1 = 0$, the line is perpendicular to the normals of the plane and is therefore parallel to the plane. Furthermore, the point $(3, -1, -2)$ lies on the line and in the plane. Therefore the line lies entirely in the plane.

712. EXERCISES

1. Complete the proof of Theorem I, § 710. *Hint:* Find a point (x_0, y_0, z_0) whose coordinates satisfy (2), and hence put (2) into the form of (1).
2. Prove Theorems II and III, § 710.
3. Find the equation of the plane through the point $(-3, 5, 1)$ perpendicular to a line with direction numbers $4, 1, -3$.
4. Find the equation of the plane through the point $(1, 2, -4)$ perpendicular to the radius vector of the point.
5. Find the equation of the plane through the point $(1, -1, 6)$ parallel to the plane $2x + 5y - z + 9 = 0$.
6. Find the equation of the plane that is the perpendicular bisector of the segment $(4, 3, -9)(8, -1, 5)$.
7. Find the equation of the plane through the points $(4, -3, 2)$ and $(1, 1, -5)$ and parallel to the z-axis. *Hint:* Find the equation of the line in the xy-plane that passes through the points $(4, -3)$ and $(1, 1)$.

In Exercises 8–11, find the equation of the plane through the three given points.

8. $(0, 0, 0), (1, 1, -1), (0, 2, 1)$.
9. $(-3, 0, 0), (-6, -2, 0), (3, 0, 2)$.
10. $(1, 1, 6), (2, -1, -1), (5, -2, 3)$.
11. $(2, 1, 3), (6, 3, 5), (-2, 2, 8)$.
12. By using direction numbers of vectors orthogonal to both vectors $(x_1 - x_3, y_1 - y_3, z_1 - z_3)$ and $(x_2 - x_3, y_2 - y_3, z_2 - z_3)$, show that the equation of the plane through the three noncollinear points $(x_i, y_i, z_i), i = 1, 2, 3$, can be written:

$$\begin{vmatrix} x - x_3 & y - y_3 & z - z_3 \\ x_1 - x_3 & y_1 - y_3 & z_1 - z_3 \\ x_2 - x_3 & y_2 - y_3 & z_2 - z_3 \end{vmatrix} = \begin{vmatrix} x & y & z & 1 \\ x_1 & y_1 & z_1 & 1 \\ x_2 & y_2 & z_2 & 1 \\ x_3 & y_3 & z_3 & 1 \end{vmatrix} = 0.$$

✓ **13.** Let $\vec{u} = (\lambda, \mu, \nu)$ be a unit vector normal to a plane Π. Let the equation of Π be written in the form
$$\lambda x + \mu y + \nu z + d = 0. \quad (1)$$
Show that the *directed distance from* Π *to* any point p_0: (x_0, y_0, z_0) (the direction being that of $\vec{u}$) is
$$\lambda x_0 + \mu y_0 + \nu z_0 + d. \quad (2)$$
Hint: Let p_1: (x_1, y_1, z_1) be any point in Π, and find the component of $\overrightarrow{p_1 p_0}$ in the direction of $\vec{u}$.

14. Prove that the nonnegative distance between the plane $ax + by + cz + d = 0$ and the point p_0: (x_0, y_0, z_0) is
$$\left| \frac{ax_0 + by_0 + cz_0 + d}{\sqrt{a^2 + b^2 + c^2}} \right|. \quad (3)$$
(Cf. Ex. 13.)

15. Prove that the distance between the parallel planes
$$ax + by + cz + d_1 = 0, \quad ax + by + cz + d_2 = 0$$
is
$$\left| \frac{d_2 - d_1}{\sqrt{a^2 + b^2 + c^2}} \right|.$$
(Cf. Exs. 13–14.)

16. Prove that if $\Pi_1 = \Pi_1(x, y, z)$ and $\Pi_2 = \Pi_2(x, y, z)$ are two linear expressions in x, y, and z, and if $\Pi_1 = 0$ and $\Pi_2 = 0$ are two intersecting planes, then the family of planes
$$k_1 \Pi_1 + k_2 \Pi_2 = 0, \quad (4)$$
where k_1 and k_2 are parameters or arbitrary constants not both zero, is the family of all planes passing through the line of intersection of the planes $\Pi_1 = 0$ and $\Pi_2 = 0$.

In Exercises 17–22, find the equation of the plane that passes through the line of intersection of the planes
$$x - y + 2z + 5 = 0 \quad \text{and} \quad 2x + 3y - z - 1 = 0$$
and that satisfies the given condition. (Cf. Ex. 16.)

✓ **17.** It passes through the origin.
18. It passes through the point $(1, -1, 4)$.
✓ **19.** It is parallel to the z-axis.
20. It is perpendicular to the plane $x + 2y - 2z = 0$.
✓ **21.** It is parallel to the segment $(1, 1, -1)$ $(3, 5, -3)$.
22. It has equal y and z intercepts.
23. Show that the following lines are parallel:
$$x + y - 2z + 5 = 0, \quad 2x + 5y + 2z - 6 = 0;$$
$$2x + y - 6z - 1 = 0, \quad 3x + 4y - 4z = 0.$$
24. Show that the following lines are perpendicular:
$$2x + y - 4z + 10 = 0, \quad x + 5y + 7z - 6 = 0;$$
$$5x - y - 3z + 18 = 0, \quad 3x - 4y + 5z + 1 = 0.$$
25. Show that the following three lines are identical:
$$\frac{x-1}{1} = \frac{y+9}{-2} = \frac{z-28}{5}; \quad x = -4 + t, \quad y = 1 - 2t, \quad z = 3 + 5t;$$
$$7x + y - z + 30 = 0, \quad 4x - 3y - 2z + 25 = 0.$$

In Exercises 26–29, write equations in symmetric form and in parametric form for the line through the given points.

26. $(5, 1, 8), (7, 4, 5)$.
27. $(-3, 2, 0), (1, 1, 1)$.
28. $(3, 4, 1), (3, 7, 7)$.
29. $(7, -9, -2), (20, -9, -2)$.

In Exercises 30–33, write equations in symmetric form and in parametric form for the line through the given point perpendicular to the given plane.

30. $(3, 5, 6);\ 5x + 8y - z + 4 = 0$.
31. $(-2, 1, 0);\ x - 3y - 2z - 5 = 0$.
32. $(0, 5, 1);\ 6x + 2y - 7 = 0$.
33. $(1, -3, -4);\ 3y = 10$.

In Exercises 34–36, write the equation of the plane through the point $(5, -1, 0)$ perpendicular to the given line.

34. $\dfrac{x-2}{3} = \dfrac{y}{-2} = \dfrac{z+5}{-6}$.
35. $\dfrac{x+3}{4} = \dfrac{y+7}{0} = \dfrac{z-1}{-5}$.
36. $3x - y - 3z - 16 = 0,\ 3x + y - 2z - 5 = 0$.
37. Write the equations of the three planes passing through the line $\dfrac{x-3}{-5} = \dfrac{y+2}{6} = \dfrac{z}{1}$, each of which is parallel to a coordinate axis.

In Exercises 38–39, show that the two lines intersect, and find the point of intersection.

38. $\dfrac{x+2}{5} = \dfrac{y-2}{3} = \dfrac{z+2}{0};\ \dfrac{x+3}{2} = \dfrac{y-5}{0} = \dfrac{z-7}{-3}$.
39. $x = 3 + 5t, y = 5 + 8t, z = -t;\ x = 10 + 6t, y = -7 - 2t, z = 7 + 3t$.

In Exercises 40–41, find the point where the given line intersects the given plane.

40. $x = -9 + 5t, y = -17 + 8t, z = 12 - 6t;\ 5x - 3y + 6z - 8 = 0$.
41. $\dfrac{x+8}{2} = \dfrac{y-8}{-4} = \dfrac{z+2}{1};\ 4x + y + z + 11 = 0$.

In Exercises 42–43, find the (orthogonal) projection of the given point on the given plane.

42. $(13, 4, -27);\ 3x + y - 7z + 4 = 0$.
43. $(-10, 13, 3);\ 6x - 5y - 2z + 1 = 0$.

In Exercises 44–45, show that the two lines intersect, and find the equation of the plane containing them.

44. $\dfrac{x-1}{2} = \dfrac{y+4}{5} = \dfrac{z-3}{-3};\ \dfrac{x+3}{3} = \dfrac{y-9}{-4} = \dfrac{z+14}{7}$.
45. $2x + 2y + 3z + 3 = 0,\ x + 2y + 2z + 1 = 0;$
$x - y + z + 3 = 0,\ 2x + y + 8z + 9 = 0$.

In Exercises 46–47, write the equation of the plane containing the first line and parallel to the second.

46. $x = -y = z;\ \dfrac{x-5}{1} = \dfrac{y+2}{11} = \dfrac{z+5}{-3}$.
47. $\dfrac{x-4}{1} = \dfrac{y-5}{2} = \dfrac{z+1}{3};\ \dfrac{x+13}{2} = \dfrac{y+10}{4} = \dfrac{z+3}{-5}$.
*48. Find the distance between the point $(2, -1, 7)$ and the line $\dfrac{x-5}{1} = \dfrac{y-2}{-2} = \dfrac{z+1}{1}$. *Hint:* Find the magnitude of the vector from the point $(5, 2, -1)$ to the point $(2, -1, 7)$, and the sine of the angle between this vector and the given line.

★49. Find the distance between the lines $\dfrac{x+5}{3} = \dfrac{y-3}{2} = \dfrac{z}{4}$ and $\dfrac{x-1}{1} = \dfrac{y+2}{0} = \dfrac{z+3}{4}$. *Hint:* Find the parallel planes containing these two lines.

713. SURFACES. SECTIONS, TRACES, INTERCEPTS

For present purposes the word *surface* will be used to mean the graph of an equation of the form $f(x, y, z) = 0$, where it will be assumed that the function $f(x, y, z)$ is sufficiently well-behaved to exclude some of the pathological anomalies that are otherwise possible. For further discussion, see §§ 1811–1823, and the author's *Solid Analytic Geometry* referred to in § 701.

The curve of intersection of a surface and a plane is called a **section** of that surface by the plane. The section of a surface by a coordinate plane is called the **trace** of the surface in that plane. An equation of a trace of a surface in the two variables of the coordinate plane of the trace is easily obtained by setting the third variable equal to zero in the given equation of the surface.

If a coordinate axis intersects a surface, such a point of intersection (or the appropriate coordinate) is called an **intercept**. The intercepts of a surface on any axis can be found by setting equal to zero the variables corresponding to the other axes. Sections of a surface are very useful in determining its general shape or in drawing a sketch of it.

Example. Find the traces and intercepts of the surface
$$x^2 + y^2 - z^2 - 2x = 0.$$

Solution. By equating to zero in turn x, y, and z, we obtain the equations of the traces in the *yz*-, *xz*-, and *xy*-planes:
$$y^2 - z^2 = 0, \quad x^2 - z^2 - 2x = 0, \quad x^2 + y^2 - 2x = 0.$$
These traces are two straight lines, a hyperbola, and a circle, respectively. The *x*-intercepts are 0 and 2, and the *y*- and *z*-intercepts are both zero.

714. SPHERES

Using the formula for the distance between two points (§ 701), we see that the equation

(1) $$(x - x_0)^2 + (y - y_0)^2 + (z - z_0)^2 = R^2$$

is satisfied by the coordinates of those points and only those points $p: (x, y, z)$ that are at a distance R from the point $p_0: (x_0, y_0, z_0)$. It is therefore an equation of the sphere with center at p_0 and radius equal to R.

If equation (1) is expanded and its terms are collected, it assumes the form

(2) $$x^2 + y^2 + z^2 + 2px + 2qy + 2rz + d = 0.$$

On the other hand, if we start with any equation of the form (2), by completing squares we can write it in the form

(3) $$(x + p)^2 + (y + q)^2 + (z + r)^2 = s,$$

where $s = p^2 + q^2 + r^2 - d$. If $s > 0$, (3) is obviously an equation of a **real sphere**. If $s = 0$, equation (3) is satisfied by the coordinates of only one real point, $(-p, -q, -r)$. In this case the real graph consists of this one point and is called a **point sphere**. If $s < 0$, equation (3) has only imaginary solutions and is called an **imaginary sphere**. In any case, we call the surface (2) a **sphere** and the real point $(-p, -q, -r)$ its **center**.

Example. Describe the graph of the equation

$$4x^2 + 4y^2 + 4z^2 - 16x + 24y + 8z + d = 0,$$

where d is in turn equal to 47, 56, and 65.

Solution. We divide each member of the equation by 4 and complete squares to obtain the equivalent equation

$$(x - 2)^2 + (y + 3)^2 + (z + 1)^2 = \frac{56 - d}{4}.$$

Regardless of the value of d the graph is a sphere with center $(2, -3, -1)$. If $d = 47$, the sphere is real with radius equal to $\frac{3}{2}$. If $d = 56$, the sphere is a point sphere whose only real point is its center. If $d = 65$, the sphere is imaginary.

715. CYLINDERS

A *cylinder* is any surface generated by a line moving parallel to a fixed line. A more precise definition is the following:

Definition. *Let C be a plane curve and let F denote the family of all lines through points of C perpendicular to the given plane. The surface consisting of all points of these lines is called a* **cylinder**. *Each line of the family F is called a* **ruling** *or* **generator**, *and the curve C is called a* **directrix**. *A cylinder is said to be* **parallel** *to its rulings and* **perpendicular** *to any plane perpendicular to them.*

Since the normal sections of a cylinder cut by planes perpendicular to the rulings are congruent, there is no ambiguity in speaking of *the* directrix. (Cf. Fig. 706.)

If the directrix is one or more straight lines, the cylinder is one or more planes. Otherwise the cylinder is a curved surface, usually named after the directrix. For example, if the directrix is a circle, the cylinder is a right circular cylinder. Similarly, we speak of elliptic, hyperbolic, and parabolic cylinders.

Although a cylinder need not be parallel to a coordinate axis, we shall consider in this chapter only those cylinders that are, and shall assume also that the directrix is in a coordinate plane. The principal theorem for such cylinders is the following:

Theorem. *A surface whose equation has a missing variable is a cylinder parallel to the axis of the missing variable. The curve corresponding to this equation, as an equation in the other two variables, is the directrix located in their coordinate plane.*

Proof. For definiteness let
$$(1) \qquad f(x, y, z) = f(x, y) = 0$$
be an equation in which z is missing. This equation, regarded either as an equation in three variables or as an equation in two variables, will in general impose restrictions on x and y, but none on z. If (x_1, y_1, z_1) is any point on

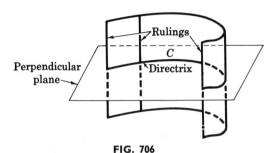

FIG. 706

the graph of (1), then (x_1, y_1, z_2), where z_2 has any value whatsoever, is also on the graph. Therefore, corresponding to any point (x_1, y_1) on the curve
$$(2) \qquad f(x, y) = 0, \quad z = 0,$$
the line $x = x_1, y = y_1$, parallel to the z-axis, lies in the surface (1), which must therefore be a cylinder parallel to the z-axis. The curve (2) is a directrix of the surface (1).

If two variables are missing, the cylinder is parallel to the axes of both missing variables and consists of a collection of planes.

Example 1. Discuss and sketch the surface
$$x^2 + 4y^2 = 16.$$
Solution. The surface is an elliptic cylinder parallel to the z-axis. The normal sections are ellipses with semiaxes 4 and 2 with centers on the z-axis. The z-axis is called the *axis* of the cylinder.

Example 2. Discuss the surface $z^4 = 1$.
Solution. The surface consists of two real horizontal planes, $z = 1$ and $z = -1$.

Example 3. Discuss the graph of the equation $z = \sin x$.
Solution. The surface is a "sinusoidal" cylinder parallel to the y-axis, extending above and below the xy-plane.

716. SURFACES OF REVOLUTION

A surface S is called a **surface of revolution** in case there is a line Λ such that every section of S perpendicular to Λ is a circle with center on Λ. Such a surface can be thought of as being generated by a plane curve, called a **generatrix**, revolved about Λ as axis of revolution. Particularly well adapted to algebraic treatment is a surface of revolution having a coordinate axis as axis of revolution.

Figure 707 illustrates the surface obtained by revolving about the x-axis a portion of the graph of a function $f(x)$ and the graphs of the equation $y = f(x)$ and those of the related equations $y = -f(x)$, $y = |f(x)|$, and $y^2 = (f(x))^2$. It is readily seen that the surfaces obtained by revolving about the

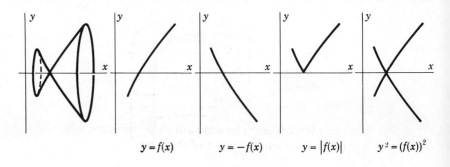

FIG. 707

x-axis the curves shown in Figure 707 are all the same, and that therefore it is not quite reasonable to speak of *the* generatrix, although it is sometimes convenient to do so.

We shall obtain the equation of the surface of revolution by considering the last curve of Figure 707, which can be described as the locus of a point in the xy-plane moving so that the square of its distance from the x-axis is equal to the square of $f(x)$, where x is the directed distance of the point from the y-axis. As the curve is revolved about the x-axis, this point remains at a constant distance from the x-axis and also at a constant distance from the yz-plane. The surface of revolution can therefore be described as the locus of a point in space moving so that the square of its distance from the x-axis is equal to the square of $f(x)$, where x is now the directed distance of the point from the yz-plane. Since the square of the distance between the point (x, y, z) and the x-axis is $y^2 + z^2$, the equation of the surface of revolution is

$$y^2 + z^2 = (f(x))^2.$$

In other words, to find the equation of the surface obtained by revolving the curve $y = f(x)$ about the x-axis, we square each side of the equation $y = f(x)$ and replace y^2 by $y^2 + z^2$. The student will undoubtedly observe with pleasure that this is one occasion when he can square both sides of an equation with impunity—there is no danger that anything extraneous will be introduced in the process.

It is obvious that the graph of any equation in which y and z appear only in the form $y^2 + z^2$ is a surface of revolution about the x-axis. We can now say, conversely, that any such surface is the graph of such an equation.

A similar procedure is followed in finding an equation of a surface obtained by revolving a curve in any coordinate plane about a coordinate axis lying in it. The results are tabulated below, not as a substitute for separate analyses of individual problems, but to emphasize the naturalness of the formulas.

Curve in the	revolved about	replace	by
xy-plane	x-axis	y^2	$y^2 + z^2$
	y-axis	x^2	$x^2 + z^2$
xz-plane	x-axis	z^2	$y^2 + z^2$
	z-axis	x^2	$x^2 + y^2$
yz-plane	y-axis	z^2	$x^2 + z^2$
	z-axis	y^2	$x^2 + y^2$

A check on one's work is given by comparing the generatrix with the trace of the surface of revolution in the plane of the generatrix.

Example 1. Write the equation of the surface of revolution obtained by revolving the curve
$$x = z^2, \quad y = 0$$
(*a*) about the x-axis; (*b*) about the z-axis.

Solution. (*a*) Replace z^2 by $y^2 + z^2$. The equation is $x = y^2 + z^2$. (*b*) Square both members of the equation $x = z^2$ and replace x^2 by $x^2 + y^2$. The equation is $x^2 + y^2 = z^4$.

Example 2. Show that $x^2 + y^2 = z^2$ is the equation of a right circular cone.

Solution. The surface is obtained by revolving about the z-axis (*i*) the curve $x^2 = z^2$ or $x = z$ or $x = -z$ in the xz-plane, or (*ii*) the curve $y^2 = z^2$ or $y = z$ or $y = -z$ in the yz-plane. The surface obtained by revolving one of two intersecting lines about the other is a right circular cone.

717. EXERCISES

In Exercises 1–4, find an equation of the trace of the given surface in the specified coordinate plane. Identify the trace.

1. $x^2 + y^2 + z^2 - 2xz + 5z - 4 = 0$; xy-plane.
2. $xy + xz + yz = 1$; xz-plane.
3. $x^2 + 4y^2 + z^2 + 4xy - 2xz - 2x - 4y + z + 1 = 0$; xy-plane.
4. $x^2 + xy - 3xz - 2 = 0$; yz-plane.

In Exercises 5–6, find the intercepts of the given surface on the specified coordinate axis.

5. $x^2 + 3y^2 + 5xz - 2x + y - 3 = 0$; x-axis.
6. $3x^2 - z^2 + xy - 8yz + y - 3z - 2 = 0$; y-axis.

In Exercises 7–8, write the equation of the sphere with the given center and radius.

7. $(2, -5, 1)$; 6. **8.** $(3, 0, -4)$; 2.

In Exercises 9–10, write the equation of the sphere having the first point as center and passing through the second point.

9. $(4, 1, 3)$, $(2, -1, 2)$. **10.** $(-2, -1, 5)$, $(8, -7, 3)$.

In Exercises 11–14, find the center, and state whether the sphere is real, a point sphere, or imaginary. If it is real find the radius.

11. $x^2 + y^2 + z^2 - 6x + 2y + 4z - 11 = 0$.
12. $x^2 + y^2 + z^2 + 2x + 14y - 10z + 94 = 0$.
13. $2x^2 + 2y^2 + 2z^2 + 6x - 8y - 2z + 13 = 0$.
14. $9x^2 + 9y^2 + 9z^2 - 6x + 30y - 18z + 31 = 0$.

In Exercises 15–16, find the equation of the sphere through the four given points. *Hint:* Substitute four times in (2), § 714.

15. $(0, 0, 0)$, $(1, 0, 0)$, $(0, 1, 0)$, $(0, 0, 1)$.
16. $(3, -2, 1)$, $(5, -2, 2)$, $(3, -1, 4)$, $(4, -1, 5)$.

17. Prove that the plane tangent to the real sphere

(1) $$x^2 + y^2 + z^2 + 2px + 2qy + 2rz + d = 0$$

at the point (x_1, y_1, z_1) has the equation

$$x_1 x + y_1 y + z_1 z + p(x_1 + x) + q(y_1 + y) + r(z_1 + z) + d = 0.$$

In Exercises 18–19, find the equation of the plane tangent to the given sphere at the given point. (Cf. Ex. 17.)

18. $x^2 + y^2 + z^2 - 6x - 2y - 4z - 7 = 0$; $(5, 0, 6)$.
19. $x^2 + y^2 + z^2 - 10x - 12z + 40 = 0$; $(3, 1, 2)$.

In Exercises 20–22, describe the graph of the system of equations.

20. $x^2 + y^2 + z^2 - 8x - 2y + 4z - 79 = 0$; $7x + 4y - 4z + 41 = 0$.
21. $x^2 + y^2 + z^2 - 2x - 10y - 12z + 26 = 0$,
$x^2 + y^2 + z^2 + 4x + 8y - 6z - 40 = 0$.
22. $x^2 + y^2 + z^2 + 8x - 2y - 12z + 37 = 0$,
$x^2 + y^2 + z^2 - 10x + 2y - 23 = 0$.

In Exercises 23–24, find the equation of the sphere through the circle of intersection of the two given spheres and through the specified point.

*23. $x^2 + y^2 + z^2 + 2x + 8y - 4z - 20 = 0$,
$x^2 + y^2 + z^2 - 10x - 8y - 2z = 0$; $(1, 3, -1)$.
*24. $x^2 + y^2 + z^2 - 6x + 4y + z + 1 = 0$,
$x^2 + y^2 + z^2 - 2x + 3y + 6z + 1 = 0$; $(3, -2, -1)$.

In Exercises 25–30, show that the surface is a cylinder parallel to a coordinate axis. Name this coordinate axis and sketch the surface.

25. $x^2 + y^2 = 9$. **26.** $x^2 + 4z = 16$.
27. $yz = 2$. **28.** $x^2 + y^2 = 4x$.
29. $x^2 = z^2$. **30.** $|y| + |z| = 1$.

In Exercises 31–34, write an equation of the right circular cylinder with the given radius and axis. Sketch the surface.

31. 3; x-axis. **32.** 5; z-axis.

33. 2; $x = 0, z = 2$. 34. 5; $x = 3, y = 4$.

In Exercises 35–38, sketch the surface.

35. $z(x^2 + y^2 - 1) = 0$. 36. $z(x^2 + y^2) = 0$.
37. $x^2 - y^2 = 0$. 38. $x^3 - y^3 = 0$.

In Exercises 39–42, sketch the curve.

39. $y = 4x$, $x^2 + z^2 = 16$.
40. $x + y + z = 1$, $(x - 1)^2 + (y - 1)^2 = 1$.
41. $x^2 + z^2 = 1$, $y^2 + z^2 = 1$.
42. $x^2 + z^2 = 4$, $y^2 + z^2 = 9$.

In Exercises 43–48, write an equation of the surface obtained by revolving the given curve about the specified axis. Draw a figure in each case.

43. $x^2 + 2y^2 = 8, z = 0$; (a) x-axis; (b) y-axis.
44. $4x^2 - 9z^2 = 5, y = 0$; (a) x-axis; (b) z-axis.
45. $6y^2 + 6z^2 = 7, x = 0$; (a) y-axis; (b) z-axis.
46. $2x + 3y = 6, z = 0$; (a) x-axis; (b) y-axis.
47. $y = 2, x = 0$; (a) y-axis; (b) z-axis.
48. $x^{\frac{2}{3}} + z^{\frac{2}{3}} = 1, y = 0$; (a) x-axis; (b) z-axis.

In Exercises 49–52, state which coordinate axis is the axis of revolution for the surface, and write equations of a generatrix in the specified coordinate plane. Draw a figure.

49. $x^2 + y^2 + z = 2$; xz-plane. 50. $x^2 - 4y^2 - 4z^2 = 8$; xy-plane.
51. $x^2 - 4y^2 - 4z^2 = 0$; xz-plane.
52. $x^2 + z^2 = 4y^4 - 4y^3 + 5y^2 - 2y + 1$; yz-plane.
53. Find an equation of the torus obtained by revolving about the y-axis the circle in the xy-plane with center $(a, 0, 0)$ and radius b, where $b < a$.
★54. Formulate a rule for finding an equation of the surface obtained by revolving a curve in the xy-plane about the line $x = a, z = 0$.

718. THE STANDARD QUADRIC SURFACES

The graph of any equation of the second degree:

(1) $ax^2 + bx^2 + cz^2 + 2fyz + 2gxy + 2hxy + 2px + 2qy + 2rz + d = 0$,

where not all of the coefficients a, b, c, f, g, and h vanish, is called a **quadric surface** or **conicoid**. It can be shown (cf. the author's *Solid Analytic Geometry* already referred to) that except for cylinders (§ 715), degenerate cases (such as point spheres and pairs of planes), and imaginary surfaces (cf. § 714) every quadric surface is one of the standard types listed below (with a suitable choice of coordinate axes). The letters a, b, and c used in the accompanying formulas are assumed to be positive.

Ellipsoid.

(2) $$\frac{x^2}{a^2} + \frac{y^2}{b^2} + \frac{z^2}{c^2} = 1.$$

This surface is symmetrical with respect to each coordinate plane and axis and the origin. (Cf. Fig. 708.) The section of the surface by the plane

$z = k$ is a real ellipse if $|k| < c$, a point (actually two imaginary lines intersecting in a real point) if $|k| = c$, and an imaginary ellipse if $|k| > c$. This can be seen if equation (2) is written in the form

$$\frac{x^2}{a^2} + \frac{y^2}{b^2} = \frac{c^2 - z^2}{c^2}.$$

Therefore the surface lies between the tangent planes $z = \pm c$. Similarly, it is bounded by the planes $x = \pm a$ and $y = \pm b$. The line segments on the axes with the intercepts as end points, or their lengths, $2a$, $2b$, and $2c$, are called the **axes** of the ellipsoid, and a, b, and c are called the **semiaxes**.

If the axes are all equal the ellipsoid is a sphere.

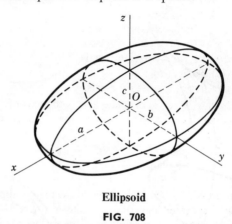

Ellipsoid

FIG. 708

If two axes are equal and different from the third, the ellipsoid is called a **spheroid**, **oblate** if the third axis is shorter than the first two (like the earth), **prolate** if the third axis is longer than the first two (like a football).

Hyperboloid of one sheet.

(3) $$\frac{x^2}{a^2} + \frac{y^2}{b^2} - \frac{z^2}{c^2} = 1.$$

This surface is symmetrical with respect to each coordinate plane and axis and the origin. (Cf. Fig. 709.) Each section by a plane parallel to the xy-plane is a real ellipse, and each section by a plane parallel to either of the other two coordinate planes is a hyperbola or a degenerate hyperbola consisting of two straight lines. (Which of these sections are degenerate?) The surface is in one connected piece and accordingly is called a hyperboloid of *one sheet*.

Transverse and **conjugate axes** and **semiaxes** are defined by analogy with the terminology of plane analytic geometry. There are two transverse axes and one conjugate axis. The transverse semiaxes are a and b, and the conjugate semiaxis is c.

Hyperboloid of two sheets.

(4) $$\frac{x^2}{a^2} + \frac{y^2}{b^2} - \frac{z^2}{c^2} = -1.$$

This surface is symmetrical with respect to each coordinate plane and axis and the origin. (See Fig. 709.) Each section by a plane $z = k$ is a real ellipse if $|k| > c$, a point if $|k| = c$, and an imaginary ellipse if $|k| < c$. Each section by a plane parallel to one of the other coordinate planes is a hyperbola. The surface consists of two separate pieces and accordingly is called a hyperboloid of *two sheets*.

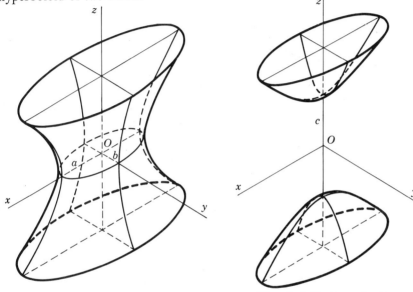

Hyperboloid of one sheet   Hyperboloid of two sheets

FIG. 709

Transverse and **conjugate axes** and **semiaxes** are defined as for the hyperboloid of one sheet. In this case there are one transverse axis and two conjugate axes. The transverse semiaxis is c and the conjugate semiaxes are a and b. Notice that the hyperboloids (3) and (4) might appropriately be called *conjugate* since a transverse axis of one is a conjugate axis of the other.

Quadric cone.

(5) $$\frac{x^2}{a^2} + \frac{y^2}{b^2} - \frac{z^2}{c^2} = 0.$$

This surface is symmetrical with respect to each coordinate plane and axis and the origin. (Cf. Fig. 710.) If $p_1 : (x_1, y_1, z_1)$ is any point on the surface, different from the origin O, then any point on the line Op_1 is also on the surface, since the coordinates of such a point have the form kx_1, ky_1, and kz_1, which satisfy (5) whenever x_1, y_1, and z_1 do. Therefore the surface

(5) is a cone, according to the general definition of a cone as the surface generated by a moving line that passes through a fixed point, called the **vertex**. The vertex of (5) is the origin.

Each section of (5) by a plane parallel to the xz- or yz- plane is a hyperbola (degenerate if the section is the trace of the surface in either of these coordinate planes). Each section by a plane parallel to the xy-plane is an ellipse (degenerating to a point, or two imaginary intersecting lines, if the section is the trace in the xy-plane).

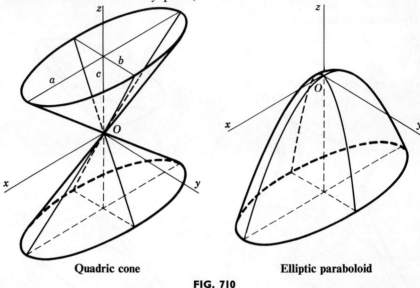

Quadric cone **Elliptic paraboloid**

FIG. 710

The cone (5) is called the **asymptotic cone** of either of the hyperboloids (3) or (4).

Elliptic paraboloid.

(6) $$\frac{x^2}{a^2} + \frac{y^2}{b^2} + 2z = 0.$$

This surface is symmetrical with respect to the xz-plane, the yz-plane, and the z-axis. (Cf. Fig. 710.) The origin is called the **vertex** of the paraboloid (6). The section of the surface by the plane $z = k$ is a real ellipse if $k < 0$, a point ellipse if $k = 0$, and an imaginary ellipse if $k > 0$. Sections by planes parallel to the other coordinate planes are parabolas with vertical axes, opening downward.

Hyperbolic paraboloid.

(7) $$\frac{x^2}{a^2} - \frac{y^2}{b^2} + 2z = 0.$$

This is a saddle-shaped surface symmetrical with respect to the xz-plane, the yz-plane, and the z-axis. (Cf. Fig. 711.) The origin is called the **vertex** of the paraboloid (7). The section of the surface by the plane $z = k$ is a hyperbola, which is degenerate if $k = 0$. Sections by planes parallel to the other coordinate planes are parabolas with vertical axes, opening upward or downward.

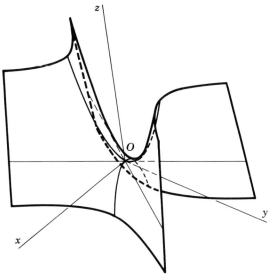

Hyperbolic paraboloid
FIG. 711

Example. Show that the surface
(8) $$z = xy$$
is a hyperbolic paraboloid.

Solution. A rotation of $-45°$ about the origin in the xy-plane (about the z-axis in space), with equations:
(9) $$x = \frac{x' + y'}{\sqrt{2}}, \quad y = \frac{-x' + y'}{\sqrt{2}}$$
transforms equation (8) to
(10) $$x'^2 - y'^2 + 2z = 0.$$
This is equation (7) in the new coordinate system, with $a = b = 1$.

719. EXERCISES

1. Show that the sections of the ellipsoid (2), § 718, by planes parallel to a coordinate plane are similar ellipses (ellipses having the same eccentricity), decreasing in size as their distances from the origin increase.

2. Prove that a spheroid is a surface of revolution obtained by revolving an ellipse about an axis—about the minor axis for an oblate spheroid, and about the major axis for a prolate spheroid.

In Exercises 3 and 4, determine the semiaxes of the ellipsoid. Draw a figure.

3. $4x^2 + y^2 + 9z^2 = 36.$ **4.** $x^2 + 2y^2 + 3z^2 = 6.$

In Exercises 5–8, show that the ellipsoid is a spheroid. Determine whether the spheroid is oblate or prolate, and name the axis of revolution. Draw a figure.

5. $4x^2 + 9y^2 + 4z^2 = 36$.
6. $9x^2 + 9y^2 + 4z^2 = 36$.
7. $25x^2 + y^2 + z^2 = 75$.
8. $x^2 + 25y^2 + 25z^2 = 25$.

9. Show that the sections of (3), § 718, by planes parallel to the xy-plane are similar ellipses, increasing in size as their distances from the origin increase. (Cf. Ex. 1.)

10. Show that the sections of (3), § 718, by planes parallel to, say, the xz-plane are hyperbolas (or degenerate hyperbolas) whose projections on the xz-plane have the same asymptotes. Are these projections similar? Which sections are degenerate, and what are their projections? (Cf. Ex. 9.)

11. Prove that a hyperboloid of one sheet with equal transverse axes is the surface of revolution obtained by revolving a hyperbola about its conjugate axis.

12. Formulate and prove statements for the hyperboloid (4), § 718, similar to those of Exercises 1, 9, and 10.

13. Prove that a hyperboloid of two sheets with equal conjugate axes is the surface of revolution obtained by revolving a hyperbola about its transverse axis.

In Exercises 14–17, determine whether the hyperboloid is one of one sheet or one of two sheets and give the transverse and conjugate semiaxes in each case.

14. $4x^2 + y^2 - 9z^2 = 36$.
15. $x^2 - 4y^2 + 9z^2 + 36 = 0$.
16. $x^2 + 1 = y^2 + z^2$.
17. $z^2 - 1 = x^2 + y^2$.

18. Formulate and prove statements for the quadric cone (5), § 718, similar to those of Exercises 1, 9, and 10.

19. Show that if $a = b$, (5), § 718, is a right circular cone.

20. Show that the traces in the xz-plane or in the yz-plane of the surfaces (3), (4), and (5), § 718, are a pair of conjugate hyperbolas and their asymptotes.

21. Show that no two of the surfaces (3), (4), and (5), § 718, have a point in common.

22. Prove that the parabolic sections of (6), § 718, by planes parallel to the xz- (or yz-) plane are congruent.

23. Discuss the surface (6), § 718, if $a = b$.

24. Discuss similarity of hyperbolic sections of (7), § 718, by planes parallel to the xy-plane.

25. Prove that the parabolic sections of (7), § 718, by planes parallel to the yz-plane are congruent parabolas opening upward and that the parabolic sections by planes parallel to the xz-plane are congruent parabolas opening downward.

In Exercises 26–49, name the surface. If the surface is a surface of revolution, name the axis of revolution. Draw a figure.

26. $3x^2 + y^2 + 4z^2 = 0$.
27. $x^2 + 9z^2 = 9$.
28. $y^2 - 9z^2 = 81$.
29. $4y^2 + 5 = 0$.
30. $9x^2 - 4y^2 + 9z^2 = 36$.
31. $x^2 + 4y^2 = 2z$.
32. $5y^2 = 8$.
33. $4x^2 = y^2 + z^2$.
34. $y^2 + 2z^2 + 4 = 0$.
35. $x^2 + z^2 = y$.
36. $4x^2 + 4y^2 + z^2 = 16$.
37. $4x = 4y^2 - z^2$.
38. $3x^2 + 4y^2 + z^2 = -12$.
39. $z^2 = 0$.
40. $4y^2 + 9z^2 = 0$.
41. $4x^2 + 4y^2 + 4z^2 = 25$.
42. $x^2 - 4y^2 - 4z^2 = 4$.
43. $x^2 = 4y$.

44. $x^2 - 16y^2 = 0$.
46. $4y^2 + 4z^2 = 9$.
48. $x^2 + y^2 + z^2 + 2xy + 2xz + 2yz = 1$.
45. $x = (y - 3z)(y + 3z)$.
47. $x^2 + y^2 + z^2 = x$.
49. $y = z^2 - 4z + 3$.

In Exercises 50–53, write an equation of the described locus of a point p. Name the locus.

50. The point p is equidistant from the point $(0, 0, -1)$ and the plane $z = 1$.

51. The point p is equidistant from the z-axis and the xy-plane.

52. The distance of p from the origin is equal to twice its distance from the xy-plane.

53. The sum of the squares of the distances of p from the origin and the xy-plane is equal to 4.

In Exercises 54–55, determine the nature of the described locus of a point p. Name the surface.

54. The sum of the squares of the distances of p from a line and a fixed point on the line is constant.

55. The sum of the squares of the distances of p from a line and a plane perpendicular to the line is constant.

8

Arcs and Curves

801. DUHAMEL'S PRINCIPLE FOR INTEGRALS

The definite integral is defined as the limit of sums of a certain type. For the product of two functions $f(x)$ and $g(x)$ these sums take the form $\sum_{i=1}^{n} f(x_i)g(x_i)\,\Delta x_i$. In practice one is occasionally faced with the problem of evaluating a limit of sums of the form $\sum_{i=1}^{n} f(x_i)g(x_i')\,\Delta x_i$ which resemble the standard approximating sums of a definite integral but which differ from the latter in that the two given functions are evaluated at possibly distinct values of the independent variable in the ith subinterval. The fact that this limit exists and is equal to the definite integral $\int_a^b f(x)g(x)\,dx$ is known as **Bliss's Theorem** (Theorem XV, § 401).

In this section we state a useful theorem of which Bliss's Theorem is a simple special case (where $\phi(x, y) = xy$). In the following section a proof is given under the more restrictive assumption that $f(t)$ and $g(t)$ are continuous. (For a general proof, and extensions to more than two variables, cf. *RV*, Exs. 20, 21, 28, 30, § 1110.)

Theorem. Duhamel's Principle for Integrals. *Let $f(t)$ and $g(t)$ be integrable on $[a, b]$ and let $\phi(x, y)$ be everywhere continuous. Then, in the sense of § 401, the limit of the sum $\sum_{i=1}^{n} \phi(f(t_i), g(t_i'))\,\Delta t_i$, where $a_{i-1} \leq t_i \leq a_i$ and $a_{i-1} \leq t_i' \leq a_i$, as the norm of the net $\mathfrak{N} : a = a_0, a_1, \cdots, a_n = b$ tends toward zero, exists and is equal to the definite integral $\int_a^b \phi(f(t), g(t))\,dt$, which also exists:*

(1) $$\lim_{|\mathfrak{N}| \to 0} \sum_{i=1}^{n} \phi(f(t_i), g(t_i'))\,\Delta t_i = \int_a^b \phi(f(t), g(t))\,dt.$$

*802. A PROOF WITH CONTINUITY HYPOTHESES

We give here a proof of Duhamel's Principle under the additional assumption that $f(t)$ and $g(t)$ are continuous for $a \leq t \leq b$ (cf. *RV*, Ex. 28, § 1110).

Since $f(t)$ and $g(t)$ are continuous on a closed interval they are bounded there (Theorem I, § 213), so that there exists a number K such that for $a \leq t \leq b$, $|f(t)| \leq K$ and $|g(t)| \leq K$. Since $\phi(x, y)$ is assumed to be everywhere continuous it is, in particular, continuous on the closed square $A : -K \leq x \leq K, -K \leq y \leq K$. Consequently, by the Theorem of § 617, $\phi(x, y)$ is uniformly continuous on A. Corresponding to a preassigned positive number ϵ, we can therefore find a positive number η such that $\sqrt{(x_2 - x_1)^2 + (y_2 - y_1)^2} < \eta$ implies $|\phi(x_2, y_2) - \phi(x_1, y_1)| < \epsilon/2(b - a)$ (in which case, of course, $|y_2 - y_1| < \eta$ implies $|\phi(x_1, y_2) - \phi(x_1, y_1)| < \epsilon/2(b - a)$). We now require the positive number δ to be so small that whenever the norm of a net on the closed interval $[a, b]$ is less than δ, then any sum of the form $\sum_{i=1}^{n} \phi(f(t_i), g(t_i)) \Delta t_i$ approximates the integral $\int_a^b \phi(f(t), g(t)) \, dt$ within $\epsilon/2$. (The integral exists since the integrand is a continuous function of the variable t. Cf. Ex. 25, § 611.) Then, since $g(t)$ is uniformly continuous on $[a, b]$ (cf. § 224), we can (and do) further require that δ be so small that when the norm of a net is less than δ, and $a_{i-1} \leq t_i \leq a_i$ and $a_{i-1} \leq t_i' \leq a_i$, then $|g(t_i) - g(t_i')| < \eta$. Owing to the properties of the numbers η and δ, whenever the norm of the net is less than δ we have

(1) $$\left| \sum \phi(f(t_i), g(t_i)) \Delta t_i - \int_a^b \phi(f(t), g(t)) \, dt \right| < \frac{\epsilon}{2},$$

and

(2) $|\sum \phi(f(t_i), g(t_i')) \Delta t_i - \sum \phi(f(t_i), g(t_i)) \Delta t_i|$
$$\leq \sum |\phi(f(t_i), g(t_i')) - \phi(f(t_i), g(t_i))| \Delta t_i < \sum \frac{\epsilon}{2(b-a)} \Delta t_i = \frac{\epsilon}{2}.$$

By the triangle inequality,

$$\left| \sum \phi(f(t_i), g(t_i')) \Delta t_i - \int_a^b \phi(f(t), g(t)) \, dt \right| < \epsilon.$$

Since this statement is equivalent to (1), § 801, the proof is complete.

803. ARCS AND CURVES

Definition. *An **arc** is a continuous image of a closed interval.* For example, an arc in E_2 is the set of all points (x, y) whose coordinates are given by two continuous functions of a real variable,

(1) $$x = f(t), \quad y = g(t),$$

where $a \leq t \leq b$. The variable t is called a **parameter**, the representation (1) is a **parametrization** of the arc, and the functions $f(t)$ and $g(t)$ are **parametrization functions**. *A **closed curve** is an arc for which the images of the endpoints, a and b, of the defining interval are identical;* in E_2, $f(a) = f(b), g(a) = g(b)$. *A **simple arc** is a continuous one-to-one image of a closed interval. A **simple closed curve** is a closed curve for which the* *end-points of the defining interval are the only two distinct points that map into the same point.* (Cf. Fig. 801.)

A similar definition applies to arcs and curves in higher-dimensional spaces.

The single word **curve** will be used to mean the same as *arc* except that the defining interval need not be closed or bounded.

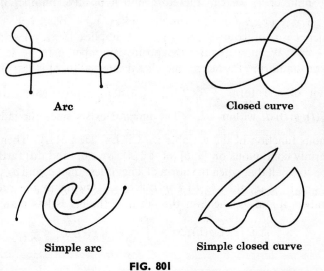

FIG. 801

An arc can be thought of as the path traced out by a particle in motion during a closed interval of time. In a closed curve the particle returns to its initial position. In a simple arc it never repeats an earlier position, and in a simple closed curve it repeats only its initial position. In any case, an arc is a set of points, and this set of points constitutes the same arc no matter how many times any particular portions of that set may be traced out by the moving particle.

NOTE 1. Since a closed interval is compact, the mapping of the defining interval onto a simple arc has a continuous inverse. Thus, *a simple arc could be defined as a homeomorph of a closed interval.* (Cf. Theorem II and the Note, § 612.) In a similar way, since in the definition of a closed curve the end-points of the defining interval can be considered as being brought together and identified, *a simple closed curve could be defined as a homeomorph of the circumference of a circle.*

NOTE 2. Since any two closed intervals are homeomorphic (Note, § 612), it is immaterial what the defining interval in the preceding Definition is. For simplicity, it can be taken to be the closed unit interval [0, 1].

NOTE 3. It is easy to see that the parametrization functions f and g of the above Definition are not unique for a given arc or curve. An example in E_2 is the straight line segment from $(0, 0)$ to $(1, 1)$, given by the three distinct parametrizations $x = t, y = t$; $x = t^2, y = t^2$; and $x = 1 - t, y = 1 - t$; $0 \leq t \leq 1$.

↙NOTE 4. In 1890 Peano (cf. § 101) made public a remarkable discovery about certain arcs and curves. He had found that, whatever one's intuitions might say to the contrary, it is possible for an arc or curve to "fill space"—that is, to pass at least once through every point of a higher dimensional set, like a closed square, or a "cube" in a Euclidean space of any number of dimensions. For a discussion of an arc that completely fills the closed unit square $A: 0 \leq x \leq 1, 0 \leq y \leq 1$ in E_2, see *RV*, § 1104.

804. ARC LENGTH

The length of an arc or curve is defined in terms of lengths of approximating inscribed polygons obtained by joining a finite sequence of points, each to the next, by straight line segments. As will be disclosed in succeeding discussion, such a definition necessitates a formulation based on a particular parametrization. Otherwise, for a given finite set of points on an arc, it would be meaningless to speak of joining one of these points to the "next." The parametrization tells us, for instance, which way to proceed from a point of intersection of a self-crossing arc.

Definition. *Let C be an arc in E_2 with parametrization $x = f(t)$, $y = g(t)$, $a \leq t \leq b$, and let $\mathfrak{N}: a_0 < a_1 < \cdots < a_n = b$ be a net on $[a, b]$, the corresponding points of C being $p_0, p_1, \cdots, p_n$. Let $L_\mathfrak{N}$ denote the length $d(p_0, p_1) + d(p_1, p_2) + \cdots + d(p_{n-1}, p_n)$ of the **inscribed polygon** $[p_0 p_1 \cdots p_n]$. Then the **length** L of the arc C for the parametrization $x = f(t)$, $y = g(t)$ is the (finite or infinite) least upper bound, for all nets $\mathfrak{N}$ on $[a, b]$:*

(1) $L \equiv \sup L_\mathfrak{N} = \sup [d(p_0, p_1) + \cdots + d(p_{n-1}, p_n)].$

*In case the polygonal lengths $L_\mathfrak{N}$ are bounded, so that (1) is finite, C is called **rectifiable**. Otherwise it is **nonrectifiable** and has **infinite length**.*

A similar definition applies to higher-dimensional spaces.

As illustrated in Example 2, below, an arc may have infinitely many lengths, depending on the parametrization.† However, for *simple* arcs and *simple* closed curves, it can be demonstrated that *arc length is independent of the parametrization.* (For a proof, see *RV*, § 1107.)

Example 1. Show that the simple arc

$$x = t, y = t \sin \frac{1}{t} \text{ for } 0 < t \leq 1, x = 0, y = 0 \text{ for } t = 0,$$

is not rectifiable. (Cf. Fig. 303.)

Solution. For any given positive integer n, choose for t the values

$$t_0 = 0, t_1 = \frac{2}{(2n-1)\pi}, t_2 = \frac{2}{(2n-3)\pi}, \cdots, t_{n-2} = \frac{2}{3\pi}, t_{n-1} = \frac{2}{\pi}, t_n = 1.$$

† It is possible to formulate a concept of arc length independent of parametrization by defining it to be the greatest lower bound of the lengths L given by all possible parametrizations.

The intermediate points of the corresponding inscribed polygon are points of tangency of the graph of $y = x \sin \dfrac{1}{x}$ with the straight lines $y = \pm x$ (cf. Fig. 303). The length L of the polygon $[p_0 p_1 \cdots p_n]$ is certainly greater than the sum of the absolute values of the ordinates at the points $p_1, p_2, \cdots, p_{n-1}$, or

$$\frac{2}{\pi} + \frac{2}{3\pi} + \frac{2}{5\pi} + \cdots + \frac{2}{(2n-1)\pi}.$$

By the divergence of the harmonic series (Theorem IV, § 1108), these sums are unbounded. Therefore, the lengths L are unbounded. (Cf. Ex. 25, § 807.)

*Example 2. Find the length of the arc C in E_2 whose points (x, y) satisfy the relations $0 \leq x \leq 1$, $y = 0$, given by the parametrization functions $x = f(t), y = 0$,

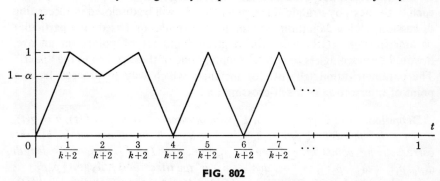

FIG. 802

where the graph of $f(t)$, as shown in Figure 802, is made up of straight line segments. The interval $0 \leq t \leq 1$ is broken up into $k + 2$ equal parts, where k is a positive odd integer, and α is an arbitrary number such that $0 \leq \alpha \leq 1$. The arc C, in other words, is traced out by a particle moving as follows: first across the arc; then a retracing of amount α; then to the right-hand end of the arc; then back and forth across the arc an arbitrary number of times.

Solution. It is not hard to see that the arc length of C, according to the Definition, is equal to the total straight-line distance covered by the moving point. In other words, it is equal to $1 + 2\alpha + 2 + 2 + \cdots + 2 = k + 2\alpha$. Therefore the length of the arc C can be any real number L such that $L \geq 1$. Clearly the smallest of all possible lengths given by *all* parametrizations is $L = 1$.

805. INTEGRAL FORM FOR ARC LENGTH

It is our purpose in this section to show that if enough more than mere continuity is demanded of the parametrization functions, the length of a rectifiable arc can be expressed as a definite integral. The conditions of the following theorem are not the most general possible (cf. RV, Exs. 23, 29, § 1110). They have been chosen for their simplicity, and are sufficiently general for most purposes. The theorem is formulated for an arc in the Euclidean plane, but an analogous statement, with an analogous proof, applies to higher-dimensional spaces (cf. § 809, and Ex. 23, § 807).

INTEGRAL FORM FOR ARC LENGTH

Theorem I. *Let C be an arc with parametrization functions $x = f(t)$, $y = g(t)$ which, together with their first derivatives, are continuous on the closed interval $[a, b]$. Then C is rectifiable and its length L is given by the formula*

$$(1) \qquad L = \int_a^b \sqrt{(f'(t))^2 + (g'(t))^2}\, dt.$$

Proof. Since the integrand is continuous, the integral (1) exists. Let its value be I. If $\epsilon > 0$ we wish first to find $\delta > 0$ such that

$$(2) \qquad |\mathfrak{N}| < \delta \quad \text{implies} \quad |L_\mathfrak{N} - I| < \epsilon.$$

Such a number δ is provided by Duhamel's Principle (§ 801) applied to the function $\phi(x, y) \equiv \sqrt{x^2 + y^2}$, since

$$L_\mathfrak{N} = \sum_{i=1}^n d(p_{i-1}, p_i) = \sum_{i=1}^n \sqrt{[f(a_i) - f(a_{i-1})]^2 + [g(a_i) - g(a_{i-1})]^2}$$

$$= \sum_{i=1}^n \sqrt{[f'(t_i)]^2 + [g'(t_i')]^2}\, \Delta t_i.$$

The last expression, above, is given by the Law of the Mean (§ 305) applied to the two functions $f(t)$ and $g(t)$ on the interval $[a_{i-1}, a_i]$. Finally, since (2) implies that the inequalities

$$(3) \qquad I - \epsilon < L_\mathfrak{N} < I + \epsilon$$

hold for all polygonal lengths for which $|\mathfrak{N}| < \delta$, and since for any given net $\mathfrak{N}$ the effect of interpolating additional points is to increase (or leave unchanged) the corresponding polygonal length, we can conclude that the supremum of $L_\mathfrak{N}$, for all possible nets, satisfies the inequalities:

$$(4) \qquad I - \epsilon < \sup L_\mathfrak{N} \leq I + \epsilon.$$

(Give a proof of (4) in Ex. 20, § 807.) By the arbitrariness of ϵ this means that $L = I$, and the proof is complete.

If a variable upper limit t (in place of b) is used, we shall denote the variable arc length from the fixed point $p(a)$ to the variable point $p(t)$ by the letter s:

$$(5) \qquad s = s(t) \equiv \int_a^t \sqrt{(f'(u))^2 + (g'(u))^2}\, du.$$

Here, of course, a new letter is used for the variable of integration.

The arc length s thus becomes a differentiable function of the parameter t and, by Theorem I, § 405, $\dfrac{ds}{dt} = \sqrt{(f'(t))^2 + (g'(t))^2}$, or

$$(6) \qquad \left(\frac{ds}{dt}\right)^2 = \left(\frac{dx}{dt}\right)^2 + \left(\frac{dy}{dt}\right)^2.$$

In terms of differentials this takes the form

$$(7) \qquad ds^2 = dx^2 + dy^2,$$

a relation which shows how ds is associated with the tangent to the curve. (Cf. Fig. 803.)

A point on a curve $x = f(t)$, $y = g(t)$ at which $f'(t) = g'(t) = 0$ is called a **singular point**. Any other point of the curve is a **regular point**. For example, for the semicubical parabola $x = t^2$, $y = t^3$, the origin is a singular point and all other points are regular.

NOTE 1. The examples $x = t$, $y = t$ and $x = t^3$, $y = t^3$, $-1 \leq t \leq 1$, show that an arc may have a singular point for one parametrization and fail to have one for another.

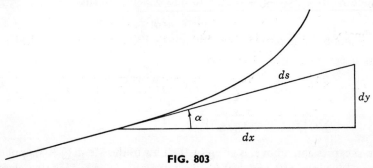

FIG. 803

NOTE 2. A point where a curve crosses itself should not be confused with a singular point. For example, the folium of Descartes has no singular points in the parametrization given in (3), § 319: $x = 3at/(t^3 + 1)$, $y = 3at^2/(t^3 + 1)$.

Under the assumption that a curve C has no singular points, the function $s(t)$ is seen to be a differentiable function of t, with a positive derivative so that (§ 306) s is a strictly increasing differentiable function of t, and t is a strictly increasing differentiable function of s. This permits the use of the arc length itself as a parameter:

(8) $$x = F(s), \quad y = G(s).$$

In this case, formula (6) shows that

(9) $$\left(\frac{dx}{ds}\right)^2 + \left(\frac{dy}{ds}\right)^2 = 1.$$

In fact, if α is the inclination of the tangent line to C, it is clear from Figure 803 that $\frac{dx}{ds} = \cos \alpha$ and $\frac{dy}{ds} = \sin \alpha$, so that equation (9) becomes an expression of a familiar trigonometric identity.

A curve with continuously differentiable parametrization functions and without singular points is called a **smooth curve**. For such a curve the continuity of the functions $\frac{dx}{dt}$, $\frac{dy}{dt}$, $\frac{ds}{dt}$, and $\frac{dt}{ds}$, and hence of the functions $\frac{dx}{ds}$ and $\frac{dy}{ds}$, implies the continuity of the inclination α as a function of s. In other words, *a smooth curve C has a continuously turning tangent.*

§ 805]   INTEGRAL FORM FOR ARC LENGTH   245

If the parameter is the variable x, formula (1) for arc length becomes the familiar one of a first course in calculus,

$$(10) \qquad L = \int_a^b \sqrt{1 + \left(\frac{dy}{dx}\right)^2}\, dx,$$

a similar formula holding in case the parameter is the variable y.

If a curve is given by polar coordinates, results similar to those just given can be obtained. Suppose ρ and θ are continuous functions of a parameter t, and have continuous derivatives with respect to t. Then the rectangular coordinates x and y, by virtue of the relations

$$(11) \qquad x = \rho \cos \theta, \quad y = \rho \sin \theta,$$

are also continuously differentiable as functions of t:

$$(12) \qquad \frac{dx}{dt} = -\rho \sin \theta \frac{d\theta}{dt} + \cos \theta \frac{d\rho}{dt}, \quad \frac{dy}{dt} = \rho \cos \theta \frac{d\theta}{dt} + \sin \theta \frac{d\rho}{dt}.$$

Squaring the various members of (12), adding, and combining terms, we have

$$(13) \qquad \left(\frac{dx}{dt}\right)^2 + \left(\frac{dy}{dt}\right)^2 = \left(\frac{d\rho}{dt}\right)^2 + \rho^2 \left(\frac{d\theta}{dt}\right)^2.$$

We now substitute in (1) to obtain

$$(14) \qquad L = \int_a^b \sqrt{\left(\frac{d\rho}{dt}\right)^2 + \rho^2 \left(\frac{d\theta}{dt}\right)^2}\, dt$$

and, if the parameter is θ,

$$(15) \qquad L = \int_{\theta_1}^{\theta_2} \sqrt{\rho^2 + \left(\frac{d\rho}{d\theta}\right)^2}\, d\theta.$$

In differential form the relation between arc length and polar coordinates is

$$(16) \qquad ds^2 = d\rho^2 + (\rho\, d\theta)^2.$$

Notice that, as with rectangular coordinates, the differential of arc length can be represented as the hypotenuse of a right triangle, this hypotenuse being tangent to the arc. (Cf. Fig. 804.)

An important fact about smooth curves is given in the following theorem:

Theorem II. *If p is fixed point on a smooth curve C and q is a variable point on C that tends toward p as a limit, the quotient of the length of the chord $\overline{pq}$ to the length of the arc $\widetilde{pq}$ tends toward 1 as a limit:*

$$\lim_{q \to p} \frac{\overline{pq}}{\widetilde{pq}} = 1.$$

(See Fig. 805.)

Proof. Since $\widetilde{pq} = \Delta s$ and $\overline{pq}^2 = \Delta x^2 + \Delta y^2$, we wish to show that
$$\lim_{\Delta s \to 0} \left(\frac{\overline{pq}}{\widetilde{pq}} \right)^2 = \lim_{\Delta s \to 0} \frac{\Delta x^2 + \Delta y^2}{\Delta s^2} = 1.$$
But this reduces immediately to (9).

All of the results of this section have their three-dimensional (indeed, n-dimensional) analogues (cf. § 809), which we shall feel free to use, as if they were formally established.

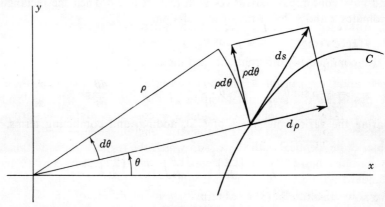

FIG. 804

Example. Express the coordinates of the points of the semi-cubical parabola
$$x = t^2, \quad y = t^3,$$
for $t \geqq 0$, in terms of arc length as parameter.

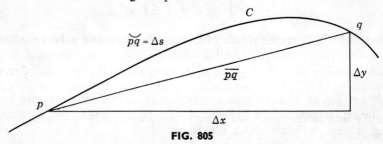

FIG. 805

First Solution. Taking the origin ($t = 0$) as the starting point, we use formula (5):
$$s = \int_0^t \sqrt{(2u)^2 + (3u^2)^2} \, du = \int_0^t u\sqrt{4 + 9u^2} \, du$$
$$= \left[\frac{(4 + 9u^2)^{\frac{3}{2}}}{27} \right]_0^t = \frac{(4 + 9t^2)^{\frac{3}{2}} - 8}{27}.$$

Solving for t in terms of s gives
$$t = \left[\frac{(27s + 8)^{\frac{2}{3}} - 4}{9} \right]^{\frac{1}{2}},$$

whence
$$x = \frac{(27s+8)^{2/3}-4}{9}, \quad y = \frac{[(27s+8)^{2/3}-4]^{3/2}}{27}.$$

Second Solution. Elimination of the parameter t gives the equation $y = x^{3/2}$, so that (10) becomes
$$s = \int_0^x \sqrt{1 + \tfrac{9}{4}u}\, du = \frac{8(1+\tfrac{9}{4}x)^{3/2}-8}{27}.$$

Solving for x in terms of s gives the previous result.

*806. REMARK CONCERNING THE TRIGONOMETRIC FUNCTIONS

In §§ 503 and 504 the basic trigonometric functions, $\sin t$ and $\cos t$, were defined, and their principal elementary properties were established. Among these properties is the parametrization of the unit circle (Ex. 13, § 504), $x^2 + y^2 = 1$, by means of the equations $x = \cos t$, $y = \sin t$. We are now in a position to show that the trigonometric functions defined analytically in Chapter 5 are identical with those with which the reader is already familiar on an intuitive basis. For this purpose we *define* the measure α of an angle in standard position to be the arc length of that portion of the unit circle enclosed by it, account being taken of sign and multiples of 2π. Complete consistency between the intuitive and analytic definitions is obtained by demonstrating that the arc length of the curve $x = \cos t$, $y = \sin t$, for t between 0 and α, is α. But this is merely a simple application of the integral form for arc length:

$$\int_0^\alpha \sqrt{(-\sin t)^2 + (\cos t)^2}\, dt = \int_0^\alpha 1 \cdot dt = \alpha.$$

The reconciliation of the two definitions is now completed as follows: In the first place, the traditional method of defining the sine and cosine functions is equivalent to considering the point where the unit circle cuts the terminal side of the angle α in standard position, the coordinates of this point being *by definition* $\cos \alpha$ and $\sin \alpha$. The analytic method defines these quantities in a different way, but the point whose coordinates are the quantities $\cos \alpha$ and $\sin \alpha$ of § 504 has been shown to have the property of being the point on the unit circle whose radius vector is the terminal side of the angle α in standard position.

807. EXERCISES

In Exercises 1–10, find the length of the given curve between the specified points or values of the parameter. The constant a is positive.

1. $y = x^3$; (0, 0) and (2, 8).
2. $y = \sqrt{x}$; (0, 0) and (4, 2).
3. $\rho = a \cos \theta$; $\theta = -\pi/2$ and $\pi/2$.

4. $\rho = k\theta, k > 0$; $(0, 0)$ and $(\alpha, k\alpha)$.
5. $x = 5 \cos t, y = 5 \sin t$; $t = 0$ and $2n\pi$.
6. $x = a \cos^3 t, y = a \sin^3 t$; $t = 0$ and 2π.
7. $x^{\frac{2}{3}} + y^{\frac{2}{3}} = a^{\frac{2}{3}}$; as a simple closed curve.
8. $y = a \cosh \dfrac{x}{a}$; $x = 0$ and $x = b > 0$.
9. $\rho = 2/(1 + \cos \theta)$; $\theta = 0$ and $\pi/2$.
10. $x = e^t \cos t, y = e^t \sin t$; $t = 0$ and $\pi/2$.

In Exercises 11–14, express x and y, or ρ and θ, in terms of arc length as parameter, distances being measured from the specified point or value of the parameter.

11. $x = r \cos \theta, y = r \sin \theta$, r a positive constant, $\theta \geq 0$; $\theta = 0$.
12. $x = \theta - \sin \theta, y = 1 - \cos \theta, 0 \leq \theta \leq 2\pi$; $\theta = 0$.
13. $x = \cos^3 t, y = \sin^3 t, 0 \leq t \leq \pi/2$; $t = 0$.
14. $\rho = e^\theta, \theta \geq 0$; $\theta = 0$.
15. Criticize the following derivation of the arc length of the cardioid, $\rho = 1 + \cos \theta$: Since $\rho' = -\sin \theta$, $\rho^2 + \rho'^2 = 2 + 2 \cos \theta = 4 \cos^2(\theta/2)$. Hence formula (12) gives $\int_0^{2\pi} 2 \cos (\theta/2) d\theta = 4 \sin (\theta/2) \Big]_0^{2\pi} = 0$.
16. Prove that every polygon is an arc.
17. Let A be a set in E_2 shaped like a letter E, consisting of the segments joining the following pairs of points: $(0, 4)$ and $(3, 4)$; $(0, 2)$ and $(2, 2)$; $(0, 0)$ and $(3, 0)$; and $(0, 0)$ and $(0, 4)$. Find a specific parametrization that yields A as an arc. Find its arc length under this parametrization. (Cf. Exs. 18, 27.)
18. Find a specific parametrization that yields the set A of Exercise 17 as a closed curve. Find its arc length under this parametrization. (Cf. Exs. 17, 27.)
19. Prove that any set that can be parametrized as an arc can also be parametrized as a closed curve, and conversely.
20. Prove (4), § 805.
21. State Duhamel's Principle for a function of n variables.
*22. Prove Duhamel's Principle for a function of n variables each of which is a continuous function of the parameter.
*23. Derive the integral form

$$L = \int_a^b \sqrt{\sum_{i=1}^n (f_i'(t))^2}\, dt$$

for an arc $x_i = f_i(t), i = 1, 2, \cdots, n$, in E_n, where the derivatives are continuous.
*24. Show that the total arc length of the ellipse $x = a \sin \theta, y = b \cos \theta$, where $a > b$, is

$$4a \int_0^{\frac{\pi}{2}} \sqrt{1 - k^2 \sin^2 \theta}\, d\theta,$$

where k is the eccentricity of the ellipse. The integral is an elliptic integral of the second kind (cf. (4), § 511).
*25. Show that the arc $y = x^2 \sin (1/x)$ ($y = 0$ when $x = 0$) is rectifiable on $[0, 1]$. (Cf. Example 1, § 804.)
*26. Prove that the circumference of a circle of radius r is $2\pi r$, and hence justify the definition of π given in § 504. Explain how, if a circle is not considered as a *simple* closed curve, its circumference can be any number $L \geq 2\pi r$.
*27. Show that the set A of Exercise 17 cannot be parametrized as a *simple* arc.

808. CYLINDRICAL AND SPHERICAL COORDINATES

Many problems in three dimensions can be more easily attacked with the aid of coordinate systems that are not rectangular. We introduce, in this section, two of the most important, and discuss arc length in terms of them. In later sections we shall put them to additional service. We refer each of the new systems of coordinates to a basic rectangular system.

I. Cylindrical coordinates

In this system a point is specified by the polar coordinates (ρ, θ) of its projection on the xy-plane and its directed distance z from the xy-plane. (Cf. Fig. 806.) With the restrictions

$$\rho \geq 0,$$
$$0 \leq \theta < 2\pi,$$

any point not on the z-axis has a unique representation (ρ, θ, z).

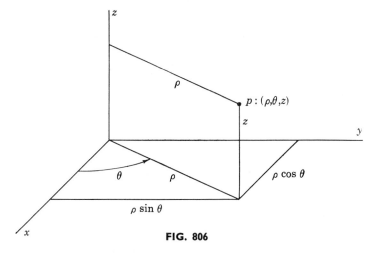

FIG. 806

If the cylindrical coordinates of a point are given, its rectangular coordinates can be obtained by means of the relations

(1)
$$\begin{cases} x = \rho \cos \theta, \\ y = \rho \sin \theta, \\ z = z. \end{cases}$$

Conversely, the cylindrical coordinates of a point are given in terms of the rectangular coordinates by the relations

(2)
$$\begin{cases} \rho^2 = x^2 + y^2, \\ \tan \theta = y/x, \\ z = z, \end{cases}$$

the quadrant of θ being determined by the signs of x and y.

These relations can also be used to transform an equation of a surface in one coordinate system into an equation in the other coordinate system. For example, by means of relations (2), the equation $z = \rho^2$ becomes $z = x^2 + y^2$.

II. Spherical coordinates

The spherical coordinates (r, ϕ, θ) of a point p, as indicated in Figure 807, are (i) its distance r from the origin O, (ii) the angle ϕ between its radius

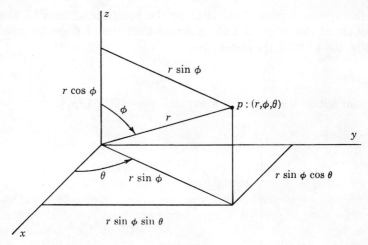

FIG. 807

vector and the z-axis, and (iii) the angular polar coordinate of θ of its projection on the xy-plane, subject to the restrictions

$$r \geq 0,$$
$$0 \leq \phi \leq \pi,$$
$$0 \leq \theta < 2\pi.$$

Again the coordinates of any point not on the z-axis are uniquely determined.

The equations of transformation are

(3)
$$\begin{cases} x = r \sin \phi \cos \theta, & r^2 = x^2 + y^2 + z^2, \\ y = r \sin \phi \sin \theta, & \cos \phi = \dfrac{z}{\sqrt{x^2 + y^2 + z^2}}, \\ z = r \cos \phi, & \tan \theta = \dfrac{y}{x}. \end{cases}$$

As an example, the equation

$$x^2 + y^2 + z^2 - 2z = 0$$

becomes $r^2 - 2r \cos \phi = 0$, or $r = 2 \cos \phi$.

809. ARC LENGTH IN RECTANGULAR, CYLINDRICAL, AND SPHERICAL COORDINATES

If the treatment of arc length for a plane curve in rectangular coordinates is paralleled for three dimensions, formulas (1), (5)–(9), § 805, have the following analogues, where $x = f(t)$, $y = g(t)$, $z = h(t)$ are the parametric equations of the given arc:

$$(1) \qquad L = \int_a^b \sqrt{(f'(t))^2 + (g'(t))^2 + (h'(t))^2}\, dt,$$

$$(2) \qquad s = s(t) = \int_a^t \sqrt{(f'(u))^2 + (g'(u))^2 + (h'(u))^2}\, du,$$

$$(3) \qquad \left(\frac{ds}{dt}\right)^2 = \left(\frac{dx}{dt}\right)^2 + \left(\frac{dy}{dt}\right)^2 + \left(\frac{dz}{dt}\right)^2,$$

$$(4) \qquad ds^2 = dx^2 + dy^2 + dz^2,$$

$$(5) \qquad x = F(s),\quad y = G(s),\quad z = H(s),$$

$$(6) \qquad \left(\frac{dx}{ds}\right)^2 + \left(\frac{dy}{ds}\right)^2 + \left(\frac{dz}{ds}\right)^2 = 1.$$

Details of these derivations are requested in Exercise 11, § 810.

If the procedure used in § 805 for deriving a formula for arc length in terms of polar coordinates is followed here, we have only to consider formulas (1) and (3), § 808, as defining relationships among coordinates all of which are considered as functions of a parameter t. We differentiate those two sets of formulas with respect to t, square, and add, obtaining for the case of cylindrical coordinates

$$(7) \qquad \left(\frac{ds}{dt}\right)^2 = \left(\frac{d\rho}{dt}\right)^2 + \rho^2\left(\frac{d\theta}{dt}\right)^2 + \left(\frac{dz}{dt}\right)^2,$$

and for the case of spherical coordinates

$$(8) \qquad \left(\frac{ds}{dt}\right)^2 = \left(\frac{dr}{dt}\right)^2 + r^2\left(\frac{d\phi}{dt}\right)^2 + r^2 \sin^2 \phi \left(\frac{d\theta}{dt}\right)^2.$$

(Cf. Ex. 11, § 810.) The corresponding formulas for arc length are

$$(9) \qquad L = \int_a^b \sqrt{\left(\frac{d\rho}{dt}\right)^2 + \rho^2\left(\frac{d\theta}{dt}\right)^2 + \left(\frac{dz}{dt}\right)^2}\, dt,$$

$$(10) \qquad L = \int_a^b \sqrt{\left(\frac{dr}{dt}\right)^2 + r^2\left(\frac{d\phi}{dt}\right)^2 + r^2 \sin^2 \phi \left(\frac{d\theta}{dt}\right)^2}\, dt.$$

In Exercise 12, § 810, the student is asked to discuss formulas (7) and (8) in differential form.

810. EXERCISES

1. Write the rectangular coordinates of the points whose cylindrical coordinates are $(3, 0, 2)$ and $(6, \pi/3, -1)$.

2. Write the cylindrical coordinates of the points whose rectangular coordinates are $(-4, 0, 1)$ and $(6, 2\sqrt{3}, 0)$.

3. Write the rectangular coordinates of the points whose spherical coordinates are $(7, \pi/2, \pi)$ and $(12, 5\pi/6, 2\pi/3)$.

4. Write the spherical coordinates of the points whose rectangular coordinates are $(0, -5, 0)$ and $(1, 1, 1)$.

5. Describe the surfaces $\rho = $ constant, $\theta = $ constant, and $\rho = \sin\theta$.

6. Describe the surfaces $r = $ constant, $\phi = $ constant, and $f(r, \phi) = 0$.

7. Transform from cylindrical to rectangular coordinates: $z = \pm\rho$.

8. Transform from rectangular to cylindrical coordinates: $x^2 + y^2 = x$.

9. Transform from spherical to rectangular coordinates: $r = 3$.

10. Transform from rectangular to spherical coordinates: $x^2 + y^2 - 3z^2 = 0$.

11. Give the details in the derivation of formulas (1)–(10), § 809.

12. Write formulas (7) and (8), § 809, in differential form, and discuss them intuitively. *Hint:* See Figures 1011 and 1012.

In Exercises 13–16, find the length of the given arc between the specified points or values of the parameter. All constants are positive.

13. $x = a\cos\lambda t$, $y = a\sin\lambda t$, $z = \mu t$; $t = t_1$ and t_2.

14. $\rho = a$, $\theta = \lambda t$, $z = \mu t$; $t = t_1$ and t_2.

15. $x = at\cos\lambda t$, $y = at\sin\lambda t$, $z = \mu t$; $t = 0$ and $t > 0$.

16. $r = vt$, $\phi = \alpha$, $\theta = \lambda t$; $t = 0$ and $t > 0$.

★17. State and prove the analogue of Theorem II, § 805, for three dimensions.

811. CURVATURE AND RADIUS OF CURVATURE IN TWO DIMENSIONS

A measure of the rapidity with which a curve in E_2, or its tangent line, is turning at a particular point on the curve is given by the rate of change of the

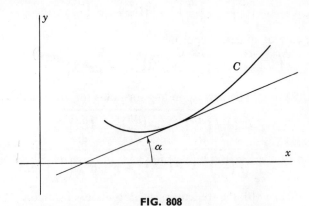

FIG. 808

angle made by the tangent line with some fixed direction (which can be taken for convenience to be that of a coordinate axis) with respect to the arc length, measured from some fixed point on the curve. The absolute value of this rate of change (if it exists) is called the **curvature** of the curve at the particular

§811]   CURVATURE AND RADIUS OF CURVATURE   253

point. That is, if α is the inclination of the tangent line to a curve C (Fig. 808), and if s is its arc length, the curvature K of C at any particular point is defined to be

(1) $$K \equiv \left|\frac{d\alpha}{ds}\right|,$$

evaluated at that point.

Assume that the curve C is given parametrically by functions $x = x(t)$ and $y = y(t)$ possessing second derivatives, and let differentiation with respect to the parameter t be indicated by primes. Then, by standard differentiation formulas of calculus, and (6), § 805, at any regular point where C has a finite slope:

$$\frac{d\alpha}{ds} = \frac{d}{dt}\left(\arctan \frac{dy}{dx}\right)\frac{dt}{ds} = \frac{\left(\arctan \frac{y'}{x'}\right)'}{s'}$$

$$= \frac{1}{1+(y'/x')^2} \cdot \frac{x'y'' - y'x''}{x'^2} \cdot \frac{1}{\sqrt{x'^2 + y'^2}},$$

so that the curvature is given by the formula

(2) $$K = \frac{|x'y'' - y'x''|}{(x'^2 + y'^2)^{\frac{3}{2}}}.$$

Owing to the symmetry between x and y in formula (2), we know that at a point with a vertical tangent line, where $x' = 0$, an interchange of the roles of the coordinate axes yields the same formula (2). If the student is interested he can carry through the derivation by considering $\alpha = \operatorname{arccot}\frac{dx}{dy}$. (Cf. Ex. 25, § 814.)

If the parameter is arc length, then, by virtue of (9), § 805,

(3) $$K = |x'y'' - y'x''|.$$

If the curve is given by an equation of the form $y = f(x)$, so that x is the parameter, then (since $x' = 1$ and $x'' = 0$),

(4) $$K = \frac{|y''|}{(1+y'^2)^{\frac{3}{2}}},$$

a similar formula holding in case the parameter is y.

If the curve is given in polar coordinates, and if θ is the parameter, then substitution of $x = \rho \cos \theta$ and $y = \rho \sin \theta$ in (2) gives (check the details):

(5) $$K = \frac{|\rho^2 + 2\rho'^2 - \rho\rho''|}{(\rho^2 + \rho'^2)^{\frac{3}{2}}}.$$

NOTE. The formulas above presuppose that the point in question is not a singular point; that is that x' and y' are not both zero there (§ 805), and that the derivatives involved exist. Throughout the reminder of this chapter the existence and continuity of all derivatives used will be assumed without specific statement. In particular *only smooth curves will be considered*.

The **radius of curvature** R of a curve at a given point is defined to be the reciprocal of the curvature at that point,

(6) $$R \equiv \frac{1}{K}.$$

(If the curvature is zero, the radius of curvature is said to be *infinite*.) Formulas for radius of curvature are given by taking reciprocals of (2)–(5).

Justification for the terminology introduced in this section resides in the two facts that *the curvature of a straight line is zero* and *the radius of curvature of a circle is its radius*. (Cf. Exs. 9–10, § 814.)

812. CIRCLE OF CURVATURE

The **center of curvature** of a curve at a given point p on the curve is the point q on the normal to the curve at p, on the concave side of the curve,† whose distance from p is the radius of curvature R of the curve at p. (Cf. Fig. 809.) The **circle of curvature** of the curve at p is the circle with center q and radius R. Its tangent at p is the same as that of the given curve at p, and its curvature is that of the given curve at p.

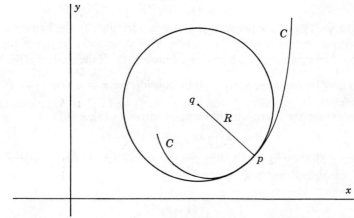

FIG. 809

In the next section we shall derive formulas for the coordinates of the center of curvature, but we illustrate by an example a standard method of obtaining the center of curvature without any further use of formulas.

Example. Find the center of curvature of the curve $y = x^3$ at the point $(1, 1)$.

Solution. At the point p: $(1, 1)$, $y' = 3$ and $y'' = 6$, and therefore $R = \frac{5}{3}\sqrt{10}$. Since the slope of the curve at p is 3, the slope of the normal there is $-\frac{1}{3}$, and

---
† Concavity is determined by the sign of $d\alpha/ds$ or, in other terms, of $x'y'' - y'x''$, together with the direction of increasing arc length.

(cf. Fig. 810) a triangle similar to the one whose hypotenuse is the radius of the circle of curvature would have sides 1 and -3, and hypotenuse $\sqrt{10}$. Since the triangle desired is larger in the ratio of $\frac{5}{3}$, the x-coordinate of the center must be 5 units to the left of 1, and the y-coordinate must be $\frac{5}{3}$ units above 1. That is, the center of curvature is $(-4, 2\frac{2}{3})$.

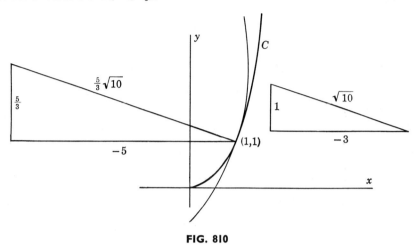

FIG. 810

*813. EVOLUTES AND INVOLUTES

In this section we shall obtain formulas for the coordinates of the center of curvature of a smooth curve, C, $x = x(t)$, $y = y(t)$, at a point where the curvature exists and is nonzero.

If the circle of curvature Γ has center (α, β) and radius R, its equation is

(1) $$(x - \alpha)^2 + (y - \beta)^2 = R^2.$$

Let x and y, for this circle Γ, be functions of a parameter (undesignated), and denote differentiation with respect to this parameter by primes. Since (1) becomes an identity in this parameter, we obtain by differentiation the following two equations:

(2) $$\begin{cases} x'(x - \alpha) + y'(y - \beta) = 0, \\ x''(x - \alpha) + y''(y - \beta) = -(x'^2 + y'^2), \end{cases}$$

true for all points of Γ.

A rather curious and most significant fact is the following: Let x and y be considered as the *original* functions $x(t)$ and $y(t)$ defining the curve C, let differentiation *with respect to the parameter t* be designated by primes, and let (α, β) be the center of curvature and R the radius of curvature at the point p at which x and y and their derivatives are evaluated. *Then equations* (1) *and* (2) *hold.*

Equation (1) is obvious. We shall establish (2) by first proving it for the special case where the parameter for the curve C is x. This can always be done unless C has a vertical tangent at (x, y). (For this latter special case the parameter can be chosen to be y, and the derivation is entirely similar to the one that follows below; the student is asked in Exercise 25, § 814, to carry through the details.) With x as the parameter, then, equations (2) take the form

(3)
$$\begin{cases} (x - \alpha) + \dfrac{dy}{dx}(y - \beta) = 0, \\ \dfrac{d^2y}{dx^2}(y - \beta) = -\left[1 + \left(\dfrac{dy}{dx}\right)^2\right]. \end{cases}$$

The reasons that equations (3) hold are threefold: In the first place, they hold in case y as a function of x is defined by the circle of curvature Γ. In the second place, since C and Γ have a common tangent at the point (x, y), the derivative $\dfrac{dy}{dx}$, when evaluated at that point, is the same for C as it is for Γ. In the third place, since C and Γ have the same curvature and the same concavity at (x, y), the second derivative $\dfrac{d^2y}{dx^2}$, evaluated at that point, is the same for C as it is for Γ (cf. (4), § 811). Therefore equations (3) hold for C.

Finally, if we substitute

$$\frac{dy}{dx} = \frac{y'}{x'} \quad \text{and} \quad \frac{d^2y}{dx^2} = \frac{1}{x'}\left(\frac{y'}{x'}\right)' = \frac{x'y'' - y'x''}{x'^3}$$

in (3) we obtain equations (2), and the proof is complete.

Solving (2) as a system of two linear equations in the two unknowns $(x - \alpha)$ and $(y - \beta)$, and rearranging terms, we find for the coordinates of the center of curvature:

(4)
$$\alpha = x - \frac{y'(x'^2 + y'^2)}{x'y'' - y'x''}, \quad \beta = y + \frac{x'(x'^2 + y'^2)}{x'y'' - y'x''}.$$

If the point p of the curve, just considered, is regarded as variable and moving along the curve, the circle of curvature can be thought of as rolling along the curve (with variable radius) and its center becomes a variable point with variable coordinates (α, β), tracing out a curve whose equations are given parametrically by (4). Such a curve is called an *evolute*, according to the following definition:

Definition. *The locus of centers of curvature of the points of a given curve is called the **evolute** of that curve. If a curve C_1 is the evolute of a curve C_2, then C_2 is called an **involute** of C_1.*

NOTE. The evolute of a curve is, by definition, unique. However, as is shown in Exercise 29, § 814, a curve may have infinitely many involutes.

It has been stated that equations (4) give the evolute of a given curve parametrically. In some cases the parameter can be eliminated to give a

§ 814] EXERCISES 257

single equation in α and β as the equation of the evolute. Often, elimination of the parameter involves using the original equation of the curve.

Example. Find the evolute of the ellipse $\dfrac{x^2}{a^2} + \dfrac{y^2}{b^2} = 1$.

First Solution. Letting x be the parameter, we have by implicit differentiation,

$$\frac{dy}{dx} = -\frac{b^2 x}{a^2 y}, \quad \frac{d^2 y}{dx^2} = -\frac{b^2(b^2 x^2 + a^2 y^2)}{a^4 y^3} = -\frac{b^4}{a^2 y^3},$$

so that equations (4) become

$$\alpha = \frac{(a^2 - b^2)x^3}{a^4}, \quad \beta = -\frac{(a^2 - b^2)y^3}{b^4},$$

and elimination of x and y, by solving here and substituting in the original equation, gives $(a\alpha)^{\frac{2}{3}} + (b\beta)^{\frac{2}{3}} = (a^2 - b^2)^{\frac{2}{3}}$. (Cf. Fig. 811.)

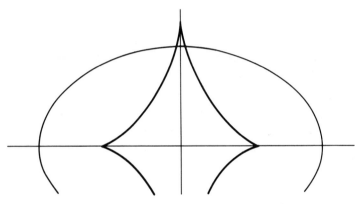

FIG. 811

Second Solution. Transform the equation to parametric form: $x = a \cos t$, $y = b \sin t$. Then substitution in (4) gives

$$\alpha = a \cos t - \frac{b \cos t(a^2 \sin^2 t + b^2 \cos^2 t)}{ab} = \frac{(a^2 - b^2)}{a} \cos^3 t,$$

$$\beta = b \sin t + \frac{-a \sin t(a^2 \sin^2 t + b^2 \cos^2 t)}{ab} = \frac{(a^2 - b^2)}{-b} \sin^3 t.$$

Elimination of the parameter t by means of a well-known trigonometric indentity gives the answer above.

814. EXERCISES

In Exercises 1–8, find the curvature, radius of curvature, and center of curvature of the given curve at the specified point. Draw a figure and show the circle of curvature.

1. $y = x^2$; $(0, 0)$.
2. $y = x^2$; $(1, 1)$.
3. $y^3 = x^2$; $(1, 1)$.
4. $y^2 = x^3 + 8$; $(1, 3)$.
5. $x = 3t^2$, $y = 3t - t^3$; $t = 1$.
6. $x = t - \sin t$, $y = 1 - \cos t$; $t = \pi$.
7. $\rho = 1 - \cos \theta$; $\theta = \pi$.
8. $\rho^2 = \cos 2\theta$; $\theta = 0$.

9. Prove that the curvature of any straight line is identically zero. State and prove a converse. Include the case of a vertical straight line.

10. Prove that the radius of curvature of any circle is its radius. *Hint:* Write its equations in parametric form $x = a + r\cos\theta$, $y = b + r\sin\theta$.

11. Show that the point of maximum curvature of a parabola $y = ax^2 + bx + c$ occurs at the vertex.

12. Show that the points of maximum curvature of $y = x^3 - 3x$ do not occur at the maximum and minimum points.

13. Show that the radii of curvature at the ends of the axes of the ellipse $b^2 x^2 + a^2 y^2 = a^2 b^2$ are b^2/a and a^2/b.

14. Find the minimum value of the radius of curvature of the curve $y = e^x$. Check your answer by working the same problem for the inverse function $y = \ln x$.

15. Show that the "circle of curvature" of $y = x^4$ at the origin is the x-axis.

16. The bowl of a goblet is a part of the surface obtained by revolving about the y-axis the curve $y^3 = x^4$. If a spherical pellet is dropped into the goblet, show that this pellet cannot touch the bottom of the container.

★17. Show that the evolute of the cycloid, $x = a(\theta - \sin\theta)$, $y = a(1 - \cos\theta)$, is again a cycloid, which can be obtained by translating the original cycloid.

★18. Demonstrate the unreliability of one's intuitions in the following "reasoning": A wheel is rolling on a straight line. At any instant the point at the top of the wheel is rotating instantaneously about the point at the bottom of the wheel. "Therefore" the cycloid of Exercise 17 has radius of curvature at its maximum point equal to the diameter of the generating circle. (Show that this radius of curvature is actually equal to *twice* the diameter.)

In Exercises 19–24, find the center of curvature at a general point. Draw a figure showing the evolute of the given curve.

★19. $y^2 = 2px$.   **★20.** $x^2 - y^2 = 1$.

★21. $x = t^2, y = t^3$.   **★22.** $x = 3t^2, y = 3t - t^3$.

★23. The parabola $\sqrt{x} + \sqrt{y} = 1$, $0 \le x \le 1$, $0 \le y \le 1$.

★24. The hypocycloid $4x = 3\cos\theta + \cos 3\theta$, $4y = 3\sin\theta - \sin 3\theta$.

★25. Prove the validity of formulas (2), §§ 811 and 813, in the case of a vertical tangent ($x' = 0$).

★26. The **tractrix** is the curve defined by the differential equation and initial condition:

(1) $$\frac{dy}{dx} = -\frac{y}{\sqrt{a^2 - y^2}}, \quad \lim_{x \to 0+} y = a; \quad 0 < x, 0 < y < a.$$

Show that the tractrix can be regarded as a "curve of pursuit," describing the path traced out by a point $p: (x, y)$ that starts at the point $(0, a)$ and moves steadily toward a fugitive point q on the x-axis while remaining a constant distance from it (cf. Fig. 812). Show that the tractrix (1) has the equation

$$x = -\sqrt{a^2 - y^2} + a \ln[(a + \sqrt{a^2 - y^2})/y],$$

and that its evolute is the (half-) catenary $y = a \cosh(x/a)$. For an interesting property of the tractrix, see Exercise 16, § 1912.

★27. Show that the center of curvature of a curve at a point p on the curve is the limiting position of the point of intersection of the normal lines to the curve at p and at a neighboring point q on the curve, as q approaches p along the curve.

§ 814] EXERCISES 259

Hint: Choose a rectangular coordinate system so that p is the origin and the positive half of the y-axis is normal to the curve on its concave side. (Cf. Fig. 813.) The radius of curvature at p is $1/y''$, evaluated at $x = 0$. The y-intercept of the normal at q is $y + x/y'$ which by l'Hospital's Rule, tends toward $1/y''$, as x approaches zero.

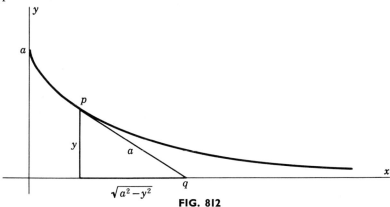

FIG. 812

★28. Let a circle be drawn through three points p_0, p_1, and p_2 of a curve, and let p_1 and p_2 approach p_0 along the curve. The circle in the limiting position of this circle (if it exists) is called the **osculating circle** of the curve at the point p_0. Show that the osculating circle is identical with the circle of curvature. *Hint:* Choose

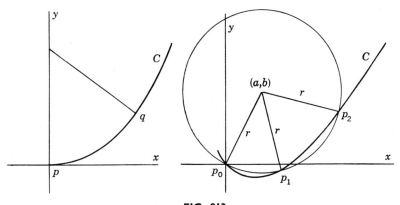

FIG. 813

a coordinate system as in Ex. 27, let p_i have coordinates (x_i, y_i), $i = 0, 1, 2$, $(x_0 = y_0 = 0)$, and let $y = f(x)$ be the equation of the curve near the origin. (Cf. Fig. 813.) Let the circle through p_0, p_1, and p_2 have center (a, b) and radius r, and let $F(x) = (x - a)^2 + (y - b)^2 - r^2$, where $y = f(x)$. Then $F(0) = F(x_1) = F(x_2) = 0$. Therefore, by Rolle's Theorem there exist x_3 between the smallest and the middle one of 0, x_1, and x_2, and x_4 between the middle one and the largest of 0, x_1, and x_2 such that $F'(x_3) = F'(x_4) = 0$, and, again, there exists x_5 between x_3 and x_4 such that $F''(x_5) = 0$. Show that $F''(x) = 2 + 2y'^2 + 2(y - b)y''$, and since $x_5 \to 0$, the facts that $y \to 0$ and $y' \to 0$ imply $b \to 1/y''$, evaluated at $x = 0$. Then use the form of $F'(x)$ to show that $a \to 0$.

★29. *The unwinding string.* Suppose a string is thought of as being first laid out along a curve C and then unwound, as indicated in Figure 814. What is the curve Γ described by a point on the string? Show that the given curve is the evolute of this new curve, and therefore that, although any curve has by definition, only

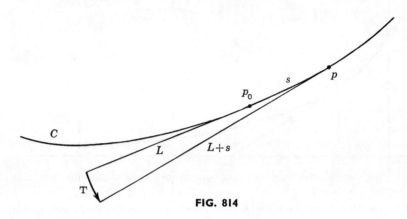

FIG. 814

one evolute, it may have (since there are infinitely many points on the string) infinitely many involutes. *Hint:* Let the given curve C have parametric equations $x = f(s)$, $y = g(s)$, where s is the arc length from a fixed point p_0 on C. Since a tangent line segment of unit length at p, in the appropriate direction, has horizontal and vertical projections equal to $-f'$ and $-g'$, respectively, and the string, in being unwound from tangency at p_0 to tangency at p, increases its length from an initial length L by an amount equal to the arc length s between p_0 and p to a new length $L + s$, the parametric equations of Γ are $x = f - (L + s)f'$, $y = g - (L + s)g'$. Equations (4), § 813, applied to Γ, become $\alpha = f$, $\beta = g$, as requested. (Good use must be made of the identity $f'^2 + g'^2 = 1$.)

★30. Find the equations of the spiral involute obtained by "unwinding a string" (cf. Ex. 29) from the unit circle, starting from the point $(1, 0)$. Then verify that the unit circle is the evolute of this involute.

★31. With suitable differentiability assumptions, prove that the normal to a curve at any point is tangent to the evolute at the corresponding point. *Hint:* Differentiate formulas (4), § 813, and show that $\beta':\alpha' = -x':y'$.

★32. Prove that if arc length s is the parameter, then
$$K = \sqrt{x''^2 + y''^2}.$$

9

Partial Differentiation

901. PARTIAL DERIVATIVES

When a function of several real variables is considered as dependent on just one of these variables, the others being held constant, it may possess a derivative with respect to this one variable. Such a derivative (if it exists) is called a *partial derivative*, according to the following definition, which is formulated for simplicity in terms of a function of two variables:

Definition. *Let* $u = f(x, y)$ *be defined in a region R of the Euclidean plane. At a point* (x, y) *of R the* **partial derivatives** *of u with respect to x and of u with respect to y, written*

(1)
$$\begin{cases} \dfrac{\partial u}{\partial x} = \dfrac{\partial f}{\partial x} = u_x = f_x = f_x(x, y) = f_1(x, y), \\[2mm] \dfrac{\partial u}{\partial y} = \dfrac{\partial f}{\partial y} = u_y = f_y = f_y(x, y) = f_2(x, y), \end{cases}$$

respectively, are the limits of the difference quotients:

(2)
$$\begin{cases} f_1(x, y) \equiv \lim_{h \to 0} \dfrac{f(x+h, y) - f(x, y)}{h}, \\[2mm] f_2(x, y) \equiv \lim_{k \to 0} \dfrac{f(x, y+k) - f(x, y)}{k}. \end{cases}$$

The derivatives of the preceding Definition, in distinction to the derivatives *of* derivatives treated in the following section, are called **partial derivatives of the first order** or, for simplicity, **first partial derivatives**.

NOTE 1. We shall discuss later (cf. § 917) relative merits and weaknesses of the notations of (1). Let us point out at this time, however, that one advantage of the numerical subscript notation f_1, f_2 is that it stresses the *form* of the function rather than the particular letters used. For example, whether we write $f(x, y, z) = xy^2x^3$ or $f(\xi, \eta, \zeta) = \xi \eta^2 \zeta^3, f_1(5, 3, 2) = 72$.

NOTE 2. For a geometrical interpretation of first partial derivatives, see Exercise 17, § 904, and its accompanying Figure 901.

Example 1. The first partial derivatives of $u = \sin(x + 2y - 3z)$ are $u_x = \cos(x + 2y - 3z)$, $u_y = 2\cos(x + 2y - 3z)$, and $u_z = -3\cos(x + 2y - 3z)$.

Example 2. Direct substitution shows that $u = \dfrac{Ax^2 + By^2}{Cx^2 + Dy^2}$ is a solution of the *partial differential equation* $x\dfrac{\partial u}{\partial x} + y\dfrac{\partial u}{\partial y} = 0$.

902. PARTIAL DERIVATIVES OF HIGHER ORDER

If a function of several variables has first partial derivatives, these derivatives (which are again functions of these same several variables) may in turn have first partial derivatives. These are called **partial derivatives of the second order**, or **second partial derivatives**, of the original function. Notation for the case of a function $u = f(x, y, z)$ of three variables, and differentiation with respect to x or y is:

$$\frac{\partial}{\partial x}\left(\frac{\partial u}{\partial x}\right) = \frac{\partial^2 u}{\partial x^2} = \frac{\partial^2 f}{\partial x^2} = u_{xx} = f_{xx}(x, y, z) = f_{11}(x, y, z),$$

$$\frac{\partial}{\partial y}\left(\frac{\partial u}{\partial x}\right) = \frac{\partial^2 u}{\partial y \partial x} = \frac{\partial^2 f}{\partial y \partial x} = u_{yx} = f_{yx}(x, y, z) = f_{21}(x, y, z),$$

$$\frac{\partial}{\partial x}\left(\frac{\partial u}{\partial y}\right) = \frac{\partial^2 u}{\partial x \partial y} = \frac{\partial^2 f}{\partial x \partial y} = u_{xy} = f_{xy}(x, y, z) = f_{12}(x, y, z),$$

$$\frac{\partial}{\partial y}\left(\frac{\partial u}{\partial y}\right) = \frac{\partial^2 u}{\partial y^2} = \frac{\partial^2 f}{\partial y^2} = u_{yy} = f_{yy}(x, y, z) = f_{22}(x, y, z).$$

Partial derivatives of order higher than the second, and for functions of any number of variables, are defined and denoted similarly. For example,

$$\frac{\partial}{\partial x}\left(\frac{\partial^2 u}{\partial x^2}\right) = \frac{\partial^3 u}{\partial x^3}, \quad \frac{\partial}{\partial y}\left(\frac{\partial^2 u}{\partial y \partial x}\right) = \frac{\partial^3 u}{\partial y^2 \partial x}, \quad \frac{\partial}{\partial y}\left(\frac{\partial^2 u}{\partial x \partial z}\right) = \frac{\partial^3 u}{\partial y \partial x \partial z}.$$

Partial derivatives in which more than one of the independent variables is a variable of differentiation are called **mixed partial derivatives**. Thus $\dfrac{\partial^2 u}{\partial x \partial y}$ is a mixed partial derivative of second order.

It might appear at first that for a function f of two variables, x and y, there are two mixed partial derivatives of the second order, f_{xy} and f_{yx}, and six of the third order, $f_{xxy}, f_{xyx}, f_{yxx}, f_{xyy}, f_{yxy},$ and f_{yyx}, with a corresponding increase in the number as the order increases, or as more independent variables are present. However, as is shown in the following section, in case the mixed partial derivatives involved are *continuous*, certain equality relations exist among them. For example, for a function f of x and y, $f_{xy} = f_{yx}, f_{xxy} = f_{xyx} = f_{yxx}$, and $f_{xyy} = f_{yxy} = f_{yyx}$, so that there are only one mixed partial derivative

§ 903] EQUALITY OF MIXED PARTIAL DERIVATIVES

of the second order and two of the third order. In general, in the presence of continuity, two higher-order mixed partial derivatives are equal whenever the number of times each independent variable is a variable of differentiation is the same for the two. For example, $f_{zyzxzx} = f_{xxyzzz}$. This general fact is a consequence of the simpler fact that $f_{xy} = f_{yx}$, since this latter equality states that two adjacent subscript letters can be interchanged, and any permutation of subscripts can be obtained by a sequence of adjacent interchanges. Furthermore, in a proof of the relation $f_{xy} = f_{yx}$ the presence of variables other than x and y is irrelevant since they are held fixed. Thus everything depends on the basic fact that $f_{xy} = f_{yx}$ for functions of *two* variables. In the following section this relation is stated with precision, and proved in a slightly more general form than that indicated above.

*903. EQUALITY OF MIXED PARTIAL DERIVATIVES

Theorem. *Assume that (i) the function f of the two real variables x and y, and its two first partial derivatives f_x and f_y, exist in a region R, and (ii) the mixed partial derivative f_{xy} exists in R and is continuous at the point (x_0, y_0) of R. Then the mixed partial derivative f_{yx} exists at (x_0, y_0) and is equal to f_{xy} at that point.*

Proof. Choose a small square neighborhood of (x_0, y_0) lying in R, and choose h and k so that their absolute values are less than half the length of a side of this neighborhood. Let $A(h, k)$ and $\psi(y)$ be defined by the equations

(1) $\quad A(h, k) \equiv f(x_0 + h, y_0 + k) - f(x_0 + h, y_0)$
$$- f(x_0, y_0 + k) + f(x_0, y_0),$$

(2) $\quad \psi(y) \equiv f(x_0 + h, y) - f(x_0, y),$

with y restricted to lie appropriately near y_0. (The function ψ also depends on h, but it is its dependence on y that concerns us here.) Then $A(h, k) = \psi(y_0 + k) - \psi(y_0)$, and since $f_y(x, y)$ exists in the chosen neighborhood of (x_0, y_0), $\psi(y)$ is differentiable for y near y_0 (and is therefore also continuous for such y). Thus, by the Law of the Mean for functions of one variable (§ 305),

(3) $\quad A(h, k) = \psi(y_0 + k) - \psi(y_0) = k\psi'(y_0 + \theta_1 k)$

for some θ_1 such that $0 < \theta_1 < 1$ (θ_1 depending on h and k).

This equation can be written

(4) $\quad A(h, k) = k[f_y(x_0 + h, y_0 + \theta_1 k) - f_y(x_0, y_0 + \theta_1 k)].$

Again, by the Law of the Mean, applied to $f_y(x, y_0 + \theta_1 k)$ considered as a function of the single variable x, the quantity in brackets, on the right in (4), can be written $h f_{xy}(x_0 + \theta_2 h, y_0 + \theta_1 k)$, for some θ_2 such that $0 < \theta_2 < 1$, so that

(5) $\quad A(h, k) = hk f_{xy}(x_0 + \theta_2 h, y_0 + \theta_1 k).$

Therefore, for $hk \neq 0$,

(6) $$\frac{A}{hk} = f_{xy}(x_0 + \theta_2 h, y_0 + \theta_1 k).$$

We now start afresh with the function

(7) $$\phi(x) \equiv f(x, y_0 + k) - f(x, y_0),$$

for x near x_0. For $hk \neq 0$,

(8) $$\frac{A}{hk} = \frac{1}{k} \frac{\phi(x_0 + h) - \phi(x_0)}{h},$$

and since $\phi'(x_0)$ exists, it follows that for $k \neq 0$, $\lim_{h \to 0} \frac{A}{hk}$ exists and is equal to $\frac{1}{k} \phi'(x_0)$, or

(9) $$\lim_{h \to 0} \frac{A}{hk} = \frac{1}{k} [f_x(x_0, y_0 + k) - f_x(x_0, y_0)].$$

By (6), and the continuity of f_{xy} at (x_0, y_0), $\lim_{(h,k) \to (0,0)} \frac{A}{hk}$ $(hk \neq 0)$ exists and is equal to $f_{xy}(x_0, y_0)$. Therefore (Note 1, § 607)

(10) $$\lim_{k \to 0} \lim_{h \to 0} \frac{A}{hk}$$

exists and is equal to $f_{xy}(x_0, y_0)$. Finally, by (9), this iterated limit is, by definition, $f_{yx}(x_0, y_0)$.

904. EXERCISES

In Exercises 1–4, find the first partial derivatives of the given function with respect to each of the independent variables.

1. $u = (x - y) \sin(x + y)$.
2. $u = xe^y + ye^x$.
3. $f(x, y, z) = \sqrt{x^2 + y^2 + z^2}$.
4. $x = r \sin \phi \cos \theta$.

In Exercises 5–8, find all second-order partial derivatives of each function, and verify the appropriate equalities among the mixed partial derivatives.

5. $f(x, y) = Ax^2 + 2Bxy + Cy^2$.
6. $z = \dfrac{x}{x + y}$.
7. $u = xe^y + y \sin z$.
8. $u = z \operatorname{Arctan} \dfrac{y}{x}$.

In Exercises 9–16, show that the function u satisfies the given differential equation.

9. $u = \dfrac{xy}{x + y}$; $x \dfrac{\partial u}{\partial x} + y \dfrac{\partial u}{\partial y} = u$.
10. $u = Ax^3 + Bx^2 y + Cxy^2 + Dy^3$; $xu_x + yu_y + zu_z = 3u$.
11. $u = (x^2 + y^2 + z^2)^{-\frac{1}{2}}$; $u_x^2 + u_y^2 + u_z^2 = u^4$.
12. $u = \operatorname{Arctan} \dfrac{y}{x}$; $xu_x + yu_y = 0$.
13. $u = \operatorname{Arctan} \dfrac{y}{x}$; $xu_y - yu_x = 1$.

14. $u = \dfrac{xy}{x+y}$; $x^2 \dfrac{\partial^2 u}{\partial x^2} + 2xy \dfrac{\partial^2 u}{\partial x \, \partial y} + y^2 \dfrac{\partial^2 u}{\partial y^2} = 0.$

15. $u = ce^{-n^2 t} \sin nx$; $\dfrac{\partial^2 u}{\partial x^2} = \dfrac{\partial u}{\partial t}.$

16. $u = c \sin akx \cos kt$; $\dfrac{\partial^2 u}{\partial x^2} = a^2 \dfrac{\partial^2 u}{\partial t^2}.$

17. Interpret the first partial derivatives of $z = f(x, y)$ as slopes of curves formed by cutting the surface $z = f(x, y)$ by planes of the form $x = a$ and $y = b$. (Cf. Fig. 901.)

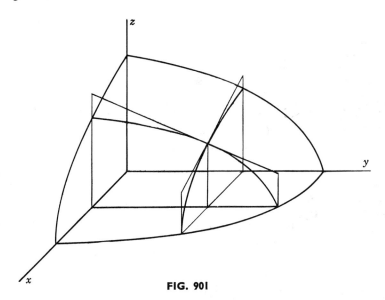

FIG. 901

18. As a consequence of the Theorem of § 903, assuming that $f(x, y, z)$ and all of its partial derivatives to the fourth order are continuous (in a given region) prove that $f_{xxy} = f_{yxx}$ and $f_{yxzx} = f_{xxyz}.$

A function $u(x, y)$ is called **harmonic** if and only if it satisfies **Laplace's equation**

$$(1) \qquad \dfrac{\partial^2 u}{\partial x^2} + \dfrac{\partial^2 u}{\partial y^2} = 0.$$

In Exercises 19–22, show that u is harmonic.

19. $u = x^3 - 3xy^2.$

20. $u = e^x \sin y.$

21. $u = \ln(x^2 + y^2).$

22. $u = \operatorname{Arctan} \dfrac{y}{x}.$

23. Assuming existence and continuity of derivatives whenever desired, show that if two functions, u and v, satisfy the **Cauchy-Riemann differential equations**

$$(2) \qquad \dfrac{\partial u}{\partial x} = \dfrac{\partial v}{\partial y}, \; \dfrac{\partial u}{\partial y} = -\dfrac{\partial v}{\partial x},$$

then both functions are harmonic. (Cf. Exs. 19–22.)

★24. If $f(x, y)$ is a given function, define $g(x, y) \equiv f(y, x)$. Show that
$$g_x(x, y) = f_x(y, x), \quad \text{or} \quad g_1(x, y) = f_2(y, x),$$
$$g_y(x, y) = f_y(y, x), \quad \text{or} \quad g_2(x, y) = f_1(y, x).$$
Illustrate by simple (but not completely trivial) examples.

★25. Let $f(x,y) \equiv xy\dfrac{x^2 - y^2}{x^2 + y^2}$, for $x^2 + y^2 \neq 0, f(0, 0) \equiv 0$. Show that $f_{xy}(0, 0) = 1$ and $f_{yx}(0, 0) = -1$. Why does this not invalidate the Theorem of § 903?

★26. Show that the function
$$f(x, y) \equiv \frac{xy}{x^2 + y^2}, \quad \text{for} \quad x^2 + y^2 \neq 0, \quad f(0, 0) \equiv 0,$$
considered in §§ 606–608, has first partial derivatives everywhere, although it is not everywhere continuous. In other words, show that although the existence of a derivative implies continuity of a function of one variable (§ 302), this fact does not extend to functions of more than one variable. (Cf. Exs. 27–28.)

★27. Prove that if $f(x, y)$ has *bounded* first partial derivatives in a region it must be continuous there. (Cf. Exs. 26, 28.) *Hints:* Apply the Law of the Mean (§ 305) to each bracketed part of
$$f(x_0 + h, y_0 + k) - f(x_0, y_0) = [f(x_0 + h, y_0 + k) - f(x_0, y_0 + k)]$$
$$+ [f(x_0, y_0 + k) - f(x_0, y_0)].$$

★28. Assume that $f(x, y)$ possesses first partial derivatives $f_1(x, y)$ and $f_2(x, y)$ in a region R and that f_1 and f_2 are continuous at a point p of R. Prove that f is continuous in some neighborhood of p. (Cf. Ex. 27.)

★29. Give an example of a function $f(x, y)$ for which $f_{xy}(x, y)$ is identically 0 but for which $f_{yx}(x, y)$ never exists. *Hint:* Let $f(x, y)$ be a nondifferentiable function of x alone.

905. THE FUNDAMENTAL INCREMENT FORMULA

If $u = f(x)$ is a differentiable function of the single variable x, equation (4), § 302:

(1) $$\Delta y = \frac{dy}{dx} \Delta x + \epsilon \Delta x,$$

where $\epsilon \to 0$ as $\Delta x \to 0$, expresses an increment Δy of the dependent variable, in terms of the derivative, plus a correction term involving an infinitesimal. It is the purpose of this section to obtain a corresponding formula for the increment of a dependent variable that is a function of several independent variables. For simplicity the details are presented for the case of a function of two independent variables. The case of three or more independent variables is given as an exercise (Ex. 11, § 909).

Assume that $u = f(x, y)$ possesses first partial derivatives $f_1(x, y)$ and $f_2(x, y)$ in a region R, and that f_1 and f_2 are continuous at some point (x_0, y_0) of R. Furthermore, take a square neighborhood of (x_0, y_0) that lies in R and let Δx and Δy be numerically less than half the length of a side of this

§ 905]   THE FUNDAMENTAL INCREMENT FORMULA   267

neighborhood, so that the point $(x_0 + \Delta x, y_0 + \Delta y)$ lies in R. Then the increment Δu of the dependent variable can be written as follows:

(2)
$$\begin{aligned}\Delta u &= f(x_0 + \Delta x, y_0 + \Delta y) - f(x_0, y_0) \\ &= [f(x_0 + \Delta x, y_0 + \Delta y) - f(x_0, y_0 + \Delta y)] \\ &\quad + [f(x_0, y_0 + \Delta y) - f(x_0, y_0)].\end{aligned}$$

We now apply the Law of the Mean (§ 305) to each of the bracketed quantities in (2) and obtain

(3) $\quad \Delta u = f_1(x_0 + \theta_1 \Delta x, y_0 + \Delta y) \Delta x + f_2(x_0, y_0 + \theta_2 \Delta y) \Delta y,$

where θ_1 and θ_2 are quantities between 0 and 1 that depend on Δx and Δy.

Continuity of the first partial derivatives is now used. Since $(x_0 + \theta_1 \Delta x, y_0 + \Delta y) \to (x_0, y_0)$ and $(x_0, y_0 + \theta_2 \Delta y) \to (x_0, y_0)$, as $\Delta x \to 0$ and $\Delta y \to 0$, $f_1(x_0 + \theta_1 \Delta x, y_0 + \Delta y) \to f_1(x_0, y_0)$ and $f_2(x_0, y_0 + \theta_2 \Delta y) \to f_2(x_0, y_0)$. In other words,

(4) $\quad \begin{cases} f_1(x_0 + \theta_1 \Delta x, y_0 + \Delta y) = f_1(x_0, y_0) + \epsilon_1, \\ f_2(x_0, y_0 + \theta_2 \Delta y) = f_2(x_0, y_0) + \epsilon_2, \end{cases}$

where ϵ_1 and ϵ_2 are infinitesimals (tending toward zero as Δx and Δy tend toward zero). Substitution of (4) in (3) gives the Fundamental Increment Formula:

Theorem. Fundamental Increment Formula. *If $u = f(x, y)$ possesses first partial derivatives in a region R, and if these first partial derivatives are continuous at some point (x_0, y_0) of R, then (for increments Δx and Δy sufficiently small numerically) the increment Δu of the independent variable u can be written*

(5) $$\Delta u = \frac{\partial u}{\partial x} \Delta x + \frac{\partial u}{\partial y} \Delta y + \epsilon_1 \Delta x + \epsilon_2 \Delta y,$$

where the partial derivatives are evaluated at the point (x_0, y_0) and ϵ_1 and ϵ_2 are infinitesmals, tending toward zero with Δx and Δy.

The Fundamental Increment Formula for a function u of several variables $x, y, z, \cdots$ takes the form (cf. Ex. 11, § 909):

(6) $$\Delta u = \frac{\partial u}{\partial x} \Delta x + \frac{\partial u}{\partial y} \Delta y + \frac{\partial u}{\partial z} \Delta z + \cdots + \epsilon_1 \Delta x + \epsilon_2 \Delta y + \epsilon_3 \Delta z + \cdots,$$

where $\epsilon_1, \epsilon_2, \epsilon_3, \cdots$ are infinitesimal functions of the increments $\Delta x, \Delta y, \Delta z, \cdots$.

Whenever the increment of a function u of several variables can be expressed in the form (6), the function u is said to be **differentiable** at the particular point at which the first partial derivatives are evaluated. If a function possesses first partial derivatives in a neighborhood of a point and if these first partial derivatives are continuous at the point, the function is said to be **continuously differentiable** at the point. The preceding Theorem can now be restated: *continuous differentiability implies differentiability.*

Note 1. From the form of (6) we can infer that *differentiability of a function at a point implies its continuity there.* (Cf. Ex. 27, § 904.)

Note 2. Mere *existence* of the first partial derivatives of a function does not imply its differentiability. (Cf. Ex. 26, § 904, and Note 1.)

Note 3. *Differentiability does not imply continuous differentiability.* (Cf. the Example, below.)

Example. Define $\phi(x) \equiv x^2 \sin(1/x)$, $x \neq 0$, $\phi(0) \equiv 0$ (Example 3, § 303) and let $f(x, y) \equiv \phi(x) + \phi(y)$. Then $f_1(x, y) = \phi'(x)$ and $f_2(x, y) = \phi'(y)$ are both discontinuous at the origin. On the other hand, f is differentiable there, since formula (5) becomes $\Delta u = 0 \cdot \Delta x + 0 \cdot \Delta y + [\Delta x \sin(1/\Delta x)] \Delta x + [\Delta y \sin(1/\Delta y)] \Delta y$, and the quantities in brackets are infinitesimals.

906. DIFFERENTIALS

By analogy with the case of a function of one variable (§ 311), we define the **differential**, or **total differential**, of a differentiable function $u = f(x, y)$ of two variables:

(1) $$du \equiv \frac{\partial u}{\partial x} dx + \frac{\partial u}{\partial y} dy,$$

where dx and dy are two new independent variables. The differential du, then, is a function of *four* independent variables: x and y (the coordinates of the point at which the partial derivatives are evaluated) and dx and dy. These latter quantities, dx and dy, are also called **differentials**. Again (cf. § 312) it is convenient, for most purposes, to identify these differentials with actual increments:

(2) $$dx = \Delta x, \quad dy = \Delta y.$$

For a function $u = f(x, y, z, \cdots)$ of any number of variables, formulas similar to (1) and (2) hold by definition:

(3) $$du = \frac{\partial u}{\partial x} dx + \frac{\partial u}{\partial y} dy + \frac{\partial u}{\partial z} dz + \cdots,$$

(4) $$dx = \Delta x, \quad dy = \Delta y, \quad dz = \Delta z, \cdots.$$

The independence of the variables $dx, dy, dz, \cdots$ in (3) has the following consequence: Whenever the differential of a function u of several variables $x, y, z, \cdots$ is represented as a *linear combination* (cf. § 1602) of the differentials $dx, dy, dz, \cdots$,

(5) $$du = P\, dx + Q\, dy + R\, dz + \cdots,$$

the coefficients must be the first partial derivatives of u:

(6) $$P = \frac{\partial u}{\partial x}, \quad Q = \frac{\partial u}{\partial y}, \quad R = \frac{\partial u}{\partial z}, \cdots.$$

This can be seen by equating the right-hand members of (3) and (5), setting first $dx = 1$, $dy = dz = \cdots = 0$, then $dx = 0$, $dy = 1$, $dz = \cdots = 0$, and so forth.

§ 907] CHANGE OF VARIABLES. THE CHAIN RULE 269

With the notation of differentials the Fundamental Increment Formula (§ 905) becomes

(7) $$\Delta u = du + \epsilon_1 \Delta x + \epsilon_2 \Delta y + \cdots.$$

(Cf. (1), § 312: $\Delta y = dy + \epsilon \Delta x$.) It follows that a differential is a "good approximation" to an increment, for a function of several variables, in the same sense that it is for a function of one variable. (Cf. § 913.)

Example. For the function $u = x^3 y^2$, find Δu, du, and the infinitesimals ϵ_1 and ϵ_2 of (7).

Solution. The differential is $du = 3x^2 y^2 \, dx + 2x^3 y \, dy$. The increment is $\Delta u = (x + \Delta x)^3 (y + \Delta y)^2 - x^3 y^2$, which can be written

(8) $$\begin{aligned} \Delta u = du &+ (3xy^2 \, \Delta x^2 + 6x^2 y \, \Delta x \, \Delta y + x^3 \, \Delta y^2) \\ &+ (y^2 \, \Delta x^3 + 6xy \, \Delta x^2 \, \Delta y + 3x^2 \, \Delta x \, \Delta y^2) \\ &+ (2y \, \Delta x^3 \, \Delta y + 3x \, \Delta x^2 \, \Delta y^2) + \Delta x^3 \, \Delta y^2. \end{aligned}$$

The infinitesimals ϵ_1 and ϵ_2 are not uniquely determined, since many distinct groupings of terms on the right in (8) can be used. According to one such grouping,

$$\epsilon_1 = \epsilon_1(\Delta x, \Delta y) = 3xy^2 \, \Delta x + 6x^2 y \, \Delta y + y^2 \, \Delta x^2 + 2y \, \Delta x^2 \, \Delta y + \Delta x^2 \, \Delta y^2,$$
$$\epsilon_2 = \epsilon_2(\Delta x, \Delta y) = x^3 \, \Delta y + 6xy \, \Delta x^2 + 3x^2 \, \Delta x \, \Delta y + 3x \, \Delta x^2 \, \Delta y.$$

NOTE. It can be shown (cf. Ex. 12, § 911) that the differential dz of a function $z = f(x, y)$ of two variables is associated with the tangent plane to the surface $z = f(x, y)$ in much the same way that the differential dy of a function $y = f(x)$ of one variable is associated with the tangent line to the curve $y = f(x)$.

907. CHANGE OF VARIABLES. THE CHAIN RULE

It is our aim in this section to discuss the following question: *If u is a differentiable function of several variables $x, y, z, \cdots$, each of which is a differentiable function of several variables $r, s, t, \cdots$, how can the first partial derivatives of u with respect to the new variables be expressed in terms of the partial derivatives of the given functions?*

The simplest case was discussed in § 302: If u is a differentiable function of a single variable x which in turn is a differentiable function of a single variable t, then u becomes a differentiable function of t, and $du/dt = (du/dx)(dx/dt)$. The partial derivatives referred to above are generalizations of the ordinary derivatives of this special case.

We proceed to the next simplest case. Let u be a differentiable function of several variables. For simplicity let us assume that u is a function $f(x, y)$ of *two* independent variables. Furthermore, assume that each of these two variables is a differentiable function of a single variable t. Allowing t to change by an increment $\Delta t \, (\neq 0)$, we express, with the aid of the fundamental increment formula (5), § 905, the difference quotient as follows:

(1) $$\frac{\Delta u}{\Delta t} = \frac{\partial u}{\partial x} \frac{\Delta x}{\Delta t} + \frac{\partial u}{\partial y} \frac{\Delta y}{\Delta t} + \epsilon_1 \frac{\Delta x}{\Delta t} + \epsilon_2 \frac{\Delta y}{\Delta t}.$$

Considering u as a function of the single variable t, we let Δt tend toward zero, and take limits of both members of (1). Since x and y are differentiable functions of t, and since ϵ_1 and ϵ_2 are both infinitesimals (as Δt tends toward zero so do Δx and Δy, and consequently so do ϵ_1 and ϵ_2), we conclude by taking the limit of the right-hand member of (1) that u is a differentiable function of t and its derivative is given by the formula

(2) $$\boxed{\frac{du}{dt} = \frac{\partial u}{\partial x}\frac{dx}{dt} + \frac{\partial u}{\partial y}\frac{dy}{dt}.}$$

A similar result (for similar reasons) is valid for a function u of more than two variables:

(3) $$\frac{du}{dt} = \frac{\partial u}{\partial x}\frac{dx}{dt} + \frac{\partial u}{\partial y}\frac{dy}{dt} + \frac{\partial u}{\partial z}\frac{dz}{dt} + \cdots.$$

The form of (3) shows that if u as a function of $x, y, z, \cdots$ and $x, y, z, \cdots$ as functions of t are all *continuously* differentiable, then u as a function of t is too.

Still more generally (give a proof in Exercise 12, § 909), if $u = f(x, y, z, \cdots)$ is a differentiable function of several variables each of which is a differentiable function of several variables $r, s, t, \cdots$, then u, as a function of these new independent variables, is differentiable and, since formula (3) can be applied to each of the independent variables separately, the following equations, known as the general **chain rule** are valid:

(4) $$\begin{cases} \dfrac{\partial u}{\partial r} = \dfrac{\partial u}{\partial x}\dfrac{\partial x}{\partial r} + \dfrac{\partial u}{\partial y}\dfrac{\partial y}{\partial r} + \dfrac{\partial u}{\partial z}\dfrac{\partial z}{\partial r} + \cdots, \\[4pt] \dfrac{\partial u}{\partial s} = \dfrac{\partial u}{\partial x}\dfrac{\partial x}{\partial s} + \dfrac{\partial u}{\partial y}\dfrac{\partial y}{\partial s} + \dfrac{\partial u}{\partial z}\dfrac{\partial z}{\partial s} + \cdots, \\[4pt] \dfrac{\partial u}{\partial t} = \dfrac{\partial u}{\partial x}\dfrac{\partial x}{\partial t} + \dfrac{\partial u}{\partial y}\dfrac{\partial y}{\partial t} + \dfrac{\partial u}{\partial z}\dfrac{\partial z}{\partial t} + \cdots, \\ \cdots \end{cases}$$

In case $u = f(x)$ is a differentiable function of a single variable which is a differentiable function of several variables, $r, s, t, \cdots$, formulas (4) take the form

(5) $$\begin{cases} \dfrac{\partial u}{\partial r} = \dfrac{du}{dx}\dfrac{\partial x}{\partial r}, \\[4pt] \dfrac{\partial u}{\partial s} = \dfrac{du}{dx}\dfrac{\partial x}{\partial s}, \\ \cdots \end{cases}$$

As before, if the given functions are *continuously* differentiable, so is the resulting composite function.

In § 311 it was pointed out that one consequence of the chain rule is that the formula $dy = (dy/dx)\, dx$ is valid whether x is the independent variable or x is dependent on a third variable. A similar statement is true for functions of several variables.

§ 907] CHANGE OF VARIABLES. THE CHAIN RULE

Theorem. *If u is a differentiable function $f(x, y, z, \cdots)$ of several variables, then*

(6) $$du = \frac{\partial u}{\partial x} dx + \frac{\partial u}{\partial y} dy + \frac{\partial u}{\partial z} dz + \cdots,$$

whether the variables $x, y, z, \cdots$ are considered as independent variables or as differentiable functions of other variables $r, s, t, \cdots$.

Proof. The problem is to show that the right-hand member of (6), when expressed in terms of the new variables, reduces to $\frac{\partial u}{\partial r} dr + \frac{\partial u}{\partial s} ds + \cdots$, which we know to be the value of the left-hand member when u is considered as a function of the new variables. To achieve this result we substitute for $dx, dy, \cdots$, in (6), the expressions $\frac{\partial x}{\partial r} dr + \frac{\partial x}{\partial s} ds + \cdots$, $\frac{\partial y}{\partial r} dr + \frac{\partial y}{\partial s} ds + \cdots, \cdots$, expand the products, and then rearrange terms so that those involving dr, those involving $ds, \cdots$, are separately assembled. (It would be well to check the mechanical details!) If dr is factored from those terms involving it, its coefficient (better check this!) will be found to be $\frac{\partial u}{\partial x} \frac{\partial x}{\partial r} + \frac{\partial u}{\partial y} \frac{\partial y}{\partial r} + \cdots$, or $\frac{\partial u}{\partial r}$ (formulas (4)). Since the coefficients of the remaining differentials, $ds, dt, \cdots$ are found similarly to be $\frac{\partial u}{\partial s}, \frac{\partial u}{\partial t}, \cdots$, the proof is complete.

NOTE. The student should be on guard against treating a partial symbol ∂x, or the like, as if it had a meaning of its own similar to that of a total differential dx. For example, in the first equation of (4), "cancellation" of the symbols $\partial x, \partial y, \partial z, \cdots$ would lead to the ridiculous conclusion

$$\frac{\partial u}{\partial r} = \frac{\partial u}{\partial r} + \frac{\partial u}{\partial r} + \frac{\partial u}{\partial r} + \cdots.$$

Example 1. If $u = f(x, y) = xy$, then $du = x\, dy + y\, dx$, whether x or y are independent variables, differentiable functions of a new independent variable (this is the differential form for the formula for differentiating a product of two functions), or differentiable functions of several new variables.

Example 2. If $u = e^{xy}$ and $x = r + s$, $y = r - 2s$, then $\frac{\partial u}{\partial x} = ye^{xy}$, $\frac{\partial u}{\partial y} = xe^{xy}$, $\frac{\partial x}{\partial r} = \frac{\partial x}{\partial s} = \frac{\partial y}{\partial r} = 1$, $\frac{\partial y}{\partial s} = -2$. Therefore $\frac{\partial u}{\partial r} = ye^{xy} \cdot 1 + xe^{xy} \cdot 1 = e^{r^2 - rs - 2s^2}$. $(2r - s)$, and $\frac{\partial u}{\partial s} = ye^{xy} \cdot 1 + xe^{xy}(-2) = e^{r^2 - rs - 2s^2} \cdot (-r - 4s)$. This result is easily checked by direct substitution before differentiation.

Example 3. Prove that if u is a function of x and y in which x and y occur only in the combination xy, then $x \frac{\partial u}{\partial x} = y \frac{\partial u}{\partial y}$ (assuming differentiability).

Solution. Let $u = f(t)$, where $t = xy$. Then by (5), $\dfrac{\partial u}{\partial x} = \dfrac{du}{dt}\dfrac{\partial t}{\partial x} = y\dfrac{du}{dt}$; $\dfrac{\partial u}{\partial y} = \dfrac{du}{dt}\dfrac{\partial t}{\partial y} = x\dfrac{du}{dt}$. Hence $x\dfrac{\partial u}{\partial x} = xy\dfrac{du}{dt} = y\dfrac{\partial u}{\partial y}$.

*908. HOMOGENEOUS FUNCTIONS. EULER'S THEOREM

For simplicity, statements in this section will be made in terms of functions of two variables. Extensions to several variables are immediate.

Definition. *A function $f(x, y)$ is **homogeneous** of degree n in a region R if and only if for all x, y, and positive λ† such that both (x, y) and $(\lambda x, \lambda y)$ are in R,*

(1) $$f(\lambda x, \lambda y) = \lambda^n f(x, y).$$

Example 1. $f(x, y) \equiv 2x^3 - 8xy^2$ is homogeneous of degree 3, since $f(\lambda x, \lambda y) = 2\lambda^3 x^3 - 8\lambda^3 xy^2 = \lambda^3 f(x, y)$. The region R may be taken to be the entire plane.

Example 2. $f(x, y) \equiv (x^2 + 4y^2)^{-\frac{1}{3}}$ is homogeneous of degree $-\frac{2}{3}$ in the plane (except at the origin).

Example 3. $f(x, y) \equiv x^3 + xy$ is not homogeneous of any degree, since $f(\lambda x, \lambda y)/f(x, y)$ is not independent of x and y.

Theorem. Euler's Theorem. *If $f(x, y)$ is continuously differentiable and homogeneous of degree n in a region R, then in R*

(2) $$xf_1(x, y) + yf_2(x, y) = nf(x, y).$$

Proof. Since R is open, it follows that for any fixed point (x_0, y_0) in R, $(\lambda x_0, \lambda y_0)$ is in R for all λ sufficiently near 1 (why?). For such λ we define a function $g(\lambda)$, making use of (1):

(3) $$g(\lambda) \equiv f(\lambda x_0, \lambda y_0) \equiv \lambda^n f(x_0, y_0).$$

Applying the chain rule to (3), where the intermediate variables are the two functions of λ, λx_0 and λy_0, we have

$$g'(\lambda) = x_0 f_1(\lambda x_0, \lambda y_0) + y_0 f_2(\lambda x_0, \lambda y_0) = n\lambda^{n-1} f(x_0, y_0).$$

With $\lambda = 1$, equation (2) is obtained for the fixed but arbitrary point (x_0, y_0) of R.

For a partial converse of Euler's theorem see Exercise 31, § 909.

909. EXERCISES

1. Find $\dfrac{du}{dt}$ by the chain rule. Check by substitution before differentiation.

$u = \ln(1 + x^2 + y^2)$, $x = \cos t$, $y = \sin t$;
$u = e^x \sin y$, $x = \ln t$, $y = \text{Arcsin } 3t$.

† Restricting λ to positive values means that $f(x, y)$ is being studied at points (x, y) and $(\lambda x, \lambda y)$ on the *same* side of the origin. It permits inclusion of such functions as $\sqrt{x^2 + y^2}$.

EXERCISES

2. Find $\dfrac{\partial u}{\partial r}$ and $\dfrac{\partial u}{\partial s}$ by the chain rule. Check by substitution before differentiation.

$$u = x^3, \quad x = re^s;$$
$$u = x^2 + y^2 + z^2, \quad x = r\cos s, \quad y = r\sin s, \quad z = r.$$

In Exercises 3–6, show that the given partial differential equation must be satisfied by a differentiable function of the variables x and y, if these variables occur only in the combination specified.

3. $x + y$; $\quad \dfrac{\partial u}{\partial x} - \dfrac{\partial u}{\partial y} = 0.$

4. $x - y$; $\quad \dfrac{\partial u}{\partial x} + \dfrac{\partial u}{\partial y} = 0.$

5. y/x; $\quad x\dfrac{\partial u}{\partial x} + y\dfrac{\partial u}{\partial y} = 0.$

6. $x^2 + y^2$; $\quad y\dfrac{\partial u}{\partial x} - x\dfrac{\partial u}{\partial y} = 0.$

In Exercises 7–8, show that the given partial differential equation must be satisfied by a differentiable function of the variables x, y, and z, if these variables occur only in the combinations specified.

7. $x + y + z$; $\quad \dfrac{\partial u}{\partial x} = \dfrac{\partial u}{\partial y} = \dfrac{\partial u}{\partial z}.$

8. $x - y, y - z, z - x$; $\quad \dfrac{\partial u}{\partial x} + \dfrac{\partial u}{\partial y} + \dfrac{\partial u}{\partial z} = 0.$

In Exercises 9–10, express in terms of the individual functions and their derivatives.

9. $\dfrac{d}{dx} f(g(x) + h(x)).$

10. $\dfrac{d}{dx} f(g(x)h(x)).$

11. Derive the Fundamental Increment Formula (6), § 905, for a function of several variables, with a careful statement of hypotheses.

12. Prove the statement preceding (4), § 907, that if u is a differentiable function of $x, y, z, \cdots$ each of which is a differentiable function of $r, s, t, \cdots$, then u is a differentiable function of $r, s, t, \cdots$. *Hint:* Letting $\epsilon_i, \zeta_i, \eta_k, \cdots$ denote infinitesimals, write

$$\Delta u = \dfrac{\partial u}{\partial x} \Delta x + \dfrac{\partial u}{\partial y} \Delta y + \cdots + \epsilon_1 \Delta x + \epsilon_2 \Delta y + \cdots$$

$$= \dfrac{\partial u}{\partial x}\left[\dfrac{\partial x}{\partial r}\Delta r + \dfrac{\partial x}{\partial s}\Delta s + \cdots + \zeta_1 \Delta r + \zeta_2 \Delta s + \cdots\right]$$

$$+ \dfrac{\partial u}{\partial y}\left[\dfrac{\partial y}{\partial r}\Delta r + \dfrac{\partial y}{\partial s}\Delta s + \cdots + \eta_1 \Delta r + \eta_2 \Delta s + \cdots\right]$$

$$+ \cdots$$

and reassemble terms in the manner suggested in the proof of the Theorem of § 907.

13. Use the chain rule to derive the formula for differentiating u^v, where u and v are differentiable functions of x. Hence differentiate x^x and $x^{x^x} \equiv x^{(x^x)}$. *Hint:* Use logarithms.

14. Show that $d(x^{y^z}) = x^{y^z-1}y^z\, dx + (\ln x)x^{y^z}y^{z-1}z\, dy + (\ln x)(\ln y)x^{y^z}y^z\, dz.$ (Cf. Ex. 13.)

15. Express the derivative of the determinant

$$\begin{vmatrix} a(x) & b(x) & c(x) \\ d(y) & e(y) & f(y) \\ g(z) & h(z) & k(z) \end{vmatrix},$$

where x, y, and z are functions of a variable t, as a sum of three determinants obtained by differentiating a row at a time. Generalize to nth-order determinants, and thus obtain a formula for differentiating a determinant all of whose elements a_{ij} are functions of a single variable t: $a_{ij} = a_{ij}(t)$.

16. Prove that if $u = f(x, y)$ is harmonic (cf, Exs. 19–22, § 904), then so is
$$\phi(x, y) \equiv f\left(\frac{x}{x^2 + y^2}, \frac{y}{x^2 + y^2}\right).$$

*17. Assuming appropriate continuity conditions, prove that if $u = f(x, y)$, $x = \phi(r, s)$, and $y = \psi(r, s)$, then

$$\frac{\partial^2 u}{\partial r^2} = f_1\phi_{11} + f_2\psi_{11} + \phi_1^2 f_{11} + 2\phi_1\psi_1 f_{12} + \psi_1^2 f_{22},$$

$$\frac{\partial^2 u}{\partial r \partial s} = f_1\phi_{12} + f_2\psi_{12} + \phi_1\phi_2 f_{11} + (\phi_1\psi_2 + \phi_2\psi_1)f_{12} + \psi_1\psi_2 f_{22},$$

$$\frac{\partial^2 u}{\partial s^2} = f_1\phi_{22} + f_2\psi_{22} + \phi_2^2 f_{11} + 2\phi_2\psi_2 f_{12} + \psi_2^2 f_{22}.$$

Make up two examples suitable for verifying these formulas.

*18. Obtain formulas similar to those of Exercise 17 for $u = f(x)$, $x = \phi(r, s)$, and verify them for two examples.

*19. Prove that if $f(x, y, z) = g(u)$, where $u = x^2 + y^2 + z^2$, then $f_{xx} + f_{yy} + f_{zz}$ is a function of u only.

In Exercises 20–23, show that the given function is homogeneous. Specify the region and the degree. Verify Euler's theorem in each case.

*20. $\dfrac{xy}{x^2 + y^2}$.

*21. $\sqrt{x^2 - xy}$.

*22. $\operatorname{Arctan} \dfrac{y}{x}$.

*23. $z \ln\left(1 + \dfrac{y}{x}\right)$.

In Exercises 24–25, use Euler's theorem to show that the given function is not homogeneous in any region.

*24. e^{xy}.

*25. $\ln(1 + x + y)$.

*26. Prove that if $f(x, y)$ is continuously differentiable and homogeneous of degree n in a region R in the right half-plane, then $f(x, y) \equiv x^n f(1, y/x)$. In particular, show that if $n = 0$, the value of $f(x, y)$ depends only on the ratio of the independent variables—that is, only on the polar angle θ.

*27. Define homogeneity, and establish Euler's theorem, for a function of several variables.

*28. Prove that the first partial derivatives of a homogeneous function of degree n are homogeneous functions of degree $n - 1$.

*29. Using appropriate continuity assumptions, establish for a homogeneous function of degree n the formula
$$x^2 f_{11}(x, y) + 2xy f_{12}(x, y) + y^2 f_{22}(x, y) = n(n - 1)f(x, y).$$
(Cf. Ex. 28.)

*30. Generalize Exercise 29, both as to the number of variables and as to the order of the derivatives.

*31. Define a function $f(x, y)$ to be **locally homogeneous** in a region R in case, for every point of R, the Definition, § 908, applies for all values of λ in a suitable neighborhood of $\lambda = 1$. Prove the following extension and converse of Euler's theorem: *If $f(x, y)$ is continuously differentiable in a region R, then it is locally homogeneous of degree n there if and only if $xf_1(x, y) + yf_2(x, y) = nf(x, y)$ there.* *Hint:* Assume the preceding equation holds throughout R, let (x, y) be any point in R, and let I be an open interval containing the number 1 and such that $\lambda \in I$ implies $(\lambda x, \lambda y) \in R$. With x and y fixed, define $g(\lambda) \equiv f(\lambda x, \lambda y)$. Show that $g(\lambda)$ satisfies the equation $\lambda g'(\lambda) = ng(\lambda)$, so that $g(\lambda)$ has the form $g(\lambda) = c\lambda^n$. Substitute $\lambda = 1$ to evaluate c.

★32. Specify conditions on a region so that the condition $xf_1(x, y) + yf_2(x, y) = nf(x, y)$ is equivalent to homogeneity of degree n there. (Cf. Exs. 31, 33.)

★33. Let R be the region consisting of all points (x, y) such that $x > 0$, with the exception of the half-line $x = 1$, $y \geq 0$. Define $f(x, y) \equiv 0$ except when $x > 1$ and $y > 0$. Define $f(x, y) \equiv y^4/x^4$ if $x > 1$ and $y > 0$. Prove that f is locally homogeneous of degree 0 in R, but not homogeneous there. (Cf. Exs. 26, 31, 32.)

★910. DIRECTIONAL DERIVATIVES. TANGENTS AND NORMALS

We shall consider now a few applications of the preceding sections to three-dimensional Euclidean space. These same ideas find their analogues in the plane and in higher demensions (cf. Exs. 9, 11, § 911).

Let a smooth space curve C (§ 805) be given. That is, C is prescribed by parametrization functions $x(t)$, $y(t)$, $z(t)$ having continuous first derivatives not all of which vanish at the same point (C has no singular points). Then, since Δx, Δy, Δz represent direction numbers of a secant line of C, so do the quotients $\Delta x/\Delta t$, $\Delta y/\Delta t$, $\Delta z/\Delta t$. Taking the limits of these quotients we see that the derivatives $x'(t)$, $y'(t)$, $z'(t)$ are direction numbers of the tangent line to the curve C. In particular, if the parameter is taken to be arc length s, measured from some fixed point of the curve, the derivatives dx/ds, dy/ds, dz/ds become direction *cosines* of the tangent line, oriented in the direction of increasing arc length (formula (6), § 809):

(1) $$\frac{dx}{ds} = \cos \alpha, \quad \frac{dy}{ds} = \cos \beta, \quad \frac{dz}{ds} = \cos \gamma$$

In other words, if C is given by functions $x(s)$, $y(s)$, $z(s)$, the quantities (1) are the components of the *unit tangent vector*.

We now address ourselves to the following questions: If $u = f(x, y, z)$ is a differentiable function defined in a region R of space, how does u change as the point (x, y, z) moves in a certain direction? In which direction does u change most rapidly? How fast? Just what do these questions really mean?

In the first place, by "rate of change" of u we shall mean *rate of change with respect to distance*. Accordingly, we shall consider the behavior of the function $u = f(x, y, z)$ as the point (x, y, z) is constrained to move along a smooth curve C lying in R, thinking of u as a function of arc length s:

(2) $$u(s) \equiv f(x(s), y(s), z(s)).$$

The rate of change of u with respect to s takes the form (from the chain rule):

$$(3) \quad \frac{du}{ds} = \frac{df}{ds} = f_1(x, y, z)\frac{dx}{ds} + f_2(x, y, z)\frac{dy}{ds} + f_3(x, y, z)\frac{dz}{ds}$$

$$= \frac{\partial u}{\partial x}\cos\alpha + \frac{\partial u}{\partial y}\cos\beta + \frac{\partial u}{\partial z}\cos\gamma.$$

It is now evident that the rate of change of u is the scalar product of two vectors (§ 707). The first of these, called the **gradient** of the function f, has components

$$(4) \quad f_1(x, y, z), \quad f_2(x, y, z), \quad f_3(x, y, z),$$

and depends only on the function f and the point (x, y, z). The second is the unit tangent vector (1) of the curve C and depends only on the *direction* of the curve at the specified point. Because of this last fact, the derivative (3) is called the **directional derivative** of u in the given direction. With the aid of § 706 we have the important result:

Theorem I. *The directional derivative of a function $f(x, y, z)$ at a given point and in a given direction is equal to the component of the gradient of that function, evaluated at the given point, in the given direction.*

Corollary. *The maximum and minimum values of the directional derivative of a function $f(x, y, z)$, at a given point, are the magnitude of its gradient and the negative of this magnitude, evaluated at that point, respectively. The function $f(x, y, z)$ has its greatest and least rates of change in the direction of the gradient and in the opposite direction, respectively.*

If the smooth curve C through the given point is chosen to be a straight line parallel to a coordinate axis, and having the same orientation, Δs becomes Δx, Δy, or Δz, and the directional derivative of $u = f(x, y, z)$ reduces to one of the first partial derivatives of f. In other words, *the directional derivative of a differentiable function is a generalization of a first partial derivative.*

For present purposes (cf. Note 1, § 1812) we shall define a **surface** as the set of all points (x, y, z) whose coordinates satisfy an equation of the form $f(x, y, z) = 0$, where f is differentiable. A **smooth surface** is a surface for which f has continuous first partial derivatives not all of which vanish identically. Any point of a surface at which these three first partial derivatives all vanish is called a **singular point**. Any other point of the surface is a **regular point**.

Consider now a surface S with equation $f(x, y, z) = 0$, let C be any smooth curve lying wholly in the surface, and let $p: (x_0, y_0, z_0)$ be any point of C that is a regular point of S (it is also a regular point of C). Then $u(s) \equiv f(x(s), y(s), z(s))$ vanishes identically, so that the directional derivative of f also vanishes identically and, in particular, at the point p:

$$(5) \quad \frac{du}{ds} = f_1\frac{dx}{ds} + f_2\frac{dy}{ds} + f_3\frac{dz}{ds} = 0,$$

all derivatives being evaluated at p. By means of the perpendicularity condition for two lines ($\cos \theta = 0$) we conclude that the gradient, evaluated at p, is perpendicular or normal to *all* smooth curves lying in the surface and passing through p. For this reason we define the **normal direction** to a surface $f(x, y, z) = 0$ at a regular point to be that of the gradient of f evaluated at the point. In other words, *the line normal to a surface $f(x, y, z) = 0$ at a regular point has as a set of direction numbers the three first partial derivatives of f evaluated at the point.*

If $f(x, y, z)$ is a differentiable function and if $p: (x_0, y_0, z_0)$ is a point at which f is defined, by considering the surface whose equation is $F(x, y, z) \equiv f(x, y, z) - f(x_0, y_0, z_0) = 0$, and agreeing that the zero vector is normal to all vectors and to all surfaces we can infer immediately the truth of the following theorem:

Theorem II. *The gradient of a differentiable function $f(x, y, z)$, evaluated at any point (x_0, y_0, z_0), is normal to the surface $f(x, y, z) = f(x_0, y_0, z_0)$ at that point.*

Motivated by this theorem and the Corollary to Theorem I we call the maximum directional derivative of a differentiable function $u = f(x, y, z)$ at a point p the **normal derivative** of f at p, and denote it $\dfrac{du}{dn}$ or $\dfrac{df}{dn}$. We have, then,

(6) $$\frac{du}{dn} = \frac{df}{dn} = \sqrt{f_1^2 + f_2^2 + f_3^2}.$$

In conclusion, we assemble some equations arising from the foregoing discussion, pertaining to a curve $C: (x(t), y(t), z(t))$ at a regular point $p: (x_0, y_0, z_0) = (x(t_0), y(t_0), z(t_0))$ and a surface $S: f(x, y, z) = 0$ at a regular point $p: (x_0, y_0, z_0)$ (cf. §§ 710, 711):

Normal plane and tangent line to the curve C:

(7) $$x'(t_0)(x - x_0) + y'(t_0)(y - y_0) + z'(t_0)(z - z_0) = 0,$$

(8) $$\frac{x - x_0}{x'(t_0)} = \frac{y - y_0}{y'(t_0)} = \frac{z - z_0}{z'(t_0)}.$$

Tangent plane and normal line to the surface S:

(9) $$f_1(x_0, y_0, z_0)(x - x_0) + f_2(x_0, y_0, z_0)(y - y_0) + f_3(x_0, y_0, z_0)(z - z_0) = 0,$$

(10) $$\frac{x - x_0}{f_1(x_0, y_0, z_0)} = \frac{y - y_0}{f_2(x_0, y_0, z_0)} = \frac{z - z_0}{f_3(x_0, y_0, z_0)}.$$

Example. Discuss the gradient of the function $u = x^2 + y^2 + z^2$. In particular, find its normal derivative, and its directional derivative in a direction normal to the paraboloid $z = 3x^2 + y^2$ at the point $(1, 1, 4)$, directed inward.

Solution. Let $f(x, y, z) \equiv x^2 + y^2 + z^2$. The gradient of f has components $2x, 2y, 2z$. At any point, then, it has the same direction as the radius vector of that point, and twice the magnitude. It always points directly away from the origin, and is normal to the sphere with center at the origin passing through the point. The function f is increasing most rapidly in that direction, and has a normal derivative of $2\sqrt{x^2 + y^2 + z^2}$. Normals to the paraboloid have direction numbers $6x, 2y, -1$, which become at the given point $6, 2, -1$. Converting these to direction cosines we have $-6/\sqrt{41}, -2/\sqrt{41}, 1/\sqrt{41}$ and, by equation (3) for the directional derivative,

$$\frac{du}{ds} = \left[2\left(-\frac{6}{\sqrt{41}}\right) + 2\left(-\frac{2}{\sqrt{41}}\right) + 8\left(\frac{1}{\sqrt{41}}\right)\right] = -\frac{8}{\sqrt{41}} = -1.249.$$

The normal derivative at this same point is $2\sqrt{18} = 6\sqrt{2} = 8.485$.

*911. EXERCISES

In Exercises 1–2, find the equations of the tangent line and normal plane to the given curve at the specified point.

*1. $x = a \cos t, y = b \sin t, z = ct, t = \pi/2$.
*2. $x = at \cos t, y = bt \sin t, z = ct, t = \pi/2$.

In Exercises 3–4, find the equations of the normal line and tangent plane to the given surface at the specified point.

*3. $z = e^x \sin y$; $(0, 0, 0)$.
*4. $Ax^2 + By^2 + Cz^2 + 2Dyz + 2Exz + 2Fxy = G$; (x_1, y_1, z_1).

In Exercises 5–6, prove that the surfaces intersect orthogonally (that is, their normals are perpendicular).

*5. $x^2 + y^2 + z^2 - 2x + 4y - 8z = 0$,
$3x^2 - y^2 + 2z^2 - 6x - 4y - 16z + 31 = 0$.

*6. $\dfrac{x^2}{a^2 - r^2} + \dfrac{y^2}{b^2 - r^2} + \dfrac{z^2}{c^2 - r^2} = 1$,

$\dfrac{x^2}{a^2 - s^2} + \dfrac{y^2}{b^2 - s^2} + \dfrac{z^2}{c^2 - s^2} = 1, \quad \dfrac{x^2}{a^2 - t^2} + \dfrac{y^2}{b^2 - t^2} + \dfrac{z^2}{c^2 - t^2} = 1,$

where $0 < r < a < s < b < t < c$.

In Exercises 7–8, discuss the gradient of the given function. Find the normal derivative, and the directional derivative at the given point in the given direction.

*7. $u = xyz$; $(4, -2, 3)$; $\cos \alpha : \cos \beta : \cos \gamma = 2 : -1 : -2$ directed upward.
*8. $u = x^2 + yz$; $(-3, 1, -2)$; direction making equal acute angles with the positive coordinate axes.

*9. Discuss the gradient and directional derivative of a function $f(x, y)$ of two variables. Show that the gradient of $f(x, y)$ at the point (x_0, y_0) is normal to the curve $f(x, y) = f(x_0, y_0)$. Derive the formulas $\dfrac{du}{ds} = \dfrac{\partial u}{\partial x} \cos \alpha + \dfrac{\partial u}{\partial y} \cos \beta$ and $\dfrac{du}{dn} = \sqrt{\left(\dfrac{\partial u}{\partial x}\right)^2 + \left(\dfrac{\partial u}{\partial y}\right)^2}$.

*10. Find the normal derivative of the function $u = x^2 - y^2$ at the point $(2, 1)$, and the value of the directional derivative of this function at the same point in the direction of a line making an angle of $120°$ with the positive x-axis (directed upward to the left). (Cf. Ex. 9.)

*11. Discuss the gradient and directional derivative of a function f of n variables.
*12. Show that for a linear function z of two variables x and y, the differential dz and the increment Δz are the same. Hence obtain a relation between the differential of a more general function $z(x, y)$ and the tangent plane to the surface $z = z(x, y)$.

912. THE LAW OF THE MEAN

The Law of the Mean for a function of one variable (§ 305) can be expressed by the formula
(1) $$f(a + h) = f(a) + f'(a + \theta h)h,$$
where $0 < \theta < 1$, the assumptions on the function $f(x)$ being continuity on the closed interval $[a, a + h]$ (or $[a + h, a]$ if $h < 0$), and differentiability for points between a and $a + h$.

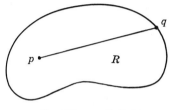 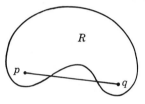

Conditions satisfied   Conditions not satisfied

FIG. 902

For a function of several variables the Law of the Mean takes the following form (which we state for simplicity for a function of *two* variables):

Theorem I. Law of the Mean. *Let $f(x, y)$ be continuous in a region or closed region R and differentiable in the interior of R, and let $p: (a, b)$ and $q: (a + h, b + k)$ be any two distinct points of R such that all points $r: (a + \theta h, b + \theta k)$, where $0 < \theta < 1$, of the straight line segment I joining p and q (r not equal to p or q) belong to the interior of R (cf. Fig. 902). Then there exists a number θ such that $0 < \theta < 1$ and*
(2) $$f(a + h, b + k) = f(a, b) + f_1(a + \theta h, b + \theta k)h + f_2(a + \theta h, b + \theta k)k.$$
That is, there exists a point r in the interior of I such that for the function $u = f(x, y)$ the increment Δu for the points p and q is equal to the differential du evaluated at r, with $dx = h$ and $dy = k$.

Proof. Let the function $\phi(t)$ of a single real variable t be defined on the unit interval $0 \leq t \leq 1$:
(3) $$\phi(t) = f(a + th, b + tk).$$
Since $\phi(t)$ is continuous on $[0, 1]$ and differentiable in $(0, 1)$, the Law of the Mean (1) can be applied (with $a = 0$, $h = 1$):
(4) $$\phi(1) = \phi(0) + \phi'(\theta).$$
By the chain rule, (2) is the same as (4).

An important consequence of the Law of the Mean is the following (cf. Theorem I, § 306):

Theorem II. *If a function is continuous in a region or closed region R and differentiable in the interior, and if its first partial derivatives vanish throughout the interior, then the function is constant in R.*

Proof. Assume the contrary. That is, assume that the function f assumes distinct values at two points of R. Then, by Theorem IV, § 610, there are two points in the interior of R at which f has distinct values. Join these two points by a broken line segment lying in R. Then there must be consecutive vertices p and q of this broken line segment such that $f(p) \neq f(q)$. But this is a contradiction to the Law of the Mean (2), since the differential of f vanishes identically between p and q.

If we apply the Extended Law of the Mean (§ 307) to the function $\phi(t) \equiv f(a + th, b + tk)$ (assuming appropriate continuity and differentiability), we obtain for the case $n = 2$:

(5) $$\phi(1) = \phi(0) + \phi'(0) + \frac{\phi''(\theta)}{2!}, \quad 0 < \theta < 1,$$

or, in terms of the function $f(x, y)$:

(6) $$f(a + h, b + k) = f(a, b) + f_1(a, b)h + f_2(a, b)k$$
$$+ \tfrac{1}{2}[f_{11}(a + \theta h, b + \theta k)h^2$$
$$+ 2f_{12}(a + \theta h, b + \theta k)hk + f_{22}(a + \theta h, b + \theta k)k^2].$$

It is helpful at this stage to use a more compressed notation, showing the binomial behavior of this expansion, with the substituted values of x and y indicated by subscripts:

(7) $$f(a + h, b + k) = f(a, b) + \left[\left(h\frac{\partial}{\partial x} + k\frac{\partial}{\partial y}\right)f(x, y)\right]_{\substack{x=a \\ y=b}}$$
$$+ \frac{1}{2}\left[\left(h\frac{\partial}{\partial x} + k\frac{\partial}{\partial y}\right)^2 f(x, y)\right]_{\substack{x=a+\theta h \\ y=b+\theta k}}.$$

The formula for the **Extended Law of the Mean** in general, for a function of two variables (under suitable continuity and differentiability conditions), can be written:

(8) $$f(a + h, b + k) = f(a, b) + \sum_{i=1}^{n-1} \frac{1}{i!}\left[\left(h\frac{\partial}{\partial x} + k\frac{\partial}{\partial y}\right)^i f(x, y)\right]_{\substack{x=a \\ y=b}} + R_n,$$

where $$R_n = \frac{1}{n!}\left[\left(h\frac{\partial}{\partial x} + k\frac{\partial}{\partial y}\right)^n f(x, y)\right]_{\substack{x=a+\theta h \\ y=b+\theta k}}, \quad 0 < \theta < 1.$$

(The student is asked to establish (8) in Exercise 27, § 916, and to extend the law of the mean to several variables in Exercises 28, 30, § 916.) The Extended Law of the Mean (8) is also called **Taylor's Formula with a Remainder**.

913. APPROXIMATIONS BY DIFFERENTIALS

As in the case of a single variable (§ 312), the Extended Law of the Mean ($n = 2$) provides a measure of the accuracy with which a differential approximates an increment. Let us first recall the formula for a function of one variable ((5), § 312):

(1) $$\Delta u - du = \tfrac{1}{2} f''(a + \theta h) h^2.$$

For a function $u = f(x, y)$ this takes the form ((6), § 912):

(2) $$\Delta u - du = \tfrac{1}{2}[f_{11}h^2 + 2f_{12}hk + f_{22}k^2]_{\substack{x=a+\theta h,\\ y=b+\theta k}},$$

the differential du being evaluated at $x = a$, $y = b$, with $dx = h$, $dy = k$.

If B is a bound for the second partial derivatives of f, for points near (a, b), that is, if $|f_{ij}| \leq B$, $i, j = 1, 2$, for $|x - a| + |y - b| \leq \delta$, then $|h| + |k| \leq \delta$ implies

$$|f_{11}h^2 + 2f_{12}hk + f_{22}k^2| \leq B(|h| + |k|)^2 \leq B\delta^2,$$

so that

(3) $$|\Delta u - du| \leq \tfrac{1}{2}B\delta^2.$$

Example 1. A racing car is clocked over a 15-mile course at 5 minutes. However, there is a possible error in the measurement of the course length of as much as 0.03 miles and an error in the measurement of the time of as much as 0.02 minutes. Approximately how accurate is the estimate of 3 miles per minute, or 180 miles per hour, for the average speed?

Solution. Let the course length, measured time, and average speed be denoted by x, t, and v, respectively, with x in miles and t in minutes. Then $v = x/t$ and

$$dv = \frac{1}{t}\,dx - \frac{x}{t^2}\,dt.$$

Upon making use of the values $x = 15$ and $t = 5$ and the estimates $|dx| \leq 0.03$ and $|dt| \leq 0.02$, we have $|dv| \leq (0.2)(0.03) + (0.6)(0.02) = 0.018$. Therefore the computed average speed should be accurate to within 0.018 miles per minute, or approximately one mile per hour.

Since the second-order partial derivatives of the function $v = x/t$ are 0, $-1/t^2$, and $2x/t^3$, and since $|1/t^2| \leq 1/(4.98)^2$ and $|2x/t^3| \leq 30.06/(4.98)^3$, we have, for the right-hand member of (3):

$$\tfrac{1}{2}B\delta^2 \leq \frac{1}{2} \cdot \frac{30.06}{(4.98)^3} \cdot (0.05)^2 < \frac{1}{2} \cdot \frac{32}{120} \cdot \frac{1}{400} = \frac{0.02}{60}.$$

Therefore, the above error estimate is accurate to within 0.02 miles per hour.

When one is considering *relative errors*, as distinguished from the errors themselves, it is often helpful to find the differential of the natural logarithm of a function. This is a consequence of the particular form of the differential of $\ln x$: $d(\ln x) = dx/x$.

Example 2. A water tank is formed by the bottom half of a right circular cylinder with a horizontal axis, closed at the ends by (vertical) plane semicircles.

The radius of this cylinder is to be determined by measuring the length l of the tank and the volume V that it contains. If the quantities l and V are measured with errors up to 1% and 3%, respectively, determine approximately, by means of differentials, the maximum relative error in the computed radius.

Solution. Since $V = \frac{1}{2}\pi r^2 l$, $r = (V/\frac{1}{2}\pi l)^{\frac{1}{2}}$ and
$$\ln r = \tfrac{1}{2} \ln V - \tfrac{1}{2} \ln \tfrac{1}{2}\pi - \tfrac{1}{2} \ln l.$$
Hence $\dfrac{dr}{r} = \dfrac{dV}{2V} - \dfrac{dl}{2l}$ and, with the data above,
$$\left|\frac{dr}{r}\right| \leq \frac{1}{2}\left|\frac{dV}{V}\right| + \frac{1}{2}\left|\frac{dl}{l}\right| \leq \frac{1}{2}(0.01 + 0.03) = 0.02.$$

In other words, the maximum relative error in the determination of the radius is approximately 2%.

We shall not make any attempt to use the inequality (3).

914. MAXIMA AND MINIMA

If $f(x, y)$ is differentiable in a neighborhood of a point (a, b) at which it has a relative **extremum** (that is, a relative maximum or minimum value; cf. § 309, Theorem II, § 610), it is clear from elementary considerations (cf. § 309) that the following conditions are *necessary*:

(1) $\qquad\qquad f_1(a, b) = 0, \quad f_2(a, b) = 0.$

It is our purpose in this section to establish a set of conditions that are *sufficient* for a maximum or minimum.† A simple example, like $z = xy$ or $z = x^2 - y^2$ each of which has as graph a hyperbolic paraboloid with a saddle-like shape near the origin, shows that conditions (1) are not sufficient for an extremum. A slightly subtler example is $z = x^2 + 4xy + y^2$. For this surface the two traces $z = x^2$ and $z = y^2$ in the coordinate planes $y = 0$ and $x = 0$ are parabolas opening upward, so that one might expect the origin to be a minimum point. However, for values of x and y arbitrarily near 0 and such that $y = -x$, z is negative, and the origin is *not* an extreme point. This surface is again a hyperbolic paraboloid, as we see by noting the character of the sections $z = x^2 + 4xy + y^2 = $ constant. The fact that these sections are hyperbolas is disclosed by the negativeness of the determinant

(2) $\qquad\qquad AC - B^2 = \begin{vmatrix} A & B \\ B & C \end{vmatrix}$

associated with the quadratic expression $Ax^2 + 2Bxy + Cy^2$, where $A = C = 1$, $B = 2$.

The discussion of the preceding paragraph contains the nucleus of a method for approaching the problem of extrema with the aid of the Law of the Mean. Assume that $u = f(x, y)$ has continuous second partial derivatives in a neighborhood of (a, b), and that the first partial derivatives vanish at

† The method of *Lagrange multipliers* is presented in § 928.

that point. Then the Extended Law of the Mean for $n = 2$ ((6), § 912) can be written

(3) $\quad \Delta u = f(a + h, b + k) - f(a, b)$
$= \tfrac{1}{2}[f_{11}(a + \theta h, b + \theta k)h^2 + 2f_{12}(a + \theta h, b + \theta k)hk$
$+ f_{22}(a + \theta h, b + \theta k)k^2],$

where $0 < \theta < 1$.

There are three possibilities that present themselves, corresponding to inequalities involving Δu for values of h and k near 0:

(4) $\quad \begin{cases} f \text{ has a maximum value at } (a, b): \Delta u \leq 0, \\ f \text{ has a minimum value at } (a, b): \Delta u \geq 0, \\ f \text{ has neither: } \Delta u \text{ changes sign.}^\dagger \end{cases}$

In order to study the signs of Δu, we write

(5) $\quad \begin{cases} f_{11} \Delta u = \tfrac{1}{2}[(f_{11}h + f_{12}k)^2 + (f_{11}f_{22} - f_{12}^2)k^2], \\ f_{22} \Delta u = \tfrac{1}{2}[(f_{11}f_{22} - f_{12}^2)h^2 + (f_{12}h + f_{22}k)^2], \end{cases}$

where the partial derivatives are evaluated at the point $x = a + \theta h$, $y = b + \theta k$. We are now ready to formulate a set of sufficient conditions for maxima and minima:

Theorem. *Let $f(x, y)$ be a function with continuous second partial derivatives in a neighborhood of the point (a, b), and assume that at that point the first partial derivatives vanish:* $f_1(a, b) = f_2(a, b) = 0$. *Let*

(6) $\quad D(x, y) \equiv f_{11}(x, y)f_{22}(x, y) - [f_{12}(x, y)]^2, \quad D \equiv D(a, b).$

Then

(i) *if $D > 0$, f has an extreme value at (a, b);*
(ii) *if $D > 0$, this extreme value is a maximum or a minimum according as $f_{11}(a, b)$ and $f_{22}(a, b)$ are both negative or both positive;*
(iii) *if $D < 0$, f has neither a maximum nor a minimum at (a, b);*
(iv) *if $D = 0$, any of the preceding alternatives are possible.*

Proof. We observe first that if $D > 0$, $f_{11}(a, b)$ and $f_{22}(a, b)$ must both be nonzero and of the same sign. We infer from (5), and continuity, that for values of h and k sufficiently near 0, but not both equal to 0, $D(x, y)$ is positive and the three quantities $\Delta u, f_{11}$, and f_{22} are all nonzero and of the same sign, and conclusions (i) and (ii) follow from (4). (Cf. Note 4, § 209.)

If $D < 0$ the conclusion is less immediate. If $f_{11}(a, b) \neq 0$ we can use (5) to find pairs of points arbitrarily near (a, b) at which $f_{11} \Delta u$, and hence Δu

---

† In case Δu changes sign, the point (a, b) is said to be a **saddle point** of the surface $z = f(x, y)$. Thus, the origin is a saddle point of the hyperbolic paraboloid (7), § 718. For a description of a particularly interesting type of saddle point surrounded by three "peaks" and three "valleys," called a **monkey saddle** point, see D. Hilbert and S. Cohn-Vossen, *Geometry and the Imagination* (New York, Chelsea Publishing Company, 1952), page 191.

itself, have opposite signs. For example, if $k = 0$ and $|h|$ is small, $f_{11} \Delta u = \frac{1}{2}(f_{11}h)^2$ is positive; and if $h = -[f_{12}(a, b)/f_{11}(a, b)]k$ and $|k|$ is small, $f_{11} \Delta u$ is negative (check the details). A similar argument applies if $f_{22}(a, b) \neq 0$. If $D < 0$ and $f_{11}(a, b) = f_{22}(a, b) = 0$, then let $h = k$ and $h = -k$, in turn, directly in equation (3) (how does this show (iii)?).

Finally, the examples $\pm (x^4 + y^4)$ and $x^4 - y^4$ show that if $D = 0$ no conclusion can be drawn.

NOTE 1. The routine procedure in solving an extremum problem is: (i) find the **critical points** (a, b) such that $f_1(a, b) = f_2(a, b) = 0$; (ii) test the function for each critical point. In practice it is often simpler to analyze a given function directly instead of its second derivatives. Techniques such as completing squares, squaring variables, etc., may suggest themselves. (Cf. Ex. 17, § 310. Also cf. §§ 925–929.

NOTE 2. For functions of more than two variables the attack on the problem of sufficiency conditions for extrema is again the extended law of the mean. However, the analysis of the signs of Δu becomes much more difficult, and depends on the subject of *positive definite quadratic forms*. For a reference, see G. Birkhoff and S. MacLane, *A Survey of Modern Algebra* (New York, the Macmillan Company, 1953).

NOTE 3. The example $f(x, y) \equiv (x^2 - y)(2x^2 - y)$ shows that it is possible for a function to have a critical point (the origin) at which it has neither a relative maximum nor a relative minimum but which is such that whenever the values of the variables are linearly related in a neighborhood of that critical point (that is, for any straight line through the origin) the function has a relative extremum (minimum in this case). (The student should check these statements.) (Cf. Example 2, § 606.)

Example 1. Find the minimum value of the function
$$f(x, y) \equiv x^2 + 5y^2 - 6x + 10y + 6.$$
First Solution. Completion of squares gives
$$f(x, y) = (x - 3)^2 + 5(y + 1)^2 - 8,$$
with a minimum value of -8 at $x = 3$, $y = -1$.

Second Solution. By (1), the only critical point is $(3, -1)$. The corresponding value of f is -8. Since $f_{11} = 2, f_{12} = 0, f_{22} = 10$, $D > 0$ and $(3, -1)$ is a minimum point.

Example 2. Test $f(x, y) \equiv x^2 + xy$ for extrema.

First Solution. The only critical point is $(0, 0)$. Since $f(0, 0) = 0, f(\epsilon, 0) = \epsilon^2 > 0$, and $f(\epsilon, -2\epsilon) = -\epsilon^2 < 0$, the function has no extrema.

Second Solution. At the origin $D = -1 < 0$, so that (as before) the function has no extrema.

Example 3. Find the dimensions and the volume of the largest rectangular parallelepiped that has faces in the coordinate planes and one vertex in the ellipsoid

(7)
$$\frac{x^2}{a^2} + \frac{y^2}{b^2} + \frac{z^2}{c^2} = 1.$$

Solution. The problem is to maximize the function $V = xyz$ subject to the constraint (7). This is equivalent to finding the values of x, y, z subject to (7) that maximize $x^2 y^2 z^2$. Still more simply, if we let $u \equiv x^2/a^2$, $v \equiv y^2/b^2$, and $w \equiv z^2/c^2$, we seek positive numbers u, v, w subject to the constraint

(8) $$u + v + w = 1$$

that maximize $a^2 b^2 c^2 uvw$ or, equivalently, that maximize uvw. Symmetry considerations suggest the solution $u = v = w = \frac{1}{3}$. In analytic terms, we wish to maximize the function of u and v (u and v positive with sum less than 1):

$$\phi(u, v) \equiv uv(1 - u - v) = uv - u^2 v - uv^2.$$

The critical points for $\phi(u, v)$ are $(0, 0)$ and $(\frac{1}{3}, \frac{1}{3})$, the first clearly inappropriate. At the point $(\frac{1}{3}, \frac{1}{3})$, $\phi_{11}\phi_{22} - \phi_{12}^2 = \frac{1}{3} > 0$ and $\phi_{11} < 0$. Therefore ϕ has a maximum value there. The original parallelepiped requested therefore corresponds to $u = v = w = \frac{1}{3}$, and has dimensions $x = a/\sqrt{3}$, $y = b/\sqrt{3}$, $z = c/\sqrt{3}$ and volume $abc/3\sqrt{3}$.

NOTE 4. If extreme values are sought for a function on a restricted set that is not open, it is necessary to examine its values at all points of its domain of definition that are not interior points. Examples 4 and 5, below, illustrate the general ideas. For further discussion, see sections 925–929.

Example 4. Let R be the closed rectangle: $|x| \leq 3$, $|y| \leq 1$. Find the extrema on R of the functions (a) $x^2 + 4xy + y^2$; (b) $x^2 + 2xy + y^2$; (c) $x^2 + xy + y^2$.

Solution. (a): If $f(x, y) \equiv x^2 + 4xy + y^2$, $f_1 = 2x + 4y$, $f_2 = 4x + 2y$, $f_{11} = f_{22} = 2, f_{12} = 4$, and $D(x, y) = -12$. Therefore there can be no relative extremum, and *a fortiori* no absolute extremum, in the interior of R. We test the upper edge of R by setting $y = 1$: $x^2 + 4x + 1$ has a minimum value of -3 at $x = -2$ and a maximum value of 22 at $x = 3$. A similar checking of the remaining three sides of R reveals that these two values of f are the extrema on R.

(b): The fact that $f(x, y) = (x + y)^2$ is a perfect square implies that the minimum value is 0 (whenever $x + y = 0$). The maximum value is attained at a point that maximizes $|x + y|$, and is equal to 16 at $(3, 1)$ and $(-3, -1)$.

(c): Since the first partial derivatives of $f(x, y) = x^2 + xy + y^2$ vanish at the origin, and $f_{11}f_{22} - f_{12}^2 = 3$ and $f_{12} = 2 > 0$, the minimum value of 0 is attained at the origin. As above, the maximum value is attained at $(3, 1)$ and $(-3, -1)$, and is equal to 13.

Example 5. Maximize $x^2 + xy$ if $\dfrac{x^2}{a^2} + \dfrac{y^2}{b^2} \leq 1$ $(a > 0, b > 0)$.

Solution. By Example 2, the maximum value must be attained on the ellipse itself. One procedure (of several) is to parametrize the ellipse: $x = a \cos \theta$, $y = b \sin \theta$, and seek the maximum value of the resulting function of θ:

$$\phi(\theta) \equiv a^2 \cos^2 \theta + ab \sin \theta \cos \theta.$$

From elementary symmetry considerations it is evident that this maximum is attained for a value of θ between 0 and $\frac{1}{2}\pi$, corresponding to a first quadrant point on the ellipse. Routine methods for maximizing a function of a single real variable give $\tan 2\theta = b/a$, so that $\sin 2\theta = b/\sqrt{a^2 + b^2}$ and $\cos 2\theta = a/\sqrt{a^2 + b^2}$. The desired maximum is $\frac{1}{2}a^2 + \frac{1}{2}a\sqrt{a^2 + b^2}$.

915. EXERCISES

1. Find a value of θ between 0 and 1 satisfying (2), § 912, for the function $x^2 + xy$ with $a = 3$, $b = -5$, $h = -1$, $k = 2$.

2. Write out the equation (2), § 912, for the function $e^x \sin y$ with $a = 0$, $b = 0$, $h = 1$, $k = \pi/6$.

3. Find a value of θ between 0 and 1 satisfying (6), § 912, for the function $x^2 y$ with $a = 1$, $b = 1$, $h = 1$, $k = 2$.

4. Write out equation (6), § 912, for the function $x^2 \ln y$, with $a = 0$, $b = 1$, $h = 1$, $k = e - 1$.

5. The legs of a right triangle are measured to be 20 and 30 feet, with maximum errors of 1 inch. Find approximately the maximum possible error in calculating the area of the triangle. Estimate the accuracy of the preceding computation.

6. Do Exercise 5 for the hypotenuse, instead of the area.

7. Obtain linear approximation formulas for values of x and y near 0:

(a) $\sqrt{\dfrac{1+x}{1+y}}$;   (b) $\text{Arctan}\,\dfrac{y}{1+x}$.

8. Discuss for extrema, considering all possible values of a: $z = x^3 - 3axy + y^3$.

9. Find the maximum and minimum values of $(ax^2 + by^2)e^{-(x^2+y^2)}$ for $0 \leq |a| \leq b$. Consider all possible cases.

10. Discuss the behavior of each function near the origin:
(a) $23x^2y^7 - 17x^4y + 3x^2 - xy + 2y^2$;
(b) $-7y^{10} - 11x^3y^3 - x^2 + 5xy - 3y^2$;
(c) $x^4 - 2x^2y^2 + y^4 + x^2 - 6xy + 9y^2$.

*11. Discuss the behavior of each function near the origin:
(a) $x^{10} - x^7y^7 + y^{10}$;
(b) $x^{10} - x^5y^5 + y^{10}$;
(c) $x^{10} - x^3y^3 + y^{10}$.

Hint: Do (b) first, and compare the others with it.

*12. Discuss the behavior of each function near the origin:
(a) $x^{10} - 3x^7y^7 + y^{10}$;
(b) $x^{10} - 3x^5y^5 + y^{10}$;
(c) $x^{10} - 3x^3y^3 + y^{10}$.

*13. For each part find the minimum value of the given function, and the points (x, y) that give this minimum:
(a) $|x + 2y - 5| + |x + 2y + 1|$;
(b) $|x - 2| + \sqrt{x^2 + y^2}$;
(c) $\sqrt{(x-1)^2 + y^2} + \sqrt{(x+1)^2 + y^2}$.

*14. Prove that the function
$$e^{\frac{x^2+y^2}{x}} + \frac{5}{x} + \frac{11}{y} + \sin xy^2$$
has an absolute minimum value in the open first quadrant: $x > 0$, $y > 0$. (Do not attempt to find this minimum value!)

*15. Prove that the function
$$x^2 + xy + y^2 + \frac{1}{x} + \frac{1}{y}$$

has an absolute minimum value of $3\sqrt[4]{3}$ in the open first quadrant: $x > 0$, $y > 0$, and that this minimum value is attained at a unique point.

Exercises 16–18 are concerned with a set of n points in E_2: p_1:(x_1, y_1), p_2:(x_2, y_2), $\cdots$, p_n:(x_n, y_n). They illustrate in different ways the principle of **least squares** (cf. § 1605).

*16. Show that the point p:(x, y) that minimizes the sum of the squares of the distances

(1) $$[d(p, p_1)]^2 + [d(p, p_2)]^2 + \cdots + [d(p, p_n)]^2$$

is the **mean center**, with coordinates

$$x = \frac{x_1 + x_2 + \cdots + x_n}{n}, \quad y = \frac{y_1 + y_2 + \cdots + y_n}{n}.$$

*17. For any line L let d_i be the distance between L and p_i, $i = 1, 2, \cdots, n$. Find the line L for which the sum of the squares $d_1^2 + d_2^2 + \cdots + d_n^2$ is a minimum. *Hint:* Translate axes so that the mean center (Ex. 16) is at the origin, and write the equation of L in normal form: $x \cos \theta + y \sin \theta - p = 0$, with θ and p as parameters.

*18. Assuming that the x-coordinates, $x_1, x_2, \cdots, x_n$ are not all equal, show that there is one and only one line $y = mx + b$ that minimizes the sum of the squares of the "vertical distances":

(2) $$(mx_1 + b - y_1)^2 + (mx_2 + b - y_2)^2 + \cdots + (mx_n + b - y_n)^2,$$

and that the numbers m and b are the (unique) solution of the pair of equations:

(3) $$(\Sigma x_i^2)m + (\Sigma x_i)b = \Sigma x_i y_i,$$
$$(\Sigma x_i)m + nb = \Sigma y_i.$$

Hint: The minimum of the function $\Sigma(x - x_i)^2$ is zero if and only if the numbers $x_1, x_2, \cdots, x_n$ are identical. The equations (3) are the **normal equations** for the given set of points, and the resulting line $y = mx + b$ is the **line of regression of y on x**. This is the line of "best fit" in the sense that if $d_i = y_i - y$ is the "error" made by replacing each given point (x_i, y_i) by the computed point (x_i, y) on the line, then *this* is the line for which the sum of the squares of these errors, Σd_i^2, is a minimum.

19. Find the extrema of $x^2y - xy^2$ on the closed square: $0 \leq x \leq 1$, $0 \leq y \leq 1$.

20. Find the extrema of $x^2y - xy^2$ on the closed quarter-disk: $x \geq 0$, $y \geq 0$, $x^2 + y^2 \leq 1$.

*21. Minimize $|x + y|$ if $\dfrac{x^2}{a^2} - \dfrac{y^2}{b^2} \geq 1$ ($a > 0$, $b > 0$).

*22. Find the extrema of $e^{x+y} \sin(x - y)$ on the closed half-plane: (a) $x + y \geq 1$; (b) $x + y \leq 1$.

*23. Find the extrema of $e^{xy}(x + y)$ on the closed half-plane $y \geq a$. Consider all cases.

24. Prove that a linear function $ax + by + c$ with domain a simple closed polygon and its "inside" (cf. § 1027) attains its extreme values at vertices of the polygon.

25. Find the extrema of $-3x + 5y + 10$ on the triangle with vertices $(2, 1)$, $(1, -1)$, $(0, 3)$. (Cf. Ex. 24.)

*26. The four points $(1, 0)$, $(0, 1)$, $(1, 1)$, and $(2, 2)$ are the vertices of a (closed) quadrilateral in three distinct ways. Show that any extremum of any linear function is the same for these three quadrilaterals, and that no such extremum can be attained at the point (1, 1) unless the linear function is constant.

27. Establish formula (8), § 912.

28. Generalize the Law of the Mean (Theorem I, § 912) to a function of several variables.

29. Write out the form taken on by equation (2), § 912, when that equation is extended to a function of three variables and applied to the function xyz with $a = 0, b = 0, c = 0, h = 1, k = 2, l = 3$. (Cf. Ex. 28.)

★30. Generalize the Extended Law of the Mean (formula (8), § 912) to a function of several variables.

★31. Obtain for a function of 3 variables the formula corresponding to (2), § 913.

★32. Show by the following example that a function $f(x, y)$ may be continuously differentiable throughout a region R and such that its first partial derivative $f_1(x, y)$ vanishes identically there, yet not depend solely on y. Hence show the inadequacy of the following reasoning in proving Theorem II, § 912: "If $f_1(x, y) = 0$, f does not depend on x, and if $f_2(x, y) = 0$, f does not depend on y; therefore f is a constant." Show that if R is a region in E_2 such that for every line parallel to the x-axis and meeting R the points that this line has in common with R constitute an interval, finite or infinite, then the existence and identical vanishing of the partial derivative f_1 of a function f implies that f is a function of y alone. *Example:* Let R consist of the plane with the points (x, y) such that $x = 0, y \geq 0$ deleted, and let $f(x, y) \equiv 0$ except in the first quadrant, where $f(x, y) \equiv y^4$.

★33. Show that the Law of the Mean (2), § 912, if the condition $(a + \theta h, b + \theta k) \in I(R)$ is omitted, fails for the function of Exercise 32, by taking $a = -1, b = 1, h = 2, k = 0$.

★34. Say what you can about a function $f(x, y)$ all of whose second partial derivatives vanish identically in a region R. *Hint:* Show that $f_1(x, y) = a$ and $f_2(x, y) = b$, identically, and consider the function $f(x, y) - ax - by$.

★35. Let R be a region in E_2 such that for every line parallel to a coordinate axis and meeting R the points that this line has in common with R constitute an interval, finite or infinite. Say what you can about a function $f(x, y)$ having the following property:

(a) $f_{12}(x, y) = f_{21}(x, y) = 0$ identically in R;
(b) $f_{11}(x, y) = f_{22}(x, y) = 0$ identically in R;
(c) $f_1(x, y) = f_2(x, y) = x$ identically in R.
(Cf. Ex. 32.)

916. DIFFERENTIATION OF AN IMPLICIT FUNCTION

It is often desirable or necessary to treat functions that are not given explicitly, but instead are defined by or involved implicitly in some functional relation. For example, it may be convenient to study properties of the function $y = \sqrt{1 - x^2}$ by means of the equation $x^2 + y^2 = 1$. On the other hand, the inverse function of $x + e^x$ is of necessity not written explicitly, but is *defined* implicitly by the equation $y + e^y = x$. We shall be faced, then, with the study of certain properties of a function $y = \phi(x)$ determined by an equation in the two variables x and y, which can be written in the form

$f(x, y) = 0$. More generally, we shall investigate functions of several variables which are implicitly specified. The question of *existence* of a function satisfying a given equation is treated in § 932, and will not be pursued here. Our concern in this section is to establish and use certain relationships between the derivative (or partial derivatives if there is more than one independent variable) of a function implicitly defined and the partial derivatives of the functions appearing in the defining equation.

A basic fact of the greatest importance is that when a variable is defined, by a given equation, as a function of the remaining variables the given equation, considered as a relation in these remaining variables, is an *identity*. For example, when $\sqrt{1 - x^2}$ is substituted for y in the equation $x^2 + y^2 = 1$, this equation reduces to the identity $x^2 + (1 - x^2) = 1$. Similarly, if $y = \phi(x)$ is the function defined by the equation $y + e^y = x$, then the equation $\phi(x) + e^{\phi(x)} = x$ is true for all x by definition. In this section we shall use the expression "is defined by" to mean "satisfies identically."

We are now ready to derive the basic differentiation formulas for functions assumed to exist (cf. § 932 for their existence). The ranges of the variables are unspecified for reasons of expediency.

Theorem I. *Let $y = \phi(x)$ be a differentiable function defined by the equation $f(x, y) = 0$, where $f(x, y)$ is differentiable and $f_2(x, y)$ does not vanish. Then*

(1) $$\frac{dy}{dx} = -\frac{f_x}{f_y} = -\frac{f_1(x, \phi(x))}{f_2(x, \phi(x))}.$$

Proof. Let $u = f(x, y)$, and let x and y be considered as functions of the new independent variable t: $x = t$, $y = \phi(t)$. Then u, as a function of t, is differentiable and, by the chain rule,

(2) $$\frac{du}{dt} = \frac{\partial f}{\partial x}\frac{dx}{dt} + \frac{\partial f}{\partial y}\frac{dy}{dt}.$$

As a function of t alone, however, u is identically zero, so that the left-hand member of (2) vanishes identically. Using the fact that $dx/dt = 1$, solving (2) for dy/dt, and relabeling the variable t with the letter x, we obtain (1).

NOTE. A mnemonic device for formula (1) is the following: Write the quotient $-f_x/f_y$ in the form $-(\partial f/\partial x)/(\partial f/\partial y)$, "cancel" ∂f, and obtain $-\partial y/\partial x$. This, of course, is nonsense (cf. the Note, § 907), but the point is that it tells us where the subscripts x and y belong. (The minus sign must be remembered separately.) This device is a little like one that serves to recall the change of base formula for logarithms (Ex. 11, § 502), $\log_a x = (\log_a b)(\log_b x)$: "cancel" the b's and shift the subscript a.

The proof of the next theorem is nearly identical in method with that of Theorem I, and is requested in Exercise 10, § 918.

Theorem II. Let $u = \phi(x, y, z, \cdots)$ be a differentiable function of the k variables $x, y, z, \cdots$ defined by the equation $f(x, y, z, \cdots, u) = 0$, where the function $f(x, y, z, \cdots, u)$ of the $k + 1$ variables $x, y, z, \cdots, u$ is differentiable and $f_{k+1}(x, y, z, \cdots, u)$ does not vanish. Then

(3)
$$\begin{cases} \dfrac{\partial u}{\partial x} = -\dfrac{f_x}{f_u} = -\dfrac{f_1(x, y, z, \cdots, \phi(x, y, z, \cdots))}{f_{k+1}(x, y, z, \cdots, \phi(x, y, z, \cdots))}, \\ \dfrac{\partial u}{\partial y} = -\dfrac{f_y}{f_u} = -\dfrac{f_2(x, y, z, \cdots, \phi(x, y, z, \cdots))}{f_{k+1}(x, y, z, \cdots, \phi(x, y, z, \cdots))}, \\ \cdots \end{cases}$$

Example 1. Find the slope of the hyperbola
$$x^2 - 4xy - 3y^2 = 9$$
at the point $(2, -1)$.

First Solution. If $f(x, y) \equiv x^2 - 4xy - 3y^2 - 9$, y is defined as a function of x by the equation $f(x, y) = 0$. By (1),
$$\frac{dy}{dx} = -\frac{f_x}{f_y} = -\frac{2x - 4y}{-4x - 6y} = \frac{x - 2y}{2x + 3y} = \frac{4}{1} = 4.$$

Second Solution. Differentiate with respect to x:
$$2x - 4y - 4x\frac{dy}{dx} - 6y\frac{dy}{dx} = 0,$$
and solve for dy/dx.

Example 2. Find $\dfrac{\partial z}{\partial y}$ if $x^2 y - 8xyz = yz + z^3$.

First Solution. Let $f(x, y, z) \equiv 8xyz + yz + z^3 - x^2 y$. Then, by (3):
$$\frac{\partial z}{\partial y} = -\frac{\partial f/\partial y}{\partial f/\partial z} = -\frac{8xz + z - x^2}{8xy + y + 3z^2}.$$

Second Solution. Solve
$$x^2 - 8xz - 8xy\frac{\partial z}{\partial y} = z + y\frac{\partial z}{\partial y} + 3z^2\frac{\partial z}{\partial y}.$$

917. SOME NOTATIONAL PITFALLS

Suppose u is a given function of three variables x, y, and z, and that z is a given function of the two variables x and y. What does $\dfrac{\partial u}{\partial x}$ mean? This meaning depends on whether u is being considered as a function of all three variables or as a function of only the two variables x and y. The notation $\dfrac{\partial u}{\partial x}$ does not specify which. This ambiguity becomes particularly confusing if we recklessly apply the chain rule, using the symbol $\dfrac{\partial u}{\partial x}$ twice:

(1)
$$\frac{\partial u}{\partial x} = \frac{\partial u}{\partial x} + \frac{\partial u}{\partial z}\frac{\partial z}{\partial x},$$

where in the left-hand member u is a function of x and y, and in the right-hand member u is a function of x, y, and z. Such a muddled equation as (1) should be avoided. One of the simplest means of getting away from it is to use functional notation and to be more generous in the use of letters. To be precise, for the question above let $u = f(x, y, z)$ be the given function of three variables. Following the notation of § 907 and observing that a change of variables is taking place, we can let r and s denote the new variables and represent the old variables in terms of the new by the formulas

(2) $$x = r, \quad y = s, \quad z = \phi(r, s).$$

Finally, if we write $g(r, s) \equiv f(r, s, \phi(r, s))$, the chain rule gives $\dfrac{\partial g}{\partial r} = \dfrac{\partial f}{\partial x}\dfrac{\partial x}{\partial r} + \dfrac{\partial f}{\partial y}\dfrac{\partial y}{\partial r} + \dfrac{\partial f}{\partial z}\dfrac{\partial z}{\partial r}$ which, by (2), reduces to $\dfrac{\partial g}{\partial r} = \dfrac{\partial f}{\partial x} + \dfrac{\partial f}{\partial z}\dfrac{\partial z}{\partial r}.$

Finally, upon restoration of the notation x and y for the new variables, equation (1) becomes

(3) $$\frac{\partial g}{\partial x} = \frac{\partial f}{\partial x} + \frac{\partial f}{\partial z}\frac{\partial z}{\partial x}.$$

In this example the purpose of using two new letters r and s is to clarify the possibly disturbing fact that in the original formulation of the problem the two new variables x and y are identical with two of the original variables. With practice and sophistication the student will learn to dispense with such auxiliary variables as r and s, and proceed immediately to equation (3).

The numerical subscript notation for partial derivatives (§ 901) is also suitable for resolving such notational problems as that just discussed. Equation (3) becomes

(4) $$g_1(x, y) = f_1(x, y, \phi(x, y)) + f_3(x, y, \phi(x, y))\phi_1(x, y).$$

A similar problem is the following: Suppose z is defined as a function of x and y by means of the equation

(4) $$z = f(x, y, z).$$

What is $\dfrac{\partial z}{\partial x}$? It should be clear, in spite of the deceptive equality of (4), that this is *not* the same as $\dfrac{\partial f}{\partial x}$. One method of answering the question is the following: Define the function $F(x, y, z)$:

(5) $$F(x, y, z) \equiv f(x, y, z) - z,$$

and write (4) in the form

(6) $$F(x, y, z) = 0.$$

Then, by (3), § 916, $\dfrac{\partial z}{\partial x} = -\dfrac{F_x}{F_z} = -\dfrac{f_x}{f_z - 1} = \dfrac{f_x}{1 - f_z}.$ More simply, perhaps, considering z as a function of x and y, we can differentiate (4) directly:

(7) $$\frac{\partial z}{\partial x} = \frac{\partial f}{\partial x} + \frac{\partial f}{\partial z}\frac{\partial z}{\partial x},$$

and solve for $\dfrac{\partial z}{\partial x}$, with the same result as before.

Example. If $u = x^2y + y^2z + z^2x$ and if z is defined implicitly as a function of x and y by the equation $x^2 + yz + z^3 = 0$, find $\dfrac{\partial u}{\partial x}$, where u is considered as a function of x and y alone.

Solution. Let $f(x, y, z) \equiv x^2y + y^2z + z^2x$, $g(x, y, z) \equiv x^2 + yz + z^3$, $z = \phi(x, y)$ be the function determined by $g(x, y, z) = 0$, and $\psi(x, y) \equiv f(x, y, \phi(x, y))$. Then

$$\psi_1(x, y) = f_1 + f_3\phi_1 = f_1 + f_3\left(-\frac{g_1}{g_3}\right) = 2xy + z^2 - \frac{2x(y^2 + 2zx)}{y + 3z^2}$$

$$= \frac{6xyz^2 - 4x^2z + yz^2 + 3z^4}{y + 3z^2}.$$

918. EXERCISES 9/9

In these Exercises assume differentiability wherever appropriate.

In Exercises 1–4, find $\dfrac{dy}{dx}$.

1. $x^2 + xy + 3y^2 = 17$.
2. $x^{\frac{2}{3}} + y^{\frac{2}{3}} = a^{\frac{2}{3}}$.
3. $x \sin y + y \sin x = b$.
4. $x = ay + be^{-cy}$.

In Exercises 5–8, find $\dfrac{\partial z}{\partial x}$ and $\dfrac{\partial z}{\partial y}$.

5. $Ayz + Bzx + Cxy = D$.
6. $x^3 + y^3 + z^3 = 3axyz$.
7. $x = ze^{yz}$.
8. $xe^z = y \sin z$.

9. If the equation $f(x, y, z) = 0$ defines each variable as a differentiable function of the other two establish the formulas

$$\frac{\partial z}{\partial x} = -\frac{f_1}{f_3}, \frac{\partial z}{\partial y} = -\frac{f_2}{f_3}, \frac{\partial y}{\partial x} = -\frac{f_1}{f_2}, \frac{\partial y}{\partial z} = -\frac{f_3}{f_2}, \frac{\partial x}{\partial y} = -\frac{f_2}{f_1}, \frac{\partial x}{\partial z} = -\frac{f_3}{f_1},$$

where the variable being differentiated is considered as a function of the other two. In what sense are $\dfrac{\partial z}{\partial x}$ and $\dfrac{\partial x}{\partial z}$ reciprocals? Is it true that $\dfrac{\partial x}{\partial y}\dfrac{\partial y}{\partial z}\dfrac{\partial z}{\partial x} = -1$? Generalize to the case of n variables.

10. Prove Theorem II, § 916.

In Exercises 11–14, u is defined as a function of x, y, and z by the first equation, and z is defined as a function of x and y by the second equation. Considering u, as a consequence, as a function of x and y, find $\dfrac{\partial u}{\partial x}$ and $\dfrac{\partial u}{\partial y}$.

11. $x^2 + y^3 + z^4 + u^5 = 1$, $x + y^2 + z^3 = 1$.
12. $u = x^3 + y^3 + z^3 + u^3$, $z = x^2 + y^2 + z^2$.
13. $u = x + y + z + e^u$, $z = x + y + \sin z$.
14. $f(x, y, z, u) = 0$, $g(x, y, z) = 0$.

15. Examine Theorems 1 and 2, § 916, in the linear case, $f(x, y) = ax + by$, $f(x, y, \cdots, u) = ax + by + \cdots + qu$. Indicate the significance of the non-vanishing of the appropriate derivative.

16. If $z = f(x, y, z)$, find $\dfrac{\partial x}{\partial y}, \dfrac{\partial y}{\partial z}, \dfrac{\partial z}{\partial x}$.

17. If $f(x, y) = g(y, z)$, find $\dfrac{\partial x}{\partial y}, \dfrac{\partial y}{\partial z}, \dfrac{\partial z}{\partial x}$.

18. If $x^2 + y^2 = f(x, y, z)$, find $\dfrac{\partial x}{\partial y}, \dfrac{\partial y}{\partial z}, \dfrac{\partial z}{\partial x}$.

19. If $z = f(x + y + z, xyz)$, find $\dfrac{\partial x}{\partial y}, \dfrac{\partial y}{\partial z}, \dfrac{\partial z}{\partial x}$.

20. If $g(x) = f(x, y, g(y))$, find $\dfrac{dy}{dx}$.

21. Show that if y is defined as a function of x by the equation $f(x, y) = 0$, then
$$\frac{d^2y}{dx^2} = -\frac{f_1^2 f_{22} - 2f_1 f_2 f_{12} + f_2^2 f_{11}}{f_2^3}.$$

22. Show that if z is defined as a function of x and y by the equation $f(x, y, z) = 0$, then
$$\frac{\partial^2 z}{\partial x \, \partial y} = -\frac{f_1 f_2 f_{33} - f_1 f_3 f_{23} - f_2 f_3 f_{13} + f_3^2 f_{12}}{f_3^3}.$$

919. ENVELOPE OF A FAMILY OF PLANE CURVES

Definition. *An envelope of a family of plane curves is a curve that has the following two properties:* (i) <u>at every one of its points it is tangent to at least one curve of the family</u>; (ii) <u>it is tangent to every curve of the family at at least one point.</u> *The envelope of a family of curves is the totality* (union) *of its envelopes.*

Example 1. The family of curves $y = (x - t)^2$, where t is a parameter, is the set of parabolas obtained by translating the parabola $y = x^2$ in the direction of the x-axis. The x-axis clearly satisfies the definition above, and is therefore an envelope. It appears obvious (intuitively) that it is *the* envelope, since there is no other curve that is an envelope. The point on the envelope that corresponds to the curve $y = (x - t)^2$, for any fixed t, is given by $x = t$, $y = 0$.

Example 2. The circles of the family $(x - t)^2 + y^2 = 1$, have centers on the x-axis and radii equal to 1. The line $y = 1$ is an envelope, as is the line $y = -1$. The envelope consists of the two lines $y^2 = 1$. The points on the envelope that correspond to the curve $(x - t)^2 + y^2 = 1$, for any fixed t, are given by $x = t$, $y = \pm 1$.

It is our purpose in this section to find necessary conditions which a well-behaved envelope of a well-behaved family of curves must satisfy.

Theorem. *Let*

(1) $$f(x, y, t) = 0$$

be the equation of a family F of curves, where t is a parameter, and let

(2) $$x = \phi(t), \quad y = \psi(t)$$

be the parametric equations of an envelope E of F. *Assume furthermore that f, ϕ, and ψ, are continuously differentiable and that for each value of t the curve* (1) *and E have a common tangent at the point* $(\phi(t), \psi(t))$. *Then the coordinates* (2) *of each point of E must satisfy, in addition to equation* (1), *the equation*

(3) $$f_3(x, y, t) = 0.$$

Proof. Since (1) and E have a common tangent at the point $(\phi(t), \psi(t))$, this point certainly lies on the curve (1), for each t. Therefore the equation $f(\phi(t), \psi(t), t) = 0$ is an *identity* in t and, by the chain rule,

(4) $$f_1(\phi(t), \psi(t), t)\phi'(t) + f_2(\phi(t), \psi(t), t)\psi'(t) + f_3(\phi(t), \psi(t), t) = 0.$$

Since our objective is to establish the vanishing of the third term of (4), it is sufficient to show that the sum of the first two terms of (4) is zero. This is trivial unless at least one of these terms is nonzero. We assume for definiteness that both $f_2(\phi(t), \psi(t), t)$ and $\psi'(t)$ are nonzero, for the particular value of t considered, and use the fact that (1) and E have a common tangent at $(\phi(t), \psi(t))$. In the first place (§ 916), the slope of (1) at $(\phi(t), \psi(t))$ is equal to $-f_1(\phi(t), \psi(t), t)/f_2(\phi(t), \psi(t), t)$, which is finite. Therefore the tangent to E at $(\phi(t), \psi(t))$ is not vertical, and $\phi'(t) \neq 0$. Its slope is therefore $\psi'(t)/\phi'(t)$. The equality $-f_1/f_2 = \psi'/\phi'$ is equivalent to the desired vanishing of $f_1\phi' + f_2\psi'$.

NOTE. In simple cases, since equations (1) and (3) are both satisfied by the coordinates of points of the envelope, the envelope can often be found by eliminating t from that pair of equations.

We now consider the two examples given above, and others, in the light of the Theorem and Note.

Example 1 (again). Let $f(x, y, t) \equiv (x - t)^2 - y$. Then $f_3(x, y, t) = -2(x - t)$. The coordinates of points on an envelope must satisfy identically the two equations

$$x - t = 0, \quad (x - t)^2 = y.$$

In other words, $y = 0$, and only points on the x-axis are on the envelope. Therefore (with the aid of the preceding discussion) we know that the x-axis is *the* envelope.

Example 2 (again). Let $f(x, y, t) \equiv (x - t)^2 + y^2 - 1$. Then $f_3(x, y, t) = -2(x - t)$. As in Example 1, the two equations (1) and (3) lead to the envelope. In this case its equation is $y^2 = 1$.

Example 3. Find the envelope of the family of parabolas obtained by arbitrarily translating the parabola $y = x^2$ so that the vertex lies on the line $y = 2x$.

Solution. The family has the equation $f(x, y, t) = (x - t)^2 - (y - 2t) = 0$. Then $f_3(x, y, t) = -2x + 2t + 2$. Elimination of t between equations (1) and (3) leads to the equation $y = 2x - 1$. This line must therefore contain the envelope. Direct verification shows that it *is* the envelope.

Example 4. Discuss the envelope of the family of curves $f(x, y, t) \equiv (x - t)^2 - y^3 = 0$, obtained by translating parallel to the x-axis the semicubical parabola $y^3 = x^2$.

Solution. The procedure used in Example 1 yields the same result: $y = 0$. However, the x-axis is clearly *not* an envelope. In fact, there is *no* envelope. This example shows the importance of checking the result of eliminating t from equations (1) and (3).

Example 5. Discuss the envelope of the family of curves $\sqrt[3]{y} = x - t$.

Solution. If $f(x, y, t) \equiv \sqrt[3]{y} - x + t$, equation (3) becomes $1 = 0$. This inconsistent equation might lead us to believe that there is *no* envelope. However, the conditions of the preceding Theorem are not satisfied: $f(x, y, t)$ is not a differentiable function, since $f_2(x, y, t) = (\frac{1}{3})y^{-\frac{2}{3}}$ becomes infinite on the x-axis.

There are two courses open to us now: (*i*) examine the points where $f(x, y, t)$ fails to be differentiable, and (*ii*) rewrite the original equation. If we choose the latter alternative, and define $f(x, y, t) \equiv y - (x - t)^3 = 0$, we have $f_3(x, y, t) = 3(x - t)^2$. The result, as in preceding examples, is that the x-axis is the envelope.

920. EXERCISES 9/10

In Exercises 1–6, find the envelope of the given family of curves. Draw a figure.
1. The lines $x \cos \phi + y \sin \phi = 3$.
2. The lines $y = tx + t^2$.
3. The circles $(x - t)^2 + y^2 = 2t^2$.
4. The circles $(x - t)^2 + y^2 = t$.
5. The parabolas $x^2 = t(y - t)$.
6. The parabolas $x^2 = t^2(y - t)$.
7. Explain the absence of envelope for the family of hyperbolas $4(x - 3t)^2 - 9(y - 2t)^2 = 36$.

8. Find the envelope of the family obtained by translating parallel to the x-axis the curve $y^2 = x^2(2 - x)$ (cf. Fig. 317, p. 105). Check your answer by maximizing y.

9. Translate the parabola $y = x^2$ by moving its vertex along the parabola $y = x^2$. Find the envelope of the resulting family.

10. A variable circle moves so that it is always tangent to the x-axis and its center remains on the parabola $y = x^2$. Find the envelope.

11. A line segment of fixed length a moves so that its end-points are on the two coordinate axes. Find the envelope. *Hint:* Denote the intercepts $a \cos \theta$ and $a \sin \theta$.

12. A line moves so that its intercepts a and b have a constant positive sum, $a + b = c > 0$. Find the envelope.

13. The ellipse $\dfrac{x^2}{a^2} + \dfrac{y^2}{b^2} = 1$ varies so that its area is constant. Find the envelope.

14. The ellipse $\dfrac{x^2}{a^2} + \dfrac{y^2}{b^2} = 1$ varies so that the sum of its semiaxes is a constant c. Find the envelope.

15. The ellipse $\dfrac{x^2}{a^2} + \dfrac{y^2}{b^2} = 1$ varies so that the distance $c = \sqrt{a^2 + b^2}$ between the ends of the axes is constant. Find the envelope.

16. A variable circle moves so that it always passes through the origin and its center lies on the hyperbola $x^2 - y^2 = 1$. Find the envelope. *Hint:* Parametrize the hyperbola: $x = \cosh t,\ y = \sinh t$.

17. Prove that any smooth curve (§ 805) $x = \phi(t),\ y = \psi(t)$ is an envelope of its tangents. Prove furthermore that if the curvature of the given curve (§ 811) exists and is not zero, the curve is *the* envelope of its tangents. *Hint:* Equations (1) and (3), § 919, become $(y - \psi)\phi' = (x - \phi)\psi'$ and $(y - \psi)\phi'' = (x - \phi)\psi''$, respectively. Show that these together with $\phi'\psi'' \neq \phi''\psi'$ imply $x = \phi,\ y = \psi$, for each t.

In Exercises 18–20, verify the statements of Exercise 17 for the given curve.

18. $x = a \cos \theta,\ y = b \sin \theta$.

19. $x^2 + y^2 = 4$.   **20.** $y = x^3$.

*21. Let $x = \phi(t),\ y = \psi(t)$ be a smooth curve with nonzero curvature (§§ 805, 811). Prove that the envelope of its normals is its evolute (§ 813).

In Exercises 22–24, find the evolute requested in the specified Exercise of § 814, as the envelope of the normals of the given curve. (Cf. Ex. 21.)

*22. Exercise 19.   ***23.** Exercise 21.   ***24.** Exercise 22.

921. SEVERAL FUNCTIONS DEFINED IMPLICITLY. JACOBIANS

In § 916 we considered some properties of a function $u = \phi(x, y, z, \cdots)$ defined implicitly by an equation of the form $f(x, y, z, \cdots, u) = 0$. The problem of an implicit function becomes particularly simple in the linear case (cf. Ex. 15, § 918). As with linear expressions, the case of more general functions defined implicitly extends to the study of functions defined by *systems* of equations.

Let us review some important facts regarding systems of linear equations.† The most important single theorem is Cramer's Rule, which deals with a system of n linear equations in n unknowns. If the determinant D of such a system is nonzero, Cramer's Rule assures us that there is precisely one solution and specifies the form of this solution by means of quotients of determinants. If the determinant D is zero or if the number m of equations and the number n of unknowns are unequal it is possible that the system may have infinitely many solutions or none at all. It is reasonable to expect, however (in the absence of other information), that if $m < n$ there are infinitely many solutions and if $m > n$ there are no solutions. Under any circumstances regarding m and n, whenever a system of linear equations has at least one solution it is possible to obtain all solutions by suitably selecting

---

† For a treatment of systems of linear equations see M. Bôcher, *Introduction to Higher Algebra* (New York, The Macmillan Company, 1935), N. Conkwright, *Introduction to the Theory of Equations* (New York, Ginn and Company, 1941), G. Birkhoff and S. MacLane, *A Survey of Modern Algebra* (New York, The Macmillan Company, 1953), and D. C. Murdoch, *Linear Algebra for Undergraduates* (New York, John Wiley and Sons, 1957).

§ 921] SEVERAL FUNCTIONS DEFINED IMPLICITLY 297

k of the equations and k of the unknowns ($k \leq m$, $k \leq n$), and solving these k equations for these k unknowns uniquely in terms of the remaining $n - k$ unknowns.

Because of considerations such as those of the preceding paragraph we shall restrict our discussion of systems of equations to those for which the number of equations and the number of variables considered to be defined by the system as dependent on the remaining variables are equal.

We now proceed to the study of a system of equations of the form

(1)
$$\begin{cases} f^{(1)}(x_1, x_2, \cdots, x_m, u_1, u_2, \cdots, u_n) = 0, \\ f^{(2)}(x_1, x_2, \cdots, x_m, u_1, u_2, \cdots, u_n) = 0, \\ \quad \cdot \\ \quad \cdot \\ \quad \cdot \\ f^{(n)}(x_1, x_2, \cdots, x_m, u_1, u_2, \cdots, u_n) = 0, \end{cases}$$

where the functions are specified by superscripts instead of subscripts to avoid confusion with subscript notation for partial derivatives. The variables $u_1, \cdots, u_n$ are considered to be defined by (1) as functions of $x_1, \cdots, x_m$. The question of *existence* of such solutions is treated in § 932. In this section we shall deduce certain consequences of their existence, under appropriate conditions. For simplicity the region under consideration will not be specified here, and the details of the discussion will be carried through for the case $n = 2$. Since the method is essentially the same for all n, no loss of generality is suffered.

Accordingly, we consider a system of two equations

(2)
$$\begin{cases} f(x, y, \cdots, u, v) = 0, \\ g(x, y, \cdots, u, v) = 0, \end{cases}$$

where f and g are differentiable. Let $u = \phi(x, y, \cdots)$ and $v = \psi(x, y, \cdots)$ be two differentiable functions of the remaining variables that satisfy the two equations (2) identically. Our objective is to express the first partial derivatives of ϕ and ψ in terms of those of f and g. We apply the chain rule to (2), differentiating first with respect to x:

(3)
$$\begin{cases} \dfrac{\partial f}{\partial x} + \dfrac{\partial f}{\partial u}\dfrac{\partial u}{\partial x} + \dfrac{\partial f}{\partial v}\dfrac{\partial v}{\partial x} = 0, \\ \dfrac{\partial g}{\partial x} + \dfrac{\partial g}{\partial u}\dfrac{\partial u}{\partial x} + \dfrac{\partial g}{\partial v}\dfrac{\partial v}{\partial x} = 0, \end{cases}$$

where a compressed notation has been adopted. The second term of the first equation of (3), for example, could have been written more completely

$$f_u(x, y, \cdots, \phi(x, y, \cdots), \psi(x, y, \cdots)) \cdot \phi_x(x, y, \cdots).$$

Since system (3) may be regarded as a system of two linear equations in the two unknowns $\dfrac{\partial u}{\partial x}$ and $\dfrac{\partial v}{\partial x}$, we may solve for these two unknowns by Cramer's Rule:

$$(4) \quad \frac{\partial u}{\partial x} = - \frac{\begin{vmatrix} \dfrac{\partial f}{\partial x} & \dfrac{\partial f}{\partial v} \\ \dfrac{\partial g}{\partial x} & \dfrac{\partial g}{\partial v} \end{vmatrix}}{\begin{vmatrix} \dfrac{\partial f}{\partial u} & \dfrac{\partial f}{\partial v} \\ \dfrac{\partial g}{\partial u} & \dfrac{\partial g}{\partial v} \end{vmatrix}}, \quad \frac{\partial v}{\partial x} = - \frac{\begin{vmatrix} \dfrac{\partial f}{\partial u} & \dfrac{\partial f}{\partial x} \\ \dfrac{\partial g}{\partial u} & \dfrac{\partial g}{\partial x} \end{vmatrix}}{\begin{vmatrix} \dfrac{\partial f}{\partial u} & \dfrac{\partial f}{\partial v} \\ \dfrac{\partial g}{\partial u} & \dfrac{\partial g}{\partial v} \end{vmatrix}},$$

where, to be sure, the denominator determinant must be different from zero.

Determinants of the type appearing in (4) are of great importance in pure and applied mathematics. They are called *functional determinants* or *Jacobians*, after the German mathematician C. G. J. Jacobi (1804–1851). Their precise definition and a more concise notation for them follow.

Definition. *If the variables $u_1, \cdots, u_n$ are differentiable functions of the variables $x_1, \cdots, x_n$ (and possibly of more variables $x_{n+1}, \cdots, x_m$ as well):*

$$(5) \quad \begin{cases} u_1 = f^{(1)}(x_1, x_2, \cdots, x_n, x_{n+1}, \cdots, x_m), \\ u_2 = f^{(2)}(x_1, x_2, \cdots, x_n, x_{n+1}, \cdots, x_m), \\ \quad \vdots \\ u_n = f^{(n)}(x_1, x_2, \cdots, x_n, x_{n+1}, \cdots, x_m), \end{cases}$$

the Jacobian or functional determinant of $u_1, u_2, \cdots, u_n$ with respect to $x_1, x_2, \cdots, x_n$ is defined as an nth-order determinant, and denoted, as follows:

$$(6) \quad \frac{\partial(u_1, \cdots, u_n)}{\partial(x_1, \cdots, x_n)} = \frac{\partial(f^{(1)}, \cdots, f^{(n)})}{\partial(x_1, \cdots, x_n)} \equiv \begin{vmatrix} \dfrac{\partial f^{(1)}}{\partial x_1} & \cdots & \dfrac{\partial f^{(1)}}{\partial x_n} \\ \vdots & & \vdots \\ \dfrac{\partial f^{(n)}}{\partial x_1} & \cdots & \dfrac{\partial f^{(n)}}{\partial x_n} \end{vmatrix}.$$

With this notation equation (4) takes the form

$$(7) \quad \frac{\partial u}{\partial x} = - \frac{\dfrac{\partial(f, g)}{\partial(x, v)}}{\dfrac{\partial(f, g)}{\partial(u, v)}}, \quad \frac{\partial v}{\partial x} = - \frac{\dfrac{\partial(f, g)}{\partial(u, x)}}{\dfrac{\partial(f, g)}{\partial(u, v)}}.$$

§ 921] SEVERAL FUNCTIONS DEFINED IMPLICITLY

In an entirely similar way, again under the assumption that the Jacobian $\dfrac{\partial(f, g)}{\partial(u, v)}$ is nonzero, we have for the partial derivatives with respect to y

$$(8) \qquad \frac{\partial u}{\partial y} = -\frac{\dfrac{\partial(f, g)}{\partial(y, v)}}{\dfrac{\partial(f, g)}{\partial(u, v)}}, \quad \frac{\partial v}{\partial y} = -\frac{\dfrac{\partial(f, g)}{\partial(u, y)}}{\dfrac{\partial(f, g)}{\partial(u, v)}},$$

with analogous formulas holding for the partial derivatives with respect to any of the other variables of which u and v are considered to be functions.

The student should observe the resemblance between formulas (7) and (8), and formula (1) § 916, where $n = 1$. In the simplest case $n = 1$, the Jacobian becomes a partial derivative, and therefore may be regarded as a generalization of a partial derivative. Other similarities will become evident in the sequel.

Notational pitfall. The notation $\dfrac{\partial u}{\partial x}$, appearing in (7) is ambiguous, in that it tells only that u is considered as one dependent variable, but it does *not* identify the other. Suppose y, instead of v, were the other dependent variable. Then the formula for $\dfrac{\partial u}{\partial x}$ would be $-\dfrac{\partial(f, g)}{\partial(x, y)} \Big/ \dfrac{\partial(f, g)}{\partial(u, y)}$. This is a different function, even though the notation $\partial u/\partial x$ is the same! One standard notational device to clarify this ambiguity is to specify the other *independent* variables (*the ones being held fixed*) by subscripts. For example, if the only variables involved in (2) were x, y, u, and v, then the partial derivatives $\partial u/\partial x$ just considered would be denoted in formula (7), $\left(\dfrac{\partial u}{\partial x}\right)_y$; in the instance immediately above, $\left(\dfrac{\partial u}{\partial x}\right)_v$. A pertinent question in this connection is this: Under what circumstances are $\partial u/\partial x$ and $\partial x/\partial u$ reciprocals? The answer is simple: When the *variables held constant* are the same for both derivatives. (Cf. Ex. 9, § 918, Ex. 13, § 924.) With the subscript notation just introduced and where the only variables involved are x, y, u, and v we can say, then, that $(\partial u/\partial x)_y$ and $(\partial x/\partial u)_y$ are reciprocals; so are $(\partial u/\partial x)_v$ and $(\partial x/\partial u)_v$.

We formalize the more general case with a theorem (whose proof is requested in Exercise 14, § 924):

Theorem. *Let the functions* $f^{(i)}(x_1, \cdots, x_m, u_1, \cdots, u_n)$, $i = 1, \cdots, n$, *of system* (1) *be differentiable, let* $u_j = u_j(x_1, \cdots, x_m)$, $j = 1, \cdots, n$, *be differentiable functions that satisfy system* (1) *identically in the variables* $x_1, \cdots, x_m$, *and assume that the Jacobian* $\dfrac{\partial(f^{(1)}, \cdots, f^{(n)})}{\partial(u_1, \cdots, u_n)}$, *called the **Jacobian of the system**, be nonzero for all values of the variables concerned.*

Then the partial derivatives of the functions u_j are given by the formula

(9) $$\frac{\partial u_j}{\partial x_k} = -\frac{\dfrac{\partial(f^{(1)}, f^{(2)}, \cdots, f^{(j)}, \cdots, f^{(n)})}{\partial(u_1, u_2, \cdots, x_k, \cdots, u_n)}}{\dfrac{\partial(f^{(1)}, f^{(2)}, \cdots, f^{(j)}, \cdots, f^{(n)})}{\partial(u_1, u_2, \cdots, u_j, \cdots, u_n)}},$$

where the numerator Jacobian is obtained from the denominator Jacobian by formally replacing u_j by x_k.

922. COORDINATE TRANSFORMATIONS. INVERSE TRANSFORMATIONS

An algebraic fact which we shall need in this section comes from the theory of determinants (see the references given in § 921; also cf. Ex. 27, § 1705). This states that the product of two nth-order determinants can be written as an nth-order determinant obtained as follows: Let A and B be nth-order determinants whose general elements (ith row and jth column) are a_{ij} and b_{ij}, respectively, and define C to be the nth order determinant whose general element is

(1) $$c_{ij} = \sum_{k=1}^{n} a_{ik} b_{kj} = a_{i1} b_{1j} + a_{i2} b_{2j} + \cdots + a_{in} b_{nj}.$$

Then $AB = C$.

We omit the proof, but illustrate with a simple example, which the student should check:

Example 1. $\begin{vmatrix} 3 & -1 \\ 5 & 6 \end{vmatrix} \cdot \begin{vmatrix} 1 & 4 \\ -2 & 7 \end{vmatrix} = \begin{vmatrix} 5 & 5 \\ -7 & 62 \end{vmatrix}$. (This says that the product of 23 and 15 is 345.)

One simple consequence of this fact about the multiplication of determinants should be noted: If both factors, A and B, are nonzero, so is their product AB.

We consider now a system of the form

(2) $$\begin{cases} x = x(u, v), \\ y = y(u, v), \end{cases}$$

where $x(u, v)$ and $y(u, v)$ are continuously differentiable functions of the variables u and v. The system (2) can be thought of as defining a *point transformation* T, carrying certain points from the uv-plane into points of the xy-plane (cf. § 612). Under suitably restrictive conditions on the functions and on the sets considered, this transformation will establish a *one-to-one correspondence* between two sets, and the *inverse transformation* T^{-1}, which carries the points in the xy-plane back into points in the uv-plane, will be defined. (Such sufficiency conditions are treated in § 1229.) In the case of a one-to-one correspondence, the system (2) can be considered alternatively as reassigning to a point (u, v) (or (x, y)) a new pair of coordinates (x, y) (or (u, v)). When so considered the transformation is called a **coordinate transformation**. In either case the Jacobian $\dfrac{\partial(x,y)}{\partial(u,v)}$ is called the **Jacobian of the transformation** (2).

NOTE. Since a coordinate transformation is commonly used for transforming an equation in x and y (or $x, y, \cdots$) into an equation in other variables, and since such a transformation of an equation is usually achieved by substituting for $x, y, \cdots$ their expressions in terms of the new variables, it is customary to represent a coordinate transformation in the form (2), instead of writing the new variables in terms of $x, y, \cdots$, as was done in § 612.

A simple and familiar example may help to clarify the situation:

Example 2. Consider the relation between rectangular and polar coordinates in the plane, and write

(3) $\qquad \begin{cases} x = u \cos v, \\ y = u \sin v, \end{cases}$

where u and v represent the polar coordinates ρ and θ, respectively. The student is already familiar with (3) as a coordinate transformation. Let us consider it as a point transformation. Figure 903 illustrates a one-to-one correspondence between two regions, S in the uv-plane and R in the xy-plane. If the region S were

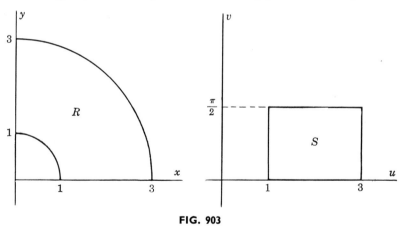

FIG. 903

extended upward until v became greater than 2π, the correspondence would no longer be one-to-one, since the partial annulus R would extend around the origin and overlap itself. Similarly, if u were allowed to be zero, the v-axis in the uv-plane would correspond to the origin in the xy-plane, and the transformation (3) would be many-to-one instead of one-to-one. The Jacobian of the transformation is

$\begin{vmatrix} \cos v & -u \sin v \\ \sin v & u \cos v \end{vmatrix} = u$, which vanishes only on the v-axis in the uv-plane (the origin in the xy-plane).

Now suppose we have *two* transformations,

(4) $\quad \begin{cases} x = x(u, v), \\ y = y(u, v), \end{cases} \quad \begin{cases} u = u(r, s), \\ v = v(r, s), \end{cases}$

where the functions are differentiable, so that upon applying the transformations in succession, we have x and y expressed in terms of r and s:

(5) $\quad \begin{cases} x = \phi(r, s) \equiv x(u(r, s), v(r, s)), \\ y = \psi(r, s) \equiv y(u(r, s), v(r, s)). \end{cases}$†

Since this *composition of transformations* has such a close similarity to the *composition of functions* (we can speak of a transformation of a transformation instead of a function of a function) it might be expected that we should have some sort of chain rule involving Jacobians, corresponding to the familiar formula $\dfrac{dy}{dx} = \dfrac{dy}{du}\dfrac{du}{dx}$ for functions. Let us see.

Applying the chain rule for partial differentiation to (5), we have:

(6) $\quad \begin{cases} \dfrac{\partial x}{\partial r} = \dfrac{\partial x}{\partial u}\dfrac{\partial u}{\partial r} + \dfrac{\partial x}{\partial v}\dfrac{\partial v}{\partial r}, & \dfrac{\partial x}{\partial s} = \dfrac{\partial x}{\partial u}\dfrac{\partial u}{\partial s} + \dfrac{\partial x}{\partial v}\dfrac{\partial v}{\partial s}, \\[1em] \dfrac{\partial y}{\partial r} = \dfrac{\partial y}{\partial u}\dfrac{\partial u}{\partial r} + \dfrac{\partial y}{\partial v}\dfrac{\partial v}{\partial r}, & \dfrac{\partial y}{\partial s} = \dfrac{\partial y}{\partial u}\dfrac{\partial u}{\partial s} + \dfrac{\partial y}{\partial v}\dfrac{\partial v}{\partial s}. \end{cases}$

This is the spot where we use the multiplication theorem for determinants. We observe, from (6):

$$\begin{vmatrix} \dfrac{\partial x}{\partial u} & \dfrac{\partial x}{\partial v} \\ \dfrac{\partial y}{\partial u} & \dfrac{\partial y}{\partial v} \end{vmatrix} \cdot \begin{vmatrix} \dfrac{\partial u}{\partial r} & \dfrac{\partial u}{\partial s} \\ \dfrac{\partial v}{\partial r} & \dfrac{\partial v}{\partial s} \end{vmatrix} = \begin{vmatrix} \dfrac{\partial x}{\partial r} & \dfrac{\partial x}{\partial s} \\ \dfrac{\partial y}{\partial r} & \dfrac{\partial y}{\partial s} \end{vmatrix},$$

a relation which we shall call the **chain rule for transformations** and which can be written more concisely:

(7) $\quad \dfrac{\partial(x, y)}{\partial(r, s)} = \dfrac{\partial(x, y)}{\partial(u, v)}\dfrac{\partial(u, v)}{\partial(r, s)}.$

In the form (7), the chain rule bears a very close resemblance to the chain rule for functions of one variable, and can be easily remembered by virtue of the apparent "cancellation" of the symbol $\partial(u, v)$ (which, of course, is meaningless by itself). Observe that we can conclude from (7) that if both Jacobians of the transformations (4) that are combined by composition are nonzero, then the Jacobian of the resultant transformation (5) is also nonzero.

† Observe that the notation of (4) is functional notation, and that since x is *not* the same function of r and s as it is of u and v, we must *not* write $x = x(r, s)$!

An important special case of (7) is that in which the second of the two transformations (4) is the inverse of the first:

(8) $\quad \begin{cases} x = x(u, v), \\ y = y(u, v), \end{cases} \begin{cases} u = u(x, y), \\ v = v(x, y), \end{cases}$

where the system (5) reduces to $x = \phi(r, s) = r$ and $y = \psi(r, s) = s$. The resultant Jacobian is then the Jacobian of the **identity transformation** that leaves every point invariant:

$$\frac{\partial(x, y)}{\partial(r, s)} = \begin{vmatrix} 1 & 0 \\ 0 & 1 \end{vmatrix} = 1.$$

The chain rule (7) then reduces to

(9) $\quad \dfrac{\partial(x, y)}{\partial(u, v)} \dfrac{\partial(u, v)}{\partial(x, y)} = 1,$

stating that *the Jacobians of a transformation and its inverse are reciprocals*. This corresponds to the familiar fact for functions of one variable that dy/dx and dx/dy are reciprocals. From (9) we infer that *whenever a transformation and its inverse exist and are prescribed by differentiable functions, both the transformation and its inverse have nonzero Jacobians*.

Since the functions $u(x, y)$ and $v(x, y)$ of the inverse transformation are defined implicitly by a system of equations

$$f(x, y, u, v) \equiv x - x(u, v) = 0 \quad \text{and} \quad g(x, y, u, v) \equiv y - y(u, v) = 0,$$

their first partial derivatives can be obtained by formulas (7) and (8), § 921:

$$\frac{\partial u}{\partial x} = -\frac{\dfrac{\partial(f, g)}{\partial(x, v)}}{\dfrac{\partial(f, g)}{\partial(u, v)}} = \frac{\dfrac{\partial y}{\partial v}}{\dfrac{\partial(x, y)}{\partial(u, v)}}, \quad \frac{\partial v}{\partial x} = -\frac{\dfrac{\partial(f, g)}{\partial(u, x)}}{\dfrac{\partial(f, g)}{\partial(u, v)}} = \frac{-\dfrac{\partial y}{\partial u}}{\dfrac{\partial(x, y)}{\partial(u, v)}},$$

$$\frac{\partial u}{\partial y} = -\frac{\dfrac{\partial(f, g)}{\partial(y, v)}}{\dfrac{\partial(f, g)}{\partial(u, v)}} = \frac{-\dfrac{\partial x}{\partial v}}{\dfrac{\partial(x, y)}{\partial(u, v)}}, \quad \frac{\partial v}{\partial y} = -\frac{\dfrac{\partial(f, g)}{\partial(u, y)}}{\dfrac{\partial(f, g)}{\partial(u, v)}} = \frac{\dfrac{\partial x}{\partial u}}{\dfrac{\partial(x, y)}{\partial(u, v)}}.$$

For the general case of the chain rule for transformations from one n-dimensional space to another (that is, where there are n dependent and n independent variables), the details are completely equivalent to those given above, and will be omitted here (cf. Ex. 15, § 924). We merely note in passing the particular forms and values of the Jacobians for a change of coordinates in three-dimensional space from rectangular coordinates to cylindrical and spherical coordinates (§ 808):

(10) Cylindrical: $\quad \dfrac{\partial(x, y, z)}{\partial(\rho, \theta, z)} = \begin{vmatrix} \cos \theta & -\rho \sin \theta & 0 \\ \sin \theta & \rho \cos \theta & 0 \\ 0 & 0 & 1 \end{vmatrix} = \rho.$

(11) Spherical:

$$\frac{\partial(x, y, z)}{\partial(r, \phi, \theta)} = \begin{vmatrix} \sin\phi\cos\theta & r\cos\phi\cos\theta & -r\sin\phi\sin\theta \\ \sin\phi\sin\theta & r\cos\phi\sin\theta & r\sin\phi\cos\theta \\ \cos\phi & -r\sin\phi & 0 \end{vmatrix} = r^2 \sin\phi.$$

Example 3. Write Laplace's equation $\dfrac{\partial^2 u}{\partial x^2} + \dfrac{\partial^2 u}{\partial y^2} = 0$ (cf. Exs. 19–22, § 904) in polar coordinates.

Solution. Since $x = \rho\cos\theta$ and $y = \rho\sin\theta$,

(12) $\begin{cases} \dfrac{\partial u}{\partial \rho} = \dfrac{\partial u}{\partial x}\cos\theta + \dfrac{\partial u}{\partial y}\sin\theta, \\ \dfrac{\partial u}{\partial \theta} = -\dfrac{\partial u}{\partial x}\rho\sin\theta + \dfrac{\partial u}{\partial y}\rho\cos\theta. \end{cases}$

Solving (12) for $\dfrac{\partial u}{\partial x}$ and $\dfrac{\partial u}{\partial y}$, we have

(13) $\begin{cases} \dfrac{\partial u}{\partial x} = \dfrac{\partial u}{\partial \rho}\cos\theta - \dfrac{\partial u}{\partial \theta}\dfrac{\sin\theta}{\rho}, \\ \dfrac{\partial u}{\partial y} = \dfrac{\partial u}{\partial \rho}\sin\theta + \dfrac{\partial u}{\partial \theta}\dfrac{\cos\theta}{\rho}. \end{cases}$

Putting each equation of (13) to double duty we get

$$\frac{\partial^2 u}{\partial x^2} = \frac{\partial}{\partial x}\left[\frac{\partial u}{\partial x}\right] = \frac{\partial}{\partial \rho}\left[\frac{\partial u}{\partial \rho}\cos\theta - \frac{\partial u}{\partial \theta}\frac{\sin\theta}{\rho}\right]\cos\theta - \frac{\partial}{\partial \theta}\left[\frac{\partial u}{\partial \rho}\cos\theta - \frac{\partial u}{\partial \theta}\frac{\sin\theta}{\rho}\right]\frac{\sin\theta}{\rho}$$

$$= \left[\frac{\partial^2 u}{\partial \rho^2}\cos\theta + \frac{\partial u}{\partial \theta}\frac{\sin\theta}{\rho^2} - \frac{\partial^2 u}{\partial \rho\,\partial\theta}\frac{\sin\theta}{\rho}\right]\cos\theta$$

$$- \left[-\frac{\partial u}{\partial \rho}\sin\theta + \frac{\partial^2 u}{\partial \theta\,\partial\rho}\cos\theta - \frac{\partial u}{\partial \theta}\frac{\cos\theta}{\rho} - \frac{\partial^2 u}{\partial \theta^2}\frac{\sin\theta}{\rho}\right]\frac{\sin\theta}{\rho}.$$

Similarly,

$$\frac{\partial^2 u}{\partial y^2} = \left[\frac{\partial^2 u}{\partial \rho^2}\sin\theta - \frac{\partial u}{\partial \theta}\frac{\cos\theta}{\rho^2} + \frac{\partial^2 u}{\partial \rho\,\partial\theta}\frac{\cos\theta}{\rho}\right]\sin\theta$$

$$+ \left[\frac{\partial u}{\partial \rho}\cos\theta + \frac{\partial^2 u}{\partial \theta\,\partial\rho}\sin\theta - \frac{\partial u}{\partial \theta}\frac{\sin\theta}{\rho} + \frac{\partial^2 u}{\partial \theta^2}\frac{\cos\theta}{\rho}\right]\frac{\cos\theta}{\rho}.$$

Addition of these expressions gives

(14) $\quad \dfrac{\partial^2 u}{\partial x^2} + \dfrac{\partial^2 u}{\partial y^2} = \dfrac{\partial^2 u}{\partial \rho^2} + \dfrac{1}{\rho^2}\dfrac{\partial^2 u}{\partial \theta^2} + \dfrac{1}{\rho}\dfrac{\partial u}{\partial \rho} = 0.$

Alternatively, Laplace's equation can be written

(15) $\quad \rho\dfrac{\partial}{\partial \rho}\left(\rho\dfrac{\partial u}{\partial \rho}\right) + \dfrac{\partial^2 u}{\partial \theta^2} = 0.$

923. FUNCTIONAL DEPENDENCE

Definition. *If* $u = u(x, y, \cdots)$, $v = v(x, y, \cdots)$, $\cdots$ *are m differentiable functions of the n variables* $x, y, \cdots$, *and if these m functions satisfy, identically*

in the variables $x, y, \cdots$ in some region R, an equation of the form
(1) $$F(u, v, \cdots) = 0,$$
where F is differentiable for the appropriate values of the m variables $u, v, \cdots$ and where not all of its m first partial derivatives vanish simultaneously, then the functions $u, v, \cdots$ are said to be **functionally dependent** in R.

Our principal objective in this section will be to establish an important *necessary* condition for functional dependence. In considering such a *consequence* of functional dependence we shall concern ourselves only with the case $m \leq n$ because, under ordinary circumstances, if $m > n$ functional dependence normally holds. For example, the two functions of one variable $u = e^x$ and $v = \sin x$ are related by the equation $\sin \ln u - v = 0$. Similarly, if $m = 3$ and $n = 2$ we can ordinarily solve the first two of the equations $u = f(x, y)$, $v = g(x, y)$, $w = h(x, y)$ for x and y in terms of u and v and by substituting the results in the third, obtain a relation between u, v, and w.

Theorem. *If m functions of n variables, where $m \leq n$, are functionally dependent in a region R, then every mth-order Jacobian of these m functions with respect to m of the n variables vanishes identically in R.*

Proof. For definiteness and simplicity we give the details for the case of *two* functions $u = u(x, y, \cdots)$ and $v = v(x, y, \cdots)$ of the n variables $x, y, \cdots$, where $n \geq 2$. The method is general (cf. Ex. 16, § 924). We select an arbitrary pair of independent variables, r and s, and establish the identical vanishing of the Jacobian $\partial(u, v)/\partial(r, s)$. To achieve this we use the chain rule for the function $F(u, v)$ which by hypothesis vanishes identically in the variables $x, y, \cdots$ and whose two first partial derivatives do not vanish simultaneously. The chain rule, then, for the variables r and s, gives the two equations:

(2) $$\begin{cases} \dfrac{\partial F}{\partial u} \dfrac{\partial u}{\partial r} + \dfrac{\partial F}{\partial v} \dfrac{\partial v}{\partial r} = 0, \\[1em] \dfrac{\partial F}{\partial u} \dfrac{\partial u}{\partial s} + \dfrac{\partial F}{\partial v} \dfrac{\partial v}{\partial s} = 0. \end{cases}$$

If the Jacobian $\partial(u, v)/\partial(r, s)$ were different from zero, at some point of R, Cramer's Rule would state that the only values of $\partial F/\partial u$ and $\partial F/\partial v$ that satisfy system (2) are zero. This contradicts the assumption that $\partial F/\partial u$ and $\partial F/\partial v$ never vanish simultaneously. Therefore the Jacobian $\partial(u, v)/\partial(r, s)$ must vanish identically.

The theorem just established provides a means of establishing the functional *independence* of a set of functions: if the Jacobians specified in the theorem are not all identically zero, then the functions are *not* functionally related. However, the substantial and difficult theorem is the converse of this one. In § 934 such a converse will be established. We should observe, however, that Theorem II, § 912, is a simple special case of the converse: If there is just *one* function u, the Jacobians of our theorem reduce to the first partial derivatives of that function. If these all vanish, the function is dependent (by itself) since it is a constant and satisfies an equation of the form $F(u) \equiv u - c = 0$.

NOTE. Functional dependence should not be confused with *linear dependence* (which means that one of the functions is a linear combination $c_1 f_1 + c_2 f_2 + \cdots$ of the others). For example, $\sin x$ and $\cos x$ are functionally dependent ($\sin^2 x + \cos^2 x = 1$) and linearly independent. (Cf. §§ 704, 1602; also *RV*, § 1533.)

924. EXERCISES

In these Exercises assume differentiability wherever appropriate.

In Exercises 1–2, if the variables u and v are defined implicitly as functions of the remaining variables, find their first (partial) derivatives.

1. $x^2 + 2uv = 1$, $x^3 - u^3 + v^3 = 1$.
2. $u + e^v = x + y$, $e^u + v = x - y$.

In Exercises 3–4, if the variables u, v, and w are defined implicitly as functions of the remaining variables, find their first (partial) derivatives.

3. $u + v + w = x$, $uv + uw + vw = x^2$, $uvw = 1$.
4. $u + \sin v = x$, $v + \sin w = y$, $w + \sin u = 1$.

In Exercises 5–6, find the indicated first partial derivatives, where the notation is that of the Notational Pitfall, § 921.

5. If $x + y + z + u = 1$, $x^2 + y^2 + z^2 + u^2 = 1$, find $\left(\dfrac{\partial u}{\partial x}\right)_y$, $\left(\dfrac{\partial u}{\partial x}\right)_z$, $\left(\dfrac{\partial x}{\partial u}\right)_y$, $\left(\dfrac{\partial x}{\partial u}\right)_z$.

6. If $x + y + z + u + v = 1$, $x^2 + y^2 + z^2 + u^2 + v^2 = 1$, find $\left(\dfrac{\partial u}{\partial x}\right)_{yz}$, $\left(\dfrac{\partial u}{\partial x}\right)_{yv}$, $\left(\dfrac{\partial u}{\partial x}\right)_{zv}$.

In Exercises 7–9, verify that the Jacobians of the direct and inverse transformations are reciprocals, for the designated Exercise of § 618.

7. Exercise 1.   **8.** Exercise 2.   **9.** Exercise 5.

10. Find the Jacobian of the transformation of Exercise 4, § 618. Explain why, if the transformation is equal to its own inverse, its Jacobian is not equal to its own reciprocal and therefore have the value ± 1.

In Exercises 11–12, show that the given sets of functions are not functionally dependent.

11. $u = \sin x + \sin y$, $v = \sin(x + y)$.
12. $u = e^x + \ln y + xyz$, $v = \ln x + e^y + xyz$.

13. Prove the statement in the Notational Pitfall, § 921, about the reciprocal nature of $\partial u/\partial x$ and $\partial x/\partial u$.

14. Prove the Theorem of § 921.

15. State and prove the chain rule for transformations in n-dimensional spaces (§ 922).

16. Prove the Theorem, § 923, for arbitrary m and n.

*17. Let a space curve be given as the intersection of two surfaces $f(x, y, z) = g(x, y, z) = 0$. Show that the three Jacobians

(1) $$\frac{\partial(f,g)}{\partial(y, z)}, \quad \frac{\partial(f,g)}{\partial(z, x)}, \quad \frac{\partial(f,g)}{\partial(x, y)}$$

(assuming they are not all zero) are a set of direction numbers for the tangent to the curve. *Hint:* The tangent line is "tangent" to both surfaces.

*18. Let a surface $f(x, y, z) = 0$ have its coordinates prescribed parametrically: $x(u, v), y(u, v), z(u, v)$. Show that the three Jacobians

(2) $$\frac{\partial(y, z)}{\partial(u, v)}, \quad \frac{\partial(z, x)}{\partial(u, v)}, \quad \frac{\partial(x, y)}{\partial(u, v)}$$

(assuming they are not all zero) are a set of direction numbers for the normal to the surface. *Hint:* As an equation in u and v, $f(x, y, z) = 0$ is an identity. Show that the Jacobians (2) are proportional to f_1, f_2, f_3.

*19. If $u = f(x, y, z)$, and if z is defined as a function of x and y by the equation $g(x, y, z) = 0$, show that u as a function of x and y has first partials

$$\frac{\partial u}{\partial x} = \frac{\frac{\partial(f,g)}{\partial(x, z)}}{\frac{\partial g}{\partial z}}, \quad \frac{\partial u}{\partial y} = \frac{\frac{\partial(f,g)}{\partial(y, z)}}{\frac{\partial g}{\partial z}}.$$

*20. If $w = f(x, y, z, u, v)$, and if u and v are defined as functions of x, y, z by the equations $g(x, y, z, u, v) = h(x, y, z, u, v) = 0$, show that w as a function of x, y, z has first partials

$$\frac{\partial w}{\partial x} = \frac{\frac{\partial(f,g,h)}{\partial(x, u, v)}}{\frac{\partial(g, h)}{\partial(u, v)}}, \quad \frac{\partial w}{\partial y} = \frac{\frac{\partial(f,g,h)}{\partial(y, u, v)}}{\frac{\partial(g, h)}{\partial(u, v)}}, \quad \frac{\partial w}{\partial z} = \frac{\frac{\partial(f,g,h)}{\partial(z, u, v)}}{\frac{\partial(g, h)}{\partial(u, v)}}.$$

*21. Let u and v be functions of a single variable x which is a function of two variables r and s. Show that the Jacobian of u and v with respect to r and s vanishes identically.

*22. Let u and v be functions of three variables x, y, z, each of which is a function of the two variables r and s. Establish the following chain rule:

$$\frac{\partial(u, v)}{\partial(r, s)} = \frac{\partial(u, v)}{\partial(y, z)}\frac{\partial(y, z)}{\partial(r, s)} + \frac{\partial(u, v)}{\partial(z, x)}\frac{\partial(z, x)}{\partial(r, s)} + \frac{\partial(u, v)}{\partial(x, y)}\frac{\partial(x, y)}{\partial(r, s)}.$$

Hint: The complete expansion of the left-hand side consists of 18 terms, while that on the right contains 24 terms. However, 6 of these cancel out and the rest match up.

*24. If $u = f(x, y)$, and if x and y are defined as functions of y and z and of x and z, respectively, by the equation $g(x, y, z) = 0$, show that (cf. the Notational Pitfall, § 921):

$$\left(\frac{\partial u}{\partial z}\right)_x - \left(\frac{\partial u}{\partial z}\right)_y = \left(\frac{\partial u}{\partial y}\right)_z \left(\frac{\partial y}{\partial z}\right)_x = -\left(\frac{\partial u}{\partial x}\right)_z \left(\frac{\partial x}{\partial z}\right)_y.$$

*24. If $x = \rho \cos \theta$, $y = \rho \sin \theta$, show that

$$\left(\frac{\partial u}{\partial x}\right)^2 + \left(\frac{\partial u}{\partial y}\right)^2 = \left(\frac{\partial u}{\partial \rho}\right)^2 + \frac{1}{\rho^2}\left(\frac{\partial u}{\partial \theta}\right)^2.$$

*25. Show that the Cauchy-Riemann differential equations

$$\frac{\partial u}{\partial x} = \frac{\partial v}{\partial y}, \quad \frac{\partial u}{\partial y} = -\frac{\partial v}{\partial x},$$

(cf. Ex. 23, § 904) can be expressed as follows in polar coordinates:

$$\frac{\partial u}{\partial \rho} = \frac{1}{\rho}\frac{\partial v}{\partial \theta}, \quad \frac{\partial v}{\partial \rho} = -\frac{1}{\rho}\frac{\partial u}{\partial \theta}.$$

*26. Show that Laplace's equation in three rectangular coordinates,

$$\frac{\partial^2 u}{\partial x^2} + \frac{\partial^2 u}{\partial y^2} + \frac{\partial^2 u}{\partial z^2} = 0,$$

can be written as follows in cylindrical coordinates:

$$\frac{\partial^2 u}{\partial \rho^2} + \frac{1}{\rho^2}\frac{\partial^2 u}{\partial \theta^2} + \frac{1}{\rho}\frac{\partial u}{\partial \rho} + \frac{\partial^2 u}{\partial z^2} = 0.$$

(Cf. Example 1, § 1718).

*27. Show that Laplace's equation in three rectangular coordinates,

$$\frac{\partial^2 u}{\partial x^2} + \frac{\partial^2 u}{\partial y^2} + \frac{\partial^2 u}{\partial z^2} = 0,$$

can be written as follows in spherical coordinates:

$$\frac{1}{r^2 \sin^2 \phi}\left[\sin^2 \phi \frac{\partial}{\partial r}\left(r^2 \frac{\partial u}{\partial r}\right) + \sin \phi \frac{\partial}{\partial \phi}\left(\sin \phi \frac{\partial u}{\partial \phi}\right) + \frac{\partial^2 u}{\partial \theta^2}\right] = 0.$$

Hints: By solving the three equations resulting from differentiating $x = r \sin \phi \cos \theta$, $y = r \sin \phi \sin \theta$, $z = r \cos \phi$, obtain

(3) $\begin{cases} \dfrac{\partial u}{\partial x} = \dfrac{\partial u}{\partial r}\sin\phi\cos\theta + \dfrac{\partial u}{\partial \phi}\dfrac{\cos\phi\cos\theta}{r} - \dfrac{\partial u}{\partial \theta}\dfrac{\sin\theta}{r\sin\phi}, \\ \dfrac{\partial u}{\partial y} = \dfrac{\partial u}{\partial r}\sin\phi\sin\theta + \dfrac{\partial u}{\partial \phi}\dfrac{\cos\phi\sin\theta}{r} + \dfrac{\partial u}{\partial \theta}\dfrac{\cos\theta}{r\sin\phi}, \\ \dfrac{\partial u}{\partial z} = \dfrac{\partial u}{\partial r}\cos\phi - \dfrac{\partial u}{\partial \phi}\dfrac{\sin\phi}{r}. \end{cases}$

Then, following the model of Example 3, § 922, put each equation of (3) to double service, and add. This is rather tedious, but it is finite—41 terms to combine. (Cf. Example 2, § 1718.)

925. EXTREMA WITH ONE CONSTRAINT. TWO VARIABLES

In a first course in calculus a student meets and learns to solve problems of the following types:

Example 1. Find the rectangle of maximum area that can be inscribed in the ellipse

(1) $$\frac{x^2}{a^2} + \frac{y^2}{b^2} = 1,$$

assuming that the sides of the rectangle are parallel to the coordinate axes.

Example 2. A playpen is to be constructed alongside a house in the shape of a rectangle, the wall of the house serving as one side, a piece of fencing forming the other three sides. (*a*) What is the maximum area for a given length of fencing? (*b*) What are the most economical dimensions for a given area?

Example 3. What are the relative dimensions of a right circular cylinder if (*a*) the volume is maximized for a given total surface area; (*b*) the total surface area is minimized for a given volume?

Example 4. (*a*) How large a right circular cylinder can be fitted inside a right circular cone? (*b*) How small a right circular cone can be fitted around a right circular cylinder? In both cases assume that the axes of symmetry coincide.

Each of these problems involves two variables and two functions of these variables. One of these functions is to be maximized or minimized, and the other is to be held fixed. The restriction imposed by this latter condition takes the form of an equation, and is called a **constraint** on the variables. The total problem is called an **extremal problem with constraint.**

Our first objective in this section will be to examine the first of the preceding Examples in several of the many ways in which one may attack an extremal problem with constraint. Some of these methods are appropriate for only certain types of problems. Others are more generally applicable. After obtaining some rather general techniques, we shall apply these to the remaining three Examples, given above, by way of illustration. Another method, that of *Lagrange multipliers*, is treated in § 928.

First Solution, Example 1. Explicit solving. Let (x, y) be a point in the closed first quadrant and on the ellipse (1), and form the rectangle R with vertices $(x, \pm y)$, $(-x, \pm y)$. Since the area of R is $4xy$, the problem is to maximize the function xy subject to the constraint (1), in the closed first quadrant. We know at the outset that the maximum sought does exist, since xy is continuous on a compact set, and this maximum is attained at some point in the open quadrant. By solving for y in (1) $\left(y = \dfrac{b}{a} \sqrt{a^2 - x^2} \right)$, substituting in the expression xy, and discarding the unnecessary positive constant b/a, we can reformulate our problem as that of maximizing the function

(2) $$x \sqrt{a^2 - x^2},$$

for $0 < x < a$. Routine differentiation of (2) reveals that between 0 and a there is one and only one critical value of x: $x = \frac{1}{2}\sqrt{2}\, a$. Without any further test, then, we know that the desired maximum is provided by $x = \frac{1}{2}\sqrt{2}\, a$. Further substitution gives $y = \frac{1}{2}\sqrt{2}\, b$, and the maximum possible area of R is $2ab$. Notice that this is just half the area of the circumscribed rectangle with sides $x = \pm a$, $y = \pm b$. Instead of solving for y in terms of x we could, of course, have solved for x in terms of y.

Second Solution, Example 1. Implicit solving. Instead of solving explicitly for y in terms of x (say) we may consider y defined implicitly by (1) as a function of x. Implicit differentiation gives:

$$\frac{x}{a^2} + \frac{y}{b^2} y' = 0, \quad y' = -\frac{b^2 x}{a^2 y}.$$

Equating to 0 the derivative with respect to x of the function xy as a function of x, under the assumption that an extremum is under consideration, we get:

$$xy' + y = \frac{a^2y^2 - b^2x^2}{a^2y} = 0.$$

Therefore the two terms on the left of (1) are equal to each other and thus to $\frac{1}{2}$, and $x = \frac{1}{2}\sqrt{2}\,a$, $y = \frac{1}{2}\sqrt{2}\,b$, as before.

Third Solution, Example 1. *New parameter.* Let the ellipse (1) be parametrized: $x = a\cos\theta$, $y = b\sin\theta$. The problem, then, is to maximize $ab\cos\theta\sin\theta$ or, equivalently, $\sin 2\theta$, for $0 < \theta < \frac{1}{2}\pi$. The maximum is attained at $\theta = \frac{1}{4}\pi$.

Fourth Solution, Example 1. *Undesignated parameter.* Let x and y both be differentiable functions of an unnamed parameter, and satisfy (1). If primes indicate differentiation with respect to this new parameter, we have simultaneously:

(3) $$\begin{cases} yx' + xy' = 0, \\ \dfrac{x}{a^2}x' + \dfrac{y}{b^2}y' = 0, \end{cases}$$

for reasons outlined above. Assuming that x' and y' do not both vanish at the point (x, y) where xy is maximum (that is, this is a regular point of the ellipse), we can eliminate x' and y'. The previously found solution is again obtained.

Fifth Solution, Example 1. *Determinant.* Elimination of x' and y' from (3) is simplest in terms of the vanishing of their coefficient determinant:

(4) $$\begin{vmatrix} y & x \\ \dfrac{x}{a^2} & \dfrac{y}{b^2} \end{vmatrix} = 0.$$

Sixth Solution, Example 1. *Differentials.* This is the same as the Fourth Solution, except that dx and dy are used in place of x' and y', respectively.

Seventh Solution, Example 1. *Simplifying substitutions.* If we let $u \equiv x/a$ and $v \equiv v/b$, we have the relation $u^2 + v^2 = 1$, and wish to maximize the function $abuv$ or, equivalently, uv. The details of the work have hereby been simplified. A further simplification can be achieved by squaring uv. With the notation $s = u^2$ and $t = v^2$, we seek a maximum for st for positive s and t such that $s + t = 1$. This is the classical problem of maximizing the area of a rectangle with given perimeter. Since the solution is a square, $s = t$, and the solution found above is repeated.

Eighth Solution, Example 1. *Interchange roles.* It is especially clear in the fourth and fifth solutions that if, instead of finding an extremum for xy subject to the constraint that $(x^2/a^2) + (y^2/b^2)$ is a constant, we seek an extreme value for $(x^2/a^2) + (y^2/b^2)$ subject to the constraint that xy is a constant, the resultant relation (4) between x and y is the same. If this technique is employed with the notation of the substitution $u = x/a$, $v = y/b$ of the Seventh Solution, we find that we are looking for the point of the first-quadrant portion of the hyperbola $uv = 1$ that is nearest the origin. This is clearly on the quadrant-bisecting line $u = v$. Therefore $x/a = y/b$, as before.

The special form of equations (3) and (4) suggests a single coordinated attack on the general problem:

EXTREMA WITH ONE CONSTRAINT

Theorem. *Let $f(x, y)$ and $g(x, y)$ be differentiable in a neighborhood $N_{(a,b)}$ of the point (a, b). Assume that one of these functions has a relative extreme value at (a, b) subject to a constraint on the variables obtained by equating the other function to a constant in the neighborhood $N_{(a,b)}$. Finally, assume that this equation of constraint defines one of the two variables x or y as a differentiable function of the other in a neighborhood of the corresponding number b or a. In short:*

(5) $\quad \begin{cases} f(x, y) \text{ is extreme,} \\ g(x, y) \text{ is constant,} \end{cases}$ or $\begin{cases} f(x, y) \text{ is constant,} \\ g(x, y) \text{ is extreme,} \end{cases}$

at (a, b). Then the Jacobian of f and g with respect to x and y must vanish at (a, b):

(6) $\quad \left[\dfrac{\partial(f, g)}{\partial(x, y)}\right]_{\substack{x=a \\ y=b}} = \begin{vmatrix} f_1(a, b) & f_2(a, b) \\ g_1(a, b) & g_2(a, b) \end{vmatrix} = 0.$

Proof. Assume for definiteness that $f(x, y)$ has an extreme value subject to the constraint

(7) $\quad\quad\quad\quad\quad g(x, y) = k,$

which defines y as a differentiable function $\phi(x)$ of x near $x = a$. Then, by implicit differentiation, since (7) must hold identically:

(8) $\quad\quad\quad\quad g_1(a, b) + g_2(a, b)\phi'(a) = 0.$

Furthermore, since $f(x, y)$, as a function of x, has a relative extremum at $x = a$, its derivative with respect to x must vanish there:

(9) $\quad\quad\quad\quad f_1(a, b) + f_2(a, b)\phi'(a) = 0.$

Multiplication of (8) and (9) by $f_2(a, b)$ and $g_2(a, b)$, respectively, and subsequent subtraction, give (6).

NOTE 1. If $f_1(a, b)$ and $f_2(a, b)$ are not both zero, and if $g_1(a, b)$ and $g_2(a, b)$ are not both zero, then conclusion (6) can be expressed by the statement that the two curves $f(x, y) = f(a, b)$ and $g(x, y) = g(a, b)$ are tangent to each other at (a, b).

By the preceding Theorem, the two parts of each of Examples 2 and 3 lead to identical relations between the variables of the form (9). We shall carry the solutions of these two Examples only this far.

Solution, Example 2. Let the dimensions of the rectangle be x parallel to the house wall and y perpendicular to the wall. Letting $f(x, y) = xy$ and $g(x, y) = x + 2y$, we have from (6):

$$\begin{vmatrix} y & x \\ 1 & 2 \end{vmatrix} = 0, \quad \text{or} \quad x = 2y.$$

Solution, Example 3. If r and h denote the base radius and altitude, respectively, and if $f(r, h) = \pi r^2 h$ and $g(r, h) = 2\pi r^2 + 2\pi r h$, (6) gives

$$2\pi^2 r \begin{vmatrix} 2h & r \\ 2r + h & r \end{vmatrix} = 0, \quad \text{or} \quad h = 2r.$$

Solution, Example 4. (a) If R and H denote the base radius and altitude, respectively, of the given cone, and if r and h denote the base radius and altitude, respectively, of the inscribed cylinder, we seek a maximum value of $f(r, h) = \pi r^2 h$ subject to the constraint
$$g(r, h) \equiv \frac{r}{R} + \frac{h}{H} = 1.$$
Equation (6) is
$$\pi r \begin{vmatrix} 2h & r \\ \dfrac{1}{R} & \dfrac{1}{H} \end{vmatrix} = 0.$$
Therefore $\dfrac{r}{R} = \dfrac{2h}{H} = \dfrac{2}{3}$, $r = \dfrac{2}{3} R$, $h = \dfrac{1}{3} H$. (b) In this case the equation of constraint is the same, except that r and h are constants and R and H are the variables. The function $F(R, H) \equiv \pi R^2 H$ is to be minimized. Equation (6) is
$$\pi R \begin{vmatrix} 2H & R \\ -\dfrac{r}{R^2} & -\dfrac{h}{H^2} \end{vmatrix} = 0,$$
with the same solution as before: $R = \dfrac{3}{2} r$, $H = 3h$.

NOTE 2. In the First Solution, Example 1, we indicated how it is possible to dispense with a formal test for maxima or minima of the type presented in §§ 309 and 914. It should be appreciated, however, that some argument is logically necessary, and that equation (6) is only a *necessary* and not a *sufficient* condition for an extremum. As an illustration we present one more example.

Example 5. Find the point of the parabola
(10) $$2x^2 + 2y = 3$$
that is nearest the origin.

Solution. We seek a minimum for the function $x^2 + y^2$ (the *square* of the distance to be minimized) subject to the constraint (10). Equation (6) reduces to
(11) $$x(2y - 1) = 0.$$
The simultaneous solutions of (10) and (11) are $x = 0$, $y = \frac{3}{2}$, and $x = \pm 1$, $y = \frac{1}{2}$, with corresponding distances $\sqrt{x^2 + y^2}$ of $\frac{3}{2}$ and $\frac{1}{2}\sqrt{5}$, respectively. There are thus *two* nearest points $(\pm 1, \frac{1}{2})$, at a distance from the origin of $\frac{1}{2}\sqrt{5}$. The significance of the point $(0, \frac{3}{2})$ (which is at a greater distance of $\frac{3}{2}$) becomes evident if one expresses the square of the distance of (x, y) from the origin in terms of x as a parameter:
$$x^2 + y^2 = x^2 + (\tfrac{3}{2} - x^2)^2 = x^4 - 2x^2 + \tfrac{9}{4}.$$

926. EXTREMA WITH ONE CONSTRAINT. MORE THAN TWO VARIABLES

The analogue of the Theorem of § 925, in case there are three variables instead of two, can be phrased as follows:

Theorem. *Let $f(x, y, z)$ and $g(x, y, z)$ be differentiable in a neighborhood $N_{(a,b,c)}$ of the point (a, b, c). Assume that one of these functions has a relative* extreme value at (a, b, c) subject to a constraint on the variables obtained by equating the other function to a constant in the neighborhood $N_{(a,b,c)}$. Finally, assume that this equation of constraint defines one of the three variables as a differentiable function of the other two in a neighborhood of the corresponding point (b, c), (c, a), or (a, b). In short:

(1) $\begin{cases} f(x, y, z) \text{ is extreme,} \\ g(x, y, z) \text{ is constant,} \end{cases}$ or $\begin{cases} f(x, y, z) \text{ is constant,} \\ g(x, y, z) \text{ is extreme,} \end{cases}$

at (a, b, c). Then the gradients (normals) of f and g at the point (a, b, c) are collinear:

(2) $f_1(a,b,c) : f_2(a,b,c) : f_3(a,b,c) = g_1(a,b,c) : g_2(a,b,c) : g_3(a,b,c)$.

Equivalently, the three Jacobians $\partial(f,g)/\partial(y,z)$, $\partial(f,g)/\partial(z,x)$, and $\partial(f,g)/\partial(x,y)$ all vanish at (a, b, c):

(3) $\begin{vmatrix} f_2 & f_3 \\ g_2 & g_3 \end{vmatrix} = \begin{vmatrix} f_3 & f_1 \\ g_3 & g_1 \end{vmatrix} = \begin{vmatrix} f_1 & f_2 \\ g_1 & g_2 \end{vmatrix} = 0.$

Proof. Assume for definiteness that $f(x, y, z)$ has an extreme value subject to the constraint

(4) $g(x, y, z) = k,$

which defines z as a differentiable function $\phi(x, y)$ of x and y near (a, b). Then, by implicit differentiation:

(5) $\begin{cases} g_1(a, b, c) + g_3(a, b, c)\phi_1(a, b) = 0, \\ g_2(a, b, c) + g_3(a, b, c)\phi_2(a, b) = 0. \end{cases}$

Furthermore, since $f(x, y, z)$, as a function of x and y, has a relative extremum at (a, b), its first partial derivatives must vanish there:

(6) $\begin{cases} f_1(a, b, c) + f_3(a, b, c)\phi_1(a, b) = 0, \\ f_2(a, b, c) + f_3(a, b, c)\phi_2(a, b) = 0. \end{cases}$

Equations (5) and (6) together imply (3) and, by § 707, (2) as well. (For example, multiply through the first equation of (5) by $f_3(a, b, c)$, and the first equation of (6) by $g_3(a, b, c)$, and subtract the results.)

NOTE. If the gradients of f and g at (a, b, c) are both nonzero vectors, that is, if the point (a, b, c) is a regular point for the two surfaces $f(x, y, z) = f(a, b, c)$ and $g(x, y, z) = g(a, b, c)$, conclusion (2) can be expressed by the statement that these two surfaces are tangent to each other at (a, b, c).

The case of two functions of more than three variables, $f(x, y, z, \cdots)$ and $g(x, y, z, \cdots)$, one function extreme and one function constant, is entirely similar to the preceding case (cf. Ex. 12, § 929). The analogue of (3) is the vanishing of all possible 2 by 2 Jacobian determinants formed by taking arbitrary pairs of columns of the **Jacobian matrix**:

(7) $\begin{pmatrix} f_1 & f_2 & f_3 & \cdots \\ g_1 & g_2 & g_3 & \cdots \end{pmatrix}.$

Whenever a two-rowed matrix has this property it is said to be of **rank less than 2**.†

Example 1. A plane Π passing through the point $p_1: (x_1, y_1, z_1)$ in the first octant has three positive intercepts on the coordinate axes, and therefore cuts off a tetrahedron from the first octant. Find the minimum volume of this tetrahedron.

Solution. From the intercept form of the plane Π, (7), § 709, where r, s, and t are its x-, y-, and z-intercepts, respectively, we have

$$(8) \qquad \frac{x_1}{r} + \frac{y_1}{s} + \frac{z_1}{t} = 1.$$

Letting rst be the function to be minimized, we have the statement of proportionality, from (2):

$$st : tr : rs = -\frac{x_1}{r^2} : -\frac{y_1}{s^2} : -\frac{z_1}{t^2}.$$

We infer from this that the three terms on the left of (8) are all equal to each other, and consequently to $\frac{1}{3}$. The volume of the smallest tetrahedron is therefore

$$\tfrac{1}{6}(3x_1)(3y_1)(3z_1) = \tfrac{9}{2}x_1 y_1 z_1.$$

Example 2. Maximize the function $e^x y z^2$, if x, y, and z are nonnegative numbers whose sum is (a) 100; (b) 1.

Solution. If $f(x, y, z) \equiv e^x y z^2$ and $g(x, y, z) \equiv x + y + z$, the collinear vectors (2) are

$$(e^x y z^2, e^x z^2, 2e^x y z) \quad \text{and} \quad (1, 1, 1).$$

Consequently, since it is obvious that the maximum value of f cannot occur when either y or z is zero, we conclude that $y = 1$ and $z = 2$. For part (a) this gives $x = 97$, and the maximum value of f as $4e^{97}$. For part (b), however, it is clear that this method is invalid since it leads to a point outside the domain under consideration, which is the closed triangle in which the plane $x + y + z = 1$ cuts the closed first octant. This means that the maximum value of f, for this triangle, must be attained on one of its three edges. Since neither y nor z can vanish, we conclude that $x = 0$, and proceed to maximize yz^2 subject to the constraint $y + z = 1$. This extremal problem gives $y = \frac{1}{3}$ and $z = \frac{2}{3}$, so that the solution to part (b) of the original problem is $e^0 \tfrac{1}{3}(\tfrac{2}{3})^2 = \tfrac{4}{27}$.

927. EXTREMA WITH MORE THAN ONE CONSTRAINT

The problem of finding an extreme value for a function whose variables are subjected to more than one constraint is similar to those already considered. We state a general theorem for n variables constrained by $m - 1$ equations, and prove it for the case $m = 3$, $n = 4$. Proofs for the special case $m = n = 3$ and for the general theorem are requested in Exercises 13 and 14, § 929.

Theorem. *Let* $\{f^{(i)}(x_1, x_2, \cdots, x_n)\}$, $i = 1, 2, \cdots, m$, *be* m *functions differentiable in a neighborhood* N_p *of the point* $p: (a_1, a_2, \cdots, a_n)$, *where*

† For a discussion of *rank of a matrix*, see any good book on the theory of matrices, such as one of the references cited in the footnote, page 296.

$m \leq n$. Assume that one of these functions has a relative extremum at p subject to the constraints on the variables obtained by equating the remaining $m - 1$ functions to constants in N_p. Finally, assume that these $m - 1$ equations of constraint define $m - 1$ of the n variables $x_1, x_2, \cdots, x_n$ as differentiable functions of the other $n - m + 1$ variables in a neighborhood of a point in E_{n-m+1} whose coordinates are those coordinates of p that are appropriate to the $n - m + 1$ variables that are being considered as independent. Then the m by n Jacobian matrix

(1)
$$\begin{pmatrix} f_1^{(1)} & f_2^{(1)} & f_3^{(1)} & \cdots & f_n^{(1)} \\ f_1^{(2)} & f_2^{(2)} & f_3^{(2)} & \cdots & f_n^{(2)} \\ \cdots \cdots \cdots & & & \cdots \\ f_1^{(m)} & f_2^{(m)} & f_3^{(m)} & \cdots & f_n^{(m)} \end{pmatrix}$$

must have rank less than m; that is, every m by m Jacobian determinant made up of any m columns of (1) must vanish.

Proof for $m = 3$, $n = 4$. For simplicity and definiteness we let the three given functions and the four variables be denoted $f(x, y, u, v)$, $g(x, y, u, v)$, and $h(x, y, u, v)$, differentiable in a neighborhood of $p: (a, b, c, d)$, and assume that f is extreme at p subject to equations of constraint obtained by setting g and h equal to constants. Finally, we assume that these two equations define u and v as functions of x and y. Our objective is to establish the vanishing of the four Jacobians, evaluated at p:

(2) $$\frac{\partial(f, g, h)}{\partial(x, y, u)} = \frac{\partial(f, g, h)}{\partial(x, y, v)} = \frac{\partial(f, g, h)}{\partial(x, u, v)} = \frac{\partial(f, g, h)}{\partial(y, u, v)} = 0.$$

For reasons amplified in recent discussion:

(3) $$\begin{cases} f_1 + f_3 u_1 + f_4 v_1 = 0, \\ g_1 + g_3 u_1 + g_4 v_1 = 0, \\ h_1 + h_3 u_1 + h_4 v_1 = 0, \end{cases}$$

(4) $$\begin{cases} f_2 + f_3 u_2 + f_4 v_2 = 0, \\ g_2 + g_3 u_2 + g_4 v_2 = 0, \\ h_2 + h_3 u_2 + h_4 v_2 = 0, \end{cases}$$

all quantities evaluated at p. Equations (3) and (4) imply immediately the vanishing of the final two Jacobians of (2). (For example, equations (3) can be regarded as a system of three homogeneous linear equations satisfied by the three "variables" 1, u_1, and v_1, and since these three quantities are not all zero their coefficient determinant must vanish.) As representative of the remaining two, the first Jacobian of (2) can be shown to vanish as follows: In the first place, if $v_2 = 0$, equations (4) show that the second column of $\partial(f, g, h)/\partial(x, y, u)$ is a multiple of the third, and the Jacobian vanishes. On the other hand, if $v_2 \neq 0$, equations (4) can all be multiplied through by $-v_1/v_2$, the results being added to the corresponding equations of (3). The three equations thus produced imply the vanishing of the first Jacobian of (2), as desired.

Example 1. Find the highest and lowest points of the ellipse formed by the intersection of the plane $x + y + z = 1$ and the ellipsoid $16x^2 + 4y^2 + z^2 = 16$.

Solution. The problem is to find the extreme values of z subject to the two given equations of constraint. (These extreme values exist since the ellipse is a compact set; cf. Ex. 30, § 605.) The matrix (1) is square, with only one vanishing Jacobian:

$$\begin{vmatrix} 0 & 0 & 1 \\ 1 & 1 & 1 \\ 32x & 8y & 2z \end{vmatrix} = 0,$$

or $y = 4x$. Elimination of y and then x gives $21z^2 - 32z - 64 = 0$. The maximum and minimum values of z are therefore $8/3$ and $-8/7$, respectively.

★Example 2. Find the maximum value of the determinant

(5) $$D = \begin{vmatrix} x_{11} & x_{12} & x_{13} \\ x_{21} & x_{22} & x_{23} \\ x_{31} & x_{32} & x_{33} \end{vmatrix},$$

subject to the three constraints

(6) $$x_{i1}^2 + x_{i2}^2 + x_{i3}^3 = a_i^2, \quad \text{where} \quad a_i > 0, i = 1, 2, 3.$$

Solution. The matrix (1) becomes the 4 by 9 matrix

(7) $$\begin{pmatrix} X_{11} & X_{12} & X_{13} & X_{21} & X_{22} & X_{23} & X_{31} & X_{32} & X_{33} \\ x_{11} & x_{12} & x_{13} & 0 & 0 & 0 & 0 & 0 & 0 \\ 0 & 0 & 0 & x_{21} & x_{22} & x_{23} & 0 & 0 & 0 \\ 0 & 0 & 0 & 0 & 0 & 0 & x_{31} & x_{32} & x_{33} \end{pmatrix},$$

where X_{ij} is the cofactor of x_{ij} in D, $i, j = 1, 2, 3$, except that factors of 2 have been canceled from the last three rows for simplicity (cf. Ex. 15, § 909). Since the rank of (7) must be less than 4, the theory of simultaneous systems of linear equations† assures us that there exist four constants, k, m_1, m_2, and m_3, not all zero, such that k times the first row of (7) is equal to m_1 times the second row plus m_2 times the third row plus m_3 times the fourth row. Since the maximum value of D is positive, and certainly *nonzero*, the number k, above, cannot equal 0 (for otherwise all *four* numbers k, m_1, m_2, and m_3 would be forced to vanish). We therefore can (and do) assume without loss of generality that $k = 1$. We can conclude from the form of (7):

the row of cofactors from the first row of $D = m_1$ times the first row,
the row of cofactors from the second row of $D = m_2$ times the second row,
the row of cofactors from the third row of $D = m_3$ times the third row.

Expansion of D according to the elements of each of the three rows gives:

$$D = m_1 \sum_{j=1}^{3} x_{1j}^2 = m_1 a_1^2 = m_2 a_2^2 = m_3 a_3^2,$$

† See any of the references cited in the footnote, page 296.

and since D is nonzero, m_1, m_2, and m_3 are all nonzero. Since the cofactors from any row of D form a vector orthogonal to both of the other row-vectors, we can conclude that the three row-vectors of D are mutually orthogonal. Therefore (§ 922),

$$D^2 = \begin{vmatrix} x_{11} & x_{12} & x_{13} \\ x_{21} & x_{22} & x_{23} \\ x_{31} & x_{32} & x_{33} \end{vmatrix} \cdot \begin{vmatrix} x_{11} & x_{21} & x_{31} \\ x_{12} & x_{22} & x_{32} \\ x_{13} & x_{23} & x_{33} \end{vmatrix} = \begin{vmatrix} a_1^2 & 0 & 0 \\ 0 & a_2^2 & 0 \\ 0 & 0 & a_3^2 \end{vmatrix},$$

and $D = a_1 a_2 a_3$.

⋆NOTE. From the result of Example 2 we can infer immediately a useful inequality. This states that *no third order determinant can exceed numerically the product of the magnitude of its row-vectors.* With the notation of (5):

(8) $\quad |D| \leq \sqrt{x_{11}^2 + x_{12}^2 + x_{13}^2} \sqrt{x_{21}^2 + x_{22}^2 + x_{23}^2} \sqrt{x_{31}^2 + x_{32}^2 + x_{33}^2}.$

If M is any upper bound for the absolute values of the elements of D:

$$|x_{ij}| \leq M, \quad i, j = 1, 2, 3,$$

then the inequality (8) simplifies:

(9) $\quad |D| \leq \sqrt{3^3} M^3.$

Both the general result of Example 2 and the inequalities (8) and (9) generalize in a fairly direct and simple manner to determinants of order n. In particular, the extension of (9) to nth-order determinants is an important theorem in the theory of integral equations, due to the French mathematician Hadamard (1865–): *If D is an nth-order determinant all of whose elements x_{ij} satisfy the inequality $|x_{ij}| \leq M, i, j = 1, 2, \cdots, n$, then*

(10) $\quad |D| \leq \sqrt{n^n} M^n$

(Give proofs of the preceding statements in Ex. 15, § 929.)

928. LAGRANGE MULTIPLIERS

The French mathematician Lagrange (1736–1813) gave a useful procedural technique for solving extremal problems with constraints. The idea is to add to the extremal function constant multiples of the constraint functions, and to equate to zero all first partial derivatives of the result. The constant factors by which the constraint functions are multiplied are called **Lagrange multipliers.** If the Lagrange multipliers are eliminated from the equations got by setting all first partial derivatives equal to zero, the results are identical with those obtained in the preceding three sections. In order to guarantee the *existence* of Lagrange multipliers, for any given extremal problem, we must impose an added hypothesis on those previously given. The precise statements are detailed below for the cases of two variables and one constraint (Theorem I) and four variables and two constraints (Theorem II). Two other special cases are requested in Exercise 16, § 929.

Theorem I. *Let $f(x, y)$ and $g(x, y)$ be differentiable in a neighborhood $N_{(a,b)}$ of (a, b). Assume that f has a relative extremum there subject to a constraint on x and y of the form $g(x, y) = k$ that defines y as a differentiable* function $\phi(x)$ of x in a neighborhood of $x = a$. Furthermore, assume that $g_2(a, b) \neq 0$. Then there must exist a number λ (called a **Lagrange multiplier**) such that

(1) $\qquad f_1(a, b) + \lambda g_1(a, b) = f_2(a, b) + \lambda g_2(a, b) = 0.$

Proof. By (6), § 925, λ can be taken to be $-f_2(a, b)/g_2(a, b)$.

Theorem II. *Let $f(x, y, u, v)$, $g(x, y, u, v)$, and $h(x, y, u, v)$ be differentiable in a neighborhood N_p of p: (a, b, c, d). Assume that f has a relative extremum there subject to two constraints on the variables of the form $g(x, y, u, v) = k_1$ and $h(x, y, u, v) = k_2$ that define u and v as differentiable functions of x and y in a neighborhood of (a, b). Furthermore, assume that the Jacobian $\partial(g, h)/\partial(u, v)$ is nonzero at p. Then there must exist two constants λ and μ (called **Lagrange multipliers**) such that at p:*

(2) $\qquad f_1 + \lambda g_1 + \mu h_1 = f_2 + \lambda g_2 + \mu h_2$
$\qquad\qquad = f_3 + \lambda g_3 + \mu h_3 = f_4 + \lambda g_4 + \mu h_4 = 0.$

Proof. Let λ and μ be defined by the last two of equations (2). Then

$$\lambda = -\frac{\begin{vmatrix} f_3 & h_3 \\ f_4 & h_4 \end{vmatrix}}{\begin{vmatrix} g_3 & h_3 \\ g_4 & h_4 \end{vmatrix}}, \quad \mu = -\frac{\begin{vmatrix} g_3 & f_3 \\ g_4 & f_4 \end{vmatrix}}{\begin{vmatrix} g_3 & h_3 \\ g_4 & h_4 \end{vmatrix}},$$

and

$$\begin{vmatrix} g_3 & h_3 \\ g_4 & h_4 \end{vmatrix} (f_1 + \lambda g_1 + \mu h_1) = \begin{vmatrix} f_1 & g_1 & h_1 \\ f_3 & g_3 & h_3 \\ f_4 & g_4 & h_4 \end{vmatrix},$$

$$\begin{vmatrix} g_3 & h_3 \\ g_4 & h_4 \end{vmatrix} (f_2 + \lambda g_2 + \mu h_2) = \begin{vmatrix} f_2 & g_2 & h_2 \\ f_3 & g_3 & h_3 \\ f_4 & g_4 & h_4 \end{vmatrix}.$$

By (2), § 927, the two third-order determinants vanish, and equations (2) are established.

Example 1. Show that the lengths of the semiaxes of the ellipse

(3) $\qquad\qquad ax^2 + 2bxy + cy^2 = 1$

are the reciprocals of the square roots of the roots λ_1 and λ_2 of the "characteristic equation" of the matrix $\begin{pmatrix} a & b \\ b & c \end{pmatrix}$ of the "quadratic form" $ax^2 + 2bxy + cy^2$:

(4) $\qquad\qquad \begin{vmatrix} a - \lambda & b \\ b & c - \lambda \end{vmatrix} = 0.$

Show that the equation of the axis corresponding to $\lambda = \lambda_i$ is $(a - \lambda_i)x + by = 0$ (or $bx + (c - \lambda_i)y = 0$), $i = 1, 2$.

Solution. From the symmetry of (3) with respect to (0, 0), we know that the center of the ellipse is at the origin and that the semiaxes are the extreme values of $\sqrt{x^2 + y^2}$. We maximize and minimize $f(x, y) = x^2 + y^2$ subject to the constraint $g(x, y) = -(ax^2 + 2bxy + cy^2) = -1$. Since the extreme values of f are positive numbers, we can use a reciprocal to denote the Lagrange multiplier, and equate to zero the first partial derivatives of $x^2 + y^2 - \frac{1}{\lambda}(ax^2 + 2bxy + cy^2)$:

(5) $$2\left(x - \frac{ax + by}{\lambda}\right) = 2\left(y - \frac{bx + cy}{\lambda}\right) = 0.$$

If equations (5) are rewritten

(6) $$\begin{cases} (a - \lambda)x + by = 0, \\ bx + (c - \lambda)y = 0, \end{cases}$$

the values of λ are determined as the roots of (4), since equations (6) have nontrivial solutions for x and y if and only if their coefficient determinant vanishes. For each of these values of λ, we have from (3) and (5):

(7) $$1 = x(ax + by) + y(bx + cy) = \lambda(x^2 + y^2),$$

so that the values of λ are the reciprocals of the extreme values of $x^2 + y^2$; that is, the values of λ are the reciprocals of the squares of the semiaxes of (3). The final statement to be established is a direct consequence of equations (6). (Why?) (Cf. Exs. 21–23, § 929.)

Example 2. Maximize the expression

(8) $$\sum_{i=1}^{n} a_i x_i,$$

where $a_1, a_2, \cdots, a_n$ are n given positive constants, if

(9) $$\sum_{i=1}^{n} x_i^2 \leq 1.$$

First Solution. It is clear that the maximum value of (8) exists, since the set in E_n described by (9) is compact, and that this maximum is given by nonnegative values of $x_1, \cdots, x_n$ subject to *equality* in (9). We equate to zero all n first partial derivatives of the expression $\Sigma a_i x_i - \frac{1}{2}\lambda(\Sigma x_i^2)$:

$$a_1 - \lambda x_1 = a_2 - \lambda x_2 = \cdots = a_n - \lambda x_n = 0.$$

Solving for the x's and substituting in the equality (9), we find $x_i = a_i/\lambda$, $i = 1, \cdots, n$, where $\lambda = \sqrt{\Sigma a_i^2}$. Consequently, the maximum value of (8) is $\sqrt{\Sigma a_i^2}$.

★*Second Solution.* By Schwarz's inequality (Ex. 32, § 107):

$$\Sigma a_i x_i \leq \sqrt{\Sigma a_i^2} \sqrt{\Sigma x_i^2} \leq \sqrt{\Sigma a_i^2}.$$

Thus, $\sqrt{\Sigma a_1^2}$ is an upper bound for (8). Substitution of $x_i = a_i/\sqrt{\Sigma a_i^2}$, $i = 1, \cdots, n$, shows that this number is attained as a maximum value.

Example 3. Minimize the expression $\sum_{i=1}^{n} x_i^2$ subject to the constraint $\sum_{i=1}^{n} a_i x_i = 1$, where $a_i > 0$, $i = 1, 2, \cdots, n$.

Solution. As with Example 2, $a_i = \lambda x_i$, $i = 1, \cdots, n$. Substitution in the equation of constraint gives $\lambda = \Sigma a_i^2$. Therefore the minimum value of Σx_i^2 is $1/\Sigma a_i^2$.

929. EXERCISES

1. Find the maximum perimeter of a rectangle that can be inscribed in the ellipse $b^2x^2 + a^2y^2 = a^2b^2$, assuming the sides parallel to the coordinate axes.

2. Find the point of the first quadrant part of the curve $x^2y = 2$ that is nearest the origin.

3. Find the maximum volume for a rectangular parallelepiped that can be inscribed in the ellipsoid $(x^2/a^2) + (y^2/b^2) + (z^2/c^2) = 1$, assuming the edges parallel to the coordinate axes.

4. Maximize xyz, if $xy + 2xz + 3yz = 72$.

5. Show that the product of three positive numbers whose sum is prescribed is a maximum if and only if these numbers are equal.

6. Find the minimum distance between the origin $(0, 0, 0)$ and a point on (a) the hyperbolic paraboloid $2z = x^2 - y^2 + 1$; (b) the elliptic paraboloid $2z = x^2 + y^2 - 10$.

7. Find the extreme values of $x^2 + y^2 + z^2$ if $\frac{x^2}{1^2} + \frac{y^2}{2^2} + \frac{z^2}{3^2} = 1$. Show that equations (3), § 926, give the three results 1, 4, and 9. Show therefore that those equations give only necessary and not sufficient conditions for an extremum. Give a geometrical explanation in terms of tangency to the given ellipsoid of spheres $x^2 + y^2 + z^2 = 1, 4,$ and 9.

8. Find the highest point on the curve of intersection of the sphere $x^2 + y^2 + z^2 = 30$ and the cone $x^2 + 2y^2 - z^2 = 0$. (That is, maximize z.)

9. Find the extreme distances from the origin for points on the curve:
$$3x^2 + 2y^2 + z^2 = 30, \quad z^2 = 2xy.$$

10. If $x, y, z, u, v,$ and w are six positive numbers such that
$$\frac{x}{u} + \frac{y}{v} + \frac{z}{w} = 1,$$
show that (a) $27xyz \leq uvw$; (b) $\sqrt{x} + \sqrt{y} + \sqrt{z} \leq \sqrt{u + v + w}$. *Hints:* For (a) maximize xyz; for (b) minimize $u + v + w$.

11. Three numbers $x, y,$ and z such that $x \geq 2, y \geq 2,$ and $z \geq 2$ are related by the equation $x + y + z = 12$. With these restrictions maximize each of the following functions: (a) x^3y^2z; (b) xy/z; (c) $(x + y)/(y + z)$; (d) $xy/(y + z)$.

12. State and prove a theorem that generalizes the Theorem of § 926 to the case of two functions of n variables, one function extreme and the other function constant.

13. Prove the Theorem of § 927 for the case $m = n = 3$.

14. Prove the Theorem of § 927 in general.

*15. Prove the analogue of inequality (8), § 927, for nth-order determinants, and thus derive the inequality of Hadamard, (10), § 927.

16. State and prove the analogues of the two theorems of § 928 for the cases of three variables and either one or two constraints.

17. Prove that the shortest line segment from a point to a curve without singular points (alternatively, between two curves without singular points) is perpendicular to the curve (both curves).

18. Obtain the formula for the distance between a point and a plane given in (3), Ex. 14, § 711, by minimizing the square of the distance between that point and an arbitrary point of the plane.

§ 930] DIFFERENTIATION UNDER THE INTEGRAL SIGN 321

19. Prove that the maximum value of the product $x_1 x_2 \cdots x_n$ of n positive numbers whose sum has a prescribed value:
$$x_1 + x_2 + \cdots + x_n = a$$
is equal to a^n/n^n. Conclude that the geometric mean of n positive numbers can never exceed their arithmetic mean:
$$\sqrt[n]{x_1 x_2 \cdots x_n} \leq \frac{x_1 + x_2 + \cdots + x_n}{n}.$$

20. Maximize $\sum_{i=1}^{n} a_i x_i$, where $a_i > 0$, $i = 1, 2, \cdots, n$, subject to the constraint $\sum_{i=1}^{n} x_i = 1$, with $x_i \geq 0$, $i = 1, 2, \cdots, n$.

*21. Use the method of Example 1, § 928, to find the equations of the axes of the ellipse $8x^2 + 4xy + 5y^2 = 36$, and the simplified equation resulting from the elimination of the xy-term by rotation of coordinate axes.

*22. Show that the lengths of the semiaxes of the ellipsoid
(1) $\quad ax^2 + by^2 + cz^2 + 2fyz + 2gxz + 2hxy = 1$
are the reciprocals of the square roots of the roots of the equation
$$\begin{vmatrix} a - \lambda & h & g \\ h & b - \lambda & f \\ g & f & c - \lambda \end{vmatrix} = 0.$$
Use this method to find the lengths of the semiaxes of the ellipsoid $5x^2 + 3y^2 + 3z^2 + 2xy - 2xz - 2yz - 12 = 0$. (Cf. Example 1, § 928; Exs. 7, 21, 23.)

*23. Let the ellipsoid (1) of Exercise 22 be cut by the plane $Ax + By + Cz = 0$ in an ellipse. Show that the semiaxes of this ellipse are the reciprocals of the square roots of the roots of the equation
$$\begin{vmatrix} A & B & C & 0 \\ a - \lambda & h & g & A \\ h & b - \lambda & f & B \\ g & f & c - \lambda & C \end{vmatrix} = 0.$$

Hints: Obtain a system of four homogeneous linear equations in the unknowns x, y, z, and μ by combining with the equation of the given plane the results of setting equal to zero the first partial derivatives with respect to x, y, z of the function
$$(x^2 + y^2 + z^2) - \frac{1}{\lambda}(ax^2 + by^2 + cz^2 + 2fyz + 2gxz + 2hxy) \\ + 2\mu(Ax + By + Cz):$$
$$\begin{cases} (a - \lambda)x + hy + gz - \lambda A \mu = 0, \\ hx + (b - \lambda)y + fz - \lambda B \mu = 0, \\ gx + fy + (c - \lambda)z - \lambda C \mu = 0. \end{cases}$$
Multiply these three equations by x, y, and z, respectively, add, and use the given equations of constraint. (Cf. Example 1, § 928; Ex. 22.)

*930. DIFFERENTIATION UNDER THE INTEGRAL SIGN. LEIBNITZ'S RULE

In this section we shall study functions of the form
(1) $$F(x) \equiv \int_{\phi(x)}^{\psi(x)} f(x, y)\, dy,$$

and investigate their continuity and differentiability properties. We shall first assume that ϕ and ψ are constants, and then permit them to be more general functions of x. It should be appreciated that (1) is a function of x alone, and not of y.

Theorem I. *If $f(x, y)$ is continuous on the closed rectangle A: $a \leq x \leq b$, $c \leq y \leq d$, then $F(x) \equiv \int_c^d f(x, y)\, dy$ exists and is continuous for $a \leq x \leq b$.* (Cf. Ex. 7, § 931.)

Proof. Existence follows from continuity of $f(x, y)$. Since A is compact $f(x, y)$ is uniformly continuous on A (§ 617). Therefore, if ϵ is an arbitrary positive number, we can find a positive number δ such that whenever $|x_1 - x_2| < \delta$ and $|y_1 - y_2| < \delta$ (and *a fortiori* when $|x_1 - x_2| < \delta$ and $y_1 = y_2$), $|f(x_2, y_2) - f(x_1, y_1)| < \epsilon/(d - c)$. Accordingly, we let $|x_1 - x_2| < \delta$ and find (cf. Ex. 2, § 402):

$$|F(x_2) - F(x_1)| \leq \int_c^d |f(x_2, y) - f(x_1, y)|\, dy < \int_c^d \frac{\epsilon}{d - c}\, dy = \epsilon.$$

In discussing the differentiability of $F(x)$, defined in (1), we shall consider the domain of definition of the function $f(x, y)$ to be limited to the rectangle A: $a \leq x \leq b$, $c \leq y \leq d$. The partial derivative $f_1(x, y)$, therefore, when considered along the vertical sides of A, $x = a$, $x = b$, is given by a one-sided limit. For example:

$$f_1(a, c) \equiv \lim_{h \to 0+} [f(a + h, c) - f(a, c)]/h.$$

Theorem II. Leibnitz's Rule. *If $f(x, y)$ and $f_1(x, y)$ exist and are continuous on the closed rectangle A: $a \leq x \leq b$, $c \leq y \leq d$, then the function $F(x) \equiv \int_c^d f(x, y)\, dy$ is differentiable for $a \leq x \leq b$, and*

(2) $$F'(x) = \int_c^d f_1(x, y)\, dy.$$

(Cf. Ex. 7, § 931.)

Proof. Since A is compact, $f_1(x, y)$ is uniformly continuous there. Let a positive number ϵ be given, and choose $\delta > 0$ so that $|x_1 - x_2| < \delta$ and $|y_1 - y_2| < \delta$ imply $|f_1(x_2, y_2) - f_1(x_1, y_1)| < \epsilon/(d - c)$ (and then let $y_1 = y_2$). Letting h be an arbitrary number such that $0 < |h| \leq \delta$, and letting x_0 be an arbitrary point of the interval $[a, b]$, we form a difference which we wish to show is numerically less than ϵ (once this is established, the proof is complete):

(3) $$\left| \frac{F(x_0 + h) - F(x_0)}{h} - \int_c^d f_1(x_0, y)\, dy \right|.$$

§ 930] DIFFERENTIATION UNDER THE INTEGRAL SIGN

Using the definition of $F(x)$ and the law of the mean for a function of one variable (the θ depending on y), we have for (3):

$$\left| \frac{1}{h} \int_c^d [f(x_0 + h, y) - f(x_0, y)] \, dy - \int_c^d f_1(x_0, y) \, dy \right|$$

$$= \left| \int_c^d f_1(x_0 + \theta h, y) \, dy - \int_c^d f_1(x_0, y) \, dy \right|$$

$$\leq \int_c^d |f_1(x_0 + \theta h, y) - f_1(x_0, y)| \, dy < \int_c^d \frac{\epsilon}{d - c} \, dy = \epsilon.$$

NOTE 1. The function $f_1(x_0 + \theta(y)h, y)$, in spite of the unknown nature of $\theta(y)$, is an integrable function of y since it is equal to $\frac{1}{h}[f(x_0 + h, y) - f(x_0, y)]$, a continuous function of y. Consequently, $f_1(x_0 + \theta h, y) - f_1(x_0, y)$ is also integrable, and all integrals appearing in the preceding proof do exist.

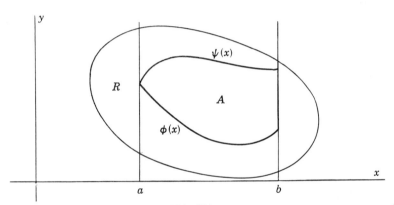

FIG. 904

Example 1. Let $F(x) = \int_1^2 \sin(xe^y) \, dy$. Then $F'(x) = \int_1^2 e^y \cos(xe^y) \, dy$. In this case the resulting integration can be performed explicitly:

$$F'(x) = \frac{1}{x} \int_1^2 \cos(xe^y) \, d(xe^y) = \frac{1}{x} [\sin(xe^y)]_1^2 = \frac{\sin(e^2 x) - \sin(ex)}{x}.$$

Theorems I and II can be generalized as follows (for hints on the proof, cf. Exs. 8, 9, § 931; for generalizations, cf. Exs. 10, 11, § 931):

Theorem III. Let $\phi(x)$ and $\psi(x)$ be continuous for $a \leq x \leq b$ and for such x let $\phi(x) \leq \psi(x)$. Let A designate the set of all points such that $a \leq x \leq b$ and $\phi(x) \leq y \leq \psi(x)$. If $f(x, y)$ is continuous in a region R containing A, then $F(x) = \int_{\phi(x)}^{\psi(x)} f(x, y) \, dy$ is continuous for $a \leq x \leq b$. (Cf. Fig. 904.)

Theorem IV. General Form of Leibnitz's Rule. *Under the assumptions of Theorem III and the additional hypotheses that $\phi(x)$ and $\psi(x)$ are continuously differentiable for $a \leq x \leq b$, and that $f_1(x, y)$ exists and is continuous in R, the function $F(x) \equiv \int_{\phi(x)}^{\psi(x)} f(x, y)\, dy$ is continuously differentiable, and*

(4) $\quad F'(x) = f(x, \psi(x))\psi'(x) - f(x, \phi(x))\phi'(x) + \int_{\phi(x)}^{\psi(x)} f_1(x, y)\, dy.$

A special case of Theorem IV was given in Exercise 2, § 407.

NOTE 2. The preceding theorems state that under suitable continuity conditions two limiting operations can be interchanged. By Theorem I the limit of an integral is the integral of the limit, and by Theorem II the derivative of an integral is the integral of the derivative. (Cf. Theorem II, § 1009, and the Note, § 1408.)

The extension of Leibnitz's Rule (Theorem II) to improper integrals is presented in Chapter 14.

Example 2. If $F(x) \equiv \int_{x^2}^{x^3} \tan(xy^2)\, dy$,

$$F'(x) = 3x^2 \tan(x^7) - 2x \tan(x^5) + \int_{x^2}^{x^3} y^2 \sec^2(xy^2)\, dy.$$

★931. EXERCISES

In Exercises 1–2, evaluate the indicated derivative.

★1. $\dfrac{d}{dx} \displaystyle\int_0^x \dfrac{\sin xy}{y}\, dy.$ 　　★2. $\dfrac{d}{dx} \displaystyle\int_{-x}^x \dfrac{1 - e^{-xy}}{y}\, dy.$

In Exercises 3–4, verify Leibnitz's Rule.

★3. $\displaystyle\int_x^{x^3} (x^2 + y^2)\, dy.$ 　　★4. $\displaystyle\int_0^{\frac{x}{2}} \sqrt{x^2 - y^2}\, dy.$

★5. Define $I_n(x) \equiv \displaystyle\int_a^x (x - t)^{n-1} f(t)\, dt, n = 1, 2, \cdots$. Prove that

$$\frac{d^n I_n}{dx^n} = (n - 1)!\, f(x).$$

★6. Show that if $y(x)$ satisfies the "integral equation"

$$y(x) = 4 \int_0^x (t - x) y(t)\, dt - \int_0^x (t - x) f(t)\, dt,$$

then $y(x)$ satisfies the differential equation and boundary conditions

$$\frac{d^2 y}{dx^2} + 4y = f(x),$$

$$y(0) = 0, \quad y'(0) = 0.$$

★7. Prove that Theorems I and II, § 930, remain valid if the interval $[a, b]$ for x is replaced by an arbitrary interval, finite or infinite, open or closed or neither.

★8. Prove Theorem III, § 930. *Hints:* For a given $x_0 \in [a, b]$ let $\eta > 0$ be such that the rectangle B: $|x - x_0| \leq \eta$, $\phi(x_0) - \eta \leq y \leq \psi(x_0) + \eta$ is contained in R, and let γ be a constant between $\phi(x_0)$ and $\psi(x_0)$. First prove that $G(x, u) \equiv \int_\gamma^u f(x, y)\, dy$ is continuous in B. Then write $\int_u^v f(x, y)\, dy = G(x, v) - G(x, u)$.

*9. Prove Theorem IV, § 930. *Hints:* Proceed as in the hints of Ex. 8, and prove that $G(x, u)$ is continuously differentiable. Then use the chain rule.

*10. Same as Exercise 7 for Theorems III and IV, § 930.

*11. Extend the Theorems of § 930 to functions of several variables (the integration applying to only one).

*932. THE IMPLICIT FUNCTION THEOREM

In this section we study conditions that guarantee the existence of an implicitly defined function, when one equation is given, and an implicitly defined set of functions, when a system of equations is given. We start with the simplest case, that of a function of one variable, $y(x)$ defined by an equation in two variables, $f(x, y) = 0$. To illustrate some of the difficulties that may arise we consider first a few simple examples.

Example 1. If the given equation is $f(x, y) \equiv x - e^y = 0$, the function $f(x, y)$ has all of the continuity and differentiability conditions one could desire, for *all* real values of x and y. This equation defines a single-valued function $y = \phi(x) \equiv \ln x$, satisfying the given equation identically, but this function is defined only for *positive x*.

Example 2. The function $f(x, y) \equiv 2x - \sin x$ is well-behaved (in some senses at least), but the equation $f(x, y) = 0$ defines y as a function of x only in the sense of a completely unrestricted range!

Example 3. Let $f(x, y) \equiv x^4 - y^2$. Then the equation $f(x, y) = 0$ defines y as a *double-valued* function of x ($x \neq 0$). It also defines y as a *single-valued* function of x, but it does so in many ways. Four of these are $y = x^2$, $y = -x^2$, $y = x|x|$, and $y = -x|x|$, and if discontinuities are permitted, many more solutions are possible (for example, $y = x^2$ if x is rational, and $y = -x^2$ if x is irrational). However, if we choose a point on the original curve *not the origin*, there will be determined *in a suitably restricted neighborhood* of this point the graph of a unique function $y = \phi(x)$ satisfying the identity $f(x, \phi(x)) = 0$. (Cf. the Note below.)

We shall state and prove two special cases of the Implicit Function Theorem. The general case, and its proof, should then be clear (cf. Ex. 2, § 935). For an explicit statement and proof of the general theorem, see C. Carathéodory, *Variationsrechnung und Partielle Differentialgleichungen Erster Ordnung* (Leipzig, B. G. Teubner, 1935), pp. 9–12.

The details in the two proofs that follow are designed for generalization and are therefore not necessarily the most economical for either specific theorem. Notice how hypothesis *(iii)*, in the simplest case (Theorem I), precludes the situation of Example 2, where the variable y does not appear, and the difficulty associated with the origin in Example 3, where uniqueness fails. (Cf. § 601 for point set notation.)

Theorem I. Implicit Function Theorem. *If*

(i) $f(x, y)$ *and* $f_2(x, y)$ *are continuous in a neighborhood* $N_{(a,b)}$ *of the point* (a, b) *(in E_2)*,

(ii) $f(a, b) = 0$,

(iii) $f_2(a, b) \neq 0$,

then there exist neighborhoods N_a of a and N_b of b (in E_1) and a function $\phi(x)$ defined in N_a such that

(iv) *for all x in the neighborhood N_a, $\phi(x) \in N_b$, $(x, \phi(x)) \in N_{(a,b)}$, and $f(x, \phi(x)) = 0$,*

(v) $\phi(x)$ *is uniquely determined by (iv); that is, if $\Phi(x)$ is defined in N_a and has properties (iv), then $\Phi(x) = \phi(x)$ throughout N_a,*

(vi) $\phi(a) = b$,

(vii) $\phi(x)$ *is continuous in N_a,*

(viii) *if $f_1(x, y)$ exists and is continuous in $N_{(a,b)}$, then $\phi'(x)$ exists and is continuous in N_a, and*

$$\phi'(x) = -\frac{f_1(x, \phi(x))}{f_2(x, \phi(x))}.$$

Proof. By continuity and (iii), there exist closed neighborhoods of the points a and b:

$$A: |x - a| \leq \alpha, \quad B: |y - b| \leq \beta,$$

such that $x \in A$ and $y \in B$ imply $(x, y) \in N_{(a,b)}$ and $f_2(x, y) \neq 0$. (Cf. Note 4, § 209.)

We first establish uniqueness or, more precisely, that for each $x \in A$ there is *at most* one $y \in B$ such that $f(x, y) = 0$. For, if y' and y'' were *two* such y, for some x, then by the Law of the Mean (§ 305):

$$0 = f(x, y'') - f(x, y') = f_2(x, y_1)(y'' - y'),$$

where y_1 is between y' and y'' and therefore belongs to B. This contradicts the assumption that for $x \in A$ and $y \in B$, $f_2(x, y) \neq 0$.

We now denote by C the compact two-point set of numbers η such that $|\eta - b| = \beta$ (that is, $\eta = b \pm \beta$), and define the nonnegative function

(1) $$F(x, y) \equiv (f(x, y))^2.$$

Then, since $C \subset B$, and by the uniqueness just established, $\eta \in C$ implies $f(a, \eta) \neq 0$. Therefore there exists a positive number ϵ such that for $\eta \in C$,

(2) $$\begin{cases} F(a, \eta) \geq \epsilon, \\ F(a, b) = 0. \end{cases}$$

We infer from (2) and the continuity of $F(x, y)$, as a function of x, at the three values of y: $y = b$, $y = \eta \in C$, the existence of a positive number $\delta \leq \alpha$ such that $|x - a| \leq \delta$ implies

(3) $$\begin{cases} F(x, \eta) > \tfrac{1}{2}\epsilon, \\ F(x, b) < \tfrac{1}{2}\epsilon. \end{cases}$$

§ 932]   THE IMPLICIT FUNCTION THEOREM   327

If N_a denotes the neighborhood of a: $|x - a| < \delta$, and D its closure, we can infer from (3) that for each fixed $x \in D$, $F(x, y)$ as a function of y on the compact set B has a minimum value attained at an *interior* point $\bar{y}$ of B. By Theorem I, § 309,
$$F_2(x, \bar{y}) = 2f(x, \bar{y})f_2(x, \bar{y}) = 0,$$
and hence $f(x, \bar{y}) = 0$.

With the uniqueness already established as the first part of this proof, and with N_b defined as the interior of B, we now have proved the existence of a function $y = \bar{y} = \phi(x)$ satisfying conclusions (iv), (v), and (vi) of the theorem.

To prove continuity of $\phi(x)$, we write, with x and $x + \Delta x$ in N_a, and with $\phi \equiv \phi(x)$ and $\Delta\phi \equiv \phi(x + \Delta x) - \phi(x)$:
$$0 = f(x + \Delta x, \phi + \Delta\phi) - f(x, \phi)$$
$$= [f(x + \Delta x, \phi + \Delta\phi) - f(x, \phi + \Delta\phi)] + [f(x, \phi + \Delta\phi) - f(x, \phi)].$$
By the Law of the Mean, we can write this last bracketed expression as $f_2(x, y_1) \Delta\phi$, where y_1 is between ϕ and $\phi + \Delta\phi$, and solve for $\Delta\phi$:

(4) $$\Delta\phi = -\frac{f(x + \Delta x, \phi + \Delta\phi) - f(x, \phi + \Delta\phi)}{f_2(x, y_1)}.$$

By the uniform continuity of $f(x, y)$ on the closed rectangle: $x \in D$, $y \in B$, we can conclude that the numerator on the right of (4) tends to 0 with Δx, while the denominator is bounded from 0. Therefore $\Delta\phi \to 0$ and ϕ is continuous.

Finally, the formula of (viii) follows from (4) by one further application of the Law of the Mean (to the numerator of the fraction). The continuity of $\phi'(x)$ is implied by its representation as a quotient of continuous functions.

Theorem II. Implicit Function Theorem. *Let*

(5) $$\begin{cases} f(x, y, z, u, v) = 0, \\ g(x, y, z, u, v) = 0, \end{cases}$$

be a system of two equations to be solved for the two unknowns u and v in terms of the remaining variables. If

(i) *f, g, f_4, f_5, g_4, and g_5 are continuous in a neighborhood $N_{(a,b,c,d,e)}$ of the point (a, b, c, d, e) (in E_5),*
(ii) *$f(a, b, c, d, e) = g(a, b, c, d, e) = 0$,*
(iii) *the Jacobian $J(x, y, z, u, v) = \partial(f, g)/\partial(u, v) = f_4 g_5 - f_5 g_4$ is nonzero at (a, b, c, d, e),*

then there exist neighborhoods $N_{(a,b,c)}$ of $(a, b, c,)$ (in E_3) and $N_{(d,e)}$ of (d, e) (in E_2) and functions $u = \phi(x, y, z)$, $v = \psi(x, y, z)$ defined in $N_{(a,b,c)}$ such that

(iv) *for all (x, y, z) in the neighborhood $N_{(a,b,c)}$, $(\phi(x, y, z), \psi(x, y, z)) \in N_{(d,e)}$, $(x, y, z, \phi(x, y, z), \psi(x, y, z)) \in N_{(a,b,c,d,e)}$, and $f(x, y, z, \phi(x, y, z), \psi(x, y, z)) = g(x, y, z, \phi(x, y, z), \psi(x, y, z)) = 0,$*

(v) ϕ and ψ are uniquely determined by (iv); that is, if Φ and Ψ are defined in $N_{(a,b,c)}$ and have properties (iv), then $\Phi = \phi$, $\Psi = \psi$ throughout $N_{(a,b,c)}$,

(vi) $\phi(a, b, c) = d$, $\psi(a, b, c) = e$,

(vii) ϕ and ψ are continuous in $N_{(a,b,c)}$,

(viii) if $f_1, f_2, f_3, g_1, g_2,$ and g_3 exist and are continuous in $N_{(a,b,c,d,e)}$, then ϕ and ψ have continuous first derivatives, which are given by formulas of type (7) and (8), § 921.

Proof. By continuity and (iii), there exist closed neighborhoods of the points (a, b, c) and (d, e):

$$A: (x - a)^2 + (y - b)^2 + (z - c)^2 \leq \alpha^2, \quad B: (u - d)^2 + (v - e)^2 \leq \beta^2,$$

such that $(x, y, z) \in A$, $(u_i, v_i) \in B$, $i = 1, 2$, imply $(x, y, z, u_i, v_i) \in N_{(a,b,c,d,e)}$, $i = 1, 2$, and

(6) $\quad \begin{vmatrix} f_4(x, y, z, u_1, v_1) & f_5(x, y, z, u_1, v_1) \\ g_4(x, y, z, u_2, v_2) & g_5(x, y, z, u_2, v_2) \end{vmatrix} \neq 0.$

(The determinant $qt - rs$ is a continuous function of the four variables q, r, s, t.)

In order to prove that for any $(x, y, z) \in A$ there is *at most* one $(u, v) \in B$ such that

(7) $\quad F(x, y, z, u, v) \equiv [f(x, y, z, u, v)]^2 + [g(x, y, z, u, v)]^2$

vanishes, we assume there exist *two* such points (u', v') and (u'', v''), for some $(x, y, z) \in A$. The Law of the Mean (§ 912), applied to the two equations

(8) $\quad f(x, y, z, u'', v'') - f(x, y, z, u', v')$
$$= g(x, y, z, u'', v'') - g(x, y, z, u', v') = 0,$$

gives, for suitable points (u_1, v_1) and (u_2, v_2) of the straight line segment joining (u', v') and (u'', v'') (and hence belonging to the convex closed neighborhood B—cf. Ex. 33, § 618, Definition II, § 1807):

(9) $\quad \begin{cases} f_4(x, y, z, u_1, v_1)(u'' - u') + f_5(x, y, z, u_1, v_1)(v'' - v') = 0, \\ g_4(x, y, z, u_2, v_2)(u'' - u') + g_5(x, y, z, u_2, v_2)(v'' - v') = 0. \end{cases}$

Since $(u'' - u')^2 + (v'' - v')^2 \neq 0$, this contradicts (6).

We now denote by C the compact set of all points (η, ζ) in E_2 such that

(10) $\quad (\eta - d)^2 + (\zeta - e)^2 = \beta^2.$

Then, since $C \subset B$, and by the uniqueness just established, $(\eta, \zeta) \in C$ implies $F(a, b, c, \eta, \zeta) > 0$. Therefore, by the compactness of C, there exists a positive number ϵ such that for all $(\eta, \zeta) \in C$:

(11) $\quad \begin{cases} F(a, b, c, \eta, \zeta) \geq \epsilon, \\ F(a, b, c, d, e) = 0. \end{cases}$

We infer from (11) and the continuity of $F(x, y, z, u, v)$ (Ex. 1, § 935) the existence of a positive number $\delta \leq \alpha$ such that if D is the closed neighborhood (with interior $N_{(a,b,c)}$) defined by

(12) $$D: (x - a)^2 + (y - b)^2 + (z - c)^2 \leq \delta^2,$$

then $(x, y, z) \in D$ and $(\eta, \zeta) \in C$ imply

(13) $$\begin{cases} F(x, y, z, \eta, \zeta) > \tfrac{1}{2}\epsilon, \\ F(x, y, z, d, e) < \tfrac{1}{2}\epsilon. \end{cases}$$

Therefore, for each fixed $(x, y, z) \in D$, $F(x, y, z, u, v)$ as a function of (u, v) on the compact set B has a minimum value attained at an *interior* point $(\bar{u}, \bar{v})$ of B. By § 914, $F_4(x, y, z, \bar{u}, \bar{v}) = F_5(x, y, z, \bar{u}, \bar{v}) = 0$ and

(14) $$\begin{cases} f(x, y, z, \bar{u}, \bar{v}) f_4(x, y, z, \bar{u}, \bar{v}) + g(x, y, z, \bar{u}, \bar{v}) g_4(x, y, z, \bar{u}, \bar{v}) = 0, \\ f(x, y, z, \bar{u}, \bar{v}) f_5(x, y, z, \bar{u}, \bar{v}) + g(x, y, z, \bar{u}, \bar{v}) g_5(x, y, z, \bar{u}, \bar{v}) = 0. \end{cases}$$

Since $J(x, y, z, \bar{u}, \bar{v}) \neq 0$ for $(x, y, z) \in A$ and $(\bar{u}, \bar{v}) \in B$, it follows that

(15) $$F(x, y, z, \bar{u}, \bar{v}) = f(x, y, z, \bar{u}, \bar{v}) = g(x, y, z, \bar{u}, \bar{v}) = 0.$$

With the uniqueness already established as the first part of this proof, and with $N_{(d,e)}$ defined as the interior of B, we now have proved the existence of two functions

(16) $$u = \bar{u} = \phi(x, y, z), \quad v = \bar{v} = \psi(x, y, z)$$

satisfying conclusions (iv), (v), and (vi) of the theorem.

To prove continuity of ϕ and ψ, we write

$$\begin{aligned} 0 &= [f(x + \Delta x, y + \Delta y, z + \Delta z, \phi + \Delta\phi, \psi + \Delta\psi) \\ &\quad - f(x, y, z, \phi + \Delta\phi, \psi + \Delta\psi)] \\ &\quad + [f(x, y, z, \phi + \Delta\phi, \psi + \Delta\psi) - f(x, y, z, \phi, \psi + \Delta\psi)] \\ &\quad + [f(x, y, z, \phi, \psi + \Delta\psi) - f(x, y, z, \phi, \psi)] \\ &= [f(x + \Delta x, y + \Delta y, z + \Delta z, \phi + \Delta\phi, \psi + \Delta\psi) \\ &\quad - f(x, y, z, \phi + \Delta\phi, \psi + \Delta\psi)] \\ &\quad + f_4(x, y, z, u_1, \psi + \Delta\psi)\,\Delta\phi + f_5(x, y, z, \phi, v_2)\,\Delta\psi, \end{aligned}$$

where u_1 and v_2 lie between ϕ and $\phi + \Delta\phi$, and ψ and $\psi + \Delta\psi$, respectively, with a similar expression for g. By (6), Cramer's Rule gives $\Delta\phi$ and $\Delta\psi$ as quotients of determinants, and these quotients $\to 0$ as $\Delta x^2 + \Delta y^2 + \Delta z^2 \to 0$, by uniform continuity of f on the compact set $(x, y, z) \in D$, $(u, v) \in B$.

Finally, the formulas of (viii) follow from the preceding application of Cramer's Rule if one sets in turn $\Delta y = \Delta z = 0$, $\Delta x = \Delta z = 0$, and $\Delta x = \Delta y = 0$, with suitable use of the Law of the Mean. The continuity of the partial derivatives is implied by their representation as quotients of continuous functions.

NOTE. An alternative formulation of the Implicit Function Theorem is obtained, as a Corollary to the Theorem as stated above, by expressing the uniqueness of the solution in terms of *continuity*. Specifically, in the statement of Theorem I (II) all reference to the neighborhood $N_b(N_{(d,e)})$ may be omitted if the assumption of properties (*vi*) and (*vii*) for the function Φ (functions Φ and Ψ) of part (*v*) is made. The reason for this, for Theorem II to be precise, is that if ϕ, Φ, ψ, and Ψ are *all* continuous in $N_{(a,b,c)}$ and if $\phi(a, b, c) = \Phi(a, b, c) = d$ and $\psi(a, b, c) = \Psi(a, b, c) = e$, then the membership $(\phi(x, y, z), \psi(x, y, z)) \in N_{(d,e)}$ implies the membership $(\Phi(x, y, z), \Psi(x, y, z)) \in N_{(d,e)}$, for all points (x, y, z) of $N_{(a,b,c)}$. Detailed hints are given in Exercise 3, § 935. In Example 3, above, for any point on the original curve, not the origin, a unique *continuous* solution exists if the values of x alone, without regard to y, are sufficiently restricted.

*933. EXISTENCE THEOREM FOR INVERSE TRANSFORMATIONS

We shall restrict the details of our discussion in this section to transformations from a two-dimensional space E_2 to a two-dimensional space E_2. The general existence theorem for n dimensions, and its proof, are completely analogous to those for the case $n = 2$, and need not be given explicitly. (Cf. Ex. 4, § 935.)

Consider a transformation, then, of the form

(1) $$\begin{cases} x = x(u, v), \\ y = y(u, v). \end{cases}$$

Certainly it is not *always* possible to solve such a system for u and v in terms of x and y. For example, the system $x = u + v$, $y = 2u + 2v$ has no solution at all unless $2x = y$, and then not a unique solution in any sense. We now state a theorem that guarantees, under certain conditions, the existence of an inverse transformation.

Theorem. *If*

(i) *$x(u, v)$ and $y(u, v)$ are continuously differentiable in a neighborhood of the point (c, d),*
(ii) *the Jacobian $J = \partial(x, y)/\partial(u, v)$ is nonzero at (c, d),*

then there exist a neighborhood $N_{(a,b)}$ of the point $(a, b) \equiv (x(c, d), y(c, d))$ (in E_2) and functions $u = u(x, y)$, $v = v(x, y)$, defined in $N_{(a,b)}$ and such that

(iii) *the following two identities hold throughout $N_{(a,b)}$:*
$$x(u(x, y), v(x, y)) = x,$$
$$y(u(x, y), v(x, y)) = y,$$
(iv) *$u(a, b) = c$ and $v(a, b) = d$,*
(v) *$u(x, y)$ and $v(x, y)$ are continuous in $N_{(a,b)}$,*
(vi) *if $\phi(x, y)$ and $\psi(x, y)$ (in place of $u(x, y)$ and $v(x, y)$, respectively) have properties (iii), (iv), (v), then $\phi(x, y) = u(x, y)$ and $\psi(x, y) = v(x, y)$ in $N_{(a,b)}$,*
(vii) *$u(x, y)$ and $v(x, y)$ are continuously differentiable in $N_{(a,b)}$, with partial derivatives*
$$u_x = y_v/J, \quad u_y = -x_v/J, \quad v_x = -y_u/J, \quad v_y = x_u/J,$$

§ 934] CONDITIONS FOR FUNCTIONAL DEPENDENCE

(viii) *the set B of all points in the uv-plane that are images of points of $N_{(a,b)}$ under the inverse transformation is an open set, and the correspondence between points of B and points of $N_{(a,b)}$ is one-to-one.*

Proof. We define the system

(2) $$\begin{cases} f(x, y, u, v) \equiv x - x(u, v) = 0, \\ g(x, y, u, v) \equiv y - y(u, v) = 0, \end{cases}$$

and apply the Implicit Function Theorem for a system of two equations, which is applicable since the two Jacobians $\partial(f, g)/\partial(u, v)$ and $\partial(x, y)/\partial(u, v)$ are identical. That theorem, in the form of the Note, § 932, gives (*iii*), (*iv*), (*v*), (*vi*), and part of (*vii*) immediately, while the formulas of (*vii*) were obtained in § 922. Conclusion (*viii*) is somewhat more difficult, but not much. Notice that the part of the theorem already established states, in part, that the point (a, b) into which (c, d) is carried by the transformation is completely surrounded by a neighborhood $N_{(a,b)}$ of points that are images of points in the original neighborhood of (c, d), under the given transformation. This fact, applied to the inverse transformation at an arbitrary point of B, establishes the desired openness. The fact that the transformation is one-to-one follows from the fact that all of the transformation functions considered are single-valued.

The argument used in proving conclusion (*viii*) also establishes the following:

Corollary. *If the transformation functions* (1) *are continuously differentiable over an open set B in the uv-plane, and if the Jacobian $\partial(x, y)/\partial(u, v)$ of this transformation does not vanish in B, then the image A of B by this transformation is an open set in the xy-plane. In other words, if we define an* **open mapping** *defined on a region, to be one that always maps open sets onto open sets, then any transformation of the type just considered is open.* (Cf. Ex. 5, § 935.)

NOTE 1. A transformation satisfying the conditions of the Corollary need not be one-to-one. Example: $x = u \cos v$, $y = u \sin v$, B the rectangle $1 < u < 3$, $0 < v < 3\pi$.

NOTE 2. If the Jacobian $\partial(x, y)/\partial(u, v)$ vanishes, the set A of the Corollary need not be open. Example: $x = u \cos v$, $y = u \sin v$, B the rectangle $-1 < u < 1$, $0 < v < \pi/2$.

*934. SUFFICIENCY CONDITIONS FOR FUNCTIONAL DEPENDENCE

A set of necessary conditions for functional dependence was obtained in § 923. In a sense which we shall discuss in this section these conditions, which involve the identical vanishing of certain Jacobians, are also sufficient. The simplest case, that of *one* function, was established in Theorem II, § 912, and alluded to at the end of § 923. In the present section we shall give full details for only one case, that of two functions of two variables. (Cf. Exs. 6, 7, § 935.) We first look at a simple example:

Example. Let $u = f(x, y) \equiv x + y$, $v = g(x, y) \equiv 2x + 2y$. The Jacobian $\dfrac{\partial(u, v)}{\partial(x, y)} = \begin{vmatrix} 1 & 1 \\ 2 & 2 \end{vmatrix} = 0$. The two functions, u and v, are related by the equation $v = 2u$. The pair of equations $u = x + y$, $v = 2x + 2y$ defines a transformation that carries the entire xy-plane onto the line $v = 2u$ in the uv-plane. The transformation is, in this sense, *degenerate*.

Theorem. *If*

(i) *the functions $u = f(x, y)$ and $v = g(x, y)$ are continuously differentiable in a region R in the xy-plane,*
(ii) *the Jacobian $\partial(u, v)/\partial(x, y)$ vanishes identically in R,*
(iii) *(a, b) is a point of R where not all first partial derivatives u_x, u_y, v_x, v_y vanish,*

then there exist a neighborhood $N_{(c,d)}$ of the point $(c, d) \equiv (f(a, b), g(a, b))$ and a function $F(u, v)$ defined in $N_{(c,d)}$ such that

(iv) *F is continuously differentiable in $N_{(c,d)}$,*
(v) *at least one of the two partial derivatives F_u and F_v is nonzero throughout $N_{(c,d)}$,*
(vi) *there exists a neighborhood $N_{(a,b)}$ in which $f(x, y)$ and $g(x, y)$ are functionally dependent by means of F:*

$$F(u, v) = F(f(x, y), g(x, y)) = 0.$$

Proof. The *idea* of the proof is that if we can solve for one of the variables (say y) in one of the equations (say the first) and substitute the result in the other, then the other independent variable (x in this case) drops out, leaving a relationship between u and v. For details, then, assume for definiteness that u_y is nonzero at (a, b): $f_2(a, b) \neq 0$. The Implicit Function Theorem for a single equation guarantees that we can solve the equation $u = f(x, y)$ for y in the sense that there exists a neighborhood $N_{(a,c)}$ of the point (a, c), and a function $y = \phi(x, u)$, defined and continuously differentiable in $N_{(a,c)}$ such that $b = \phi(a, c)$ and, for all (x, u) in $N_{(a,c)}$, $(x, \phi(x, u)) \in R$ and $f(x, \phi(x, u)) = u$. Furthermore,

$$\frac{\partial \phi}{\partial x} = -\frac{f_1(x, \phi(x, u))}{f_2(x, \phi(x, u))}, \quad \frac{\partial \phi}{\partial u} = \frac{1}{f_2(x, \phi(x, u))}.$$

We define a new function in $N_{(a,c)}$: $\psi(x, u) \equiv g(x, \phi(x, u))$. In $N_{(a,c)}$ this function is continuously differentiable, and

$$\frac{\partial \psi}{\partial x} = \frac{\partial g}{\partial x} + \frac{\partial g}{\partial y}\frac{\partial \phi}{\partial x} = \frac{f_2 g_1 - f_1 g_2}{f_2} = \frac{1}{f_2}\frac{\partial(u, v)}{\partial(x, y)} = 0,$$

because of the vanishing of the Jacobian, for all (x, u) in $N_{(a,c)}$. This means that for any x sufficiently near a, $\psi(x, u)$ is actually only a function of u

(cf. Ex. 32, § 915), defined and differentiable for u in a neighborhood N_c of c: $G(u) \equiv g(x, \phi(x, u))$. An interpolated comment is now in order: if (x, y) is sufficiently near (a, b), $f_2(x, y)$ is of constant sign, and hence, since *for any such fixed x the two functions $u = f(x, y)$ and $y = \phi(x, u)$ can be considered as inverse functions*, the relation $\phi(x, f(x, y)) = y$ (as well as the relation $f(x, \phi(x, u)) = u$) is satisfied. We now conclude that if (x, y) is in a sufficiently restricted neighborhood $N_{(a,b)}$ of (a, b), $u = f(x, y)$ will be in N_c, so that, when $\phi(x, u)$ is replaced by y and u by $f(x, y)$, the equation defining G becomes $G(f(x, y)) = g(x, \phi(x, f(x, y))) = g(x, y)$, which must hold in $N_{(a,b)}$. Finally, we let $F(u, v) \equiv G(u) - v$, and the proof is complete.

NOTE 1. Hypothesis (*iii*) does not impose much, for if the four partial derivatives all vanished throughout R (or any subregion of R), both functions u and v would be constant (Theorem II, § 912) and therefore trivially dependent in R (or the subregion).

NOTE 2. The proof just given indicates what happens geometrically, and shows that the Example comes close to being representative. The equation $v = G(u)$ determines a curve C in the uv-plane having the property that an entire neighborhood of the point (a, b) is mapped by the transformation $u = f(x, y)$, $v = g(x, y)$ onto a part of C. In case u and v are constants (cf. Note 1), the transformation carries a two-dimensional set into a single point. In other words, a *degenerate* transformation (one with identically vanishing Jacobian) is associated with a *lowering* of dimension.

The case of functional dependence of two functions of more than two variables is entirely analogous. For example, if $u = f(x, y, z)$ and $v = g(x, y, z)$ have identically vanishing Jacobians but have partial derivatives that are not all zero at a particular point (say $f_3 \neq 0$), then the details of the preceding proof would be altered essentially only by solving for $z = \phi(x, y, u)$, such that $f(x, y, \phi(x, y, u)) = u$, where $\partial \phi / \partial x = -f_1/f_3$ and $\partial \phi / \partial y = -f_2/f_3$, and by defining $\psi(x, y, u) \equiv g(x, y, \phi(x, y, u))$. Then the vanishing of the Jacobians implies that ψ is a function of u alone. From the point of view of a transformation, taken in Note 2, the degeneracy again causes a lowering of dimensions, and again to less than the two-dimensionality normally expected. (Cf. Ex. 6, § 935.)

For the general case of n functions, hypothesis (*iii*) takes the form of a requirement that not all $(n-1)$th-order minors of the Jacobians under consideration vanish at the particular point specified. (Cf. Ex. 7, § 935.)

*935. EXERCISES

*1. Prove the existence of a positive number δ as claimed for the implication involving (12) and (13), § 932. *Hint:* If there were no such δ, then there would be a sequence $\{(x_n, y_n, z_n, \eta_n, \zeta_n)\}$ such that $(x_n - a)^2 + (y_n - b)^2 + (z_n - c)^2 \leq 1/n$, $(\eta_n, \zeta_n) \in C$, and $F(x_n, y_n, z_n, \eta_n, \zeta_n) \leq \frac{1}{2}\epsilon$. Choose a convergent subsequence which $\to (a, b, c, \bar{\eta}, \bar{\zeta})$.

★2. State and prove a form of the Implicit Function Theorem appropriate for solving the system of equations $f(x, y, u, v, w) = g(x, y, u, v, w) = h(x, y, u, v, w) = 0$ for u, v, and w in terms of x and y.

★3. Prove the statements of the Note, § 932. *Hints:* For Theorem II, § 932, consider a point (λ, μ, ν) nearest (a, b, c) which is a limit point of points (x, y, z) for which the points $(\Phi(x, y, z), \Psi(x, y, z))$ and $(\phi(x, y, z), \psi(x, y, z))$ are distinct. Then $(\Phi(\lambda, \mu, \nu), \Psi(\lambda, \mu, \nu)) = (\phi(\lambda, \mu, \nu), \psi(\lambda, \mu, \nu)) \in N_{(d,e)}$, and the two mappings must agree in a neighborhood of (λ, μ, ν). (Contradiction.)

★4. State and prove an existence theorem for the inverse of a transformation from E_3 into E_3.

★5. State and prove the analogue of the Corollary, § 933, for three dimensions.

★6. State and prove the theorem on functional dependence of two functions of three variables, alluded to in the penultimate paragraph of § 934.

★7. State and prove a theorem on functional dependence of three functions of three variables.

In Exercises 8–11, use the principle of the Theorem of § 934 to establish functional dependence for the given functions. Then find an explicit relation between them.

★8. $u = \dfrac{x^2 - y^2}{x^2 + y^2}$, $v = \dfrac{2xy}{x^2 + y^2}$.

★9. $u = \dfrac{xe^z}{y^2} + \dfrac{y^2 e^{-z}}{x}$, $v = \dfrac{xe^z}{y^2} - \dfrac{y^2 e^{-z}}{x}$.

★10. $u = xe^y \sin z$, $v = xe^y \cos z$, $w = x^2 e^{2y}$.

★11. $u = t^2 + x^2 + y^2 + z^2$, $v = tx + ty + tz + xy + xz + yz$, $w = t + x + y + z$.

10

Multiple Integrals

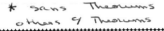

1001. INTRODUCTION

The Riemann integral of a function of one variable is defined over an interval $[a, b]$. Since the set over which the function is integrated is always a closed interval, conditions for integrability concern only the function itself, and not its domain. For example, all continuous functions are integrable over $[a, b]$.

In higher-dimensional spaces, however, one is interested in integrating functions over more general sets than rectangles (and other higher-dimensional analogues of intervals). Integrability conditions, therefore, involve not only the function but the set over which it is integrated. A function may fail to be integrable either by having "too many" discontinuities or by being integrated over "too pathological" a set. Not even a constant function is integrable if the set over which it is integrated is too irregular.

We shall start by defining the double integral in terms of rectangular subdivisions or *nets*. This permits a simple definition of *area* of a set and, in terms of area, a second formulation of the double integral, based on more general subdivisions than rectangles.

Multiple integrals of functions of more than two variables are treated similarly, but we shall not attempt to present all of the details. Triple integrals are discussed specifically in later sections.

For a more complete treatment see the author's *Real Variables*.

1002. DOUBLE INTEGRALS

We begin by defining the double integral of a function $f(x, y)$ over a closed rectangle R, $a \leq x \leq b$, $c \leq y \leq d$, assuming that f is defined there. A **net** $\mathfrak{N}$ on R is a finite system of lines parallel to the coordinate axes, at abscissas $a = a_0 < a_1 < a_2 < \cdots < a_p = b$ and at ordinates $c = c_0 < c_1 < c_2 < \cdots < c_q = d$ (cf. Fig. 1001). The rectangle R is thus cut by a net into a finite number of smaller closed rectangles $R_1, R_2, \cdots, R_n$ $(n = pq)$, with areas $\Delta A_1, \Delta A_2, \cdots, \Delta A_n$ (where the area of a rectangle is *defined* to be the product of its width and its length). The **norm** $|\mathfrak{N}|$ of a net $\mathfrak{N}$ is the greatest length of the diagonals of the rectangles $R_1, R_2, \cdots, R_n$. Let (x_i, y_i) be an arbitrary point of R_i, $i = 1, 2, \cdots, n$, and form the sum

(1) $$\sum_{i=1}^{n} f(x_i, y_i) \Delta A_i = f(x_1, y_1) \Delta A_1 + \cdots + f(x_n, y_n) \Delta A_n.$$

The integral of f is defined as the limit of the sums (1):

Definition I. *A function $f(x, y)$, defined on the closed rectangle R: $a \leq x \leq b$, $c \leq y \leq d$, **is integrable** there if and only if the limit*

(2) $$\lim_{|\mathfrak{N}| \to 0} \sum_{i=1}^{n} f(x_i, y_i) \Delta A_i,$$

in the sense of Definition I, § 401, and with the preceding notation, exists and

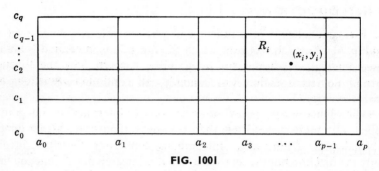

FIG. 1001

*is finite. In case of integrability the **integral** or **double integral** of f over R is the limit* (2):

(3) $$\iint_R f(x, y)\, dA \equiv \lim_{|\mathfrak{N}| \to 0} \sum_{i=1}^{n} f(x_i, y_i) \Delta A_i.$$

Integration over an arbitrary (nonempty) bounded set is defined as follows:

Definition II. *If a function $f(x, y)$ is defined on a bounded set S, let the function $f_S(x, y)$ be defined to be equal to $f(x, y)$ if (x, y) is a point of S, and zero otherwise, and let R be any closed rectangle, with sides parallel to the coordinate axes, containing S. Then f is **integrable on** (or **over**) S if and only if f_S is integrable on R, and* (*in case of integrability*) *the **integral** or **double integral** of f over S is defined by the equation*

(4) $$\iint_S f(x, y)\, dA \equiv \iint_R f_S(x, y)\, dA.$$

NOTE. Definition II is independent of the choice of the rectangle R (cf. Ex. 17, § 1010).

As in the case of a function of one variable (§ 401), every integrable function is bounded and its integral is unique (cf. Ex. 18, § 1010, Theorem II, § 1006).

Under suitable restrictions on the function $f(x,y)$, the integral of Definition I exists, and under further restrictions on the set S the integral of Definition II exists (Theorem II, § 1003, Theorem XI, § 1006).

1003. AREA

Area has at this point been defined only for rectangles. We now extend its definition:

Definition I. *A bounded set S **has area** if and only if the function $f(x, y) \equiv 1$ is integrable over S. In case S has area, its **area** is*

$$A(S) \equiv \iint_S 1\, dA = \iint_S dA.$$

NOTE 1. The concepts of *having area* and of *area* are independent of the rectangular coordinate system. (Cf. Ex. 26, § 1010.)

NOTE 2. For rectangles the new definition of area is consistent with the original. That is, according to Definition I, the area of any rectangle is equal to the product of its width and its length. (Cf. Note 1.)

NOTE 3. If $f(x)$ is integrable on $[a, b]$ and if $f(x) \geq 0$ there, then the integral $\int_a^b f(x)\, dx$ can be interpreted as the area under the curve $y = f(x)$, above the x-axis, between $x = a$ and $x = b$. (Cf. Note 2, § 1008.)

★NOTE 4. The concept defined above as *area*, when extended to Euclidean spaces of arbitrary dimension, is known as **content** (or **Jordan content**, after the French mathematician C. Jordan (1838–1922)). In E_3 it is known as **volume**. Its most valuable generalization is *Lebesgue measure*. Any finite set in any Euclidean space clearly has content zero (and *a fortiori* measure zero). Any subset of a set of content zero is a set of content zero. The set of rational numbers of the unit interval [0, 1], as a set in E_1, has no Jordan content, zero or otherwise, by virtue of the nonintegrability of the function of Example 6, § 201, discussed in § 401. This same set, when considered as a set in E_2, has content or area zero (cf. Ex. 19, § 1010).

It is convenient to have a simple method of establishing whether or not a given set has area. We shall show in Theorem IV, § 1005, that a necessary and sufficient condition for the existence of area for a bounded set is that *its frontier have zero area.* By means of this result we shall be able to prove the following theorem (cf. § 1005):

Theorem I. *Any bounded set whose frontier consists of a finite number of rectifiable simple arcs has area.*

In terms of area we can now state a convenient two-dimensional analogue of the integrability of a continuous function of one variable over a closed interval (Theorem VIII, § 401):

Theorem II. *A function $f(x, y)$ bounded and continuous on a set with area is integrable over that set.*

This theorem is a corollary of Theorem XI, § 1006.

Some of the most important properties of area are contained in the theorem:

*****Theorem III.** *Any subset of a set with zero area is a set with zero area. If S_1 and S_2 are bounded sets with area, and if their intersection has zero area, then their union has area and*

(1) $$A(S_1 \cup S_2) = A(S_1) + A(S_2).$$

In general, if $S_1, S_2, \cdots, S_n$ are bounded sets with area such that every pair has an intersection of zero area, then

(2) $$A(S_1 \cup S_2 \cup \cdots \cup S_n) = A(S_1) + A(S_2) + \cdots + A(S_n).$$

(Cf. § 601 for point set notation.)

The proof is left to the reader in Exercise 20, § 1010. (Also cf. *RV*, Theorem III, § 1303.)

We conclude this section with an example of a bounded set without area.

Example. Let S consist of the rational points in the unit square $0 \leq x \leq 1$, $0 \leq y \leq 1$ (that is, both coordinates are rational; cf. Example 10, § 602). Let $f(x, y) \equiv 1$, and let $\mathfrak{N}$ be an arbitrary net over the unit square. Then in every rectangle R_i of $\mathfrak{N}$ there are points where $f_S(x, y) = 1$ and points where $f_S(x, y) = 0$. Therefore the sums $\sum_{i=1}^{n} f_S(x_i, y_i) \Delta A_i$ may have values ranging from 0 (in case *all* of the points (x_i, y_i) are chosen so that $f_S(x_i, y_i) = 0$) to 1 (in case *all* of the points are chosen so that $f_S(x_i, y_i) = 1$). This is true however small the norm of the net may be, so that a limit cannot exist. Therefore f_S is not integrable over the unit square, f is not integrable over S, and S does not have area. The frontier of S is the closed unit square, whose area is equal to 1.

1004. SECOND FORMULATION OF THE DOUBLE INTEGRAL

In § 1002 the double integral of a function over a closed rectangle was formulated by means of cutting this rectangle up into smaller closed rectangles, where any two of these rectangles have at most one side in common. A very natural extension of this process (sometimes informally called the **cracked china** definition of the double integral) is a **partition** Π of a bounded set S with area into a finite number of closed **pieces** $S_1, S_2, \cdots, S_n$, each with area, such that any two of these pieces have at most a set of zero area in common. (Cf. Fig. 1002.) The **diameter** $\delta(A)$ of á compact set A is the maximum distance between any two of its points (cf. Fig. 1002, and Ex. 18, § 618). The **norm** $|\Pi|$ of a partition Π is the greatest of the diameters

§ 1005] INNER AND OUTER AREA. AREA CRITERION   339

$\delta(S_1), \cdots, \delta(S_n)$ of its pieces. If (x_i, y_i) is an arbitrary point of S_i and if ΔA_i denotes its area, we can form sums similar to (1), § 1002:

(1) $$\sum_{i=1}^{n} f(x_i, y_i) \Delta A_i = f(x_1, y_1) \Delta A_1 + \cdots + f(x_n, y_n) \Delta A_n.$$

The matter of special importance is the existence of the limit of these sums:

Theorem. *With the preceding notation, if $f(x, y)$ is integrable over S, the limit of the sums (1), as the norm of the partition tends toward zero, exists and is equal to the integral of f over S:*

(2) $$\lim_{|\Pi| \to 0} \sum_{i=1}^{n} f(x_i, y_i) \Delta A_i = \iint_S f(x, y) \, dA.$$

A proof of a simple but useful special case is given in § 1007. For a proof of the general theorem, see *RV*, § 1307.

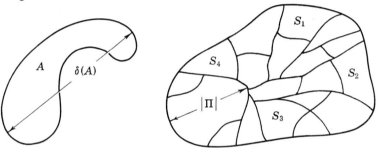

FIG. 1002

★1005. INNER AND OUTER AREA. CRITERION FOR AREA

Let S be a bounded set contained in a closed rectangle R with sides parallel to the coordinate axes, let $f(x, y) \equiv 1$ in S, let a net $\mathfrak{N}$ be imposed on R, and consider the sum

(1) $$\sum_{i=1}^{n} f_S(x_i, y_i) \Delta A_i,$$

whose limit, as $|\mathfrak{N}| \to 0$, defines the area $A(S)$. Our immediate goal is to determine the extreme values of (1), for a fixed $\mathfrak{N}$, as the points (x_i, y_i) vary.

Accordingly, we look at an arbitrary term of (1) and observe that there are three possibilities: (*i*) if the rectangle R_i is completely contained in S, then $f_S(x_i, y_i) \Delta A_i = \Delta A_i$ for all choices of points (x_i, y_i); (*ii*) if R_i intersects both S and its complement, then $f_S(x_i, y_i) \Delta A_i = \Delta A_i$ or 0 according as (x_i, y_i) belongs to S or to the complement; (*iii*) if R_i has no points in common with S, then $f_S(x_i, y_i) \Delta A_i = 0$. Therefore (cf. Fig. 1003) the smallest possible value of (1), obtained by choosing the points (x_i, y_i) outside S whenever possible, is the total area of the rectangles of $\mathfrak{N}$ contained in S (doubly cross-hatched in Fig. 1003), while the largest possible value, obtained

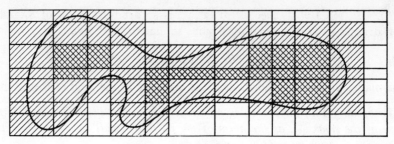

FIG. 1003

by choosing (x_i, y_i) in S whenever possible, is the total area of the rectangles of $\mathfrak{N}$ that intersect S (singly or doubly cross-hatched in Fig. 1003). That is, if we define:

(2) $\qquad a(\mathfrak{N}) \equiv$ total area of all $R_i \subset S$,

(3) $\qquad A(\mathfrak{N}) \equiv$ total area of all R_i intersecting S,

then

(4) $$a(\mathfrak{N}) \leq \sum_{i=1}^{n} f_S(x_i, y_i) \Delta A_i \leq A(\mathfrak{N}).$$

In terms of (2) and (3) we define two important concepts:

Definition. *The **inner area** and **outer area** of a bounded set S, denoted $\underline{A}(S)$ and $\bar{A}(S)$, respectively, are defined:*

$$\underline{A}(S) \equiv \sup_{\mathfrak{N}} a(\mathfrak{N}), \quad \bar{A}(S) \equiv \inf_{\mathfrak{N}} A(\mathfrak{N}),$$

formed with respect to all nets $\mathfrak{N}$ over a closed rectangle containing S.

Example. The set of the Example, § 1003, has inner area 0 and outer area 1.

We observe in passing that whether a bounded set has area or not, it always has both inner and outer area. A limit formulation for inner and outer area follows:

Theorem I. *If S is a bounded set, the limits of $a(\mathfrak{N})$ and $A(\mathfrak{N})$, as $|\mathfrak{N}| \to 0$, exist and equal the inner and outer area, respectively:*

$$\lim_{|\mathfrak{N}| \to 0} a(\mathfrak{N}) = \underline{A}(S), \quad \lim_{|\mathfrak{N}| \to 0} A(\mathfrak{N}) = \bar{A}(S).$$

Outline of Proof. We shall indicate only the proof that $a(\mathfrak{N}) \to \underline{A} \equiv \underline{A}(S)$. If $\epsilon > 0$, first choose a net $\mathfrak{N}_1$ such that $a(\mathfrak{N}_1) > \underline{A} - \frac{\epsilon}{2}$. Then let $\delta > 0$ be such that $|\mathfrak{N}| < \delta$ implies that the total area of those rectangles of $\mathfrak{N}$ that intersect the lines of $\mathfrak{N}_1$ is less than $\epsilon/2$. It follows that $a(\mathfrak{N}) > \underline{A} - \epsilon$. Complete details are requested in Exercise 21, § 1010.

§ 1005] INNER AND OUTER AREA. AREA CRITERION 341

Proofs of the following two theorems are left to the reader (Ex. 22, § 1010):

Theorem II. *If S is a bounded set,*

(5) $$\underline{A}(S) \leq \bar{A}(S),$$

equality holding if and only if S has area. In case of equality, $A(S) = \underline{A}(S) = \bar{A}(S)$.

Theorem III. *A bounded set has zero area if and only if its outer area is zero; that is, if and only if corresponding to $\epsilon > 0$ there exists a net $\mathfrak{N}$ such that $A(\mathfrak{N}) < \epsilon$.*

Since $A(\mathfrak{N}) - a(\mathfrak{N})$, for a given net $\mathfrak{N}$, is the total area of the rectangles of $\mathfrak{N}$ that intersect both S and its complement, and since the frontier $F(S)$ of S consists of points p such that every neighborhood of p intersects both S and its complement, the following theorem should seem plausible:

Theorem IV. *A bounded set has area if and only if its frontier has zero area.*

Proof. (i) Assume that S has area, and let the containing rectangle R be sufficiently large to contain the closure of S in its interior. If ϵ is a given positive number, choose a net $\mathfrak{N}$ such that $A(\mathfrak{N}) - a(\mathfrak{N}) < \epsilon$, and consider the rectangles $\{R_{i_k}\}$ that intersect both S and S'. We shall show that every point of $F(S)$ belongs to at least one R_{i_k}, so that $\bar{A}(F(S)) < \epsilon$, and consequently $A(F(S)) = 0$. If $p \in F(S)$, then three possibilities present themselves. 1: p is in the interior of some R_j of $\mathfrak{N}$; then R_j is one of $\{R_{i_k}\}$. 2: p is a point of a common edge of exactly two rectangles R_j and R_l; then at least one of R_j and R_l must be one of $\{R_{i_k}\}$. 3: p is a common vertex of four rectangles of $\mathfrak{N}$; then at least one of these four must be one of $\{R_{i_k}\}$. (These last three parts follow readily by the indirect method of proof.)

(ii) Assume that $A(F(S)) = 0$, let $\epsilon > 0$, and choose a net $\mathfrak{N}$ such that the total area of all rectangles $\{R_{i_k}\}$ of $\mathfrak{N}$ that intersect $F(S)$ is less than ϵ. We shall show that every rectangle of $\mathfrak{N}$ that intersects both S and S' is one of $\{R_{i_k}\}$. Assume the contrary: let R_j be a rectangle containing a point $p: (x_0, y_0)$ of S and a point $q: (x_1, y_1)$ of S', but no point of $F(S)$. For the segment $I: (x_0 + (x_1 - x_0)t, y_0 + (y_1 - y_0)t), 0 \leq t \leq 1$, joining p and q, let t_2 be the least upper bound of the values of t for points of I that belong to S, and let r be the corresponding point of I. Then $r \in F(S)$, since if $r \in S$ then r is a limit point of S', and if $r \in S'$ then r is a limit point of S. (Contradiction.)

Theorem I, § 1003, is an immediate corollary of the preceding theorem and the following:

Theorem V. *Any rectifiable simple arc has zero area.*

Proof. Let the simple arc C have length L, and let $\epsilon > 0$. Enclose C in a closed square and impose on this square a net $\mathfrak{N}$ that cuts the square into smaller squares, all of side δ, where $0 < \delta < \min\left(L, \dfrac{\epsilon}{8L}\right)$. Let $k \equiv \left[\dfrac{L}{\delta}\right] + 1$ (cf. Example 5, § 201), so that $\dfrac{L}{\delta} < k \leq \dfrac{L}{\delta} + 1$, and cut C into k pieces, each of length less than δ (this is possible since $L/k < \delta$). Each of these pieces can intersect at most 4 squares of $\mathfrak{N}$ (this is true since each has length less than the side of each square). Since the area of four squares is $4\delta^2$, the total area of the squares containing points of C is not greater than

$$4k\delta^2 \leq 4\delta^2 \left(\frac{L+\delta}{\delta}\right) < 4\delta(L+L) < 8L(\epsilon/8L) = \epsilon.$$

With the aid of Theorem III, the proof is complete.

NOTE 1. *A simple arc of infinite length may have zero area.* The graph of $y = x \sin(1/x)$ for $0 \leq x \leq 1$ is an example. (Cf. Example 1, § 804.)

NOTE 2. The following sets may fail to have area: a simple arc (and therefore a compact set); a bounded region; a compact region. (Cf. *RV*, Notes 2, 3, § 1305.)

★1006. THEOREMS ON DOUBLE INTEGRALS

In this section we consider some of the most important properties of the double integral. A few are direct extensions of theorems of §§ 401 and 403 and need virtually no change in proof. Others require an entirely new formulation. Theorem XI states sufficient conditions for integrability of a function f over a set S in terms of the continuity of f and the nature of S. All of these theorems can be immediately translated into theorems for any Euclidean space E_n, $n = 1, 2, 3, \cdots$, the word *area* being replaced by the word *content* (or, for E_3, *volume*). Discussion in this section is limited to the plane E_2.

We start by introducing a new concept, which is helpful in simplifying many statements:

Definition I. *A property $P(p)$ which for each point p of a set S is either true or false is said to hold **very nearly everywhere** in S if and only if the subset of S for which P is false is of zero area.*

Examples. The function $f(x,y) \equiv x - y$ is very nearly everywhere nonzero in the unit square $0 \leq x \leq 1$, $0 \leq y \leq 1$. The function $f(x,y) \equiv 2x^2 + 3y^2$ is very nearly everywhere positive in the unit disk $x^2 + y^2 \leq 1$. The function $f(x,y) \equiv xy/(x^2+y^2)$, $x^2 + y^2 \neq 0$, $f(0,0) \equiv 0$, is very nearly everywhere continuous in the unit disk.

NOTE 1. A well-known generalization of *very nearly everywhere* is *almost everywhere*, based on Lebesgue measure zero instead of zero area. (For the space E_1 cf. *RV*, Exs. 52–54, § 503.)

Our first theorem makes use of Definition I.

Theorem I. *A bounded function that is equal to zero very nearly everywhere on a bounded set S is integrable on S, and its integral over S is zero.*

Proof. Assume $|f(x, y)| < K$ on S, let Z be a subset of S with zero area such that $f(x, y) = 0$ on $S - Z$, and let $\epsilon > 0$. Extend f to a closed rectangle $R \supset S$ by means of the function f_S, and let $\mathfrak{N}$ be a net over R such that relative to the set Z, $A(\mathfrak{N}) < \epsilon/K$. Then, since for every point (x, y) belonging to a rectangle not contributing to $A(\mathfrak{N})$, $f_S(x, y) = 0$, we have the inequalities

$$\left| \sum_{i=1}^{n} f_S(x_i, y_i) \Delta A_i \right| \leq K \cdot A(\mathfrak{N}) < \epsilon,$$

and the proof is complete.

Theorems II-V are immediate extensions of Theorems I-IV, § 401, and have analogous proofs (cf. Ex. 23, § 1010). The set S is assumed to be bounded.

Theorem II. *If $\lim\limits_{|\mathfrak{N}| \to 0} \sum\limits_{i=1}^{n} f(x_i, y_i) \Delta A_i$ exists, the limit is unique.*

Theorem III. *If $f(x, y)$ and $g(x, y)$ are integrable over S, and if $f(x, y) \leq g(x, y)$, there, then*

$$\iint_S f(x, y) \, dA \leq \iint_S g(x, y) \, dA.$$

Theorem IV. *If $f(x, y)$ is integrable over S and if k is a constant, then $kf(x, y)$ is integrable over S and*

$$\iint_S kf(x, y) \, dA = k \iint_S f(x, y) \, dA.$$

Theorem V. *If $f(x, y)$ and $g(x, y)$ are integrable over S, then so are their sum and difference, and*

$$\iint_S [f(x, y) \pm g(x, y)] \, dA = \iint_S f(x, y) \, dA \pm \iint_S g(x, y) \, dA.$$

A similar statement holds for an arbitrary finite number of functions.

The extension of Theorem VI, § 401, is:

Theorem VI. *If $f(x, y)$ and $g(x, y)$ are bounded on a bounded set S and equal very nearly everywhere there, then the integrability of either implies that of the other, and the equality of their integrals.*

Proof. Assume that $f(x, y)$ is integrable on S and let $h(x, y) \equiv g(x, y) - f(x, y)$. Then by Theorem I, $h(x, y)$ is integrable over S, and since $g(x, y) = f(x, y) + h(x, y)$ the result follows from Theorem V.

Theorem V, § 401, takes the form:

Theorem VII. *If S_1 and S_2 are bounded sets having an intersection $Z \equiv S_1 \cap S_2$ of zero area, and if $f(x, y)$ is integrable over S_1 and integrable over S_2, then $f(x, y)$ is integrable over their union $S_1 \cup S_2$, and*

$$\iint_{S_1 \cup S_2} f(x, y)\, dA = \iint_{S_1} f(x, y)\, dA + \iint_{S_2} f(x, y)\, dA.$$

A similar result holds for any finite number of bounded sets each pair of which has an intersection of zero area.

Proof. Let R be a closed rectangle containing $S_1 \cup S_2$, and define $f_i(x, y)$ ($i = 1, 2$) over R to be equal to $f(x, y)$ if $(x, y) \in S_i$ and equal to 0 otherwise. Also define $g(x, y)$ to be equal to $f(x, y)$ if $(x, y) \in S_1 \cup S_2$ and equal to 0 otherwise. Since f_1 and f_2 are integrable over R, then (Theorem V) so is $f_1 + f_2$, and

$$\iint_R (f_1 + f_2)\, dA = \iint_R f_1\, dA + \iint_R f_2\, dA = \iint_{S_1} f\, dA + \iint_{S_2} f\, dA.$$

On the other hand, $g(x, y) = f_1(x, y) + f_2(x, y)$ except possibly on the set Z, and the result is a consequence of Theorem VI. The general result for a finite number of sets follows by mathematical induction.

Theorem VII, § 401, becomes:

Theorem VIII. *If $f(x, y)$ is constant, $f(x, y) \equiv k$, on a bounded set S with area, then $f(x, y)$ is integrable there and*

$$\iint_S f(x, y)\, dA = kA(S).$$

Proof. When $k = 1$ the theorem is true by definition, and for other values of k it follows from Theorem IV.

Sufficiency conditions for integrability can be obtained with the aid of step-functions (cf. § 403):

Definition II. *A **step-function** is a bounded function that is constant in the interior of each rectangle of some net over a closed rectangle.*

Theorem IX. *Any step-function $\sigma(x, y)$ is integrable, and if $\sigma(x, y)$ has the values $\sigma_1, \cdots, \sigma_n$ in the interiors of the rectangles $R_1, \cdots, R_n$, respectively, then*

$$\iint_R \sigma(x, y)\, dA = \sum_{i=1}^n \sigma_i \Delta A_i.$$

Proof. This is an immediate consequence of Theorems VI, VII, and VIII.

Theorem X. *A function $f(x, y)$, defined on a bounded set S, is integrable there if and only if corresponding to $\epsilon > 0$ there exist step-functions $\sigma(x, y)$ and $\tau(x, y)$ defined on a closed rectangle $R \supset S$ such that*

(1) $$\sigma(x, y) \leq f(x, y) \leq \tau(x, y)$$

on S, $\sigma(x, y) \leq \tau(x, y)$ on R, and

(2) $$\iint_R [\tau(x, y) - \sigma(x, y)]\, dA < \epsilon.$$

Proof. The proof is identical (*mutatis mutandis*) with that of Theorem II, § 403. (Give the details in Ex. 24, § 1010.)

Theorem XI. *A function $f(x, y)$, defined and bounded on a bounded set S with area, and continuous very nearly everywhere there, is integrable over S.*

Proof. As a preliminary we shall show that it can be assumed without loss of generality that S is a closed rectangle with sides parallel to the coordinate axes. This is done by showing that under the assumptions of the theorem, the extension f_S of f to a closed rectangle R containing S is continuous very nearly everywhere on R. (The integrability of f_S over R, then, implies that of f over S.) This is true since if Z is a subset of S of zero area such that f is continuous on $S - Z$, the only possible points of discontinuity of f_S are members either of Z or of $F(S)$, the frontier of S (prove this!). Since both Z and $F(S)$ have zero area, so does their union $Z \cup F(S)$ (prove this!).

In the remaining parts of the proof we shall assume that S is a closed rectangle with sides parallel to the coordinate axes.

Let Z be a subset of the rectangle S, of zero area, such that f is continuous on $S - Z$, and let $\epsilon > 0$. We seek step-functions $\sigma(x, y)$ and $\tau(x, y)$ satisfying (1) and (2) for S. The first step in this direction is to subdivide S by a net $\mathfrak{N}_0$, to be held fixed, such that the total area of the rectangles of $\mathfrak{N}_0$ that intersect Z is less than $\epsilon/4K$, where $|f(x, y)| < K$ on S. Define $\sigma(x, y)$ and $\tau(x, y)$ on *these* rectangles to be equal to $-K$ and K, respectively. Then $\sigma(x, y) < f(x, y) < \tau(x, y)$ there, and the integral $\iint [\tau(x, y) - \sigma(x, y)]\, dA$ over these rectangles is less than $2K(\epsilon/4K) = \epsilon/2$.

Finally, in each closed rectangle $R_i^{(0)}$ of $\mathfrak{N}_0$ that lies in $S - Z$, f is continuous, and therefore uniformly continuous, so that there exists a $\delta_i > 0$ such that $\sqrt{(x - x')^2 + (y - y')^2} < \delta_i$ implies $|f(x, y) - f(x', y')| < \epsilon/2A$, where $A = A(S)$. Let δ be the least of these δ_i, and adjoin further lines to those of $\mathfrak{N}_0$ to obtain a net $\mathfrak{N}$ of norm $< \delta$. For *this* net $\mathfrak{N}$, define σ and τ, where they are not already defined, as follows: on the lines of $\mathfrak{N}$, $\sigma \equiv \tau \equiv f$. In the interior of a rectangle R_i of $\mathfrak{N}$, define $\sigma(x, y) \equiv \sigma_i \equiv \inf\limits_{(x,y) \in R_i} f(x, y)$ and $\tau(x, y) \equiv \tau_i \equiv \sup\limits_{(x,y) \in R_i} f(x, y)$. Then (1) is clearly satisfied, and (2) follows from $\iint_S [\tau(x, y) - \sigma(x, y)]\, dA < \dfrac{2K\epsilon}{4K} + \dfrac{\epsilon}{2A} A = \epsilon.$ This completes the proof.

Theorem XII. *If f and g are continuous on a set S of positive area, and if $f \leq g$ there and f and g are not identically equal in $I(S)$, then*

$$\iint_S f\,dA < \iint_S g\,dA.$$

Proof. Let $h(x, y) \equiv g(x, y) - f(x, y)$, and assume $h(a, b) > 0$, where $(a, b) \in I(S)$. Let ϵ be a positive number and $N_{(a,b)} \subset S$ be a square neighborhood of (a, b) such that $h(x, y) \geq \epsilon$ in $N_{(a,b)}$ (cf. Ex. 18, § 212), and define $\phi(x, y) \equiv \epsilon$ if $(x, y) \in N_{(a,b)}$ and $\phi(x, y) \equiv 0$ otherwise. Then (Theorem III) $h \geq \phi$, and

$$\iint_S g\,dA - \iint_S f\,dA = \iint_S h\,dA \geq \iint_S \phi\,dA = \epsilon A(N_{(a,b)}) > 0.$$

Theorem XIII. Mean Value Theorem for Double Integrals. *If $f(x, y)$ is continuous on a compact region R with area, there exists a point (ξ, η) in the interior of R such that*

$$\iint_R f(x, y)\,dA = f(\xi, \eta) \cdot A(R).$$

Proof. Let $m \equiv \min_{(x,y) \in R} f(x, y)$, $M \equiv \max_{(x,y) \in R} f(x, y)$, and assume $m < M$ (otherwise f is constant). Then $m \leq f \leq M$, and f is not identically equal to either m or M in $I(R)$. Hence (Theorem XII),

$$mA(R) < \iint_R f\,dA < MA(R).$$

Therefore (Theorem IV, § 610) there exists a point $(\xi, \eta) \in I(R)$ such that $f(\xi, \eta) = \left[\iint_R f\,dA\right] / A(R)$.

*1007. PROOF OF THE SECOND FORMULATION

In this section we shall give a proof of the Theorem of § 1004:

(1) $$\lim_{|\Pi| \to 0} \sum_{i=1}^{n} f(x_i, y_i) \Delta A_i = \iint_S f(x, y)\,dA,$$

for the special case where f is continuous and every closed piece S_i of the partition Π is a *compact region*. (For the general proof, see *RV*, § 1307.)

We assume, then, that the integral on the right of (1) exists, and let $\epsilon > 0$ be given. Since every piece of the partition Π is compact, so is S itself (why?), and f is uniformly continuous there (cf. § 617). Consequently there exists a $\delta > 0$ such that whenever the distance between the points (x, y) and (ξ, η) is less than δ, $|f(x, y) - f(\xi, \eta)| < \epsilon/A(S)$, where $A(S)$ is the area of S.

By Theorems VII and XIII, § 1006,

$$\iint_S f(x, y)\, dA = \sum \iint_{S_i} f(x, y)\, dA = \sum f(\xi_i, \eta_i)\, \Delta A_i,$$

where $(\xi_i, \eta_i) \in I(S_i)$, $i = 1, 2, \cdots, n$. Therefore, if $|\Pi| < \delta$,

$$\left| \sum f(x_i, y_i)\, \Delta A_i - \iint_S f(x, y)\, dA \right| = \left| \sum f(x_i, y_i)\, \Delta A_i - \sum f(\xi_i, \eta_i)\, \Delta A_i \right|$$
$$\leq \sum |f(x_i, y_i) - f(\xi_i, \eta_i)|\, \Delta A_i < \frac{\epsilon}{A(S)} \sum \Delta A_i = \epsilon.$$

The limit (1) is thus established, and the proof is complete.

1008. ITERATED INTEGRALS, TWO VARIABLES

Under suitable conditions on the functions $f(x, y)$, $\phi(x)$, and $\psi(x)$, the integral

(1) $$\int_{\phi(x)}^{\psi(x)} f(x, y)\, dy,$$

which is a function of x (and not of y), is integrable over an interval $[a, b]$ (cf. § 930). In this case the integral

(2) $$\int_a^b \left\{ \int_{\phi(x)}^{\psi(x)} f(x, y)\, dy \right\} dx = \int_a^b \int_{\phi(x)}^{\psi(x)} f(x, y)\, dy\, dx$$

is called an **iterated integral** or, more completely, a **two-fold iterated integral**.

Similarly, a two-fold iterated integral with the order of integration reversed may sometimes be formed:

(3) $$\int_c^d \left\{ \int_{\alpha(y)}^{\beta(y)} f(x, y)\, dx \right\} dy = \int_c^d \int_{\alpha(y)}^{\beta(y)} f(x, y)\, dx\, dy.$$

Example 1. $$\int_0^1 \int_0^x (x^2 + 4xy)\, dy\, dx = \int_0^1 [x^2 y + 2xy^2]_0^x\, dx$$
$$= \int_0^1 (x^3 + 2x^3)\, dx = \int_0^1 3x^3\, dx = \tfrac{3}{4}.$$

The importance of iterated integrals lies in their use for the evaluation of double integrals. This corresponds to the evaluation of definite integrals of functions of one variable by means of indefinite integrals (the Fundamental Theorem of Integral Calculus, § 405). We state a corresponding Fundamental Theorem now, and prove it (in a more general form) in the following section.

Theorem I. Fundamental Theorem for Double Integrals. *If $f(x, y)$ is defined over a bounded set S, bounded below and above by the graphs of functions $\phi(x)$ and $\psi(x)$, respectively, and at the extreme left and right by the lines $x = a$*

and $x = b$ (that is, S consists of all (x, y) such that $a \leq x \leq b$ and $\phi(x) \leq y \leq \psi(x)$; cf. Fig. 1004), then the following relation holds whenever both integrals exist:

(4) $$\iint_S f(x, y) \, dA = \int_a^b \int_{\phi(x)}^{\psi(x)} f(x, y) \, dy \, dx.$$

A similar statement holds with an interchange in the roles of x and y (cf. Fig. 1004):

(5) $$\iint_S f(x, y) \, dA = \int_c^d \int_{\alpha(y)}^{\beta(y)} f(x, y) \, dx \, dy.$$

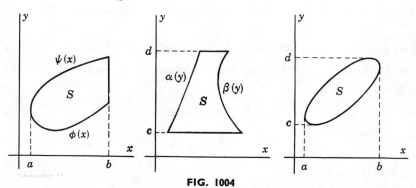

FIG. 1004

If S is bounded above and below and to the left and right by graphs of bounded functions (cf. Fig. 1004), then the double integral and the iterated integrals of (4) and (5) are all equal.

NOTE 1. With the preceding results as background, the following two alternative notations for a double integral are often used:

$$\iint_S f(x, y) \, dx \, dy = \iint_S f(x, y) \, dy \, dx.$$

Example 2. Evaluate $I = \iint_S x^2 y \, dA$ in two ways by means of iterated integrals, where S is the region between the parabola $x = y^2$ and the line $x + y = 2$.

Solution. The simpler *formulation* uses x as the first variable of integration:

$$I = \int_{-2}^{1} \int_{y^2}^{2-y} x^2 y \, dx \, dy = \int_{-2}^{1} \frac{y}{3}[x^3]_{y^2}^{2-y} \, dy = -15\tfrac{3}{40}.$$

With y as the first variable of integration, the iterated integral must be expressed as the sum of two:

$$I = \int_0^1 \int_{-\sqrt{x}}^{\sqrt{x}} x^2 y \, dy \, dx + \int_1^4 \int_{-\sqrt{x}}^{2-x} x^2 y \, dy \, dx = 0 - 15\tfrac{3}{40} = -15\tfrac{3}{40}.$$

The *evaluation* details turn out to be easier with the second formulation.

The following relation between integrals and area between curves is a simple consequence of the Fundamental Theorem of this section (cf. (xii), § 1011, for a corresponding formula for volume between surfaces):

Theorem II. *If the set S of Theorem I has area, and if $\psi(x) - \phi(x)$ is integrable over $[a, b]$, then*

(6) $$A(S) = \int_a^b [\psi(x) - \phi(x)]\, dx.$$

Proof. In formula (4), let $f(x, y) \equiv 1$.

★NOTE 2. It can be proved (cf. *RV*, § 1309) that the existence of the left-hand member of (6) implies that of the right-hand member, and that the existence of the right-hand member of (6) implies that of the left-hand member if and only if ϕ and ψ are both integrable. As a consequence, *if $f(x)$ is a nonnegative function on (a, b), and if S is the set of points (x, y) such that $a \leq x \leq b$ and $0 \leq y \leq f(x)$, then S has area if and only if $f(x)$ is integrable and, in case of existence,*

$$A(S) = \int_a^b f(x)\, dx.$$

★1009. PROOF OF THE FUNDAMENTAL THEOREM

We establish first a basic theorem, whose title is borrowed from a similar but more elegant theorem in Lebesgue Theory, proved by the Italian mathematician G. Fubini in 1910. (Cf. *RV*, § 1309, for a generalization.)

Theorem I. Fubini's Theorem. *If $f(x, y)$ is defined over the closed rectangle $R: a \leq x \leq b, c \leq y \leq d$, then*

(1) $$\iint_R f(x, y)\, dA = \int_a^b \int_c^d f(x, y)\, dy\, dx,$$

whenever both of these integrals exist.

Proof. The first step is to observe that this theorem is true for step-functions (Ex. 25, § 1010). Next assume that $f(x, y)$ is integrable over R, let $\epsilon > 0$, and form the related step-functions $\sigma(x, y)$ and $\tau(x, y)$ (§ 1006) such that $\sigma \leq f \leq \tau$, and $\iint [\tau - \sigma]\, dA < \epsilon$. The equality (1) results from the two sets of inequalities:

(2) $$\iint_R \sigma(x, y)\, dA \leq \iint_R f(x, y)\, dA \leq \iint_R \tau(x, y)\, dA,$$

(3) $$\int_a^b \int_c^d \sigma(x, y)\, dy\, dx \leq \int_a^b \int_c^d f(x, y)\, dy\, dx \leq \int_a^b \int_c^d \tau(x, y)\, dy\, dx.$$

The Fundamental Theorem for Double Integrals (Theorem I, § 1008) is a direct consequence of Fubini's Theorem, obtained by extending the domain of definition of $f(x, y)$ from the given set S of Theorem I, § 1008, to a closed rectangle R by defining $f(x, y)$ to be identically zero outside S.

An immediate corollary to Fubini's Theorem is the interchangeability of the order of integration in an iterated integral (cf. Note 2, § 930, and the Note, § 1408):

Theorem II. *If $f(x, y)$ is integrable over the closed rectangle $R: a \leq x \leq b$, $c \leq y \leq d$, then the two iterated integrals $\int_a^b \int_c^d f(x, y) \, dy \, dx$ and $\int_c^d \int_a^b f(x, y) \, dx \, dy$ are equal whenever they both exist.*

1010. EXERCISES

In Exercises 1–4, write the double integral $\iint_S f(x, y) \, dA$ in two ways in terms of iterated integrals, where S is the given set. Compute the area of S by setting $f(x, y) \equiv 1$.

1. Bounded by the third degree curves $y^2 = x^3$ and $y = x^3$.
2. Inside both the circle $x^2 + y^2 = 2ax$ and the parabola $x^2 = ay$, $a > 0$.
3. Inside the circle $x^2 + y^2 = 25$ and above the line $3x = 4y$.
4. Inside both parabolas $y = x^2$ and $10x = y^2 - 15y + 24$.

In Exercises 5–8, determine the region over which the integration extends.

5. $\int_0^1 \int_0^{x+1} f(x, y) \, dy \, dx$.
6. $\int_0^2 \int_{y^2-3}^{\frac{1}{2}y} f(x, y) \, dx \, dy$.
7. $\int_0^1 \int_0^{(1-y^3)^{\frac{3}{2}}} f(x, y) \, dx \, dy$.
8. $\int_0^2 \int_{\sqrt{x}}^{\sqrt{20-x^2}} f(x, y) \, dy \, dx$.

In Exercises 9–12, transform the given iterated integral to one (or more) where the order of integration is reversed.

9. $\int_{-1}^2 \int_{-x}^{2-x^2} f(x, y) \, dy \, dx$.
10. $\int_0^2 \int_{y^3}^{4\sqrt{2y}} f(x, y) \, dx \, dy$.
11. $\int_1^4 \int_{y/2}^y f(x, y) \, dx \, dy$.
12. $\int_0^1 \int_0^{x+e^x} f(x, y) \, dy \, dx$.

In Exercises 13–16, evaluate the double integral $\iint_S f(x, y) \, dA$ by means of an iterated integral, for the given function $f(x, y)$ and set S.

13. $f(x, y) = x^2 + xy$; S is the square: $0 \leq x \leq 1, 0 \leq y \leq 1$.
14. $f(x, y) = x + y$; S is the triangular region bounded by $x = 2y$, $x = 6$, $y = 0$.
15. $f(x, y) = xy$; S is the first quadrant portion inside the circle $x^2 + y^2 = a^2$.
16. $f(x, y) = x^2 + y^2$; S is the region bounded by $x = 0$, $y = x$, and $y = e^{-x}$.
★17. Prove that Definition II, § 1002, is independent of the rectangle R. *Hints:* Show first that it is sufficient to establish this result for rectangles R_1 and R_2 such that $R_1 \subset R_2$. It is almost trivial to show that the existence of $\iint_{R_2} f_S(x, y) \, dA$ implies that of $\iint_{R_1} f_S(x, y) \, dA$, and their equality. Proof of the reverse implication

involves constructing thin strips enclosing the edges of R_1 as midlines, and of sufficiently small total area.

*18. Prove that every integrable function $f(x, y)$ is bounded.
*19. Prove the statement in the last sentence of Note 4, § 1003.
*20. Prove Theorem III, § 1003. *Hint:* Use Theorem VII, § 1006.
*21. Complete the details of the proof of Theorem I, § 1005.
*22. Prove Theorems II and III, § 1005.
*23. Prove Theorems II–V, § 1006.
*24. Give the details of the proof of Theorem X, § 1006.
*25. Prove Fubini's Theorem, § 1009, for step-functions.
*26. Prove Notes 1 and 2, § 1003. *Hints:* (i) First prove that *any* rectangle (regardless of position) has area since it is a compact region whose frontier consists of four line segments. (ii) Then show that *area* is invariant under translations

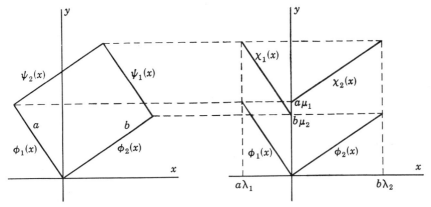

FIG. 1005

of axes. (iii) For a given rectangle R whose sides are not parallel to the coordinate axes, translate so that one vertex is at the origin and the adjacent sides have lengths a and b and direction cosines (λ_1, μ_1) and (λ_2, μ_2), respectively, where $\mu_1 > 0$ and $\mu_2 > 0$. Then (cf. Fig. 1005) by translation of the parts ψ_1 and ψ_2 of the upper curve of R to the functions χ_1 and χ_2 show that the total area of R is equal to $\left| \int_0^{a\lambda_1} (\chi_1 - \phi_1) \, dx \right| + \left| \int_0^{b\lambda_2} (\chi_2 - \phi_2) \, dx \right| = |a\lambda_1| \cdot b\mu_2 + |b\lambda_2| \cdot a\mu_1 = ab(|\lambda_1|\mu_2 + |\lambda_2|\mu_1) = ab(\mu_1^2 + \mu_2^2) = ab$. (iv) Conclude that *area* is invariant under rotations.

*27. Prove that the area of a parallelogram three of whose vertices are (x_i, y_i), $i = 1, 2, 3$, is the absolute value of the determinant

(1) $$\begin{vmatrix} x_1 & y_1 & 1 \\ x_2 & y_2 & 1 \\ x_3 & y_3 & 1 \end{vmatrix}.$$

Hints: First show that (1) is invariant under translations, so that it can be assumed that $x_3 = y_3 = 0$ and that (1) has the form

(2) $$\begin{vmatrix} x_1 & y_1 \\ x_2 & y_2 \end{vmatrix}.$$

Then show that (2) is invariant under rotations, so that it can be assumed that (2) has the form
$$\begin{vmatrix} x_1 & 0 \\ x_2 & y_2 \end{vmatrix} = x_1 y_2.$$
Then compute the area by appropriate subdivision and translation of the parts of the upper curve, as in Exercise 26.

*28. Prove that a function integrable on a bounded set is integrable on any subset that has area. (Cf. Theorem XII, § 401.)

*29. Prove that if $f(x, y)$ is integrable on a bounded set S with area, then so is $|f(x, y)|$. (Cf. Theorem XIII, § 401.)

*30. Prove that the product of two functions integrable on a bounded set with area is integrable on that set. (Cf. Theorem XIV, § 401.)

*31. Prove **Bliss's Theorem** for functions of two variables: *If $f(x, y)$ and $g(x, y)$ are integrable on a bounded set S with area, then the limit*

$$\lim_{|\mathfrak{N}| \to 0} \sum_{i=1}^{n} f_S(x_i, y_i) g_S(x_i', y_i') \Delta A_i$$

exists and is equal to $\iint_S f(x, y) g(x, y) \, dA$. (Cf. Theorem XV, § 401.)

*32. State and prove the analogue of Duhamel's Principle, § 801, for functions of two variables, assuming continuity for all functions involved, and that the set of integration is compact with area.

1011. TRIPLE INTEGRALS. VOLUME

The **triple integral** of a function $f(x, y, z)$ over a closed rectangular parallelepiped R and, more generally, over a bounded set W, is defined (if it exists) in a manner similar to that used for a double integral. We shall omit all details in the text, but call on the student for a few in the exercises of § 1012. A few of the more important facts are listed below:

(*i*) The **integral** of $f(x, y, z)$ over W (if it exists) is the same as the integral of its extension f_W over $R \supset W$, defined in terms of a **net** $\mathfrak{N}$ of planes parallel to the coordinate planes, which subdivide R into smaller closed rectangular parallelepipeds R_i of volume ΔV_i, $i = 1, 2, \cdots, n$:

(1) $$\iiint_W f \, dV = \iiint_R f_W \, dV = \lim_{|\mathfrak{N}| \to 0} \sum_{i=1}^{n} f_W(x_i, y_i, z_i) \Delta V_i.$$

(*ii*) The **volume** of W (if it exists) is defined to be the integral of the function 1 over W:

(2) $$V(W) \equiv \iiint_W 1 \, dV.$$

(*iii*) The definition of (1) is independent of the containing R, and that of (2) is independent of the rectangular coordinate system. (Cf. Ex. 14, § 1012.)

(*iv*) A bounded set has volume if and only if its frontier has zero volume. The analogue of Theorem III, § 1003, holds in E_3. (Cf. Ex. 7, § 1012.)

(v) Any compact smooth surface has zero volume. Therefore any bounded set whose frontier consists of a finite number of compact smooth surfaces has volume. (Cf. Ex. 15, § 1012.)

(vi) A second formulation of the triple integral of f over W (bounded and with volume) is given by partitions Π of W into closed pieces W_i that are compact regions with volume V_i, $i = 1, \cdots, n$:

$$\iiint_W f\,dV = \lim_{|\Pi| \to 0} \sum_{i=1}^{n} f(x_i, y_i, z_i)\,\Delta V_i.$$

(Cf. Ex. 8, § 1012.)

★(vii) The Theorems of § 1006 have analogues in E_3. In particular, step-functions in E_3 have definition and properties similar to those for step-functions in E_2. A function $f(x, y, z)$ is integrable over a bounded set W if and only if corresponding to $\epsilon > 0$ there exist step-functions $\sigma(x, y, z)$ and $\tau(x, y, z)$ defined on a closed rectangular parallelepiped $R \supset W$ such that

$\sigma \leq f \leq \tau$ on W, $\sigma \leq \tau$ on R, and $\iiint_R [\tau - \sigma]\,dV < \epsilon$. (Cf. Ex. 9, § 1012.)

★(viii) A function bounded and very nearly everywhere continuous on a bounded set with volume is integrable there. (Cf. Ex. 10, § 1012.)

(ix) Under suitable conditions a triple integral can be evaluated by means of a (three-fold) **iterated integral** of the form

$$(3) \qquad \int_{a_1}^{a_2} \int_{y_1(x)}^{y_2(x)} \int_{z_1(x,y)}^{z_2(x,y)} f(x, y, z)\,dz\,dy\,dx,$$

or one of the other 5 forms got by permuting the variables.

NOTE. Alternative notations for a triple integral, suggested by (ix) (cf. Note 1, § 1008) are:

$$\iiint_W f(x, y, z)\,dz\,dy\,dx = \iiint_W f(x, y, z)\,dx\,dy\,dz = \cdots.$$

★(x) **Fubini's Theorem.** If $f(x, y, z)$ is defined over R: $a_1 \leq x \leq a_2$, $b_1 \leq y \leq b_2$, $c_1 \leq z \leq c_2$, then

$$(4) \qquad \iiint_R f(x, y, z)\,dV = \int_{a_1}^{a_2} \int_{b_1}^{b_2} \int_{c_1}^{c_2} f(x, y, z)\,dz\,dy\,dx$$

whenever both integrals exist. (Cf. Ex. 11, § 1012.)

(xi) **Fundamental Theorem for Triple Integrals.** If $f(x, y, z)$ is defined over a bounded set W consisting of all (x, y, z) such that $a_1 \leq x \leq a_2$, $y_1(x) \leq y \leq y_2(x)$, and $z_1(x, y) \leq z \leq z_2(x, y)$, then

$$(5) \qquad \iiint_W f(x, y, z)\,dV = \int_{a_1}^{a_2} \int_{y_1(x)}^{y_2(x)} \int_{z_1(x,y)}^{z_2(x,y)} f(x, y, z)\,dz\,dy\,dx$$

whenever both integrals exist. (Cf. Fig. 1006, Ex. 12, § 1012.)

(xii) If S is a bounded set with area in the xy-plane, and if $f(x, y, z)$ is defined over a bounded set W consisting of all (x, y, z) such that $(x, y) \in S$ and $z_1(x, y) \leq z \leq z_2(x, y)$, then

(6) $$\iiint_W f(x, y, z) \, dV = \iint_S \left[\int_{z_1(x,y)}^{z_2(x,y)} f(x, y, z) \, dz \right] dA$$

whenever both integrals exist. In particular, when $f(x, y, z) = 1$,

(7) $$V(W) = \iint_S [z_2(x, y) - z_1(x, y)] \, dA$$

whenever both members exist. (Cf. Fig. 1006, Ex. 13, § 1012.)

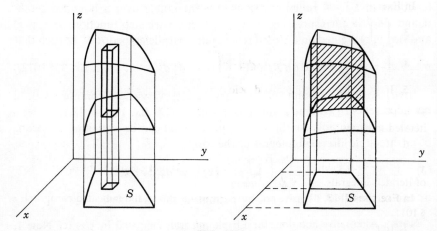

FIG. 1006

(xiii) If S is a bounded set with area in the xy-plane, if $f(x, y)$ is defined, bounded, and nonnegative over S, and if W is the set of all (x, y, z) such that $(x, y) \in S$ and $0 \leq z \leq f(x, y)$, then W has volume if and only if $f(x, y)$ is integrable over S and, in case of existence,

$$V(W) = \iint_S f(x, y) \, dA.$$

(Cf. *RV*, §§ 1309, 1311.)

Example. Write the triple integral $\iiint_W f(x, y, z) \, dV$ as an iterated integral, and compute the volume of W by setting $f(x, y, z) = 1$, where W is the region bounded by the paraboloids $z = x^2 + y^2$ and $2z = 12 - x^2 - y^2$.

Solution. The curve of intersection is the circle $x^2 + y^2 = 4$, $z = 4$. Therefore the integral can be written

$$\int_{-2}^{2} \int_{-\sqrt{4-x^2}}^{\sqrt{4-x}} \int_{x^2+y^2}^{\frac{12-x^2-y^2}{2}} f(x, y, z) \, dz \, dy \, dx.$$

If $f(x, y, z) \equiv 1$, properties of symmetry (that is, of even functions) can be used to simplify this to

$$V = 6 \int_0^2 \int_0^{\sqrt{4-x^2}} (4 - x^2 - y^2) \, dy \, dx = 4 \int_0^2 (4 - x^2)^{\frac{3}{2}} \, dx = 12\pi.$$

1012. EXERCISES

In Exercises 1 and 2, write the triple integral $\iiint_W f(x, y, z) \, dV$ as an iterated integral. Compute the volume of W by setting $f(x, y, z) \equiv 1$.

1. W is the region bounded by the plane $z = 7$ and the paraboloid $z = 23 - x^2 - y^2$.

2. W is the region inside the octahedron $|x| + |y| + |z| = 1$.

In Exercises 3 and 4, find the region of integration.

3. $\int_0^1 \int_{y^2}^{\sqrt{y}} \int_0^{9-x^2-y^2} f(x, y, z) \, dz \, dx \, dy.$

4. $\int_1^e \int_x^{x^2} \int_0^{\ln x} f(x, y, z) \, dz \, dy \, dx.$

5. If W is the region bounded below by the xy-plane and above by the sphere $x^2 + y^2 + z^2 = 2x$, express the triple integral $\iiint_W f(x, y, z) \, dV$ in six ways as iterated integrals.

6. If W is the region bounded by the planes $x = 1$, $y = 0$, $z = 0$, and $x - y - z = 0$, evaluate $\iiint_W x^3 y^2 z \, dV$ in at least 3 of the 6 possible ways in terms of iterated integrals.

In Exercises 7–13, discuss and prove the statement in the indicated paragraph of §1011.

★**7.** (iv).
★**9.** (vii).
★**11.** (x).
★**13.** (xii).

★**8.** (vi).
★**10.** (viii).
★**12.** (xi).

★**14.** Discuss and prove the statements of (iii), §1011. *Hint:* In order to prove that the volume of any rectangular parallelepiped with edges a, b, and c is abc, translate one vertex to the origin and let the neighboring vertices be $(a\lambda_1, a\mu_1, av_1)$, $(b\lambda_2, b\mu_2, bv_2)$, and $(c\lambda_3, c\mu_3, cv_3)$, where $v_i \geq 0$, $i = 1, 2, 3$. As in Exercise 26, §1010, translate the upper planes to obtain an integral of constant differences over parallelograms in the xy-plane. Use Exercise 27, §1010, to obtain

$$V = av_1 \begin{Vmatrix} b\lambda_2 & b\mu_2 \\ c\lambda_3 & c\mu_3 \end{Vmatrix} + bv_2 \begin{Vmatrix} c\lambda_3 & c\mu_3 \\ a\lambda_1 & a\mu_1 \end{Vmatrix} + cv_3 \begin{Vmatrix} a\lambda_1 & a\mu_1 \\ b\lambda_2 & b\mu_2 \end{Vmatrix}$$

$$= abc \left\{ v_1 \begin{Vmatrix} \lambda_2 & \mu_2 \\ \lambda_3 & \mu_3 \end{Vmatrix} + \cdots \right\} = abc(v_1^2 + v_2^2 + v_3^2) = abc.$$

★**15.** Prove that any compact smooth surface has zero volume ((v), §1011.) *Hints:* By use of the Heine-Borel Theorem (Ex. 31, §618) and the Implicit Function Theorem (§932) show that the problem can be reduced to showing that if $z = f(x, y)$ is continuous on a closed rectangle, its graph has zero volume.

★16. Prove that the volume of a parallelepiped four of whose noncoplanar vertices are (x_i, y_i, z_i), $i = 1, 2, 3, 4$, is the absolute value of the determinant

(1)
$$\begin{vmatrix} x_1 & y_1 & z_1 & 1 \\ x_2 & y_2 & z_2 & 1 \\ x_3 & y_3 & z_3 & 1 \\ x_4 & y_4 & z_4 & 1 \end{vmatrix}.$$

(Cf. Ex. 27, § 1010.)

★17. Justify the method used in Integral Calculus for obtaining a volume by integrating a cross-section area. Discuss in particular volumes of revolution.

★18. Justify the method of "cylindrical shells" used in Integral Calculus for obtaining volumes of revolution.

In Exercises 19–23, state and prove the three-dimensional analogue of the statement of the indicated Exercise of § 1010.

★19. Exercise 28.   **★20.** Exercise 29.   **★21.** Exercise 30.
★22. Exercise 31.   **★23.** Exercise 32.

1013. DOUBLE INTEGRALS IN POLAR COORDINATES

Frequently a problem whose solution depends upon the evaluation of a double integral is formulated most naturally in terms of polar coordinates.

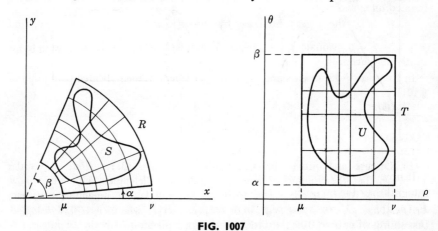

FIG. 1007

In many other cases the evaluation of a double integral in rectangular coordinates is greatly simplified by means of a change of coordinates, from rectangular to polar. Let S be a set with area contained in a **polar rectangle** R, $0 < \mu \leq \rho \leq \nu$, $\alpha \leq \theta \leq \beta$ (cf. Fig. 1007), over which is imposed a **polar net** $\mathfrak{N}$, consisting of circles $\rho = $ constant and rays $\theta = $ constant, which cut R into smaller polar rectangles $R_1, R_2, \cdots, R_n$. This corresponds to a rectangular net in the $\rho\theta$-plane (cf. Fig. 1007), which cuts the set U that corresponds to S and the set T that corresponds to R into smaller pieces. Let $f(x, y)$ be a function integrable on S, define f_S to be equal to f

§ 1013] DOUBLE INTEGRALS IN POLAR COORDINATES 357

on S and equal to 0 outside S, and let $F(\rho, \theta) \equiv f(\rho \cos \theta, \rho \sin \theta)$, $F_U(\rho, \theta) \equiv f_S(\rho \cos \theta, \rho \sin \theta)$. Then the integral of $f(x, y)$ over S can be expressed, by the second formulation of the double integral, § 1004, as the limit

(1) $$\lim_{|\mathfrak{N}|\to 0} \sum_{i=1}^{n} f_S(x_i, y_i) \Delta A_i,$$

where (x_i, y_i) is an arbitrary point and ΔA_i is the area of the polar rectangle R_i.

It can be shown (Ex. 14, § 1015) that the area of a polar rectangle $\rho_1 \leq \rho \leq \rho_2$, $\theta_1 \leq \theta \leq \theta_2$ ($\rho_1 > 0$) is equal to $\bar{\rho} \Delta \rho \Delta \theta$, where $\bar{\rho}$ is the average radius $\frac{1}{2}(\rho_1 + \rho_2)$, $\Delta \rho \equiv \rho_2 - \rho_1$, and $\Delta \theta \equiv \theta_2 - \theta_1$. The limit (1) can therefore be written in the form

(2) $$\lim_{|\mathfrak{N}|\to 0} \sum_{i=1}^{n} f_S(x_i, y_i) \bar{\rho}_i \Delta \rho_i \Delta \theta_i,$$

where $\bar{\rho}_i$, $\Delta \rho_i$, and $\Delta \theta_i$ have the meanings just prescribed, for the polar rectangle R_i, and for expediency about to be realized $f_S(x, y)$ is evaluated at a point (x_i, y_i) of R_i whose ρ-coordinate is $\bar{\rho}_i$.

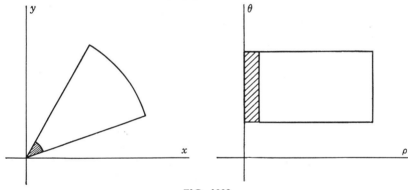

FIG. 1008

It now becomes apparent that the limit (2) can be interpreted as the double integral over the rectangle T (the variables being ρ and θ) of the *new* function $F_U(\rho, \theta) \cdot \rho$. If we write this new double integral as an iterated integral (assuming of course that the integrals exist), we obtain:

(3) $$\iint_S f(x, y)\, dA = \iint_R f_S(x, y)\, dA = \iint_T F_U(\rho, \theta) \rho\, dA$$

$$= \iint_U F(\rho, \theta) \rho\, dA = \int_\alpha^\beta \int_\mu^\nu F_U(\rho, \theta) \rho\, d\rho\, d\theta = \int_\mu^\nu \int_\alpha^\beta F_U(\rho, \theta) \rho\, d\theta\, d\rho.$$

Before formalizing our results we observe:

NOTE. The restriction that μ be *positive* may be replaced by the assumption $\mu \geq 0$. To be sure, the correspondence between R and T in Figure 1008 is no longer one-to-one (the origin in the xy-plane corresponds to the entire θ-axis in the

358   MULTIPLE INTEGRALS   [§ 1013

$\rho\theta$-plane), but if ρ is sufficiently small, the two corresponding shaded regions in Figure 1008 are both of area less than any preassigned positive number. In the limit they disappear. Therefore, equality between two integrals which exists for the two corresponding unshaded regions of Figure 1008 is also valid for the entire sector (extending to the origin) and rectangle (extending to the θ-axis).

By virtue of (3), the preceding Note, and the Fundamental Theorem, § 1008, we have:

Theorem. *If $f(x, y)$ is defined over a bounded set S, bounded toward the origin and away from the origin by the graphs of functions $\rho = \rho_1(\theta)$ and*

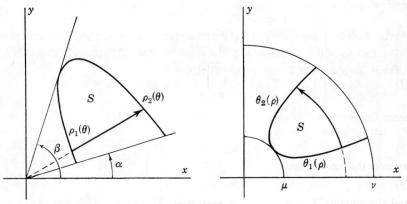

FIG. 1009

$\rho = \rho_2(\theta)$, *respectively, and by the two rays $\theta = \alpha$ and $\theta = \beta$ (cf. Fig. 1009), and if $F(\rho, \theta) \equiv f(\rho \cos \theta, \rho \sin \theta)$ then the following relation holds whenever the two integrals exist:*

(4) $$\iint_S f(x, y) \, dA = \int_\alpha^\beta \int_{\rho_1(\theta)}^{\rho_2(\theta)} F(\rho, \theta) \rho \, d\rho \, d\theta.$$

A similar statement holds for the reverse order of integration (cf. Fig. 1009):

(5) $$\iint_S f(x, y) \, dA = \int_\mu^\nu \int_{\theta_1(\rho)}^{\theta_2(\rho)} F(\rho, \theta) \rho \, d\theta \, d\rho.$$

Example 1. Find the total area inside the lemniscate $\rho^2 = a^2 \cos 2\theta$.

Solution. Integrating the function 1, and using symmetry, we have

$$A = 4 \int_0^{\frac{\pi}{4}} \int_0^{a\sqrt{\cos 2\theta}} \rho \, d\rho \, d\theta = 2a^2 \int_0^{\frac{\pi}{4}} \cos 2\theta \, d\theta = a^2.$$

Example 2. Evaluate $\int_0^2 \int_0^{\sqrt{4-x^2}} (x^2 + y^2) \, dy \, dx.$

§ 1014]   VOLUMES WITH DOUBLE INTEGRALS   359

Solution. The region of integration is bounded in the first quadrant by the coordinate planes and the circle $x^2 + y^2 = 4$. Transforming the given integral I to polar coordinates, we have

$$I = \int_0^{\frac{\pi}{2}} \int_0^2 \rho^2 \cdot \rho \, d\rho \, d\theta = \int_0^{\frac{\pi}{2}} \left[\frac{\rho^4}{4}\right]_0^2 d\theta = 2\pi.$$

Example 3. Evaluate the improper integrals

$$\int_0^{+\infty} e^{-x^2} \, dx \quad \text{and} \quad \int_{-\infty}^{+\infty} e^{-x^2} \, dx.$$

Solution. Let S_n be the closed square $0 \leq x \leq n$, $0 \leq y \leq n$, and C_n the closed quarter-circle $x \geq 0$, $y \geq 0$, $x^2 + y^2 \leq n^2$. Integrating the function $e^{-(x^2+y^2)}$ over S_n and C_n, we have

(6) $$J_n \equiv \iint_{S_n} e^{-(x^2+y^2)} \, dA = \int_0^n e^{-x^2} \, dx \int_0^n e^{-y^2} \, dy = \left[\int_0^n e^{-x^2} \, dx\right]^2,$$

(7) $$K_n \equiv \iint_{C_n} e^{-(x^2+y^2)} \, dA = \int_0^{\frac{\pi}{2}} \int_0^n e^{-\rho^2} \rho \, d\rho \, d\theta = \frac{\pi}{4}(1 - e^{-n^2}).$$

Since the integrand in (6) and (7) is positive, $J_n \uparrow$ and $K_n \uparrow$ as $n \uparrow$, and hence $J \equiv \lim J_n$ and $K \equiv \lim K_n$ exist. From the inclusions

(8) $$C_n \subset S_n \subset C_{2n}$$

we infer the inequalities

(9) $$K_n \leq J_n \leq K_{2n},$$

and the equality $J = K$. From (7) we have $K_n \to \pi/4$. Therefore $J_n \to \pi/4$. Taking square roots and using elementary properties of symmetry, we have:

(10) $$\int_0^{+\infty} e^{-x^2} \, dx = \tfrac{1}{2}\sqrt{\pi}, \quad \int_{-\infty}^{+\infty} e^{-x^2} \, dx = \sqrt{\pi}.$$

1014. VOLUMES WITH DOUBLE INTEGRALS IN POLAR COORDINATES

Combining statements of §§ 1011 and 1013 regarding volumes and the method of evaluating double integrals by means of polar coordinates, we have for the volume under the surface $z = f(x, y)$ ($f \geq 0$), with $F(\rho, \theta) \equiv f(\rho \cos \theta, \rho \sin \theta)$:

$$V = \int_\alpha^\beta \int_{\rho_1(\theta)}^{\rho_2(\theta)} F(\rho, \theta) \rho \, d\rho \, d\theta \quad \text{or} \quad \int_\mu^\nu \int_{\theta_1(\rho)}^{\theta_2(\rho)} F(\rho, \theta) \rho \, d\theta \, d\rho,$$

and for the volume between the surfaces $z = f(x, y)$ and $z = g(x, y)$ ($f \leq g$), with $F(\rho, \theta) \equiv f(\rho \cos \theta, \rho \sin \theta)$, $G(\rho, \theta) \equiv g(\rho \cos \theta, \rho \sin \theta)$:

$$\int_\alpha^\beta \int_{\rho_1(\theta)}^{\rho_2(\theta)} [G(\rho, \theta) - F(\rho, \theta)] \rho \, d\rho \, d\theta \quad \text{or} \quad \int_\mu^\nu \int_{\theta_1(\rho)}^{\theta_2(\rho)} [G(\rho, \theta) - F(\rho, \theta)] \rho \, d\rho \, d\theta,$$

where the limits of integration are determined by the region of integration.

Example 1. Find the volume bounded by the half-cone $z = 2\sqrt{5(x^2 + y^2)}$ and the paraboloid $z = 20 - 2(x^2 + y^2)$.

Solution. The two surfaces meet in the circle $x^2 + y^2 = 5$, $z = 10$. Therefore the volume can be expressed

$$V = \int_{-\sqrt{5}}^{\sqrt{5}} \int_{-\sqrt{5-x^2}}^{\sqrt{5-x^2}} [20 - 2(x^2 + y^2) - 2\sqrt{5(x^2 + y^2)}]\, dy\, dx$$

$$= \int_0^{2\pi} \int_0^{\sqrt{5}} [20 - 2\rho^2 - 2\sqrt{5}\rho]\rho\, d\rho\, d\theta = \frac{125\pi}{3}.$$

Example 2. Find the volume common to a sphere of radius a, a right circular half-cone of generating angle η, and a dihedral wedge of angle α, if the vertex of

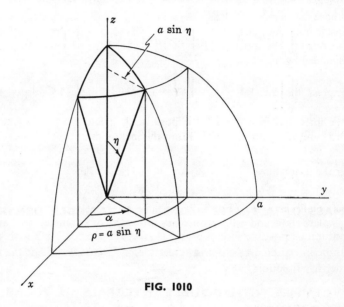

FIG. 1010

the half-cone is at the center of the sphere and the two half-planes of the wedge contain the axis of the cone.

Solution. Let the positive half of the z-axis be the axis of the half-cone, with vertex at the origin and let the half-planes be $\theta = 0$ and $\theta = \alpha$ (cf. Fig. 1010). The equations of the cone and sphere, in cylindrical coordinates (§ 808), are $z = \rho \cot \eta$ and $\rho^2 + z^2 = a^2$, respectively, so that the desired volume is
$\int_0^\alpha \int_0^{a \sin \eta} \{\sqrt{a^2 - \rho^2} - \rho \cot \eta\}\rho\, d\rho\, d\theta = \tfrac{1}{3}\alpha a^3(1 - \cos \eta)$. The special case of a hemisphere is given by $\alpha = 2\pi$, $\eta = \pi/2$: $V = \tfrac{2}{3}\pi a^3$.

1015. EXERCISES

In Exercises 1–4, evaluate the given integral by changing to polar coordinates.

1. $\int_0^a \int_0^{\sqrt{a^2 - x^2}} x^2\, dy\, dx$.

2. $\int_0^a \int_0^{\sqrt{ax - x^2}} x^2\, dy\, dx$.

3. $\int_0^1 \int_0^x (x^2 + y^2)^{\frac{3}{2}} \, dy \, dx.$

4. $\int_0^2 \int_0^{2\sqrt{4-x^2}} \sqrt{4x^2 + y^2} \, dy \, dx.$

Hint for Ex. 4: Replace the letter y by the letter z, and then substitute $z = 2y$ to obtain $4 \int_0^2 \int_0^{\sqrt{4-x^2}} \sqrt{x^2 + y^2} \, dy \, dx.$

In Exercises 5–8, find the area of the given region.

5. Inside the cardioid $\rho = 1 + \cos \theta$.
6. One loop of the four-leaved rose $\rho = a \sin 2\theta$.
7. Inside the circle $\rho = 2a \cos \theta$, outside the circle $\rho = a$.
8. Inside the small loop of the limaçon $\rho = 1 + 2 \cos \theta$.

In Exercises 9–12, find the volume of the given region.

9. Above the plane $z = 0$, below the cone $z = \rho$, inside the cylinder $\rho = 2a \cos \theta$.
10. Inside both the sphere $x^2 + y^2 + z^2 = a^2$ and the cylinder $\rho = a \cos \theta$.
11. Bounded by the sphere $x^2 + y^2 + z^2 = 20$ and the paraboloid $z = x^2 + y^2$.
12. Bounded by the plane $x + y + z = \frac{1}{2}$ and the paraboloid $z = x^2 + y^2$.

13. Express the integral $\int_0^{\frac{\pi}{4}} \int_0^{a \sec \theta} F(\rho, \theta) \rho \, d\rho \, d\theta$ by means of iterated integrals where the order of integration is reversed.

14. Show that the area of the polar rectangle $\mu \leq \rho \leq \nu$, $\alpha \leq \theta \leq \beta$ ($\mu \geq 0$) is equal to $\frac{1}{2}(\mu + \nu)(\nu - \mu)(\beta - \alpha)$, by means of rectangular coordinates. *Hint:* First show that the formula is correct when $\mu = 0$, $\beta = \pi/2$, and $\beta - \alpha$ is an acute angle.

1016. MASS OF A PLANE REGION OF VARIABLE DENSITY

A continuous positive function $\delta(x, y)$ over a bounded set S with positive area can be interpreted as a density function.† The **mass** of S is the integral of the density function:

(1) $$M(S) \equiv \iint_S \delta(x, y) \, dA.$$

Justification for these definitions lies primarily in two facts: (i) If the density is constant over S, the mass is equal to the product of the area and the density: $M(S) = \delta \cdot A(S)$. In other words, *constant density is mass per unit area.* (ii) The ratio of the mass of a compact region R to its area is equal, by the Mean Value Theorem for double integrals (Theorem XIII, § 1006), to an *average density* $\delta(\xi, \eta)$ evaluated at a point (ξ, n) of R. Therefore, by permitting R to shrink to any fixed point p of R, we conclude that *density at a point can be regarded as a limit of mass per unit area.*

Example. Find the mass of a circle of radius a if its density is k times the distance from the center.

† Such a set in E_2 with a density function is sometimes called a **lamina**. It is a mathematical idealization of a portion of a thin metal sheet of variable density or thickness.

Solution. With the aid of polar coordinates we have
$$M(S) = k \iint_S \sqrt{x^2 + y^2}\, dA = k \int_0^{2\pi} \int_0^a \rho^2\, d\rho\, d\theta = \frac{2k\pi a^3}{3}.$$

1017. MOMENTS AND CENTROID OF A PLANE REGION

For a bounded set S with positive area, and a density function $\delta(x, y)$, we have the following *definitions*, whose justification (Ex. 13, § 1018) lies in a comparison of the approximating sums for the double integrals with the discrete case of a finite collection of point masses:

First moment with respect to the y-axis: $M_y \equiv \iint_S \delta(x, y) x\, dA.$

First moment with respect to the x-axis: $M_x \equiv \iint_S \delta(x, y) y\, dA.$

Second moment with respect to the y-axis: $I_y \equiv \iint_S \delta(x, y) x^2\, dA.$

Second moment with respect to the x-axis: $I_x \equiv \iint_S \delta(x, y) y^2\, dA.$

Polar second moment with respect to the origin: $I_0 \equiv \iint_S \delta(x, y)(x^2 + y^2)\, dA.$

Centroid: $(\bar{x}, \bar{y}) \equiv \left(\dfrac{M_y}{M}, \dfrac{M_x}{M} \right).$

NOTE 1. In case of a density function δ identically equal to 1, the concepts defined above are purely geometric in character. However, they retain considerable interest and practicality in many fields, including Engineering and Statistics. The formulas for centroid become
$$(\bar{x}, \bar{y}) = \left(\frac{M_y}{A}, \frac{M_x}{A} \right).$$

NOTE 2. With a variable density function associated with a mass distribution, second moments are also called **moments of inertia** (this explains the notation I), and the centroid is also called the **center of mass**.

NOTE 3. The following relation always obtains:
$$I_0 = I_x + I_y.$$

More generally, first and second moments of a set S can be defined with respect to an arbitrary straight line L:
$$M_L \equiv \iint_S \delta(x, y) D(x, y)\, dA,$$
$$I_L \equiv \iint_S \delta(x, y) [D(x, y)]^2\, dA,$$

where $D(x, y)$ is either of the two possible directed (perpendicular) distances from the line L to the point (x, y).

Similarly the polar second moment can be extended to an arbitrary fixed point $p_0:(x_0, y_0)$:

$$I_{p_0} \equiv \iint_S \delta(x, y)[(x - x_0)^2 + (y - y_0)^2]\, dA.$$

NOTE 4. The centroid has the important property that for any line L through the centroid the first moment M_L is zero. This means that the centroid can be uniquely defined as the point that has this property, and is therefore independent of the coordinate system. (Cf. Ex. 14, § 1018.)

NOTE 5. The centroid can also be defined as the point p that minimizes the polar moment I_p. (Cf. Ex. 15, § 1018.)

Example. Find M_y, M_x, I_y, I_x, I_0, and the centroid for the quarter-circle bounded in the first quadrant by the coordinate axes and the circle $x^2 + y^2 = a^2$. (Assume constant density $\delta = 1$.)

Solution. By symmetry and Note 3,

$$M_y = M_x = \iint_S x\, dA = \int_0^{\frac{\pi}{2}} \int_0^a \rho^2 \cos\theta\, d\rho\, d\theta = \frac{a^3}{3},$$

$$I_y = I_x = \frac{1}{2} I_0 = \frac{1}{2} \int_0^{\frac{\pi}{2}} \int_0^a \rho^3\, d\rho\, d\theta = \frac{\pi a^4}{16}.$$

The centroid is $\left(\dfrac{4a}{3\pi}, \dfrac{4a}{3\pi}\right)$.

1018. EXERCISES

In Exercises 1–4, find the centroid of the given region. Assume $\delta = 1$.

1. Bounded in the first quadrant by the parabola $y^2 = 2px$ and the lines $x = b$ and $y = 0$.
2. First quadrant portion inside the hypocycloid $x^{\frac{2}{3}} + y^{\frac{2}{3}} = a^{\frac{2}{3}}$.
3. Inside the circle $x^2 + y^2 = 2ay$ and outside the circle $x^2 + y^2 = a^2$.
4. First-quadrant loop of the four-leaved rose $\rho = a \sin 2\theta$.

In Exercises 5–8, find the centroid of the given region with the given density function.

5. The square $0 \le x \le 1$, $0 \le y \le 1$; $\delta(x, y) = kx$.
6. First quadrant portion inside the circle $x^2 + y^2 = a^2$; $\delta(x, y) = k(x^2 + y^2)$.
7. Inside the circle $x^2 + y^2 = 2ax$; $\delta(x, y) = k\sqrt{x^2 + y^2}$.
8. First-quadrant loop of the four-leaved rose $\rho = a \sin 2\theta$; $\delta(\rho, \theta) = k\rho$.

In Exercises 9–12, find the specified second moment for the given set and density function.

9. I_x; region bounded by the ellipse $b^2x^2 + a^2y^2 = a^2b^2$; $\delta = 1$.
10. I_y; inside the lemniscate $\rho^2 = a^2 \cos 2\theta$; $\delta = 1$.
11. I_y; triangle with vertices $(0, 0)$, (h, a), $(h, a + b)$, $h > 0$, $b > 0$; $\delta = kx$.
12. I_0; inside the cardioid $\rho = a(1 + \cos\theta)$, $a > 0$; $\delta = k\rho$.
13. Give the details in the justification of the definitions of § 1017, called for in the first paragraph, § 1017.
14. Prove the statements of Note 4, § 1017, paying special attention to uniqueness.

15. Let $(\bar{x}, \bar{y})$ be the centroid of a bounded set S with area and a density function, and let $I_{x=\bar{x}}$, $I_{y=\bar{y}}$, and $I_{(\bar{x},\bar{y})}$ denote the second moments of S with respect to the lines $x = \bar{x}$ and $y = \bar{y}$, and the point $(\bar{x}, \bar{y})$, respectively. Establish the formulas

$$I_y = I_{x=\bar{x}} + M(S)\bar{x}^2,$$
$$I_x = I_{y=\bar{y}} + M(S)\bar{y}^2,$$
$$I_0 = I_{(\bar{x},\bar{y})} + M(S)(\bar{x}^2 + \bar{y}^2).$$

Hence prove Note 5, § 1017.

16. Show that the centroid of a triangle (of constant density) is the point of intersection of its medians.

1019. TRIPLE INTEGRALS, CYLINDRICAL COORDINATES

In order to evaluate a triple integral in terms of cylindrical coordinates (cf. § 808), we proceed as with double integrals in polar coordinates, and use the second formulation for the triple integral (cf. (*vi*), § 1011), with a

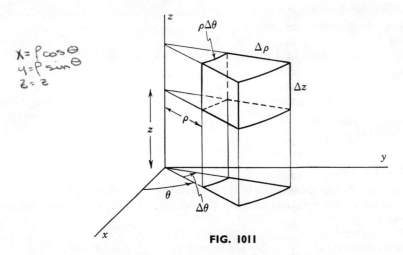

FIG. 1011

decomposition of a master "polar rectangular parallelepiped" into smaller ones. A typical element of volume is shown in Figure 1011. The actual volume of this element of volume is equal (by § 1013) to $\bar{\rho}\,\Delta z\,\Delta\rho\,\Delta\theta$, where $\bar{\rho}$ is the average of the two extreme values of ρ. Therefore (cf. § 1013), the triple integral of a function $F(\rho, \theta, z) = f(x, y, z)$ over a bounded set W, expressed as the triple integral of the extended function F_W over a polar rectangular parallelepiped R containing W (cf. § 1011), can be written

$$(1) \quad \iiint_W f\,dV = \lim_{|\Re|\to 0} \sum_{i=1}^n f_W(x_i, y_i, z_i)\,\Delta V_i$$
$$= \lim_{|\Re|\to 0} \sum_{i=1}^n F_W{}^*(\bar{\rho}_i, \theta_i, z_i)\bar{\rho}_i\,\Delta z_i\,\Delta\rho_i\,\Delta\theta_i$$
$$= \int_\alpha^\beta \int_\mu^\nu \int_c^d F_W{}^*(\rho, \theta, z)\,dz\,d\rho\,d\theta,$$

§ 1020] TRIPLE INTEGRALS, SPHERICAL COORDINATES   365

the iterated integral being extended over the rectangular parallelepiped in the $\rho\theta z$-space that corresponds to R, W^* being the set that corresponds to W.

By the Fundamental Theorem for triple integrals (cf. (xi), § 1011), the iterated integral of (1) can be written with six orders of integration. For a suitably defined set W, one of these integrals becomes

(2) $$\iiint_W f \, dV = \int_\alpha^\beta \int_{\rho_1(\theta)}^{\rho_2(\theta)} \int_{z_1(\rho,\theta)}^{z_2(\rho,\theta)} F(\rho, \theta, z)\rho \, dz \, d\rho \, d\theta.$$

NOTE. The assumption $\rho > 0$ can be replaced by the weaker inequality $\rho \geq 0$, for the same reasons in E_3 that apply in E_2. (Cf. the Note, § 1013).

Example. Use cylindrical coordinates to compute the volume of a right circular cone of base radius a and altitude h.

Solution. Let the equation of the cone be $az = hp$. Then the desired volume is

$$\int_0^{2\pi} \int_0^a \int_{\frac{h\rho}{a}}^h \rho \, dz \, d\rho \, d\theta = \frac{h}{a}\int_0^{2\pi}\int_0^a \rho(a - \rho) \, d\rho \, d\theta = \frac{\pi a^2 h}{3}.$$

1020. TRIPLE INTEGRALS, SPHERICAL COORDINATES

Following the pattern used for cylindrical coordinates in § 1019, we seek suitable specifications for the second formulation of the triple integral in

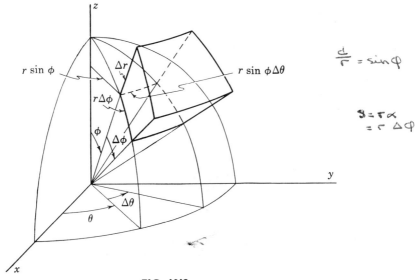

FIG. 1012

spherical coordinates. The type of "*polar rectangular parallelepiped*" used in this case is one that is bounded by two spheres with centers at the origin ($r = $ constant), two half-cones with axes on the z-axis ($\phi = $ constant), and two half-planes through the z-axis ($\theta = $ constant) (cf. Fig. 1012). If the increments are small it appears that the volume of this *element of volume* should be approximately equal to the product of the lengths of the three mutually orthogonal "sides," Δr, $r\,\Delta\phi$, and $r\sin\phi\,\Delta\theta$. That is, we should expect to have (approximately) $\Delta V = r^2 \sin\phi\,\Delta r\,\Delta\phi\,\Delta\theta$.

We shall now find the *precise* value of the element of volume. In Example 2, § 1014, the volume common to a sphere of radius a, a right circular half-cone of generating angle η, and a dihedral wedge of angle α, if the vertex of the half-cone is at the center of the sphere and the two half-planes of the wedge contain the axis of the cone, was found to be $\frac{1}{3}\alpha a^3(1 - \cos\eta)$. Therefore (as we see by subtraction) the volume bounded by such a wedge, between two such concentric spheres of radius a and b, $a < b$, and two such half-cones of generating angles η and ζ, $\eta < \zeta$, is $\frac{1}{3}\alpha(b^3 - a^3)(\cos\eta - \cos\zeta)$. By the Law of the Mean (§ 305), $(b^3 - a^3) = (b - a) \cdot 3\bar{r}^2$ where $a < \bar{r} < b$, and $(\cos\eta - \cos\zeta) = (\zeta - \eta) \cdot \sin\bar{\phi}$ where $\eta < \bar{\phi} < \zeta$. Therefore, with the notation indicated in Figure 1012, the element of volume is exactly equal to

(1) $$\Delta V = \bar{r}^2 \sin\bar{\phi}\,\Delta r\,\Delta\phi\,\Delta\theta,$$

where $\bar{r}$ is between r and $r + \Delta r$, and $\bar{\phi}$ is between ϕ and $\phi + \Delta\phi$. Therefore the approximating sums for $\iiint_W f(x, y, z)\,dV$ can be written $\sum_{i=1}^{n} f_W(x_i, y_i, z_i)\bar{r}_i^2 \sin\bar{\phi}_i\,\Delta r_i\,\Delta\phi_i\,\Delta\theta_i$, where $x_i = \bar{r}_i \sin\bar{\phi}_i \cos\theta_i$, $y_i = \bar{r}_i \sin\bar{\phi}_i \sin\theta_i$, $z_i = \bar{r}_i \cos\bar{\phi}_i$, and the iterated integral takes the form (for one of the six possible orders of integration):

(2) $$\iiint_W f\,dV = \int_\alpha^\beta \int_{\phi_1(\theta)}^{\phi_2(\theta)} \int_{r_1(\phi,\theta)}^{r_2(\phi,\theta)} F(r, \phi, \theta) r^2 \sin\phi\,dr\,d\phi\,d\theta.$$

NOTE. As with the variable ρ in polar coordinates in E_2 (and also in E_3) so with the variable r in spherical coordinates, the assumption $r > 0$ can be replaced by the weaker inequality $r \geq 0$. (Cf. the Notes, §§ 1013, 1019.)

Example. Evaluate $\int_{-a}^{a} \int_{-\sqrt{a^2-x^2}}^{\sqrt{a^2-x^2}} \int_{a-\sqrt{a^2-x^2-y^2}}^{a+\sqrt{a^2-x^2-y^2}} \sqrt{x^2 + y^2 + z^2}\,dz\,dy\,dx$.

Solution. The region of integration is the interior of the sphere $x^2 + y^2 + z^2 = 2az$. In spherical coordinates this is $r = 2a\cos\phi$, and the integral becomes

$$\int_0^{2\pi} \int_0^{\frac{\pi}{2}} \int_0^{2a\cos\phi} r^3 \sin\phi\,dr\,d\phi\,d\theta = \frac{8\pi a^4}{5}.$$

1021. MASS, MOMENTS, AND CENTROID OF A SPACE REGION

As in the two-dimensional case (§ 1016), if $\delta(x, y, z)$ is a continuous positive function over a compact region W with volume, the **mass** is given by integration:

(1) $$M \equiv \iiint_W \delta(x, y, z)\,dV.$$

§ 1021] CENTROID OF A SPACE REGION

The first moment of W is defined with respect to any plane, and the second moment (or moment of inertia) with respect to any plane, line or point. We give here only typical formulas for coordinate planes and axes and the origin (cf. § 1017):

First moment with respect to the yz-plane: $\quad M_{yz} \equiv \iiint_W \delta(x, y, z) x \, dV.$

Second moment with respect to the yz-plane: $\quad I_{yz} \equiv \iiint_W \delta(x, y, z) x^2 \, dV.$

Second moment with respect to the x-axis: $\quad I_x \equiv \iiint_W \delta(x, y, z)(y^2 + z^2) \, dV.$

Polar second moment with respect to the origin:
$$I_0 \equiv \iiint_W \delta(x, y, z)(x^2 + y^2 + z^2) \, dV.$$

Centroid: $(\bar{x}, \bar{y}, \bar{z}) \equiv \left(\dfrac{M_{yz}}{M}, \dfrac{M_{zx}}{M}, \dfrac{M_{xy}}{M} \right).$

NOTE 1. The second moments satisfy such relations as
$$I_{xy} + I_{xz} = I_x,$$
$$I_{xy} + I_{xz} + I_{yz} = I_0,$$
$$I_x + I_y + I_z = 2I_0.$$

NOTE 2. Although the preceding formulas are expressed in terms of rectangular coordinates, they are immediately available for cylindrical and spherical coordinates.

Example 1. Find the centroid of the portion of the sphere $x^2 + y^2 + z^2 \leq a^2$ in the first octant, assuming constant density.

Solution. There is no loss of generality in assuming $\delta = 1$, and since $\bar{x} = \bar{y} = \bar{z}$, we need compute only
$$M_{xy} = \int_0^{\frac{\pi}{2}} \int_0^{\frac{\pi}{2}} \int_0^a (r \cos \phi) r^2 \sin \phi \, dr \, d\phi \, d\theta = \frac{\pi a^4}{16}.$$
Since $V = \dfrac{\pi a^3}{6}$, the centroid has coordinates $\bar{x} = \bar{y} = \bar{z} = \frac{3}{8} a$.

Example 2. Find the moment of inertia I_L of a right circular cylinder of base radius a, altitude h, and density proportional to the distance from the axis of the cylinder, with respect to a line L parallel to the axis of the cylinder and at a distance b from it.

Solution. Let the cylinder be described by the inequalities $0 \leq \rho \leq a, 0 \leq z \leq h$, and the line L by $x = b, y = 0$. If the density is $\delta = k\rho$,
$$I_L = \int_0^{2\pi} \int_0^a \int_0^h k\rho[(x-b)^2 + y^2]\rho \, dz \, d\rho \, d\theta$$
$$= \int_0^{2\pi} \int_0^a \int_0^h k\rho^2(\rho^2 + b^2) \, dz \, d\rho \, d\theta + 0$$
$$= 2\pi k a^3 h \left(\frac{a^2}{5} + \frac{b^2}{3} \right).$$

If M is the mass of the cylinder and I_z its moment of inertia with respect to its axis, we have
$$M = \frac{2\pi k a^3 h}{3} \quad \text{and} \quad I_z = \frac{2\pi k a^5 h}{5} = \frac{3}{5} a^2 M,$$
so that
$$I_L = I_z + b^2 M = (\tfrac{3}{5} a^2 + b^2) M.$$

Example 3. Find the moment of inertia of a sphere of constant density with respect to a diameter.

Solution. Let the sphere be $x^2 + y^2 + z^2 \leq a^2$, with density k. Then by symmetry and Note 1,
$$I_x = I_y = I_z = \frac{2}{3} I_0 = \frac{2}{3} \int_0^{2\pi} \int_0^{\pi} \int_0^a k r^4 \sin\phi \, dr \, d\phi \, d\theta = \frac{8\pi k a^5}{15}.$$
Since the mass of the sphere is $\dfrac{4\pi k a^3}{3}$, the answer can be expressed $\dfrac{2}{5} a^2 M$.

1022. EXERCISES

1. Find the volume bounded by the sphere $x^2 + y^2 + z^2 = 8$ and the paraboloid $4z = x^2 + y^2 + 4$.

2. Find the volume and centroid of the region bounded by the hyperboloid $x^2 + y^2 - z^2 + a^2 = 0$ and the cone $2x^2 + 2y^2 - z^2 = 0$, for $z \geq 0$.

3. Find the volume of the region bounded by the sphere $x^2 + y^2 + z^2 = 13$ and the cone $x^2 + y^2 = (z - 1)^2$, for $z \geq 1$.

4. Find the moment of inertia of a homogeneous solid right circular cylinder (base radius a, altitude h, density k) with respect to its axis.

5. Find the centroid of the wedge-like region in the first octant bounded by the coordinate planes $x = 0$ and $z = 0$, the plane $z = by$ ($b > 0$), and the cylinder $x^2 + y^2 = a^2$.

6. Find the moment of inertia of a homogeneous right circular cone (base radius a, altitude h, density k) with respect to its axis.

7. Find the centroid and the moment of inertia with respect to its axis of symmetry of a homogeneous hemisphere ($\delta = k$).

8. Find M and $\bar{z}$ for the region bounded by the sphere $r = a$ and the half-cone $\phi = \alpha$ ($0 < \alpha \leq \tfrac{1}{2}\pi$), if the density is constant ($\delta = k$).

9. Find M, $\bar{z}$, and I_z for the hemisphere $0 \leq r \leq a$, $0 \leq \phi \leq \tfrac{1}{2}\pi$, if $\delta = k\rho$.

10. Find the mass and the moment of inertia of a sphere of radius a with respect to a diameter if the density is proportional to the distance from the center ($\delta = kr$).

11. Let W be the rectangular parallelepiped $0 \leq x \leq a$, $0 \leq y \leq b$, $0 \leq z \leq c$, with constant density k. Find M, I_{yz}, I_x, and I_0.

12. Find the centroid and I_z for the tetrahedron with vertices $(0, 0, 0)$, $(a, 0, 0)$, $(0, b, 0)$, $(0, 0, c)$, and constant density. *Hint*: For I_z let z be the final variable of integration.

13. Find the centroid of the homogeneous hemispherical shell $a \leq r \leq b$, $0 \leq \phi \leq \tfrac{1}{2}\pi$.

14. The density of a cube is proportional to the distance from one edge. Find its mass and its moment of inertia with respect to that edge. (Let the edges $= a$ and the constant of proportionality $= k$.)

15. Discuss the *Theorem of Pappus* (due to the Greek geometer Pappus, who lived during the second half of the third century A.D.): *Let S be a plane region in E_2*

§ 1023] MASS, MOMENTS, AND CENTROID OF AN ARC   369

lying entirely to the right of the y-axis, with centroid $(\bar{x}, \bar{y})$. Let W be the solid region in E_3 obtained by revolving A about the y-axis. Then
$$V(W) = 2\pi \bar{x} A(S).$$
Extend this to the formula $M(W) = 2\pi \bar{x} M(S)$, assuming a variable density for S. (Cf. Note 3, § 1814.)

16. Use the theorem of Pappus (Ex. 15) to determine the centroid of a plane semicircle, and the volume of a solid torus obtained by revolving the circle $(x - b)^2 + y^2 \leq a^2$ ($a < b$) about the y-axis. (Cf. Example 2, § 1814.)

17. Let a solid torus (doughnut) of constant density be generated by revolving the circle $(x - b)^2 + y^2 \leq a^2$ ($a < b$) about the y-axis. Prove that the moment of inertia of this torus with respect to its axis of revolution is $\frac{1}{4}(3a^2 + 4b^2)M$, where M is its mass. (Cf. Ex. 16.)

18. Let I_c be the moment of inertia of a body of mass M with respect to a line L_c through its centroid, and let L be a line parallel to L_c and at a distance b from it. If I_L denotes the moment of inertia of the body with respect to the line L, show that $I_L = I_c + b^2 M$. Use this formula to check the result of Example 2, § 1021.

19. State and prove the analogue for E_3 of Note 4, § 1017.

20. State and prove the analogue for E_3 of Note 5, § 1017.

1023. MASS, MOMENTS, AND CENTROID OF AN ARC

The concept of a mass determined by a density $\delta(s)$ as a function of arc length s on a rectifiable arc is similar to the two- and three-dimensional mass-distribution analogues. We state typical formulas for mass, moments, and centroid for a plane curve. Similar formulas hold for a space curve.

Mass $M \equiv \int_{s_1}^{s_2} \delta(s) \, ds.$

First moment with respect to the y-axis: $M_y \equiv \int_{s_1}^{s_2} \delta(s) x \, ds.$

Second moment with respect to the y-axis: $I_y \equiv \int_{s_1}^{s_2} \delta(s) x^2 \, ds.$

Polar second moment with respect to the origin: $I_0 \equiv \int_{s_1}^{s_2} \delta(s)(x^2 + y^2) \, ds.$

Centroid: $(\bar{x}, \bar{y}) \equiv \left(\dfrac{M_y}{M}, \dfrac{M_x}{M} \right).$

Example. Find the centroid of a homogeneous semicircular thin rod.

Solution. Let the rod be described in polar coordinates by $\rho = a$, $0 \leq \theta \leq \pi$, with $\delta = k$. Then $s = a\theta$, so that
$$M = \int_0^\pi k(a d\theta) = \pi a k,$$
$$M_x = \int_0^\pi k(a \sin \theta)(a d\theta) = 2a^2 k.$$
Therefore the centroid is $\left(0, \dfrac{2a}{\pi} \right).$

1024. ATTRACTION

According to the Newtonian law of gravitational attraction, the force between two point masses, m_1 and m_2, is equal to Gm_1m_2/d^2, where G is the universal constant of gravitation and d is the distance between the two masses. Since force is a *vector*, total forces between masses that are not point masses, but are distributed through a region, in one, two, or three dimensions, are found by determining their *components*. After the component of the force is found for an *element of mass*, the total force component results from integration over the region. This principle is illustrated in the Examples below.

NOTE. There is one theoretical difficulty that should be mentioned, although we shall not discuss it in detail. Suppose (for definiteness) we are interested in setting up the magnitude of the force between an element of mass, expressed in

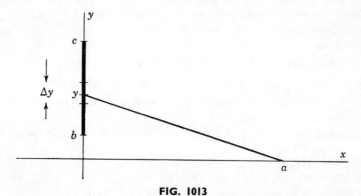

FIG. 1013

spherical coordinates and a unit mass at a distance d. In § 1020 we saw that the element of volume ΔV is given *precisely* by an expression of the form $\bar{r}^2 \sin \bar{\phi} \, \Delta r \, \Delta \phi \, \Delta \theta$, where $\bar{r}$ and $\bar{\phi}$ are appropriately chosen intermediate values of the variables r and ϕ. If the mass of a region is determined by a continuous *variable* density, $\delta(r, \phi, \theta)$, then the element of mass, ΔM, is given by the product $\delta(r', \phi', \theta') \, \Delta V$, where (r', ϕ', θ') is a point in the element of volume whose existence is guaranteed by the Mean Value Theorem (Theorem XIII, § 1006) applied to the function $\delta(r, \phi, \theta)$. But r' and ϕ' are certainly not in general the same as $\bar{r}$ and $\bar{\phi}$! Furthermore, the distance $d(r, \phi, \theta)$ must be evaluated at some intermediate point (r'', ϕ'', θ''), which cannot be expected to be the same as either of the first two points. After the appropriate component is formed, the total force is represented as the limit of a sum which is of the type involved in a multiple integral except that the functions whose product forms the integrand are evaluated at *different points* instead of the *same point*. This whole problem can be resolved without too much difficulty almost precisely as in the proof of Bliss's Theorem (Ex. 5, § 404), which is a special case of Duhamel's Principle for Integrals (§ 801). (Cf. Exs. 31, 32, § 1010; Exs. 22, 23, § 1012.) In Exercise 10, § 1025, hints are

§ 1024] ATTRACTION 371

given for the details of a proof assuming continuity of all functions involved, and simple conditions on the region of integration.

In the Examples below we avoid all of these considerations, and use just *one* representative point from each element of volume.

Example 1. Find the attractive force between a rod of uniform density and a unit mass (not a part of the rod).

Solution. Let the density and length of the rod be $\delta = k$ and L, respectively. Establish a coordinate system, as in Figure 1013, with the rod between $(0, b)$ and $(0, c)$, and the unit mass at $(a, 0)$, with $b < c$ and $a \geq 0$. The distance between the unit mass at $(a, 0)$ and the element of mass of the rod, $k\,\Delta y$, at $(0, y)$ is $\sqrt{a^2 + y^2}$. Therefore the magnitude of the x-component of the element of force is $\dfrac{Ga}{(a^2 + y^2)^{\frac{3}{2}}} k\,\Delta y$, and that of the y-component is $\dfrac{Gy}{(a^2 + y^2)^{\frac{3}{2}}} k\,\Delta y$. Upon integration, we find

$$|F_x| = Gak \int_b^c \frac{dy}{(a^2 + y^2)^{\frac{3}{2}}} = \frac{Gk}{a}\left\{\frac{c}{\sqrt{a^2 + c^2}} - \frac{b}{\sqrt{a^2 + b^2}}\right\},$$

$$|F_y| = Gk \int_b^c \frac{y\,dy}{(a^2 + y^2)^{\frac{3}{2}}} = Gk\left\{\frac{1}{\sqrt{a^2 + b^2}} - \frac{1}{\sqrt{a^2 + c^2}}\right\}.$$

From these components the direction and magnitude of the force vector can be determined. Special cases are of interest. One is: $a = 0, 0 < b < c$. Then $|F_y| = \dfrac{Gk}{bc}(c - b) = \dfrac{GM}{bc}$. Another is obtained by letting $b = 0$ and $c \to +\infty$. Then $|F_x| = |F_y| = Gk/a$. Finally, let $c \to +\infty$ and $b \to -\infty$. Then $|F_x| = 2Gk/a$.

Example 2. Find the attractive force between a homogeneous plane semicircular lamina and a unit mass situated on the axis of symmetry of the lamina and on the circumference (extended).

First Solution. Let the lamina be bounded above by the circle $\rho = 2a \sin \theta$ and below by the horizontal line $y = \rho \sin \theta = a$, the unit mass being located at the origin, and let the density be $\delta = k$. Since the force between an element of area at (ρ, θ) and the unit mass is $Gk\rho\,d\rho\,d\theta/\rho^2$, the total force is obtained by integrating the y-component of this:

$$F = Gk \int_{\frac{\pi}{4}}^{\frac{3\pi}{4}} \int_{a \csc \theta}^{2a \sin \theta} \frac{\sin \theta \, d\rho \, d\theta}{\rho} = 2Gk \int_{\frac{\pi}{4}}^{\frac{\pi}{2}} \sin \theta \ln (2 \sin^2 \theta) \, d\theta.$$

Integration by parts gives the result $2Gk[\ln(3 + 2\sqrt{2}) - \sqrt{2}]$. We notice (as we should expect) that this result is independent of the radius a. Computation shows that this force is equal (approximately) to $MG/(1.501a)^2$. In other words, the attraction is the same as it would be if the entire mass were concentrated at the point $\rho = 1.501a$, $\theta = \tfrac{1}{2}\pi$.

Second Solution. Let the lamina be the right-hand half of the circle $x^2 + y^2 = a^2$, and consider it to be cut into vertical strips of thickness Δx. Then, by the result of Example 1, the force on the unit mass at $(-a, 0)$ contributed by the strip at abscissa x is (approximately) equal to

$2Gk \, \Delta x \sqrt{a^2 - x^2}/(x + a)\sqrt{2a^2 + 2ax} = \sqrt{2Gk} \, \Delta x \sqrt{a - x}/\sqrt{a}(a + x)$. Therefore the total force is $\dfrac{\sqrt{2Gk}}{\sqrt{a}} \displaystyle\int_0^a \dfrac{\sqrt{a - x}}{a + x} dx$, which is equal to the result given by the first solution.

Example 3. Show that a solid sphere whose density is a function of the distance from the center of the sphere only, attracts a particle outside the sphere as if all of its mass were concentrated at its center.

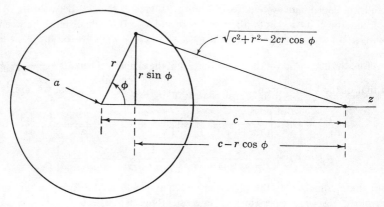

FIG. 1014

Solution. Center the sphere, of radius a, at the origin, and let a unit mass be located on the positive half of the z-axis, a distance $c > a$ from the origin (cf. Fig. 1014). Only the component of the force in the z-direction is of interest. This is represented by the integral

(1) $$F_z = \int_0^{2\pi} \int_0^{\pi} \int_0^a \dfrac{G\delta(r)(c - r \cos \phi)}{(c^2 + r^2 - 2cr \cos \phi)^{\frac{3}{2}}} r^2 \sin \phi \, dr \, d\phi \, d\theta.$$

Since the limits of integration in (1) are constants, the order of integration is immaterial. Integrating first with respect to θ and then with respect to ϕ, we find

(2) $$F_z = \dfrac{4\pi G}{c^2} \int_0^a r^2 \delta(r) \, dr = \dfrac{GM}{c^2},$$

where M is the mass of the sphere:

(3) $$M = \int_0^{2\pi} \int_0^{\pi} \int_0^a \delta(r) r^2 \sin \phi \, dr \, d\phi \, d\theta = 4\pi \int_0^a r^2 \delta(r) \, dr.$$

Formula (2) is that sought.

1025. EXERCISES

In Exercises 1–6, find the attractive force between the given set with constant density k and a unit mass at the given point.

1. The circular wire $x^2 + y^2 = a^2$, $z = 0$; $(0, 0, h)$.

2. The circular disk $x^2 + y^2 \leq a^2$, $z = 0$; $(0, 0, h)$.
3. The rectangular lamina $|x| \leq a$, $|y| \leq b$; $(c, 0)$ $(c > a)$.
4. The solid cylinder $x^2 + y^2 \leq a^2$, $0 \leq z \leq h$; $(0, 0, c)$ $(c > h)$.
5. The solid hemisphere $r \leq 2a \cos \phi$, $z \geq a$; $(0, 0, 0)$.
6. The solid cone $\phi \leq \alpha = \text{Arctan } (a/h)$, $0 < z \leq h$; $(0, 0, 0)$. Discuss the convergence of any improper integral concerned.

7. Show that the attractive force between two solid spheres, for each of which the density is a function of the distance from the center of the sphere only, is the same as if the total mass of each were concentrated at its center.

8. Show that the attractive force exerted by a spherical shell on a point mass inside the shell, if the density is a function of the distance from the center of the sphere only, is zero.

9. Discuss the attraction between the sphere of Example 3, § 1024, and a point mass inside the sphere. Show that if the sphere is homogeneous the attraction varies directly as the distance of the point mass from the center of the sphere. Discuss the behavior of any improper integrals concerned.

*10. Let R be a closed rectangular parallelepiped with faces parallel to the coordinate planes, and let $f(x, y, z)$, $g(x, y, z)$, and $h(x, y, z)$ be continuous on R. If $\mathfrak{N}$ is a net that subdivides R into pieces $S_1, \cdots, S_n$ of volume $\Delta V_1, \cdots, \Delta V_n$, and if (x_i, y_i, z_i), (x_i', y_i', z_i'), and (x_i'', y_i'', z_i'') are arbitrary points of S_i, $i = 1, \cdots, n$, show that

$$\lim_{|\mathfrak{N}| \to 0} \sum_{i=1}^{n} f(x_i, y_i, z_i) g(x_i', y_i', z_i') h(x_i'', y_i'', z_i'') \Delta V_i$$
$$= \iiint_R f(x, y, z) g(x, y, z) h(x, y, z) \, dV.$$

(Cf. Ex. 5, § 404; also §§ 801, 802, Exs. 31, 32, § 1010.) *Hint:* Write $f(p)g(p')g(p'') - f(p)g(p)h(p) = f(p)g(p')h(p'') - f(p)g(p')h(p) + f(p)g(p')h(p) - f(p)g(p)h(p)$.

1026. JACOBIANS AND TRANSFORMATIONS OF MULTIPLE INTEGRALS

The question to which we now address ourselves is the following: "What is the effect on a multiple integral

$$\iint_R \cdots \int f(x, y, \cdots) \, dx \, dy \cdots$$

of a change of variables, from $x, y, \cdots$ to $u, v, \cdots$?"

This question has been partially answered (§ 406) for the case of an integral of a function of a single variable: If $f(x)$ is continuous and if $x(u)$ is continuously differentiable, then

(1) $$\int_a^b f(x) \, dx = \int_c^d f(x(u)) \frac{dx}{du} \, du,$$

where $a = x(c)$ and $b = x(d)$.

It has also been answered to some extent for transformations to polar coordinates (§ 1013), cylindrical coordinates (§ 1019), and spherical coordinates (§ 1020), with the formulas

(2) $$\iint_S f(x, y) \, dA = \int_\alpha^\beta \int_\mu^\nu f_S(x(\rho, \theta), y(\rho, \theta))\rho \, d\rho \, d\theta,$$

(3) $$\iiint_W f(x, y, z) \, dV = \int_\alpha^\beta \int_\mu^\nu \int_c^d f_{W^*}(x(\rho, \theta, z), y(\rho, \theta, z), z)\rho \, dz \, d\rho \, d\theta,$$

(4) $$\iiint_W f(x, y, z) \, dV$$
$$= \int_\alpha^\beta \int_{\phi_1}^{\phi_2} \int_{r_1}^{r_2} f_{W^*}(x(r, \phi, \theta), y(r, \phi, \theta), z(r, \phi, \theta))r^2 \sin \phi \, dr \, d\phi \, d\theta.$$

In each of these cases the transformation of the multiple integral is achieved by (i) substituting for x, y, and z in the integrand their expressions in terms of the new variables, and (ii) introducing an extra factor in the new integrand $\left(\dfrac{dx}{du}, \rho, \rho, \text{ and } r^2 \sin \phi, \text{ respectively}\right)$. We notice that, in analogy with (1) where the extra factor is the *derivative* of the original variable with respect to the new, the extra factor in (2), (3), and (4) is the *Jacobian* of the original variables with respect to the new (cf. § 922):

In (2): $\dfrac{\partial(x, y)}{\partial(\rho, \theta)} = \rho,$

in (3): $\dfrac{\partial(x, y, z)}{\partial(\rho, \theta, z)} = \rho,$

in (4): $\dfrac{\partial(x, y, z)}{\partial(r, \phi, \theta)} = r^2 \sin \phi.$

All of these transformation formulas, (1)–(4), are special instances of a general rule, involving Jacobians. The statement of this rule, called the *Transformation Theorem*, is given below for the case of two variables. A proof of the Transformation Theorem is very difficult and will not be attempted in this book. (For a proof, and references, see RV, § 1328.) A general discussion based on intuitive ideas is given in the following section (§ 1027).

Theorem. Transformation Theorem for Double Integrals. *If*

(i) *T is a one-to-one transformation between open sets Ω in the uv-plane and $\Psi = T(\Omega)$ in the xy-plane, defined by continuously differentiable functions $x = x(u, v)$ and $y = y(u, v)$;*

(ii) *the Jacobian $J(u, v) = \dfrac{\partial(x, y)}{\partial(u, v))}$ is nonzero throughout Ω;*

(iii) R and S are bounded sets in the xy-plane and uv-plane, respectively, whose closures are subsets of Ψ and Ω, respectively ($\bar{R} \subset \Psi$, $\bar{S} \subset \Omega$), such that $R = T(S)$;

(iv) $f(x, y)$ is defined over R and $g(u, v) \equiv f(x(u, v), y(u, v))$;

then

(v) R has area if and only if S has area;

(vi) assuming R and S have area, f is integrable over R if and only if $g|J|$ is integrable over S, and in case of integrability

(5) $$\iint_R f(x, y)\, dA = \iint_S g(u, v)\, |J(u, v)|\, dA.$$

(Cf. Fig. 1015.)

A similar theorem holds for multiple integrals of any number of variables.

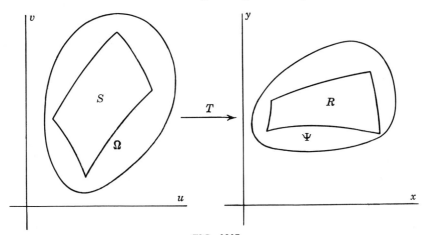

FIG. 1015

1027. GENERAL DISCUSSION

In this section we hope to appeal to the reader's intuition, the goal being to explain why the appearance of a Jacobian in the transformation of a multiple integral should be expected, and to develop a feeling for the geometric significance of the Jacobian in its present role. No attempt at mathematical rigor will be made, but it is hoped that the present discussion may suggest ideas that point toward a proof.

Let us examine the transformation T prescribed (in the Theorem, § 1026) by the functions $x(u, v)$ and $y(u, v)$, from the point of view of the *element of area dA* in the xy-plane that corresponds to positive differential increments du and dv in the variables u and v. (In the case of polar coordinates ρ and θ, for instance, $dA = \rho\, d\rho\, d\theta$.)

Accordingly, let S be the closed rectangle in the uv-plane with sides parallel to the coordinate axes and opposite vertices (u_0, v_0) and $(u_0 + du, v_0 + dv)$, and let R be the corresponding set ($R = T(S)$) in the xy-plane (cf. Fig. 1016). The secant vector joining the points in the xy-plane that correspond to (u_0, v_0) and $(u_0 + du, v_0)$ (shown by a broken line arrow from 1 to 2 in Figure 1016) has components

$$x(u_0 + du, v_0) - x(u_0, v_0) \quad \text{and} \quad y(u_0 + du, v_0) - y(u_0, v_0)$$

or, by the Law of the Mean (§ 912),

$$x_1(u_0 + \theta_1\, du, v_0)\, du \quad \text{and} \quad y_1(u_0 + \theta_2\, du, v_0)\, du,$$

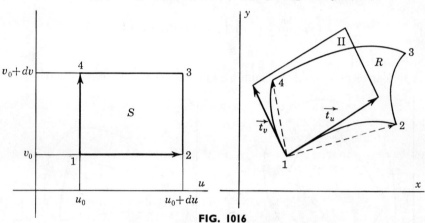

FIG. 1016

where $0 < \theta_1 < 1$ and $0 < \theta_2 < 1$. This vector approximates the corresponding *tangent* differential vector $\vec{t}_u$ with components

(1) $$\frac{\partial x}{\partial u}\, du \quad \text{and} \quad \frac{\partial y}{\partial u}\, du,$$

evaluated at the point (u_0, v_0).

Similarly, the segment in the xy-plane that corresponds to the one joining (u_0, v_0) to $(u_0, v_0 + dv)$ (shown by a broken line arrow from 1 to 4 in Fig. 1016) is approximately equal to the tangent differential vector $\vec{t}_v$ with components

(2) $$\frac{\partial x}{\partial v}\, dv \quad \text{and} \quad \frac{\partial y}{\partial v}\, dv,$$

evaluated at the point (u_0, v_0).

The set $R = T(S)$, therefore, is approximated by the parallelogram Π determined by the vectors (1) and (2). Using a formula from Plane Analytic Geometry (cf. Ex. 27, § 1010), we have, for the area of Π:

(3) $$A(\Pi) = \text{absolute value of } \begin{vmatrix} \dfrac{\partial x}{\partial u}\, du & \dfrac{\partial y}{\partial u}\, du \\ \dfrac{\partial x}{\partial v}\, dv & \dfrac{\partial y}{\partial v}\, dv \end{vmatrix} = \left| \dfrac{\partial(x, y)}{\partial(u, v)} \right| du\, dv.$$

Before taking the next step, let us recall that all of the differential expressions that we have encountered in earlier sections (§§ 312, 805, 809, 906) have been associated with a *tangent* line to a curve (or tangent plane to a surface) rather than with the curve (or surface) itself. With this background in mind it seems appropriate now to *define* the **differential element of area**, corresponding to a point (u_0, v_0) and differentials du and dv, by the equation

(4) $$dA \equiv |J(u, v)|\, du\, dv,$$

where $J(u, v)$ is the Jacobian $\partial(x, y)/\partial(u, v)$ of the transformation.

The relation (4) leads directly to the formula for change of variables in the Transformation Theorem, § 1026:

(5) $$\iint_R f(x, y)\, dA = \iint_S f(x(u, v), y(u, v))|J(u, v)|\, du\, dv.$$

The mapping T transforms a rectangle of area $du\, dv$ into a set that is approximated by one of area $|J|\, du\, dv$. Expressed slightly differently, this says that the transformation produces a change of area near any point by multiplying that area by a scale factor of proportionality, this factor being the magnitude of the Jacobian of the transformation evaluated at the point. Thus, whenever $|J| > 1$ a magnification or stretching takes place, and whenever $|J| < 1$ a diminution or shrinking is produced—the magnitude of this stretching or shrinking being equal to $|J|$. It should be appreciated, however, that this change of scale near a point is not uniform, but depends on the direction. For example, the transformation $x = 3u$, $y = \frac{1}{2}v$ stretches by a factor of 3 in the x-direction and shrinks by a factor of $\frac{1}{2}$ in the y-direction, the net effect on the area being an increase by a factor of $\frac{3}{2}$.

The next question is the significance of the *sign* of the Jacobian. This rests on the meaning of the sign of the determinant that gives the area of a parallelogram. It is thus a matter of Plane Analytic Geometry to show that a *positive* Jacobian corresponds to a *preservation* of orientation (that is, the smallest-angle rotation of $\vec{t}_u$ into the direction of $\vec{t}_v$, in Figure 1016, is counterclockwise), and a *negative* Jacobian corresponds to a *reversal* of orientation.

In three (or more) dimensions, an analysis similar to the preceding is possible. The student is asked in Exercise 11, § 1028, to carry out the details for E_3. In this case *three* tangent vectors determine an approximating *parallelepiped*, whose volume is $|J(u, v, w)|\, du\, dv\, dw$ (cf. Ex. 16, § 1012). The sign of the Jacobian again has the significance that it is positive or negative according as the orientation is preserved or reversed. (Cf. § 1703.)

NOTE 1. The assumptions (*i*) and (*ii*) of the Transformation Theorem, § 1026, that the transformation T is one-to-one and that the Jacobian never vanishes can be relaxed to the extent of permitting exceptional behavior in either regard on a set of zero area. A precise statement in given in *RV*, Exercise 12, § 1330. This principle was discussed in the Note, § 1013, for polar coordinates, and is illustrated again in Example 3, below.

NOTE 2. In practice, in order to determine the image of a closed region under a given transformation it is often sufficient to find the image of the frontier only. Complete statements follow in Theorems I and II (for proofs, cf. *RV*, Exs. 14 and 15, § 1330), with the aid of the Definition:

Definition. *A **Jordan curve** in E_2 is a simple closed curve C such that its complement C' consists of two disjoint regions, one bounded and one unbounded, the frontier of each of which is C. The bounded region is called the **inside** of C, and the unbounded region is called the **outside** of C.† Similar formulations hold in higher dimensions.*

Theorem I. *A region R whose frontier is a Jordan curve C is either the inside of C or the outside of C. Any such region is the interior of its closure.*

Theorem II. *Let R be a bounded region with a Jordan curve C as its frontier, and let M be a mapping that is continuous on its closure $\bar{R}$. If the image of R, under M, is an open set S, and that of C is a Jordan curve D, then S is the inside of D.*

A similar statement holds in higher dimensions.

We shall now illustrate by means of a few examples some typical simplifications of multiple integrals that can be effected through a change of variables.

1. *Simplification of the domain of integration.*

Example 1. Simplify $\iint_R f(x, y) \, dA$, where R is the interior of the parallelogram with vertices $(0, 0)$, $(1, 2)$, $(-4, 3)$, and $(-3, 5)$.

Solution. The linear transformation that maps $(0, 0) \to (0, 0)$, $(1, 2) \to (1, 0)$, and $(-4, 3) \to (0, 1)$ is
$$\begin{cases} x = u - 4v, \\ y = 2u + 3v. \end{cases}$$
Since $\partial(x, y)/\partial(u, v) = 11$,
$$\iint_R f(x, y) \, dA = 11 \int_0^1 \int_0^1 f(u - 4v, 2u + 3v) \, dv \, du.$$

Example 2. Let R be the region bounded in the first quadrant by the hyperbolas $x^2 - y^2 = a$, $x^2 - y^2 = b$, $2xy = c$, $2xy = d$, where $0 < a < b$, $0 < c < d$. By means of the transformation
$$\begin{cases} u = x^2 - y^2 \\ v = 2xy, \end{cases}$$

† The **Jordan curve theorem** states that every simple closed curve in E_2 is a Jordan curve. For a proof see M. H. A. Newman, *Elements of the Topology of Plane Sets of Points* (Cambridge University Press, 1951).

since $\partial(u, v)/\partial(x, y) = 4(x^2 + y^2) = 4\sqrt{u^2 + v^2}$,

$$\iint_R f(x, y)\, dA = \frac{1}{4} \int_a^b \int_c^d \frac{g(u, v)}{\sqrt{u^2 + v^2}}\, dv\, du.$$

If $f(x, y) = x^2 + y^2$, the integral has the value $\frac{1}{4}(b - a)(d - c)$.

2. *Simplification of the integrand. Separation of variables.*

Example 3. Evaluate $\displaystyle\int_0^1 \int_0^{1-x} e^{\frac{y-x}{y+x}}\, dy\, dx$.

Solution. Let R be the interior of the triangle in the xy-plane with vertices $(0, 0)$, $(1, 0)$, $(0, 1)$, and let S be the interior of the square in the uv-plane with vertices $(0, 0)$, $(1, 0)$, $(0, 1)$, and $(1, 1)$, and map R onto S by the transformation

$$\begin{cases} u = x + y, \\ v = y/(x + y), \end{cases} \qquad \begin{cases} x = u - uv, \\ y = uv. \end{cases}$$

Then $\partial(x, y)/\partial(u, v) = u$, and

$$\iint_R e^{\frac{y-x}{y+x}}\, dy\, dx = \iint_S u e^{2v-1}\, dv\, du = \int_0^1 u\, du \int_0^1 e^{2v-1}\, dv$$
$$= \tfrac{1}{4}(e - e^{-1}) = \tfrac{1}{2} \sinh 1.$$

The mapping is not one-to-one on the frontiers of the sets, and the Jacobian vanishes on the v-axis, which corresponds to the point $x = 0$, $y = 0$. However, the result is valid by Note 1.

1028. EXERCISES

1. Evaluate $\displaystyle\iint_R \sqrt{x + y}\, dA$, where R is the parallelogram bounded by the lines $x + y = 0$, $x + y = 1$, $2x - 3y = 0$, $2x - 3y = 4$.

2. Evaluate $\displaystyle\int_0^1 \int_0^{1-x} \frac{1}{(1 + x^2 + y^2)^2}\, dy\, dx$ by means of a suitable rotation of axes.

3. Evaluate $\displaystyle\iint_R xy\, dA$, where R is the region in the first quadrant bounded by the hyperbolas $x^2 - y^2 = a$ and $x^2 - y^2 = b$, where $0 < a < b$, and the circles $x^2 + y^2 = c$ and $x^2 + y^2 = d$, where $b < c < d$.

4. Evaluate $\displaystyle\iint_R (x - y)^4 e^{x+y}\, dA$, where R is the square with vertices $(1, 0)$, $(2, 1)$, $(1, 2)$, and $(0, 1)$.

5. Evaluate $\displaystyle\int_0^1 \int_0^x \sqrt{x^2 + y^2}\, dy\, dx$ by means of the transformation $x = u$, $y = uv$.

*6. Compute $\displaystyle\iint_R (ax + by)^n\, dA$, where R is the unit disk $x^2 + y^2 \leq 1$, by using a suitable rotation ($a^2 + b^2 > 0$, n a positive integer). (Cf. Ex. 36, § 416.)

★7. Evaluate $\int_0^1 \int_0^{1-x} y \ln(1 - x - y) \, dy \, dx$, using the transformation $x = u - uv$, $y = uv$.

★8. Show that the mapping
$$x = u - uv, \quad y = uv - uvw, \quad z = uvw$$
transforms the interior of the tetrahedron bounded by the planes $x = 0$, $y = 0$, $z = 0$, $x + y + z = 1$ into the interior of the cube $0 < u < 1$, $0 < v < 1$, $0 < w < 1$. Use this transformation to compute the coordinates of the centroid of the tetrahedron.

9. Explain why the transformation equation (1), § 1026, for a function of one variable has no absolute value signs.

10. State the analogue of the Transformation Theorem, § 1026, for functions of n variables.

11. Carry out the detailed discussion requested in the paragraph preceding Note 1, § 1027.

11

Infinite Series of Constants

1101. BASIC DEFINITIONS

If $a_1, a_2, \cdots, a_n, \cdots$ is a sequence of numbers, the expression

(1) $$\sum_{n=1}^{+\infty} a_n = a_1 + a_2 + \cdots + a_n + \cdots$$

(also written Σa_n if there is no possible misinterpretation) is called an **infinite series** or, in brief, a **series**. The numbers $a_1, a_2, \cdots, a_n, \cdots$ are called **terms**, and the number a_n is called the **nth term** or **general term**.

The sequence $\{S_n\}$, where

$$S_1 \equiv a_1,\ S_2 \equiv a_1 + a_2, \cdots,\ S_n \equiv a_1 + a_2 + \cdots + a_n,$$

is called the **sequence of partial sums** of the series (1). An infinite series **converges** or **diverges** according as its sequence of partial sums converges or diverges. In case of convergence, the series (1) is said to have a **sum** equal to the limit of the sequence of partial sums, and if $S \equiv \lim_{n \to +\infty} S_n$, we write

(2) $$\Sigma a_n = \sum_{n=1}^{+\infty} a_n = a_1 + \cdots + a_n + \cdots = S,$$

and say that the series **converges to S**.

In case $\lim_{n \to +\infty} S_n$ exists and is infinite the series Σa_n, although divergent, is said to have an **infinite sum**:

(3) $$\Sigma a_n = \sum_{n=1}^{+\infty} a_n = a_1 + \cdots + a_n + \cdots = +\infty, -\infty, \text{ or } \infty.$$

and the series is said to **diverge to** $+\infty, -\infty,$ or ∞.

Example 1. $\dfrac{1}{2} + \dfrac{1}{4} + \cdots + \dfrac{1}{2^n} + \cdots = 1$, since the sum of the first n terms is

$$S_n = \frac{1}{2} + \frac{1}{4} + \cdots + \frac{1}{2^n} = 1 - \frac{1}{2^n},$$

and $S_n \to 1$ (cf. Example 3, § 202).

Example 2. The series whose general term is $2n$ diverges to $+\infty$:
$$2 + 4 + 6 + \cdots + 2n + \cdots = +\infty.$$

Example 3. The series
$$1 - 1 + 1 - 1 + \cdots + (-1)^{n+1} + \cdots$$
diverges since its sequence of partial sums is
$$1, 0, 1, 0, 1, 0, 1, \cdots.$$
This sequence diverges since, if it converged to S, every subsequence would also converge to S, and S would equal both 0 and 1, in contradiction to the uniqueness of the limit of a sequence. (Cf. § 204; also Example 4, § 202.)

1102. THREE ELEMENTARY THEOREMS

Direct consequences of Theorems I, II, and X, § 204, are the following:

Theorem I. *The alteration of a finite number of terms of a series has no effect on convergence or divergence* (*although it will in general change the sum in case of convergence*).

Theorem II. *If a series converges or has an infinite sum, this sum is unique.*

Theorem III. *Multiplication of the terms of a series by a nonzero constant k does not affect convergence or divergence. If the original series converges, the new series converges to k times the sum of the original, for any constant k:*

(1) $$\sum_{n=1}^{+\infty} k a_n = k \sum_{n=1}^{+\infty} a_n.$$

(Cf. Ex. 27, § 1107.)

1103. A NECESSARY CONDITION FOR CONVERGENCE

Theorem. *If a series converges, its general term tends toward zero as n becomes infinite. Equivalently, if the general term of a series does not tend toward zero as n becomes infinite, the series diverges.*

Proof. Assume the series, $a_1 + a_2 + \cdots$, converges to S. Then $S_n = S_{n-1} + a_n$, or $a_n = S_n - S_{n-1}$. Since the sequence of partial sums converges to S, both S_n and S_{n-1} tend toward S as n becomes infinite. Hence (Theorem VII, § 204),
$$\lim_{n \to +\infty} a_n = \lim_{n \to +\infty} (S_n - S_{n-1}) = S - S = 0.$$

Example. Show that the series $2 + 4 + \cdots + 2n + \cdots$ and $0 + 1 + 0 + 1 + \cdots$ diverge.

Solution. In neither case does the general term tend toward zero, although the second series contains a subsequence converging to zero.

NOTE. The condition of the theorem is necessary but *not sufficient* for convergence. It will be shown later (Theorem IV, § 1108) that the *harmonic series* $1 + \frac{1}{2} + \frac{1}{3} + \cdots$ diverges, although the general term, $\frac{1}{n}$, converges to zero.

1104. THE GEOMETRIC SERIES

The series

(1) $$a + ar + ar^2 + \cdots + ar^{n-1} + \cdots$$

is known as a **geometric series**.

Theorem. *The geometric series* (1), *where* $a \neq 0$, *converges if* $|r| < 1$, *with* $\dfrac{a}{1-r}$ *as its sum, and diverges if* $|r| \geq 1$.

Proof. By use of the formula of algebra for the sum of the terms of a finite geometric progression, the sum of the first n terms of (1) can be written

$$S_n = \frac{a - ar^n}{1-r} = \frac{a}{1-r} - \frac{a}{1-r} r^n.$$

If $|r| < 1$, $\lim\limits_{n \to +\infty} r^n = 0$ (Theorem XIII, § 204), and therefore

$$\lim_{n \to +\infty} S_n = \frac{a}{1-r} - \frac{a}{1-r} \lim_{n \to +\infty} r^n = \frac{a}{1-r}.$$

If $|r| \geq 1$, $|ar^{n-1}| = |a| \cdot |r|^{n-1} \geq |a|$, and the general term of (1) does not tend toward zero. Therefore, by the Theorem of § 1103, the series (1) diverges.

Example. Determine for each of the following series whether the series converges or diverges. In case of convergence, find the sum:
(a) $\frac{1}{2} + \frac{1}{4} + \frac{1}{8} + \frac{1}{16} + \cdots$; (b) $2 + 3 + \frac{9}{2} + \frac{27}{4} + \cdots$;
(c) $3 - 2 + \frac{4}{3} - \frac{8}{9} + \cdots$.

Solution. (a) The series converges since $r = \frac{1}{2} < 1$, and the sum is $\dfrac{\frac{1}{2}}{1 - \frac{1}{2}} = 1$.
(b) The series diverges since $r = \frac{3}{2} > 1$. (c) The series converges since $r = -\frac{2}{3}$, and $|r| < 1$. The sum is $\dfrac{3}{1 - (-\frac{2}{3})} = \dfrac{9}{5}$.

1105. POSITIVE SERIES

For a **positive series** (a series whose terms are positive) or, more generally, a **nonnegative series** (a series whose terms are nonnegative), there is a simple criterion for convergence.

Theorem. *The series* $\sum\limits_{n=1}^{+\infty} a_n$, *where* $a_n \geq 0$, *converges if and only if the sequence of partial sums is bounded; that is, if and only if there is a number A such that* $a_1 + \cdots + a_n \leq A$ *for all values of n. If this condition is satisfied, the sum of the series is less than or equal to A. A series of nonnegative terms either converges, or diverges to $+\infty$.*

Proof. Since $S_{n+1} = S_n + a_n$, the condition $a_n \geq 0$ is equivalent to the condition $S_n \uparrow$, that is, that the sequence $\{S_n\}$ is monotonically increasing.

The Theorem of this section is therefore a direct consequence of the Note following Theorem XIV, § 204.

1106. THE INTEGRAL TEST

A convenient test for establishing convergence or divergence for certain series of positive terms makes use of the technique of integration.

Theorem. Integral Test. *If Σa_n is a positive series and if $f(x)$ is a positive monotonically decreasing function, defined for $x \geq a$, where a is some real*

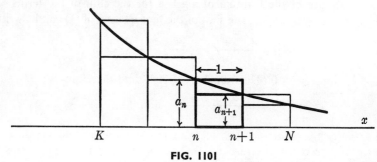

FIG. 1101

number, and if (for integral values of n) $f(n) = a_n$ for $n > a$, then the improper integral

(1) $$\int_a^{+\infty} f(x)\, dx$$

and the infinite series

(2) $$\sum_{n=1}^{+\infty} a_n$$

either both converge, or both diverge to $+\infty$.

Proof. Let us observe that both (1) and (2) always exist, in the finite or $+\infty$ sense. (Why? Supply details in Ex. 28, § 1107.) We wish to show that the finiteness of either implies the finiteness of the other. With the aid of results given in §§ 412, 1102, and 1105, we have only to show that if K is a fixed positive integer $\geq a$ and if N is an arbitrary integer $> K$, then the boundedness (as a function of N) of either

(3) $$I(N) \equiv \int_K^N f(x)\, dx$$

or

(4) $$J(N) \equiv \sum_{n=K}^N a_n$$

implies that of the other. (Cf. Ex. 29, § 1107). Let us look at the geometric representations of (3) and (4) given in Figure 1101. Each term of (4) is the area of a rectangle of width 1 and height a_n. This suggests that if (3) is written as the sum

$$(5) \quad I(N) = \int_K^N f(x)\, dx = \sum_{n=K}^{N-1} \int_n^{n+1} f(x)\, dx,$$

then each term of (5) is trapped between successive terms of (4). Indeed, for the interval $n \leq x \leq n+1$, since $f(x)$ is monotonically decreasing,

$$a_n = f(n) \geq f(x) \geq f(n+1) = a_{n+1},$$

so that

$$a_n = \int_n^{n+1} a_n\, dx \geq \int_n^{n+1} f(x)\, dx \geq \int_n^{n+1} a_{n+1}\, dx = a_{n+1}.$$

Therefore

$$\sum_{n=K}^{N-1} a_n \geq \int_K^N f(x)\, dx \geq \sum_{n=K}^{N-1} a_{n+1}.$$

Consequently (in the notation of (3) and (4)), since

$$\sum_{n=K}^{N-1} a_{n+1} = \sum_{n=K+1}^{N} a_n = J(N) - a_K,$$

$$J(N-1) \geq I(N) \geq J(N) - a_K.$$

These two inequalities show that the boundedness of either $J(N)$ or $I(N)$ implies that of the other (Ex. 30, § 1107) and the proof is complete.

Example 1. Show that the **p-series**

$$\sum_{n=1}^{+\infty} \frac{1}{n^p}$$

converges if $p > 1$ and diverges if $p \leq 1$.

Solution. This follows immediately from Example 3, § 413, if $p \geq 0$. If $p < 0$, the conclusion is trivial.

Example 2. Show that $\sum_{n=2}^{+\infty} \frac{1}{n(\ln n)^p}$ converges if and only if $p > 1$.

Solution. The integral $\int_2^{+\infty} \frac{dx}{x(\ln x)^p}$ converges if and only if $p > 1$. (Ex. 28, § 416.)

1107. EXERCISES

In Exercises 1–4, find an expression for the sum of the first n terms of the series. Thus find the sum of the series, if it exists.

1. $\dfrac{1}{1 \cdot 2} + \dfrac{1}{2 \cdot 3} + \dfrac{1}{3 \cdot 4} + \cdots$. *Hint:* $\dfrac{1}{n(n+1)} = \dfrac{1}{n} - \dfrac{1}{n+1}$.

2. $\dfrac{1}{1 \cdot 3} + \dfrac{1}{3 \cdot 5} + \dfrac{1}{5 \cdot 7} + \cdots$.

3. $\frac{1}{3} + \frac{2}{9} + \frac{4}{27} + \frac{8}{81} + \cdots$.
4. $1 + 2 + 3 + 4 + \cdots$.

In Exercises 5–8, write down the first four terms and the general term of the series whose sequence of partial sums is given.

5. $2, 2\frac{1}{2}, 2\frac{2}{3}, 2\frac{3}{4}, \cdots$.
6. $1, 3, 1, 3, 1, 3, \cdots$.
7. $1.3, 0.91, 1.027, 0.9919, \cdots, 1 - (-.3)^n, \cdots$.
8. $2, 1.3, 1.09, 1.027, \cdots, 1 + (.3)^{n-1}, \cdots$.

In Exercises 9–12, determine whether the given geometric series is convergent or not, and find its sum in case of convergence.

9. $12 - 8 + \cdots$.
10. $8 - 12 + \cdots$.
11. $0.27 + 0.027 + \cdots$.
12. $101 - 100 + \cdots$.

In Exercises 13–16, find as a quotient of integers the rational number represented by the repeating decimal, by means of geometric series.

13. $0.5555\cdots$.   *Hint:* $0.5555\cdots$ means $0.5 + 0.05 + 0.005 + \cdots$.
14. $3.285555\cdots$.   *Hint:* Write the number $3.28 + 0.005555\cdots$.
15. $6.1727272\cdots$.
16. $0.428571428571428571\cdots$.

In Exercises 17–26, establish convergence or divergence by the integral test of § 1106.

17. $\sum \dfrac{1}{2n+1}$.
18. $\sum \dfrac{n}{n^2+3}$.

19. $\sum \dfrac{1}{n^2+4}$.
20. $\sum \dfrac{n}{e^n}$.

21. $\displaystyle\sum_{n=3}^{+\infty} \dfrac{1}{n^2-4}$.
22. $\displaystyle\sum_{n=4}^{+\infty} \dfrac{1}{\sqrt{n^2-9}}$.

23. $\sum \dfrac{\ln n}{n}$.
24. $\displaystyle\sum_{n=3}^{+\infty} \dfrac{1}{n(\ln n)(\ln \ln n)^2}$.

25. $\sum \dfrac{1}{(n+1)^3}$.
26. $\sum \dfrac{1}{n^3+1}$.

27. Establish formula (1) of Theorem III, § 1102, for the case of an infinite sum ($+\infty$, $-\infty$, or ∞).

28. Explain why both (1) and (2), § 1106, must exist, as either finite or infinite limits. (Cf. Theorem X, § 401.)

29. For the Theorem of § 1106 prove that it is sufficient to show that the boundedness of either (3) or (4) implies that of the other. Explain why we may use K in place of a in (3) and K in place of 1 in (4).

30. Supply the details needed in the last sentence of the proof of § 1106.

31. Prove that $\displaystyle\sum_{n=3}^{+\infty} \dfrac{1}{n(\ln n)(\ln \ln n)^p}$ converges if and only if $p > 1$. More generally, prove that $\displaystyle\sum_{n=N}^{+\infty} \dfrac{1}{n(\ln n)(\ln \ln n)\cdots(\ln \ln \cdots \ln n)^p}$, where N is sufficiently large for $(\ln \ln \cdots \ln n)$ to be defined, converges if and only if $p > 1$. (Cf. Ex. 29, § 416.)

*32. Indicate by a graph how you could define a function $f(x)$ having the properties: $f(x)$ is positive, continuous, defined for $x \geq 0$, $\displaystyle\int_0^{+\infty} f(x)\,dx$ converges, and $\displaystyle\sum_{n=1}^{+\infty} f(n)$ diverges. (Cf. Ex. 42, § 416.)

*33. Indicate by a graph how you could define a function $f(x)$ having the properties: $f(x)$ is positive, continuous, defined for $x \geq 0$, $\displaystyle\int_0^{+\infty} f(x)\,dx$ diverges, and $\displaystyle\sum_{n=1}^{+\infty} f(n)$ converges.

1108. COMPARISON TESTS. DOMINANCE

In establishing the integral test of § 1106 we have already made use of the fact that a series of positive terms converges if and only if its sequence of partial sums is *bounded*. In order to prove convergence for a series of positive terms it is necessary to show only that the terms *approach zero fast enough* to keep the partial sums bounded. One of the simplest and most useful ways of doing this is to compare the terms of the given series with those of a series whose behavior is known. The ideas are similar to those relevant to comparison tests for improper integrals (§ 415).

Definition I. *The statement that a series Σb_n* **dominates** *a series Σa_n means that $|a_n| \leq b_n$ for every positive integer n.*

NOTE 1. Any dominating series consists automatically of nonnegative terms, although a dominated series may have negative terms.

Theorem I. Comparison Test. *Any nonnegative series dominated by a convergent series converges. Equivalently, any series that dominates a divergent nonnegative series diverges.*

Proof. We shall prove only the first form of the statement (cf. Ex. 21, § 1111). Assume $0 \leq a_n \leq b_n$ and that Σb_n converges. The convergence of Σb_n implies the boundedness of the partial sums $\sum_{n=1}^{N} b_n$, which implies the boundedness of the partial sums $\sum_{n=1}^{N} a_n$. By the Theorem of § 1105, the proof is complete.

Example 1. Prove that the series
$$1 + \frac{1}{2^2} + \frac{1}{3^3} + \cdots + \frac{1}{n^n} + \cdots$$
converges.

Solution. Since $\frac{1}{n^n} \leq \frac{1}{2^n}$, for $n > 1$, the given series is dominated by a convergent geometric series (except for the first term).

Example 2. Prove that the series
$$\frac{1}{2} + \frac{1}{3} + \frac{1}{2^2} + \frac{1}{3^2} + \frac{1}{2^3} + \frac{1}{3^3} + \cdots$$
converges.

Solution. This series is dominated by
$$\frac{1}{2} + \frac{1}{2} + \frac{1}{2^2} + \frac{1}{2^2} + \frac{1}{2^3} + \frac{1}{2^3} + \cdots,$$
whose partial sums are bounded by twice the sum of the convergent geometric series $\Sigma \frac{1}{2^n}$.

Let us adapt the "big O" notation of § 415 to the present topic:†

Definition II. *If $\{a_n\}$ and $\{b_n\}$ are two sequences, the notation $a_n = O(b_n)$ (read "a sub n is big O of b sub n"), means that there exist a positive integer N and a positive number K such that $n > N$ implies $|a_n| \leq K b_n$. If simultaneously $a_n = O(b_n)$ and $b_n = O(a_n)$, the two sequences (and also the two series Σa_n and Σb_n) are said to be of the **same order of magnitude** at $+\infty$.*

NOTE 2. As in § 415, the "big O" notation is also used in the following sense: $a_n = b_n + O(c_n)$ means $a_n - b_n = O(c_n)$.

Big O and order of magnitude relationships are usually established by taking limits:

Theorem II. *Let $\{a_n\}$ and $\{b_n\}$ be sequences of positive terms and assume that $\lim_{n \to +\infty} \dfrac{a_n}{b_n} = L$ exists, in a finite or infinite sense. Then:*
- *(i) $0 \leq L < +\infty$ implies $a_n = O(b_n)$,*
- *(ii) $0 < L \leq +\infty$ implies $b_n = O(a_n)$,*
- *(iii) $0 < L < +\infty$ implies that $\{a_n\}$ and $\{b_n\}$ are of the same order of magnitude at $+\infty$.*

(Give the details of the proof in Ex. 22, § 1111. Cf. Theorem II, § 415.)

Theorem III. *If Σa_n and Σb_n are series of nonnegative terms and if $a_n = O(b_n)$, then the convergence of Σb_n implies that of Σa_n. Equivalently, the divergence of Σa_n implies that of Σb_n.*

Proof. The convergence of $\sum_{n=N+1}^{+\infty} K b_n$ implies that of $\sum_{n=N+1}^{+\infty} a_n$ by Theorem I.

Corollary. *Two positive series that have the same order of magnitude either both converge or both diverge.*

Because of the usefulness of the *p*-series, $\Sigma \dfrac{1}{n^p}$, given in Example 1, § 1106, as a "test series," we restate here the facts established in that Example.

Theorem IV. *The **p-series** $\sum_{n=1}^{+\infty} \dfrac{1}{n^p}$ converges if $p > 1$ and diverges if $p \leq 1$. As a special case, the **harmonic series**, $\sum_{n=1}^{+\infty} \dfrac{1}{n}$, diverges.*‡

---

† The "little o" notation of § 415 takes the form for sequences: $a_n = o(b_n)$ means that there exist a positive integer N and a sequence of nonnegative terms K_n such that $K_n \to 0$ and such that $n > N$ implies $|a_n| \leq K_n b_n$. Obviously $a_n = o(b_n)$ implies $a_n = O(b_n)$.

‡ The function defined by the convergent *p*-series is called the **Riemann Zeta-function**:

$$(1) \quad \zeta(z) \equiv \sum_{n=1}^{+\infty} \frac{1}{n^z}.$$

This function is defined for complex values of z, by (1) if the real part of $z > 1$, and by other means (infinite products, improper integrals) for other values of z. Its values have been tabulated (cf. Eugen Jahnke and Fritz Emde, *Tables of Functions*, New York, Dover Publications, 1943). For further comments, see the Note, § 1120.

(Cf. Ex. 31, § 1107, for a sequence of test series similar to the *p*-series, each converging more slowly than the preceding.)

The technique of the comparison test for positive series, particularly in the form of Theorem III, consists usually of four steps:
 (*i*) get a "feeling" for how rapidly the terms of the given series are approaching zero;
 (*ii*) construct a new test series of positive terms which dominates, is dominated by, or is of the same order of magnitude as the original series;
 (*iii*) establish the necessary inequalities, or the limit of the quotient of the general terms;
 (*iv*) infer convergence if the dominating series converges, and divergence if the dominated series diverges.

In practice, one frequently postpones applying a comparison test to a given series until after one has tried one of the more automatic tests (like the ratio test) given in subsequent sections.

Example 3. Test for convergence or divergence:
$$\frac{1}{1 \cdot 3} + \frac{2}{5 \cdot 7} + \frac{3}{9 \cdot 11} + \frac{4}{13 \cdot 15} + \cdots.$$

Solution. The general term is $\dfrac{n}{(4n-3)(4n-1)}$, which evidently is of the same order of magnitude at $+\infty$ as $\dfrac{1}{n}$ (the limit of the quotient is $\tfrac{1}{16}$). Since the test series $\Sigma \dfrac{1}{n}$ diverges, so does the given series.

Example 4. Prove that if $a_n = O(c_n)$ and $b_n = O(d_n)$, then $a_n + b_n = O(c_n + d_n)$ and $a_n b_n = O(c_n d_n)$. Prove that if $a_n = O(c_n)$ and b_n and d_n are sequences of positive numbers which are of the same order of magnitude at $+\infty$, then $\dfrac{a_n}{b_n} = O\left(\dfrac{c_n}{d_n}\right)$.

Solution. For the first part, if $|a_n| \leq K_1 c_n$ and $|b_n| \leq K_2 d_n$, then $|a_n + b_n| \leq \max(K_1, K_2)(c_n + d_n)$ and $|a_n b_n| \leq K_1 K_2 c_n d_n$. For the second part, since $1/b_n = O(1/d_n)$, the result follows from the product form of the first part.

Example 5. Test for convergence or divergence:
$$\frac{\ln 5}{1 \cdot 2} + \frac{\sqrt{3} \ln 9}{2 \cdot 3} + \frac{\sqrt{5} \ln 13}{3 \cdot 4} + \cdots.$$

Solution. The general term is $\dfrac{\sqrt{2n-1}\, \ln(4n+1)}{n(n+1)}$. Inasmuch as $\sqrt{2n-1} = O(n^{\frac{1}{2}})$ and $\ln(4n+1) = O(n^p)$ for every $p > 0$ (by l'Hospital's Rule), the numerator $= O(n^q)$ for every $q > \tfrac{1}{2}$ (cf. Example 4). Since the denominator is of the same order of magnitude as n^2 at $+\infty$, the general term $= O(n^r)$ for every $r > -\tfrac{3}{2}$. Finally, since Σn^r converges for every $r < -1$, we have only to choose r between $-\tfrac{3}{2}$ and -1 to establish the convergence of the given series.

1109. THE RATIO TEST

One of the most practical routine tests for convergence of a positive series makes use of the ratio of consecutive terms.

Theorem. Ratio Test. *Let Σa_n be a positive series, and define the* **test ratio**
$$r_n \equiv \frac{a_{n+1}}{a_n}.$$
Assume that the limit of this test ratio exists:
$$\lim_{n \to +\infty} r_n = \rho,$$
where $0 \leq \rho \leq +\infty$. Then

(i) *if $0 \leq \rho < 1$, Σa_n converges;*
(ii) *if $1 < \rho \leq +\infty$, Σa_n diverges;*
(iii) *if $\rho = 1$, Σa_n may either converge or diverge, and the test fails.*

Proof of (i). Assume $\rho < 1$, and let r be any number such that $\rho < r < 1$. Since $\lim_{n \to +\infty} r_n = \rho < r$, we may choose a neighborhood of ρ that excludes r. Since every r_n, for n greater than or equal to some N, lies in this neighborhood of ρ, we have:
$$n \geq N \quad \text{implies} \quad r_n < r.$$
This gives the following sequence of inequalities:

$$\frac{a_{N+1}}{a_N} < r, \quad \text{or} \quad a_{N+1} < r a_N,$$

$$\frac{a_{N+2}}{a_{N+1}} < r, \quad \text{or} \quad a_{N+2} < r a_{N+1} < r^2 a_N,$$

$$\frac{a_{N+3}}{a_{N+2}} < r, \quad \text{or} \quad a_{N+3} < r a_{N+2} < r^3 a_N,$$
$$\cdots \qquad \qquad \cdots$$

Thus, each term of the series

(1) $$a_{N+1} + a_{N+2} + a_{N+3} + \cdots$$

is less than the corresponding term of the series

(2) $$r a_N + r^2 a_N + r^3 a_N + \cdots.$$

But the series (2) is a geometric series with common ratio $r < 1$, and therefore converges. Since (2) dominates (1), the latter series also converges. Therefore (Theorem I, § 1102), the original series Σa_n converges.

Proof of (ii). By reasoning analogous to that employed above, we see that since $\lim_{n \to +\infty} r_n > 1$, whether the limit is finite or infinite, there must be a number N such that $r_n \geq 1$ whenever $n \geq N$. In other words,

(3) $$n \geq N \quad \text{implies} \quad \frac{a_{n+1}}{a_n} \geq 1, \quad \text{or} \quad a_{n+1} \geq a_n.$$

The inequalities (3) state that beyond the first N terms, each term is at least as large as the preceding term. Since these terms are positive, the limit

THE RATIO TEST

of the general term cannot be 0 (take $\epsilon = a_N > 0$), and therefore (§ 1103) the series Σa_n diverges.

Proof of (iii). For any *p*-series the test ratio is $r_n = \dfrac{a_{n+1}}{a_n} = \left(\dfrac{n}{n+1}\right)^p$, and since the function x^p is continuous at $x = 1$ (Ex. 7, § 502),

$$\lim_{n \to +\infty} r_n = \lim_{n \to +\infty} \left(\frac{n}{n+1}\right)^p = \left[\lim_{n \to +\infty} \frac{n}{n+1}\right]^p = 1^p = 1.$$

If $p > 1$ the *p*-series converges and if $p \leq 1$ the *p*-series diverges, but in either case $\rho = 1$.

NOTE 1. For convergence it is important that the *limit* of the test ratio be less than 1. It is not sufficient that the test ratio itself be always less than 1. This is shown by the harmonic series, which diverges, whereas the test ratio $n/(n+1)$ is always less than 1. However, if an inequality of the form $\dfrac{a_{n+1}}{a_n} \leq r < 1$ holds for all sufficiently large n (whether the limit ρ exists or not), the series Σa_n converges. (Cf. Ex. 28, § 1111.)

NOTE 2. For divergence it is sufficient that the test ratio itself be greater than 1. In fact, if $r_n \geq 1$ for all sufficiently large n, the series Σa_n diverges, since this inequality is the inequality (3) upon which the proof of (*ii*) rests.

NOTE 3. The ratio test may fail, not only by the equality of ρ and 1, but by the failure of the limit ρ to exist, finitely or infinitely. For example, the series of Example 2, § 1108, converges, although the test ratio $r_n = a_{n+1}/a_n$ has no limit, since $r_{2n-1} = 2^n/3^n \to 0$ and $r_{2n} = 3^n/2^{n+1} \to +\infty$.

NOTE 4. The ratio test provides a simple proof that certain limits are zero. That is, if $\lim\limits_{n \to +\infty} a_{n+1}/a_n$, where $a_n > 0$ for all n, exists and is less than 1, then $a_n \to 0$. This is true because a_n is the general term of a convergent series.

Example 1. Prove that $\lim\limits_{n \to +\infty} \dfrac{x^n}{n!} = 0$, for every real number x.

Solution. Without loss of generality we can assume $x > 0$ (take absolute values). Then
$$\frac{x^{n+1}}{(n+1)!} \div \frac{x^n}{n!} = \frac{x}{n+1} \to 0.$$

Example 2. Use the ratio test to establish convergence of the series
$$1 + \frac{1}{1!} + \frac{1}{2!} + \frac{1}{3!} + \cdots + \frac{1}{(n-1)!} + \cdots.$$

Solution. Since $a_n = \dfrac{1}{(n-1)!}$, $a_{n+1} = \dfrac{1}{n!}$ and
$$r_n = \frac{a_{n+1}}{a_n} = \frac{(n-1)!}{n!} = \frac{1}{n}.$$
Therefore $\rho = \lim\limits_{n \to +\infty} r_n = 0 < 1$.

Example 3. Use the ratio test to establish convergence of the series
$$\frac{1}{2} + \frac{2^2}{2^2} + \frac{3^2}{2^3} + \frac{4^2}{2^4} + \cdots + \frac{n^2}{2^n} + \cdots.$$

Solution. Since $a_n = \dfrac{n^2}{2^n}$, $a_{n+1} = \dfrac{(n+1)^2}{2^{n+1}}$ and

$$r_n = \frac{a_{n+1}}{a_n} = \left(\frac{n+1}{n}\right)^2 \cdot \frac{2^n}{2^{n+1}} = \frac{1}{2}\left(\frac{n+1}{n}\right)^2.$$

Therefore $\rho = \lim\limits_{n\to +\infty} r_n = \dfrac{1}{2} \lim\limits_{n\to +\infty} \left(1 + \dfrac{1}{n}\right)^2 = \dfrac{1}{2} < 1$.

Example 4. Test for convergence or divergence:

$$\frac{1}{3} + \frac{2!}{3^2} + \frac{3!}{3^3} + \cdots + \frac{n!}{3^n} + \cdots.$$

Solution. Since $a_n = \dfrac{n!}{3^n}$, $a_{n+1} = \dfrac{(n+1)!}{3^{n+1}}$ and

$$r_n = \frac{(n+1)!}{n!} \cdot \frac{3^n}{3^{n+1}} = \frac{n+1}{3}.$$

Therefore $\rho = \lim\limits_{n\to +\infty} \dfrac{n+1}{3} = +\infty$, and the series diverges.

NOTE 5. Experience teaches us that if the general term of a series involves the index n either exponentially or factorially (as in the preceding Examples) the ratio test can be expected to answer the question of convergence or divergence, while if the index n is involved only algebraically or logarithmically (as in the p-series and Examples 3 and 5, § 1108), the ratio test can be expected to fail.

1110. THE ROOT TEST

A test somewhat similar to the ratio test is the following, whose proof is requested in Exercise 23, § 1111:

Theorem. Root Test. *Let Σa_n be a nonnegative series, and assume the existence of the following limit:*

$$\lim_{n\to +\infty} \sqrt[n]{a_n} = \sigma,$$

where $0 \leq \sigma \leq +\infty$. Then
  (i) *if $0 \leq \sigma < 1$, Σa_n converges;*
  (ii) *if $1 < \sigma \leq +\infty$, Σa_n diverges;*
  (iii) *if $\sigma = 1$, Σa_n may either converge or diverge, and the test fails.*

NOTE 1. The inequality $\sqrt[n]{a_n} < 1$, for all n, is not sufficient for convergence, although an inequality of the form $\sqrt[n]{a_n} \leq r < 1$, for $n > N$, does guarantee convergence. (Cf. Ex. 29, § 1111, Note 1, § 1109.)

NOTE 2. For divergence it is sufficient to have an inequality of the form $\sqrt[n]{a_n} \geq 1$, for $n > N$, since this precludes $\lim\limits_{n\to +\infty} a_n = 0$.

NOTE 3. The ratio test is usually easier to apply than the root test, but the latter is more powerful. (Cf. Exs. 31–32, § 1111.)

Example. Use the root test to establish convergence of the series

$$1 + \frac{2}{2^1} + \frac{3}{2^2} + \frac{4}{2^3} + \cdots.$$

Solution. Since $\sqrt[n]{a_n} = \sqrt[n]{\dfrac{n}{2^{n-1}}} = \dfrac{\sqrt[n]{n}}{2^{\frac{n-1}{n}}}$, the problem of finding $\lim\limits_{n\to+\infty} \sqrt[n]{a_n}$ can be reduced to that of finding the two limits $\lim\limits_{n\to+\infty} \sqrt[n]{n}$ and $\lim\limits_{n\to+\infty} 2^{1-\frac{1}{n}}$. The second of these is not indeterminate, owing to the continuity of the function 2^x at $x = 1$, and has the value $2^1 = 2$. To evaluate $\lim\limits_{n\to+\infty} \sqrt[n]{n}$, or $\lim\limits_{x\to+\infty} x^{\frac{1}{x}}$, let $y = x^{\frac{1}{x}}$, and take logarithms (cf. § 317): $\ln y = \dfrac{\ln x}{x}$. Then, by l'Hospital's Rule (§ 316), $\lim\limits_{x\to+\infty} \ln y = \lim\limits_{x\to+\infty} \dfrac{1/x}{1} = 0$, and $y \to e^0 = 1$. Therefore $\lim\limits_{n\to+\infty} \sqrt[n]{a_n} = \tfrac{1}{2} < 1$, and Σa_n converges.

1111. EXERCISES

In Exercises 1–20, establish convergence or divergence of the given series.

1. $\dfrac{1}{3} + \dfrac{\sqrt{2}}{5} + \dfrac{\sqrt{3}}{7} + \dfrac{\sqrt{4}}{9} + \cdots$.

2. $\dfrac{1}{\sqrt{1\cdot 2}} + \dfrac{1}{\sqrt{2\cdot 3}} + \dfrac{1}{\sqrt{3\cdot 4}} + \dfrac{1}{\sqrt{4\cdot 5}} + \cdots$.

3. $\dfrac{\sqrt{3}}{2\cdot 4} + \dfrac{\sqrt{5}}{4\cdot 6} + \dfrac{\sqrt{7}}{6\cdot 8} + \dfrac{\sqrt{9}}{8\cdot 10} + \cdots$.

4. $\dfrac{\sqrt{2}+1}{3^3-1} + \dfrac{\sqrt{3}+1}{4^3-1} + \dfrac{\sqrt{4}+1}{5^3-1} + \dfrac{\sqrt{5}+1}{6^3-1} + \cdots$.

5. $\dfrac{1}{3} + \dfrac{1\cdot 2}{3\cdot 5} + \dfrac{1\cdot 2\cdot 3}{3\cdot 5\cdot 7} + \dfrac{1\cdot 2\cdot 3\cdot 4}{3\cdot 5\cdot 7\cdot 9} + \cdots$.

6. $\dfrac{1!}{2^5} + \dfrac{2!}{2^6} + \dfrac{3!}{2^7} + \dfrac{4!}{2^8} + \cdots$.

7. $\dfrac{1}{\ln 2} + \dfrac{1}{\ln 3} + \dfrac{1}{\ln 4} + \dfrac{1}{\ln 5} + \cdots$.

8. $\dfrac{2!}{4!} + \dfrac{3!}{5!} + \dfrac{4!}{6!} + \dfrac{5!}{7!} + \cdots$.

9. $\sum\limits_{n=1}^{+\infty} \dfrac{\sqrt{n}}{n^2+4}$.

10. $\sum\limits_{n=1}^{+\infty} \dfrac{\ln n}{n\sqrt[3]{n}+1}$.

11. $\sum\limits_{n=1}^{+\infty} \dfrac{n^4}{n!}$.

12. $\sum\limits_{n=1}^{+\infty} \dfrac{3^{2n-1}}{n^2+1}$.

13. $\sum\limits_{n=1}^{+\infty} e^{-n^2}$.

14. $\sum\limits_{n=2}^{+\infty} \dfrac{1}{(\ln n)^n}$.

15. $\sum\limits_{n=1}^{+\infty} \dfrac{\sqrt{n+1}-\sqrt{n}}{n^\alpha}$.

16. $\sum\limits_{n=1}^{+\infty} \dfrac{n!}{n^n}$.

17. $\sum\limits_{n=2}^{+\infty} \dfrac{1}{(\ln n)^\alpha}$.

18. $\sum\limits_{n=1}^{+\infty} \dfrac{1}{1+a^n}, \alpha > -1$.

19. $\sum\limits_{n=1}^{+\infty} (\sqrt[n]{n}-1)^n$.

20. $\sum\limits_{n=1}^{+\infty} r^n |\sin n\alpha|, \alpha > 0, r > 0$.

21. Prove the comparison test (Theorem I, § 1108) for divergence.
22. Prove Theorem II, § 1108.
23. Prove the root test, § 1110.
24. Prove that if $a_n \geq 0$ and there exists a number $k > 1$ such that $\lim\limits_{n \to +\infty} n^k a_n$ exists and is finite, then Σa_n converges.
25. Prove that if $a_n \geq 0$ and $\lim\limits_{n \to +\infty} n a_n$ exists and is positive, then Σa_n diverges.

*26. Prove that any series of the form $\sum\limits_{n=1}^{+\infty} \dfrac{d_n}{10^n}$, where $d_n = 0, 1, 2, \cdots, 9$, converges. Hence show that any decimal expansion $0.d_1 d_2 \cdots$ represents some real number r, where $0 \leq r \leq 1$.

*27. Prove the converse of Exercise 26: If $0 \leq r \leq 1$, then there exists a decimal expansion $0.d_1 d_2 \cdots$ representing r. Show that this decimal expansion is unique unless r is positive and representable by a (unique) terminating decimal (cf. Ex. 5, § 114), in which case r is also representable by a (unique) decimal composed, from some point on, of repeating 9's.

*28. Establish the following form of the ratio test: The positive series Σa_n converges if $\varlimsup\limits_{n \to +\infty} \dfrac{a_{n+1}}{a_n} < 1$, and diverges if $\varliminf\limits_{n \to +\infty} \dfrac{a_{n+1}}{a_n} > 1$. (Cf. Ex. 11, § 219.)

*29. Establish the following form of the root test: If Σa_n is a nonnegative series and if $\sigma \equiv \varlimsup\limits_{n \to +\infty} \sqrt[n]{a_n}$, then Σa_n converges if $\sigma < 1$ and diverges if $\sigma > 1$. (Cf. Ex. 28.)

*30. Apply Exercises 28 and 29 to the series $\dfrac{1}{2} + \dfrac{1}{3} + \dfrac{1}{2^2} + \dfrac{1}{3^2} + \cdots$ of Example 2, § 1108, and show that $\varlimsup \dfrac{a_{n+1}}{a_n} = +\infty$, $\varliminf \dfrac{a_{n+1}}{a_n} = 0$, $\varlimsup \sqrt[n]{a_n} = 1/\sqrt{2}$, and $\varliminf \sqrt[n]{a_n} = 1/\sqrt{3}$. Thus show that the ratio test of § 1109 and that of Exercise 28 both fail, that the root test of § 1110 fails, and that the root test of Exercise 29 succeeds in establishing convergence of the given series.

*31. Prove that if $\lim\limits_{n \to +\infty} \dfrac{a_{n+1}}{a_n}$ exists, then $\lim\limits_{n \to +\infty} \sqrt[n]{a_n}$ also exists and is equal to it. *Hints:* For the case $\lim\limits_{n \to +\infty} \dfrac{a_{n+1}}{a_n} = L$, where $0 < L < +\infty$, let α and β be arbitrary numbers such that $0 < \alpha < L < \beta$. Then for $n \geq $ some N, $\alpha a_n < a_{n+1} < \beta a_n$. Hence

$$\alpha a_N < a_{N+1} < \beta a_N$$
$$\alpha^2 a_N < \alpha a_{N+1} < a_{N+2} < \beta a_{N+1} < \beta^2 a_N$$
$$\cdot \quad \cdot \quad \cdot$$
$$\alpha^p a_N < a_{N+p} < \beta^p a_N.$$

Thus, for $n > N$,
$$\alpha^n \dfrac{a_N}{\alpha^N} < a_n < \beta^n \dfrac{a_N}{\beta^N},$$

and
$$\alpha \leq \left\{ \begin{array}{c} \varliminf\limits_{n \to +\infty} \sqrt[n]{a_n} \\ \varlimsup\limits_{n \to +\infty} \sqrt[n]{a_n} \end{array} \right\} \leq \beta.$$

*32. The example 1, 2, 1, 2, $\cdots$ shows that $\lim\limits_{n\to+\infty} \sqrt[n]{a_n}$ may exist when $\lim\limits_{n\to+\infty} \dfrac{a_{n+1}}{a_n}$ does not. Find an example of a convergent positive series Σa_n for which this situation is also true.

*33. Prove that $\lim\limits_{n\to+\infty} \sqrt[n]{\dfrac{n^n}{n!}} = e$. (Cf. Ex. 31.)

*1112. MORE REFINED TESTS

The tests discussed in preceding sections are those most commonly used in practice. There are occasions, however, when such a useful test as the ratio test fails, and it is extremely difficult to devise an appropriate test series for the comparison test. We give now some sharper criteria which may sometimes be used in the event that $\lim\limits_{n\to+\infty} \dfrac{a_{n+1}}{a_n} = 1$.

Theorem I. *Kummer's Test.* Let Σa_n be a positive series, and let $\{p_n\}$ be a sequence of positive constants such that

$$\text{(1)} \qquad \lim_{n\to\infty}\left[p_n \frac{a_n}{a_{n+1}} - p_{n+1}\right] = L$$

exists and is positive $(0 < L \leq +\infty)$. Then Σa_n converges. If the limit (1) exists and is negative $(-\infty \leq L < 0)$ (or, more generally, if $p_n \dfrac{a_n}{a_{n+1}} - p_{n+1} \leq 0$ for $n \geq N$), and if $\Sigma \dfrac{1}{p_n}$ diverges, then Σa_n diverges.

Proof. Let r be any number such that $0 < r < L$. Then (cf. the proof of the ratio test) there must exist a positive integer N such that $n \geq N$ implies $p_n \dfrac{a_n}{a_{n+1}} - p_{n+1} > r$. For any positive integer m, then, we have the sequence of inequalities:

$$p_N a_N - p_{N+1} a_{N+1} > r a_{N+1},$$
$$p_{N+1} a_{N+1} - p_{N+2} a_{N+2} > r a_{N+2},$$
$$\cdots \cdots$$
$$p_{N+m-1} a_{N+m-1} - p_{N+m} a_{N+m} > r a_{N+m}.$$

Adding on both sides we have, because of cancellations by pairs on the left:

$$\text{(2)} \qquad p_N a_N - p_{N+m} a_{N+m} > r(a_{N+1} + \cdots + a_{N+m}).$$

Using the notation S_n for the partial sum $a_1 + \cdots + a_n$, we can write the sum in parentheses of (2) as $S_{N+m} - S_N$, and obtain by rearrangement of terms:

$$rS_{N+m} < rS_N + p_N a_N - p_{N+m} a_{N+m} < rS_N + p_N a_N.$$

Letting B denote the constant $(rS_N + p_N a_N)/r$, we infer that $S_n < B$ for $n > N$. In other words, the partial sums of Σa_n are bounded and hence (§ 1105) the series converges.

To prove the second part we infer from the inequality $p_n \dfrac{a_n}{a_{n+1}} - p_{n+1} \leq 0$, which holds for $n \geq N$, the sequence of inequalities

$$p_N a_N \leq p_{N+1} a_{N+1} \leq \cdots \leq p_n a_n,$$

for any $n > N$. Denoting by A the positive constant $p_N a_N$, we conclude from the comparison test, the inequality

$$a_n \geq A \cdot \frac{1}{p_n}, \quad \text{for } n > N,$$

and the divergence of $\Sigma \dfrac{1}{p_n}$, that Σa_n also diverges.

Theorem II. Raabe's Test. *Let Σa_n be a positive series and assume that*

$$\lim_{n \to +\infty} n\left(\frac{a_n}{a_{n+1}} - 1\right) = L$$

exists (finite or infinite). Then
  (i) if $1 < L \leq +\infty$, Σa_n converges;
  (ii) if $-\infty \leq L < 1$, Σa_n diverges;
  (iii) if $L = 1$, Σa_n may either converge or diverge, and the test fails.

Proof. The first two parts are a consequence of Kummer's Test with $p_n \equiv n$. (Cf. Ex. 5, § 1113, for further suggestions.)

In case the limit L of Raabe's Test exists and is equal to 1, a refinement is possible:

Theorem III. *Let Σa_n be a positive series and assume that*

$$\lim_{n \to +\infty} \ln n \left[n\left(\frac{a_n}{a_{n+1}} - 1\right) - 1 \right] = L$$

exists (finite or infinite). Then
  (i) if $1 < L \leq +\infty$, Σa_n converges;
  (ii) if $-\infty \leq L < 1$, Σa_n diverges;
  (iii) if $L = 1$, Σa_n may either converge or diverge, and the test fails.

Proof. The first two parts are a consequence of Kummer's Test with $p_n \equiv n \ln n$. (Cf. Ex. 9, § 1113. For a further refinement, and a proof of *(iii)*, cf. RV, Exs. 10, 11, § 713.)

Example 1. Test for convergence or divergence:

$$\left(\frac{1}{2}\right)^p + \left(\frac{1 \cdot 3}{2 \cdot 4}\right)^p + \left(\frac{1 \cdot 3 \cdot 5}{2 \cdot 4 \cdot 6}\right)^p + \cdots.$$

Solution. Since $\lim\limits_{n \to +\infty} \dfrac{a_{n+1}}{a_n} = 1$, the ratio test fails, and we turn to Raabe's test. We find

$$n\left(\frac{a_n}{a_{n+1}} - 1\right) = n\left[\left(\frac{2n+2}{2n+1}\right)^p - 1\right] = \frac{2n}{2n+1} \cdot \frac{(1+x)^p - 1}{2x},$$

where $x = (2n + 1)^{-1}$. The limit of this expression, as $n \to +\infty$, is (by l'Hospital's Rule):
$$\lim_{x \to 0} \frac{(1 + x)^p - 1}{2x} = \lim_{x \to 0} \frac{p(1 + x)^{p-1}}{2} = \frac{p}{2}.$$
Therefore the given series converges for $p > 2$ and diverges for $p < 2$. For the case $p = 2$, see Example 2.

Example 2. Test for convergence or divergence;
$$\left(\frac{1}{2}\right)^2 + \left(\frac{1 \cdot 3}{2 \cdot 4}\right)^2 + \left(\frac{1 \cdot 3 \cdot 5}{2 \cdot 4 \cdot 6}\right)^2 + \cdots.$$

Solution. Both the ratio test and Raabe's test fail (cf. Example 1). Preparing to use Theorem III, we simplify the expression
$$n\left(\frac{a_n}{a_{n+1}} - 1\right) - 1 = n\frac{4n + 3}{4n^2 + 4n + 1} - 1 = \frac{-n - 1}{(2n + 1)^2}.$$
Since $\lim_{n \to +\infty} \ln n \frac{-n - 1}{(2n + 1)^2} = 0 < 1$, the given series diverges.

*1113. EXERCISES

In Exercises 1–4, test for convergence or divergence.

*1. $\sum_{n=1}^{+\infty} \frac{2 \cdot 4 \cdot 6 \cdots 2n}{1 \cdot 3 \cdot 5 \cdots (2n + 1)}$. (Cf. Ex. 4.)

*2. $\sum_{n=1}^{+\infty} \frac{1 \cdot 3 \cdots (2n - 1)}{2 \cdot 4 \cdots 2n} \cdot \frac{1}{2n + 1}$.

*3. $\sum_{n=1}^{+\infty} \frac{1 \cdot 3 \cdots (2n - 1)}{2 \cdot 4 \cdots 2n} \cdot \frac{4n + 3}{2n + 2}$.

*4. $\sum_{n=1}^{+\infty} \left[\frac{2 \cdot 4 \cdot 6 \cdots 2n}{1 \cdot 3 \cdot 5 \cdots (2n + 1)}\right]^p$.

*5. Prove Theorem II, § 1112. *Hint:* For examples for part (*iii*), define the terms of $\sum_{n=1}^{+\infty} a_n$ inductively, with $a_1 \equiv 1$, and $\frac{a_n}{a_{n+1}} = 1 + \frac{1}{n} + \frac{k}{n \ln n}$. Then use Theorem III, § 1112.

*6. Prove that the **hypergeometric series**
$$1 + \frac{\alpha \cdot \beta}{1 \cdot \gamma} + \frac{\alpha(\alpha + 1)\beta(\beta + 1)}{1 \cdot 2 \cdot \gamma \cdot (\gamma + 1)} + \cdots$$
$$+ \frac{\alpha(\alpha + 1) \cdots (\alpha + n - 1) \cdot \beta(\beta + 1) \cdots (\beta + n - 1)}{n! \, \gamma(\gamma + 1) \cdots (\gamma + n - 1)} + \cdots$$
converges if and only if $\gamma - \alpha - \beta > 0$ (γ not 0 or a negative integer).

*7. Prove **Gauss's Test**: *If the positive series Σa_n is such that*
$$\frac{a_n}{a_{n+1}} = 1 + \frac{h}{n} + O\left(\frac{1}{n^2}\right),$$
then Σa_n converges if $h > 1$, and diverges if $h \leq 1$. Show that $O(1/n^2)$ could be replaced by the weaker condition $O(1/n^\alpha)$, where $\alpha > 1$.

*8. State and prove limit-superior and limit-inferior forms for the Theorems of § 1112. (Cf. Exs. 28–29, § 1111.)

*9. Prove the first two parts of Theorem III, § 1112. *Hint:* This is equivalent to showing (using Kummer's test with $p_n \equiv n \ln n$) that

$$\ln n \left[n \left(\frac{a_n}{a_{n+1}} - 1 \right) - 1 \right] - \left\{ n \ln n \frac{a_n}{a_{n+1}} - (n+1) \ln (n+1) \right\} \to 1,$$

or

$$(n+1) \ln (n+1) - [n \ln n + \ln n] = (n+1) \ln \left(\frac{n+1}{n} \right) \to 1.$$

1114. SERIES OF ARBITRARY TERMS

If a series has terms of one sign, as we have seen for nonnegative series, there is only one kind of divergence—to infinity—and convergence of the series is equivalent to the boundedness of the partial sums. We wish to turn our attention now to series whose terms may be either positive or negative—or zero. The behavior of such series is markedly different from that of nonnegative series, but we shall find that we can make good use of the latter to clarify the former.

1115. ALTERNATING SERIES

An **alternating series** is a series of the form

(1) $$c_1 - c_2 + c_3 - c_4 + \cdots,$$

where $c_n > 0$ for every n.

Theorem. *An alternating series* (1) *whose terms satisfy the two conditions*
 (i) $c_{n+1} < c_n$ *for every* n,
 (ii) $c_n \to 0$ *as* $n \to +\infty$,
converges. If S and S_n denote the sum, and the partial sum of the first n terms, respectively, of the series (1),

(2) $$|S_n - S| < c_{n+1}.$$

Proof. We break the proof into six parts (cf. Fig. 1102):

A. The partial sums S_{2n} (consisting of an even number of terms) form an increasing sequence.

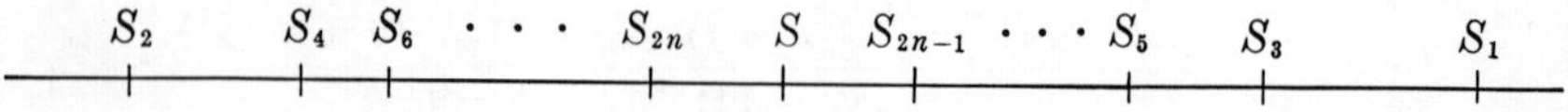

$$\underset{S_2}{|} \quad \underset{S_4}{|} \; \underset{S_6}{|} \quad \cdots \quad \underset{S_{2n}}{|} \quad \underset{S}{|} \quad \underset{S_{2n-1}}{|} \quad \cdots \quad \underset{S_5}{|} \quad \underset{S_3}{|} \quad \underset{S_1}{|}$$

FIG. 1102

B. The partial sums S_{2n-1} (consisting of an odd number of terms) form a decreasing sequence.
 C. For every m and every n, $S_{2m} < S_{2n-1}$.
 D. S exists.
 E. For every m and every n, $S_{2m} < S < S_{2n-1}$.
 F. The inequality (2) holds.

§ 1116] ABSOLUTE AND CONDITIONAL CONVERGENCE   399

A: Since $S_{2n+2} = S_{2n} + (c_{2n+1} - c_{2n+2})$, and since $c_{2n+2} < c_{2n+1}$, $S_{2n+2} > S_{2n}$.

B: Since $S_{2n+1} = S_{2n-1} - (c_{2n} - c_{2n+1})$, and since $c_{2n+1} < c_{2n}$, $S_{2n+1} < S_{2n-1}$.

C: If $2m < 2n - 1$,
$$S_{2n-1} - S_{2m} = (c_{2m+1} - c_{2m+2}) + \cdots + (c_{2n-3} - c_{2n-2}) + c_{2n-1} > 0,$$
and $S_{2m} < S_{2n-1}$. If $2m > 2n - 1$,
$$S_{2m} - S_{2n-1} = -(c_{2n} - c_{2n+1}) - \cdots - (c_{2m-2} - c_{2m-1}) - c_{2m} < 0,$$
and $S_{2m} < S_{2n-1}$.

D: From *A*, *B*, and *C* it follows that $\{S_{2n}\}$ and $\{S_{2n-1}\}$ are bounded monotonic sequences and therefore converge. We need only show that their limits are equal. But this is true, since
$$\lim_{n \to +\infty} S_{2n+1} - \lim_{n \to +\infty} S_{2n} = \lim_{n \to +\infty} (S_{2n+1} - S_{2n}) = \lim_{n \to +\infty} c_{2n+1} = 0.$$

E: By the fundamental theorem on convergence of monotonic sequences (Theorem XIV, § 204), for an arbitrary fixed n and variable m, $S_{2m} < S_{2n+1}$ implies $S \leq S_{2n+1} < S_{2n-1}$. Similarly, for an arbitrary fixed m and variable n, $S_{2m+2} < S_{2n+1}$ implies $S_{2m} < S_{2m+2} \leq S$.

F: On the one hand,
$$0 < S_{2n-1} - S < S_{2n-1} - S_{2n} = c_{2n},$$
while, on the other hand
$$0 < S - S_{2n} < S_{2n+1} - S_{2n} = c_{2n+1}.$$

Example 1. The **alternating harmonic series**
$$1 - \tfrac{1}{2} + \tfrac{1}{3} - \tfrac{1}{4} + \cdots$$
converges, since the conditions of the Theorem above are satisfied.

Example 2. Prove that the series $\sum_{n=1}^{+\infty} (-1)^n \dfrac{\ln n}{n}$ converges.

Solution. All of the conditions of the alternating series test are obvious except for the inequality $\dfrac{\ln(n+1)}{n+1} < \dfrac{\ln n}{n}$. The simplest way of establishing this is to show that $f(x) \equiv \dfrac{\ln x}{x}$ is strictly decreasing since its derivative is
$$f'(x) = \frac{1 - \ln x}{x^2} < 0, \quad \text{for } x > e.$$

1116. ABSOLUTE AND CONDITIONAL CONVERGENCE

We introduce some notation. Let $\sum_{n=1}^{+\infty} a_n$ be a given series of arbitrary terms. For every positive integer n we define p_n to be the larger of the two numbers a_n and 0, and q_n to be the larger of the two numbers $-a_n$ and 0 (in case $a_n = 0$, $p_n = q_n = 0$):

(1) $\qquad p_n \equiv \max(a_n, 0), \quad q_n \equiv \max(-a_n, 0).$

Then p_n and q_n are nonnegative numbers (at least one of them being zero) satisfying the two equations

(2) $$p_n - q_n = a_n, \quad p_n + q_n = |a_n|,$$

and the two inequalities

(3) $$0 \leq p_n \leq |a_n|, \quad 0 \leq q_n \leq |a_n|.$$

The two nonnegative series Σp_n and Σq_n are called the **nonnegative** and **nonpositive parts**, respectively, of the series Σa_n. A third nonnegative series related to the original is $\Sigma|a_n|$, called the **series of absolute values** of Σa_n.

We now assign labels to the partial sums of the series considered:

$$S_n \equiv a_1 + \cdots + a_n, \quad A_n \equiv |a_1| + \cdots + |a_n|,$$
$$P_n \equiv p_1 + \cdots + p_n, \quad Q_n \equiv q_1 + \cdots + q_n.$$

From (1) and (2) we deduce:

(4) $$P_n - Q_n = S_n, \quad P_n + Q_n = A_n,$$

and are ready to draw some conclusions about convergence.

From the first equation of (4) we can solve for either P_n or Q_n

$$(P_n = Q_n + S_n \quad \text{and} \quad Q_n = P_n - S_n),$$

and conclude (with the aid of a limit theorem) that the convergence of any two of the three series Σa_n, Σp_n, and Σq_n implies the convergence of the third. This means that if *both* the nonnegative and nonpositive parts of a series converge the series itself must also converge. It also means that if a series converges, then the nonnegative and nonpositive parts must either *both converge* or *both diverge*. (Why?) The alternating harmonic series of § 1115 is an example of a convergent series whose nonnegative and nonpositive parts both diverge.

The inequalities (3) and the second equation of (4) imply (by the comparison test and a limit theorem) that the series of absolute values, $\Sigma|a_n|$ converges if and only if *both* the nonnegative and nonpositive parts of the series converge. Furthermore, in case the series $\Sigma|a_n|$ converges, if we define $P \equiv \lim_{n \to +\infty} P_n$, $Q \equiv \lim_{n \to +\infty} Q_n$, $S \equiv \lim_{n \to +\infty} S_n$, and $A \equiv \lim_{n \to +\infty} A_n$, we have from (4), $P - Q = S$, $P + Q = A$.

We give a definition and a theorem embodying some of the results just obtained:

Definition. *A series Σa_n* **converges absolutely** *(or is* **absolutely convergent***) if and only if the series of absolute values, $\Sigma|a_n|$, converges. A series* **converges conditionally** *(or is* **conditionally convergent***) if and only if it converges and does not converge absolutely.*

Theorem I. *An absolutely convergent series is convergent. In case of absolute convergence, $|\Sigma a_n| \leq \Sigma|a_n|$.*

NOTE 1. For an absolutely convergent series, the nonnegative and nonpositive parts both converge. For a conditionally convergent series, the nonnegative and nonpositive parts both diverge.

NOTE 2. An alternative proof of Theorem I is provided by the Cauchy criterion (cf. Exs. 22–26, § 1117, for a discussion).

NOTE 3. The tests established for convergence of nonnegative series are of course immediately available for absolute convergence of arbitrary series. We state here the ratio test for arbitrary series, of nonzero terms, both because of its practicality and because the conclusion for $\rho > 1$ is not simply that the series fails to converge absolutely but that it *diverges* (cf. Ex. 12, § 1117).

Theorem II. Ratio Test. *Let Σa_n be a series of nonzero terms, and define the test ratio $r_n \equiv a_{n+1}/a_n$. Assume that the limit of the absolute value of this test ratio exists:*
$$\lim_{n \to +\infty} \left| \frac{a_{n+1}}{a_n} \right| = \rho,$$
where $0 \leq \rho \leq +\infty$. Then
  (i) *if $0 \leq \rho < 1$, Σa_n converges absolutely;*
  (ii) *if $1 < \rho \leq +\infty$, Σa_n diverges;*
  (iii) *if $\rho = 1$, Σa_n may converge absolutely, or converge conditionally, or diverge, and the test fails.*

NOTE 4. The ratio test never establishes convergence in the case of a conditionally convergent series.

Example 1. The alternating p-series,
$$1 - \frac{1}{2^p} + \frac{1}{3^p} - \frac{1}{4^p} + \cdots,$$
converges absolutely if $p > 1$, converges conditionally if $0 < p \leq 1$, and diverges if $p \leq 0$.

*Example 2. Show that the series $\dfrac{1}{2} - \dfrac{1 \cdot 3}{2 \cdot 4} + \dfrac{1 \cdot 3 \cdot 5}{2 \cdot 4 \cdot 6} - \dfrac{1 \cdot 3 \cdot 5 \cdot 7}{2 \cdot 4 \cdot 6 \cdot 8} + \cdots$ converges conditionally.

Solution. By Example 1, § 1112, the series fails to converge absolutely. In order to prove that the series converges we can use the alternating series test to reduce the problem to showing that the general term tends toward 0 (clearly $c_n \downarrow$). Letting $p = 3$ in Example 1, § 1112, we know that $c_n^3 \to 0$. Therefore $c_n \to 0$.

1117. EXERCISES

In Exercises 1–10, test for absolute convergence, conditional convergence, or divergence.

1. $\sum_{n=1}^{+\infty} \dfrac{(-1)^{n-1}}{2n+3}$.

2. $\sum_{n=1}^{+\infty} \dfrac{(-1)^n n}{n+2}$.

3. $\sum_{n=1}^{+\infty} \dfrac{(-1)^n \sqrt[3]{n^2+5}}{\sqrt[4]{n^3+n+1}}$.

4. $\sum_{n=1}^{+\infty} (-1)^{n-1} \dfrac{n^4}{(n+1)!}$.

5. $\sum_{n=1}^{+\infty} (-1)^n \dfrac{n \ln n}{e^n}$.

6. $\sum_{n=1}^{+\infty} (-1)^n \dfrac{\cos n\alpha}{n^2}$.

7. $e^{-x}\cos x + e^{-2x}\cos 2x + e^{-3x}\cos 3x + \cdots$.

8. $1 + r\cos\theta + r^2\cos 2\theta + r^3\cos 3\theta + \cdots$.

9. $\dfrac{1}{2(\ln 2)^p} - \dfrac{1}{3(\ln 3)^p} + \dfrac{1}{4(\ln 4)^p} - \cdots$.

★10. $\left(\dfrac{1}{2}\right)^p - \left(\dfrac{1\cdot 3}{2\cdot 4}\right)^p + \left(\dfrac{1\cdot 3\cdot 5}{2\cdot 4\cdot 6}\right)^p - \cdots$.

11. Prove that if the condition (*i*) of § 1115 is replaced by $c_{n+1} \leq c_n$, the conclusion is altered only by replacing (2), § 1115, by $|S_n - S| \leq c_{n+1}$.

12. Prove the ratio test (Theorem II) of § 1116.

13. Show by three counterexamples that each of the three conditions of the alternating series test, § 1115, is needed in the statement of that test (that is, the alternating of signs, the decreasing nature of c_n, and the limit of c_n being 0).

★14. Prove the **Schwarz** (or **Cauchy**) and the **Minkowski inequalities** for series: If $\sum_{n=1}^{+\infty} a_n^2$ and $\sum_{n=1}^{+\infty} b_n^2$ converge, then so do $\sum_{n=1}^{+\infty} a_n b_n$ and $\sum_{n=1}^{+\infty} (a_n + b_n)^2$, and

$$\left[\sum_{n=1}^{+\infty} a_n b_n\right]^2 \leq \sum_{n=1}^{+\infty} a_n^2 \sum_{n=1}^{+\infty} b_n^2,$$

$$\left[\sum_{n=1}^{+\infty} (a_n + b_n)^2\right]^{\frac{1}{2}} \leq \left[\sum_{n=1}^{+\infty} a_n^2\right]^{\frac{1}{2}} + \left[\sum_{n=1}^{+\infty} b_n^2\right]^{\frac{1}{2}}.$$

(Cf. Exs. 32–33, § 107, Exs. 20–21, § 402.)

★15. Prove that if Σa_n^2 is a convergent series, then $\Sigma |a_n|/n$ is also convergent. (Cf. Ex. 14.)

★16. If $\{s_n\}$ is a given sequence, define $\sigma_n \equiv \dfrac{1}{n}(s_1 + s_2 + \cdots + s_n)$. Prove that if $\{s_n\}$ converges to 0 then $\{\sigma_n\}$ also converges to 0. (Cf. Ex. 17.) *Hint:* Let m be a positive integer $< n$, and write

$$\sigma_n = \dfrac{1}{n}(s_1 + \cdots + s_m) + \dfrac{1}{n}(s_{m+1} + \cdots + s_n).$$

If $\epsilon > 0$, first choose m so large that whenever $k > m$, $|s_k| < \tfrac{1}{2}\epsilon$. Holding m fixed, choose $N > m$ and so large that $|s_1 + \cdots + s_m|/N < \tfrac{1}{2}\epsilon$. Then choose $n > N$, and consider separately the two groups of terms of σ_n, given above.

★17. With the notation of Exercise 16, prove that if $\{s_n\}$ converges, then $\{\sigma_n\}$ also converges and has the same limit. Show by the example $0, 1, 0, 1, \cdots$ that the convergence of $\{\sigma_n\}$ does not imply that of $\{s_n\}$. Can you find a divergent sequence $\{s_n\}$ such that $\sigma_n \to 0$? *Hint:* Assume $s_n \to l$, let $t_n \equiv s_n - l$, and use the result of Exercise 16.

★18. With the notation of Exercise 16, prove that if $\lim_{n \to +\infty} s_n = +\infty$, then $\lim_{n \to +\infty} \sigma_n = +\infty$. Show by the example $0, 1, 0, 2, 0, 3, \cdots$ that the reverse implication is not valid. *Hint:* If B is a given positive number, first choose m so large that whenever $k > m$, $s_k > 3B$. Then choose $N > 3m$ and so large that $|s_1 + \cdots + s_m|/N < B$. Then follow the hint of Exercise 16.

★19. Show by the example $1, -1, 2, -2, 3, -3, \cdots$ that with the notation of Exercise 16, $\lim_{n \to +\infty} s_n = \infty$ does not imply $\lim_{n \to +\infty} \sigma_n = \infty$. Can you find an example where $\lim_{n \to +\infty} s_n = \infty$ and $\lim_{n \to +\infty} \sigma_n = 0$?

★20. Let $\sum_{n=1}^{+\infty} a_n$ be a given series, with partial sums $S_n \equiv a_1 + \cdots + a_n$. Define the sequence of *arithmetic means*

$$\sigma_n \equiv \frac{S_1 + \cdots + S_n}{n}.$$

The series Σa_n is said to be **summable** *by Cesàro's method of arithmetic means of order* 1 (for short, summable $(C, 1)$) if and only if $\lim_{n \to +\infty} \sigma_n$ exists and is finite. Show that Exercises 16–19 prove that summability $(C, 1)$ is a generalization of convergence: that any convergent series is summable $(C, 1)$ with $\lim_{n \to +\infty} \sigma_n = \lim_{n \to +\infty} S_n$, that for nonnegative series summability $(C, 1)$ is identical with convergence, and that there are divergent series (whose terms are not of one sign) that are summable $(C, 1)$. *Hint:* Consider $1 - 1 + 1 - 1 + \cdots$.

★21. A series Σa_n is summable by *Hölder's method of arithmetic means of order* 2 (for short, summable $(H, 2)$) if and only if

$$\lim_{n \to +\infty} \frac{\sigma_1 + \cdots + \sigma_n}{n}$$

(in the notation of Ex. 20) exists and is finite. Show that summability $(C, 1)$ implies summability $(H, 2)$, but not conversely. Generalize to summability (H, r).

★22. Prove the **Cauchy criterion** for convergence of an infinite series: *An infinite series $\sum_{n=1}^{+\infty} a_n$ converges if and only if corresponding to $\epsilon > 0$ there exists a number N such that $n > m > N$ implies $|a_m + a_{m+1} + \cdots + a_n| < \epsilon$.* (Cf. § 218.)

★23. Prove that an infinite series $\sum_{n=1}^{+\infty} a_n$ converges if and only if corresponding to $\epsilon > 0$ there exists a number N such that $n > N$ and $p > 0$ imply $|a_n + a_{n+1} + \cdots + a_{n+p}| < \epsilon$. (Cf. Ex. 22.)

★24. Prove that an infinite series $\sum_{n=1}^{+\infty} a_n$ converges if and only if corresponding to $\epsilon > 0$ there exists a positive integer N such that $n > N$ implies $|a_N + a_{N+1} + \cdots + a_n| < \epsilon$. (Cf. Ex. 2, § 219.)

★25. Show by an example that the condition of Exercise 23 is not equivalent to the following: $\lim_{n \to +\infty} (a_n + \cdots + a_{n+p}) = 0$ for every $p > 0$. (Cf. Ex. 4, § 219, Ex. 43, § 1304.)

★26. Use the Cauchy criterion of Exercise 22 to prove that an absolutely convergent series is convergent.

★27. Prove the following **Abel test**: *If the partial sums of a series $\sum_{n=1}^{+\infty} a_n$ are bounded and if $\{b_n\}$ is a monotonically decreasing sequence of nonnegative numbers whose limit is 0, then $\Sigma a_n b_n$ converges.* Use this fact to establish the convergence in the alternating series test as phrased in Exercise 11. *Hint:* Let $S_n \equiv a_1 + \cdots + a_n$, and assume $|S_n| < K$ for all n. Then

$$\left| \sum_{i=m}^{n} a_i b_i \right| = \left| \sum_{i=m}^{n} (S_i - S_{i-1})b_i \right| = \left| \sum_{i=m}^{n-1} S_i(b_i - b_{i+1}) + S_n b_n - S_{m-1} b_m \right|$$

$$\leq K \left[\sum_{i=m}^{n-1} (b_i - b_{i+1}) + b_n + b_m \right] = 2K b_m.$$

★**28.** Prove the following **Abel test**: *If Σa_n converges and if $\{b_n\}$ is a bounded monotonic sequence, then $\Sigma a_n b_n$ converges.* *Hint:* Assume for definiteness that $b_n \downarrow$, let $b_n \to b$, write $a_n b_n = a_n(b_n - b) + a_n b$, and use Exercise 27.

1118. GROUPINGS AND REARRANGEMENTS

A series Σb_n is said to arise from a given series Σa_n by **grouping of terms** (or by the **introduction of parentheses**) if every b_n is the sum of a finite number of consecutive terms of Σa_n, and every pair of terms a_m and a_n, where $m < n$, appear as terms in a unique pair of terms b_p and b_q, respectively, where $p \leq q$. For example, the grouping

$$(a_1 + a_2) + (a_3) + (a_4 + a_5) + (a_6) + \cdots$$

gives rise to the series Σb_n, where $b_1 = a_1 + a_2$, $b_2 = a_3$, $b_3 = a_4 + a_5$, $b_4 = a_6, \cdots$.

Theorem I. *Any series arising from a convergent series by grouping of terms is convergent, and has the same sum as the original series.*

Proof. The partial sums of the new series form a subsequence of the partial sums of the original.

NOTE. The example
$$(2 - 1\tfrac{1}{2}) + (1\tfrac{1}{3} - 1\tfrac{1}{4}) + (1\tfrac{1}{5} - 1\tfrac{1}{6}) + \cdots$$
shows that grouping of terms may convert a divergent series into a convergent series. Equivalently, removal of parentheses may destroy convergence.

A series Σb_n is said to arise from a given series Σa_n by **rearrangement (of terms)** if there exists a one-to-one correspondence between the terms of Σa_n and those of Σb_n such that whenever a_m and b_n correspond, $a_m = b_n$. For example, the series

$$\tfrac{1}{4} + 1 + \tfrac{1}{16} + \tfrac{1}{9} + \tfrac{1}{36} + \tfrac{1}{25} \cdots$$

is a rearrangement of the p-series, with $p = 2$.

Theorem II. Dirichlet's Theorem. *Any series arising from an absolutely convergent series by rearrangement of terms is absolutely convergent, and has the same sum as the original series.*

Proof. We prove first that the theorem is true for nonnegative series. Let Σa_n be a given nonnegative series, convergent with sum A and let Σb_n be any rearrangement. If B_n is any partial sum of Σb_n, the terms of B_n consist of a finite number of terms of Σa_n, and therefore form a part of some partial sum A_m of Σa_n. Since the terms are assumed to be nonnegative, $B_n \leq A_m$, and hence, $B_n \leq A$. Therefore the partial sums of the nonnegative series Σb_n are bounded, and Σb_n converges to a sum $B \leq A$. Since Σa_n is a rearrangement of Σb_n, the symmetric relation $A \leq B$ also holds, and $A = B$.

If Σa_n is absolutely convergent, and if Σb_n is any rearrangement, the nonnegative and nonpositive parts of Σb_n (cf. § 1116) are rearrangements of the nonnegative and nonpositive parts of Σa_n, respectively. Since these latter both converge, say with sums P and Q, respectively, their rearrangements will also both converge, with sums P and Q, respectively, by the preceding paragraph. Finally, $\Sigma a_n = P - Q = \Sigma b_n$, and the proof is complete.

*Theorem III. *The terms of any conditionally convergent series can be rearranged to give either a divergent series or a conditionally convergent series whose sum is an arbitrary preassigned number.*

*Proof. We prove one case, and leave the rest as an exercise. (Ex. 2, § 1121.) Let Σa_n be a conditionally convergent series, with divergent nonnegative and nonpositive parts Σp_n and Σq_n, respectively, and let c be an arbitrary real number. Let the rearrangement be determined as follows: first put down terms $p_1 + p_2 + \cdots + p_{m_1}$ until the partial sum first exceeds c. Then attach terms $-q_1 - q_2 - q_3 - \cdots - q_{n_1}$ until the total partial sum first falls short of c. Then attach terms $p_{m_1+1} + \cdots + p_{m_2}$ until the total partial sum first exceeds c. Then terms $-q_{n_1+1} - \cdots - q_{n_2}$, etc. Each of these steps is possible because of the divergence of Σp_n and Σq_n. The resulting rearrangement of Σa_n converges to c since $p_n \to 0$ and $q_n \to 0$.

Example. Rearrange the terms of the series obtained by doubling all terms of the alternating harmonic series so that the resulting series is the alternating harmonic series, thus convincing the unwary that $2 = 1$.

Solution. Write the terms

$2[1 - \frac{1}{2} + \frac{1}{3} - \frac{1}{4} + \frac{1}{5} - \frac{1}{6} + \frac{1}{7} - \frac{1}{8} + \frac{1}{9} - \cdots]$
$= 2 - 1 + \frac{2}{3} - \frac{1}{2} + \frac{2}{5} - \frac{1}{3} + \frac{2}{7} - \frac{1}{4} + \frac{2}{9} - \cdots$
$= (2 - 1) - \frac{1}{2} + (\frac{2}{3} - \frac{1}{3}) - \frac{1}{4} + (\frac{2}{5} - \frac{1}{5}) - \frac{1}{6} + \cdots$
$= 1 - \frac{1}{2} + \frac{1}{3} - \frac{1}{4} + \frac{1}{5} - \frac{1}{6} + \frac{1}{7} - \frac{1}{8} + \frac{1}{9} - \cdots.$

1119. ADDITION, SUBTRACTION, AND MULTIPLICATION OF SERIES

Definition I. *If Σa_n and Σb_n are two series, their **sum** Σc_n and **difference** Σd_n are series defined by the equations*

(1) $\qquad c_n \equiv a_n + b_n, \quad d_n \equiv a_n - b_n.$

Theorem I. *The sum and difference of two convergent series, $\Sigma a_n = A$ and $\Sigma b_n = B$, converge to $A + B$ and $A - B$, respectively. The sum and difference of two absolutely convergent series are absolutely convergent.*

The proof is left as an exercise (Ex. 3, § 1121).

The product of two series is a more difficult matter. The definition of a product series is motivated by the form of the product of polynomials or, more generally, power series (treated in Chapter 12):

$(a_0 + a_1 x + a_2 x^2 + \cdots)(b_0 + b_1 x + b_2 x^2 + \cdots)$
$= a_0 b_0 + (a_0 b_1 + a_1 b_0) x + (a_0 b_2 + a_1 b_1 + a_2 b_0) x^2 + \cdots.$

For convenience we revise slightly our notation for an infinite series, letting the terms have subscripts 0, 1, 2, $\cdots$, and write

$$\Sigma a_n = \sum_{n=0}^{+\infty} a_n \text{ and } \Sigma b_n = \sum_{n=0}^{+\infty} b_n.$$

Definition II. *If* $\sum_{n=0}^{+\infty} a_n$ *and* $\sum_{n=0}^{+\infty} b_n$ *are two series, their* **product** $\sum_{n=0}^{+\infty} c_n$ *is defined:*

(2) $\quad c_0 = a_0 b_0, \ c_1 = a_0 b_1 + a_1 b_0, \cdots,$

$$c_n = \sum_{k=0}^{n} a_k b_{n-k} = a_0 b_n + a_1 b_{n-1} + \cdots + a_n b_0, \cdots,$$

$\cdots\cdots\cdots\cdots\cdots\cdots\cdots\cdots\cdots$

The basic questions are these: If Σa_n and Σb_n converge, with sums A and B, and if Σc_n is their product series, does Σc_n converge? If Σc_n converges to C is $C = AB$? If Σc_n does not necessarily converge, what conditions on Σa_n and Σb_n guarantee convergence of Σc_n?

The answers, in brief, are: The convergence of Σa_n and Σb_n does not guarantee convergence of Σc_n (Ex. 5, § 1121). If Σc_n does converge, then $C = AB$. (This result is due to Abel. Cf. Ex. 20, § 1311.) If both Σa_n and Σb_n converge, and if *one* of them converges absolutely, then Σc_n converges (to AB). (This result is due to Mertens. For a proof, cf. *RV*, Ex. 20, § 721.) If *both* Σa_n and Σb_n converge absolutely, then Σc_n converges absolutely (to AB). (This is our next theorem.)

Theorem II. *The product series of two absolutely convergent series is absolutely convergent. Its sum is the product of their sums.*

Proof. Let $\sum_{n=0}^{+\infty} a_n$ and $\sum_{n=0}^{+\infty} b_n$ be the given absolutely convergent series, let $\sum_{n=0}^{+\infty} c_n$ be their product series, and define the series $\sum_{n=0}^{+\infty} d_n$ to be

$$a_0 b_0 + a_0 b_1 + a_1 b_0 + a_0 b_2 + a_1 b_1 + a_2 b_0$$
$$+ a_0 b_3 + a_1 b_2 + a_2 b_1 + a_3 b_0 + a_0 b_4 + \cdots,$$

the terms following along the diagonal lines suggested in Figure 1103. Furthermore, let

$$A_n \equiv a_0 + a_1 + \cdots + a_n, \quad B_n \equiv b_0 + b_1 + \cdots + b_n,$$
$$C_n \equiv c_0 + c_1 + \cdots + c_n, \quad D_n \equiv d_0 + d_1 + \cdots + d_n.$$

Observe that every c_n is obtained by the grouping of $(n + 1)$ terms of the series Σd_n, that every C_n is a partial sum of the series Σd_n with terms occupying a triangle in the upper left-hand corner of Figure 1103, and that every $A_n B_n$ is a sum (not a strict "partial sum") of certain terms of the series Σd_n occupying a square in the upper left-hand corner of Figure 1103.

We prove the theorem first for the case of nonnegative series Σa_n and Σb_n. In this case, since every finite set of terms from Σd_n is located in some square in the upper left-hand corner of Figure 1103, *every D_m is less than or equal to some $A_n B_n$*. That is, if $A \equiv \lim_{n \to +\infty} A_n$, and $B \equiv \lim_{n \to +\infty} B_n$, the inequality

$D_m \leq AB$ holds for every m. Therefore the series Σd_n converges, the limit $D \equiv \lim_{n \to +\infty} D_n$ is finite, and $D \leq AB$. On the other hand, since $A_n B_n$ is a sum of terms of the convergent series Σd_n, the inequality $A_n B_n \leq D$ holds for every n, and hence $AB \leq D$. Thus $D = AB$. Finally, since $\{C_n\}$ is a subsequence of $\{D_n\}$ (resulting from introducing parentheses in the series Σd_n), $C \equiv \lim_{n \to +\infty} C_n = D = AB$.

If the given series, Σa_n and Σb_n, have terms of arbitrary sign, the conclusion sought is a consequence of Dirichlet's Theorem (§ 1118): Since, by the preceding paragraph, the series Σd_n converges absolutely, any rearrangement

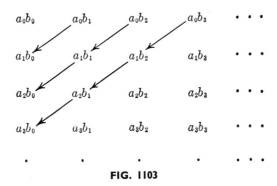

FIG. 1103

converges absolutely to the same sum. The sequence $\{C_n\}$ is a subsequence of the sequence of partial sums of Σd_n, and the sequence $\{A_n B_n\}$ is a subsequence of the sequence of partial sums of an appropriate rearrangement of Σd_n. Therefore $\lim_{n \to +\infty} C_n = \lim_{n \to +\infty} A_n B_n = \lim_{n \to +\infty} D_n$.

*1120. SOME AIDS TO COMPUTATION

The only techniques available from the preceding sections of this chapter for evaluating series are (i) the sum $a/(1-r)$ of a geometric series and (ii) the estimate $|S_n - S| < c_{n+1}$ for an alternating series. We give in this section some further means of estimating the sum of a convergent series, with illustrative examples. These will be available for computation work presented in the next chapter.

If a nonnegative series is dominated by a convergent geometric series, the formula $a/(1-r)$ provides an estimate for the sum. A useful formulation of this method makes use of the test ratio (give the proof in Ex. 6, § 1121):

Theorem I. *If $a_n > 0$, $S_n \equiv a_1 + \cdots + a_n$, $S \equiv \sum_{n=1}^{+\infty} a_n$, $r_n \equiv \dfrac{a_{n+1}}{a_n}$, $r_n \downarrow$, and $r_n \to \rho < 1$, then for any n for which $r_n < 1$,*

(1) $$S_n + \frac{a_{n+1}}{1 - \rho} \leq S \leq S_n + \frac{a_{n+1}}{1 - r_{n+1}}.$$

Example 1. Compute the sum of the series

$$1 + \frac{1}{3} + \frac{1}{2! \cdot 5} + \frac{1}{3! \cdot 7} + \cdots + \frac{1}{n!(2n+1)} + \cdots,$$

to three decimal places. $\left(\text{The sum is} \int_0^1 e^{x^2}\, dx. \text{ Cf. Example 4, § 1210.}\right)$

Solution. The sum S_6 is between 1.4625 and 1.4626. Since $r_n \to \rho = 0$, we estimate:

$$\frac{a_7}{1-0} = 0.000107, \quad \text{and} \quad \frac{a_7}{1-r_7} = \frac{0.000107}{1-0.121} < 0.00013.$$

Therefore, from (1), S must lie between 1.4626 and 1.4628. Its value to three decimal places is therefore 1.463.

If a nonnegative series is not dominated by a convergent geometric series $\left(\text{for example, if } \dfrac{a_{n+1}}{a_n} \to 1\right)$, Theorem I cannot be used. However, in this case the process of integration (cf. the integral test, § 1106) can sometimes be used, as expressed in the following theorem (give the proof in Ex. 7, § 1121):

Theorem II. *If $f(x)$ is a positive monotonically decreasing function, for $x \geq a$, if $\sum_{n=1}^{+\infty} a_n$ is a convergent positive series, with $f(n) = a_n$ for $n > a$, if $S \equiv \Sigma a_n$, and if $S_n \equiv a_1 + \cdots + a_n$, then for $n > a$,*

(6) $$S_n + \int_{n+1}^{+\infty} f(x)\, dx \leq S \leq S_n + \int_n^{+\infty} f(x)\, dx.$$

A much sharper estimate is provided by the following theorem (for a proof see *RV*, Ex. 21, § 721):

Theorem III. *Under the hypotheses of Theorem II and the additional assumption that $f''(x)$ is a positive monotonically decreasing function for $x \geq a$, then for $n > a$,*

(7) $$S_n + \frac{a_{n+1}}{2} + \int_{n+1}^{+\infty} f(x)\, dx - \frac{f'(n+2)}{12}$$
$$< S < S_n + \frac{a_{n+1}}{2} + \int_{n+1}^{+\infty} f(x)\, dx - \frac{f'(n)}{12}.$$

Example 2. Estimate the sum of the p-series with $p = 2$, using 10 terms.

Solution. With the aid of a table of reciprocals we find that $S_{10} = 1.549768$. With $f(x) \equiv x^{-2}$, $\int_{11}^{+\infty} f(x)\, dx = \frac{1}{11} = 0.090909$ and $\int_{10}^{+\infty} f(x)\, dx = \frac{1}{10} = 0.100000$. Thus the estimate of Theorem II places the sum S between 1.640 and 1.650, with an accuracy of one digit in the second decimal place.

Further computations give $\frac{1}{2} a_{11} = 0.004132$, $-\frac{1}{12} f'(12) = \frac{1}{6 \cdot 12^3} = 0.000096$, and $-\frac{1}{12} f'(10) = \frac{1}{6 \cdot 10^3} = 0.000167$. Thus the estimate of Theorem III places

§ 1120]   SOME AIDS TO COMPUTATION   409

the sum S between 1.6449 and 1.6450, with an accuracy of one digit in the fourth decimal place.

NOTE. The sum in Example 2 can be specifically evaluated by means of the techniques of Fourier series (cf. Example 2, § 1608; Example 1, § 1609; also Example 2, § 1609; Exs. 4–6, 20, § 1612; Ex. 5, § 1624):

$$\frac{\pi^2}{6} = 1 + \frac{1}{2^2} + \frac{1}{3^2} + \frac{1}{4^2} + \cdots.$$

If an alternating series converges slowly, the estimate given in the alternating series test (§ 1115) is very crude unless an excessively large number of terms is used. For example, to compute the value of the alternating harmonic series to four decimal places would require at least ten thousand terms! This particular series happens to converge to ln 2 (cf. § 1207), and this fortuitous circumstance permits a simpler and more speedy evaluation (cf. Example 4, § 1213). However, this series will be used in the following example to illustrate a technique frequently useful in evaluating slowly converging alternating series:

Example 3. Evaluate the alternating harmonic series,

$$S = 1 - \tfrac{1}{2} + \tfrac{1}{3} - \tfrac{1}{4} + \cdots$$

to four decimal places.

Solution. We start by evaluating the sum of the first 10 terms: $S_{10} = 0.645635$. We wish to estimate the remainder:

$$x \equiv \tfrac{1}{11} - \tfrac{1}{12} + \tfrac{1}{13} - \tfrac{1}{14} + \cdots.$$

If we double, remove parentheses (the student should justify this step), and introduce parentheses, we find

$$2x = (\tfrac{1}{11} + \tfrac{1}{11}) - (\tfrac{1}{12} + \tfrac{1}{12}) + (\tfrac{1}{13} + \tfrac{1}{13}) - \cdots$$

$$= \tfrac{1}{11} + \tfrac{1}{11} - \tfrac{1}{12} - \tfrac{1}{12} + \tfrac{1}{13} + \tfrac{1}{13} - \cdots$$

$$= \tfrac{1}{11} + (\tfrac{1}{11} - \tfrac{1}{12}) - (\tfrac{1}{12} - \tfrac{1}{13}) + (\tfrac{1}{13} - \tfrac{1}{14}) - \cdots$$

$$= \frac{1}{11} + \frac{1}{11 \cdot 12} - \frac{1}{12 \cdot 13} + \frac{1}{13 \cdot 14} - \cdots.$$

Again doubling, and removing and introducing parentheses, we have

$$4x = \frac{2}{11} + \frac{1}{11 \cdot 12} + \frac{2}{11 \cdot 12 \cdot 13} - \frac{2}{12 \cdot 13 \cdot 14} + \cdots,$$

or

$$2x = \frac{25}{11 \cdot 24} + \frac{1}{11 \cdot 12 \cdot 13} - \frac{1}{12 \cdot 13 \cdot 14} + \cdots.$$

Once more:

$$4x = \frac{25}{11 \cdot 12} + \frac{1}{11 \cdot 12 \cdot 13} + \frac{3}{11 \cdot 12 \cdot 13 \cdot 14} - \frac{3}{12 \cdot 13 \cdot 14 \cdot 15} + \cdots.$$

The sum of the first two terms of this series is 0.189977, and the remainder is less than the term $3/11 \cdot 12 \cdot 13 \cdot 14 < 0.000126$. Therefore x is between 0.04749, and 0.04753, and S is between 0.69312 and 0.69317. An estimate to four places is 0.6931+. (The actual value to five places is 0.69315.)

1121. EXERCISES

1. Prove that any series arising from a divergent nonnegative series by grouping of terms is divergent. Equivalently, if the introduction of parentheses into a nonnegative series produces a convergent series, the original series is convergent.

2. Can the terms of a conditionally convergent series be rearranged to give a series whose partial sums (*i*) tend toward $+\infty$? (*ii*) tend toward $-\infty$? (*iii*) tend toward ∞ but neither $+\infty$ nor $-\infty$? (*iv*) are bounded and have no limit?

3. Prove Theorem I, § 1119.

*4. Prove that if Σa_n and Σb_n are nonnegative series with sums A and B, respectively, and if Σc_n is their product series, with sum C, then $C = AB$ under all circumstances of convergence or divergence, with the usual conventions about infinity $((+\infty) \cdot (+\infty) = +\infty$, (positive number) $\cdot (+\infty) = +\infty)$ and the additional convention $0 \cdot (+\infty) = 0$.

*5. Prove that if $\sum_{n=0}^{+\infty} a_n$ and $\sum_{n=0}^{+\infty} b_n$ are both the series

$$1 - \frac{1}{\sqrt{2}} + \frac{1}{\sqrt{3}} - \frac{1}{\sqrt{4}} + \cdots$$

and if $\sum_{n=0}^{+\infty} c_n$ is their product series, then Σa_n and Σb_n converge, while Σc_n diverges.

Hint: Show that $|c_n| \geq 1$.

*6. Prove Theorem I, § 1120.

*7. Prove Theorem II, § 1120.

*8. Prove the commutative law for product series: The product series of Σa_n and Σb_n is the same as the product series of Σb_n and Σa_n.

*9. Prove the associative law for product series: Let Σd_n be the product series of Σa_n and Σb_n, and let Σe_n be the product series of Σb_n and Σc_n. Then the product series of Σa_n and Σe_n is the same as the product series of Σd_n and Σc_n.

*10. Prove the distributive law for multiplying and adding series: The product series of Σa_n and $\Sigma(b_n + c_n)$ is the sum of the product series of Σa_n and Σb_n and the product series of Σa_n and Σc_n.

*11. Using the evaluation of the series Σn^{-2} given in the Note, § 1120, show that

$$\frac{\pi^2}{8} = 1 + \frac{1}{3^2} + \frac{1}{5^2} + \frac{1}{7^2} + \cdots.$$

*12. Using the evaluation of the series Σn^{-2} given in the Note, § 1120, show that

$$\frac{\pi^2}{12} = 1 - \frac{1}{2^2} + \frac{1}{3^2} - \frac{1}{4^2} + \cdots.$$

In Exercises 13–18, compute to four significant digits.

*13. $(e =) \quad 1 + \frac{1}{1!} + \frac{1}{2!} + \frac{1}{3!} + \cdots.$

*14. $(\ln 2 =) \quad \frac{1}{2} + \frac{1}{2 \cdot 2^2} + \frac{1}{3 \cdot 2^3} + \frac{1}{4 \cdot 2^4} + \cdots.$

*15. $(\zeta(3) =) \quad 1 + \frac{1}{2^3} + \frac{1}{3^3} + \frac{1}{4^3} + \cdots.$

★16. $\dfrac{5}{2^2 \cdot 3^2} + \dfrac{9}{4^2 \cdot 5^2} + \dfrac{13}{6^2 \cdot 7^2} + \dfrac{17}{8^2 \cdot 9^2} + \cdots.$

★17. $\left(\dfrac{\pi}{4} = \right)\quad 1 - \dfrac{1}{3} + \dfrac{1}{5} - \dfrac{1}{7} + \cdots.$

★18. $\left(\ln \dfrac{3}{2} = \right)\quad \dfrac{1}{2} - \dfrac{1}{2 \cdot 2^2} + \dfrac{1}{3 \cdot 2^3} - \dfrac{1}{4 \cdot 2^4} + \cdots.$

12

Power Series

1201. INTERVAL OF CONVERGENCE

A series of the form

(1) $$\sum a_n x^n = \sum_{n=0}^{+\infty} a_n x^n = a_0 + a_1 x + a_2 x^2 + \cdots$$

is called a **power series in x**. A series of the form

(2) $$\sum a_n (x-a)^n = \sum_{n=0}^{+\infty} a_n (x-a)^n = a_0 + a_1(x-a) + \cdots$$

is called a **power series in $(x-a)$**. More generally, a series of the form

(3) $$\sum a_n [u(x)]^n = \sum_{n=0}^{+\infty} a_n [u(x)]^n = a_0 + a_1 u(x) + \cdots,$$

where $u(x)$ is a function of x, is called a **power series in $u(x)$**. It is principally series (1) or (2) that will be of interest in this chapter, and the single expression *power series* will be used to mean either series (1) or series (2).

A power series is an example of a series of *functions*. For any fixed value of x the series becomes a series of *constants*; but convergence or divergence of this series of constants depends, ordinarily, on the value of x. One is frequently interested in the question, "For what values of x does a given power series converge?" The answer is fairly simple: The values of x for which a power series converges always form an interval, which may degenerate to a single point, or encompass all real numbers, or be a finite interval, open, closed, or half-open. To prove this result we formulate it for simplicity for the power series $\sum a_n x^n$, first establishing a lemma:

Lemma. *If a power series $\sum a_n x^n$ converges for $x = x_1$ and if $|x_2| < |x_1|$, then the series converges absolutely for $x = x_2$.*

Proof. Assume $\sum a_n x_1^n$ converges. Then $\lim_{n \to +\infty} a_n x_1^n = 0$. Therefore the sequence $\{a_n x_1^n\}$, being convergent, is bounded. Let $|a_n x_1^n| < K$ for all n. If $|x_2| < |x_1|$, we may write

(4) $$|a_n x_2^n| = |a_n x_1^n| \cdot \left|\frac{x_2}{x_1}\right|^n < K r^n, \quad \text{where } 0 \leq r < 1.$$

The series $\Sigma K r^n$ is a convergent geometric series. Therefore, by comparison, $\Sigma |a_n x_2{}^n|$ converges, and $\Sigma a_n x_2{}^n$ converges absolutely.

What this lemma says, in part, is that *any* point of convergence of the power series $\Sigma a_n x^n$ is at least as close to 0 as *any* point of divergence.

Theorem I. *Let S be the set of points x for which a power series $\Sigma a_n x^n$ converges. Then either (i) S consists only of the point $x = 0$, or (ii) S consists of all real numbers, or (iii) S is an interval of one of the following forms: $(-R, R)$, $[-R, R]$, $(-R, R]$, or $[-R, R)$, where R is a positive real number.*

Proof. If neither (i) nor (ii) holds, then (by the preceding lemma) the series $\Sigma a_n x^n$ must converge for some positive number x_C and diverge for some larger positive number x_D. That is, S contains some positive numbers and is bounded above. Let R be defined to be the least upper bound of S. Then (iii) is a consequence of the Lemma. (Why?)

NOTE 1. The statements just established for the power series (1) apply to the power series (2), the only change being that the interval of convergence has the point $x = a$ instead of the point $x = 0$ as midpoint.

The set S of points x for which a power series $\Sigma a_n (x - a)^n$ converges is called the **interval of convergence**, and the number R of Theorem I (and Note 1) is called the **radius of convergence**† of $\Sigma a_n (x - a)^n$. In case (i) of Theorem I we define $R \equiv 0$, and in case (ii) we write $R \equiv +\infty$, so that in general $0 \leq R \leq +\infty$.

A further consequence of the Lemma is the Theorem:

Theorem II. *At any point interior to the interval of convergence of a power series the convergence is absolute.*

NOTE 2. All of the eventualities stated in Theorem I exist (Ex. 13, § 1202).

NOTE 3. At an end-point of the interval of convergence a power series may diverge, converge conditionally, or converge absolutely. At the two end-points all combinations are possible, except that if a power series converges absolutely at one end-point it must also converge absolutely at the other end-point. (Cf. Ex. 14, § 1202.)

The usual procedure in determining the interval of convergence of a power series is to start with the ratio test, although this may fail (cf. Example 3). The success of the ratio test depends on the existence of the limit of the ratio of successive coefficients. We specify this relationship:

Theorem III. *The radius of convergence of the power series $\Sigma a_n (x - a)^n$ is*

(5) $$R = \lim_{n \to +\infty} \left| \frac{a_n}{a_{n+1}} \right|,$$

where $0 \leq R \leq +\infty$, provided this limit exists.

† R is called the *radius of convergence* because of the analogous situation with complex numbers, where the *interval* of convergence is replaced by a *circle* of convergence whose radius is R. (Cf. Theorem VI, § 1511.)

Proof. We give the details for $0 < R < +\infty$. (Cf. Ex. 15, § 1202.) The test ratio for the series $\Sigma a_n(x-a)^n$ is $\dfrac{a_{n+1}(x-a)}{a_n}$, whose absolute value has the limit $\lim\limits_{n \to +\infty} \left|\dfrac{a_{n+1}}{a_n}\right| \cdot |x-a| = \dfrac{|x-a|}{R}$. Therefore, by the ratio test, $\Sigma a_n(x-a)^n$ converges absolutely if $|x-a| < R$ and diverges if $|x-a| > R$.

After the radius of convergence has been found, the end-points of the interval of convergence should be tested. For such points the ratio test cannot give any information, for if $\dfrac{a_{n+1}}{a_n} R$ had a limit for $R > 0$, the limit (5) would exist, and the absolute value of the test ratio must have the limit 1. To test the end-points one is forced to use some type of comparison test, or a refined test of the type discussed in § 1112. (The root test also fails at the end-points—cf. Ex. 16, § 1202.)

*NOTE 4. A universally valid formula for the radius of convergence of a power series $\Sigma a_n(x-a)^n$ makes use of *limit superior* (Ex. 11, § 220):

$$R = \dfrac{1}{\overline{\lim\limits_{n \to +\infty}} \sqrt[n]{|a_n|}},$$

with the conventions $1/0 = +\infty$, $1/+\infty = 0$. (Give the proof in Ex. 17, § 1202, and cf. Ex. 16, § 1202.)

Example 1. Determine the interval of convergence for the series

$$1 + x^2 + \dfrac{x^4}{2!} + \dfrac{x^6}{3!} + \cdots.$$

(Cf. § 1207.)

Solution. This should be treated as a power series in powers of x^2. The radius of convergence is
$$R = \lim_{n \to +\infty} \dfrac{(n+1)!}{n!} = \lim_{n \to +\infty} (n+1) = +\infty.$$
Therefore the interval of convergence is $(-\infty, +\infty)$.

Example 2. Determine the interval of convergence for the series

$$(x-1) - \dfrac{(x-1)^2}{2} + \dfrac{(x-1)^3}{3} - \dfrac{(x-1)^4}{4} + \cdots.$$

(Cf. § 1207.)

Solution. The radius of convergence is $R = \lim\limits_{n \to +\infty} \dfrac{n+1}{n} = 1$, and the midpoint of the interval of convergence is $x = 1$. We test for convergence at the end-points of the interval, $x = 2$ and $x = 0$. The value $x = 2$ gives the convergent alternating harmonic series and $x = 0$ gives minus the divergent harmonic series. The interval of convergence is $(0, 2]$.

Example 3. Determine the interval of convergence for the series

$$\dfrac{1}{2} + \dfrac{x}{3} + \dfrac{x^2}{2^2} + \dfrac{x^3}{3^2} + \dfrac{x^4}{2^3} + \dfrac{x^5}{3^3} + \dfrac{x^6}{2^4} + \dfrac{x^7}{3^4} + \cdots.$$

(Cf. Example 2, § 1108.)

Solution. The limit (5) does not exist. However, the intervals of convergence of the two series
$$\frac{1}{2} + \frac{x^2}{2^2} + \frac{x^4}{2^3} + \cdots \quad \text{and} \quad \frac{x}{3} + \frac{x^3}{3^2} + \frac{x^5}{3^3} + \cdots$$
are $(-\sqrt{2}, \sqrt{2})$ and $(-\sqrt{3}, \sqrt{3})$, respectively. Therefore the given series converges absolutely for $|x| < \sqrt{2}$ and diverges for $|x| \geq \sqrt{2}$. The interval of convergence is $(-\sqrt{2}, \sqrt{2})$.

Example 4. Determine the values of x for which the series
$$xe^{-x} + 2x^2 e^{-2x} + 3x^3 e^{-3x} + 4x^4 e^{-4x} + \cdots$$
converges.

Solution. Either the ratio test or the root test shows that the series converges absolutely for $|xe^{-x}| < 1$ (otherwise it diverges). The inequality $|xe^{-x}| < 1$ is equivalent to $|x| < e^x$, and is satisfied for all nonnegative x. To determine the negative values of x which satisfy this inequality, we let $\alpha = -x$, and solve the equation $\alpha = e^{-\alpha}$ ($\alpha = 0.567$, approximately). Then the given series converges if and only if $x > -\alpha$, and the convergence is absolute.

1202. EXERCISES

In Exercises 1–10, determine the interval of convergence, and specify the nature of any convergence at each end-point of the interval of convergence.

1. $1 - \dfrac{2x}{1!} + \dfrac{(2x)^2}{2!} - \dfrac{(2x)^3}{3!} + \cdots$.

2. $x - \dfrac{x^3}{3} + \dfrac{x^5}{5} - \dfrac{x^7}{7} + \cdots$.

3. $1 + \dfrac{x^2}{2!} + \dfrac{x^4}{4!} + \dfrac{x^6}{6!} + \cdots$.

4. $1 + x + 2!\, x^2 + 3!\, x^3 + 4!\, x^4 + \cdots$.

5. $(x+1) - \dfrac{(x+1)^2}{4} + \dfrac{(x+1)^3}{9} - \dfrac{(x+1)^4}{16} + \cdots$.

6. $(x-2) + \dfrac{(x-2)^3}{3!} + \dfrac{(x-2)^5}{5!} + \dfrac{(x-2)^7}{7!} + \cdots$.

7. $\dfrac{(\ln 2)(x-5)}{\sqrt{2}} + \dfrac{(\ln 3)(x-5)^2}{\sqrt{3}} + \dfrac{(\ln 4)(x-5)^3}{\sqrt{4}} + \cdots$.

8. $(x-1) + \dfrac{(x-1)^3}{3} + \dfrac{(x-1)^5}{5} + \dfrac{(x-1)^7}{7} + \cdots$.

★9. $1 - \dfrac{1}{2}x + \dfrac{1 \cdot 3}{2 \cdot 4} x^2 - \dfrac{1 \cdot 3 \cdot 5}{2 \cdot 4 \cdot 6} x^3 + \dfrac{1 \cdot 3 \cdot 5 \cdot 7}{2 \cdot 4 \cdot 6 \cdot 8} x^4 - \cdots$.

★10. $x + \dfrac{x^3}{6} + \dfrac{1 \cdot 3}{2 \cdot 4} \cdot \dfrac{x^5}{5} + \dfrac{1 \cdot 3 \cdot 5 \, x^7}{2 \cdot 4 \cdot 6 \, 7} + \cdots$.

11. Determine the values of x for which the series
$$\frac{1}{x-3} + \frac{1}{2(x-3)^2} + \frac{1}{3(x-3)^3} + \frac{1}{4(x-3)^4} + \cdots$$
converges, and specify the type of convergence.

12. Determine the values of x for which the series
$$\sin x - \frac{\sin^3 x}{3} + \frac{\sin^5 x}{5} - \frac{\sin^7 x}{7} + \cdots$$
converges, and specify the type of convergence.

13. Prove Note 2, § 1201.

14. Prove Note 3, § 1201.

15. Prove Theorem III, § 1201, for the cases $R = 0$ and $R = +\infty$.

16. Prove Note 4, § 1201, for the case where $\lim\limits_{n \to +\infty} \sqrt[n]{|a_n|}$ exists (replacing $\overline{\lim}$ by, lim).

★17. Prove Note 4, § 1201. (Cf. Ex. 29, § 1111.)

★18. Apply Note 4, § 1201, to Example 3, § 1201. (Cf. Ex. 30, § 1111.)

★19. Show by examples that Theorem III, § 1201, cannot be generalized in the manner of Note 4, § 1201, by the use of limits superior or inferior. (Cf. Exs. 28–30, § 1111.)

1203. TAYLOR SERIES

We propose to discuss in this section some formal procedures, nearly all of which need justification and will be discussed in future sections. The purpose of this discussion is to motivate an important formula, and raise some questions.

Let us suppose that a power series $\Sigma a_n(x - a)^n$ has a positive radius of convergence ($0 < R \leq +\infty$), and let $f(x)$ be the function defined by this series wherever it converges. That is,

(1) $\qquad f(x) = a_0 + a_1(x - a) + a_2(x - a)^2 + \cdots.$

We now differentiate term-by-term, as if the infinite series were simply a finite sum:
$$f'(x) = a_1 + 2a_2(x - a) + 3a_3(x - a)^2 + \cdots.$$
Again: $\quad f''(x) = 2a_2 + 2 \cdot 3a_3(x - a) + 3 \cdot 4a_4(x - a)^2 + \cdots.$

And so forth:
$$f'''(x) = 3! \, a_3 + 2 \cdot 3 \cdot 4a_4(x - a) + 3 \cdot 4 \cdot 5a_5(x - a)^2 + \cdots,$$
$$\cdots \cdots \cdots \cdots \cdots \cdots$$

Upon substitution of $x = a$, we have:
$$f(a) = a_0, f'(a) = a_1, f''(a) = 2! \, a_2, f'''(a) = 3! \, a_3, \cdots,$$
or, if we solve for the coefficients a_n:

(2) $\qquad a_0 = f(a), \; a_1 = f'(a), \; a_2 = \dfrac{f''(a)}{2!}, \cdots, a_n = \dfrac{f^{(n)}(a)}{n!}, \cdots.$

This suggests that if a power series $\Sigma a_n(x - a)^n$ converges to a function $f(x)$, then the coefficients of the power series should be determined by the values of that function and its successive derivatives according to equations (2). In other words, we should expect:

(3) $\quad f(x) = f(a) + f'(a)(x - a) + \dfrac{f''(a)}{2!}(x - a)^2 + \cdots$
$$+ \dfrac{f^{(n)}(a)}{n!}(x - a)^n + \cdots,$$

and, in particular, for $a = 0$:

(4) $$f(x) = f(0) + f'(0)x + \cdots + \frac{f^{(n)}(0)}{n!} x^n + \cdots .$$

Now suppose $f(x)$ has derivatives of all orders, at least in a neighborhood of the point $x = a$. Then $f^{(n)}(a)$ is defined for every n, and the series (3) exists. Regardless of any question of convergence or (in case of convergence) equality in (3), we *define* the series $\sum_{n=0}^{+\infty} \frac{f^{(n)}(a)}{n!}(x-a)^n$† on the right-hand side of (3) as the **Taylor series for the function $f(x)$ at $x = a$**. The particular case $\sum_{n=0}^{+\infty} \frac{f^{(n)}(0)}{n!} x^n$ given in (4) is known as the **Maclaurin series for $f(x)$**.

We ask two questions:

(i) For a given function $f(x)$, is the Taylor series expansion (3) universally valid in some neighborhood of $x = a$, and if not, what criteria are there for the relation (3) to be true?

The answer to this question is a major concern of the remaining sections of this chapter.

(ii) For a given power series $\Sigma a_n(x-a)^n$, converging to a function $f(x)$ in an interval with positive radius of convergence, is the Taylor series equation (3) true?

We answer this second question affirmatively, now, but defer the proof to the next chapter (Exs. 10, 13, § 1311):

Theorem I. *If $\sum_{n=0}^{+\infty} a_n(x-a)^n$ has a positive radius of convergence, and if $f(x) \equiv \Sigma a_n(x-a)^n$ in the interval of convergence of the series, then throughout that interval $f(x)$ is continuous; throughout the interior of that interval $f(x)$ has (continuous) derivatives of all orders, and relations (2) hold: $a_n = f^{(n)}(a)/n!$, for $n = 0, 1, 2, \cdots$. The given series is the Taylor series for the function $f(x)$ at $x = a$.*

One immediate consequence of this theorem is the uniqueness of a power series $\Sigma a_n(x-a)^n$ converging to a given function:

Theorem II. Uniqueness Theorem. *If a function $f(x)$ is equal to the sum of a power series $\Sigma a_n(x-a)^n$ in a neighborhood of $x = a$, and if $f(x)$ is also equal to the sum of a power series $\Sigma b_n(x-a)^n$ in a neighborhood of $x = a$, then these two power series are identical, coefficient by coefficient: $a_n = b_n$, $n = 0, 1, 2, \cdots$.*

Proof. Each power series is the Taylor series for $f(x)$ at $x = a$.

† For convenience in notation we define $f^{(0)}(x) \equiv f(x)$, and recall that $0! \equiv 1$. Although for $x = a$ and $n = 0$ an indeterminacy 0^0 develops, let us agree that *in this instance* 0^0 shall be defined to be 1.

1204. TAYLOR'S FORMULA WITH A REMAINDER

In § 307 we obtained from the Extended Law of the Mean a formulation for expanding a function $f(x)$ in terms closely related to the Taylor series discussed in the preceding section. Let us repeat the formula of Note 3, § 307, with a slight change in notation, in the following Definition and Theorem I:

Definition. *If $f^{(n)}(x)$ exists at every point of an interval I containing the point $x = a$, then the **Taylor's Formula with a Remainder** for the function $f(x)$, for any point x of I, is*

$$(1) \quad f(x) = f(a) + f'(a)(x - a) + \frac{f''(a)}{2!}(x - a)^2 + \cdots$$
$$+ \frac{f^{(n-1)}(a)}{(n-1)!}(x - a)^{n-1} + R_n(x).$$

*The quantity $R_n(x)$ is called the **remainder after n terms**.*

The principal substance of Note 3, § 307, is an explicit evaluation of the remainder $R_n(x)$:

Theorem I. Lagrange Form of the Remainder. *Let n be a fixed positive integer. If $f^{(n)}(x)$ exists at every point of an interval I (open, closed, or half-open) containing $x = a$, and if x is any point of I, then there exists a point ξ_n between a and x ($\xi_n = a$ if $x = a$) such that the remainder after n terms, in Taylor's Formula with a Remainder, is*

$$(2) \quad R_n(x) = \frac{f^{(n)}(\xi_n)}{n!}(x - a)^n.$$

The principal purpose of this section is to obtain two other forms of the remainder $R_n(x)$. We first establish an integral form of the remainder (Theorem II), from which both the Lagrange form and the Cauchy form (Theorem III) can be derived immediately (cf. Ex. 1, § 1206).

Assuming continuity of all of the derivatives involved, we start with the obvious identity $\int_0^{x-a} f'(x - t)\, dt = f(x) - f(a)$, and integrate by parts, repeatedly:

$$f(x) - f(a) = \int_0^{x-a} f'(x - t)\, dt = \left[t f'(x - t)\right]_0^{x-a} + \int_0^{x-a} t f''(x - t)\, dt$$
$$= f'(a)(x - a) + \int_0^{x-a} f''(x - t)\, d\left(\frac{t^2}{2!}\right)$$
$$= f'(a)(x - a) + \left[\frac{t^2}{2!} f''(x - t)\right]_0^{x-a} + \int_0^{x-a} \frac{t^2}{2!} f'''(x - t)\, dt$$
$$= f'(a)(x - a) + \frac{f''(a)}{2!}(x - a)^2 + \int_0^{x-a} f'''(x - t)\, d\left(\frac{t^3}{3!}\right)$$
$$= \cdots \cdots.$$

Iteration of this process an appropriate number of times leads to the formula written out in the following theorem:

Theorem II. Integral Form of the Remainder. *Let n be a fixed positive integer. If $f^{(n)}(x)$ exists and is continuous throughout an interval I containing the point $x = a$, and if x is any point of I, then the remainder after n terms, in Taylor's Formula with a Remainder, can be written*

(3) $$R_n(x) = \frac{1}{(n-1)!} \int_0^{x-a} t^{n-1} f^{(n)}(x - t)\, dt.$$

Suppose for the moment that $x > a$. Then $x - a > 0$ and, by the First Mean Value Theorem for Integrals (Theorem XI, § 401) there exists a number η_n such that $0 < \eta_n < x - a$ and, from (3),

$$R_n(x) = \frac{1}{(n-1)!} \eta_n^{n-1} f^{(n)}(x - \eta_n) \cdot (x - a).$$

In other words, there exists a number $\xi_n \equiv x - \eta_n$ between a and x such that

$$R_n(x) = \frac{1}{(n-1)!} (x - \xi_n)^{n-1} f^{(n)}(\xi_n) \cdot (x - a).$$

On the other hand, if $x \leqq a$, we obtain the same result by reversing the inequalities or replacing them by equalities. We have the conclusion:

Theorem III. Cauchy Form of the Remainder. *Let n be a fixed positive integer. If $f^{(n)}(x)$ exists and is continuous throughout an interval I containing the point $x = a$, and if x is any point of I, then there exists a point ξ_n between a and x ($\xi_n = a$ if $x = a$) such that the remainder after n terms, in Taylor's Formula with a Remainder, is*

(4) $$R_n(x) = \frac{(x - a)(x - \xi_n)^{n-1}}{(n-1)!} f^{(n)}(\xi_n).$$

1205. EXPANSIONS OF FUNCTIONS

Definition. *A series of functions $\Sigma u_n(x)$ **represents** a function $f(x)$ on a certain set A if and only if for every point x of the set A the series $\Sigma u_n(x)$ converges to the value of the function $f(x)$ at that point.*

Immediately after the question of *convergence* of the Taylor series of a function comes the question, "Does the Taylor series of a given function *represent* the function throughout the interval of convergence?" The clue to the answer lies in Taylor's Formula with a Remainder. To clarify the situation we introduce the notation of partial sums (which now depend on x):

(1) $$S_n(x) \equiv f(a) + f'(a)(x - a) + \cdots + \frac{f^{(n-1)}(a)}{(n-1)!} (x - a)^{n-1}.$$

Taylor's Formula with a Remainder now assumes the form
(2) $$f(x) = S_n(x) + R_n(x),$$
or
(3) $$R_n(x) = f(x) - S_n(x).$$
Immediately from (3) and the definition of the sum of a series, we have the theorem:

Theorem. *If I is an interval containing $x = a$ at each point of which $f(x)$ and all of its derivatives exist, and if $x = x_0$ is a point of I, then the Taylor series for $f(x)$ at $x = a$ represents $f(x)$ at the point x_0 if and only if*
$$\lim_{n \to +\infty} R_n(x_0) = 0.$$

Determining for a particular function whether its Taylor series at $x = a$ represents the function for some particular $x = x_0$, reduces, then, to determining whether $R_n(x_0) \to 0$. Techniques for doing this vary with the function concerned. In the following section we shall make use of different forms of the remainder, in order to show that certain specific expansions represent the given functions. To make any general statement specifying conditions under which a given function is represented by its Taylor series is extremely difficult. We can say that some functions, such as polynomials, e^x, and $\sin x$ (cf. Example 1, below, and Examples 5 and 6, § 1208) are *always* represented by *all* of their Taylor series, and that other functions, such as $\ln x$ and $\tan x$, are represented by Taylor series for only *parts* of their domains. Example 2, below, gives an extreme case of a function possessing a Taylor series that converges everywhere but represents the function only at one point!

Example 1. Prove that every polynomial is everywhere represented by all of its Taylor series.

Solution. If $f(x)$ is a polynomial, $f^{(n)}(x)$ exists for all n and x, and $f^{(n)}(x) = 0$ if n is greater than the degree of the polynomial. Therefore, if the Lagrange form of the remainder is used, $R_n(x)$ is identically zero for sufficiently large n, and $\lim_{n \to +\infty} R_n(x) = 0$.

Example 2. Show that the function $f(x)$ defined to be 0 when $x = 0$, and otherwise $f(x) \equiv e^{-1/x^2}$, is not represented by its Maclaurin series, although the function has derivatives of all orders everywhere.

Solution. The function has (continuous) derivatives of all orders at every point except possibly $x = 0$. At $x = 0$, $f^{(n)}(x) = 0$ for every $n = 0, 1, 2, \cdots$ (cf. Ex. 52, § 320), so that $f^{(n)}(x)$ exists and is continuous for every $n = 0, 1, 2, \cdots$ and every x. The Maclaurin series for $f(x)$ is thus
$$0 + 0 \cdot x + 0 \cdot x^2 + 0 \cdot x^3 + \cdots$$
which represents the function identically 0 everywhere, but the function $f(x)$ only at $x = 0$.

NOTE. The property that a function may have of being represented by a Taylor series is of such basic importance in analysis that it is given a special name, *analyticity*, as specified in the definition: *A function $f(x)$ is **analytic** at $x = a$ if and only if it has a Taylor series at $x = a$ that represents the function in some neighborhood of $x = a$.* For a brief treatment of analyticity for real variables, see RV, §§ 815–816. For an introduction to analytic functions of a complex variable, see § 1810 of the present text.

1206. EXERCISES

1. Derive the Lagrange form of the remainder in Taylor's formula (Theorem I, § 1204) from the integral form (Theorem II, § 1204), using the additional hypothesis of continuity of $f^{(n)}(x)$. *Hint:* Use the generalized form of the First Mean Value Theorem for Integrals (Ex. 6, § 402).

2. Show that $|x|$, $\ln x$, $\sqrt{x}$, and $\cot x$ have no Maclaurin series. Show that x^p has a Maclaurin series if and only if p is a nonnegative integer.

3. Show that if the Taylor series for $f(x)$ at $x = a$ represents $f(x)$ near a, it can be written in increment and differential notation:

$$\Delta y = dy + \frac{f''(a)}{2!} dx^2 + \frac{f'''(a)}{3!} dx^3 + \cdots.$$

Hence show that approximations by differentials (cf. § 312) are those provided by the partial sum through terms of the first degree of the Taylor series of the function.

1207. SOME MACLAURIN SERIES

In this section we derive the Maclaurin series for the five functions e^x, $\sin x$, $\cos x$, $\ln(1+x)$, and $(1+x)^m$, and (sometimes with the aid of future exercises) show that in each case the function is represented by its Maclaurin series throughout the interval of convergence.

I. The exponential function, $f(x) = e^x$. Since, for $n = 0, 1, 2, \cdots$, $f^{(n)}(x) = e^x$, $f^{(n)}(0) = 1$, and the Maclaurin series is

(1) $$1 + x + \frac{x^2}{2!} + \frac{x^3}{3!} + \cdots + \frac{x^n}{n!} + \cdots.$$

The test ratio is $\dfrac{x}{n}$, whose limit is 0. Therefore the series (1) converges absolutely for all x, and the radius of convergence is infinite: $R = +\infty$.

To show that *the series (1) represents the function e^x for all real x*, we choose an arbitrary $x \neq 0$, and use the Lagrange form of the remainder:

(2) $$R_n(x) = \frac{e^{\xi_n}}{n!} x^n,$$

where ξ_n is between 0 and x. We observe first that for a fixed x, ξ_n (although it depends on n and is not constant) satisfies the inequality $\xi_n < |x|$, and therefore e^{ξ_n} is bounded above by the constant $e^{|x|}$. Thus the problem has been reduced to showing that $\lim\limits_{n \to +\infty} \dfrac{x^n}{n!} = 0$. But $\dfrac{x^n}{n!}$ is the general term of (1), and since (1) has already been shown to converge for all x, the general term must tend toward 0.

II. The sine function, $f(x) = \sin x$. The sequence $\{f^{(n)}(x)\}$ is $\sin x$, $\cos x$, $-\sin x$, $-\cos x$, $\sin x$, $\cdots$, and the sequence $\{f^{(n)}(0)\}$ is $0, 1, 0, -1, 0, 1, \cdots$. Therefore the Maclaurin series is

(3) $\quad x - \dfrac{x^3}{3!} + \dfrac{x^5}{5!} - \dfrac{x^7}{7!} + \cdots + (-1)^{n-1} \dfrac{x^{2n-1}}{(2n-1)!} + \cdots$.

(Here we have dropped all zero terms, and used n to indicate the sequence of remaining nonzero terms, instead of the original exponent for the series $\Sigma \dfrac{f^{(n)}(0)}{n!} x^n$.) The test ratio of (3) is

$$-\dfrac{x^2}{2n(2n+1)},$$

whose limit is 0. Therefore (3) converges absolutely for all x, and the radius of convergence is infinite: $R = +\infty$.

To show that *the series* (3) *represents the function* $\sin x$ *for all real* x, we choose an arbitrary $x \neq 0$, and again use the Lagrange form of the remainder:

(4) $\quad R_n(x) = \dfrac{g_n(\xi_n)}{n!} x^n,$

where ξ_n is between 0 and x, $g_n(x)$ is $\pm \sin x$ or $\pm \cos x$, and n is now used to indicate the number of terms in the original form $\Sigma \dfrac{f^{(n)}(0)}{n!} x^n$ of the Maclaurin series. The proof that $\lim\limits_{n \to +\infty} R_n(x) = 0$ follows the same lines as the proof for e^x, with the aid of the inequality $|g_n(\xi_n)| \leq 1$.

III. The cosine function, $f(x) = \cos x$. The details of the analysis are similar to those for $\sin x$. The Maclaurin series is

(5) $\quad 1 - \dfrac{x^2}{2!} + \dfrac{x^4}{4!} - \dfrac{x^6}{6!} + \cdots + (-1)^n \dfrac{x^{2n}}{(2n)!},$

which converges absolutely for all real x ($R = +\infty$). *The series* (5) *represents the function* $\cos x$ *for all real* x.

IV. The natural logarithm, $f(x) = \ln(1 + x)$. The sequence $\{f^{(n)}(x)\}$ is

$\ln(1+x), (1+x)^{-1}, -(1+x)^{-2}, 2!(1+x)^{-3}, -3!(1+x)^{-4},$
$\cdots, (-1)^{n-1}(n-1)!(1+x)^{-n}, \cdots,$

and hence the sequence $\{f^{(n)}(0)\}$ is

$0, 1, -1, 2!, -3!, \cdots, (-1)^{n-1}(n-1)!, \cdots.$

Therefore the Maclaurin series is

(6) $\quad x - \dfrac{x^2}{2} + \dfrac{x^3}{3} - \dfrac{x^4}{4} + \cdots + (-1)^{n-1} \dfrac{x^n}{n} + \cdots.$

Theorem III, § 1201, gives the radius of convergence: $R = 1$. If $x = 1$, the series (6) is the conditionally convergent alternating harmonic series, and if $x = -1$, the series (6) is the divergent series $\Sigma - \dfrac{1}{n}$. The interval of convergence is therefore $-1 < x \leq 1$.

To show that *the series (6) represents the function* $\ln(1+x)$ *throughout the interval of convergence*, we shall derive a form of the remainder appropriate to $\ln(1+x)$ alone. Using a formula from College Algebra for the sum of a geometric progression, we have, for any $t \neq -1$:

(7) $\quad 1 - t + t^2 - t^3 + \cdots + (-t)^{n-2} = \dfrac{1 - (-t)^{n-1}}{1+t}$.

Therefore, if we solve for $\dfrac{1}{1+t}$, and integrate from 0 to x, where $-1 < x \leq 1$, we have:

$$\int_0^x \frac{dt}{1+t} = \int_0^x [1 - t + \cdots + (-1)^n t^{n-2}] \, dt + (-1)^{n-1} \int_0^x \frac{t^{n-1} \, dt}{1+t},$$

or

(8) $\quad \ln(1+x) = x - \dfrac{x^2}{2} + \dfrac{x^3}{3} - \cdots + (-1)^n \dfrac{x^{n-1}}{n-1} + R_n(x),$

where

(9) $\quad R_n(x) = (-1)^{n-1} \displaystyle\int_0^x \dfrac{t^{n-1} \, dt}{1+t}$.

If $0 \leq x \leq 1$, $|R_n(x)| \leq \displaystyle\int_0^x t^{n-1} \, dt = \dfrac{x^n}{n} \leq \dfrac{1}{n}$, and $\displaystyle\lim_{n \to +\infty} R_n(x) = 0$. If $-1 < x < 0$,

$$|R_n(x)| \leq \left| \int_0^x \left| \frac{t^{n-1}}{1+t} \right| dt \right| \leq \frac{1}{1+x} \left| \int_0^x |t^{n-1}| \, dt \right|$$

$$= \frac{1}{1+x} \left| \int_0^x t^{n-1} \, dt \right| = \frac{|x|^n}{n(1+x)} < \frac{1}{n(1+x)},$$

and $\displaystyle\lim_{n \to +\infty} R_n(x) = 0$. Therefore the series (6) converges to $\ln(1+x)$ for $-1 < x \leq 1$. In particular, if $x = 1$, an interesting special case results:

(10) $\quad \ln 2 = 1 - \dfrac{1}{2} + \dfrac{1}{3} - \dfrac{1}{4} + \cdots$.

V. The binomial function $f(x) = (1+x)^m$, where m is any real number. The sequence $\{f^{(n)}(x)\}$ is

$$(1+x)^m, \; m(1+x)^{m-1}, \; m(m-1)(1+x)^{m-2}, \cdots,$$

and the sequence $\{f^{(n)}(0)\}$ is $1, m, m(m-1), \cdots$. Therefore the Maclaurin series is the **binomial series**

(11) $\quad 1 + mx + \dfrac{m(m-1)}{2!} x^2 + \cdots + \dbinom{m}{n} x^n + \cdots,$

where

(12) $\quad \dbinom{m}{n} \equiv \dfrac{m(m-1) \cdots (m-n+1)}{n!},$

and is called the **binomial coefficient** of x^n. If m is a nonnegative integer the binomial series (11) has only a finite number of nonzero terms and hence converges to $f(x) = (1 + x)^m$ for all real x. Assume now that m is not a nonnegative integer. Since $\binom{m}{n} \big/ \binom{m}{n+1} = \dfrac{n+1}{m-n} \to -1$, we know from Theorem III, § 1201, that the radius of convergence is 1. The behavior of (11) at the end-points of the interval of convergence depends on the value of m. We state the facts here, but defer the proofs to the Exercises (Ex. 35, § 1211):

(i) $m \geq 0$: (11) converges absolutely for $x = \pm 1$.
(ii) $m \leq -1$: (11) diverges for $x = \pm 1$.
(iii) $-1 < m < 0$: (11) converges conditionally for $x = 1$, and diverges for $x = -1$.

The binomial series (11) *represents the binomial function* $(1 + x)^m$ *throughout the interval of convergence.* In proving this we shall call upon both the Lagrange and the Cauchy forms of the remainder, depending on whether $0 < x \leq 1$ or $-1 \leq x < 0$:

Assume $0 < x \leq 1$. Then the Lagrange form of the remainder is

$$(13) \qquad R_n(x) = \binom{m}{n}(1 + \xi_n)^{m-n} x^n,$$

where $0 < \xi_n < x$. Since, for $n > m$, the inequality $1 + \xi_n > 1$ implies $(1 + \xi_n)^{m-n} < 1$, $|R_n(x)| \leq \left|\binom{m}{n} x^n\right|$ for sufficiently large n. Since $\binom{m}{n} x^n$ is the general term of the binomial series (11), it must tend toward zero whenever that series converges. Therefore $\lim_{n \to +\infty} R_n(x) = 0$ for $0 < x \leq 1$ whenever (11) converges, and we conclude that the binomial series represents the binomial function throughout the interval $0 \leq x < 1$, and also at the point $x = 1$ whenever the series converges there (that is, for $m > -1$).

Assume $-1 \leq x < 0$. Then the Cauchy form of the remainder is

$$(14) \qquad R_n(x) = nx(x - \xi_n)^{n-1} \binom{m}{n}(1 + \xi_n)^{m-n},$$

where $x < \xi_n < 0$. We rewrite (14):

$$(15) \qquad R_n(x) = n\binom{m}{n} x^n (1 + \xi_n)^{m-1} \left(\dfrac{1 - \dfrac{\xi_n}{x}}{1 + \xi_n}\right)^{n-1}.$$

For $-1 \leq x < \xi_n < 0$, $0 < 1 - \dfrac{\xi_n}{x} \leq 1 + \xi_n$ (check this), so that the last factor of (15) cannot exceed 1 for $n > 1$. Also, if $m > 1$, the inequality $1 + \xi_n < 1$ implies $(1 + \xi_n)^{m-1} < 1$, while if $m < 1$ the inequality $1 + \xi_n > 1 + x$ implies $(1 + \xi_n)^{m-1} < (1 + x)^{m-1}$. In any case, then, the last two factors of (15) remain bounded as $n \to +\infty$. The problem before us has been simplified, then, to showing (under the appropriate conditions) that

$$(16) \qquad \lim_{n \to +\infty} n\binom{m}{n} x^n = 0.$$

In case $-1 < x < 0$, the relation (16) is easily established by the ratio test (check the details in Ex. 27, § 1211). Finally, if $x = -1$, the last item remaining in the proof is to show that $n\binom{m}{n} \to 0$ for $m > 0$. A technique for doing this is suggested in Exercise 36, § 1211. We conclude that the binomial series represents the binomial function throughout the interval of convergence.

1208. ELEMENTARY OPERATIONS WITH POWER SERIES

From results obtained in Chapter 11 for series of constants (§§ 1102, 1119) we have the theorem for power series (expressed here for simplicity in terms of powers of x, although similar formulations are valid for power series in powers of $(x - a)$):

Theorem. *Addition, Subtraction, and Multiplication. Let $\sum_{n=0}^{+\infty} a_n x^n$ and $\sum_{n=0}^{+\infty} b_n x^n$ be two power series representing the functions $f_1(x)$ and $f_2(x)$, respectively, within their intervals of convergence, and let γ be an arbitrary constant. Then (i) the power series $\sum \gamma a_n x^n$ represents the function $\gamma f_1(x)$ throughout the interval of convergence of $\sum a_n x^n$; (ii) the power series $\sum (a_n \pm b_n) x^n$ represents the function $f_1(x) \pm f_2(x)$ for all points common to the intervals of convergence of the two given power series; and (iii) if $c_n \equiv \sum_{k=0}^{n} a_k b_{n-k}$, $n = 0, 1, 2, \cdots$, then the power series $\sum_{n=0}^{+\infty} c_n x^n$ represents the function $f_1(x) f_2(x)$ for all points interior to both intervals of convergence of the two given power series* (cf. § 1309).

In finding the Maclaurin or Taylor series for a given function, it is well to bear in mind the import of the uniqueness theorem (Theorem II, § 1223) for power series. This means that the Maclaurin or Taylor series need not be obtained by direct substitution in the formulas defining those series. Any means that produces an appropriate power series representing the function automatically produces the Maclaurin or Taylor series.

Example 1. Since the series for e^x is $1 + x + \dfrac{x^2}{2!} + \cdots$, the series for e^{-x} is found by substituting $-x$ for x:

$$e^{-x} = 1 - x + \frac{x^2}{2!} - \frac{x^3}{3!} + \cdots.$$

The series expansion is valid for all real x, and is therefore the Maclaurin series for e^{-x}.

Example 2. The Maclaurin series for $\cos 2x$ is

$$\cos 2x = 1 - \frac{(2x)^2}{2!} + \frac{(2x)^4}{4!} - \frac{(2x)^6}{6!} + \cdots,$$

and is valid for all real x.

Example 3. The Maclaurin series for $\sinh x = \frac{1}{2}(e^x - e^{-x})$ is

$$\frac{1}{2}\left(1 + x + \frac{x^2}{2!} + \cdots\right) - \frac{1}{2}\left(1 - x + \frac{x^2}{2!} - \cdots\right) = x + \frac{x^3}{3!} + \frac{x^5}{5!} + \cdots,$$

and is valid for all real x.

Example 4. Find the Maclaurin series for $\sin\left(\dfrac{\pi}{6} + x\right)$.

Solution. Instead of proceeding in a routine manner, we expand $\sin\left(\dfrac{\pi}{6} + x\right) = \sin\dfrac{\pi}{6}\cos x + \cos\dfrac{\pi}{6}\sin x$, and obtain:

$$\frac{1}{2}\left[1 - \frac{x^2}{2!} + \frac{x^4}{4!} - \cdots\right] + \frac{\sqrt{3}}{2}\left[x - \frac{x^3}{3!} + \frac{x^5}{5!} - \cdots\right]$$

$$= \frac{1}{2} + \frac{\sqrt{3}}{2}x - \frac{1}{2}\frac{x^2}{2!} - \frac{\sqrt{3}}{2}\frac{x^3}{3!} + \frac{1}{2}\frac{x^4}{4!} + \frac{\sqrt{3}}{2}\frac{x^5}{5!} - \cdots.$$

Example 5. The Taylor series for e^x at $x = a$ is most easily obtained by writing $e^x = e^a e^{x-a}$ and expanding the second factor by means of the Maclaurin series already established:

$$e^x = e^a\left[1 + (x - a) + \frac{(x - a)^2}{2!} + \frac{(x - a)^3}{3!} + \cdots\right].$$

Example 6. The Taylor series for $\sin x$ at $x = a$ can be found by writing

$$\sin x = \sin[a + (x - a)] = \sin a \cos(x - a) + \cos a \sin(x - a)$$

$$= \sin a\left[1 - \frac{(x - a)^2}{2!} + \frac{(x - a)^4}{4!} - \cdots\right]$$

$$+ \cos a\left[(x - a) - \frac{(x - a)^3}{3!} + \frac{(x - a)^5}{5!} - \cdots\right].$$

If $a = \dfrac{\pi}{6}$, the coefficients are those of Example 4.

Example 7. The Taylor series for $\ln x$ at $x = a > 0$ can be found by writing

$$\ln x = \ln[a + (x - a)] = \ln a + \ln\left[1 + \frac{x - a}{a}\right]$$

$$= \ln a + \frac{x - a}{a} - \frac{(x - a)^2}{2a^2} + \frac{(x - a)^3}{3a^3} - \cdots.$$

This is valid for $0 < x \leq 2a$.

Example 8. The Taylor series for x^m at $x = a > 0$ can be found by writing

$$x^m = [a + (x - a)]^m = a^m\left[1 + \left(\frac{x - a}{a}\right)\right]^m$$

$$= a^m\left[1 + m\left(\frac{x - a}{a}\right) + \frac{m(m - 1)}{2!}\left(\frac{x - a}{a}\right)^2 + \cdots\right].$$

This is valid for $0 < x < 2a$; also at 0 for $m > 0$ and at $2a$ for $m > -1$.

Example 9. The Maclaurin series for $e^x \sin ax$ is found by multiplying the series:

$$\left[1 + x + \frac{x^2}{2!} + \frac{x^3}{3!} + \cdots\right]\left[ax - \frac{a^3 x^3}{3!} + \frac{a^5 x^5}{5!} - \cdots\right].$$

If we wish the terms of degree ≤ 5, we have

$$\left(1 + x + \frac{x^2}{2} + \frac{x^3}{6} + \frac{x^4}{24} + \cdots\right)\left(ax - \frac{a^3 x^3}{6} + \frac{a^5 x^5}{120} + \cdots\right)$$

$$= ax + ax^2 + \frac{a}{6}(3 - a^2)x^3 + \frac{a}{6}(1 - a^2)x^4 + \frac{a}{120}(5 - 10a^2 + a^4)x^5 + \cdots.$$

1209. SUBSTITUTION OF POWER SERIES

Sometimes it is important to obtain a power series for a composite function, where each of the constituent functions has a known power series. The most useful special case of a general theorem for such substitutions is the principal theorem of this section. We begin with a lemma:

★Lemma. *If the terms of the doubly infinite array*

(1)
$$c_{11}, c_{12}, c_{13}, \cdots$$
$$c_{21}, c_{22}, c_{23}, \cdots$$
$$c_{31}, c_{32}, c_{33}, \cdots$$
$$\cdots\cdots\cdots\cdots,$$

when arranged in any manner to form an infinite series, give an absolutely convergent series whose sum is C, then every row series $c_{m1} + c_{m2} + \cdots$ converges absolutely, and if $c_m \equiv \sum_{n=1}^{+\infty} c_{mn}$, then $\sum_{m=1}^{+\infty} c_m$ converges absolutely, with sum C.

★Proof. We first assume that every $c_{mn} \geq 0$. Then any partial sum of terms in any row is bounded by C, so that each row series converges. Furthermore, the sum of the terms in the rectangle made up of the elements of the first M rows and the first n columns is bounded by C, so that when $n \to +\infty$ we have in the limit $\sum_{m=1}^{M} c_m \leq C$, and hence $\sum_{m=1}^{+\infty} c_m \leq C$. On the other hand, if $\epsilon > 0$, there exists a finite sequence of terms of (1) whose sum exceeds $C - \epsilon$. If M is the largest index of the rows from which these terms are selected, then $\sum_{m=1}^{M} c_m$ must also exceed $C - \epsilon$. Therefore $\sum_{m=1}^{+\infty} c_m \geq C - \epsilon$ and, since ϵ is arbitrarily small, $\sum_{m=1}^{+\infty} c_m \geq C$. In combination with a preceding inequality this gives $\sum_{m=1}^{+\infty} c_m = C$.

We now remove the assumption that $c_{mn} \geq 0$, and (by splitting the entire array into nonnegative and nonpositive parts in the manner of § 1116) immediately draw every conclusion stated in the lemma, except the equality $\sum_{m=1}^{+\infty} c_m = C$. But this equality follows from the fact that for any $\epsilon > 0$ there exists a number M such that the sum of the absolute values of all terms of (1) appearing below the Mth row is less than ϵ (check the details of this carefully in Ex. 28, § 1211). This completes the proof.

Theorem I. Substitution. *Let*

(1) $\quad y = f(u) \equiv a_0 + a_1 u + a_2 u^2 + \cdots, \text{ and}$

(2) $\quad u = g(x) \equiv b_1 x + b_2 x^2 + \cdots,$

where both power series have positive radii of convergence. Then the composite function $h(x) \equiv f(g(x))$ is represented by a power series having a positive radius of convergence, obtained by substituting the entire series (2) for the quantity u in (1), expanding, and collecting terms:

$$
\begin{aligned}
(3) \quad y = h(x) &= a_0 + a_1(b_1 x + b_2 x^2 + \cdots) + a_2(b_1 x + \cdots)^2 + \cdots \\
&= a_0 + a_1 b_1 x + (a_1 b_2 + a_2 b_1^2) x^2 \\
&\quad + (a_1 b_3 + 2 a_2 b_1 b_2 + a_3 b_1^3) x^3 \\
&\quad + (a_1 b_4 + 2 a_2 b_1 b_3 + a_2 b_2^2 + 3 a_3 b_1^2 b_2) x^4 + \cdots.
\end{aligned}
$$

Proof. We first exploit the continuity of $g(x)$ at $x = 0$ (Theorem I, § 1203) and observe that if x belongs to a sufficiently small neighborhood of $x = 0$ and if u is defined in terms of x by (2), then u belongs to the interior of the interval of convergence of (1), and the expansions indicated by (1) and the first line of (3) are valid. It now remains to justify the removal of parentheses and the subsequent rearrangement in (3). For this purpose we shall insist on restricting x to so small an interval about $x = 0$ that $v \equiv \sum_{n=1}^{+\infty} |b_n x^n|$ is inside the interval of convergence of (1). Then, as a consequence of the absolute convergence of all series concerned, we can apply the preceding lemma to the double array

(4)
$$
\begin{array}{ccccc}
a_0, & 0, & 0, & 0, & \cdots \\
0, & a_1 b_1 x, & a_1 b_2 x^2, & a_1 b_3 x^3, & \cdots \\
0, & 0, & a_2 b_1^2 x^2, & 2 a_2 b_1 b_2 x^3, & \cdots \\
\cdot & \cdot & \cdot & \cdot & \cdot
\end{array}
$$

With a final appeal to § 1118 the proof is complete. (Give precise details, particularly for the last step of the proof, in Ex. 29, § 1211.)

Example 1. Find the terms of the Maclaurin series for $e^{\sin x}$, through terms of degree 5.

Solution. The series (1) and (2) are

$$y = f(u) = e^u = 1 + u + \frac{u^2}{2!} + \frac{u^3}{3!} + \frac{u^4}{4!} + \cdots,$$

$$u = g(x) = \sin x = x - \frac{x^3}{3!} + \frac{x^5}{5!} - \cdots.$$

The double array (4) becomes

1,	0,	0,	0,	0,	0,	...
0,	x,	0,	$-\dfrac{x^3}{6}$,	0,	$\dfrac{x^5}{120}$,	...
0,	0,	$\dfrac{x^2}{2}$,	0,	$-\dfrac{x^4}{6}$,	0,	...
0,	0,	0,	$\dfrac{x^3}{6}$,	0,	$-\dfrac{x^5}{12}$,	...
0,	0,	0,	0,	$\dfrac{x^4}{24}$,	0,	...
0,	0,	0,	0,	0,	$\dfrac{x^5}{120}$,	...

. . . .

Therefore the series sought is

$$1 + x + \frac{x^2}{2} - \frac{x^4}{8} - \frac{x^5}{15} + \cdots.$$

*The radius of convergence is infinite, since each basic series converges absolutely, everywhere.

Example 2. Find the terms of the Maclaurin series for $e^{\cos x}$ through terms of degree 6.

Solution. The Maclaurin series for $\cos x$ has a nonzero constant term. Therefore we write

$$e^{\cos x} = e^{1+g(x)} = e \cdot e^{g(x)},$$

where $g(x) = -\dfrac{x^2}{2} + \dfrac{x^4}{24} - \dfrac{x^6}{720} + \cdots$. We proceed as before, obtaining

$$e^{\cos x} = e\left\{1 + \left[-\frac{x^2}{2} + \frac{x^4}{24} - \cdots\right] + \frac{1}{2}\left[-\frac{x^2}{2} + \frac{x^4}{24} - \cdots\right]^2 + \cdots\right\}$$

$$= e\left\{1 + \left[-\frac{x^2}{2} + \frac{x^4}{24} - \frac{x^6}{720}\right] + \frac{1}{2}\left[\frac{x^4}{4} - \frac{x^6}{24}\right] + \frac{1}{6}\left[-\frac{x^6}{8}\right] + \cdots\right\}$$

$$= e\left(1 - \frac{x^2}{2} + \frac{x^4}{6} - \frac{31x^6}{720} + \cdots\right).$$

Before presenting more examples, let us record for future use a convenient device, which sometimes simplifies the work connected with the method of undetermined coefficients, illustrated in the second solution of Example 3, below:

Theorem II. *If* $\displaystyle\sum_{n=0}^{+\infty} a_n x^n$ *represents a function* $f(x)$ *in a neighborhood I of $x = 0$, then (i) if $f(x)$ is an even function in I the power series* $\Sigma a_n x^n$ *consists of only even degree terms, and (ii) if $f(x)$ is an odd function in I the power series* $\Sigma a_n x^n$ *consists of only odd degree terms.*

Proof. We shall give the details only for part (*i*) (cf. Ex. 32, § 1211, for (*ii*)). Since
$$f(x) = a_0 + a_1 x + a_2 x^2 + a_3 x^3 + \cdots,$$
$$f(-x) = a_0 - a_1 x + a_2 x^2 - a_3 x^3 + \cdots,$$
and therefore
$$f(x) - f(-x) = 2a_1 x + 2a_3 x^3 + 2a_5 x^5 + \cdots.$$
If $f(x)$ is even, the function $f(x) - f(-x)$ is identically 0 in I, and every coefficient in its power series expansion must vanish, by the uniqueness theorem (Theorem II, § 1203).

Example 3. Find the terms of the Maclaurin series for sec x through terms of degree 8, and ★ determine an interval within which the series converges.

First Solution. Since $\sec x = \dfrac{1}{\cos x} = \dfrac{1}{1 - g(x)}$, where $g(x) = \dfrac{x^2}{2!} - \dfrac{x^4}{4!} + \cdots$, the Maclaurin series for sec x is found by substituting the power series for $g(x)$ in the series
$$\frac{1}{1-u} = 1 + u + u^2 + u^3 + \cdots.$$
We therefore collect terms from
$$1 + \left[\frac{x^2}{2!} - \frac{x^4}{4!} + \frac{x^6}{6!} - \frac{x^8}{8!}\right] + \left[\frac{x^2}{2!} - \frac{x^4}{4!} + \frac{x^6}{6!}\right]^2 + \left[\frac{x^2}{2!} - \frac{x^4}{4!}\right]^3 + \left[\frac{x^2}{2!}\right]^4,$$
and get
$$\sec x = 1 + \frac{x^2}{2} + \frac{5x^4}{24} + \frac{61x^6}{720} + \frac{277x^8}{8064} + \cdots.$$

★The above procedures have been validated for any interval such that $\dfrac{x^2}{2!} + \dfrac{x^4}{4!} + \dfrac{x^6}{6!} + \cdots < 1$, or $\cosh x < 2$. Therefore the series found for sec x converges to sec x within (at least) the interval $(-1.3, 1.3)$. Actually, as is easily shown by the theory of analytic functions of a complex variable, the interval of convergence is $\left(-\dfrac{\pi}{2}, \dfrac{\pi}{2}\right)$. (Cf. § 1810.)

Second Solution. Since sec x is an even function and is represented by its Maclaurin series (cf. the first solution), its Maclaurin series must have the form
$$\sec x = a_0 + a_2 x^2 + a_4 x^4 + a_6 x^6 + \cdots,$$
all coefficients of odd degree terms being 0. We form the product of this series and that of cos x and have
$$1 = (a_0 + a_2 x^2 + a_4 x^4 + \cdots)\left(1 - \frac{x^2}{2!} + \frac{x^4}{4!} - \cdots\right)$$
$$= a_0 + \left(-\frac{a_0}{2!} + a_2\right) x^2 + \left(\frac{a_0}{4!} - \frac{a_2}{2!} + a_4\right) x^4$$
$$+ \left(-\frac{a_0}{6!} + \frac{a_2}{4!} - \frac{a_4}{2!} + a_6\right) x^6 + \left(\frac{a_0}{8!} - \frac{a_2}{6!} + \frac{a_4}{4!} + \frac{a_2}{2!} + a_0\right) x^8 + \cdots.$$
Equating corresponding coefficients, we have the recursion formulas $a_0 = 1$, $a_2 = \dfrac{a_0}{2!}$, $a_4 = \dfrac{a_2}{2!} - \dfrac{a_0}{4!}, \cdots$, from which we can evaluate the coefficients, one after the other. The result is the same as that of the first solution.

1210. INTEGRATION AND DIFFERENTIATION OF POWER SERIES

As useful adjuncts to the methods of the two preceding sections, we state three theorems, the second and third of which are proved in the next chapter:

Theorem I. *A power series* $\sum_{n=0}^{+\infty} a_n(x-a)^n$ *and its derived series* $\sum_{n=1}^{+\infty} na_n(x-a)^{n-1}$ *have the same radius of convergence.*

Proof. For simplicity of notation, we assume that $a = 0$. In the first place, since $|a_n| \leq n|a_n|$, for $n = 1, 2, \cdots$, if the derived series converges absolutely for some particular $x = x_0$, then $x_0(\sum na_n x_0^{n-1}) = \sum na_n x_0^n$ converges absolutely and $\sum a_n x_0^n$ also converges absolutely. That is, the radius of convergence of the derived series can be no larger than that of the original series. On the other hand, it can be no smaller, for let $0 < \alpha < \beta$, and assume that the original series converges absolutely for $x = \beta$. We shall show that the derived series converges absolutely for $x = \alpha$. This will conclude the proof (why?). Our contention follows from the fact that $na_n \alpha^n = O(|a_n \beta^n|)$ for the nonzero terms of the series, and this fact in turn follows by means of the ratio test from the limit:

$$\lim_{n \to +\infty} \frac{na_n \alpha^n}{a_n \beta^n} = \lim_{n \to +\infty} n\left(\frac{\alpha}{\beta}\right)^n = \lim_{n \to +\infty} \frac{n}{e^{n \ln \beta/\alpha}} = 0.$$

For an alternative proof, see Exercise 37, § 1211.

Theorem II. *If $f(x)$ is represented by a power series $\sum a_n(x-a)^n$ in its interval of convergence I, and if α and β are any two points belonging to I, then the series can be integrated term by term:*

$$\int_\alpha^\beta f(x)\,dx = \sum_{n=0}^{+\infty} a_n \int_\alpha^\beta (x-a)^n\,dx = \sum_{n=0}^{+\infty} \frac{a_n}{n+1}[(\beta-a)^{n+1} - (\alpha-a)^{n+1}].$$

(Cf. Exs. 11, 14, § 1311.)

Theorem III. *A function $f(x)$ represented by a power series $\sum a_n(x-a)^n$ in the interior of its interval of convergence is differentiable there, and its derivative is represented there by the derived series:*

$$f'(x) = \sum_{n=1}^{+\infty} na_n(x-a)^{n-1}.$$

(Cf. Ex. 12, § 1311.)

Example 1. The series

$$\frac{1}{1+t} = 1 - t + t^2 - t^3 + \cdots.$$

has radius of convergence $R = 1$. Therefore, for $|x| < 1$, integration from 0 to x gives
$$\ln(1+x) = x - \frac{x^2}{2} + \frac{x^3}{3} - \cdots.$$

Example 2. The series
$$(1+t)^{-\frac{1}{2}} = 1 - \frac{1}{2}t + \frac{1\cdot 3}{2\cdot 4}t^2 - \frac{1\cdot 3\cdot 5}{2\cdot 4\cdot 6}t^3 + \cdots$$
has radius of convergence $R = 1$. Therefore the same is true for
$$(1-t^2)^{-\frac{1}{2}} = 1 + \frac{1}{2}t^2 + \frac{1\cdot 3}{2\cdot 4}t^4 + \frac{1\cdot 3\cdot 5}{2\cdot 4\cdot 6}t^6 + \cdots.$$
From this, by integrating from 0 to x, where $|x| < 1$, we find
$$\operatorname{Arcsin} x = x + \frac{1}{2}\frac{x^3}{3} + \frac{1\cdot 3}{2\cdot 4}\frac{x^5}{5} + \frac{1\cdot 3\cdot 5}{2\cdot 4\cdot 6}\frac{x^7}{7} + \cdots.$$

Example 3. Find the terms of the Maclaurin series for $\tan x$ through terms of degree 9. (Cf. Ex. 39, § 1211.)

First Solution. Multiply the power series for $\sin x$ and $\sec x$ (Example 3, § 1209):
$$\tan x = \left(x - \frac{x^3}{3!} + \frac{x^5}{5!} - \cdots\right)\left(1 + \frac{x^2}{2} + \frac{5x^4}{24} + \cdots\right)$$
$$= x + \frac{x^3}{3} + \frac{2x^5}{15} + \frac{17x^7}{315} + \frac{62x^9}{2835} + \cdots.$$

Second Solution. Since $\tan x$ is an odd function its Maclaurin series (which exists and represents the function, by the first solution) has the form
$$\tan x = a_1 x + a_3 x^3 + a_5 x^5 + \cdots.$$
Furthermore, since $\cos x \tan x = \sin x$, the coefficients of the product series for $\cos x \tan x$ must be identically equal to those of the sine series:
$$a_1 = 1, \quad -\frac{a_1}{2} + a_3 = -\frac{1}{6}, \quad \frac{a_1}{24} - \frac{a_3}{2} + a_5 = \frac{1}{120}, \cdots.$$
The result of solving these recursion formulas is the same as that found in the first solution.

Third Solution. As in the second solution, let
$$\tan x = a_1 x + a_3 x^3 + a_5 x^5 + a_7 x^7 + \cdots.$$
Differentiation and use of the identity $\sec^2 x = 1 + \tan^2 x$ give
$$a_1 + 3a_3 x^2 + 5a_5 x^4 + 7a_7 x^6 + 9a_9 x^8 + \cdots$$
$$= 1 + [a_1 x + a_3 x^3 + a_5 x^5 + a_7 x^7 + \cdots]^2$$
$$= 1 + a_1^2 x^2 + 2a_1 a_3 x^4 + (2a_1 a_5 + a_3^2) x^6 + \cdots.$$
Equating corresponding coefficients produces the recursion formulas $a_1 = 1$, $3a_3 = a_1^2$, $5a_5 = 2a_1 a_3$, $\cdots$, and the result of the first solution. This is by far the shortest of the three methods.

Example 4. Find the exact sum of the series
$$\frac{1}{1!\cdot 3} + \frac{1}{2!\cdot 4} + \frac{1}{3!\cdot 5} + \cdots + \frac{1}{n!(n+2)} + \cdots.$$

Solution. Start with the series
$$e^x = 1 + x + \frac{x^2}{2!} + \frac{x^3}{3!} + \cdots.$$

Multiply by x and integrate from 0 to 1:
$$\int_0^1 xe^x\,dx = \frac{1}{2} + \frac{1}{3} + \frac{1}{2!\cdot 4} + \frac{1}{3!\cdot 5} + \cdots.$$
Since the value of the integral is 1, the original series converges to $\frac{1}{2}$.

1211. EXERCISES

In Exercises 1–12, find the Maclaurin series for the given function.

1. $\cosh x = \dfrac{e^x + e^{-x}}{2}$.
2. $\dfrac{1}{1 + x^2}$.
3. $\operatorname{Arctan} x = \displaystyle\int_0^x \dfrac{dt}{1 + t^2}$.
4. $\sqrt{e^x}$.
5. $\cos x^2$.
6. $\ln(2 + 3x)$.
7. $(1 - x^4)^{-\frac{1}{2}}$.
8. $\sqrt{4 + x}$.
9. $\ln \dfrac{1 + x}{1 - x}$.
10. $\displaystyle\int_0^x \dfrac{\sin t}{t}\,dt$.
11. $\displaystyle\int_0^x e^{-t^2}\,dt$.
12. $\displaystyle\int_0^x \sin t^2\,dt$.

In Exercises 13–18, find the terms of the Maclaurin series for the given function through terms of the specified degree.

13. $\dfrac{1}{1 + e^x}$; 5.
14. $e^x \cos x$; 5.
15. $\tanh x$; 7.
16. $e^{\tan x}$; 5.
17. $\ln \dfrac{\sin x}{x}$; 6.
18. $\ln \cos x$; 8.

In Exercises 19–22, find the Taylor series for the given function at the specified value of $x = a$.

19. $\cos x$; a.
20. $\cos x$; $\dfrac{\pi}{4}$.
21. $\ln x$; e.
22. x^m; 1.

In Exercises 23–26, find the exact sum of the infinite series.

23. $1 + 2x + 3x^2 + 4x^3 + \cdots + (n + 1)x^n + \cdots$.
24. $\dfrac{x^3}{1 \cdot 3} - \dfrac{x^5}{3 \cdot 5} + \dfrac{x^7}{5 \cdot 7} - \dfrac{x^9}{7 \cdot 9} + \cdots$.
25. $1 - 2x^2 + 3x^4 - 4x^6 + \cdots$.
26. $\dfrac{1}{1 \cdot 2 \cdot 2} - \dfrac{1}{2 \cdot 3 \cdot 2^2} + \dfrac{1}{3 \cdot 4 \cdot 2^3} - \dfrac{1}{4 \cdot 5 \cdot 2^4} + \cdots$.

27. Prove relation (16), § 1207, for $-1 < x < 0$.
*28. Check the final details of the proof of the lemma, § 1209.
*29. Give the details requested at the end of the proof of Theorem I, § 1209.
30. Prove part (ii) of Theorem II, § 1209.
*31. By reasoning analogous to that used in Part IV, § 1207, for the function $\ln(1 + x)$, show that
$$\operatorname{Arctan} x = x - \frac{x^3}{3} + \frac{x^5}{5} - \frac{x^7}{7} + \cdots,$$

for $|x| \leq 1$. In particular, derive the formula

$$\frac{\pi}{4} = 1 - \frac{1}{3} + \frac{1}{5} - \frac{1}{7} + \cdots.$$

*32. Prove that if $\quad f(x) = a_k x^k + a_{k+1} x^{k+1} + \cdots,$

where k is a nonnegative integer and $a_k \neq 0$, within some neighborhood of $x = 0$, then $f(x)$ and x^k are of the same order of magnitude as $x \to 0$. What is the order of magnitude of $1/f(x)$?

*33. Prove that if

$$f(x) = \frac{a_0 + a_1 x + a_2 x^2 + \cdots}{x^k},$$

where k is a nonnegative integer and $a_0 \neq 0$, within some deleted neighborhood of $x = 0$, then $f(x)$ and x^{-k} are of the same order of magnitude as $x \to 0$. What is the order of magnitude of $1/f(x)$?

*34. Prove that e is irrational. *Hint:* Assume $e = p/q$, where p and q are positive integers, and write $e = S + R$, where

$$S = 1 + \frac{1}{1!} + \frac{1}{2!} + \cdots + \frac{1}{q!},$$

$$R = \frac{1}{(q+1)!} + \frac{1}{(q+2)!} + \cdots,$$

and show that the integer $eq!$ has the form $Sq! + Rq!$, where $Sq!$ is an integer and $Rq!$ is not.

*35. Prove the facts given in Part V, § 1207, regarding convergence at the endpoints of the interval of convergence of the binomial series. *Hints:* Consider only values of $n > m$. Then $\left|\dfrac{a_n}{a_{n+1}}\right| = \dfrac{n+1}{n-m}$. If $m > 0$ use Raabe's test. If $m \leq -1$, $\left|\dfrac{a_n}{a_{n+1}}\right| \leq 1$ and $a_n \not\to 0$. For $-1 < m < 0$ and $x = -1$, use Raabe's test, the terms being ultimately of one sign. For $-1 < m < 0$ and $x = 1$, the series is alternating, $|a_{n+1}| < |a_n|$, and the problem is to show that $a_n \to 0$. To do this, let k be a positive integer greater than $1/(m+1)$, and show that $a_n^k \to 0$, as $n \to +\infty$, by establishing the convergence of Σa_n^k with the aid of Raabe's test.

*36. Prove that $\lim\limits_{n \to +\infty} n \binom{m}{n} = 0$, for $m > 0$. *Hint:* Use the method suggested in the hints of Exercise 35, k being a positive integer $> 1/m$.

*37. Prove Theorem I, § 1210, by using limits superior and the formula of Note 4, § 1201.

*38. The numbers B_n in the Maclaurin series

(1) $$\frac{x}{e^x - 1} = \sum_{n=0}^{+\infty} \frac{B_n x^n}{n!},$$

called **Bernoulli numbers**, play an important role in the theory of infinite series (for instance, the coefficients in the Maclaurin series for $\tan x$ are expressible in terms of the Bernoulli numbers). (Cf. Ex. 39.) By equating the function x and the product of the series (1) and the Maclaurin series for $e^x - 1$, obtain the recursive formula ($n \geq 2$):

(2) $$\binom{n}{0} B_0 + \binom{n}{1} B_1 + \binom{n}{2} B_2 + \cdots + \binom{n}{n-1} B_{n-1} = 0.$$

Evaluate $B_0, B_1, \cdots, B_{10}$. Prove that $B_{2n+1} = 0$ for $n = 1, 2, \cdots$. *Hint:* What kind of a function is $\dfrac{x}{e^x - 1} + \dfrac{1}{2}x$?

★39. Establish the series expansions, for x near 0 (cf. Ex. 38):

$$x \coth x = \sum_{n=0}^{+\infty} \frac{B_{2n}}{(2n)!}(2x)^{2n}, \qquad x \cot x = \sum_{n=0}^{+\infty} \frac{(-1)^n B_{2n}}{(2n)!}(2x)^{2n},$$

$$\tanh x = \sum_{n=1}^{+\infty} \frac{B_{2n}}{(2n)!} 2^{2n}(2^{2n} - 1)x^{2n-1}, \qquad \tan x = \sum_{n=1}^{+\infty} \frac{(-1)^{n+1} B_{2n}}{(2n)!} 2^{2n}(2^{2n} - 1)x^{2n-1},$$

$$x \csch x = \sum_{n=0}^{+\infty} \frac{B_{2n}}{(2n)!} (2 - 2^{2n})x^{2n}, \qquad x \csc x = \sum_{n=0}^{+\infty} \frac{(-1)^n B_{2n}}{(2n)!} (2 - 2^{2n})x^{2n}.$$

Hints: $x \coth x = x + 2x(e^{2x} - 1)^{-1}$, $\tanh x = 2 \coth 2x - \coth x$, $\csch x = \frac{1}{2}\coth \frac{1}{2}x - \frac{1}{2}\tanh \frac{1}{2}x$, $\tan x = \cot x - 2 \cot 2x$, and $\csc x = \frac{1}{2}\tan \frac{1}{2}x + \frac{1}{2}\cot \frac{1}{2}x$. Exploit the relationship between the Maclaurin series for $\sinh x$ and $\sin x$, and for $\cosh x$ and $\cos x$.

1212. INDETERMINATE EXPRESSIONS

It is frequently possible to find a simple evaluation of an indeterminate expression by means of Maclaurin or Taylor series. This can often be expedited by the "big O" or "order of magnitude" concepts.

Example 1. Find $\lim_{x \to 0} \dfrac{1 - \cos x}{x^2}$.

Solution. Since

$$\cos x = 1 - \frac{x^2}{2!} + \frac{x^4}{4!} - \cdots = 1 - \frac{x^2}{2} + O(x^4), \qquad \frac{1 - \cos x}{x^2} = \frac{1}{2} + O(x^2) \to \frac{1}{2}.$$

Example 2. Find $\lim_{n \to +\infty} n\{\ln (n + 1) - \ln n\}$.

Solution. We write the expression

$$\ln (n + 1) - \ln n = \ln \left(1 + \frac{1}{n}\right) = \frac{1}{n} + O\left(\frac{1}{n^2}\right).$$

Therefore $n\{\ln (n + 1) - \ln n\} = 1 + O\left(\dfrac{1}{n}\right) \to 1$.

Example 3. Find $\lim_{x \to 0} \dfrac{\ln (1 + x)}{e^{2x} - 1}$.

Solution. $\dfrac{\ln (1 + x)}{e^{2x} - 1} = \dfrac{x + O(x^2)}{2x + O(x^2)} = \dfrac{1 + O(x)}{2 + O(x)} \to \frac{1}{2}$.

1213. COMPUTATIONS

The principal techniques most commonly used for computations by means of power series have already been established. In any such computation one wishes to obtain some sort of definite range within which the sum of a series must lie. The usual tools for this are (*i*) an estimate provided by a dominating series (Theorem I, § 1120), (*ii*) an estimate provided by the integral test (Theorems II and III, § 1120), *(iii)* the alternating series estimate (Theorem, § 1115), and *(iv)* some form of the remainder in Taylor's Formula (§ 1204).

Another device is to seek a different series to represent a given quantity. For instance, some logarithms can be computed more efficiently with the series

(1) $$\ln \frac{1+x}{1-x} = 2\left[x + \frac{x^3}{3} + \frac{x^5}{5} + \frac{x^7}{7} + \cdots\right]$$

(cf. Ex. 9, § 1211) than with the Maclaurin series for $\ln(1+x)$.

Finally, one must not forget that infinite series are not the only means for computation. We include in Example 5 one illustration of the use of an approximation (Simpson's Rule) to a definite integral.

We illustrate some of these techniques in the following examples:

Example 1. Compute the sine of one radian to five decimal places.

Solution. $\sin 1 = 1 - \frac{1}{3!} + \frac{1}{5!} - \frac{1}{7!} = 0.84147$ with an error of less than $1/9! < 0.000003$.

Example. 2. Compute $\sqrt{89}$ to six decimal places.

Solution. We approximate first by 9.4, whose square is 88.36, and write
$$\sqrt{89} = \sqrt{88.36 + .64} = 9.4\sqrt{1 + 0.64/88.36}.$$

Computing $(1+x)^{\frac{1}{2}}$, where $x = 0.64/88.36$, we have, from the binomial series,
$$1 + \frac{1}{2}x - \frac{1 \cdot 1}{2 \cdot 4}x^2 = 1.00361991,$$

with an error of $R_3(x) = \frac{1 \cdot 1 \cdot 3}{2 \cdot 4 \cdot 6}\xi^3$, from (13), § 1207. This error is between 0 and 0.000000003, so that $\sqrt{1+x}$ is between 1.00361991 and 1.00361992, and $\sqrt{89}$ is between 9.4340271 and 9.4340273. To six decimal places, $\sqrt{89} = 9.434027$.

Example 3. Compute e to five decimal places.

Solution. Substitution of $x=1$ in the Maclaurin series for e^x gives
$$e = 1 + \frac{1}{2!} + \frac{1}{3!} + \cdots + \frac{1}{9!} + R_{10}(1), \quad = 2.718281 + \frac{e^\xi}{10!}$$

where $R_{10}(1)$ is between 0 and $\frac{e}{10!} < 0.000001$. Therefore e lies between 2.718281 and 2.718282, and is equal to 2.71828, to five decimal places.

Example 4. Compute $\ln 2$ by use of series (1), to five decimal places.

Solution. Solving $\frac{1+x}{1-x} = 2$ for x, we have $x = \frac{1}{3}$, and
$$\ln 2 = 2[\tfrac{1}{3} + \tfrac{1}{3}(\tfrac{1}{3})^3 + \tfrac{1}{5}(\tfrac{1}{3})^5 + \cdots] = 0.6931468 + [\tfrac{2}{13}(\tfrac{1}{3})^{13} + \cdots].$$

The remainder, in brackets, by the fact that it is dominated by a convergent geometric series with ratio $1/9$, is less than $[\tfrac{2}{13}(\tfrac{1}{3})^{13}] \div [1 - \tfrac{1}{9}]$ (cf. Theorem I, § 1120). This quantity, in turn, is less than 0.0000001. Therefore $\ln 2$ is between

0.693146 and 0.693148. To five places, ln 2 = 0.69315. (Cf. Example 3, § 1120.)

Example 5. Compute π to five significant digits.

First Solution. One method is to evaluate $\frac{\pi}{4}$ by use of the series for Arctan x (cf. Exs. 3 and 31, § 1211) with $x = 1$:

$$\frac{\pi}{4} = 1 - \frac{1}{3} + \frac{1}{5} - \frac{1}{7} + \cdots.$$

Although this series converges very slowly, its terms can be combined in the manner of Example 3, § 1120, to produce a manageable series for computation. However, if use is made of such a trigonometric identity as

$$\text{Arctan } 1 = \text{Arctan } \tfrac{1}{2} + \text{Arctan } \tfrac{1}{3},$$

two much more rapidly converging series are obtained:

$$\frac{\pi}{4} = \left[\frac{1}{2} - \frac{1}{3}\left(\frac{1}{2}\right)^3 + \frac{1}{5}\left(\frac{1}{2}\right)^5 - \cdots\right] + \left[\frac{1}{3} - \frac{1}{3}\left(\frac{1}{3}\right)^3 + \frac{1}{5}\left(\frac{1}{3}\right)^5 - \cdots\right].$$

(Complete the details in Ex. 21, § 1214.)

The student may be interested in finding other trigonometric identities which give even more rapidly converging series. Two such identities are Arctan $1 = 2$ Arctan $\tfrac{1}{3}$ + Arctan $\tfrac{1}{7}$ and Arctan $1 = 4$ Arctan $\tfrac{1}{5}$ − Arctan $\tfrac{1}{239}$.

Second Solution. Since

$$\frac{\pi}{4} = \int_0^1 \frac{dx}{1+x^2} = \text{Arctan } 1,$$

an estimate of π is given by use of Simpson's Rule (Ex. 17, § 402). For simplicity, we take $n = 10$ and complete the table:

x_k	$1 + x_k^2$	y_k	$y_k, 2y_k, 4y_k$
0	1	1.000000	1.000000
0.1	1.01	0.990099	3.960396
0.2	1.04	0.961538	1.923076
0.3	1.09	0.917431	3.669724
0.4	1.16	0.862069	1.724138
0.5	1.25	0.800000	3.200000
0.6	1.36	0.735294	1.470588
0.7	1.49	0.671141	2.684564
0.8	1.64	0.609756	1.219512
0.9	1.81	0.552486	2.209944
1.0	2.00	0.500000	0.500000
			23.561942

Therefore $\dfrac{\pi}{4} = \dfrac{23.561942}{30} = 0.785398$ (rounding off to six places to allow for previous round-off errors). Since the fourth derivative of $(1 + x^2)^{-1}$ is numerically less than 24 for $0 < x < 1$ (cf. Ex. 22, § 1409), the error estimate is less than $(24/180)(.1)^4 < 0.000013$. Thus, to four decimal places, $\pi = 3.1416$.

1214. EXERCISES

In Exercises 1–10, evaluate the limit by use of Maclaurin or Taylor series.

1. $\lim\limits_{x\to 0} \dfrac{1-\cos x}{\tan^2 x}$.

2. $\lim\limits_{x\to 0} \dfrac{\sin x - x}{\tan x - x}$.

3. $\lim\limits_{x\to 0} \dfrac{\operatorname{Arcsin} x - x - \dfrac{x^3}{6}}{\operatorname{Arctan} x - x + \dfrac{x^3}{3}}$.

4. $\lim\limits_{x\to 0} \dfrac{2 - x - 2\sqrt{1-x}}{x^2}$.

5. $\lim\limits_{x\to 0}\left[\dfrac{\sin x}{x^7} - \dfrac{1}{x^6} + \dfrac{1}{6x^4} - \dfrac{1}{120x^2}\right]$.

6. $\lim\limits_{x\to 1} \dfrac{(x-1)\ln x}{\sin^2(x-1)}$.

7. $\lim\limits_{x\to 0} \dfrac{(1+x)^{\frac{3}{2}} - (1-x)^{\frac{3}{2}}}{x}$.

8. $\lim\limits_{x\to 0} \dfrac{x\tan x}{\sqrt{1-x^2}-1}$.

9. $\lim\limits_{x\to 1} \dfrac{e^{2x} - e^2}{\ln x}$.

10. $\lim\limits_{n\to+\infty} \ln n\left[(n+1)\ln\dfrac{n+1}{n} - 1\right]$.

In Exercises 11–20, compute the given quantity to the specified number of decimal places, by use of Maclaurin series.

11. e^2, 5.
12. $\sqrt{e}$, 5.
13. $\ln 3$, 4.
14. $\cos\tfrac{1}{2}$, 4.
15. $\sqrt[3]{10}$, 5.
16. $\tan 0.1$, 5.
17. $\displaystyle\int_0^{0.5} \dfrac{\sin t}{t}\,dt$, 4.
18. $\displaystyle\int_0^{0.1} \dfrac{\ln(1+x)}{x}\,dx$, 4.
19. $\displaystyle\int_0^{0.5} \sqrt{1-x^3}\,dx$, 4.
20. $\displaystyle\int_0^1 e^{-x^2}\,dx$, 4.

21. Complete the computation details in the first solution for Example 5, § 1213.

1215. TAYLOR SERIES, SEVERAL VARIABLES

As with a function of one variable, it is possible that a function of two or more variables has an infinite series expansion. Assuming that the function $f(x, y)$ has continuous partial derivatives of all orders, the form of this expansion, called the **Taylor series of the function** and obtained from Taylor's Formula with a Remainder (formula (8), § 912), is

$$(1) \quad f(a+h, b+k) = f(a,b) + \sum_{i=1}^{+\infty} \dfrac{1}{i!}\left[\left(h\dfrac{\partial}{\partial x} + k\dfrac{\partial}{\partial y}\right)^i f(x,y)\right]_{\substack{x=a\\y=b}},$$

where h and k represent the increments $x - a$ and $y - b$, respectively. The condition that the series on the right in (1) converges to the function $f(x, y)$ is that the remainder after n terms,

$$(2) \quad R_n = \left[\left(h\dfrac{\partial}{\partial x} + k\dfrac{\partial}{\partial y}\right)^n f(x,y)\right]_{\substack{x=a+\theta h\\y=b+\theta k}}, \quad 0 < \theta < 1,$$

is an infinitesimal, as a function of n.

A similar formulation holds for a function of more than two variables.

In practice, most Taylor series are most easily obtained by combining Taylor series of functions of a single variable.

Example 1. Find the Taylor series for the function $f(x, y) = e^x \sin y$, through terms of degree 3, for $a = b = 0$.

First Solution. Evaluate:

$$f(0, 0) = e^0 \sin 0 = 0,$$

$$\left[\left(h\frac{\partial}{\partial x} + k\frac{\partial}{\partial y}\right) f(x, y)\right]_{\substack{x=0 \\ y=0}} = h \cdot 0 + k \cdot 1 = k = y - b = y,$$

$$\frac{1}{2}\left[\left(h\frac{\partial}{\partial x} + k\frac{\partial}{\partial y}\right)^2 f(x, y)\right]_{\substack{x=0 \\ y=0}} = \frac{1}{2}[h^2 \cdot 0 + 2hk \cdot 1 + k^2 \cdot 0] = hk = xy,$$

$$\frac{1}{6}\left[\left(h\frac{\partial}{\partial x} + k\frac{\partial}{\partial y}\right)^3 f(x, y)\right]_{\substack{x=0 \\ y=0}} = \frac{1}{6}[3h^2 k - k^3] = \frac{1}{6}(3x^2 y - y^3).$$

Therefore

$$e^x \sin y = y + xy + \tfrac{1}{6}(3x^2 y - y^3) + \cdots.$$

Second Solution. Form the product:

$$\left(1 + x + \frac{x^2}{2} + \cdots\right)\left(y - \frac{y^3}{6} + \cdots\right) = y + xy + \frac{1}{2}x^2 y - \frac{1}{6}y^3 + \cdots.$$

The second solution shows that the Taylor series represents the given function for all real x and y.

Example 2. Find the Taylor series for the function $f(x, y) = \dfrac{1}{xy}$, through terms of degree 2, for $a = 2, b = -1$.

First Solution. Evaluate:

$$f(a, b) = \frac{1}{2(-1)} = -\frac{1}{2},$$

$$\left[\left(h\frac{\partial}{\partial x} + k\frac{\partial}{\partial y}\right) f(x, y)\right]_{\substack{x=a \\ y=b}} = -\frac{h}{2^2(-1)} - \frac{k}{2(-1)^2} = \frac{1}{4}(x-2) - \frac{1}{2}(y+1),$$

$$\frac{1}{2}\left[\left(h\frac{\partial}{\partial x} + k\frac{\partial}{\partial y}\right)^2 f(x, y)\right]_{\substack{x=a \\ y=b}} = \frac{1}{2}\left[\frac{2h^2}{2^3(-1)} + \frac{2hk}{2^2(-1)^2} + \frac{2k^2}{2(-1)^3}\right]$$

$$= -\tfrac{1}{8}(x-2)^2 + \tfrac{1}{4}(x-2)(y+1) - \tfrac{1}{2}(y+1)^2.$$

Therefore

$$\frac{1}{xy} = -\frac{1}{2} + \frac{x-2}{4} - \frac{y+1}{2} - \frac{(x-2)^2}{8} + \frac{(x-2)(y+1)}{4} - \frac{(y+1)^2}{2} + \cdots.$$

Second Solution. Form the product of the two series:

$$\frac{1}{x} = \frac{1}{a+h} = \frac{1}{a}\cdot\frac{1}{1+\dfrac{h}{a}} = \frac{1}{2}\left(1 - \frac{x-2}{2} + \frac{(x-2)^2}{4} - \cdots\right),$$

$$\frac{1}{y} = \frac{1}{b+k} = -\frac{1}{1-k} = -(1 + (y+1) + (y+1)^2 + \cdots).$$

The second solution shows that the Taylor series represents the given function for $|x - 2| < 2$ and $|y + 1| < 1$.

1216. EXERCISES

1. Find the Taylor series for the function $f(x, y) \equiv e^{x+y}$, through terms of degree 3, for $a = 0$, $b = 0$, using three methods: by formula (1), § 1215, by expanding e^r and substituting $r = x + y$, and by multiplying two series.

2. Find the Taylor series for the function $f(x, y) \equiv \ln(xy)$, through terms of degree 2, for $a = 1$, $b = 1$, using two methods.

3. Verify the quadratic approximation formulas for values of x and y near 0:

(a) $e^x \ln(1 + y) = y + xy - \dfrac{y^2}{2}$;

(b) $\cos x \cos y = 1 - \tfrac{1}{2}(x^2 + y^2)$.

4. Show that the approximation to an increment of a function given by a differential is the same as that given by the Taylor series of the function through the terms of the first degree. (Cf. Ex. 3, § 1206.)

13

*Uniform Convergence and Limits

*1301. UNIFORM CONVERGENCE OF SEQUENCES

Let
(1) $$S_1(x), S_2(x), \cdots, S_n(x), \cdots$$
be a sequence of functions defined on a set A. We say that this sequence of functions **converges on** A in case, for every fixed x of A, the sequence of

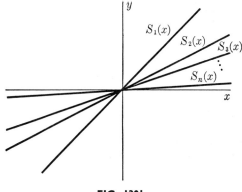

FIG. 1301

constants $\{S_n(x)\}$ converges. Assume that (1) converges on A, and define
(2) $$S(x) \equiv \lim_{n \to +\infty} S_n(x).$$
Then the rapidity with which $S_n(x)$ approaches $S(x)$ can be expected to depend (rather heavily) on the value of x.

Let us write down explicitly the analytic formulation of (2):

Corresponding to any point x in A and any $\epsilon > 0$, there exists a number $N = N(x, \epsilon)$ (dependent on both x and ϵ) such that $n > N$ implies
$$|S_n(x) - S(x)| < \epsilon.$$

Example 1. If $S_n(x) \equiv \dfrac{x}{n}$, and if $A = (-\infty, +\infty)$, then $\lim\limits_{n \to +\infty} S_n(x)$ exists and is equal to 0 for every x in A. (Cf. Fig. 1301.) If $\epsilon > 0$, the inequality

$|S_n(x) - S(x)| < \epsilon$ is equivalent to $n > |x|/\epsilon$. Therefore $N = N(x, \epsilon)$ can be defined to be $N(x, \epsilon) \equiv |x|/\epsilon$. We can see how N must depend on x. For instance, if we were asked, "How large must n be in order that $|S_n(x) - S(x)| = \dfrac{|x|}{n} <$ 0.001 ?" we should be entitled to reply, "Tell us first how large x is." If $x = 0.001$, the second term of the sequence provides the desired degree of approximation; if $x = 1$, we must proceed at least past the 1000th term; and if $x = 1000$, we must choose $n > 1{,}000{,}000$.

If it is possible to find N, in the definition of (2), as a function of ϵ alone, independent of x, then a particularly powerful type of convergence occurs.

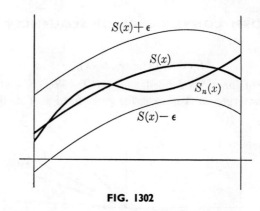

FIG. 1302

This is prescribed in the definition:

Definition. *A sequence of functions $\{S_n(x)\}$, defined on a set A,* **converges uniformly on A** *to a function $S(x)$ defined on A, with the notation*

$$S_n(x) \xrightarrow{\rightarrow} S(x),$$

if and only if corresponding to $\epsilon > 0$ there exists a number $N = N(\epsilon)$, dependent on ϵ alone and not on the point x, such that $n > N$ implies

$$|S_n(x) - S(x)| < \epsilon$$

for every x in A. (Fig. 1302.)

Let us contrast the definitions of *convergence* and *uniform convergence*. (Cf. § 224.) The most obvious distinction is that convergence is defined *at a point*, whereas uniform convergence is defined *on a set*. These concepts are also distinguished by the order in which things happen. In the case of convergence we have (*i*) the point x, (*ii*) the number $\epsilon > 0$, and (*iii*) the number N, which depends on both x and ϵ. In the case of uniform convergence we have (*i*) the number $\epsilon > 0$, (*ii*) the number N, which depends only on ϵ, and (*iii*) the point x.

§ 1301]   UNIFORM CONVERGENCE AND LIMITS

Example 2. If $S_n(x) = \dfrac{x}{n}$, and if A is the interval $(-1000, 1000)$, then the sequence $\{S_n(x)\}$ *converges uniformly to 0 on the set A.* The function $N(\epsilon)$ can be chosen: $N(\epsilon) \equiv 1000/\epsilon$. Then

$$n > N(\epsilon) \quad \text{and} \quad |x| \leq 1000$$

imply

$$|S_n(x) - S(x)| = \frac{|x|}{n} < \frac{1000}{1000/\epsilon} = \epsilon.$$

A quick and superficial reading of Example 1 might lead one to believe, since N seems to depend so thoroughly on x, that even when the set A is restricted as it is in Example 2 the convergence could not be uniform. We have seen, however, that it is. Are we really *sure*, now, that the convergence in Example 1 fails to be uniform? In order to show this, we formulate the negation of uniform convergence (give the proof in Ex. 25, § 1304):

Negation of Uniform Convergence. *A sequence of functions $\{S_n(x)\}$, defined on a set A, fails to converge uniformly on A to a function $S(x)$ defined on A if and only if there exists a positive number ϵ having the property that for any number N there exist a positive integer $n > N$ and a point x of A such that $|S_n(x) - S(x)| \geq \epsilon$.*

Example 3. Show that the convergence of Example 1 is not uniform on the interval $(-\infty, +\infty)$.

Solution. We can take $\epsilon = 1$. If N is an arbitrary number, let us choose any $n > N$, and hold this n fixed. Next we pick $x > n$. Then (for this pair n and x),

$$S_n(x) - S(x) = \frac{x}{n} > \frac{n}{n} = 1 = \epsilon.$$

NOTE. Any convergent sequence of constants (constant functions) converges uniformly on any set.

Example 4. Show that the sequence $\left\{\dfrac{x}{e^{nx}}\right\}$ converges uniformly on $[0, +\infty)$.

Solution. For a fixed value of n, the nonnegative function $f(x) = xe^{-nx}$ has a maximum value on $[0, +\infty)$ given by $x = 1/n$, and equal to $1/ne$. Therefore the sequence approaches 0 uniformly on $[0, +\infty)$.

Example 5. Show that the sequence $\left\{\dfrac{nx}{e^{nx}}\right\}$ converges uniformly on $[\alpha, +\infty)$, for any $\alpha > 0$, but not uniformly on $(0, +\infty)$.

Solution. For $x \geq \alpha$, and a fixed $n > 1/\epsilon$, the function $f(x) = nxe^{-nx}$ is monotonically decreasing, with a maximum value of $n\alpha e^{-n\alpha}$. This quantity is independent of x and approaches 0 as $n \to +\infty$. On the entire interval $(0, +\infty)$, the function $f(x) = nxe^{-nx}$ has a maximum value given by $x = 1/n$ and equal to $1/e$. The convergence is therefore not uniform.

*1302. UNIFORM CONVERGENCE OF SERIES

Let
(1) $$u_1(x) + u_2(x) + \cdots + u_n(x) + \cdots$$
be a series of functions defined on a set A, and let
$$S_n(x) \equiv u_1(x) + \cdots + u_n(x).$$
We say that this series of functions **converges on** A in case the sequence $\{S_n(x)\}$ converges on A. The series (1) **converges uniformly on** A if and only if the sequence $\{S_n(x)\}$ converges uniformly on A.

NOTE. Any convergent series of constants (constant functions) converges uniformly on any set.

Corresponding to the condition $a_n \to 0$ which is necessary for the convergence of the series of constants Σa_n, we have:

Theorem. *If the series $\Sigma u_n(x)$ converges uniformly on a set A, then the general term $u_n(x)$ converges to 0 uniformly on A.*

Proof. By the triangle inequality, if $S_n(x) \equiv u_1(x) + \cdots + u_n(x)$ and $S(x) \equiv \sum_{n=1}^{+\infty} u_n(x)$,
$$|u_n(x)| = |S_n(x) - S_{n-1}(x)| = |[S_n(x) - S(x)] + [S(x) - S_{n-1}(x)]|$$
$$\leq |S_n(x) - S(x)| + |S_{n-1}(x) - S(x)|.$$
Let $\epsilon > 0$ be given. If N is chosen such that $n > N - 1$ implies
$$|S_n(x) - S(x)| < \tfrac{1}{2}\epsilon$$
for all x in A, then $n > N$ implies $|u_n(x)| < \epsilon$ for all x in A.

Example. Show that the Maclaurin series for e^x converges uniformly on a set A if and only if A is bounded.

Solution. If the set A is bounded, it is contained in some interval of the form $[-\alpha, \alpha]$. Using the Lagrange form of the remainder in Taylor's formula for e^x at $x = a = 0$, we have (with the standard notation) for any x,
$$|S_n(x) - S(x)| = |R_n(x)| = \frac{e^{\xi_n}}{n!}|x|^n \leq \frac{e^{\alpha}}{n!}\alpha^n.$$
Since $\lim_{n \to +\infty} \dfrac{e^{\alpha}}{n!}\alpha^n = 0$ and $\dfrac{e^{\alpha}}{n!}\alpha^n$ is independent of x, the uniform convergence on A is established.

If the set A is unbounded, we can show that $\sum_{n=0}^{+\infty} \dfrac{x^n}{n!}$ fails to converge uniformly on A by showing that the general term does not approach 0 uniformly on A. This we do with the aid of the Negation of Uniform Convergence formulated in § 1301: letting ϵ be 1 and n be any fixed positive integer, we can find an x in A such that $|x|^n > n!$

*1303. DOMINANCE AND THE WEIERSTRASS M-TEST

The role of dominance in uniform convergence of series of functions is similar to that of dominance in convergence of series of constants (§ 1108).

Definition. *The statement that a series of functions $\Sigma v_n(x)$ **dominates** a series of functions $\Sigma u_n(x)$ on a set A means that all terms are defined on A and that for any x in A $|u_n(x)| \leq v_n(x)$ for every positive integer n.*

Theorem I. Comparison Test. *Any series of functions $\Sigma u_n(x)$ dominated on a set A by a series of functions $\Sigma v_n(x)$ that is uniformly convergent on A is uniformly convergent on A.*

Proof. From previous results for series of constants, we know that the series $\Sigma u_n(x)$ converges for every x in A. If $u(x) \equiv \Sigma u_n(x)$ and $v(x) \equiv \Sigma v_n(x)$, we have (cf. Theorem I, § 1116):

$$|[u_1(x) + u_2(x) + \cdots + u_n(x)] - u(x)| = |u_{n+1}(x) + u_{n+2}(x) + \cdots|$$
$$\leq v_{n+1}(x) + v_{n+2}(x) + \cdots = |[v_1(x) + v_2(x) + \cdots + v_n(x)] - v(x)|.$$

If $\epsilon > 0$ and if $N = N(\epsilon)$ is such that for $n > N$,

$$|[v_1(x) + \cdots + v_n(x)] - v(x)| < \epsilon$$

for all x in A, then $|[u_1(x) + \cdots + u_n(x)] - u(x)| < \epsilon$ for all x in A.

Corollary. *A series of functions converges uniformly on a set whenever its series of absolute values converges uniformly on that set.*

Example 1. Since the Maclaurin series for $\sin x$ and $\cos x$ are dominated on any set by the series obtained by substituting $|x|$ for x in the Maclaurin series for e^x, these series for $\sin x$ and $\cos x$ converge uniformly on any bounded set. Neither converges uniformly on an unbounded set. (Cf. the Example, § 1302.)

Since any convergent series of constants converges uniformly on any set, we have as a special case of Theorem I the extremely useful test for uniform convergence due to Weierstrass:

Theorem II. Weierstrass M-Test. *If $\Sigma u_n(x)$ is a series of functions defined on a set A, if ΣM_n is a convergent series of nonnegative constants, and if for every x of A,*

(1) $$|u_n(x)| \leq M_n, n = 1, 2, \cdots,$$

then $\Sigma u_n(x)$ converges uniformly on A.

Example 2. Prove that the series
$$1 + e^{-x} \cos x + e^{-2x} \cos 2x + \cdots$$
converges uniformly on any set that is bounded below by a positive constant.

Solution. If $\alpha > 0$ is a lower bound of the set A, then for any x in A, $|e^{-nx} \cos nx| \leq e^{-nx} \leq e^{-n\alpha}$. By the Weierstrass M-test, with $M_n \equiv e^{-n\alpha}$, the given series converges uniformly on A (the series $\Sigma e^{-n\alpha}$ is a geometric series with common ratio $e^{-\alpha} < 1$). (Cf. Ex. 26, § 1304.)

*1304. EXERCISES

In Exercises 1–10, use the Weierstrass M-test to show that the given series converges uniformly on the given set.

*1. $\sum n^2 x^n$; $[-\frac{1}{2}, \frac{1}{2}]$.

*2. $\sum \dfrac{x^n}{n^2}$; $[-1, 1]$.

*3. $\sum \dfrac{x^{2n-1}}{(2n-1)!}$; $(-1000, 2000)$.

*4. $\sum \dfrac{x^n}{n^n}$; $x = 1, 2, \cdots, K$.

*5. $\sum \dfrac{\sin nx}{n^2 + 1}$; $(-\infty, +\infty)$.

*6. $\sum \dfrac{\sin nx}{e^n}$; $(-\infty, +\infty)$.

*7. $\sum n e^{-nx}$; $[\alpha, +\infty)$, $\alpha > 0$.

*8. $\sum \dfrac{e^{nx}}{5^n}$; $(-\infty, \alpha]$, $\alpha < \ln 5$.

*9. $\sum \left(\dfrac{\ln x}{x}\right)^n$; $[1, +\infty)$.

*10. $\sum (x \ln x)^n$; $(0, 1]$.

In Exercises 11–20, show that the sequence whose general term is given converges uniformly on the first of the two given intervals, and that it converges but not uniformly on the second of the two given intervals. The letter η denotes an arbitrarily small positive number.

*11. $\sin^n x$; $[0, \frac{1}{2}\pi - \eta]$; $[0, \frac{1}{2}\pi)$.

*12. $\sqrt[n]{\sin x}$; $[\eta, \frac{1}{2}\pi]$; $(0, \frac{1}{2}\pi]$.

*13. $\dfrac{x}{x+n}$; $[0, b]$; $[0, +\infty)$.

*14. $\dfrac{x+n}{n}$; $[a, b]$; $(-\infty, +\infty)$.

*15. $\dfrac{nx}{1 + nx}$; $[\eta, +\infty)$; $(0, +\infty)$.

*16. $\dfrac{nx}{1 + n^2 x^2}$; $[\eta, +\infty)$; $(0, +\infty)$.

*17. $\dfrac{\ln(1 + nx)}{n}$; $[0, b]$; $[0, +\infty)$.

*18. $n^2 x^2 e^{-nx}$; $[\eta, +\infty)$; $(0, +\infty)$.

*19. $\dfrac{x^n}{1 + x^n}$; $[0, 1 - \eta]$; $[0, 1)$.

*20. $\dfrac{\sin nx}{1 + nx}$; $[\eta, +\infty)$; $(0, +\infty)$.

In Exercises 21–24, show that the given series converges uniformly on the first of the two given intervals, and that it converges but not uniformly on the second of the two given intervals. The letter η denotes an arbitrarily small positive number.

*21. $\sum x^n$; $[0, 1 - \eta]$; $[0, 1)$.

*22. $\sum \dfrac{x^n}{n}$; $[0, 1 - \eta]$; $[0, 1)$.

*23. $\sum \dfrac{1}{n^x}$; $[1 + \eta, +\infty)$; $(1, +\infty)$.

*24. $\sum \dfrac{nx}{e^{nx}}$; $[\eta, +\infty)$; $(0, +\infty)$.

*25. Prove the Negation of Uniform Convergence, § 1301.

*26. Prove that the convergence of the series in Example 2, § 1303, is not uniform on $(0, +\infty)$.

*27. Prove that if a sequence of functions converges uniformly on a set, then any subsequence converges uniformly on that set. Show by an example that the converse is false; that is, show that a sequence which converges nonuniformly on a set may have a uniformly convergent subsequence. (Cf. Ex. 28.)

*28. Prove that if a monotonic sequence of functions (e.g., $S_{n+1}(x) \leq S_n(x)$ for each x) converges on a set and contains a uniformly convergent subsequence, then the convergence of the original sequence is uniform. (Cf. Ex. 27.)

★29. Prove that if a sequence converges uniformly on a set A, then it converges uniformly on any set contained in A.

★30. Prove that if a sequence converges uniformly on a set A and on a set B, then it converges uniformly on their union, $A \cup B$. Show that any convergent sequence converges uniformly on every finite set. Show that if a sequence converges uniformly on an open interval (a, b) and converges at each endpoint, then the sequence converges uniformly on the closed interval $[a, b]$.

★31. Prove that if $S_n(x) \rightrightarrows S(x)$ on a set A, and if $S(x)$ is bounded on A, then the functions $S_n(x)$ are ultimately uniformly bounded on A: there exist constants K and N such that $|S_n(x)| \leq K$ for all $n > N$ and all x in A.

★32. Prove that if $S_n(x) \rightrightarrows S(x)$ on a set A, and if each $S_n(x)$ is bounded on A (there exists a sequence of constants $\{K_n\}$ such that $|S_n(x)| \leq K_n$ for each n and all x in A), then the functions $S_n(x)$ are uniformly bounded on A. (Cf. Ex. 31.)

★33. If $f(x)$ is defined on $[\frac{1}{2}, 1]$ and continuous at 1, prove that $\{x^n f(x)\}$ converges for every x of $[\frac{1}{2}, 1]$, and that this convergence is uniform if and only if $f(x)$ is bounded and $f(1) = 0$.

★34. If two sequences $\{f_n(x)\}$ and $\{g_n(x)\}$ converge uniformly on a set A, prove that their sum $\{f_n(x) + g_n(x)\}$ also converges uniformly on A. Show by an example that their product $\{f_n(x)g_n(x)\}$ need not converge uniformly on A. Prove, however, that if both original sequences are uniformly bounded (cf. Ex. 31), then the sequence $\{f_n(x)g_n(x)\}$ converges uniformly. What happens if only one of the sequences is uniformly bounded?

★35. Construct a series $\Sigma u_n(x)$ such that (i) $\Sigma u_n(x)$ converges on a set A, (ii) $u_n(x) \rightrightarrows 0$ on A, and (iii) $\Sigma u_n(x)$ does not converge uniformly on A.

★36. Prove the ratio test for uniform convergence: If $u_n(x)$ are bounded non-vanishing functions on a set A, and if there exist constants N and ρ, where $\rho < 1$, such that $\left|\dfrac{u_{n+1}(x)}{u_n(x)}\right| \leq \rho$ for all $n > N$ and all x in A, then $\Sigma u_n(x)$ converges uniformly on A.

In Exercises 37–42, find an appropriate function $N(\epsilon)$, as prescribed in the Definition of uniform convergence, § 1301, for the sequence of the given Exercise (and the set specified in that exercise).

★37. Exercise 11. **★38.** Exercise 13. **★39.** Exercise 15.

★40. Exercise 17. *Hint:* The inequality $\ln(1 + nb) < \epsilon n$ is equivalent to $1 + nb < e^{\epsilon n}$; and $1 + \epsilon n + \dfrac{\epsilon^2 n^2}{2} < e^{\epsilon n}$. Thus require $1 + nb < 1 + \dfrac{\epsilon^2 n^2}{2}$.

★41. Exercise 18. *Hint:* $n^2 x^2 e^{-nx}$ is a decreasing function of x for $x > \dfrac{2}{n}$. First require $n > \dfrac{2}{\eta}$. Then guarantee $n^2 \eta^2 < \epsilon e^{n\eta}$ (cf. Ex. 40).

★42. Exercise 19.

★43. Prove that the Cauchy condition for convergence of an infinite series Σa_n (Ex. 22, § 1117) can be expressed in the form: $a_n + a_{n+1} + \cdots + a_{n+p} \rightrightarrows 0$, uniformly in $p > 0$. (Cf. Exs. 23, 25, § 1117.)

★★44. Explain what you would mean by saying that a series of functions is **uniformly summable** $(C, 1)$ on a given set. Prove that a uniformly convergent series is uniformly summable, but not conversely. Generalize to uniform summability (H, r). Give examples. (Cf. Exs. 20–21, § 1117.)

★45. Prove the **Cauchy criterion** for uniform convergence of a sequence of functions: *A sequence $\{S_n(x)\}$ converges uniformly on a set A if and only if corresponding to $\epsilon > 0$ there exists $N = N(\epsilon)$ such that $m > N$ and $n > N$ imply $|S_m(x) - S_n(x)| < \epsilon$ for every x in A.* Hint: By § 218, $S(x) \equiv \lim_{n \to +\infty} S_n(x)$ exists. Assume that the convergence is not uniform, use the Negation of § 1301, N as prescribed above for $\frac{1}{2}\epsilon$, and the triangle inequality, for fixed $n > N$ and arbitrary $m > N$:

$$|S_n(x) - S(x)| \leq |S_n(x) - S_m(x)| + |S_m(x) - S(x)|.$$

★46. State and prove the Cauchy criterion for uniform convergence of a series of functions. (Cf. Ex. 45, above, Exs. 22–24, § 1117.)

★47. Prove the **Abel test** for uniform convergence: *If the partial sums of a series $\sum_{n=1}^{+\infty} u_n(x)$ of functions are uniformly bounded on a set A (Ex. 31), and if $\{v_n(x)\}$ is a monotonically decreasing sequence of nonnegative functions converging uniformly to 0 on A, then the series $\sum_{n=1}^{+\infty} u_n(x)v_n(x)$ converges uniformly on A.* (Cf. Ex. 27, § 1117.)

★48. Adapt and prove the Abel test of Exercise 28, § 1117, for uniform convergence.

★49. Prove the alternating series test for uniform convergence: *If $\{v_n(x)\}$ is a monotonically decreasing sequence of nonnegative functions converging uniformly to 0 on a set A, then $\sum_{n=1}^{+\infty} (-1)^{n-1} v_n(x)$ converges uniformly on A.* (Cf. Ex. 47.)

★1305. UNIFORM CONVERGENCE AND CONTINUITY

The example $S_n(x) \equiv x^{\frac{1}{2n-1}}$, $-1 \leq x \leq 1$ (Fig. 1303), shows that the limit of a sequence of continuous functions need not be continuous. The limit in this case is the signum function (Example 1, § 206), which is discontinuous at $x = 0$. As we shall see, this is possible because the convergence is not uniform. For example, let us set $\epsilon = \frac{1}{2}$. Then however large n may be there will exist a positive number x so small that the inequality $|S_n(x) - S(x)| < \epsilon$, which is equivalent to $1 - x^{\frac{1}{2n-1}} < \frac{1}{2}$, or $x^{\frac{1}{2n-1}} > \frac{1}{2}$, fails.

In case of uniform convergence we have the basic theorem:

Theorem. *If $S_n(x)$ is continuous at every point of a set A, $n = 1, 2, \cdots$, and if $S_n(x) \rightrightarrows S(x)$ on A, then $S(x)$ is continuous at every point of A.*

Proof. Let a be any point of A, and let $\epsilon > 0$. We first choose a positive integer N such that $|S_N(x) - S(x)| < \frac{1}{3}\epsilon$ for every x of A. Holding N fixed, and using the continuity of $S_N(x)$ at $x = a$, we can find a $\delta > 0$ such that $|x - a| < \delta$ implies $|S_N(x) - S_N(a)| < \frac{1}{3}\epsilon$. (The number δ apparently depends on *both* N and ϵ, but since N is determined by ϵ, δ is a function of ϵ alone—for the fixed value $x = a$.) Now we use the triangle inequality:

$$|S(x) - S(a)| \leq |S(x) - S_N(x)| + |S_N(x) - S_N(a)| + |S_N(a) - S(a)|.$$

§ 1306] UNIFORM CONVERGENCE AND INTEGRATION 449

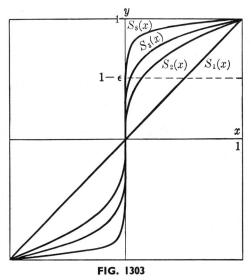

FIG. 1303

Then $|x - a| < \delta$ implies
$$|S(x) - S(a)| < \tfrac{1}{3}\epsilon + \tfrac{1}{3}\epsilon + \tfrac{1}{3}\epsilon = \epsilon,$$
and the proof is complete.

Corollary. *If $f(x) \equiv \Sigma u_n(x)$, where the series converges uniformly on a set A, and if every term of the series is continuous on A, then $f(x)$ is continuous on A.* (Ex. 19, § 1308.)

Example. Show that the function $f(x)$ defined by the series $\sum_{n=0}^{+\infty} e^{-nx} \cos nx$ of Example 2, § 1303, is continuous on the set $(0, +\infty)$—that is, for positive x.

Solution. Let $x = a > 0$ be given, and let $\alpha = \tfrac{1}{2}a$. Then the given series converges uniformly on $[\tfrac{1}{2}a, +\infty)$, by Example 2, § 1303. Therefore $f(x)$ is continuous on $[\tfrac{1}{2}a, +\infty)$ and, in particular, at $x = a$.

*1306. UNIFORM CONVERGENCE AND INTEGRATION

The example illustrated in Figure 1304 shows that the limit of the integral (of the general term of a convergent sequence of functions) need not equal the integral of the limit (function). The function $S_n(x)$ is defined to be $2n^2x$ for $0 \leq x \leq 1/2n$, $2n(1 - nx)$ for $1/2n \leq x \leq 1/n$, and 0 for $1/n \leq x \leq 1$. The limit function $S(x)$ is identically 0 for $0 \leq x \leq 1$. For every n, $\int_0^1 S_n(x)\,dx = \tfrac{1}{2}$ (the integral is the area of a triangle of altitude n and base $1/n$), but $\int_0^1 S(x)\,dx = 0$. Again, the reason that this kind of misbehavior is possible is that the convergence is not uniform (cf. Ex. 22, § 1308). (For another example of the same character, where the functions $S_n(x)$ are defined by single analytic formulas, see Ex. 25, § 1308.)

In case of uniform convergence we have the theorem:

Theorem. *If $S_n(x)$ is integrable on $[a, b]$ for $n = 1, 2, \cdots$, and if $S_n(x) \rightrightarrows S(x)$ on $[a, b]$, then $S(x)$ is integrable on $[a, b]$ and*

$$(1) \qquad \lim_{n \to +\infty} \int_a^b S_n(x)\, dx = \int_a^b \lim_{n \to +\infty} S_n(x)\, dx = \int_a^b S(x)\, dx.$$

Proof. We shall first prove that $S(x)$ is integrable on $[a, b]$. The idea is to approximate $S(x)$ by a particular $S_N(x)$, then to squeeze $S_N(x)$ between

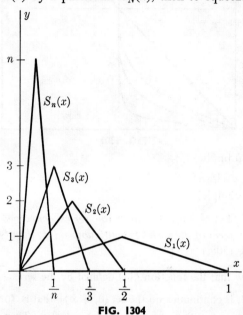

FIG. 1304

two step-functions (§ 403), and finally to construct two new step-functions that squeeze $S(x)$. Accordingly, for a given $\epsilon > 0$, we find an index N such that $|S_N(x) - S(x)| < \dfrac{\epsilon}{4(b-a)}$ for $a \leq x \leq b$. Since $S_N(x)$ is integrable on $[a, b]$, there must exist step-functions $\sigma_1(x)$ and $\tau_1(x)$ such that $\sigma_1(x) \leq S_N(x) \leq \tau_1(x)$ on $[a, b]$, and $\int_a^b [\tau_1(x) - \sigma_1(x)]\, dx < \tfrac{1}{2}\epsilon$. Define the new step-functions:

$$\sigma(x) \equiv \sigma_1(x) - \frac{\epsilon}{4(b-a)}, \quad \tau(x) \equiv \tau_1(x) + \frac{\epsilon}{4(b-a)}.$$

Then, for $a \leq x \leq b$,

$$\sigma(x) < \sigma_1(x) + [S(x) - S_N(x)] \leq S(x),$$
$$\tau(x) > \tau_1(x) + [S(x) - S_N(x)] \geq S(x),$$

and

$$\int_a^b [\tau(x) - \sigma(x)]\, dx = \int_a^b \left\{[\tau_1(x) - \sigma_1(x)] + \frac{\epsilon}{2(b-a)}\right\} dx < \tfrac{1}{2}\epsilon + \tfrac{1}{2}\epsilon = \epsilon.$$

By Theorem II, § 403, $S(x)$ is integrable on $[a, b]$.

We now wish to establish the limit (1). Since the difference between the integrals of two integrable functions is the integral of their difference, (1) is equivalent to

(2) $$\int_a^b [S_n(x) - S(x)]\, dx \to 0.$$

Finally, by Exercise 2, § 402,

$$\left| \int_a^b [S_n(x) - S(x)]\, dx \right| \leq \int_a^b |S_n(x) - S(x)|\, dx,$$

so that if n is chosen so large that $|S_n(x) - S(x)| < \epsilon/(b-a)$ on $[a, b]$, then $\left| \int_a^b [S_n(x) - S(x)]\, dx \right| < [\epsilon/(b-a)](b-a) = \epsilon$, and the proof is complete.

Corollary. *If $f(x) \equiv \Sigma u_n(x)$, where the series converges uniformly on $[a, b]$, and if every term of the series is integrable on $[a, b]$, then $f(x)$ is integrable on $[a, b]$, and the series can be integrated term by term:*

$$\int_a^b f(x)\, dx = \int_a^b \Sigma u_n(x)\, dx = \Sigma \int_a^b u_n(x)\, dx.$$

(Ex. 20, § 1308.)

Example. If $f(x)$ is the function defined by the series $\sum_{n=0}^{+\infty} e^{-nx} \cos nx$ of Example 2, § 1303, and if $0 < a < b$, then

$$\int_a^b f(x)\, dx = \sum_{n=0}^{+\infty} \int_a^b e^{-nx} \cos nx\, dx.$$

*1307. UNIFORM CONVERGENCE AND DIFFERENTIATION

The example $S_n(x) \equiv \dfrac{x}{1 + nx^2}$, $-1 \leq x \leq 1$ (Fig. 1305), shows that even with uniform convergence of differentiable functions to a differentiable function, the limit of the derivatives may not equal the derivative of the limit. Since the maximum and minimum points of $S_n(x)$ are $(n^{-\frac{1}{2}}, \frac{1}{2}n^{-\frac{1}{2}})$ and $(-n^{-\frac{1}{2}}, -\frac{1}{2}n^{-\frac{1}{2}})$, respectively, $\{S_n(x)\}$ converges uniformly to the function $S(x)$ that is identically 0 on $[-1, 1]$. However,

$$\lim_{n \to +\infty} S_n'(x) = \lim_{n \to +\infty} \frac{1 - nx^2}{(1 + nx^2)^2} = \begin{cases} 1 \text{ if } x = 0, \\ 0 \text{ if } x \neq 0, \end{cases}$$

whereas $S'(x)$ is identically 0.

The clue to the problem lies in the uniform convergence of the sequence of *derivatives*, $\{S_n'(x)\}$. We state the basic theorem, and give a proof under the additional assumption of continuity of the derivatives. For a proof of the general theorem, see *RV*, Exercise 38, § 908.

Theorem. *Assume:*
(i) *$S_n(x)$ is differentiable on $[a, b]$, for $n = 1, 2, \cdots$;*
(ii) *$\{S_n(x)\}$ converges for some point x_0 of $[a, b]$;*
(iii) *$\{S_n'(x)\}$ converges uniformly on $[a, b]$.*

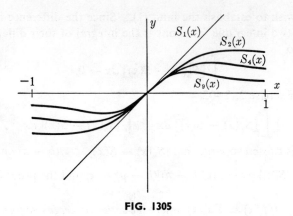

FIG. 1305

Then

(iv) $\{S_n(x)\}$ *converges uniformly on* $[a, b]$;
(v) *if* $S(x) \equiv \lim\limits_{n \to +\infty} S_n(x)$, $S'(x)$ *exists on* $[a, b]$;
(vi) $S'(x) = \lim\limits_{n \to +\infty} S_n'(x)$.

Proof if every $S_n'(x)$ *is continuous.* Under the assumption of continuity of $S_n'(x)$ we have, from the Fundamental Theorem of Integral Calculus (§ 405), $S_n(x) - S_n(x_0) = \int_{x_0}^{x} S_n'(t) \, dt$. Therefore, by the Theorem of § 1306, the sequence $\{S_n(x) - S_n(x_0)\}$ converges for every x of $[a, b]$. Therefore the limit $\lim\limits_{n \to +\infty} S_n(x)$ exists for every x of $[a, b]$ (why?), and if we define $S(x) \equiv \lim\limits_{n \to +\infty} S_n(x)$ and $T(x) \equiv \lim\limits_{n \to +\infty} S_n'(x)$, we have (§ 1306):

(1) $$S(x) = \int_{x_0}^{x} T(t) \, dt + S(x_0).$$

Since $T(x)$ is continuous on $[a, b]$ (§ 1305), the function $S(x)$ is differentiable with derivative $S'(x) = T(x)$ (Theorem I, § 405). Finally, since

$$|S_n(x) - S(x)| = \left| \int_{x_0}^{x} [S_n'(t) - S'(t)] \, dt + [S_n(x_0) - S(x_0)] \right|$$

$$\leq \left| \int_{x_0}^{x} |S_n'(t) - S'(t)| \, dt \right| + |S_n(x_0) - S(x_0)|,$$

the convergence of $\{S_n(x)\}$ is uniform (cf. Ex. 23, § 1308).

Corollary. *If* $\Sigma u_n(x)$ *is a series of differentiable functions on* $[a, b]$, *convergent at one point of* $[a, b]$, *and if the derived series* $\Sigma u_n'(x)$ *converges uniformly on* $[a, b]$, *then the original series converges uniformly on* $[a, b]$ *to a differentiable function whose derivative is represented on* $[a, b]$ *by the derived series.* (Ex. 21, § 1308.)

Example. Show that if $f(x)$ is the function defined by the series $\sum_{n=0}^{+\infty} e^{-nx} \cos nx$ of Example 2, § 1303, then

$$f'(x) = -\sum_{n=0}^{+\infty} ne^{-nx}(\cos nx + \sin nx),$$

for every $x > 0$.

Solution. Let x be a given positive number, and choose a and b such that $0 < a < x < b$. Since the original series has already been shown to converge (uniformly) on $[a, b]$, it remains only to show that the derived series converges uniformly there. This is easily done by the Weierstrass M-test, with $M_n = 2ne^{-na}$ (check the details).

*1308. EXERCISES

In Exercises 1–6, show that the convergence fails to be uniform by showing that the limit function is not continuous.

*1. $\lim_{n \to +\infty} \sin^n x$, for $0 \leq x \leq \pi$.

*2. $\lim_{n \to +\infty} e^{-nx^2}$, for $|x| \leq 1$.

*3. $\lim_{n \to +\infty} \dfrac{x^n}{1 + x^n}$, for $0 \leq x \leq 2$.

*4. $\lim_{n \to +\infty} S_n(x)$, where $S_n(x) = \dfrac{\sin nx}{nx}$ for $0 < x \leq \pi$, and $S_n(0) \equiv 1$.

*5. $(1 - x) + x(1 - x) + x^2(1 - x) + \cdots$, for $0 \leq x \leq 1$.

*6. $x^2 + \dfrac{x^2}{1 + x^2} + \dfrac{x^2}{(1 + x^2)^2} + \cdots$, for $|x| \leq 1$.

In Exercises 7–12, show that the equation is true.

*7. $\lim_{n \to +\infty} \displaystyle\int_{\frac{1}{2}\pi}^{\pi} \dfrac{\sin nx}{nx} dx = 0$.

*8. $\lim_{n \to +\infty} \displaystyle\int_{1}^{2} e^{-nx^2} dx = 0$.

*9. $\displaystyle\int_{0}^{\pi} \sum_{n=1}^{+\infty} \dfrac{\sin nx}{n^2} dx = \sum_{n=1}^{+\infty} \dfrac{2}{(2n-1)^3}$.

*10. $\displaystyle\int_{1}^{2} \sum_{n=1}^{+\infty} \dfrac{\ln nx}{n^2} dx = \sum_{n=1}^{+\infty} \dfrac{\ln 4n - 1}{n^2}$.

*11. $\displaystyle\int_{0}^{\pi} \sum_{n=1}^{+\infty} \dfrac{n \sin nx}{e^n} dx = \dfrac{2e}{e^2 - 1}$.

*12. $\displaystyle\int_{1}^{2} \sum_{n=1}^{+\infty} ne^{-nx} dx = \dfrac{e}{e^2 - 1}$.

In Exercises 13–18, show that the equation is true.

*13. $\dfrac{d}{dx} \left[\displaystyle\sum_{n=1}^{+\infty} \dfrac{x^n}{n(n+1)} \right] = \sum_{n=0}^{+\infty} \dfrac{x^n}{n+2}$, for $|x| < 1$.

*14. $\dfrac{d}{dx} \left[\displaystyle\sum_{n=1}^{+\infty} \dfrac{n}{x^n} \right] = -\sum_{n=1}^{+\infty} \dfrac{n^2}{x^{n+1}}$, for $|x| > 1$.

*15. $\dfrac{d}{dx} \left[\displaystyle\sum_{n=1}^{+\infty} \dfrac{\sin nx}{n^3} \right] = \sum_{n=1}^{+\infty} \dfrac{\cos nx}{n^2}$, for any x.

*16. $\dfrac{d}{dx} \left[\displaystyle\sum_{n=1}^{+\infty} \dfrac{\sin nx}{n^3 x} \right] = \sum_{n=1}^{+\infty} \left[\dfrac{\cos nx}{n^2 x} - \dfrac{\sin nx}{n^3 x^2} \right]$, for $x \neq 0$.

★17. $\dfrac{d}{dx}\left[\displaystyle\sum_{n=1}^{+\infty} \dfrac{1}{n^3(1+nx^2)}\right] = -2x \displaystyle\sum_{n=1}^{+\infty} \dfrac{1}{n^2(1+nx^2)^2}$, for any x.

★18. $\dfrac{d}{dx}\left[\displaystyle\sum_{n=1}^{+\infty} e^{-nx} \sin knx\right] = \displaystyle\sum_{n=1}^{+\infty} ne^{-nx}[k \cos knx - \sin knx]$, for $x > 0$.

★19. Prove the Corollary to the Theorem of § 1305.

★20. Prove the Corollary to the Theorem of § 1306.

★21. Prove the Corollary to the Theorem of § 1307.

★22. Prove that the convergence of the sequence illustrated in Figure 1304, § 1306, is not uniform.

★23. Complete the final details of the proof of § 1307.

★24. Let $S_n(x) \equiv \dfrac{1}{n} e^{-nx}$, $S(x) \equiv \lim\limits_{n \to +\infty} S_n(x)$. Show that $\lim\limits_{n \to +\infty} S_n'(0) \neq S'(0)$.

★25. Let $S_n(x) \equiv nxe^{-nx^2}$, $S(x) \equiv \lim\limits_{n \to +\infty} S_n(x)$. Show that
$$\lim_{n \to +\infty} \int_0^1 S_n(x)\, dx \neq \int_0^1 S(x)\, dx.$$

★26. Construct an example to show that it is possible to have a sequence $\{S_n(x)\}$ of continuous functions converging to a continuous function $S(x)$ nonuniformly on a closed interval $[a, b]$, but still have $\displaystyle\int_a^b S_n(x)\, dx \to \int_a^b S(x)\, dx$. *Hint:* Construct functions like those illustrated in Figure 1304, § 1306, but having uniform maximum values.

★27. Construct an example to show that it is possible to have $S_n(x) \xrightarrow{\to} S(x)$ and $S_n'(x) \to S'(x)$ nonuniformly on an interval. *Hint:* Work backwards from the result of Exercise 26, letting $\{S_n'(x)\}$ be the sequence of that exercise, and $S_n(x) \equiv \displaystyle\int_0^x S_n'(t)\, dt$.

★28. Show by an example that uniform convergence of a sequence of functions on an infinite interval is not sufficient to guarantee that the integral of the limit is the limit of the integral. (Cf. Ex. 43, § 416.) *Hint:* Consider $S_n(x) \equiv 1/n$ for $0 \leq x \leq n$, and $S_n(x) \equiv 0$ for $x > n$.

★29. Prove **Dini's Theorem:** *Any monotonic convergence of continuous functions on a closed interval (more generally, on any compact set) to a continuous function there is uniform.* *Hint:* Assume without loss of generality that for each x belonging to the closed interval A, $S_n(x)\downarrow$ and $S_n(x) \to 0$. If the convergence were *not* uniform there would exist a sequence $\{x_k\}$ of points of A such that $S_{n_k}(x_k) \geq \epsilon > 0$. Assume without loss of generality that $\{x_k\}$ converges: $x_k \to \bar{x}$. Show that for every n, $S_n(\bar{x}) \geq \epsilon$. (Alternatively, use Ex. 31, § 618.)

★30. Let $f_1(x), f_2(x), \cdots$ be a sequence of "sawtooth" functions (Fig. 1306), defined for all real x, where the graph of $f_n(x)$ is made up of line segments of slope ± 1, such that $f_n(x) = 0$ for $x = \pm m \cdot 4^{-n}$, $m = 0, 1, 2, \cdots$, and $f_n(x) = \frac{1}{2} \cdot 4^{-n}$ for $x = \frac{1}{2} \cdot 4^{-n} + m \cdot 4^{-n}$, $m = 0, 1, 2, \cdots$. Let $f(x) \equiv \displaystyle\sum_{n=1}^{+\infty} f_n(x)$. Prove that $f(x)$ is everywhere continuous and nowhere differentiable.† *Hint for nondifferentiability:* If a is any fixed point, show that for any $n = 1, 2, \cdots$ a number h_n can

† This example is modeled after one due to Van der Waerden. Cf. E. C. Titchmarsh, *Theory of Functions* (Oxford, Oxford University Press, 1932).

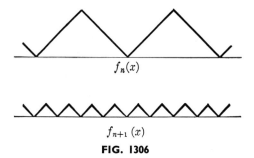

FIG. 1306

be chosen as one of the numbers 4^{-n-1} or -4^{-n-1} such that $f_n(a+h_n)-f_n(a)=\pm h_n$. Then $f_m(a+h_n)-f_m(a)$ has the value $\pm h_n$ for $m \leq n$, and otherwise vanishes. Hence the difference quotient $[f(a+h_n)-f(a)]/h_n$ is an integer of the same parity as n (even if n is even and odd if n is odd). Therefore its limit as $n \to +\infty$ cannot exist as a finite quantity.

*1309. POWER SERIES. ABEL'S THEOREM

Chapter 12 contains the statements of three important theorems (Theorem I, § 1203, Theorems II and III, § 1210) having to do with continuity, integration, and differentiation of power series. When only *interior* points of the interval of convergence are involved, these theorems are simple corollaries (cf. Exs. 10–12, § 1311) of the following basic theorem:

Theorem I. *A power series converges uniformly on any interval whose end-points lie in the interior of its interval of convergence.*

Proof. For simplicity of notation we shall assume that the series has the form $\sum\limits_{n=0}^{+\infty} a_n x^n$ (cf. Ex. 9, § 1311). Let $R > 0$ be the radius of convergence of $\Sigma a_n x^n$, and let I be an interval with end-points α and β, where $\max(|\alpha|,|\beta|) = r < R$. Choose a fixed γ such that $r < \gamma < R$, and define $M_n \equiv |a_n \gamma^n|$, $n = 0, 1, \cdots$. Since γ is interior to the interval of convergence of $\Sigma a_n x^n$, $\Sigma a_n \gamma^n$ converges absolutely. Therefore the convergent series of constants ΣM_n dominates the series $\Sigma a_n x^n$ throughout I, and by the Weierstrass M-test, the uniform convergence desired is established.

Behavior of a power series at an end-point of the interval of convergence usually involves more subtle and delicate questions than does the behavior at interior points of the interval. One of the most useful and elegant tools for treating convergence at and near an end-point is due to the Norwegian mathematician N. H. Abel (1802–1829). In this section we present the statement of Abel's Theorem, together with two corollaries (whose proofs are requested in Exercise 15, § 1311). Two other corollaries are the remaining unproved parts of Theorem I, § 1203, and Theorem II, § 1210, that are concerned with end-points of the interval of convergence (cf. Exs. 13–14, § 1311).

Theorem II. Abel's Theorem. *If $\sum_{n=0}^{+\infty} a_n$ is a convergent series of constants, then the power series $\sum_{n=0}^{+\infty} a_n x^n$ converges uniformly for $0 \leq x \leq 1$.*

Corollary I. *If a power series converges at an end-point P of its interval of convergence I, it converges uniformly on any closed interval that has P as one of its end-points and any interior point of I as its other end-point. If a power series converges at both end-points of its interval of convergence it converges uniformly throughout that interval.*

Corollary II. *If a function continuous throughout the interval of convergence of a power series is represented by that power series in the interior of that interval, then it is represented by that power series at any end-point of that interval at which the series converges.*

Example. Show that

(1) $$\ln(1+x) = x - \frac{x^2}{2} + \frac{x^3}{3} - \cdots,$$

for $-1 < x \leq 1$, by integrating $\sum_{n=0}^{+\infty} (-1)^n x^n$.

Solution. The geometric series $\sum_{n=0}^{+\infty} (-1)^n x^n$ converges for $|x| < 1$. Therefore the relation (1) is valid for $|x| < 1$ (Theorem II, § 1210). Since $\ln(1+x)$ is continuous at $x = 1$, the relation (1) is also true for $x = 1$, by Corollary II.

*1310. PROOF OF ABEL'S THEOREM

Notation. *If $u_0, u_1, u_2, \cdots, u_n, \cdots$ is a sequence of real numbers, we denote by*

(1) $$\max_{m}^{n} \left| u_m + u_{m+1} + \cdots + u_k \right|$$

the maximum of the $n - m + 1$ numbers

$$|u_m|, |u_m + u_{m+1}|, \cdots, |u_m + u_{m+1} + \cdots + u_n|.$$

Lemma. *If $\{a_n\}$ is any sequence of real numbers, and if $\{b_n\}$ is a monotonically decreasing sequence of nonnegative numbers ($b_n \downarrow$ and $b_n \geq 0$), then*

(2) $$\left| \sum_{k=0}^{n} a_k b_k \right| \leq b_0 \cdot \max_{0}^{n} \left| a_0 + a_1 + \cdots + a_k \right|$$

and

(3) $$\left| \sum_{k=m}^{n} a_k b_k \right| \leq b_m \cdot \max_{m}^{n} \left| a_m + a_{m+1} + \cdots + a_k \right|.$$

Proof. The two statements (2) and (3) are identical, except for notation. For convenience we shall prove (2), and then apply (3) in the proof of

Abel's Theorem. Letting $A_n \equiv a_0 + a_1 + \cdots + a_n$, $n = 0, 1, 2, \cdots$, we can write the left-hand member of (2) in the form

$$|A_0 b_0 + (A_1 - A_0)b_1 + (A_2 - A_1)b_2 + \cdots + (A_n - A_{n-1})b_n|$$
$$= |A_0(b_0 - b_1) + A_1(b_1 - b_2) + \cdots + A_{n-1}(b_{n-1} - b_n) + A_n b_n|.$$

By the triangle inequality and the assumptions on $\{b_n\}$, this quantity is less than or equal to

$$|A_0|(b_0 - b_1) + |A_1|(b_1 - b_2) + \cdots + |A_{n-1}|(b_{n-1} - b_n) + |A_n|b_n$$
$$\leq \{\text{maximum of } |A_0|, |A_1|, \cdots, |A_n|\} \cdot \{(b_0 - b_1) + \cdots + b_n\}$$
$$= b_0 \cdot \max_0^n \left| a_0 + \cdots + a_k \right|.$$

Proof of Abel's Theorem. By the Lemma, if $b_n \equiv x^n$,

$$\left| \sum_{k=m}^n a_k x^k \right| \leq 1 \cdot \max_m^n \left| a_m + \cdots + a_n \right|.$$

The convergence of Σa_n implies that for any $\epsilon > 0$ there exists a number N such that $n \geq m > N$ implies $|a_m + \cdots + a_n| < \epsilon$. Therefore, by the Cauchy criterion for uniform convergence (Ex. 45, § 1304), the proof is complete.

*1311. EXERCISES

In Exercises 1–6, obtain the given expansions by integration, and justify the inclusion of the end-points specified.

★1. $\ln(1 - x) = -x - \dfrac{x^2}{2} - \dfrac{x^3}{3} - \cdots$; $-1 \leq x < 1$.

★2. $\operatorname{Arctan} x = x - \dfrac{x^3}{3} + \dfrac{x^5}{5} - \cdots$; $|x| \leq 1$.

★3. $\operatorname{Arcsin} x = x + \dfrac{1}{2}\dfrac{x^3}{3} + \dfrac{1 \cdot 3}{2 \cdot 4}\dfrac{x^5}{5} + \dfrac{1 \cdot 3 \cdot 5}{2 \cdot 4 \cdot 6}\dfrac{x^7}{7} + \cdots$; $|x| \leq 1$.

★4. $\ln(x + \sqrt{1 + x^2}) = x - \dfrac{1}{2}\dfrac{x^3}{3} + \dfrac{1 \cdot 3}{2 \cdot 4}\dfrac{x^5}{5} - \dfrac{1 \cdot 3 \cdot 5}{2 \cdot 4 \cdot 6}\dfrac{x^7}{7} + \cdots$, $|x| \leq 1$.

★5. $\frac{1}{2}[x\sqrt{1 - x^2} + \operatorname{Arcsin} x]$
$$= x - \dfrac{1}{2}\dfrac{x^3}{3} - \dfrac{1 \cdot 1}{2 \cdot 4}\dfrac{x^5}{5} - \dfrac{1 \cdot 1 \cdot 3}{2 \cdot 4 \cdot 6}\dfrac{x^7}{7} - \cdots$$; $|x| \leq 1$.

★6. $\frac{1}{2}[x\sqrt{1 + x^2} + \ln(x + \sqrt{1 + x^2})]$
$$= x + \dfrac{1}{2}\dfrac{x^3}{3} - \dfrac{1 \cdot 1}{2 \cdot 4}\dfrac{x^5}{5} + \dfrac{1 \cdot 1 \cdot 3}{2 \cdot 4 \cdot 6}\dfrac{x^7}{7} - \cdots$$; $|x| \leq 1$.

★7. Show that
$$\int_0^x \ln(1 + t)\, dt = \dfrac{x^2}{1 \cdot 2} - \dfrac{x^3}{2 \cdot 3} + \dfrac{x^4}{3 \cdot 4} - \cdots$$; $|x| \leq 1$.

Hence evaluate $\dfrac{1}{1 \cdot 2} - \dfrac{1}{2 \cdot 3} + \dfrac{1}{3 \cdot 4} - \cdots$.

*8. Show that
$$\int_0^x \text{Arctan } t \, dt = \frac{x^2}{1 \cdot 2} - \frac{x^4}{3 \cdot 4} + \frac{x^6}{5 \cdot 6} - \cdots; \quad |x| \leq 1.$$
Hence evaluate $1 - \frac{1}{2} - \frac{1}{3} + \frac{1}{4} + \frac{1}{5} - \frac{1}{6} - \frac{1}{7} + + - - \cdots$.

*9. Prove Theorem I, § 1309, for $\sum a_n(x - a)^n$.

*10. Prove Theorem I, § 1203, for all points interior to the interval of convergence (cf. Ex. 13).

*11. Prove Theorem II, § 1210, if α and β are both interior to I (cf. Ex. 14).

*12. Prove Theorem III, § 1210.

*13. Prove that a function defined by a power series is continuous throughout the interval of convergence, and thus complete the proof of Theorem I, § 1203 (cf. Ex. 10).

*14. Complete the proof of Theorem II, § 1210, by permitting either α or β or both to be end-points of the interval of convergence (cf. Ex. 11).

*15. Prove Corollaries I and II of Theorem II, § 1309.

*16. Show that
$$\int_0^1 \frac{\text{Arctan } x}{x} \, dx = \int_0^1 \frac{|\ln x|}{1 + x^2} \, dx = \sum_{n=0}^{+\infty} \frac{(-1)^n}{(2n + 1)^2}.$$

*17. Show that
$$\int_0^1 \frac{\text{Arcsin } x}{x} \, dx = \int_0^1 \frac{|\ln x|}{\sqrt{1 - x^2}} \, dx = \sum_{n=0}^{+\infty} \frac{1 \cdot 3 \cdots (2n - 1)}{2 \cdot 4 \cdots 2n} \cdot \frac{1}{(2n + 1)^2}.$$

*18. From the validity of the Maclaurin series for the binomial function $(1 + x)^m$ for $|x| < 1$, infer the validity of the expansion for any end-point of the interval of convergence at which the series converges.

*19. Show that
$$\int_0^1 \frac{x^m}{1 + x} \, dx = \frac{1}{m + 1} - \frac{1}{m + 2} + \frac{1}{m + 3} - \cdots,$$
for $m > -1$.

*20. Prove that if Σa_n and Σb_n are convergent series, with sums A and B, respectively, if Σc_n is their product series, and if Σc_n converges, with sum C, then $C = AB$. *Hint:* Define the three functions $f(x) \equiv \Sigma a_n x^n$, $g(x) \equiv \Sigma b_n x^n$, and $h(x) \equiv \Sigma c_n x^n$, for $0 \leq x \leq 1$. Examine the continuity of these functions for $0 \leq x \leq 1$, and the equation $h(x) = f(x)g(x)$ for $0 \leq x < 1$.

*21. Verify the expansion of the following elliptic integral (cf. § 511):
$$\int_0^{\frac{\pi}{2}} \frac{dt}{\sqrt{1 - k^2 \sin^2 t}} = \frac{\pi}{2} \left[1 + \left(\frac{1}{2}\right)^2 k^2 + \left(\frac{1 \cdot 3}{2 \cdot 4}\right)^2 k^4 + \left(\frac{1 \cdot 3 \cdot 5}{2 \cdot 4 \cdot 6}\right)^2 k^6 + \cdots \right],$$
for $|k| \leq 1$. *Hint:* Expand by the binomial series:
$$(1 - k^2 \sin^2 t)^{-\frac{1}{2}} = 1 + \frac{1}{2} k^2 \sin^2 t + \frac{1 \cdot 3}{2 \cdot 4} k^4 \sin^4 t + \frac{1 \cdot 3 \cdot 5}{2 \cdot 4 \cdot 6} k^6 \sin^6 t + \cdots,$$
and use Wallis's formulas (Ex. 36, § 416).

*1312. FUNCTIONS DEFINED BY POWER SERIES. EXERCISES

*1. Let the function $e^x \equiv \exp(x)$ be *defined* by the series (1), § 1207, and let the function $y = \ln x$ be *defined* by the equation $x = e^y$. Prove the properties of the exponential and logarithmic functions enumerated in § 502. *Hint:* Begin by proving (*i*), (*ii*), and (*iii*) of Exercise 4, § 502, making use of the special case of (*i*): $e^x e^{-x} = 1$.

⋆2. Let the functions sin x and cos x be *defined* by the series (3) and (5), respectively, of § 1207. Prove the properties of the trigonometric functions enumerated in Exercises 4–17, § 504. *Hints:* Start with Exercise 4, § 504. After showing that cos 2 < 0, *define* $\frac{1}{2}\pi < 2$ to be the smallest positive number whose cosine is 0. Then proceed with Exercise 8, § 504.

⋆1313. UNIFORM LIMITS OF FUNCTIONS

A *sequence* $\{S_n(x)\}$ of functions of one real variable can be regarded as a *single* function of the *two* real variables x and n:

(1) $$S_n(x) = f(x, n).$$

For example, if x is restricted to the closed interval [0, 1], the function has as its domain of definition the set of all points (x, n) in the Euclidean plane such that $0 \leq x \leq 1$ and $n =$ a positive integer; this domain, therefore, consists of infinitely many horizontal line segments.

Viewed in this way, sequences of functions fall heir to the ideas of limits and iterated limits of §§ 606 and 607. Let us confess immediately, however, that whatever interest resides in this fact is principally conceptual. What is more important is to go in the other direction, and carry over the concept of *uniform convergence* to more general functions of two variables.

In order to focus our attention on a specific objective, we shall formulate things in this section in terms of a function $f(x, y)$ and a finite limit c for the variable y. Cases where y has an infinite or a one-sided limit are similar, and some are included in Exercise 1, § 1315. For many properties of uniform limits, the point x may be replaced by a point in E_n (cf. Ex. 2, § 1315).

Definition. *Let $f(x, y)$ be defined for all points x in a set A and all points y in a set B having $y = c$ as a limit point. Then $f(x, y)$ converges to a limit $f(x)$ **uniformly** for x in A, as $y \to c$, written*

(2) $$f(x, y) \underset{\to}{\to} f(x) \text{ on } A, \quad \text{as } y \to c,$$

if and only if corresponding to $\epsilon > 0$ there exists a positive number $\delta = \delta(\epsilon)$, dependent on ϵ alone and not on the point x, such that

(3) $$0 < |y - c| < \delta \quad \text{implies} \quad |f(x, y) - f(x)| < \epsilon$$

*for every x in A. The function $f(x)$ is called a **uniform limit** of $f(x, y)$.*

We formulate for future use:

Negation of Uniform Convergence. *Let $f(x, y)$ have its domain as specified in the preceding Definition. Then $f(x, y)$ fails to converge uniformly to a limit $f(x)$, for x in A, as $y \to c$ if and only if there exists a positive number ϵ having the property that for any positive number δ there exist a point x in A and a point y in B such that*

(4) $$0 < |y - c| < \delta \quad \text{and} \quad |f(x, y) - f(x)| \geq \epsilon.$$

In order to facilitate proofs of the three fundamental theorems on uniform limits, given in the next section, we establish the following relation between uniform limits in general and uniform limits of sequences:

Theorem. *With the assumptions and notation of the preceding Definition,*

(5) $\qquad\qquad f(x, y) \rightrightarrows \text{some function } f(x), \quad \text{as} \quad y \to c,$

if and only if for every sequence $\{y_n\}$ of points of B converging to c, and never equal to c,

(6) $\qquad\qquad\qquad \{f(x, y_n)\} \text{ converges uniformly.}$

In case the uniform limit of (6) *exists for every $\{y_n\}$, it is the same for every $\{y_n\}$, and equal to the uniform limit $f(x)$ of* (5).

Proof. The "only if" part of the proof should be trivial to the reader at this stage, but we advise his supplying the details.

As a first step in the "if" part of the proof, we observe that if the uniform limit of $\{f(x, y_n)\}$ exists for every $\{y_n\}$, it must be the *same* limit for every $\{y_n\}$ (cf. Ex. 1, § 223). We define $f(x)$ to be this limit, and now seek a contradiction to the assumption that $f(x, y)$ does *not* approach $f(x)$ uniformly as $y \to c$. According to the Negation of Uniform Convergence, for some positive number ϵ we can choose for every positive integer n the positive number $\delta = 1/n$, and construct sequences of points x_n of A and y_n of B such that

$$0 < |y_n - c| < 1/n \quad \text{and} \quad |f(x_n, y_n) - f(x_n)| \geq \epsilon.$$

The first of these two inequalities guarantees the convergence of $\{y_n\}$ to c, while the other prohibits the uniform convergence of the sequence $\{f(x, y_n)\}$ to the function $f(x)$.

*1314. THREE THEOREMS ON UNIFORM LIMITS

The three basic theorems of §§ 1305–1307 apply to uniform limits in general. We quote them now for the case where y has a finite limit c. (Cf. Ex. 9, § 1315.) We prove the first, and leave the other two to the exercises (Ex. 10, § 1315, for Theorem II; Ex. 11, § 1315, for Theorem III with the added assumption of continuity of every derivative $f_1(x, y)$ as a function of x).

Theorem I. *Let $f(x, y)$ be defined for all points x in a set A and all points y in a set B having $y = c$ as a limit point. If, for each point y of B, $f(x, y)$ is a continuous function of x for every point x of A, and if $f(x, y) \rightrightarrows f(x)$ on A, as $y \to c$, then $f(x)$ is continuous on A.*

Proof. Under the assumptions of the theorem, if $\{y_n\}$ is a sequence of points of B, none equal to c, converging to c, then, by the Theorem of § 1313, $\{f(x, y_n)\}$ is a sequence of functions of x converging uniformly on A to the function $f(x)$. Therefore, by the Theorem of § 1305, $f(x)$ is continuous on A.

Theorem II. Let $f(x, y)$ be defined for $a \leq x \leq b$, and all points y in a set B having $y = c$ as a limit point. If, for each point y of B, $f(x, y)$ is integrable on $[a, b]$, and if $f(x, y) \underset{\rightarrow}{\rightarrow} f(x)$ on $[a, b]$, as $y \to c$, then $f(x)$ is integrable on $[a, b]$ and

$$\lim_{y \to c} \int_a^b f(x, y)\, dx = \int_a^b \lim_{y \to c} f(x, y)\, dx = \int_a^b f(x)\, dx.$$

Theorem III. Let $f(x, y)$ be defined as in Theorem II, and assume furthermore:

(i) for each point y of B, $f(x, y)$ as a function of x has a derivative

$$f_1(x, y) \equiv \frac{d}{dx} f(x, y)$$

at every point of $[a, b]$;

(ii) $\lim_{y \to c} f(x_0, y)$ exists for some point x_0 of $[a, b]$;

(iii) $f_1(x, y) \underset{\rightarrow}{\rightarrow}$ a function $\phi(x)$ on $[a, b]$, as $y \to c$.

Then

(iv) $f(x, y) \underset{\rightarrow}{\rightarrow}$ a function $f(x)$ on $[a, b]$, as $y \to c$;
(v) $f(x)$ is differentiable on $[a, b]$;
(vi) $f'(x) = \phi(x) = \lim_{y \to c} f_1(x, y)$.

★1315. EXERCISES

★1. Formulate the definition of a uniform limit of a function $f(x, y)$ as $y \to +\infty$; as $y \to \infty$; as $y \to c+$; as $y \to c-$.

★2. Adapt and prove the Theorem of § 1313, and Theorem I, § 1314, when x is replaced by a point in E_n.

In Exercises 3–4, formulate and establish the Negation of the given statement.

★3. $f(x, y) \underset{\rightarrow}{\rightarrow} f(x)$ on A, as $y \to c+$.
★4. $S_{mn} \underset{\rightarrow}{\rightarrow} A_m$ for $m \geq M$, as $n \to +\infty$.

In Exercises 5–8, show that the specified limit is uniform on the first of the two given sets, and that the limit exists but is not uniform on the second of the two given sets. The letter η denotes an arbitrarily small positive number.

★5. $\lim_{y \to 0} \dfrac{xy}{x^2 + y^2}$; $[\eta, +\infty)$; $(0, +\infty)$.

★6. $\lim_{y \to 0+} \dfrac{xy}{1 + xy}$; $[0, b]$; $[0, +\infty)$.

★7. $\lim_{x \to 0} nx + 1$; $a \leq n \leq b$; $-\infty < n < +\infty$.

★8. $\lim_{n \to \infty} \dfrac{mn}{m^2 + n^2}$; $|m| \leq K$; $|m| < +\infty$.

★9. State the three theorems of § 1314 for the case $y \to +\infty$; for the case $y \to c+$.
★10. Prove Theorems II and III, § 1314, for the case $y \to c$; for the case $y \to \infty$.
★11. Prove Theorem III, § 1313, for the case $y \to c-$, under the assumption that every derivative $f_1(x, y)$ is a continuous function of x.

In Exercises 12–15, find $\delta(\epsilon)$ or $N(\epsilon)$ explicitly, according to the Definition, §1313, or Exercise 1, above, for the specified interval.

★12. $\lim\limits_{y\to 0} \dfrac{xy}{x^2 + y^2}$; $[\eta, +\infty)$.

★13. $\lim\limits_{y\to 0+} \dfrac{xy}{1 + xy}$; $[0, b]$.

★14. $\lim\limits_{x\to 0} nx + 1$; $a \leq n \leq b$.

★15. $\lim\limits_{n\to\infty} \dfrac{mn}{m^2 + n^2}$; $|m| \leq K$.

★16. Verify that $\lim\limits_{y\to 0} \int_1^2 \dfrac{xy}{x^2 + y^2}\, dx = \int_1^2 \lim\limits_{y\to 0} \dfrac{xy}{x^2 + y^2}\, dx$ (cf. Ex. 5).

★17. Verify that $\lim\limits_{y\to 0+} \int_0^b \dfrac{xy}{1 + xy}\, dx = \int_0^b \lim\limits_{y\to 0+} \dfrac{xy}{1 + xy}\, dx$ (cf. Ex. 6).

★18. Discuss the equation $\lim\limits_{y\to 0} \dfrac{d}{dx} \ln(x + y) = \dfrac{d}{dx}\left[\lim\limits_{y\to 0} \ln(x + y)\right]$.

★19. Discuss the equation $\lim\limits_{y\to 0} \dfrac{d}{dx} \dfrac{xy^2}{x^2 + y^2} = \dfrac{d}{dx}\left[\lim\limits_{y\to 0} \dfrac{xy^2}{x^2 + y^2}\right]$, with particular attention to the point $x = 0$.

★20. Prove the following **Cauchy criterion** for a function $f(x, y)$: *For x in a set A, $f(x, y) \rightrightarrows f(x)$, as $y \to c$, if and only if corresponding to $\epsilon > 0$ there exists $\delta = \delta(\epsilon) > 0$ such that $0 < |y' - c| < \delta$ and $0 < |y'' - c| < \delta$ imply $|f(x, y') - f(x, y'')| < \epsilon$ for every x in A.* (Cf. Ex. 45, §1304.)

★21. Assume that $f(x)$ has a continuous derivative $f'(x)$ on a closed interval I. Prove that the difference quotient $[f(x + \Delta x) - f(x)]/\Delta x$ approaches $f'(x)$ uniformly on I as $\Delta x \to 0$.

14

*Improper Integrals

*1401. INTRODUCTION. REVIEW

In Chapter 4 (§§ 412–416) improper integrals, for both finite and infinite intervals, were defined, and their elementary properties treated. The language of dominance and the "big O" and "little o" notation were introduced.

For convenience we repeat some statements and definitions given in the Exercises, § 416, formulated for a function $f(x)$ defined on a half-open interval $[a, b)$ or $[a, +\infty)$. Parentheses are used to achieve a compression of two alternative statements into one. Similar formulations hold for other intervals. Proofs of all statements of this section are left to the reader (e.g., cf. § 221, Theorem XII, § 401, and Note 2, § 401, for Theorem I, below).

We start by writing down explicitly an immediate consequence of convergence. If the function $f(x)$ is improperly integrable on the interval $[a, b)$ ($[a, +\infty)$), and $\epsilon > 0$, then there exists a number $\gamma = \gamma(\epsilon)$ of the interval (a, b) ($(a, +\infty)$) such that for any u of the interval (γ, b) ($(\gamma, +\infty)$):

(1) $$\left| \int_u^b f(x)\, dx \right| < \epsilon \quad \left(\left| \int_u^{+\infty} f(x)\, dx \right| < \epsilon \right).$$

Theorem I. Cauchy Criterion. *If $f(x)$ is integrable on $[a, \beta]$ for every β such that $a < \beta < b$ ($a < \beta$), then the improper integral*

(2) $$\int_a^b f(x)\, dx \quad \left(\int_a^{+\infty} f(x)\, dx \right)$$

converges if and only if corresponding to $\epsilon > 0$ there exists a number $\gamma = \gamma(\epsilon)$ such that $a < \gamma < b$ ($a < \gamma$) and such that whenever β_1 and β_2 belong to the interval (γ, b) ($(\gamma, +\infty)$),

(3) $$\left| \int_{\beta_1}^{\beta_2} f(x)\, dx \right| < \epsilon.$$

Theorem II. Comparison Test. *If the nonnegative function $g(x)$ dominates the function $f(x)$ (whose values are not assumed to be of one sign) on the interval $I = [a, b)$ ($[a, +\infty)$), if $f(x)$ and $g(x)$ are integrable on $[a, \beta]$ for every $\beta \in I$, and if the integral $\int_a^b g(x)\, dx$ $\left(\int_a^{+\infty} g(x)\, dx \right)$ converges, so does the integral (2).*

Definition. *If $f(x)$ is integrable on $[a, \beta]$ for every β such that $a < \beta < b$ ($a < \beta$), then the integral (2) is said to **converge absolutely** if and only if the improper integral*

(4) $$\int_a^b |f(x)|\, dx \quad \left(\int_a^{+\infty} |f(x)|\, dx\right)$$

*converges. The integral (1) **converges conditionally** if and only if it converges but does not converge absolutely.*

NOTE. From the fact that integrability of $f(x)$ on $[a, \beta]$ implies that of $|f(x)|$ there (Theorem XIII, § 401) we infer that the integral (4) always exists in a finite or infinite sense. It is a simple consequence of the Comparison Test (Theorem II) that *absolute convergence implies convergence*.

Example 1. Show that $\int_1^{+\infty} \frac{\sin x}{x^p}\, dx$ converges absolutely for $p > 1$. Show that $\int_0^{+\infty} \frac{\sin x}{x^p}\, dx$ converges absolutely for $1 < p < 2$, and diverges for $p \geq 2$. (Cf. Example 1, § 1402.)

Solution. Since for $x \geq 1$, $\sin x/x^p$ is continuous and dominated by $1/x^p$, the first statement follows from Example 3, § 414. The remaining parts are consequences of the fact that, as $x \to 0+$, $\sin x/x^p$ is of the same order of magnitude as $1/x^{p-1}$. Thus, by Example 3, § 413, $\int_0^1 \frac{\sin x}{x^p}\, dx$ converges absolutely for $p < 2$.

Example 2. Show that if $P(x)$ is a polynomial and $p > 0$, then
$$\int_0^{+\infty} P(x) e^{-px}\, dx$$
converges absolutely.

Solution. $P(x)e^{-px} = O\left(\frac{1}{x^2}\right)$, as $x \to +\infty$.

*1402. ALTERNATING INTEGRALS. ABEL'S TEST

In case it has been determined that a given improper integral fails to converge absolutely, it is still possible to have (conditional) convergence through a fortuitous distribution of positiveness and negativeness. For example, the improper integral $\int_0^{+\infty} \frac{\sin x}{x^p}\, dx$, as we shall see in Example 1, converges conditionally for $0 < p \leq 1$, thanks to the regular alternation of signs of $\sin x$. The situation is similar to that of alternating series (§ 1115) where, for example, the alternating p-series $1 - \frac{1}{2^p} + \frac{1}{3^p} - \cdots$ converges conditionally for $0 < p \leq 1$ (Example 1, § 1116). For integrals like the one just mentioned, whose integrands alternate signs over consecutive intervals, it is possible to relate convergence to that of an associated infinite series. Such a test is given in Exercise 13, § 1403.

In the present section, however, we shall establish convergence of such improper integrals (taken for definiteness for the interval $[a, +\infty)$) by means

§ 1402]   ALTERNATING INTEGRALS. ABEL'S TEST   465

of a powerful test whose essential ideas are due to the Norwegian mathematician N. H. Abel (1802–1829) (cf. Ex. 27, § 1117). The proof given here rests on a form of the Second Mean Value Theorem whose proof, as outlined in Exercises 25-26, § 418, is based on the concept of the Riemann-Stieltjes integral (§ 417). For a proof of Abel's Test that is independent of the Riemann-Stieltjes integral (but which requires more restrictive hypotheses), see the hint accompanying Exercise 9, § 1403.

Theorem I. Abel's Test. *Assume that $f(x)$ is continuous and $\phi(x)$ monotonically decreasing for $x \geq a$, that $\lim_{x \to +\infty} \phi(x) = 0$, and that $F(x) \equiv \int_a^x f(t)\, dt$ is bounded for all $x \geq a$. Then $\int_a^{+\infty} f(x)\phi(x)\, dx$ converges.*

Proof. We shall establish convergence by means of the Cauchy Criterion, § 1401. Let us observe first that for $a \leq u < v$ the integral $\int_u^v f(x)\phi(x)\, dx$ exists (cf. Theorems VIII, X, XIV, § 401). Let $\epsilon > 0$, assume $|F(x)| < M$ for all $x \geq a$, and let u and v be arbitrary numbers such that $a \leq u < v$. By the Bonnet form of the Second Mean Value Theorem (Ex. 26, § 418), for a suitable number ξ such that $u \leq \xi \leq v$, $\left|\int_u^v f(x)\phi(x)\, dx\right| = \phi(u)\left|\int_u^\xi f(x)\, dx\right|$
$= \phi(u)|F(\xi) - F(u)| \leq 2M\phi(u)$. Accordingly, if N is chosen so large that $\phi(N) < \epsilon/2M$, and if $v > u > N$, then the conditions of the Cauchy Criterion are satisfied and the desired convergence is proved.

Theorem II. *If $\phi(x)$ is monotonically decreasing for $x \geq a$ and if $\lim_{x \to +\infty} \phi(x) = 0$, then*

(1) $\int_a^{+\infty} \phi(x) \sin x\, dx$ *and* $\int_a^{+\infty} \phi(x) \cos x\, dx$

converge. The convergence is absolute or conditional according as $\int_a^{+\infty} \phi(x)\, dx$ converges or diverges.

Proof. The convergence is a corollary of Theorem I. The only nontrivial part remaining is to show that divergence of $\int_a^{+\infty} \phi(x)\, dx$ implies that of $\int_a^{+\infty} \phi(x) |\sin x|\, dx$ and $\int_a^{+\infty} \phi(x) |\cos x|\, dx$. We shall show this for the former only (proof for the other being essentially the same). Accordingly, let m and n be positive integers such that $n\pi > m\pi \geq a$. Then

$$\int_a^{n\pi} \phi(x) |\sin x|\, dx \geq \int_{m\pi}^{(m+1)\pi} \phi(x) |\sin x|\, dx + \cdots + \int_{(n-1)\pi}^{n\pi} \phi(x) |\sin x|\, dx$$

$$\geq \phi((m+1)\pi)\int_{m\pi}^{(m+1)\pi} |\sin x|\, dx + \cdots + \phi(n\pi)\int_{(n-1)\pi}^{n\pi} |\sin x|\, dx$$

$$= 2[\phi((m+1)\pi) + \cdots + \phi(n\pi)] \geq \frac{2}{\pi}\int_{(m+1)\pi}^{(n+1)\pi} \phi(x)\, dx.$$

These inequalities provide the desired implication.

Example 1. Show that $\int_0^{+\infty} \frac{\sin x}{x^p} dx$ converges conditionally for $0 < p \leq 1$. (Cf. Example 1, § 1401.)

Solution. By the solution of Example 1, § 1401, we need prove the statement only for the integral $\int_1^{+\infty} \frac{\sin x}{x^p} dx$. But for this integral the result follows from Theorem II (cf. Example 3, § 414).

Example 2. Show that $\int_0^{+\infty} \sin x^2 \, dx$ converges conditionally.

Solution. Making the substitution $x = \sqrt{y}$, $y = x^2$, we have

$$\int_0^b \sin x^2 \, dx = \frac{1}{2} \int_0^{b^2} \frac{\sin y}{\sqrt{y}} dy \quad (b > 0).$$

Therefore the result follows from Example 1 (why?). It is a curious fact that the given improper integral converges even though the integrand oscillates persistently between 1 and -1, and does not approach 0 (cf. Exs. 42, 43, § 416). The behavior that makes this phenomenon possible in this case is that the lengths of the intervals of oscillation shrink to zero. (Show this.)

Example 3. Show that $\int_0^{+\infty} \frac{\cos x}{\sqrt{x}} dx = \frac{1}{2} \int_0^{+\infty} \frac{\sin x}{x^{\frac{3}{2}}} dx$, and account for the remarkable fact that one integral converges conditionally, while the other converges absolutely.

Solution. Integration by parts gives

$$\int_a^b \frac{\cos x \, dx}{\sqrt{x}} = \int_a^b \frac{1}{\sqrt{x}} d(\sin x) = \frac{\sin x}{\sqrt{x}} \Big]_a^b + \frac{1}{2} \int_a^b \frac{\sin x}{x^{\frac{3}{2}}} dx.$$

If absolute value signs are introduced in the original integrand the process of integration by parts breaks down (for example, $\frac{d}{dx}|\sin x| \neq |\cos x|$). Thus no inference regarding absolute convergence can be drawn. The "remarkable fact" is no more remarkable than the fact that the series $\frac{1}{1 \cdot 2} + \frac{1}{3 \cdot 4} + \cdots$, obtained by pairing terms of the conditionally convergent alternating harmonic series, converges absolutely.

*1403. EXERCISES

In Exercises 1–4, determine whether the integral converges or diverges and, in case of convergence, whether it converges absolutely or conditionally.

★1. $\int_0^{+\infty} \frac{\cos 2x}{\sqrt{1 + x^3}} dx.$

★2. $\int_0^{+\infty} \frac{e^{-\frac{1}{2}x} \sin x}{x} dx.$

★3. $\int_0^{+\infty} \frac{x \sin x}{x^2 - x - 1} dx.$

★4. $\int_2^{+\infty} \frac{\cos x}{\ln x} dx.$

In Exercises 5–8, give the values of p for which the integral converges absolutely, and those for which it converges conditionally.

★5. $\displaystyle\int_0^{+\infty} \frac{\cos x}{x^p} dx.$

★6. $\displaystyle\int_0^{+\infty} \frac{\sin x}{(x+1)x^p} dx.$

★7. $\displaystyle\int_0^{+\infty} \frac{x^p \cos x}{1+x^2} dx.$

★8. $\displaystyle\int_0^{+\infty} \frac{\sin^2 x}{x^p} dx.$

★9. Prove Abel's Test (Theorem I, § 1402), under the additional assumption that $\phi(x)$ is continuously differentiable, without benefit of the Bonnet form of the Second Mean Value Theorem. *Hint:* Assume that $|F(x)| < M$, integrate by parts, and show, for $a \leq u < v$: $\left|\int_u^v f(x)\phi(x) dx\right| \leq |F(v)\phi(v)| + |F(u)\phi(u)| + \int_u^v |F(x)|\{-\phi'(x)\} dx \leq 3M\phi(u).$

★10. Find two examples to prove that in Abel's Test (Theorem I, § 1402) neither assumption $\phi(x) \to 0$ nor $\phi(x)\downarrow$ can be omitted.

★11. Prove the following alternative form of Abel's Test: *If $f(x)$ is continuous and $\phi(x)$ bounded and monotonic for $x \geq a$, and if $\int_a^{+\infty} f(x) dx$ converges, then so does $\int_a^{+\infty} f(x)\phi(x) dx.$* *Hint:* If $\phi(x)\uparrow$ and $\lim_{x\to+\infty} \phi(x) = L$, write $f(x)\phi(x) = [-f(x)][L - \phi(x)] + Lf(x)$, and apply Abel's Test.

★12. Assume that $\phi(x)$ is a positive monotonically decreasing function for $x \geq a$ and that $\int_a^{+\infty} \phi(x) dx$ converges. Prove that $\phi(x) = o(x^{-1})$ as $x \to +\infty$. *Hint:* $\int_x^{+\infty} \phi(t) dt \geq \int_x^{2x} \phi(t) dt \geq x\phi(x).$

★13. Assume that $f(x)$ is defined for $x \geq a$ and integrable on $[a, b]$ for every $b > a$, let $a = b_0 < b_1 < b_2 < \cdots, b_n \to +\infty$, and define $I_n \equiv \int_{b_{n-1}}^{b_n} f(x) dx.$ Prove that the convergence of $\int_a^{+\infty} f(x) dx$ implies that of $\sum_{n=1}^{+\infty} I_n$, and the equality of their values. Show by an example that the converse implication is false. Prove that with the further assumption that $f(x)$ does not change sign on any one of the intervals (b_{n-1}, b_n), $n = 1, 2, \cdots$, the convergence of $\sum_{n=1}^{+\infty} I_n$ implies that of $\int_a^{+\infty} f(x) dx$. Use this result to establish convergence of $\int_0^{+\infty} \frac{\sin x}{x} dx$ and $\int_0^{+\infty} \sin x^2 dx.$

★14. Show that $0 < \int_0^x \frac{\sin t}{t} dt < \pi$ for all positive x.

★1404. UNIFORM CONVERGENCE

Uniform convergence for an improper integral is similar to uniform convergence for an infinite series, and many of the consequences of uniform convergence that were established for infinite series in Chapter 13 have parallels for improper integrals, which we shall study in this chapter. Many of these are special cases of general theorems on uniform limits presented in §§ 1313–1315.

We start with the basic definition, formulated for definiteness for improper integrals on a half-open interval $[c, d)$ or an infinite interval $[c, +\infty)$, parentheses being used to express an alternative statement.

Definition. *Let $f(x, y)$ be defined for every point x of a set A and every y of the interval $I = [c, d)$ ($[c, +\infty)$), and assume that for every $x \in A$ and every $\beta \in I$, $f(x, y)$ as a function of y is (Riemann) integrable on $[c, \beta]$. Then the integral*

$$\text{(1)} \qquad \int_c^d f(x, y)\, dy \quad \left(\int_c^{+\infty} f(x, y)\, dy\right)$$

converges uniformly *to a function $F(x)$ for $x \in A$, written*

$$\text{(2)} \qquad \int_c^\beta f(x, y)\, dy \rightrightarrows F(x),\ \text{as}\ \beta \to d-\ (\beta \to +\infty),$$

if and only if corresponding to $\epsilon > 0$ there exists a number $\gamma \equiv \gamma(\epsilon)$ belonging to the interval I such that the inequality $\gamma < \beta < d$ ($\gamma < \beta$) implies

$$\text{(3)} \qquad \left|\int_c^\beta f(x, y)\, dy - F(x)\right| < \epsilon$$

*for every $x \in A$. The variable x is called a **parameter**.*

NOTE. As with infinite series (§ 1302), uniform convergence implies convergence, but not conversely. (Cf. the Example, below.)

The question before us is usually whether convergence, known to obtain, is uniform. It is therefore frequently convenient to formulate uniform convergence, and its negation, as follows (the reader may supply the proofs, which are immediate):

Theorem I. *Under the assumptions of the preceding Definition, and the further assumption that (1) converges for each $x \in A$, the convergence of (1) is uniform if and only if*

$$\text{(4)} \qquad \int_\beta^d f(x, y)\, dy \rightrightarrows 0 \quad \left(\int_\beta^{+\infty} f(x, y)\, dy \rightrightarrows 0\right),$$

as $\beta \to d-$ ($\beta \to +\infty$); in other words, if and only if corresponding to $\epsilon > 0$ there exists a number $\gamma = \gamma(\epsilon)$ belonging to the interval (c, d) $((c, +\infty))$ and such that the inequality $\gamma < \beta < d$ ($\gamma < \beta$) implies

$$\text{(5)} \qquad \left|\int_\beta^d f(x, y)\, dy\right| < \epsilon \quad \left(\left|\int_\beta^{+\infty} f(x, y)\, dy\right| < \epsilon\right),$$

for every $x \in A$.

Corollary. *Under the assumptions of the preceding Definition, if c' is an arbitrary point of I, the uniform convergence of (1) is equivalent to that of the integral obtained from (1) by replacing c by c'.*

Theorem II. Negation of uniform convergence. *Under the assumptions of Theorem I, the convergence of* (1) *fails to be uniform if and only if there exists a positive number ϵ such that corresponding to an arbitrary number γ belonging to the interval (c, d) $((c, +\infty))$ there exists a number β of the interval (γ, d) $((\gamma, +\infty))$ and a point $x \in A$ such that*

(6) $$\left| \int_\beta^d f(x, y) \, dy \right| \geq \epsilon \;\; \left(\left| \int_\beta^{+\infty} f(x, y) \, dy \right| \geq \epsilon \right).$$

Example. Show that the integral

(7) $$\int_0^{+\infty} \frac{\sin xy}{y} \, dy$$

converges uniformly for $x \geq \delta > 0$, but not uniformly for $x > 0$.

Solution. The substitution $u = xy$ gives

(8) $$\int_\beta^{+\infty} \frac{\sin xy}{y} \, dy = \int_{x\beta}^{+\infty} \frac{\sin u}{u} \, du.$$

The convergence of $\int_0^{+\infty} \frac{\sin u}{u} \, du$ means that corresponding to $\epsilon > 0$ there exists a number $\gamma > 0$ such that $\alpha > \gamma$ implies

$$\left| \int_\alpha^{+\infty} \frac{\sin u}{u} \, du \right| < \epsilon.$$

Therefore, if $x \geq \delta$ and $\beta > \gamma/\delta$, we conclude that the quantity in (8) is numerically less than ϵ, and (by (5)) uniform convergence of (7) is established.

We now let $\epsilon \equiv \frac{1}{2} \int_0^{+\infty} \frac{\sin u}{u} \, du$. (This is easily shown to be positive; see Example 2, § 1408, for a specific evaluation.) Then however large β may be, we can find a value of x sufficiently near 0 to ensure that the value of (8) is close enough to $\int_0^{+\infty} \frac{\sin u}{u} \, du$ to be greater than ϵ. By Theorem II, this shows that the convergence is not uniform for $x > 0$.

*1405. DOMINANCE AND THE WEIERSTRASS *M*-TEST

Dominance in uniform convergence of improper integrals is similar to dominance in uniform convergence of series (cf. § 1303). The formulations of this section are once more framed in terms of the intervals $[c, d)$ and $[c, +\infty)$.

Definition. *The statement that a function $g(x, y)$ **dominates** a function $f(x, y)$ for $x \in A$ and $y \in B$ means that for every $x \in A$ and $y \in B$,*

$$|f(x, y)| \leq g(x, y).$$

Theorem I. Comparison Test. *Let $f(x, y)$ and $g(x, y)$ be defined for every x of a set A and every y of the interval $I = [c, d)$ $([c, +\infty))$ and assume that for every $x \in A$ and every $\beta \in I$, $f(x, y)$ and $g(x, y)$ as functions of y are integrable* on $[c, \beta]$. Furthermore, assume that $g(x, y)$ dominates $f(x, y)$ for $x \in A$ and $y \in I$, and that

(1) $$\int_c^d g(x, y)\, dy \quad \left(\int_c^{+\infty} g(x, y)\, dy\right)$$

converges uniformly for $x \in A$. Then

(2) $$\int_c^d f(x, y)\, dy \quad \left(\int_c^{+\infty} f(x, y)\, dy\right)$$

converges uniformly for $x \in A$.

Proof. By the comparison test for improper integrals (§ 415) the integral (2) is absolutely convergent, and therefore (§ 1401) convergent, for each $x \in A$. If $\epsilon > 0$ is given, and if γ is a point of I such that the inequality $\gamma < \beta < d$ ($\gamma < \beta$) implies

(3) $$\int_\beta^d g(x, y)\, dy < \epsilon \quad \left(\int_\beta^{+\infty} g(x, y)\, dy < \epsilon\right),$$

for all $x \in A$, the appropriate corresponding inequalities for $f(x, y)$ follow from

$$\left|\int_\beta^{\beta'} f(x, y)\, dy\right| \leq \int_\beta^{\beta'} |f(x, y)|\, dy \leq \int_\beta^{\beta'} g(x, y)\, dy,$$

where $\beta' > \beta$ (let $\beta' \to d-$ or $+\infty$).

Since an integral of the form $\int_c^d M(y)\, dy$ $\left(\int_c^{+\infty} M(y)\, dy\right)$, where the integrand is independent of x, converges uniformly for x in any set A, whenever it converges at all, we have as a special case of Theorem I the analogue of the Weierstrass M-test for infinite series.

Theorem II. Weierstrass M-Test. *Let $f(x, y)$ and $M(y)$ be defined for every $x \in A$ and $y \in I = [c, d)$ ($[c, +\infty)$), and assume that for every $x \in A$ and every $\beta \in I$, $f(x, y)$ and $M(y)$ as functions of y are integrable on $[c, \beta]$. Furthermore, assume that for every $x \in A$,*

(4) $$|f(x, y)| \leq M(y),$$

and that

(5) $$\int_c^d M(y)\, dy \quad \left(\int_c^{+\infty} M(y)\, dy\right)$$

is convergent. Then

(6) $$\int_c^d f(x, y)\, dy \quad \left(\int_c^{+\infty} f(x, y)\, dy\right)$$

converges uniformly for $x \in A$.

Example. Show that $\int_0^{+\infty} e^{-y} \cos xy\, dy$ converges uniformly for all real x.

Solution. $|e^{-y} \cos xy| \leq e^{-y}$, and $\int_0^{+\infty} e^{-y}\, dy$ converges.

*1406. THE CAUCHY CRITERION AND ABEL'S TEST FOR UNIFORM CONVERGENCE

The Cauchy Criterion for uniform convergence of a sequence of functions is given, with hints on a proof, in Exercise 45, § 1304. It is restated for more general functions in Exercise 20, § 1315. We now state it, and give the proof, for uniform convergence of the improper integral

$$(1) \qquad \int_c^d f(x, y)\, dy \ \left(\int_c^{+\infty} f(x, y)\, dy \right).$$

Theorem I. Cauchy Criterion for Uniform Convergence. Let $f(x, y)$ be defined for every $x \in A$ and $y \in I = [c, d)$ $([c, +\infty))$, and assume that for every $x \in A$ and $\beta \in I$, $f(x, y)$ as a function of y is integrable on $[c, \beta]$. Then the integral (1) converges uniformly for $x \in A$ if and only if corresponding to $\epsilon > 0$ there exists a number $\gamma = \gamma(\epsilon) \in I$ such that whenever β_1 and β_2 belong to the interval (γ, d) $((\gamma, +\infty))$,

$$(2) \qquad \left| \int_{\beta_1}^{\beta_2} f(x, y)\, dy \right| < \epsilon,$$

for every $x \in A$.

Equivalently, the integral (1) fails to converge uniformly for $x \in A$ if and only if there exists a positive number ϵ such that corresponding to an arbitrary point $\gamma \in I$ there exist β_1 and β_2 of the interval (γ, d) $((\gamma, +\infty))$ and a point $x \in A$ such that

$$(3) \qquad \left| \int_{\beta_1}^{\beta_2} f(x, y)\, dy \right| \geq \epsilon.$$

Proof for $[c, +\infty)$: "Only if": Assuming uniform convergence, with $\varepsilon > 0$ given, we know there exists a number $\gamma \in (c, +\infty)$ such that $\beta > \gamma$ implies (for every $x \in A$)

$$\left| \int_\beta^{+\infty} f(x, y)\, dy \right| < \frac{\epsilon}{2}.$$

Therefore $\beta_1 > \gamma$ and $\beta_2 > \gamma$ imply (for every $x \in A$):

$$\left| \int_{\beta_1}^{\beta_2} f(x, y)\, dy \right| = \left| \int_{\beta_1}^{+\infty} f(x, y)\, dy - \int_{\beta_2}^{+\infty} f(x, y)\, dy \right|$$
$$\leq \left| \int_{\beta_1}^{+\infty} f(x, y)\, dy \right| + \left| \int_{\beta_2}^{+\infty} f(x, y)\, dy \right| < \frac{\epsilon}{2} + \frac{\epsilon}{2} = \epsilon.$$

"If": Assuming the statement following the words "if and only if," we are assured that the integral (1) converges for every $x \in A$, by the Cauchy Criterion of § 1401. Furthermore, we know by assumption that corresponding to $\epsilon > 0$ there exists $\gamma > c$ such that for all $x \in A$, and β and $\beta' > \gamma$, $\left| \int_\beta^{\beta'} f(x, y)\, dy \right| < \frac{\epsilon}{2}$. We now hold x and β fixed and take a limit as $\beta' \to +\infty$ (with assurance that this limit exists—cf. the Theorem, § 209, Ex. 16, § 212):

$$\lim_{\beta' \to +\infty} \left| \int_\beta^{\beta'} f(x, y) \, dy \right| = \left| \int_\beta^{+\infty} f(x, y) \, dy \right| \leq \frac{\epsilon}{2} < \epsilon.$$

By Theorem I, § 1404, the proof is complete.

A useful test for uniform convergence in cases when the convergence is *conditional*, is similar to the Abel test of § 1402, and to an analogous test for infinite series (Ex. 47, § 1304). We state the theorem only for the domain of integration $[c, +\infty)$, other formulations being similar.

Theorem II. Abel's Test. *For every $x \in A$, let $f(x, y)$ and $\phi(x, y)$, as functions of y, be such that $f(x, y)$ is continuous and $\phi(x, y)$ is monotonically decreasing for $y \geq c$. Assume, furthermore, that there exists a constant M such that*

(4) $$\left| \int_c^\beta f(x, y) \, dy \right| < M$$

for every $x \in A$ and every $\beta \geq c$, and that

(5) $$\phi(x, y) \underset{\to}{\to} 0, \quad \text{as} \quad y \to +\infty$$

uniformly for $x \in A$. Then the integral

(6) $$\int_c^{+\infty} f(x, y) \phi(x, y) \, dy$$

converges uniformly for $x \in A$.

Proof. We use the Cauchy Criterion of the preceding Theorem, and the ideas of the proof of Theorem I, § 1402 (including the Bonnet form of the Second Mean Value Theorem), and write ($c < \beta_1 < \beta_2$):

$$\left| \int_{\beta_1}^{\beta_2} f(x, y) \phi(x, y) \, dy \right| = \left| \phi(x, \beta_1) \int_{\beta_1}^{\xi} f(x, y) \, dy \right| \leq 2M \phi(x, \beta_1).$$

We therefore chose $\gamma = \gamma(\epsilon)$ so that $\phi(x, y) < \epsilon/2M$ for $y > \gamma$.

Example. Show that

(7) $$\int_0^{+\infty} \frac{e^{-xy} \sin ay}{y} \, dy, \, a \neq 0,$$

converges uniformly for $x \geq 0$.

Solution. By the Corollary to Theorem I, § 1404, we consider in place of (7) the corresponding integral on the interval $1 \leq y < +\infty$. Uniform convergence for $x \geq \delta > 0$ is easily established by the principle of dominance (cf. the Example, § 1405). However, for $x = 0$ the integral converges *conditionally*, and we call upon Abel's Test (Theorem II) to establish uniformity for x near 0. Accordingly, let $f(x, y) \equiv \sin ay$, $\phi(x, y) \equiv e^{-xy}/y$, $A \equiv [0, +\infty)$, and $c \equiv 1$. (Cf. the Corollary, Theorem I, § 1404.) The conditions of Theorem II are readily verified, so that the integral $\int_1^{+\infty} \frac{e^{-xy} \sin ay}{y} \, dy$ converges uniformly for $x \geq 0$.

*1407. THREE THEOREMS ON UNIFORM CONVERGENCE

The three theorems on uniform limits, § 1314, specialize to improper integrals, with few adjustments. We give the statements, together with essential features of the proofs that are peculiar to the present material. The interval of integration will be assumed for definiteness to be either $[c, d)$ or $[c, +\infty)$.

Theorem I. *Let I be an arbitrary interval (finite or infinite) and let J be the interval $[c, d)$ ($[c, +\infty)$). Assume that $f(x, y)$ is continuous for all x and y such that $x \in I$, $y \in J$, and that the integral*

$$(1) \qquad \int_c^d f(x, y)\, dy \quad \left(\int_c^{+\infty} f(x, y)\, dy \right)$$

converges uniformly for $x \in I$. Then the limit function

$$(2) \qquad F(x) \equiv \int_c^d f(x, y)\, dy \quad \left(F(x) \equiv \int_c^{+\infty} f(x, y)\, dy \right)$$

is continuous on I.

Proof. For each $\beta \in J$ the function

$$F_\beta(x) \equiv \int_c^\beta f(x, y)\, dy$$

is a continuous function of x, for $x \in I$ (Theorem I, § 930, Ex. 7, § 931), and as $\beta \to d-$ ($\beta \to +\infty$),

$$F_\beta(x) \rightrightarrows F(x),$$

for $x \in I$.

Theorem II. *Let I be a closed interval $[a, b]$ and let J be the interval $[c, d)$ ($[c, +\infty)$). For each $\beta \in J$ assume that $f(x, y)$ is integrable on the closed rectangle $x \in I$, $c \leq y \leq \beta$, and that the integrals $\int_c^\beta f(x, y)\, dy$, for $x \in I$ and $\int_a^b f(x, y)\, dx$, for $y \in J$, always exist. Furthermore, assume that the integral*

$$(1) \qquad \int_c^d f(x, y)\, dy \quad \left(\int_c^{+\infty} f(x, y)\, dy \right)$$

converges uniformly for $x \in I$. Then the limit function is integrable on I, and

$$(3) \qquad \int_a^b \int_c^d f(x, y)\, dy\, dx = \int_c^d \int_a^b f(x, y)\, dx\, dy$$

$$\left(\int_a^b \int_c^{+\infty} f(x, y)\, dy\, dx = \int_c^{+\infty} \int_a^b f(x, y)\, dx\, dy \right).$$

Proof. By the Fubini Theorem (§ 1009), $\int_a^b \int_c^\beta f(x, y)\, dy\, dx = \int_c^\beta \int_a^b f(x, y)\, dx\, dy$, and (3) follows from Theorem II, § 1314, by uniform convergence.

Theorem III. *Let I be the interval $[a, b]$ and let J be the interval $[c, d)$ ($[c, +\infty)$). Assume that $f(x, y)$ and $f_1(x, y)$ exist and are continuous for all (x, y), where $x \in I$, $y \in J$. Assume, furthermore, that*

(i) $\displaystyle\int_c^d f(x, y)\, dy \ \left(\int_c^{+\infty} f(x, y)\, dy\right)$ *converges for some value x_0 of $x \in I$;*

(ii) $\displaystyle\int_c^d f_1(x, y)\, dy \ \left(\int_c^{+\infty} f_1(x, y)\, dy\right)$ *converges uniformly for $x \in I$.*

Then

(iii) *the integral in (i) converges uniformly for $x \in I$;*
(iv) *the limit function defined by (iii) is differentiable for $x \in I$;*

(v)
$$\frac{d}{dx}\int_c^d f(x, y)\, dy = \int_c^d f_1(x, y)\, dy$$
$$\left(\frac{d}{dx}\int_c^{+\infty} f(x, y)\, dy = \int_c^{+\infty} f_1(x, y)\, dy\right).$$

Proof. For any $\beta \in J$ (by Theorem II, § 930, Ex. 7, § 931):

(4)
$$\frac{d}{dx}\int_c^\beta f(x, y)\, dy = \int_c^\beta f_1(x, y)\, dy.$$

The remainder of the proof follows from Theorem III, § 1314. (The continuity of the derivative (4) assumed in the proof of Theorem III, § 1314, as requested in Ex. 11, § 1315, is assured by Theorem I, § 930.)

The hypothesis of uniform convergence (or some substitute) is essential to each of the preceding theorems. The following Examples illustrate this fact.

Example 1. *The limit of the integral need not equal the integral of the limit.* Direct evaluation shows that
$$\int_0^{+\infty} 2xye^{-xy^2}\, dy = \begin{cases} 1 & \text{if } x > 0, \\ 0 & \text{if } x = 0. \end{cases}$$
If $x \to 0+$, the limit of the integral is 1, but since the limit of the integrand is identically 0, the integral of the limit is 0. Similarly, if $x \to +\infty$, the limit of the integral is 1 and the integral of the limit is 0.

Example 2. *The value of an iterated integral may depend on the order of integration:* If $f(x, y) \equiv (2y - 2xy^3)e^{-xy^2}$,
$$\int_0^1 \int_0^{+\infty} f(x, y)\, dx\, dy = \int_0^1 \left[2xye^{-xy^2}\right]_{x=0}^{x=+\infty} dy = \int_0^1 0\, dy = 0,$$
$$\int_0^{+\infty} \int_0^1 f(x, y)\, dy\, dx = \int_0^{+\infty} \left[y^2 e^{-xy^2}\right]_{y=0}^{y=1} dx = \int_0^{+\infty} e^{-x}\, dx = 1.$$

Example 3. *Differentiation under an integral sign may be meaningless.* Although
$$\int_0^{+\infty} \frac{\sin xy}{y}\, dy$$

converges uniformly for $x \geq \delta > 0$ (Example, § 1404), the integral of the derivative (with respect to x),
$$\int_0^{+\infty} \cos xy \, dy,$$
diverges for all x.

Example 4. *Differentiation under an integral sign may be incorrect.* The function
$$F(x) \equiv \int_0^{+\infty} x^3 e^{-x^2 y} \, dy$$
is equal to x for all x, including $x = 0$. Therefore $F'(x) = 1$ for all x, including $x = 0$. On the other hand, the integral of the derivative,
$$\int_0^{+\infty} (3x^2 - 2x^4 y) e^{-x^2 y} \, dy,$$
is equal to 1 if $x \neq 0$, and equal to 0 if $x = 0$. Thus formal differentiation gives an incorrect result for $x = 0$, although every integral considered converges (absolutely).

*1408. EVALUATION OF IMPROPER INTEGRALS

The Theorems of the preceding section can often be used to provide evaluations of interesting and important special improper integrals. We give several examples, and more are requested in § 1409.

NOTE. The three theorems of § 1407 can be thought of as statements concerning the interchange of the order of applying two operations. For example, in the event of uniform convergence, Theorem I, § 1407, implies in particular that the limit of an (improper) integral is the integral of the limit, Theorem II, § 1407, equates the values of two iterated integrals (involving an improper integral) where the order of integration is reversed, and Theorem III, § 1407, states conditions under which the derivative of an (improper) integral is the integral of the derivative. Other techniques that are frequently useful in connection with questions of interchange of limiting processes are given in the author's *Real Variables*: (*i*) iterated improper integrals (*RV*, § 1410), (*ii*) iterated suprema and improper integrals of infinite series (*RV*, § 1411), (*iii*) the Lebesgue bounded and dominated convergence theorems (*RV*, Ex. 49, § 503, Ex. 47, § 515), and (*iv*) the Moore-Osgood theorem (*RV*, § 1014). An extremely potent method for evaluating many definite integrals, proper and improper, is that of *contour integration* in the theory of Analytic Functions of a Complex Variable.

Example 1. Establish the formulas

(1) $$\int_0^{+\infty} e^{-px} \sin qx \, dx = \frac{q}{p^2 + q^2}, p > 0,$$

(2) $$\int_0^{+\infty} e^{-px} \cos qx \, dx = \frac{p}{p^2 + q^2}, p > 0.$$

Then, by integrating with respect to q, between a and b, prove:

(3) $$\int_0^{+\infty} e^{-px} \frac{\cos ax - \cos bx}{x} \, dx = \frac{1}{2} \ln \frac{p^2 + b^2}{p^2 + a^2}, p > 0,$$

(4) $$\int_0^{+\infty} e^{-px} \frac{\sin bx - \sin ax}{x} \, dx = \operatorname{Arctan} \frac{b}{p} - \operatorname{Arctan} \frac{a}{p}, p > 0,$$

and thus ($b = 0$)

(5) $$\int_0^{+\infty} e^{-px} \frac{\sin ax}{x} \, dx = \operatorname{Arctan} \frac{a}{p}, p > 0.$$

Solution. Equations (1) and (2) follow immediately from (4) and (5), § 407. The integration is justified since, for a fixed $p > 0$, the integrands in (1) and (2) are dominated by e^{-px}. Therefore the integrals (1) and (2) converge uniformly for all q, and Theorem II, § 1407, applies.

Example 2. By letting $p \to 0+$ in (3) and (5) prove:

(6) $$\int_0^{+\infty} \frac{\cos bx - \cos ax}{x} dx = \ln \left| \frac{a}{b} \right|, \; ab \neq 0.$$

(7) $$\int_0^{+\infty} \frac{\sin ax}{x} dx = \begin{cases} \tfrac{1}{2}\pi, & a > 0, \\ 0, & a = 0, \\ -\tfrac{1}{2}\pi, & a < 0. \end{cases}$$

Solution. In deriving (6) we may without loss of generality assume that both a and b are positive. In applying to (3) Abel's test for uniform convergence for $p \geq 0$, we let $f(p, x) \equiv \cos bx - \cos ax$ and $\phi(p, x) \equiv e^{-px}/x$, $A \equiv [0, +\infty)$, $c \equiv 1$ (cf. the Example, § 1406). The conditions of Theorem II, § 1406, are satisfied, so that the integral (3) converges uniformly for $p \geq 0$. Therefore, by Theorem I, § 1407, the function $F(p)$ defined by the left-hand member of (3) is continuous for $p \geq 0$. Since the right-hand member of (3) is also continuous for $p \geq 0$, and equal to the left-hand member for $p > 0$, equation (3) must also be satisfied when $p = 0$. This gives (6).

Equation (7), for $a \neq 0$, follows from (5) in similar fashion, with the help of the uniform convergence established in the Example, § 1406.

Example 3. By integrating (2) with respect to p, between positive numbers a and b, prove:

(8) $$\int_0^{+\infty} \frac{e^{-ax} - e^{-bx}}{x} \cos qx \, dx = \frac{1}{2} \ln \frac{q^2 + b^2}{q^2 + a^2}, \; a > 0, b > 0,$$

and thus:

(9) $$\int_0^{+\infty} \frac{e^{-ax} - e^{-bx}}{x} dx = \ln \frac{b}{a}, \; a > 0, b > 0.$$

Solution. The integration is justified by the uniform convergence of (2) for $p \geq \delta$, where $\delta = \min(a, b)$, which is true since the integrand is dominated by $e^{-\delta x}$, $x \geq 0$.

Example 4. Use differentiation and integration by parts to establish the formula

(10) $$\int_0^{+\infty} e^{-x^2} \cos rx \, dx = \tfrac{1}{2} \sqrt{\pi} e^{-\frac{1}{4}r^2}.$$

Solution. Let the function $y = \phi(r)$ be defined by the integral in (10). Then, by Theorem III, § 1407, and the uniform convergence of the resulting integral,

$$\phi'(r) = -\int_0^{+\infty} x e^{-x^2} \sin rx \, dx,$$

which, by integration by parts, gives

$$\frac{1}{2} \int_0^{+\infty} \sin rx \, d(e^{-x^2}) = \frac{1}{2} e^{-x^2} \sin rx \Big]_0^{+\infty} - \frac{1}{2} \int_0^{+\infty} e^{-x^2} d(\sin rx),$$

or

(11) $$\frac{dy}{dr} = -\frac{1}{2} ry.$$

§ 1408] EVALUATION OF IMPROPER INTEGRALS

This linear differential equation can be solved as follows: Since (11) is equivalent to

$$\frac{d}{dr}[ye^{\frac{1}{4}r^2}] = \left(\frac{dy}{dr} + \frac{1}{2}ry\right)e^{\frac{1}{4}r^2} = 0,$$

the solutions of (11) are all functions of the form $y = ke^{-\frac{1}{4}r^2}$. We can evaluate the constant k by setting $r = 0$ in the integral in (10), and using (10), § 1013.

Example 5. Use differentiation and a change of variable to establish the formula

(12) $$\int_0^{+\infty} e^{-x^2 - \frac{r^2}{x^2}} dx = \tfrac{1}{2}\sqrt{\pi} e^{-2|r|}.$$

Solution. Without loss of generality we shall assume that $r > 0$. Let the function $y = \phi(x)$, $r > 0$, be defined by the integral in (12) (which converges for all r since the integrand is dominated by e^{-x^2}). We shall now justify taking a derivative:

(13) $$\frac{dy}{dr} = \phi'(r) = \int_0^{+\infty} -\frac{2r}{x^2} e^{-x^2 - \frac{r^2}{x^2}} dx.$$

This is validated as follows: if B is an upper bound for pe^{-p}, for all $p > 0$, the integrand in (13) is dominated by $\frac{2B}{r} e^{-x^2}$. Therefore the integral of (13) converges uniformly for $0 < \alpha \leq r \leq \beta$, and thus (13) is valid for any positive r. The next step is to rewrite the integral in (13) by means of the substitution $x = r/u$, $dx = (-r/u^2) du$:

(14) $$\frac{dy}{dr} = \phi'(r) = -2 \int_0^{+\infty} e^{-\frac{r^2}{u^2} - u^2} du = -2\phi(r) = -2y.$$

Solving the linear differential equation (14) as in Example 4, we have $\frac{d}{dr}[ye^{2r}] = \left(\frac{dy}{dr} + 2y\right)e^{2r} = 0$, and therefore $y = ke^{-2r}$. Finally, this equation can be extended, by the uniform convergence in (12), to the range $r \geq 0$ (cf. Example 2), and the substitution of $r = 0$ gives an evaluation of k, and the formula (12).

Example 6. Using trigonometric identities and symmetry properties, show that

(15) $$\int_0^{\frac{1}{2}\pi} \ln \sin x \, dx = \int_0^{\frac{1}{2}\pi} \ln \cos x \, dx = -\tfrac{1}{2}\pi \ln 2.$$

Solution. The substitution $y = \tfrac{1}{2}\pi - x$ shows the equality of the two integrals; and the fact that $\ln \sin x = o(x^{-\frac{1}{2}})$ as $x \to 0+$ establishes the finiteness of their common value, which we shall denote by I. Adding the two integrals of (15), and using the substitution $y = 2x$, we have

$$2I = \int_0^{\frac{1}{2}\pi} \ln(\sin x \cos x) \, dx = \frac{1}{2}\int_0^{\pi} \ln \frac{\sin y}{2} \, dy.$$

The substitution $u = \pi - y$ shows that

$$\int_0^{\pi} \ln \sin y \, dy = \int_0^{\frac{1}{2}\pi} \ln \sin y \, dy + \int_{\frac{1}{2}\pi}^{\pi} \ln \sin y \, dy = 2I.$$

Therefore, from the preceding expression for $2I$, we have

$$2I = \frac{1}{2}\int_0^{\pi} \ln \sin y \, dy - \frac{1}{2}\int_0^{\pi} \ln 2 \, dy = I - \tfrac{1}{2}\pi \ln 2,$$

and the desired evaluation.

*1409. EXERCISES

*1. Show that $\int_0^{+\infty} \dfrac{\cos xy}{1+y^2}\, dy$ converges uniformly for all x.

*2. Show that $\int_0^{+\infty} \dfrac{e^{-xy}\sin ay}{y}\, dy$, for a given $x > 0$, converges uniformly for all a (cf. Example, § 1406).

In Exercises 3–8, show that the given integral converges uniformly for the first of the two given intervals, and that it converges but not uniformly for the second of the two given intervals. The letter η denotes an arbitrarily small positive number, and the letter that is not the variable of integration denotes the parameter.

*3. $\int_0^1 \dfrac{dx}{x^p}$; $(-\infty, 1-\eta]$; $(-\infty, 1)$.

*4. $\int_1^{+\infty} \dfrac{dx}{x^p}$; $[1+\eta, +\infty)$; $(1, +\infty)$.

*5. $\int_0^{+\infty} e^{-ax}\, dx$; $[\eta, +\infty)$; $(0, +\infty)$.

*6. $\int_0^{+\infty} \dfrac{\sin^2 ax}{ax^2}\, dx$; $[\eta, +\infty)$; $(0, +\infty)$.

*7. $\int_0^{+\infty} x^2 y e^{-xy}\, dy$; $[\eta, +\infty)$; $(0, +\infty)$.

*8. $\int_0^{+\infty} y^{\frac{17}{5}} e^{-xy^2}\, dy$; $[\eta, +\infty)$; $(0, +\infty)$.

*9. Show that $\int_0^{+\infty} x^2 e^{-xy}\, dy$ converges uniformly for $x > 0$ and that $\int_0^{+\infty} x e^{-xy}\, dy$ does not. Hence explain the phenomenon of Example 1, § 1407.

*10. Show that the integral for the gamma function, $\int_0^{+\infty} x^{\alpha-1} e^{-x}\, dx$ (Example 9, § 415), converges uniformly for $[\eta, b]$, but not uniformly for $[\eta, +\infty)$ or for $(0, b]$, where $0 < b < +\infty$.

In Exercises 11–20, establish the given evaluation.

*11. $\int_0^{+\infty} \dfrac{e^{-a^2 x^2} - e^{-b^2 x^2}}{x^2}\, dx = (b-a)\sqrt{\pi}$, $0 \leq a < b$.

Hint: Integrate $\int_0^{+\infty} \alpha e^{-\alpha^2 x^2}\, dx = \frac{1}{2}\sqrt{\pi}$, $\alpha > 0$.

*12. $\int_0^{+\infty} x^2 e^{-x^2}\, dx = \dfrac{\sqrt{\pi}}{4}$. *Hint:* Integrate by parts.

*13. $\int_0^{\frac{\pi}{2}} x \cot x\, dx = \dfrac{\pi \ln 2}{2}$. *Hint:* Integrate by parts.

*14. $\int_0^{+\infty} \dfrac{\sin^2 \alpha x}{x^2}\, dx = \dfrac{\pi}{2}|\alpha|$. *Hint:* Integrate by parts.

*15. $\int_0^{+\infty} \dfrac{1 - \cos \alpha x}{x^2}\, dx = \dfrac{\pi}{2}|\alpha|$. *Hint:* Use an identity.

EXERCISES

*16. $\int_0^{+\infty} e^{-\left(x-\frac{a}{x}\right)^2} dx = \begin{cases} \frac{1}{2}\sqrt{\pi}, & a \geq 0, \\ \frac{1}{2}\sqrt{\pi}e^{4a}, & a \leq 0. \end{cases}$

*17. $\int_0^1 \frac{\ln(1+x)}{1+x^2} dx = \frac{\pi \ln 2}{8}$. *Hint*: Let $x = \tan\theta$, and obtain three integrals by use of the identity $\cos(\frac{1}{4}\pi - \theta) = \frac{1}{2}\sqrt{2}(\cos\theta + \sin\theta)$.

*18. $\int_0^{+\infty} \frac{\sin\alpha x \sin x}{x} dx = \frac{1}{2}\ln\left|\frac{\alpha+1}{\alpha-1}\right|, \ |\alpha| \neq 1.$

*19. $\int_0^{+\infty} \frac{\sin\alpha x \cos x}{x} dx = \frac{\pi}{2}$ if $\alpha > 1$, $\frac{\pi}{4}$ if $\alpha = 1$, 0 if $-1 < \alpha < 1$, $-\frac{\pi}{4}$ if $\alpha = -1$, and $-\frac{\pi}{2}$ if $\alpha < -1$.

*20. $\int_0^{+\infty} \frac{\cos rx}{1+x^2} dx = \frac{1}{2}\pi e^{-|r|}$. *Hints*: Let the integral define y as a function of $r > 0$, and show: $\frac{dy}{dr} = -\int_0^{+\infty} \frac{[(1+x^2)-1]\sin rx}{x(1+x^2)} dx = -\frac{\pi}{2} + \int_0^{+\infty} \frac{\sin rx}{x(1+x^2)} dx$; $\frac{d^2y}{dr^2} = y$. Therefore $y = c_1 e^r + c_2 e^{-r}$ and $dy/dr = c_1 e^r - c_2 e^{-r}$. Show that as $r \to 0+$: $y \to \frac{1}{2}\pi$ and $dy/dr \to -\frac{1}{2}\pi$.

*21. Show that the integral $\int_0^{+\infty} \frac{x \, dy}{x^2 + y^2}$ converges to $\frac{\pi}{2} \operatorname{sgn} x$ (Example 1, § 206) for all x, uniformly on any bounded set. Since the signum function is discontinuous at $x = 0$, this seems to contradict Theorem I, § 1407. Explain.

*22. Using formula (2), § 1408, show that the nth derivative of $(1+x^2)^{-1}$ is equal to
$$(-1)^{\frac{n}{2}} \int_0^{+\infty} y^n e^{-y} \cos xy \, dy, \quad n = 0, 2, 4, \cdots,$$
$$(-1)^{\frac{n+1}{2}} \int_0^{+\infty} y^n e^{-y} \sin xy \, dy, \quad n = 1, 3, 5, \cdots.$$
Hence show, with the aid of Exercise 37, § 416:
$$\left|\frac{d^n}{dx^n} \frac{1}{1+x^2}\right| \leq n!, \quad n = 0, 1, 2, \cdots.$$
(Cf. the second solution of Example 5, § 1213.)

*23. Prove that $\lim_{x \to +\infty} \int_0^{+\infty} \frac{dy}{1+xy^7} = 0.$

*24. Prove that $\lim_{x \to +\infty} \int_0^{+\infty} e^{-xy^{11}} dy = 0.$

*25. Define
$$f(x, y) \equiv \begin{cases} y^{-2} & \text{if } 0 < x < y < 1, \\ -x^{-2} & \text{if } 0 < y < x < 1, \\ 0 & \text{otherwise.} \end{cases}$$
Show that $\int_0^1 \int_0^1 f(x,y) \, dx \, dy = 1, \int_0^1 \int_0^1 f(x,y) \, dy \, dx = -1$. Explain how this is possible when every single integral involved is proper.

*26. Show that $\int_0^{+\infty} \int_0^{+\infty} (x-y)e^{-(x-y)^2} dx \, dy$ is equal to $\frac{1}{4}\sqrt{\pi}$, while the iterated integral in the reverse order is equal to $-\frac{1}{4}\sqrt{\pi}$. Explain this phenomenon.

★27. Assume that $f(x, y)$ is continuous for $x \geq a$, $y \geq c$, and that $\int_c^{+\infty} f(x, y)\, dy$ converges uniformly for $x > a$. Prove that $\lim_{x \to a+} \int_c^{+\infty} f(x, y)\, dy = \int_c^{+\infty} f(a, y)\, dy$. whenever the right-hand member exists. *Hint:* If the integral converges at $x = a$ it converges uniformly for $x \geq a$. Use Theorem I, § 1407.

★1410. THE GAMMA FUNCTION

It was shown in Example 9, § 415, that the integral defining the **gamma function**:
$$\text{(1)} \qquad \Gamma(\alpha) \equiv \int_0^{+\infty} x^{\alpha-1} e^{-x}\, dx$$
converges for $\alpha > 0$. In this section we shall investigate a few of the properties of the gamma function.

I. *Continuity and derivatives.* For $\alpha > 0$ the gamma function is continuous and, in fact, differentiable to any prescribed order. The derivatives are given by formal differentiation under the integral sign, the first two being:
$$\text{(2)} \qquad \Gamma'(\alpha) = \int_0^{+\infty} x^{\alpha-1} e^{-x} \ln x\, dx,$$
$$\text{(3)} \qquad \Gamma''(\alpha) = \int_0^{+\infty} x^{\alpha-1} e^{-x} (\ln x)^2\, dx.$$

To prove these results, we establish the uniform convergence of the integrals
$$\text{(4)} \qquad \int_0^1 x^{\alpha-1} e^{-x} (\ln x)^n\, dx \quad \text{and} \quad \int_1^{+\infty} x^{\alpha-1} e^{-x} (\ln x)^n\, dx$$
for $0 < \delta \leq \alpha \leq \gamma$, for any fixed nonnegative integer n. For the first integral of (4) this is true since the integrand is dominated by
$$x^{\delta-1} e^{-x} |\ln x|^n = o(x^{\frac{1}{2}\delta - 1}) \quad \text{as } x \to 0+,$$
and for the second integral of (4) this is true since the integrand is dominated by
$$x^{\gamma-1} e^{-x} (\ln x)^n = o(x^{-2}) \quad \text{as } x \to +\infty.$$

II. *The functional equation.* For $\alpha > 0$:
$$\text{(5)} \qquad \Gamma(\alpha + 1) = \alpha \Gamma(\alpha).$$
We show this by integrating by parts:
$$\int_\epsilon^b x^\alpha e^{-x}\, dx = \int_\epsilon^b -x^\alpha\, d(e^{-x}) = -\left[x^\alpha e^{-x}\right]_\epsilon^b + \int_\epsilon^b \alpha x^{\alpha-1} e^{-x}\, dx.$$
The formula results from letting $\epsilon \to 0+$ and $b \to +\infty$.

III. *Relation to the factorial function.* For every nonnegative integer n:
$$\text{(6)} \qquad \Gamma(n + 1) = n!.$$
This is true for $n = 0$ (recall that 0! is *defined* to be equal to 1), since $\Gamma(1) = \int_0^{+\infty} e^{-x}\, dx = 1$. By the functional equation (5), $\Gamma(2) = \Gamma(1) = 1$,

$\Gamma(3) = 2\Gamma(2) = 2!$, $\Gamma(4) = 3\Gamma(3) = 3!$, etc. The general statement (6) is obtained by mathematical induction.

IV. *Monotonic behavior.* Since the integrands in (1) and (3) are positive, $\Gamma(\alpha) > 0$ and $\Gamma''(\alpha) > 0$ for all $\alpha > 0$, and the graph of $\Gamma(\alpha)$ therefore lies above the α-axis and is concave upward for all $\alpha > 0$. Furthermore, $\Gamma(\alpha)$ has exactly one relative minimum value which, since $\Gamma(1) = \Gamma(2) = 1$, is

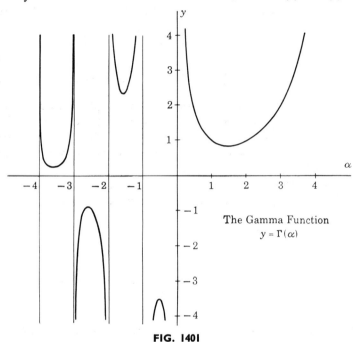

FIG. 1401

attained for some $\alpha = \alpha_0$, $1 < \alpha_0 < 2$.† Furthermore, $(\Gamma\alpha)$ must be strictly decreasing for $0 < \alpha < \alpha_0$ and strictly increasing for $\alpha > \alpha_0$. (See Fig. 1401, above, for the graph of the gamma function.)

V. *Two limits.*

(7) $$\Gamma(0+) = +\infty, \quad \Gamma(+\infty) = +\infty.$$

The first of these follows from the continuity of the gamma function at $\alpha = 1$, the fact that $\Gamma(1) = 1$, and the functional relation (5), α being permitted to $\to 0+$. The second is a consequence of (6) and the monotonic behavior of $\Gamma(\alpha)$.

VI. *Evaluation of* $\Gamma(\tfrac{1}{2})$. By means of the substitution $y = \sqrt{x}$, $x = y^2$, we find:
$$\Gamma(\tfrac{1}{2}) = \int_0^{+\infty} x^{-\tfrac{1}{2}} e^{-x}\, dx = 2\int_0^{+\infty} e^{-y^2}\, dy.$$
From Example 3, § 1013, we have
(8) $$\Gamma(\tfrac{1}{2}) = \sqrt{\pi}.$$

† The value of α_0 has been computed to be $1{\cdot}461632\cdots$, and $\Gamma(\alpha_0) = 0{\cdot}885603\cdots$.

From (5) we deduce, furthermore:

(9) $\quad \Gamma\left(\dfrac{3}{2}\right) = \dfrac{1}{2}\sqrt{\pi}, \quad \Gamma\left(\dfrac{5}{2}\right) = \dfrac{1\cdot 3}{2^2}\sqrt{\pi}, \quad \Gamma\left(\dfrac{7}{2}\right) = \dfrac{1\cdot 3\cdot 5}{2^3}\sqrt{\pi}, \cdots.$

VII. *$\Gamma(\alpha)$ for $\alpha < 0$.* Although the gamma function is defined by (1) only for $\alpha > 0$, the functional equation (5) permits extending the domain of definition to all nonintegral negative numbers, by means of the form

(10) $$\Gamma(\alpha) = \frac{\Gamma(\alpha+1)}{\alpha}.$$

For example, $\Gamma(-\tfrac{1}{2}) = \dfrac{\Gamma(\tfrac{1}{2})}{-\tfrac{1}{2}} = -2\sqrt{\pi}.$ The general character of the extended gamma function is indicated by its graph, Figure 1401. This graph has vertical asymptotes for nonpositive integral values of α. Between successive asymptotes the graph of $\Gamma(\alpha)$ alternates between the upper and lower half planes, and the successive relative maxima and minima of $\Gamma(\alpha)$ approach 0 as $\alpha \to -\infty$ (Ex. 15, § 1412).

VIII. *Other expressions for $\Gamma(\alpha)$.* By means of the indicated changes of variable in formula (1), the following two forms for $\Gamma(\alpha)$ result:

(11) $$\Gamma(\alpha) = 2\int_0^{+\infty} x^{2\alpha-1} e^{-x^2}\, dx, \quad \alpha > 0,$$

(set $x = y^2$),

(12) $$\Gamma(\alpha) = p^\alpha \int_0^{+\infty} x^{\alpha-1} e^{-px}\, dx, \quad \alpha > 0, \quad p > 0,$$

(set $x = py$).

Two other forms of the gamma function are discussed in § 1417.

IX. *Tables of the gamma function.* By virtue of the functional equation (5), any value of $\Gamma(\alpha)$ can be computed if its values are known for $1 \leq \alpha \leq 2$. Some tabulations of the gamma function are given in B. O. Peirce, *A Short Table of Integrals* (New York, Ginn and Company, 1956) and Eugen Jahnke and Fritz Emde, *Tables of Functions* (New York, Dover Publications, 1943).

*1411. THE BETA FUNCTION

It was shown in Example 8, § 415, that the integral defining the **beta function**:

(1) $$B(p,q) \equiv \int_0^1 x^{p-1}(1-x)^{q-1}\, dx$$

converges if and only if both p and q are positive. The substitution $y = 1 - x$ shows that the beta function is symmetric in p and q:

(2) $\quad B(p,q) = B(q,p), \quad p > 0, \quad q > 0.$

The two substitutions $x = \sin^2 \theta$ and $x = y/(1 + y)$ give the following two alternative expressions ($p > 0$, $q > 0$):

(3) $$B(p, q) = 2 \int_0^{\frac{\pi}{2}} \sin^{2p-1} \theta \cos^{2q-1} \theta \, d\theta,$$

(4) $$B(p, q) = \int_0^{+\infty} \frac{x^{p-1}}{(1 + x)^{p+q}} \, dx.$$

An important relation expressing the beta function in terms of the gamma function is:

(5) $$B(p, q) = \frac{\Gamma(p)\Gamma(q)}{\Gamma(p + q)}, \quad p > 0, \quad q > 0.$$

We shall now prove this.

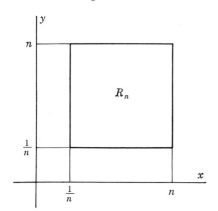 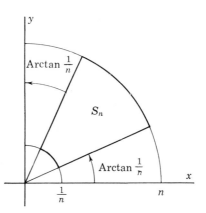

FIG. 1402

Estimates will be obtained by integration of the function

(6) $$f(x, y) \equiv 4x^{2p-1} y^{2q-1} e^{-(x^2+y^2)}$$

over the two sets in the first quadrant (ρ and θ representing polar coordinates; cf. Fig. 1402):

$$R_n : \begin{cases} \frac{1}{n} \leq x \leq n, \\ \frac{1}{n} \leq y \leq n, \end{cases} \quad S_n : \begin{cases} \frac{1}{n} \leq \rho \leq n. \\ \frac{1}{n} \leq \operatorname{Arctan} \theta \leq n. \end{cases}$$

It is a matter of elementary geometry to show the inclusions (Ex. 16, § 1412):

(7) $$R_n \subset S_{n^2}, \quad S_n \subset R_{n^3},$$

and hence, if

$$I_n \equiv \iint_{R_n} f(x, y) \, dA, \quad J_n \equiv \iint_{S_n} f(x, y) \, dA,$$

(8) $$I_n \leq J_{n^2}, \quad J_n \leq I_{n^3},$$

Since $f(x, y) \geq 0$, $I_n \uparrow$ and $J_n \uparrow$, as n increases, and by the inequalities (8),
(9) $$I \equiv \lim_{n \to +\infty} I_n = \lim_{n \to +\infty} J_n.$$
We now evaluate I_n and J_n:

$$I_n = \left\{ 2 \int_{\frac{1}{n}}^{n} x^{2p-1} e^{-x^2} dx \right\} \left\{ 2 \int_{\frac{1}{n}}^{n} y^{2q-1} e^{-y^2} dy \right\},$$

$$J_n = 4 \int_{\text{Arctan} \frac{1}{n}}^{\text{Arctan } n} \int_{\frac{1}{n}}^{n} (\rho \cos \theta)^{2p-1} (\rho \sin \theta)^{2q-1} e^{-\rho^2} \rho \, d\rho \, d\theta$$

$$= \left\{ 2 \int_{\frac{1}{n}}^{n} \rho^{2(p+q)-1} e^{-\rho^2} d\rho \right\} \left\{ 2 \int_{\text{Arctan} \frac{1}{n}}^{\text{Arctan } n} \sin^{2q-1} \theta \cos^{2p-1} \theta \, d\theta \right\}.$$

Therefore, by (9) above, formula (11), § 1410, for the gamma function, and formulas (2) and (3), above, for the beta function, we have in the limit, as $n \to +\infty$:

$$I = \Gamma(p)\Gamma(q) = \Gamma(p+q) B(p, q),$$

and formula (5), as desired.

★1412. EXERCISES

In Exercises 1–8, establish the given formula for the specified values of the variables.

★1. $\Gamma(\alpha) = \int_0^1 \left(\ln \frac{1}{x} \right)^{\alpha - 1} dx$, $\alpha > 0$.

★2. $\Gamma(\alpha) = p^\alpha \int_0^1 x^{p-1} \left(\ln \frac{1}{x} \right)^{\alpha - 1} dx$, $\alpha > 0, p > 0$.

★3. $\Gamma\left(n + \frac{1}{2}\right) = \frac{(2n)! \sqrt{\pi}}{4^n n!}$, $n = 0, 1, 2, \cdots$.

★4. $B(p+1, q) = \frac{p}{p+q} B(p, q)$, $p > 0, q > 0$.

★5. $\int_0^{+\infty} \frac{x^{\alpha - 1}}{1 + x} dx = \Gamma(\alpha) \Gamma(1 - \alpha)$, $0 < \alpha < 1$.

★6. $\int_0^1 x^{p-1}(1 - x^r)^{q-1} dx = \frac{1}{r} B\left(\frac{p}{r}, q\right)$, $p > 0, q > 0, r > 0$.

★7. $\int_0^1 \frac{x^n \, dx}{\sqrt{1 - x^2}} = \frac{\sqrt{\pi}}{2} \frac{\Gamma\left(\frac{n+1}{2}\right)}{\Gamma\left(\frac{n+2}{2}\right)}$, $n > 0$. (Cf. Ex. 6.)

★8. $\int_0^1 \frac{dx}{\sqrt{1 - x^n}} = \frac{\sqrt{\pi}}{n} \frac{\Gamma\left(\frac{1}{n}\right)}{\Gamma\left(\frac{1}{n} + \frac{1}{2}\right)}$, $n > 0$. (Cf. Ex. 6.)

In Exercises 9–12, evaluate by means of formulas already established.

★9. $\int_0^{+\infty} x^4 e^{-x^2}\, dx.$

★10. $\int_0^{+\infty} \dfrac{e^{-5x}}{\sqrt{x}}\, dx.$

★11. $\int_0^1 \left[\dfrac{1}{x}\ln\dfrac{1}{x}\right]^{\frac{1}{2}} dx.$

★12. $\int_0^1 \dfrac{x^5\, dx}{\sqrt{1-x^4}}.$

★13. Verify, by means of the substitution $u = 2\theta$ ($\alpha > 0$):

$$B(\alpha, \alpha) = 2\int_0^{\frac{\pi}{2}} \sin^{2\alpha-1}\theta \cos^{2\alpha-1}\theta\, d\theta = 2^{-2\alpha+2}\int_0^{\frac{\pi}{2}} \sin^{2\alpha-1} 2\theta\, d\theta$$

$$= 2^{-2\alpha+2}\int_0^{\frac{\pi}{2}} \sin^{2\alpha-1} u\, du = 2^{-2\alpha+1} B(\alpha, \tfrac{1}{2}),$$

and hence derive the duplication formula of A. M. Legendre (1752–1833):

$$\Gamma(2\alpha) = \dfrac{1}{\sqrt{\pi}} 2^{2\alpha-1} \Gamma(\alpha) \Gamma\left(\alpha + \dfrac{1}{2}\right).$$

★14. Show that the substitution $x = \sqrt{\tan\theta}$ leads to the following evaluation

$$\int_0^{+\infty} \dfrac{dx}{1+x^4} = \dfrac{1}{2}\int_0^{\frac{\pi}{2}} \cos^{\frac{1}{2}}\theta \sin^{-\frac{1}{2}}\theta\, d\theta = \dfrac{1}{4} B\left(\dfrac{3}{4}, \dfrac{1}{4}\right)$$

$$= \dfrac{1}{4}\Gamma\left(\dfrac{3}{4}\right)\Gamma\left(\dfrac{1}{4}\right) = \dfrac{\sqrt{\pi}}{4}\sqrt{2}\,\Gamma\left(\dfrac{1}{2}\right) = \dfrac{\pi}{2\sqrt{2}}.$$

(Cf. Ex. 13.)

★15. Prove the statements of the last sentence of VII, § 1410, regarding the graph of $\Gamma(\alpha)$ for $\alpha < 0$.

★16. Establish the inclusions (7), § 1411.

★17. Assume that the function $f(x)$ is Riemann integrable on $[\alpha, \beta]$ whenever $0 < \alpha < \beta$, that $f(x)$ is properly or improperly integrable on $[0, 1]$, and that for some real number r this function satisfies the order-of-magnitude relation at $+\infty$:

$$f(x) = O(e^{rx}) \quad \text{as} \quad x \to +\infty.$$

Prove that the integral

(1) $$F(s) \equiv \int_0^{+\infty} f(x) e^{-sx}\, dx$$

converges for $s > r$, and uniformly for $s \geq r + \eta$, where η is a fixed positive number. The function $F(s)$ defined by (1) ($s > r$) is called the **Laplace transform** of the given function $f(x)$. For an elementary treatment of the Laplace transform and its applications to differential equations, see D. V. Widder, *Advanced Calculus* (Englewood Cliffs, N.J., Prentice-Hall, 1961), chapters XIII and XIV, and D. L. Holl, C. G. Maple, and B. Vinograde, *Introduction to the Laplace Transform* (New York, Appleton-Century-Crofts, Inc., 1959). (Also cf. Exs. 16–20, § 1514.)

In Exercises 18–23, find the Laplace transform $F(s)$ of the given function $f(x)$, and state the values of s for which $F(s)$ is defined. (Cf. Ex. 17.)

★18. $f(x) = 1.$

★19. $f(x) = \sqrt{x}.$

★20. $f(x) = e^{ax}.$

★21. $f(x) = x^{p-1} e^{ax}, p > 0.$

★22. $f(x) = e^{ax} \sin bx.$

★23. $f(x) = e^{ax} \sinh bx.$

★1413. INFINITE PRODUCTS

An expression of the form

(1) $$\prod_{n=1}^{+\infty} p_n = p_1 p_2 p_3 \cdots,$$

or

(2) $$\prod_{n=1}^{+\infty} (1 + a_n) = (1 + a_1)(1 + a_2)(1 + a_3) \cdots,$$

where $1 + a_n = p_n$, $n = 1, 2, \cdots$, is called an **infinite product**. The **partial products** are

(3) $$P_n \equiv p_1 p_2 \cdots p_n = (1 + a_1)(1 + a_2) \cdots (1 + a_n), \quad n = 1, 2, \cdots.$$

For simplicity, we shall assume in this chapter that *all factors p_n are nonzero*—in other words, that a_n *is never equal to* -1. Under this assumption, the infinite product (1) or (2) is said to **converge** if and only if the limit of the partial products exists and is finite and nonzero:

(4) $$\prod_{n=1}^{+\infty} p_n \equiv \lim_{n \to +\infty} P_n = P \neq 0.$$

In all other cases the infinite product **diverges**. For example, if

(5) $$\lim_{n \to +\infty} P_n = 0,$$

the infinite product is said to **diverge to zero**. In case of convergence the number P in (4) is called the **value** of the infinite product.

Many of the theorems on infinite series have their analogues in the theory of infinite products. We shall content ourselves here with establishing three basic theorems. For a more thorough treatment of infinite products, see E. T. Whittaker and G. N. Watson, *Modern Analysis* (Cambridge University Press, 1935), or E. W. Hobson, *The Theory of Functions of a Real Variable* (Washington, D. C., The Harren Press, 1950).

Theorem I. *If the infinite product* $\prod_{n=1}^{+\infty} p_n$ *converges, the **general factor** p_n tends toward 1 as a limit:*

$$\lim_{n \to +\infty} p_n = 1.$$

Proof. $\lim_{n \to +\infty} p_n = \dfrac{\lim P_n}{\lim P_{n-1}} = \dfrac{P}{P} = 1.$

Theorem II. *Convergence of the infinite product* $\prod_{n=1}^{+\infty} p_n$ *implies convergence of the infinite series*

(6) $$\sum_{n=1}^{+\infty} \ln |p_n|.$$

The converse also holds in case $p_n > 0$ for sufficiently large n.

Proof. By Theorem I, the general factor p_n is positive for sufficiently large n, and there is no loss of generality in assuming that $p_n > 0$ for *all* n. In this case the result follows from the relations

$$S_n \equiv \sum_{k=1}^{n} \ln p_k = \ln P_n, \quad P_n = e^{S_n},$$

and the continuity of $\ln x$ and e^x.

Theorem III. *Consider the infinite product*
(7) $$(1 + a_1)(1 + a_2)(1 + a_3) \cdots$$
and the two infinite series
(8) $$a_1 + a_2 + a_3 + \cdots,$$
(9) $$a_1^2 + a_2^2 + a_3^2 + \cdots.$$

(i) *The convergence of any two of (7)–(9) implies that of the third.*
(ii) *The absolute convergence of (8) implies the convergence of both (7) and (9).*
(iii) *In case (8) converges conditionally, (7) converges or diverges to zero according as (9) converges or diverges.*

Proof. Without loss of generality we shall assume that $|a_n| < \frac{1}{2}$ for all n, since the convergence of any one of (7)–(9) implies this inequality for n sufficiently large. Taylor's Formula with a Remainder in the Lagrange form (§ 1204), applied to the function $f(x) = \ln(1 + x)$ for $n = 2$, is

$$\ln(1 + x) = x - \frac{x^2}{2(1 + \xi)^2},$$

where $0 \leq |\xi| \leq |x|$. Therefore, with the assumption $|a_n| < \frac{1}{2}$,
(10) $$c_n \equiv \ln(1 + a_n) = a_n - b_n,$$
where
(11) $$\tfrac{2}{9} a_n^2 \leq b_n \leq 2 a_n^2.$$

Since Σb_n is a positive series which, by (11), converges if and only if (9) converges, the statement (i) follows from the relation (10), thanks to Theorem II. Conclusion (ii) follows from (i) and the inequality $a_n^2 \leq \frac{1}{2}|a_n|$. Conclusion (iii) follows in part from (i) and in part from the fact that if Σa_n converges and $\Sigma a_n^2 = +\infty$, $\Sigma \ln(1 + a_n) = -\infty$ (from (10) and (11)).

Example 1. The infinite product

$$\left(1 - \frac{1}{4 \cdot 1^2}\right)\left(1 - \frac{1}{4 \cdot 2^2}\right)\left(1 - \frac{1}{4 \cdot 3^2}\right) \cdots \left(1 - \frac{1}{4n^2}\right) \cdots$$

converges, since the series $\Sigma 1/4n^2$ converges absolutely. Its value is $2/\pi$ (as can be seen by factoring each quantity in parentheses and comparing the result with (1), § 1414).

Example 2. The infinite product
$$\left(1+\frac{1}{1}\right)\left(1-\frac{1}{3}\right)\left(1+\frac{1}{3}\right)\left(1-\frac{1}{5}\right)\left(1+\frac{1}{5}\right)\cdots$$
converges, since the series $1 - \frac{1}{3} + \frac{1}{3} - \frac{1}{5} + \frac{1}{5} - \cdots$ and $1 + \frac{1}{3^2} + \frac{1}{3^2} + \frac{1}{5^2} + \frac{1}{5^2} + \cdots$ both converge. Its value is $\frac{\pi}{2}$, as shown in § 1414.

Example 3. The infinite product
$$\left(1+\frac{1}{1}\right)\left(1-\frac{1}{\sqrt{2}}\right)\left(1+\frac{1}{\sqrt{3}}\right)\left(1-\frac{1}{\sqrt{4}}\right)\cdots$$
diverges to zero, by *(iii)*, Theorem III.

*1414. WALLIS'S INFINITE PRODUCT FOR π

We wish to establish the following limit formula, due to the English mathematician J. Wallis (1616–1703):

(1)
$$\frac{\pi}{2} = \frac{2}{1} \cdot \frac{2}{3} \cdot \frac{4}{3} \cdot \frac{4}{5} \cdot \frac{6}{5} \cdot \frac{6}{7} \cdots$$
$$= \lim_{n \to +\infty} \frac{2^{4n}(n!)^4}{[(2n)!]^2(2n+1)}.$$

To this end we recall the two Wallis formulas, obtained by integration by parts and mathematical induction (cf. Ex. 36, § 416):

(2)
$$\begin{cases} I_{2n} \equiv \int_0^{\frac{\pi}{2}} \sin^{2n} x \, dx = \frac{\pi}{2} \cdot \frac{1 \cdot 3 \cdots (2n-1)}{2 \cdot 4 \cdots 2n}, \\ I_{2n+1} \equiv \int_0^{\frac{\pi}{2}} \sin^{2n+1} x \, dx = \frac{2 \cdot 4 \cdots (2n)}{3 \cdot 5 \cdots (2n+1)}, \end{cases}$$

for $n = 1, 2, \cdots$. Since, on the interval $[0, \tfrac{1}{2}\pi]$, $0 \leq \sin x \leq 1$, $I_{2n} \geq I_{2n+1} \geq I_{2n+2} > 0$, and therefore

(3)
$$1 \geq \frac{I_{2n+1}}{I_{2n}} \geq \frac{I_{2n+2}}{I_{2n}} = \frac{2n+1}{2n+2} \to 1.$$

It follows that $I_{2n+1}/I_{2n} \to 1$, and therefore

(4)
$$\frac{\pi}{2} \cdot \frac{I_{2n+1}}{I_{2n}} = \frac{2 \cdot 4 \cdots 2n}{1 \cdot 3 \cdots (2n-1)} \cdot \frac{2 \cdot 4 \cdots 2n}{3 \cdot 5 \cdots (2n+1)} \to \frac{\pi}{2}.$$

But this states that the partial products of an even number of factors of the infinite product (1) have $\tfrac{1}{2}\pi$ as a limit. Since the general factor of (1) tends toward 1, the *general* partial product also has $\tfrac{1}{2}\pi$ as a limit. The second form of (1) is obtained by multiplying numerator and denominator of (4) by $(2 \cdot 4 \cdots 2n)^2$.

*1415. EULER'S CONSTANT

A constant important in number theory and parts of the theory of analytic functions of a complex variable, known as **Euler's constant**, is defined

(1)
$$C \equiv \lim_{n \to +\infty} C_n,$$

where

(2)
$$C_n \equiv \left(1 + \frac{1}{2} + \cdots + \frac{1}{n}\right) - \ln n.$$

We shall show that C exists and is positive by establishing:

(3) $\quad C_n \downarrow,$

(4) $\quad C_n > \tfrac{1}{2}.$

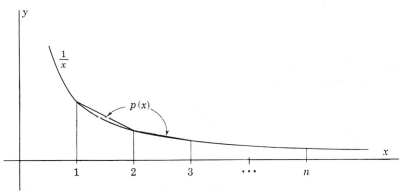

FIG. 1403

Proof of (3): The problem is to show that

$$C_n - C_{n-1} = \frac{1}{n} - \ln \frac{n}{n-1} < 0, \quad n > 1.$$

But this is equivalent to proving

$$\ln\left(1 - \frac{1}{n}\right) < -\frac{1}{n}, \quad n > 1.$$

This was established by the Law of the Mean in Example 2, § 305.

Proof of (4): On the interval $[1, n]$ define the function $p(x)$ by linear interpolation between successive pairs of points $(1, 1)$, $(2, \tfrac{1}{2})$, $(3, \tfrac{1}{3})$, $\cdots$, $\left(n, \tfrac{1}{n}\right)$ (cf. Fig. 1403). Then since the graph of $y = \tfrac{1}{x}$ is everywhere concave upward, the inequality $p(x) - \tfrac{1}{x} \geq 0$ holds throughout $[1, n]$ (why?). The resulting inequality

becomes
$$\int_1^n \left[p(x) - \frac{1}{x}\right] dx > 0$$

$$\left[\frac{1}{2} + \frac{1}{2} + \frac{1}{3} + \cdots + \frac{1}{n-1} + \frac{1}{2n}\right] - \ln n = C_n - \frac{1}{2} - \frac{1}{2n} > 0,$$

and (4) follows.

To ten decimal places, $C = 0.5772156649$.

*1416. STIRLING'S FORMULA

An important "asymptotic formula" for $n!$, due to the eighteenth century English mathematician James Stirling, gives the expression

(1) $$S_n \equiv \sqrt{2\pi n}\left(\frac{n}{e}\right)^n$$

as an "approximation" to $n!$, for large values of n, in the sense that

(2) $$\lim_{n \to +\infty} \frac{n!}{S_n} = 1.$$

It is our purpose in this section to prove the inequality for $n = 3, 4, \cdots$:

(3) $$S_n\left[1 + \frac{1}{12(n+1)}\right] < n! < S_n\left[1 + \frac{1}{12(n-2)}\right].$$

This inequality implies (2) and, in addition, gives a measure of the accuracy of S_n as an approximation for $n!$. In particular, we can infer from (3) that

(4) $$\frac{n!}{S_n} = 1 + \frac{1}{12n} + o\left(\frac{1}{n^2}\right),$$

and that for $n > 13$

(5) $$n! - S_n > \frac{(n-1)!}{13}.$$

In other words, Stirling's formula can be interpreted as a good approximation of $n!$ only in the *relative* sense of (4), whereas the actual *difference* between $n!$ and S_n grows at a rapid rate—at least as rapidly as $(n-1)!/13$.

Our derivation of (3) is based on the following trapezoidal formula for the integral of a function $f(x)$ possessing a second derivative over an interval $[a, b]$ (cf. Ex. 16, § 402):

(6) $$\int_a^b f(x)\,dx = \tfrac{1}{2}[f(a) + f(b)](b-a) - \tfrac{1}{12}f''(\xi)(b-a)^3,$$

where ξ is a suitable number between a and b.

We now consider the function $f(x) = \ln x$ on the interval $[1, n]$, integrate, and use (6) on the successive intervals $[1, 2], \cdots, [n-1, n]$:

$$\int_1^n \ln x\,dx = \int_1^2 \ln x\,dx + \int_2^3 \ln x\,dx + \cdots + \int_{n-1}^n \ln x\,dx$$

$$= \tfrac{1}{2}(\ln 1 + \ln 2) + \tfrac{1}{2}(\ln 2 + \ln 3) + \cdots \tfrac{1}{2}(\ln(n-1) + \ln n)$$

$$+ \frac{1}{12}\left[\frac{1}{\xi_2^2} + \frac{1}{\xi_3^2} + \cdots + \frac{1}{\xi_n^2}\right],$$

where $1 < \xi_2 < 2, \cdots, n-1 < \xi_n < n$. This relation can be written

(7) $$\left(n + \frac{1}{2}\right)\ln n - n + 1 - \ln n! = \frac{1}{12}\left[\frac{1}{\xi_2^2} + \cdots + \frac{1}{\xi_n^2}\right].$$

§ 1417] WEIERSTRASS'S INFINITE PRODUCT FOR $1/\Gamma(\alpha)$ 491

Since $\xi_n > n - 1$, the series $\Sigma \xi_n^{-2}$ converges. Define

$$s \equiv \sum_{k=2}^{+\infty} \frac{1}{\xi_k^2}, \quad r_n \equiv \sum_{k=n+1}^{+\infty} \frac{1}{\xi_k^2} = s - \sum_{k=2}^{n} \frac{1}{\xi_k^2}.$$

Equation (7) now becomes

(8) $\qquad \ln n! = (n + \tfrac{1}{2}) \ln n - n + 1 - s + r_n,$

or, with $\alpha \equiv e^{1-s}$:

(9) $\qquad n! = \alpha e^{r_n} \sqrt{n} \left(\frac{n}{e}\right)^n.$

Substitution of (9) in Wallis's product formula (1), § 1414, gives an evaluation of the constant α, since $r_n \to 0$:

$$\frac{\pi}{2} = \lim_{n \to +\infty} \frac{2^{4n} \alpha^4 e^{4r_n} n^2 (n/e)^{4n}}{\alpha^2 e^{2r_{2n}} \cdot 2n \cdot (2n/e)^{4n}(2n+1)}$$

$$= \lim_{n \to +\infty} \alpha^2 \frac{e^{4r_n}}{e^{2r_{2n}}} \cdot \lim_{n \to +\infty} \frac{n}{2(2n+1)} = \frac{\alpha^2}{4},$$

and $\alpha = \sqrt{2\pi}$. Therefore (9) can be written

(10) $\qquad n! = S_n e^{r_n}.$

Using the inequalities $n - 1 < \xi_n < n$, and the integral-test estimate for infinite series (Theorem II, § 1120), we have

(11) $\quad \dfrac{1}{12(n+1)} < \dfrac{1}{12}\left[\dfrac{1}{(n+1)^2} + \dfrac{1}{(n+2)^2} + \cdots \right] < r_n$

$$< \frac{1}{12}\left[\frac{1}{n^2} + \frac{1}{(n+1)^2} + \cdots \right] < \frac{1}{12(n-1)}.$$

By Taylor's Formula with a Remainder in the Lagrange form (§ 1204), if $0 < x \leq \tfrac{1}{2}$:

(12) $\qquad 1 + x < e^x = 1 + x + \dfrac{e^\xi}{2} x^2 < 1 + x + x^2,$

where $0 < \xi < x$. Therefore, with x equal to $1/12(n+1)$ and $1/12(n-1)$ in turn, we have from (11), if $n \geq 3$:

$$1 + \frac{1}{12(n+1)} < e^{r_n} < 1 + \frac{1}{12(n-1)} + \frac{1}{144(n-1)^2}$$

$$< 1 + \frac{1}{12(n-1)} + \frac{1}{12(n-1)(n-2)} = 1 + \frac{1}{12(n-2)},$$

and the derivation of (3) is complete.

★1417. WEIERSTRASS'S INFINITE PRODUCT FOR $1/\Gamma(\alpha)$

In this section we shall derive an important formula used by Weierstrass in his research on the gamma function:

(1) $\qquad \dfrac{1}{\Gamma(\alpha)} = e^{C\alpha} \alpha \prod_{n=1}^{+\infty} \left(1 + \dfrac{\alpha}{n}\right) e^{-\frac{\alpha}{n}}, \quad \alpha > 0,$

where C is Euler's constant (§ 1415). This product formula is important, in part, because it represents the (reciprocal of the) gamma function not only for $\alpha > 0$, but for all values of α (including complex numbers) except $\alpha = 0, -1, -2, \cdots$. We shall prove (1) only for real $\alpha > 0$. For a more complete discussion see E. T. Whittaker and G. N. Watson, *Modern Analysis* (Cambridge University Press, 1935), or L. V. Ahlfors, *Complex Analysis* (New York, McGraw-Hill Book Company, 1953).

We shall obtain (1) by first proving

(2) $$\Gamma(\alpha) = \lim_{n \to +\infty} \int_0^n \left(1 - \frac{x}{n}\right)^n x^{\alpha-1}\, dx, \quad \alpha > 0,$$

and then the formula due to L. Euler (1707–1783):

(3) $$\Gamma(\alpha) = \lim_{n \to +\infty} \frac{n!\, n^\alpha}{\alpha(\alpha+1)\cdots(\alpha+n)}, \quad \alpha > 0.$$

In order to prove (2), since $\Gamma(\alpha) = \lim_{n \to +\infty} \int_0^n e^{-x} x^{\alpha-1}\, dx$, we must show:

$$\lim_{n \to +\infty} \int_0^n \left[1 - e^x\left(1 - \frac{x}{n}\right)^n\right] e^{-x} x^{\alpha-1}\, dx = 0.$$

This follows from the inequalities

$$0 \le 1 - e^x\left(1 - \frac{x}{n}\right)^n \le \frac{x^2}{n}, \quad 0 \le x \le n,$$

which are consequences, in turn, of the three inequalities, established by the Law of the Mean (cf. Exs. 23, 29, § 308):

$$1 + h \le e^h, \qquad h \ge 0$$
$$1 - h \le e^{-h}, \qquad h \ge 0$$
$$1 - nh \le (1-h)^n,\ 0 \le h \le 1,\ n \ge 1,$$

as follows:

$$1 - e^x\left(1 - \frac{x}{n}\right)^n \ge 1 - e^x\left(e^{-\frac{x}{n}}\right)^n = 1 - e^x e^{-x} = 0,$$

$$1 - e^x\left(1 - \frac{x}{n}\right)^n = 1 - \left(e^{\frac{x}{n}}\right)^n\left(1 - \frac{x}{n}\right)^n \le 1 - \left(1 + \frac{x}{n}\right)^n\left(1 - \frac{x}{n}\right)^n$$

$$= 1 - \left(1 - \frac{x^2}{n^2}\right)^n \le 1 - \left(1 - n\frac{x^2}{n^2}\right) = \frac{x^2}{n}.$$

The substitution of $x = nu$ in the integral appearing in (2) gives

$$\int_0^n \left(1 - \frac{x}{n}\right)^n x^{\alpha-1}\, dx = n^\alpha \int_0^1 u^{\alpha-1}(1-u)^n\, du,$$

which reduces, by means of repeated integration by parts that reduce the exponent on $(1-u)$, to the quantity whose limit is formed in (3).

Finally, the reciprocal of (3) is equal to

$$\lim_{n \to +\infty} \alpha \left(1 + \frac{\alpha}{1}\right)\left(1 + \frac{\alpha}{2}\right) \cdots \left(1 + \frac{\alpha}{n}\right) e^{-\alpha \ln n}$$

$$= \lim_{n \to +\infty} \alpha \left(1 + \frac{\alpha}{1}\right) e^{-\alpha} \left(1 + \frac{\alpha}{2}\right) e^{-\frac{\alpha}{2}} \cdots \left(1 + \frac{\alpha}{n}\right) e^{-\frac{\alpha}{n}} e^{C_n \alpha},$$

where C_n is the quantity (2), § 1415, whose limit is equal, by definition, to Euler's constant. This fact leads immediately to the desired formula (1).

*1418. EXERCISES

*1. The infinite product $\prod_{n=1}^{+\infty} (1 + a_n)$ is said to be **absolutely convergent** if and only if the series $\sum_{n=1}^{+\infty} \ln |1 + a_n|$ is absolutely convergent. Prove that an absolutely convergent infinite product is convergent. Prove that $\prod_{n=1}^{+\infty} (1 + a_n)$ is absolutely convergent if and only if $\sum_{n=1}^{+\infty} a_n$ is absolutely convergent. *Hint:* Cf. (10) and (11), § 1413, assume $|a_n| < \frac{1}{4}$, and show that $|c_n| \leq |a_n| + |b_n| \leq 2|a_n|$, and $|a_n| \leq |c_n| + \frac{1}{2}|a_n|$.

*2. Prove that the factors of an absolutely convergent infinite product can be rearranged arbitrarily without affecting absolute convergence or the value of the product. (Cf. Ex. 1.)

*3. Use Stirling's formula to show that $\lim_{n \to +\infty} \sqrt[n]{\frac{n^n}{n!}} = e$. (Cf. Ex. 33, § 1111.)

*1419. IMPROPER MULTIPLE INTEGRALS

The definition of multiple integral, given in Chapter 10, can be extended to apply to unbounded functions or domains of integration. As might be expected, by comparison with improper integrals of functions of a single variable, this involves a process equivalent to taking a limit of a (proper) multiple integral. One important distinction between improper multiple integrals and improper single integrals is that the variety of ways in which a limit can be formed makes it impractical to formulate improper multiple integrals in such a way as to include conditional convergence (cf. § 1401). For this reason we first assume the integrand to be nonnegative (Definition I), and then, for a more general function, consider its positive and negative parts (Definition II).

Definition I. *Let $f(x, y)$ be defined and nonnegative on a set R, and integrable over every compact subset R_1 of R that has area. Then the **double integral** of f over R is defined:*

(1) $$\iint_R f(x, y) \, dA \equiv \sup_{R_1 \subset R} \iint_{R_1} f(x, y) \, dA,$$

taken over all compact subsets R_1 of R that have area. *If f or R is unbounded, the integral (1) is **improper**. If its value is finite it is **convergent**.*

Similar statements hold for higher dimensions.

Definition II. *Let $f(x, y)$ be defined on a set R, and let f^+ and f^- be its positive and negative parts, respectively:*

(2) $\qquad f^+(x, y) \equiv \max(f(x, y), 0), \quad f^-(x, y) \equiv \max(-f(x, y), 0).$

*Then f is **integrable** over R if and only if both f^+ and f^- have finite integrals (1), and*

(3) $$\iint_R f(x, y)\, dA \equiv \iint_R f^+(x, y)\, dA - \iint_R f^-(x, y)\, dA.$$

*If f or R is unbounded, the integral (3) is **improper**. Under the finiteness conditions prescribed, the integral is **convergent**.*

Similar statements hold for higher dimensions.

NOTE 1. Formulas (1) and (3) are equivalent to definitions given before in case the appropriate conditions of integrability and having area are satisfied. (Cf. Ex. 9, § 1420.)

NOTE 2. Many of the familiar properties of multiple integrals, established for previous definitions, hold for Definitions I and II. For example, if f and g are integrable over R, according to Definition II, then so is $f + g$, and

$$\iint_R (f + g)\, dA = \iint_R f\, dA + \iint_R g\, dA. \quad \text{(Cf. Ex. 10, § 1420.)}$$

A method for the evaluation of a multiple integral that is often convenient is given in the theorem (for hints on a proof, cf. Ex. 11, § 1420):

Theorem. *Assume that $\iint_R f(x, y)$ exists (finite or infinite) according to Definition I or II, and let $\{R_n\}$ be an increasing sequence of compact sets with area the union of whose interiors is the interior of R. Then*

$$\lim_{n \to +\infty} \iint_{R_n} f\, dA = \iint_R f\, dA.$$

The Transformation Theorem for multiple integrals (§ 1026) can be adapted to improper multiple integrals. For a discussion, see *RV*, § 1329.

Example 1. By Example 3, § 1013, the value of the improper double integral of $e^{-(x^2+y^2)}$ extended over the closed first quadrant is $\pi/4$.

Example 2. Determine the values of k such that

(4) $$\iint_{x^2+y^2 \leq 1} \frac{1}{(x^2 + y^2)^k}\, dA$$

converges.

Solution. The domain of integration R is the set of (x, y) such that $0 < x^2 + y^2 \leq 1$. Letting R_n be the set such that $\frac{1}{n^2} \leq x^2 + y^2 \leq 1$, we evaluate

$$\iint_{R_n} (x^2 + y^2)^{-k} \, dA = \int_0^{2\pi} \int_{\frac{1}{n}}^{1} \rho^{-2k+1} \, d\rho \, d\theta = \frac{\pi}{1-k} [1 - n^{2(k-1)}],$$

if $k \neq 1$, a formula involving $\ln n$ applying in case $k = 1$. From this it follows (from letting $n \to +\infty$) that (4) converges if and only if $k < 1$, and in case of convergence the value of (4) is $\pi/(1-k)$.

Example 3. Show that

(5) $$\iiint_{x^2+y^2+z^2 \geq 1} \frac{(x^2+y^2) \ln (x^2+y^2+z^2)}{(x^2+y^2+z^2)^k} \, dA$$

is convergent if $k > \frac{5}{2}$.

Solution. Using spherical coordinates and letting R_n be the set of points such that $1 \leq r \leq n$, we have

$$\iiint_{R_n} \frac{r^2 \sin^2 \phi \ln (r^2)}{r^{2k}} r^2 \sin \phi \, dr \, d\phi \, d\theta = \frac{16\pi}{3} \int_1^n r^{4-2k} \ln r \, dr.$$

Integration by parts establishes convergence, and the value of the integral. Thus (5) is equal to $16\pi/3(2k-5)^2$.

*1420. EXERCISES

In Exercises 1–8 determine whether the integral is convergent or divergent, and if it is convergent evaluate it.

★1. $\displaystyle\iint_{x^2+y^2 \leq 1} \frac{x^2 \, dA}{(x^2+y^2)^{\frac{3}{2}}}.$

★2. $\displaystyle\iint_{x^2+y^2 \leq 1} \frac{\ln (x^2+y^2) \, dA}{(x^2+y^2)^{\frac{1}{2}}}.$

★3. $\displaystyle\iint_{x^2+y^2 \geq 1} \frac{\ln (x^2+y^2) \, dA}{(x^2+y^2)}.$

★4. $\displaystyle\iint_{\text{Entire plane}} \frac{dA}{(1+x^2+y^2)^{\frac{3}{2}}}.$

★5. $\displaystyle\iiint_{x^2+y^2+z^2 \leq 1} \frac{x^2 \, dV}{(x^2+y^2+z^2)^2}.$

★6. $\displaystyle\iiint_{x^2+y^2+z^2 \leq 1} \frac{\ln (x^2+y^2+z^2) \, dV}{x^2+y^2+z^2}.$

★7. $\displaystyle\iiint_{x^2+y^2+z^2 \geq 1} \frac{\ln (x^2+y^2+z^2) \, dV}{(x^2+y^2+z^2)^2}.$

★8. $\displaystyle\iiint_{\text{Entire space}} \frac{dV}{(1+x^2+y^2+z^2)^{\frac{3}{2}}}.$

★9. Prove Note 1, § 1419.

★10. Prove the statement of Note 2, § 1419, regarding $\iint (f+g) \, dA$. State and prove two more theorems of similar nature. *Hints:* First establish the desired relation for nonnegative functions. Then establish and use the inequality $(f+g)^+ \leq f^+ + g^+$ and the equation $(f+g)^+ + f^- + g^- = (f+g)^- + f^+ + g^+$.

★11. Prove the Theorem, § 1419. *Hints:* First establish the result assuming $f \geq 0$ by letting R_0 be an arbitrary compact set with area lying in the *interior* of R. Show that the interiors of R_n cover R_0, and use the Heine-Borel theorem (Ex. 31, § 618).

15

Complex Variables

1501. INTRODUCTION

The role of complex variables in both pure and applied mathematics is of great and increasing importance. It is our purpose in this chapter to lay a solid groundwork for the complex number system and to introduce the reader to complex-valued functions of both real and complex variables, together with some of their uses. In most cases it will be seen that the transition from real variables to complex variables is almost immediate. A brief treatment of analytic functions of a complex variable is outlined in the Exercises of § 1810.

It is assumed that the student has already had some experience with formal algebraic manipulation of complex numbers.

1502. COMPLEX NUMBERS

Definition I. *A complex number is an ordered pair of real numbers, x and y, denoted (x, y).*

NOTE 1. Recall (cf. footnote, p. 22) that equality between two ordered pairs (x, y) and (u, v), $(x, y) = (u, v)$, means that $x = u$ and $y = v$.

NOTE 2. An alternative notation for the ordered pair (x, y) is $x + iy$. This will be treated presently (§ 1504).

Definition II. **Addition.** *The sum of two complex numbers (x, y) and (u, v) is defined:*

(1) $$(x, y) + (u, v) \equiv (x + u, y + v).$$

Definition III. **Multiplication.** *The product of two complex numbers (x, y) and (u, v) is defined:*

(2) $$(x, y)(u, v) \equiv (xu - yv, xv + yu).$$

NOTE 3. The *motivation* for these two definitions arises from the alternative notation of Note 2, where i has the property of being a "square root of -1." For example, formal multiplication of $(x + iy)$ by $(u + iv)$, with i^2 replaced by -1, leads to the expression $(xu - yv) + i(xv + yu)$.

§ 1503]   EMBEDDING OF THE REAL NUMBERS

It is a simple matter to check that both associative laws and both commutative laws hold (cf. § 102; give proofs in Ex. 1, § 1509), and that the complex numbers $(0, 0)$ and $(1, 0)$ have the additive and multiplicative properties of 0 (I(*iii*), § 102) and 1 (II(*iii*), § 102), respectively, of the real number system. In fact, the complex numbers satisfy all of the axioms of § 102. In other words:

Theorem. *The complex numbers form a field.*

Proof. The only remaining unproved assertions are the existence of the negative of a number (x, y) and the reciprocal of a nonzero number (x, y). The former is trivially seen to be $(-x, -y)$ and the latter is readily shown to be $\left(\dfrac{x}{x^2 + y^2}, -\dfrac{y}{x^2 + y^2}\right)$. (Give proofs in Ex. 1, § 1509.)

NOTE 4. It can be shown (Ex. 2, § 1509) that *the complex numbers are not an ordered field*. This means that there does not exist a set of complex numbers satisfying the four axioms of § 104 that describe "positivity."

1503. EMBEDDING OF THE REAL NUMBERS

The complex numbers that have the special form

(1) $\qquad (x, 0)$

have the following additive and multiplicative behavior:

(2) $\qquad (x, 0) + (y, 0) = (x + y, 0), \quad (x, 0)(y, 0) = (xy, 0).$

In other words, in the sense of (2), they behave in exactly the same way as the real numbers. In more technical terms, we say that the one-to-one correspondence

(3) $\qquad x \leftrightarrow (x, 0)$

between the real numbers x and the "real complex" numbers $(x, 0)$ is an **isomorphism**, meaning that the two correspondences $x \leftrightarrow (x, 0)$ and $y \leftrightarrow (y, 0)$ imply the two correspondences $x + y \leftrightarrow (x, 0) + (y, 0)$ and $xy \leftrightarrow (x, 0)(y, 0)$. If we think of the real complex numbers $(x, 0)$ as the real numbers with a new notation, we arrive at the concept of the complex number field as an **extension** of the real number field. Another way of saying the same thing is to state that by means of the correspondence (3) the real number system is **embedded** in the complex number system.

On the basis of the ideas expressed in the preceding paragraph we shall feel free henceforth to use the two words *real number* to mean either a real number x in the sense of Chapter 1 or a real complex number of the form (1), letting the context determine the distinction.

Definition. *An **imaginary number** is any complex number that is not real; that is, a complex number of the form*

$$(x, y), \quad y \neq 0.$$

A pure imaginary number is a complex number of the form

$$(0, y) \quad y \neq 0.$$

1504. THE NUMBER i

The pure imaginary number i is defined:

(1) $$i \equiv (0, 1),$$

and has the two properties:

(2) $$(0, 1)^2 = (-1, 0),$$
(3) $$(x, y) = (x, 0) + (0, 1)(y, 0).$$

We now simplify our notation in conformity with the remarks of the first paragraph of § 1503, using (1) and writing the single letter x for the ordered pair $(x, 0)$, and arrive at the familiar form for a complex number:

(4) $$z = (x, y) = x + iy, \quad \text{where} \quad i^2 = -1.$$

For any complex number $x + iy$, the **real** and **imaginary parts** are the real numbers x and y, respectively.

1505. GEOMETRICAL REPRESENTATION

Any complex number $z = x + iy$ is represented geometrically by the point (x, y) in E_2 (cf. Fig. 1501a). In this representation the x-axis is called the

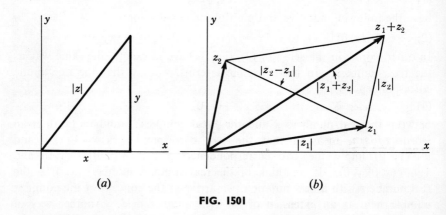

FIG. 1501

real axis and the y-axis the **imaginary axis**. The **absolute value** or **modulus** of $z = x + iy$,

(1) $$|z| \equiv \sqrt{x^2 + y^2},$$

is the distance of the *point z* from the origin O, and is the same as the magnitude of the radius vector $\overrightarrow{Oz}$.

It follows immediately from the definition of addition of complex numbers that this operation corresponds precisely to vector addition of the corresponding radius vectors, and that the absolute value of the difference $|z_2 - z_1|$

§ 1506] POLAR FORM 499

is the distance between the two points z_1 and z_2 (cf. Fig. 1501b; also, cf. § 703).

Most of the properties of absolute value for real numbers (§ 110) hold for complex numbers as well. A list of properties for complex numbers is given below, with proofs requested in Exercise 3, § 1509:

Properties of Absolute Value

I. *If z is a real complex number, $z = x + i0$, its absolute value when it is considered as a complex number is the same as its absolute value when it is considered as a real number: $|z| = |x + i0| = |x|$.*

II. $|z| \geq 0$; $|z| = 0$ if and only if $z = 0$.

III. $|z_1 z_2| = |z_1| \cdot |z_2|$.

IV. $\left|\dfrac{z_1}{z_2}\right| = \dfrac{|z_1|}{|z_2|}.$ if $z^2 \neq 0$

V. *The* **triangle inequality**† *holds:* $|z_1 + z_2| \leq |z_1| + |z_2|$.

VI. $|-z| = |z|$; $|z_1 - z_2| = |z_2 - z_1|$.

VII. $|z_1 - z_2| \leq |z_1| + |z_2|$.

VIII. $\big||z_1| - |z_2|\big| \leq |z_1 - z_2|$.

IX. *If $z = x + iy$, where x and y are real,*

$$|z| \leq |x| + |y|,$$
$$|x| \leq |z|, \quad |y| \leq |z|.$$

NOTE. Although the English mathematician Wallis (1616–1703) had done some work in representing complex numbers graphically, it was the Norwegian surveyor Wessel (1745–1818) who, in 1798, first presented the geometrical representation that is described in this section and used universally today. This representation is commonly referred to as the Argand diagram, after the French mathematician Argand (1768–1822), who published a paper on the subject in 1806.

1506. POLAR FORM

Let $z = x + iy$ be a nonzero complex number, and let $r = |z|$ be its (positive) absolute value. Then x/r and y/r are two real numbers the sum of whose squares is 1. The angle θ (which is uniquely determined if restricted to the range $0 \leq \theta < 2\pi$ (cf. Ex. 13, § 504) but need not be so restricted) such that $\cos \theta = x/r$ and $\sin \theta = y/r$ is called the **amplitude** or **argument** of the number z, and denoted $\theta = \text{amp}(z) = \arg(z)$. Since $x + iy = r(x/r + iy/r)$, it is meaningful to formulate the following definition (cf. Fig. 1502a):

Definition I. *If z is a nonzero complex number with modulus r and amplitude θ, its* **polar** *or* **trigonometric** *form is*

(1) $$z = r(\cos \theta + i \sin \theta).$$

† In geometrical terms, no side of a triangle can exceed the sum of the other two sides (cf. Fig. 1501; also cf. §§ 110, 703, 1604).

By means of the notation $e^{i\theta}$ defined by the **Euler formula**

(2) $$e^{i\theta} = \cos\theta + i\sin\theta$$

equation (1) *can be written in exponential form:*

(3) $$z = re^{i\theta}.$$

In distinction to the *polar form* $z = re^{i\theta}$, the form $z = x + iy$ in terms of the rectangular coordinates of z is sometimes called the **rectangular form** or the **Cartesian form** of z.

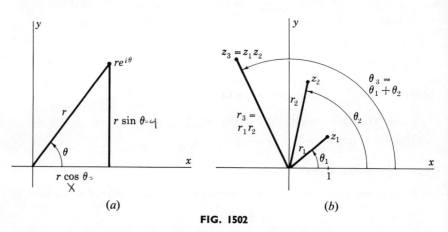

FIG. 1502

The exponential notation $e^{i\theta}$ for the expression $\cos\theta + i\sin\theta$ requires justification. This is provided by the six properties listed below. First, however, let us verify that formula (2) is at least *formally* consistent with infinite series expansions given in Chapter 12. To be precise, we substitute $i\theta$ for x in the Maclaurin series for e^x and collect the real and the imaginary terms of the result (cf. §§ 1207, 1511, Ex. 19, § 1810):

(4) $$e^{i\theta} = 1 + i\theta + \frac{(i\theta)^2}{2!} + \frac{(i\theta)^3}{3!} + \frac{(i\theta)^4}{4!} + \cdots$$
$$= \left[1 - \frac{\theta^2}{2!} + \frac{\theta^4}{4!} - \cdots\right] + i\left[\theta - \frac{\theta^3}{3!} + \frac{\theta^5}{5!} - \cdots\right]$$
$$= \cos\theta + i\sin\theta.$$

From properties of the trigonometric functions we can obtain certain properties of the function $\cos\theta + i\sin\theta$. We now list these properties, which justify the use of the exponential notation $e^{i\theta}$. Proofs are requested in Exercise 4, § 1509.

Properties of $e^{i\theta}$

I. $e^{i\theta}$ *never vanishes; in fact,* $|e^{i\theta}| = 1$.
II. $e^{i\theta_1} e^{i\theta_2} = e^{i(\theta_1 + \theta_2)}$.
III. $e^{i\theta_1}/e^{i\theta_2} = e^{i(\theta_1 - \theta_2)}$.

IV. DeMoivre's Theorem. $(e^{i\theta})^n = e^{in\theta}$, for every integer n. In particular, $(e^{i\theta})^{-1} = e^{-i\theta}$.

V. $e^{i\theta} = 1$ if and only if $\theta = 2n\pi$ for some integer n. In particular, $e^0 = 1$.

VI. $e^{i\theta}$, as a function of θ, is periodic with period 2π. That is, $e^{i(\theta+2\pi)} = e^{i\theta}$ for all θ, and if $e^{i(\theta+c)} = e^{i\theta}$, where c is real, then $c = 2n\pi$ for some integer n.

NOTE 1. If multiplication of two complex numbers is performed with the numbers in polar form, we have:
$$(r_1 e^{i\theta_1})(r_2 e^{i\theta_2}) = (r_1 r_2) e^{i(\theta_1 + \theta_2)};$$
in short: *multiply the absolute values and add the amplitudes* (cf. Fig. 1502b). For division we have, similarly,
$$\frac{r_1 e^{i\theta_1}}{r_2 e^{i\theta_2}} = \frac{r_1}{r_2} e^{i(\theta_1 - \theta_2)}, \ r_2 \neq 0;$$
in short: *divide the absolute values and subtract the amplitudes*.

NOTE 2. If r is fixed and θ variable, the equation $z = re^{i\theta}$ is a parametric equation of the circle with radius r and center at the origin. It is equivalent to the pair of equations $x = r \cos \theta, \ y = r \sin \theta$.

1507. CONJUGATES

The **conjugate** of a complex number $z = x + iy$ is defined and denoted:

(1) $$\bar{z} \equiv x - iy.$$

The following properties are immediate from the definition (give details in Ex. 5, § 1509):

Properties of Conjugates

I. A complex number is real if and only if it is equal to its conjugate: $z = \bar{z}$.

II. The sum of a complex number $z = x + iy$ and its conjugate is real, and equal to twice the real part of z: $z + \bar{z} = 2x$.

III. The product of a complex number and its conjugate is the square of its absolute value: $z\bar{z} = |z|^2$.

IV. The conjugate of $re^{i\theta}$ is $re^{-i\theta}$. In other words, a complex number and its conjugate have equal absolute values ($|\bar{z}| = |z|$) and amplitudes that are negatives of each other. They are represented by points located symmetrically with respect to the x-axis.

V. The conjugate of the sum (difference) of two complex numbers is the sum (difference) of their conjugates. This property for the sum extends to an arbitrary number of terms.

VI. The conjugate of the product (quotient) of two complex numbers is the product (quotient) of their conjugates. This property for the product extends to an arbitrary number of factors.

VII. The conjugate and reciprocal of any nonzero complex number are equal if and only if the absolute value of the number is unity: $|z| = 1$.

NOTE. Division of one complex number by another is often achieved by multiplication of numerator and denominator of a fraction by the conjugate of the denominator.

Example. Express the quotient $(1 + 2i) \div (2 - 3i)$ in rectangular form.

Solution. We use fractional form and multiply numerator and denominator by the conjugate of $2 - 3i$:

$$\frac{1+2i}{2-3i} \cdot \frac{2+3i}{2+3i} = \frac{-4+7i}{13} = -\frac{4}{13} + \frac{7}{13}i.$$

1508. ROOTS

The polar form of a complex number is especially well suited to the problem of finding square roots, cube roots, and other roots of numbers. Suppose, for example, we seek solutions for the equation

(1) $$(re^{i\theta})^n = Re^{i\alpha},$$

where $R > 0$, α and the positive integer n are given, and $r > 0$ and θ are to be determined. By the basic laws of algebra and by de Moivre's theorem (IV, § 1506), equation (1) is equivalent to $r^n e^{in\theta} = Re^{i\alpha}$. This means that $r^n = R$, and $r > 0$ is uniquely determined as the positive nth root of R:

(2) $$r = \sqrt[n]{R}.$$

We are led to the problem of finding solutions to the equation

(3) $$e^{in\theta} = e^{i\alpha}.$$

In still other terms, if we write $\theta_0 = \dfrac{\alpha}{n}$ as one determination of the amplitude θ, and write $\theta = \dfrac{\alpha}{n} + \phi$, we can express our problem as that of finding angles ϕ such that $e^{i\alpha + in\phi} = e^{i\alpha}$, or:

(4) $$(e^{i\phi})^n = e^{in\phi} = 1.$$

The problem has now been reduced to that of finding the nth roots $e^{i\phi}$ of unity, indicated by (4).

By VI, § 1506, equation (4) is satisfied if and only if $n\phi$ is some integral multiple of 2π: $n\phi = 0, \pm 2\pi, \pm 4\pi, \pm 6\pi, \cdots$, or:

(5) $$\phi = 0, \pm \frac{2\pi}{n}, \pm \frac{4\pi}{n}, \pm \frac{6\pi}{n}, \cdots.$$

A moment's reflection reveals that certain angles in the set (5) will produce the *same* complex number $e^{i\phi}$. In fact (VI, § 1506) this is true whenever these angles differ by an integral multiple of 2π. Consequently, the *complete* set of solutions of (4) is provided by any sequence of n consecutive angles from (5), the simplest being:

(6) $$\phi = 0, \frac{2\pi}{n}, \frac{4\pi}{n}, \cdots, \frac{(2n-2)\pi}{n}.$$

The n nth roots of unity, $e^{i\phi}$, for ϕ in the set (6), are represented by n points uniformly distributed on the circumference of the unit circle, starting with the point 1. In other words, these points are the vertices of a regular polygon inscribed in the unit circle and with one vertex at the point 1.

Combining this result with equations (2) and (3), we find, similarly, that there are n nth roots of any nonzero number $Re^{i\alpha}$, and that these are represented by n points uniformly distributed on the circumference of the circle $|z| = \sqrt[n]{R}$, starting with the point with amplitude α/n and having amplitudes that differ successively by $2\pi/n$. In other words, these points are the vertices of a regular polygon inscribed in the circle with center at the origin and with radius $\sqrt[n]{R}$, and with one vertex at the point $\sqrt[n]{R}e^{i\alpha/n}$. The polar form of these points is $\sqrt[n]{R}e^{i\theta}$, where $\theta = (\alpha/n) + \phi$, and ϕ is given by (6):

(7) $$\sqrt[n]{R}e^{\frac{i\alpha}{n}}, \sqrt[n]{R}e^{\frac{i(\alpha+2\pi)}{n}}, \sqrt[n]{R}e^{\frac{i(\alpha+4\pi)}{n}}, \cdots, \sqrt[n]{R}e^{\frac{i(\alpha+(2n-2)\pi)}{n}}.$$

Example 1. Find the cube roots of unity.

First Solution. We seek solutions to the equation $z^3 = 1$, or $(z-1)(z^2+z+1) = 0$. By the quadratic formula:

(8) $$z = 1, \frac{-1 \pm i\sqrt{3}}{2}.$$

Second Solution. There are three cube roots of unity, of the form $e^{i\theta}$ for $\theta = 0$, $2\pi/3$, $4\pi/3$:

(9) $$e^0 = 1, \ e^{\frac{2\pi i}{3}} = \cos\frac{2\pi}{3} + i\sin\frac{2\pi}{3} = -\frac{1}{2} + i\frac{\sqrt{3}}{2},$$

$$e^{\frac{4\pi i}{3}} = \cos\frac{4\pi}{3} + i\sin\frac{4\pi}{3} = -\frac{1}{2} - i\frac{\sqrt{3}}{2}.$$

Example 2. Find the fifth roots of $32i$.

Solution. Since $32i = 32e^{\frac{\pi i}{2}}$, (7) becomes

$$2e^{\frac{\pi i}{10}}, \ 2e^{\frac{\pi i}{2}} = 2i, \ 2e^{\frac{9\pi i}{10}}, \ 2e^{\frac{13\pi i}{10}}, \ 2e^{\frac{17\pi i}{10}}.$$

1509. EXERCISES

1. Complete the proof of the Theorem, § 1502.
2. Prove that it is impossible to define an order relation for the complex numbers relative to which they form an ordered field (Note 4, § 1502). *Hint:* $1 > 0$ (cf. Ex. 6, § 105), and $i^2 = -1$. (Cf. Ex. 5, § 105.)
3. Prove the Properties of Absolute Value, § 1505. *Hints:* For V, if $z_1 = x + iy$ and $z_2 = u + iv$, square both members of the inequality $\sqrt{(x+u)^2 + (y+v)^2} \leq \sqrt{x^2 + y^2}\sqrt{u^2 + v^2}$, simplify, and square again, ultimately obtaining $(xv - yu)^2 \geq 0$. Now reverse the logical steps and justify all operations performed. For the other properties, see the hints in the Exercises, § 111, for corresponding properties in § 110.
4. Prove the Properties of $e^{i\theta}$, § 1506.
5. Prove the Properties of Conjugates, § 1507.

6. Multiply and simplify: $(3 - i)(2 + 5i)(4 + 3i)$.

7. Divide and simplify: $\dfrac{6 + i}{5 - 8i}$.

In Exercises 8–11, find the result by means of Euler's formula.

8. $(1 + i)^{10}$.
9. $(6 - 8i)^{-18}$.
10. $(\cos 12° + i \sin 12°)^9 (\cos 18° + i \sin 18°)^{-1}$.
11. $(\cos 35° + i \sin 35°)^{16}(\cos 200° - i \sin 200°)^{-5}$.

In Exercises 12–15, find the requested roots.

12. Cube roots of -1.   **13.** Fifth roots of -32.
14. Fourth roots of $-5i$.   **15.** Fifth roots of $-\sqrt{3} + i$.

16. Using algebraic methods only, without appeal to trigonometry, prove that there are exactly two square roots of -1, that is, two numbers z such that $z^2 = -1$. More generally, prove that any nonzero complex number that has a square root has exactly two square roots.

17. Derive the quadratic formula
$$z = \frac{-b \pm \sqrt{b^2 - 4ac}}{2a}$$
for the roots of any quadratic equation $az^2 + bz + c = 0$, where a, b, and c are arbitrary complex numbers and $a \neq 0$, where $\sqrt{b^2 - 4ac}$ denotes either of the two square roots of the discriminant $b^2 - 4ac$ in case this $\neq 0$. In particular, if a, b, and c are real and the discriminant is negative, obtain the result $z = [-b \pm i\sqrt{4ac - b^2}]/2a$.

18. Prove that in the triangle inequality (V, § 1505), equality holds if and only if z_1 and z_2 lie on a ray (half-line) issuing from the origin.

19. Extend Property III, § 1505, to n terms:
$$|z_1 z_2 \cdots z_n| = |z_1| \cdot |z_2| \cdot \cdots \cdot |z_n|,$$
and prove the triangle inequality for n terms:
$$|z_1 + z_2 + \cdots + z_n| \leq |z_1| + |z_2| + \cdots + |z_n|.$$

*20. A complex number of the form $\alpha + i\beta$, where α and β are (real) integers, is called a **Gaussian integer**. A Gaussian integer a is **composite** if and only if it can be factored in the form $a = bc$, where b and c are Gaussian integers both of which are distinct from ± 1 and $\pm i$; otherwise it is **prime**. Show that as Gaussian integers 2 is composite and 3 is prime.

*21. Let $\sqrt{-5}$ denote either of the two square roots of -5, and let F be the set of all numbers of the form $p + q\sqrt{-5}$, where p and q are (real) integers. Prove that F has the following properties: (i) F contains the (real) integers; (ii) if a and b belong to F, then so do $a + b$, $a - b$, and ab; (iii) if a and b belong to F and if $ab = 0$, then either $a = 0$ or $b = 0$. Prove that in the two factorizations of 6:

(1) $\qquad\qquad 2 \cdot 3 = (1 + \sqrt{-5})(1 - \sqrt{-5})$

all four factors are prime in the sense that if any one of them is the product of two members of F, then one of these latter two factors must be ± 1. Hence prove that in the number system F a prime number may be a factor of the product of two numbers without being a factor of either of the two numbers. It also follows that the unique factorization of arithmetic (cf. RV, Ex. 29, § 107) fails for F.

*22. Let $f(z)$ be a real polynomial, that is, a polynomial with real coefficients. Prove that if a is any complex number, then $\overline{f(a)} = f(\bar{a})$.

*23. Prove that if $f(z)$ is a real rational function, that is, $f(z)$ is a quotient of real polynomials, then whenever a is a complex number for which $f(a)$ is defined, $\overline{f(a)} = f(\bar{a})$. (Cf. Ex. 22.)

*24. Prove that if $f(z)$ is a real polynomial, and a a complex number, then $f(a) = 0$ if and only if $f(\bar{a}) = 0$. That is, a complex number is a zero of a real polynomial if and only if its conjugate is. (Cf. Ex. 22.)

*25. Prove that if $f(z)$ is a real polynomial, and $a = \alpha + i\beta$ is an imaginary number ($\beta \neq 0$), then a is a zero of $f(z)$ ($f(a) = 0$) if and only if the real quadratic function $(z - a)(z - \bar{a}) = z^2 - 2\alpha z + \alpha^2 + \beta^2$ is a factor of $f(z)$. (Cf. Exs. 22–24.) *Hint:* Write $f(z)$ in the form $Q(z)(z^2 - 2\alpha z + \alpha^2 + \beta^2) + cz + d$.

*26. If $z = x + iy$ is an arbitrary complex number the **exponential function** e^z is defined:

(2) $$e^z = e^x e^{iy} = e^x(\cos y + i \sin y).$$

Prove the following properties (cf. Exs. 6, 19, § 1810):
  (*i*) The definition (2) is consistent with those already given for the cases where z is real or pure imaginary (§§ 502, 1506).
  (*ii*) e^z never vanishes; $|e^{x+iy}| = e^x$.
  (*iii*) $e^z = 1$ if and only if $z = 2n\pi i$ for some integer n.
  (*iv*) e^z is periodic with period $2\pi i$; that is, $e^{z+2\pi i} = e^z$, for all z, and if $e^{z+c} = e^z$ then $c = 2n\pi i$ for some integer n.
  (*v*) $e^{z_1} e^{z_2} = e^{z_1+z_2}$.
  (*vi*) $e^{z_1}/e^{z_2} = e^{z_1-z_2}$.
  (*vii*) $(e^z)^n = e^{nz}$ for every integer n.

*27. The **logarithmic function** $w = \log z$ ($z \neq 0$) is the inverse of the exponential function. That is, $w = \log z$ is defined by the equation $z = e^w$. Prove that $\log z$ is an infinitely-many-valued function, whose values, for a fixed z, differ by multiples of $2\pi i$. Prove that

(3) $$\log z = \ln |z| + i \operatorname{amp}(z),$$

where amp (z) is the amplitude of z. Find $\log i$.

*28. If a and b are arbitrary complex numbers ($a \neq 0$), the **power** a^b is defined by the equation

(4) $$a^b = e^{b \log a}.$$

Prove that a^b is single-valued if b is an integer (and in agreement with the usual definition for integral powers), finitely-multiple-valued if b is a nonintegral rational number, and infinitely-multiple-valued otherwise. Show that the values of i^i are all real and positive, and that the smallest value greater than 1 is approximately 111 (to the nearest integer). Examine the laws of exponents for a^b. Compare a^b with e^b when $a = e$.

*29. The **trigonometric functions** of a complex variable are defined by means of the two basic defining equations (cf. Ex. 26):

(5) $$\cos z = \frac{e^{iz} + e^{-iz}}{2}, \quad \sin z = \frac{e^{iz} - e^{-iz}}{2i}.$$

Prove the following properties (cf. Exs. 6, 19, § 1810):
  (*i*) These definitions are consistent with those already given when z is real (§ 504).

(ii) $\sin^2 z + \cos^2 z = 1$
$\sin(z_1 \pm z_2) = \sin z_1 \cos z_2 \pm \cos z_1 \sin z_2$
$\cos(z_1 \pm z_2) = \cos z_1 \cos z_2 \mp \sin z_1 \sin z_2$
(iii) $\sin z$ and $\cos z$ are periodic functions with period 2π; that is, $\sin(z + 2\pi) = \sin z$ and $\cos(z + 2\pi) = \cos z$ for all z, and if $\sin(z + c) = \sin z$ or if $\cos(z + c) = \cos z$ for all z then $c = 2n\pi$.

Find $\sin i$, $\cos i$, a number whose cosine is 4, and a number whose sine is 2. *Hint for* $\cos z = 4$: Let $e^{iz} = s$, and solve the equation $s + s^{-1} = 8$.

1510. LIMITS AND CONTINUITY

Because of the properties of absolute values of complex numbers listed in § 1505, and their close resemblance to analogous properties for real numbers, it is possible to define in a unified manner the concepts of limit and continuity, whether the variables are real or complex. The form of the definitions is the same, although in the geometrical interpretation for the complex case dimensions are doubled (cf. Chapter 6). We give here a few sample formulations. The function $f(z)$ will be assumed to be either a real-valued function of a real variable, a real-valued function of complex variable, a complex-valued function of a real variable, or a complex-valued function of a complex variable. (Cf. §§ 1513, 1810 for differentiation properties.) The assumption of the first paragraph of § 206 is still in effect: whenever a limit of a function is concerned, it will be implicitly assumed that *the quantities symbolized exist for at least some values of the independent variable neighboring the limiting value of that variable*. For example, in Definition I, below, it is assumed that a is a limit point of the domain of definition of the function $f(z)$.

Definition I. *The function $f(z)$ has the limit L as z approaches a, written*
$$\lim_{z \to a} f(z) = L, \quad \text{or} \quad f(z) \to L \text{ as } z \to a,$$
if and only if corresponding to an arbitrary positive number ϵ there exists a positive number δ such that $0 < |z - a| < \delta$ implies $|f(z) - L| < \epsilon$, for values of z for which $f(z)$ is defined.

Definition II. *The function $f(z)$ has the limit L as z becomes infinite, written*
$$f(\infty) = \lim_{z \to \infty} f(z) = L, \quad \text{or} \quad f(z) \to L \text{ as } z \to \infty,$$
if and only if corresponding to an arbitrary positive number ϵ there exists a number $N = N(\epsilon)$ such that $|z| > N$ implies $|f(z) - L| < \epsilon$, for values of z for which $f(z)$ is defined.

Definition III. *The function $f(z)$ is* **continuous** *at $z = a$ if and only if it is defined at $z = a$ and corresponding to an arbitrary positive number ϵ there exists a positive number δ such that $|z - a| < \delta$ implies $|f(z) - f(a)| < \epsilon$, for values of z for which $f(z)$ is defined.*

The limit theorems I–II, § 207, and the continuity theorems of §§ 209 and 211 carry over almost precisely. Proofs are mere restatements of those given earlier (Exs. 1–3, § 1512).

Relations between continuity involving complex variables and continuity involving real variables are stated in the following theorems (give proofs in Exs. 4–7, § 1512, with the aid of properties of absolute value, § 1506; in Exs. 8–9, § 1512, corresponding theorems in limit form are requested):

Theorem I. *If $w = f(z)$ is a real- or complex-valued function of the complex variable z, and if $z = x + iy$, where x and y are real, then w is a continuous function of z if and only if w is a continuous function of the two real variables x and y.*

Theorem II. *If $w = f(z)$ is a complex-valued function of the real or complex variable z, and if $w = u + iv$, where u and v are real, then w is a continuous function of z if and only if u and v are continuous functions of z.*

Theorem III. *The real-valued function of a real or complex variable*
$$u = |z|$$
is continuous.

Theorem IV. *If $f(z)$ is a real- or complex-valued continuous function of the real or complex variable z, then $|f(z)|$ is a real-valued continuous function of z.*

1511. SEQUENCES AND SERIES

It is a matter of routine translation from the language of real numbers into that of complex numbers to extend to the complex domain such basic notions as the convergence of a sequence or series (cf. Ex. 10, § 1512). We give one definition, and mention only a few of the more important facts, asking for proofs in the exercises (Exs. 11–19, § 1512). (Cf. Exs. 16, 17, § 1810.)

Definition. *The sequence $\{z_n\}$ has the **limit** a, written $\lim_{n \to +\infty} z_n = a$, or $z_n \to a$ as $n \to +\infty$, if and only if corresponding to an arbitrary positive number ϵ there exists a number $N = N(\epsilon)$ such that $n > N$ implies $|z_n - a| < \epsilon$. In case $\lim_{n \to +\infty} z_n = a$, z_n is said to **converge** to a, and is **convergent**. A sequence $\{z_n\}$ is **bounded** if and only if its sequence of absolute values $\{|z_n|\}$ is bounded.*

Theorem I. *If $\{z_n\}$ is a sequence of complex numbers, and if $z_n = x_n + iy_n$, where x_n and y_n are real, $n = 1, 2, \cdots$, then the sequence $\{z_n\}$ converges if and only if both sequences $\{x_n\}$ and $\{y_n\}$ converge. In case of convergence,*
$$\lim_{n \to +\infty} z_n = \lim_{n \to +\infty} x_n + i \lim_{n \to +\infty} y_n.$$

Theorem II. *Theorems I–XI and XIII, § 204, hold for complex sequences.*

***Theorem III. Fundamental Theorem on Bounded Sequences.** *Every bounded sequence of complex numbers contains a convergent subsequence.* (Cf. § 217.)

***Theorem IV. Cauchy Criterion.** *A sequence $\{z_n\}$ of complex numbers converges if and only if it is a Cauchy sequence; that is, if and only if corresponding to an arbitrary positive number ϵ there exists a number N such that $m > N$ and $n > N$ together imply $|z_m - z_n| < \epsilon$.*

Theorem V. *The following theorems on series, from Chapter 11, hold for series of complex numbers: Theorems I–III, § 1102, Theorem, § 1103, Theorem, § 1104, Theorems I–II, § 1116, Theorems I–II, § 1118, Theorems I–II, § 1119.*

For power series (Chapter 12) the following theorem holds for complex numbers (cf. Theorems I–II, § 1201; also Exs. 16, 17, § 1810):

Theorem VI. *Let S be the set of points z for which a power series $\Sigma a_n(z - a)^n$ converges. Then either (i) S consists only of the point $z = a$, or (ii) S consists of all complex numbers, or (iii) S is a disk with center at $z = a$ and positive* **radius of convergence** *R and containing all points such that $|z - a| < R$ and possibly containing in addition some or all of the circumference $|z - a| = R$. At any point interior to this* **circle of convergence** *the convergence is absolute.*

Theorem VII. *The formulas of Theorem III, § 1201, and Note 4, § 1201, for the radius of convergence of a power series both extend to the complex case.*

Concerning tests for uniform convergence and continuity of uniform limits we have:

***Theorem VIII.** *The following theorems from Chapter 13 hold for complex sequences and series: Theorems I–II and the Corollary, § 1303, Theorem and Corollary, § 1305.*

The analogue of Theorem I, § 1309, is (cf. Exs. 16, 17, § 1810):

***Theorem IX.** *A power series converges uniformly on any closed disk that lies in the interior of its circle of convergence. Consequently, any function represented by a power series throughout its circle of convergence is continuous throughout the interior of that circle.*

Finally, it should be remarked that the operations of integration and differentiation can be defined for complex-valued functions of a complex variable, and applied within the interior of the circle of convergence of a power series (cf. the Theorems of §§ 1203, 1210). Abel's Theorem (Theorem II, § 1309) can also be extended to the complex case. These matters belong properly to the subject of Analytic Functions of a Complex Variable, a brief introduction to which is offered in the Exercises of § 1810.

1512. EXERCISES

In Exercises 1–3, prove the indicated theorems, permitting the variables to be either real or complex.
1. Theorems I–VI, § 207.   2. Theorem, § 209.   3. Theorems, § 211.

In Exercises 4–7, prove the indicated theorem of § 1510.
4. Theorem I.   5. Theorem II.   6. Theorem III.   7. Theorem IV.

In Exercises 8–9, state and prove a theorem similar to the one specified, but involving limits instead of continuity.
8. Theorem I, § 1510.   9. Theorem II, § 1510.

10. Extend to the complex domain the definition of convergence of a series, § 1101.

In Exercises 11–19, prove the indicated theorem of § 1511.
11. Theorem I.   12. Theorem II.   *13. Theorem III.
*14. Theorem IV.   15. Theorem V.   16. Theorem VI.
17. Theorem VII.   *18. Theorem VIII.   *19. Theorem IX.

In Exercises 20–26, prove the given statement. (Cf. Exs. 16–20, 31, 35, § 208.)
20. If k is a constant and $\lim_{z \to a} f(z)$ exists, $\lim_{z \to a} kf(z)$ exists and is equal to $k \lim_{z \to a} f(z)$.
21. $\lim_{z \to a} z = a$.
22. If n is a positive integer, $\lim_{z \to a} z^n = a^n$.
23. Any polynomial $f(z)$ is continuous.
24. Any rational function $f(z) = g(z)/h(z)$, where $g(z)$ and $h(z)$ are polynomials, is continuous except where $h(z)$ vanishes.
25. $\lim_{z \to \infty} \frac{1}{z} = 0$.
26. If $f(z)$ is a nonconstant polynomial, with complex coefficients, $\lim_{z \to \infty} \frac{1}{f(z)} = 0$.
   Hint: Write $\frac{1}{a_0 z^n + \cdots + a_n}$ in the form $\frac{1/z^n}{a_0 + a_1/z + \cdots + a_n/z^n}$.
27. Formulate a definition for $\lim_{z \to a} f(z) = \infty$.
28. Formulate a definition for $\lim_{z \to \infty} f(z) = \infty$.
29. Formulate and prove a complex form for Exercise 31, § 208. In particular, show that if $f(z)$ is a nonconstant polynomial, $\lim_{z \to \infty} f(z) = \infty$. (Cf. Ex. 26.)
30. Formulate and prove a complex form for Exercise 35, § 208. (Cf. Exs. 26, 29.)
31. Formulate and prove complex forms for Exercises 46, 49, and 50, § 208.
*32. Prove that the functions of a complex variable z:
$$e^z, \quad \cos z, \quad \sin z,$$
defined in Exercises 26 and 29, § 1509, are represented by the Maclaurin series (1), (5), and (3) of § 1207, respectively, for all values of z.

1513. COMPLEX-VALUED FUNCTIONS OF A REAL VARIABLE

For some purposes it is fruitful to consider functions of a real variable whose values are complex. Instances of such uses are the derivation of certain formulas of integration (Example 1), the solving of linear differential equations (Example 2), the Fourier transform (Exs. 16–20, § 1514), and Fourier series (Ex. 18, § 1607).

Let $w = u + iv = f(x) = \phi(x) + i\psi(x)$ be a function of the real variable x, where $u = \phi(x)$ and $v = \psi(x)$ are real-valued. We consider the differentiability of such a complex-valued function of x.

Definition I. *The function $f(x)$ is **differentiable** at x if and only if* $\lim\limits_{h \to 0} \dfrac{f(x+h) - f(x)}{h}$ *exists (and is finite). This limit is called the **derivative** of $f(x)$ at the point x, and is written $f'(x)$.*

Theorem I. *The function $f(x) = \phi(x) + i\psi(x)$ is differentiable at the point x if and only if both functions $\phi(x)$ and $\psi(x)$ are differentiable there. In case of differentiability*

(1) $$f'(x) = \phi'(x) + i\psi'(x).$$

Proof. The two required implications, as well as equation (1), follow from the equation

(2) $$\frac{f(x+h) - f(x)}{h} = \frac{\phi(x+h) - \phi(x)}{h} + i \frac{\psi(x+h) - \psi(x)}{h}$$

and from the inequalities relating the absolute value of a complex number $W = U + iV$ and those of its real and imaginary parts (Property IX, § 1505):

(3) $$|W| \leq |U| + |V|,$$
(4) $$|U| \leq |W|, \quad |V| \leq |W|.$$

Specifically, (for the "if" part of the proof) assume that ϕ' and ψ' exist, let (2) be more concisely written $L = M + iN$, and let $U \equiv M - \phi'$, $V \equiv N - \psi'$, and $W \equiv U + iV = L - (\phi' + i\psi')$. Then, as $h \to 0$, $U \to 0$ and $V \to 0$, and therefore, by (3), $W \to 0$. Consequently $L \to \phi' + i\psi'$, and f' exists and satisfies (1). On the other hand (for the "only if" part of the proof) assume that f' exists, write it in the form $f' = r + is$, and again denote (2) by $L = M + iN$; let $U \equiv M - r$, $V \equiv N - s$, and $W \equiv U + iV = L - f'$. Then, as $h \to 0$, $W \to 0$ and therefore, by (4), $U \to 0$ and $V \to 0$. Consequently $M \to r$ and $N \to s$, and ϕ' and ψ' both exist and satisfy (1).

The following properties of complex-valued differentiable functions are now easily established (cf. Exs. 1–6, § 1514). For simplicity of statement we shall assume that the independent variable x is restricted to an interval.

Theorem II. *The sum of any finite number of differentiable functions is differentiable, and the derivative of the sum is the sum of the derivatives.*

Theorem III. *The derivative of a constant is zero.*

Theorem IV. *The product of two differentiable functions f and g is differentiable, and its derivative is given by the formula $(fg)' = fg' + gf'$.*

Theorem V. *A constant times a differentiable function is differentiable, and its derivative is the constant times the derivative of the given function.*

Theorem VI. *A function with an identically vanishing derivative throughout an interval must be constant in that interval.*

Theorem VII. *Two differentiable functions whose derivatives are equal throughout an interval must differ by a constant in that interval.*

NOTE. In Chapter 3 (Theorems I and II, § 306) the real-valued analogues of Theorems VI and VII are consequences of the Law of the Mean (§ 305). In the present section this is not the case, since the Law of the Mean (formulated precisely as stated in Theorem III, § 305, with $f(x)$ complex-valued) is false, as shown by the example $f(x) \equiv \cos x + i \sin x$ on any interval $[a, b]$ $(a < b)$. Indeed, if there were a ξ between a and b such that formula (2), § 305, held, then:

$$(\cos b + i \sin b) - (\cos a + i \sin a) = (-\sin \xi + i \cos \xi)(b - a).$$

Equating the moduli of the two members, and using a pair of trigonometric identities, we have:

$$4 \sin^2 \frac{b+a}{2} \sin^2 \frac{b-a}{2} + 4 \cos^2 \frac{b+a}{2} \sin^2 \frac{b-a}{2} = (b-a)^2,$$

or

$$\sin^2 \frac{b-a}{2} = \left(\frac{b-a}{2}\right)^2.$$

By Example 1, § 305 (and elementary properties of sin x), this is impossible, and we have the desired contradiction.

A particular complex-valued function of a real variable that is of great importance is the complex exponential function (cf. Ex. 26, § 1509):

(5) $$e^{(\alpha + i\beta)x} \equiv e^{\alpha x} e^{i\beta x} = e^{\alpha x}(\cos \beta x + i \sin \beta x),$$

where α, β, and x are real. The principal interest in this function at present lies in its differentiability and the form of its derivative:

(6) $$\frac{d}{dx}(e^{(\alpha+i\beta)x}) = (\alpha + i\beta)e^{(\alpha+i\beta)x}.$$

The verification of (6) consists of writing down and comparing the two members. To be precise, the left-hand member of (6) is equal, by Theorem I, to

$$\frac{d}{dx}(e^{\alpha x} \cos \beta x) + i \frac{d}{dx}(e^{\alpha x} \sin \beta x)$$

$$= e^{\alpha x}[(\alpha \cos \beta x - \beta \sin \beta x) + i(\alpha \sin \beta x + \beta \cos \beta x)].$$

The right-hand member of (6) is equal to

$$e^{\alpha x}[(\alpha + i\beta)(\cos \beta x + i \sin \beta x)],$$

which, when expanded, is the expression just obtained for the left-hand member.

We close this section with one last definition and theorem (give a proof of Theorem VIII in Ex. 7, § 1514):

Definition II. *Let $f(x)$ be a complex-valued function of the real variable x on the interval $[a, b]$. Then $f(x)$ is **integrable** on $[a, b]$ if and only if the limit*

(7) $$\lim_{|\mathfrak{N}| \to 0} \sum_{i=1}^{n} f(x_i) \Delta x_i$$

exists in the sense of § 401. In case the limit (7) exists its value is denoted $\int_a^b f(x)\, dx.$

Theorem VIII. *Let $f(x) = \phi(x) + i\psi(x)$, where $\phi(x)$ and $\psi(x)$ are real-valued, be defined on the interval $[a, b]$. Then $f(x)$ is integrable there if and only if both $\phi(x)$ and $\psi(x)$ are integrable there. In case of integrability,*

(8) $$\int_a^b f(x)\, dx = \int_a^b \phi(x)\, dx + i \int_a^b \psi(x)\, dx.$$

Example 1. Derive the integration formulas

(9) $$\int e^{\alpha x} \sin \beta x\, dx = \frac{e^{\alpha x}}{\alpha^2 + \beta^2}(\alpha \sin \beta x - \beta \cos \beta x) + C,$$

(10) $$\int e^{\alpha x} \cos \beta x\, dx = \frac{e^{\alpha x}}{\alpha^2 + \beta^2}(\beta \sin \beta x + \alpha \cos \beta x) + C,$$

by use of the complex exponential function (cf. Ex. 11, § 407).

Solution. The derivative of $e^{(\alpha + i\beta)x}$ is, by (5):

(11) $$\frac{d}{dx} e^{(\alpha + i\beta)x} = (\alpha + i\beta) e^{(\alpha + i\beta)x}.$$

Dividing both members of (11) by $\alpha + i\beta$, we have:

(12) $$\frac{d}{dx}\left[\frac{\alpha - i\beta}{\alpha^2 + \beta^2}(e^{\alpha x} \cos \beta x + i e^{\alpha x} \sin \beta x)\right] = e^{\alpha x}(\cos \beta x + i \sin \beta x).$$

Equating the real parts of the two sides of (12) gives (10), and equating the imaginary parts gives (9).

Example 2. Assume that the roots of the equation

(13) $$r^2 + br + c = 0,$$

where b and c are complex constants, are distinct complex numbers r_1 and r_2. Prove that every function of the form

(14) $$z = c_1 e^{r_1 x} + c_2 e^{r_2 x},$$

where c_1 and c_2 are arbitrary complex constants, is a solution of the differential equation

(15) $$\frac{d^2 z}{dx^2} + b \frac{dz}{dx} + cz = 0.$$

Conversely, prove that every solution of (15) must have the form (14).

Prove that in the case where b and c are real and r_1 and r_2 are (conjugate) imaginary numbers: $r_1 = \alpha + i\beta, r_2 = \alpha - i\beta$, where $\beta \neq 0$, every function of the form

(16) $$e^{\alpha x}(A \cos \beta x + B \sin \beta x),$$

where A and B are arbitrary real numbers, is a solution of (15). Conversely, prove that every real solution of (15) must have the form (16).

Solution. For the first part it is a routine matter of differentiation according to the rules of this section, and subsequent substitution, to verify that any function of the form (14) satisfies (15). Assuming now that $z(x)$ is a solution of (15), we write that equation in factored form $\left(\dfrac{d}{dx} - r_1\right)\left(\dfrac{d}{dx} - r_2\right)z = 0$. (This is motivated by the factorization of the left-hand member of (13) as the product $(r - r_1)(r - r_2)$.) If a new function $w(x)$ is defined:

(17) $$w(x) \equiv \frac{dz}{dx} - r_2 z,$$

then w satisfies the "first-order" differential equation

(18) $$\frac{dw}{dx} - r_1 w = 0.$$

Multiplication of (18) by the nonzero "integrating factor" $e^{-r_1 x}$ permits the rewriting of it in the form:

$$\frac{d}{dx}[we^{-r_1 x}] = e^{-r_1 x}\left[\frac{dw}{dx} - r_1 w\right] = 0.$$

Hence, by Theorem VI, the quantity $we^{-r_1 x}$ must be a constant, and w must have the form $ce^{r_1 x}$. Therefore, by (17), the original function $z(x)$ must satisfy the first-order differential equation

(19) $$\frac{dz}{dx} - r_2 z = ce^{r_1 x}.$$

We again multiply through by an integrating factor, this time $e^{-r_2 x}$ in order to write (19) as follows:

$$\frac{d}{dx}\left[ze^{-r_2 x} - \frac{ce^{(r_1 - r_2)x}}{r_1 - r_2}\right] = 0,$$

since $r_1 - r_2 \neq 0$. As before, the quantity in brackets must be a constant c_2, and it follows that $z(x)$ must have the form (14), where $c_1 = c/(r_1 - r_2)$.

For the second part it is again a formal procedure to verify that (16) is a solution of (15). Assume, finally, that $z(x)$ is a real solution of (15), which we know from the preceding paragraph must have the form (14) or, with all quantities expressed in terms of real and imaginary parts:

(20) $$\begin{aligned}z(x) &= (\gamma + i\delta)e^{\alpha x}e^{i\beta x} + (\lambda + i\mu)e^{\alpha x}e^{-i\beta x} \\ &= e^{\alpha x}[(\gamma + \lambda)\cos \beta x + (-\delta + \mu)\sin \beta x] \\ &\quad + ie^{\alpha x}[(\delta + \mu)\cos \beta x + (\gamma - \lambda)\sin \beta x].\end{aligned}$$

Since this function $z(x)$ is assumed to be real-valued, the last bracketed quantity must vanish identically, and $z(x)$ is shown to be of the specified form (16), where $A = \gamma + \lambda$ and $B = -\delta + \mu$.

1514. EXERCISES

In Exercises 1–7, prove the indicated theorem of § 1513.
1. Theorem II. 2. Theorem III. 3. Theorem IV.
4. Theorem V. 5. Theorem VI. 6. Theorem VII.
7. Theorem VIII.

8. If the functions sin ax and cos ax, where a is an arbitrary complex number and x a real variable, are defined (cf. Ex. 29, § 1509):

$$(1) \qquad \cos ax \equiv \frac{e^{iax} + e^{-iax}}{2}, \qquad \sin ax \equiv \frac{e^{iax} - e^{-iax}}{2i},$$

prove that $\dfrac{d(\sin ax)}{dx} = a \cos ax$, and $\dfrac{d(\cos ax)}{dx} = -a \sin ax$.

9. Adapt and prove the two theorems of § 405 for the case of complex-valued functions of a real variable.

10. Prove that if δ and γ are arbitrary real numbers, a an arbitrary complex number, and x a real variable, then

$$\int_{\delta}^{\gamma} e^{ax}\, dx = \frac{e^{a\gamma} - e^{a\delta}}{a}.$$

11. By squaring and cubing $e^{i\theta} = \cos\theta + i\sin\theta$, obtain the formulas (for real θ):

$\cos 2\theta = \cos^2\theta - \sin^2\theta, \qquad \sin 2\theta = 2\sin\theta\cos\theta,$

$\cos 3\theta = \cos^3\theta - 3\cos\theta\sin^2\theta, \qquad \sin 3\theta = 3\cos^2\theta\sin\theta - \sin^3\theta.$

12. Prove that for any positive integer n, and real θ, $\cos n\theta$ is equal to a polynomial in $\cos\theta$, and $\sin n\theta$ is equal to $\sin\theta$ times a polynomial in $\cos\theta$. (Cf. Ex. 11.)

★13. Assume that the roots of the equation $r^2 + br + c = 0$, where b and c are complex constants, are *equal* complex numbers $r_1 = r_2$. Prove that every function of the form

$$(2) \qquad z = c_1 e^{r_1 x} + c_2 x e^{r_2 x},$$

where c_1 and c_2 are arbitrary complex constants, is a solution of the differential equation (15), § 1513. Conversely, prove that every solution of (15), § 1513, must have the form (2).

★14. From the reduction formula of Exercise 7, § 411, deduce those of Exercises 8 and 9, § 411.

★15. Assume that $f(x)$ is a complex-valued function of a real variable x that is integrable over every finite interval. Give a definition for the improper integral

$$(3) \qquad \int_{-\infty}^{+\infty} f(x)\, dx.$$

Define absolute convergence of (3), and prove that if $f(x)$ is dominated by a real-valued function $g(x)$ for which $\int_{-\infty}^{+\infty} g(x)\, dx$ converges, then (3) converges absolutely.

★16. The **Fourier transform** of a function f of one real variable is the function g of one real variable defined by the equation

$$(4) \qquad g(s) = \frac{1}{\sqrt{2\pi}} \int_{-\infty}^{+\infty} f(t) e^{-ist}\, dt,$$

whenever this integral exists. Prove that if (3) converges absolutely, then so does (4).

★17. The **Fourier cosine transform** of a function f of one real variable is the function C of one real variable defined by the equation

$$(5) \qquad C(s) = \sqrt{\frac{2}{\pi}} \int_{-\infty}^{+\infty} f(t) \cos st\, dt,$$

§ 1515]   THE FUNDAMENTAL THEOREM OF ALGEBRA   515

whenever this integral exists. Similarly, the **Fourier sine transform** of f is the function S defined by the equation

(6) $$S(s) = \sqrt{\frac{2}{\pi}} \int_{-\infty}^{+\infty} f(t) \sin st \, dt,$$

whenever this integral exists. Show that under assumptions of absolute convergence of (3), if f is an *even* real-valued function, then its Fourier transform (4) is equal to its Fourier cosine transform (5), and if f is an *odd* real-valued function, then its Fourier transform (4) is equal to $-i$ times its Fourier sine transform (6).

18. Find the Fourier transform of the function $f(t)$ that is equal to 1 for $0 < t < \alpha$ and equal to 0 otherwise. (Cf. Ex. 16.)

19. Find the Fourier transform of the function $f(t)$ that is equal to $e^{-\alpha t}$ for $t > 0$ and equal to 0 otherwise ($\alpha > 0$). (Cf. Ex. 16.)

*** 20.** Find the Fourier transform of the function $f(t) \equiv e^{-t^2}$. (Cf. Ex. 16; also, Example 4, § 1408.)

NOTE. For an elementary treatment of Fourier transforms, see R. V. Churchill, *Fourier Series and Boundary Value Problems* (New York, McGraw-Hill Book Co., 1941).

*1515. THE FUNDAMENTAL THEOREM OF ALGEBRA

The theorem of this section was first proved in 1799 by the German mathematician Gauss (1777–1855). At the present time this theorem is generally regarded as one of the basic theorems in the theory of analytic functions of a complex variable, rather than one of algebra. (For a proof based on techniques of analytic functions, cf. Ex. 16, § 1810.) The proof given below was suggested by Cauchy (French; 1789–1857), and presented by L. E. Dickson in the Appendix of his *First Course in the Theory of Equations* (New York, John Wiley and Sons, Inc., 1922).

In this section the word *number* will mean *complex number* (which may in particular be real) and the word *polynomial* will mean a function of the form $a_0 z^n + a_1 z^{n-1} + \cdots + a_{n-1} z + a_n$, where the coefficients a_i are complex numbers. In general, a **zero** of a function $f(z)$ is a number z_0 such that $f(z_0) = 0$.

Theorem. Fundamental Theorem of Algebra. *Any nonconstant polynomial has at least one zero.*

Proof. We shall obtain a contradiction to the assumption (which we now make) that there exists a nonconstant polynomial $f(z)$ without zeros. By Exercise 24, § 1512, then, the function $g(z) \equiv 1/f(z)$ is continuous for all values of z. Therefore, by Theorems I and IV, § 1510, the real-valued function of two real variables, $h(x, y) \equiv |g(x + iy)|$, is everywhere continuous. Since $h(0, 0)$ is a positive number, there exists, by Exercise 26, § 1512, a closed disk D, with center at the origin, such that for all points outside D, $h(x, y) < h(0, 0)$. By Theorem II, § 610, the function $h(x, y)$ assumes a maximum value, for all points (x, y) belonging to D, at some point (x_0, y_0) of D, and by the inequality of the preceding sentence this maximum value is not exceeded by $h(x, y)$ at any point in the plane. That is, $h(x_0, y_0)$ is an *absolute* maximum value of $h(x, y)$. The desired contradiction follows from the following lemma:

Lemma. *If $f(z)$ is a nonconstant polynomial and if $f(z_0) \neq 0$, there exists a number z_1 such that $|f(z_1)| < |f(z_0)|$.*

Proof of Lemma. We may assume without loss of generality that $z_0 = 0$, for $g(z) \equiv f(z + z_0)$ is a polynomial such that $g(0) = f(z_0) \neq 0$, so that if z_1 is a complex number such that $|g(z_1)| < |g(0)|$, then $z_1 + z_0$ is a complex number such that $|f(z_1 + z_0)| < |f(z_0)|$. Accordingly, we assume $z_0 = 0$, and write $f(z)$ in the form

$$f(z) = b_0 + b_k z^k + b_{k+1} z^{k+1} + \cdots + b_n z^n,$$

where $f(0) = b_0 \neq 0$, and $b_k \neq 0$, k being the smallest exponent of a power of z whose coefficient is not zero. (Not *all* nonconstant powers of z have zero coefficients, since $f(z)$ is not a constant.) Write the complex number $-b_0/b_k$ in the form $-b_0/b_k = \rho e^{i\psi}$, let $\theta = \psi/k$, and let $z_1 = re^{i\theta}$, where r is to be determined. Then $b_k z_1^k = b_k r^k e^{i\psi} = -b_0 r^k/\rho$. Therefore

$$f(z_1) = [b_0 - b_0 r^k/\rho] + [b_{k+1} r^{k+1} e^{i(k+1)\theta} + \cdots + b_n r^n e^{in\theta}],$$

and, by the triangle inequality and other facts established for absolute values (§ 1505),

$$|f(z_1)| \leq \frac{|b_0|}{\rho} \cdot |\rho - r^k| + |b_{k+1}| \cdot r^{k+1} + \cdots + |b_n| \cdot r^n$$

$$= \frac{|b_0|}{\rho} \cdot |\rho - r^k| + r^k[|b_{k+1}| \cdot r + \cdots + |b_n| r^{n-k}].$$

Now choose r so small that the quantity in brackets, immediately above, is less than $|b_0|/\rho$, and also so small that $r^k < \rho$. Then

$$|f(z_1)| < \frac{|b_0|}{\rho}(\rho - r^k) + r^k \cdot \frac{|b_0|}{\rho} = |b_0|.$$

This is the desired inequality, and the proof is complete.

16

Fourier Series ~~Read~~

1601. INTRODUCTION

The concept of *infinite series* dates as far back as the ancient Greeks. (For example, Archimedes (287–212 B.C.) summed a geometric series to compute the area of a parabolic segment.) As early as the latter half of the seventeenth century A.D., power series expansions of functions were being investigated. (For example, Newton in 1676 wrote down certain binomial series for fractional exponents, and Leibnitz, late in the century, quoted the familiar although previously published Maclaurin series for e^x, $\ln(1 + x)$, $\sin x$, and $\text{Arctan } x$. In 1715 Brook Taylor (1685–1731, English) published his famous formula for expanding a given (analytic) function in a power series (cf. Chapter 12).

By the middle of the eighteenth century it became important to study the possibility of representing a function by series other than power series. This question arose in connection with the problem of the vibrating string (cf. § 1625). Following the initial attacks on this problem by d'Alembert (1717–1783, French), in 1747, and Euler (1707–1783, Swiss), in 1748, Daniel Bernoulli (1700–1782, Swiss) showed, in 1753, that the mathematical conditions imposed by physical considerations were at least formally satisfied by functions defined by infinite series whose terms involved sines and cosines of integral multiples of prescribed variables. Lagrange (1736–1813, French) continued to pursue the implied problem of representing a given function in terms of sines and cosines. Fourier (1758–1830, French), in his book, *Théorie analytique de la Chaleur* (1822), which contained results of his studies on the conduction of heat (cf. § 1626), announced the amazing result that an "arbitrary" function could be expanded in a series with general term $a_n \sin nx$. Some of the "proofs" given by Fourier were lacking in rigor. Indeed, as we shall see, he claimed far too much, since only certain functions can be expanded in the form which he prescribed. However, the extent of "arbitrariness" that can be permitted is nevertheless remarkable.

Some of the difficulties that beset the early investigators of what became known as *Fourier series* lay in the lack of precision applied at that time to such concepts as *function* and *convergence*. For some of these mathematicians, for example, the word *function* was restricted to a single analytic expression, and was not used even for a function with a broken line graph. The first rigorous proof for a fairly extensive class of functions that a Fourier series actually represents a given function was published in 1829 by Dirichlet (1805–1859, German). Such researches called for more carefully formulated statements than had traditionally been demanded. Aside from the enormous importance of Fourier series as a technique for solving "boundary value" problems (in such applied areas as vibration and heat conduction), the purely theoretical studies in this subject have had a total effect on the general theory of functions of a real variable, and on the theory of sets, that is incalculable.

In this chapter we shall present only a few of the more important portions of the classical theory of Fourier series, together with examples of their application to boundary value problems. For a more complete introduction to the subject, and further references, see the author's *Real Variables*, Chapter 15.

1602. LINEAR FUNCTION SPACES

For conceptual purposes it is often helpful, when considering a family of functions having a common domain of definition, to regard this family as a "space," and the individual functions as "points" of this space. Under these circumstances, such a space of functions is called a **function space**. A simple example is the set of all functions defined and continuous on the interval $[0, 1]$.

One of the most important methods of combining functions in a function space is by means of a *linear combination* (cf. § 704):

Definition I. *A **linear combination** of a set $f_1, f_2, \cdots, f_n$ of functions defined on a set D is a function of the form*

$$\alpha_1 f_1 + \alpha_2 f_2 + \cdots + \alpha_n f_n, \tag{1}$$

where $\alpha_1, \alpha_2, \cdots, \alpha_n$ are real numbers.

For many purposes it is important for a function space to be "closed under finite linear combinations, according to the following definition:

Definition II. *A function space S is a **linear space** if and only if whenever $f_1, f_2, \cdots, f_n$ are members of S and $\alpha_1, \alpha_2, \cdots, \alpha_n$ are arbitrary real numbers, the linear combination $\alpha_1 f_1 + \alpha_2 f_2 + \cdots + \alpha_n f_n$ is also a member of S.*

Theorem. *A function space S is a linear space if and only if it is closed under (i) addition, and (ii) multiplication by a constant; that is, if and only if (i) whenever f and g are members of S, so is $f + g$, and (ii) whenever f is a member of S, so is αf for any real number α.*

Proof. The "only if" part is trivial (let $n = 2$, $\alpha_1 = \alpha_2 = 1$, and $\alpha_1 = \alpha$, $\alpha_2 = 0$, in turn). The "if" part follows by induction from the fact that if $\alpha_1 f_1 \in S$ and $\alpha_2 f_2 \in S$, then $\alpha_1 f_1 + \alpha_2 f_2 \in S$.

§ 1603]   PERIODIC FUNCTIONS. THE SPACE $R_{2\pi}$ 519

Examples. Each of the following sets of functions, assumed to be single-valued, real-valued, and defined over a common closed interval $I = [a, b]$, is a linear space. Verification of this statement for each example is easy and is left to the reader (Ex. 7, § 1607). Other examples are given in §§ 1603 and 1608.
1. RV: all real-valued functions.
2. B: all bounded functions. $B \subseteq RV$, etc.
3. RI: all Riemann integrable functions.
4. SC: all sectionally continuous functions (cf. § 408).
5. C: all continuous functions.
6. C^n: all functions with a continuous nth derivative, where n is a fixed positive integer.
7. C^∞: all "infinitely differentiable" functions; that is, each function has derivatives of all orders.
8. Π: all polynomials.
9. Γ: all constant-valued functions.

NOTE 1. Each space of the preceding list contains all spaces that follow it in the list, and therefore is contained in all that precede it (Ex. 7, § 1607).

NOTE 2. An example of a function space that is not a linear space is the set of all monotonic functions on a given closed interval. For example, the two functions $\sin x - 2x$ and $2x$ are separately monotonic on the interval $[0, 2\pi]$, but their sum is not.

NOTE 3. The function space of all real-valued functions whose common domain of definition consists of the positive integers $1, 2, \cdots, n$ is identical with the set of all ordered n-tuples of real numbers, that is, the set of all points in the Euclidean space of n dimensions. In other words, *Euclidean spaces are special cases of function spaces*. Indeed, they are *linear* function spaces.

NOTE 4. The *functions* of a linear function space can be regarded as generalizations of *vectors* in a Euclidean space, the combination (1) corresponding to a linear combination of vectors. For example, if the domain of definition of the members of S consists of the integers 1 and 2, the sum $f + g$ of the two plane vectors f and g reduces to their resultant, obtained by the familiar completing of a parallelogram. For these reasons, linear spaces are alternatively called **vector spaces** and their members are called **vectors**. The coefficients $\alpha_1, \alpha_2, \cdots, \alpha_n$ of Definition I are called **scalars** to distinguish them from vectors. The concept of linear or vector space can be abstracted, and defined without regard to a function space. (Cf. P. R. Halmos, *Finite-Dimensional Vector Spaces* (Princeton, D. Van Nostrand Co., 1958).)

1603. PERIODIC FUNCTIONS. THE SPACE $R_{2\pi}$

A concept that will be of special importance in this chapter is prescribed:

Definition I. *A real-valued function $f(x)$ of the real variable x is **periodic** with period p, where p is a fixed real number, if and only if $f(x + p)$ is defined whenever $f(x)$ is and the equation*

(1) $$f(x + p) = f(x)$$

is an identity, true for all x of the domain of $f(x)$.

NOTE. It can be proved (cf. *RV*, § 1507) that any nonconstant continuous periodic real-valued function with domain $(-\infty, +\infty)$ has a smallest positive period and that all of its periods are (integral) multiples of that one. For example, the periods of $\sin 2x$ are the multiples of π; in other words, $\sin 2x$ is periodic with period $n\pi$ for every integer n.

It is simple to show that the set of all (real-valued) periodic functions with domain $(-\infty, +\infty)$ and period 2π is a linear space. The same is true if there is imposed the further restriction that these functions be Riemann integrable on every finite interval. The resulting linear space $R_{2\pi}$ can be equivalently defined (Ex. 9, § 1607):

Definition II. *The linear space $R_{2\pi}$ consists of all periodic functions with domain $(-\infty, +\infty)$ and period 2π that are integrable on the closed interval $[-\pi, \pi]$.*

NOTE. The space $R_{2\pi}$ can be thought of as being generated by the functions f that are defined and Riemann integrable on $[-\pi, \pi]$ with equal values at the end-points: $f(-\pi) = f(\pi)$. Any such function can be **extended periodically** to a periodic function with domain $(-\infty, +\infty)$ and period 2π by means of the formula
(2) $$f(x) \equiv f(x + 2n\pi),$$
where, for any x, n is an integer such that $-\pi \leq x + 2n\pi \leq \pi$ (cf. Ex. 10, § 1607). In place of the interval $[-\pi, \pi]$ any closed interval $[a, a + 2\pi]$ of length 2π could have been used.

1604. INNER PRODUCT. ORTHOGONALITY. DISTANCE

In the space $R_{2\pi}$ it is possible to define an *inner product* that has many properties closely analogous to those of the inner product of vectors in E_3 (cf. § 707):

Definition I. *If f and $g \in R_{2\pi}$, their **inner product** (f, g) is defined:*

(1) $$(f, g) \equiv \int_{-\pi}^{\pi} f(x)g(x)\, dx.$$

NOTE 1. Since the product of two integrable functions is integrable (Theorem XIV, § 401) the inner product (1) always exists.

NOTE 2. The inner product (1) reduces to the scalar product of two vectors in case f and g have a finite domain and the integral in (1) is replaced by a finite sum.

The following properties of inner product are simple consequences of the definition and elementary properties of the Riemann integral (Ex. 11, § 1607):

Theorem I. (*i*) *The inner product is **symmetric**:*
(2) $$(f, g) = (g, f).$$
(*ii*) *The inner product is **linear** in each member:*
(3) $$\begin{cases} (\alpha_1 f_1 + \cdots + \alpha_n f_n, g) = \alpha_1(f_1, g) + \cdots + \alpha_n(f_n, g), \\ (f, \beta_1 g_1 + \cdots + \beta_n g_n) = \beta_1(f, g_1) + \cdots + \beta_n(f, g_n). \end{cases}$$

§ 1604] INNER PRODUCT. ORTHOGONALITY. DISTANCE

Taking as point of departure the criterion for orthogonality of two vectors in terms of the vanishing of their scalar product (Note 3, § 707), we introduce a definition:

Definition II. *Two members of $R_{2\pi}$, f and g, are* **orthogonal** *if and only if their inner product vanishes:*

(4) $\qquad f \perp g \quad \text{if and only if} \quad (f, g) = 0.$†

An **orthogonal family** *of functions is a collection of functions every two distinct members of which are orthogonal.*

Continuing the analogy further, we define a new concept, which corresponds to the *magnitude* of a vector:

Definition III. *The* **norm** *of a function $f \in R_{2\pi}$ is defined:*

(5) $\qquad \|f\| \equiv (f, f)^{\frac{1}{2}} = \left\{ \int_{-\pi}^{\pi} [f(x)]^2 \, dx \right\}^{\frac{1}{2}}.$ (length of vector)

The following properties of norm, for a function, are left for the reader to establish (Ex. 12, § 1607):

Theorem II. (i) *The norm is defined for every member of $R_{2\pi}$, and is nonnegative:*

(6) $\qquad \|f\| \geq 0.$

(ii) *For the function $0 = 0(x)$ that is identically zero the norm is zero:*

(7) $\qquad \|0\| = 0.$

(iii) *The* **triangle inequality** *holds, for all f and g of $R_{2\pi}$*

(8) $\qquad \|f + g\| \leq \|f\| + \|g\|.$

(iv) *The norm is* **positive-homogeneous**; *that is, if f is integrable and α is a real number,*

(9) $\qquad \|\alpha f\| = |\alpha| \cdot \|f\|.$

If S is a linear subspace of $R_{2\pi}$ such that all members of S except for the one that is identically zero have positive norm:

(10) $\qquad \|f\| > 0 \quad \text{if} \quad f \neq 0,$

then S is called a **normed linear space**, and the norm is called **strictly positive**.

NOTE 3. Throughout the remainder of this chapter, unless a specific statement to the contrary is made, the word *norm* will be used only in the sense of equation (5). However, it should be remarked that the terms *norm* and *normed linear space* are susceptible of considerable generalization. Indeed, *any* linear space for all points of which a "norm" having the five preceding properties, (6)–(10), exists is called a **normed linear space**. Euclidean spaces are examples (cf. Note 4, below).

† More generally, two functions f and g defined on any interval $[a, b]$ are said to be **orthogonal** on that interval if and only if $\int_a^b f(x)g(x) \, dx = 0$.

NOTE 4. The **Schwarz inequality** (Ex. 20, § 402) can be written:
$$|(f,g)| \leq \|f\| \cdot \|g\|,$$
which takes the form for vectors (u_1, u_2, u_3) and (v_1, v_2, v_3) in E_3:
$$|u_1v_1 + u_2v_2 + u_3v_3| \leq \sqrt{u_1^2 + u_2^2 + u_3^2}\sqrt{v_1^2 + v_2^2 + v_3^2}.$$
If (u_1, u_2, u_3) and (v_1, v_2, v_3) are nonzero vectors with an angle θ between them, this last inequality can be written
$$\frac{|u_1v_1 + u_2v_2 + u_3v_3|}{\sqrt{u_1^2 + u_2^2 + u_3^2}\sqrt{v_1^2 + v_2^2 + v_3^2}} \leq 1 \quad \text{or} \quad |\cos \theta| \leq 1.$$
Thus the Schwarz inequality in function space can be considered as a generalization of the statement that the cosine of an angle never exceeds 1 numerically. For the significance of equality in the Schwarz inequality, see Exercise 14, § 1607.

Formula (8) is the Minkowski inequality of Exercise 21, § 402.

In terms of the *norm* of a function, we are now able to define the concept of *distance* between members of $R_{2\pi}$:

Definition IV. *The **distance** between two functions f and $g \in R_{2\pi}$ is the norm of their difference:*
$$(11) \qquad d(f, g) \equiv \|f - g\| = \left\{\int_{-\pi}^{\pi} [f(x) - g(x)]^2 \, dx\right\}^{\frac{1}{2}}.$$

We can also speak of $d(f, g)$ as the *distance from f to g*, or the *distance of g from f*.

From the properties of norm we have almost immediately (give the proof in Ex. 13, § 1607):

Theorem III. *Distance, as prescribed in Definition IV, has the following four properties:*

*(i) Distance is defined and real-valued for every pair, f and g, of members of $R_{2\pi}$, and is **nonnegative***
$$(12) \qquad d(f, g) \geq 0.$$

(ii) Every member f of $R_{2\pi}$ is at a zero distance from itself:
$$(13) \qquad d(f, f) = 0.$$

*(iii) Distance is **symmetric**, for all f and g of $R_{2\pi}$:*
$$(14) \qquad d(f, g) = d(g, f).$$

*(iv) Distance satisfies the **triangle inequality**, for all f, g, and h of $R_{2\pi}$:*
$$(15) \qquad d(f, h) \leq d(f, g) + d(g, h).$$

A space with a distance function satisfying properties (i)–(iv) is called a **distance space** or a **pseudo-metric space**. If a distance space S satisfies the additional property:

*(v) Distance is **strictly positive**, for unequal members, f and g, of S:*
$$(16) \qquad d(f, g) > 0 \quad \text{if} \quad f \neq g,$$
it is called a **metric space**.

NOTE 5. The distance (11), for a subspace S of $R_{2\pi}$, is strictly positive if and only if the norm (5) is strictly positive.

§ 1604] INNER PRODUCT. ORTHOGONALITY. DISTANCE

Definition V. *If a function with nonzero norm is divided by plus or minus its norm, the resulting function has unit norm (and therefore corresponds to a vector of unit length):*

(17)
$$\left\| \frac{f}{\pm \|f\|} \right\| = 1.$$

*In this case the function f is said to be **normalized**, and either function $\pm f/\|f\|$ is called a **result of normalizing** f. An **orthonormal family** of functions is an orthogonal family each member of which has norm equal to unity.*

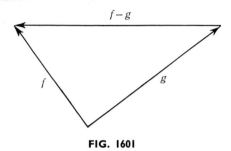

FIG. 1601

Example 1. Prove the **Pythagorean theorem**: *If f and g are orthogonal members of $R_{2\pi}$, then*

(18)
$$\|f - g\|^2 = \|f\|^2 + \|g\|^2.$$

(Cf. Fig. 1601, and Ex. 16, § 1607.)

Solution. By Theorem I,
$$\|f - g\|^2 = (f - g, f - g) = (f, f) - 2(f, g) + (g, g)$$
$$= (f, f) + (g, g) = \|f\|^2 + \|g\|^2.$$

Example 2. The functions

(19)
$$\begin{cases} 1, & \cos x, \quad \cos 2x, \quad \cos 3x, \quad \cdots, \\ & \sin x, \quad \sin 2x, \quad \sin 3x, \quad \cdots \end{cases}$$

are all members of $R_{2\pi}$, and therefore so are their linear combinations. Show that they form an orthogonal family of functions of nonzero norm. Normalize these functions and thus obtain an orthonormal family.

Solution. Standard integration procedures (check the details) give

$$\int_{-\pi}^{\pi} 1 \, dx = 2\pi,$$

$$\int_{-\pi}^{\pi} \cos^2 nx \, dx = \int_{-\pi}^{\pi} \sin^2 nx \, dx = \pi, \quad n = 1, 2, \cdots,$$

$$\int_{-\pi}^{\pi} \cos nx \, dx = \int_{-\pi}^{\pi} \sin nx \, dx = 0, \quad n = 1, 2, \cdots,$$

$$\int_{-\pi}^{\pi} \cos mx \cos nx \, dx = \int_{-\pi}^{\pi} \sin mx \sin nx \, dx = 0,$$
$$m = 1, 2, \cdots, \quad n = 1, 2, \cdots, \quad m \neq n,$$

$$\int_{-\pi}^{\pi} \cos mx \sin nx \, dx = 0, \quad m = 1, 2, \cdots, \quad n = 1, 2, \cdots.$$

Therefore the family (19) is orthogonal, the norm of the first number is $\sqrt{2\pi}$, and the norm of every other member is $\sqrt{\pi}$. A corresponding orthonormal family is

(20)
$$\begin{cases} \dfrac{1}{\sqrt{2\pi}}, \dfrac{\cos x}{\sqrt{\pi}}, \dfrac{\cos 2x}{\sqrt{\pi}}, \dfrac{\cos 3x}{\sqrt{\pi}}, \cdots, \\ \dfrac{\sin x}{\sqrt{\pi}}, \dfrac{\sin 2x}{\sqrt{\pi}}, \dfrac{\sin 3x}{\sqrt{\pi}}, \cdots. \end{cases}$$

1605. LEAST SQUARES. FOURIER COEFFICIENTS

Let $\Psi = (\psi_1, \psi_2, \cdots, \psi_n)$ be a finite orthonormal family of functions belonging to the space $R_{2\pi}$, and let S be the linear function space of all linear combinations $\psi = \beta_1 \psi_1 + \cdots + \beta_n \psi_n$ of the members of Ψ. If f is an arbitrary member of $R_{2\pi}$, does there exist a point ψ of S whose distance from f (as defined in (11), § 1604) is a minimum? If so, is ψ uniquely determined and in what way? It is to these questions that we now address ourselves.

Our first objective, then, is to minimize the quantity

(1) $$[d(f, \psi)]^2 = \|f - (\beta_1 \psi_1 + \cdots + \beta_n \psi_n)\|^2$$

as a function of the unknowns $\beta_1, \cdots, \beta_n$. We expand the right-hand member of (1) as the inner product $(f - \Sigma \beta_i \psi_i, f - \Sigma \beta_i \psi_i)$, and make use of the linearity properties of inner products stated in Theorem I, § 1604, together with the facts that $(\psi_i, \psi_j) = 0$, $i \neq j$, and $(\psi_i, \psi_i) = 1$, $i, j = 1, 2, \cdots, n$:

(2) $$[d(f, \psi)]^2 = \|f\|^2 + (\beta_1^2 + \beta_2^2 + \cdots + \beta_n^2)$$
$$- 2[\beta_1(f, \psi_1) + \beta_2(f, \psi_2) + \cdots + \beta_n(f, \psi_n)]$$
$$= \left\{ \sum_{i=1}^n [\beta_i - (f, \psi_i)]^2 \right\} + \left\{ \|f\|^2 - \sum_{i=1}^n (f, \psi_i)^2 \right\}.$$

The answers to the questions raised in the first paragraph are now accessible. Since the unknown β's occur only in the sum of squares $\Sigma[\beta_i - (f, \psi_i)]^2$, we infer that $d(f, \psi)$ is least when every term of that sum of squares vanishes. In other words, the minimum distance does exist and is given by a unique linear combination $\psi = \beta_1 \psi_1 + \cdots + \beta_n \psi_n$, where

(3) $$\beta_i = (f, \psi_i), i = 1, 2, \cdots, n.$$

The problem just solved can be thought of as that of finding the "best" approximation to a given function by means of a linear combination of the functions of a given finite orthonormal set—best in the sense of distance in the space $R_{2\pi}$. This criterion of "best approximation" is also called that of **least squares**, since the distance formula is based on squaring.

We are ready for a basic definition and theorem (summarizing the preceding remarks):

Definition. *Let* $\Phi = (\phi_1, \phi_2, \cdots)$ *be an orthonormal family (finite or infinite) of functions* $\phi_1, \phi_2, \cdots$ *belonging to* $R_{2\pi}$ *and let f be a given member* of $R_{2\pi}$. Then the **Fourier coefficients** of f with respect to the orthonormal family Φ are the numbers $\alpha_1, \alpha_2, \cdots$, defined:

(4) $$\alpha_n \equiv (f, \phi_n) = \int_{-\pi}^{\pi} f(x)\phi_n(x)\, dx.$$

Theorem. *Let $\Psi = (\psi_1, \psi_2, \cdots, \psi_n)$ be a finite subset of an orthonormal family $\Phi = (\phi_1, \phi_2, \cdots)$ of $R_{2\pi}$, and let f be a given function of $R_{2\pi}$. Then the linear combination $\beta_1\psi_1 + \cdots + \beta_n\psi_n$ of the members of Ψ that best approximates f, in the sense of the distance in $R_{2\pi}$, is uniquely given in case the coefficients $\beta_1, \cdots, \beta_n$ are the Fourier coefficients of f with respect to the corresponding members of Φ. The coefficient of any particular member of Ψ is independent of the remaining members of Ψ (or whether any other members of Φ are present at all).*

NOTE 1. In E_3, if Φ is the family of three unit vectors $\vec{i} = (1, 0, 0), \vec{j} = (0, 1, 0)$, and $\vec{k} = (0, 0, 1)$, and if $\vec{v}$ is a given vector with components v_1, v_2, and v_3:

$$\vec{v} = (v_1, v_2, v_3),$$

then the "Fourier coefficients" of $\vec{v}$ are its three components in the directions of the coordinate axes:

$$v_1 \cdot 1 + v_2 \cdot 0 + v_3 \cdot 0, \quad v_1 \cdot 0 + v_2 \cdot 1 + v_3 \cdot 0, \quad v_1 \cdot 0 + v_2 \cdot 0 + v_3 \cdot 1.$$

If Ψ consists of all *three* vectors of the orthonormal family Φ, the linear combination of $\vec{i}, \vec{j}$, and $\vec{k}$ that best approximates $\vec{v}$ has the three Fourier coefficients v_1, v_2, and v_3:

$$v_1\vec{i} + v_2\vec{j} + v_3\vec{k}.$$

In this case, the linear combination approximates $\vec{v}$ so well that it is equal to it. If Ψ consists of the *two* unit vectors $\vec{i}$ and $\vec{j}$, then the linear combination of these two vectors that best approximates $\vec{v}$ has the same Fourier coefficients (as far as they go):

$$v_1\vec{i} + v_2\vec{j}.$$

Finally, if Ψ consists of the *single* unit vector $\vec{i}$, the best approximation is

$$v_1\vec{i}.$$

NOTE 2. If S is the linear space of all linear combinations of the finite orthonormal set $\Psi = (\psi_1, \psi_2, \cdots, \psi_n)$, and if ψ is the member of S that minimizes the distance $d(f, \psi) = \|f - \psi\|$, then ψ can be thought of as the foot of a perpendicular dropped from f to S or, equivalently, as the *orthogonal projection* of f on S. This is true since the "vector" $f - \psi$ is orthogonal to every ψ_i, $i = 1, 2, \cdots, n$, and hence to every member of S. (Discuss the details in Ex. 17, § 1607.)

1606. FOURIER SERIES

Definition. *Let $\Phi = (\phi_1, \phi_2, \cdots)$ be the orthonormal family of Example 2, § 1604, belonging to the space $R_{2\pi}$:*

(1) $$\phi_1(x) \equiv \frac{1}{\sqrt{2\pi}}, \quad \phi_2(x) \equiv \frac{\cos x}{\sqrt{\pi}}, \quad \phi_3(x) \equiv \frac{\sin x}{\sqrt{\pi}},$$

$$\phi_4(x) \equiv \frac{\cos 2x}{\sqrt{\pi}}, \quad \phi_5(x) \equiv \frac{\sin 2x}{\sqrt{\pi}}, \cdots,$$

and let $f \in R_{2\pi}$. Then the **Fourier series**† of f is the series whose general term is $\alpha_n \phi_n$, where α_n is the Fourier coefficient $\alpha_n = (f, \phi_n)$. We say that f is *expanded* in a Fourier series, and write

(2) $$f \sim \alpha_1 \phi_1 + \alpha_2 \phi_2 + \cdots$$

without regard to any question of convergence.‡

Evaluation of the Fourier coefficients $\alpha_1, \alpha_2, \cdots$ gives:

$$\alpha_1 = (f, \phi_1) = \frac{1}{\sqrt{2\pi}} \int_{-\pi}^{\pi} f(x)\,dx,$$

$$\alpha_{2n} = (f, \phi_{2n}) = \frac{1}{\sqrt{\pi}} \int_{-\pi}^{\pi} f(x) \cos nx\,dx, \quad n = 1, 2, \cdots,$$

$$\alpha_{2n+1} = (f, \phi_{2n+1}) = \frac{1}{\sqrt{\pi}} \int_{-\pi}^{\pi} f(x) \sin nx\,dx, \quad n = 1, 2, \cdots.$$

Substitution of these coefficients in (2) leads to the following form for writing the Fourier series of $f(x)$:

(3) $$f(x) \sim \tfrac{1}{2}a_0 + a_1 \cos x + b_1 \sin x + a_2 \cos 2x + b_2 \sin 2x + \cdots,$$

where

(4) $$\begin{cases} a_n = \dfrac{1}{\pi} \displaystyle\int_{-\pi}^{\pi} f(x) \cos nx\,dx, & n = 0, 1, 2, \cdots, \\ b_n = \dfrac{1}{\pi} \displaystyle\int_{-\pi}^{\pi} f(x) \sin nx\,dx, & n = 1, 2, \cdots. \end{cases}$$

(Denoting the constant term by $\tfrac{1}{2}a_0$ rather than a_0 permits an economy in formulas (4), by making the formula for a_n valid when $n = 0$.)

Any series of the form

(5) $$\frac{A_0}{2} + A_1 \cos x + B_1 \sin x + A_2 \cos 2x + B_2 \sin 2x + \cdots$$

is called a **trigonometric series**, but only when the coefficients are obtained in terms of a given function by formulas of the form (4) is the series (5) a Fourier series. (Cf. Example 3, § 1609.)

NOTE 1. If a series of the form

$$\alpha_1 \phi_1 + \alpha_2 \phi_2 + \alpha_3 \phi_3 + \cdots$$

converges *uniformly* on the interval $[-\pi, \pi]$, and if a function f is *defined* to be the sum of that series (so that (2) becomes an equality by definition), then the coefficients $\alpha_1, \alpha_2, \alpha_3, \cdots$ are the Fourier coefficients of f (so that the relation (2) remains valid *as printed*). This is true since the formal procedure of multiplying on both sides of the *equality* (2) by the quantity ϕ_n and then integrating term by term is valid, and produces the desired relation $\alpha_n = (f, \phi_n)$. (Check the details.)

† If a *general* orthonormal family, in place of the particular family (1), is used, the resulting series is sometimes called the **generalized Fourier series** with respect to that orthonormal family. (For a brief treatment of generalized Fourier series, cf. *RV*, Chapter 15.)
‡ See §§ 1608, 1617, 1621 for statements regarding the manner in which the right-hand member of (2) can "represent" the function f.

Example 1. Find the Fourier series of the function $f \in R_{2\pi}$, defined for the interval $[-\pi, \pi]$:
$$f(x) = \begin{cases} x, & -\pi < x < \pi, \\ 0, & x = -\pi, \pi. \end{cases}$$

Solution. Substitution in formulas (4) gives $a_n = 0$, $n = 0, 1, 2, \cdots$ (since $f(x)$ is an odd function), and

$$b_n = \frac{1}{\pi} \int_{-\pi}^{\pi} x \sin nx \, dx = \frac{2}{\pi} \int_0^{\pi} x \, d\left(-\frac{\cos nx}{n}\right)$$

$$= -\frac{2x \cos nx}{n\pi}\bigg]_0^{\pi} = \begin{cases} 2/n, & n = 1, 3, 5, \cdots, \\ -2/n, & n = 2, 4, 6, \cdots. \end{cases}$$

Therefore,
$$f(x) \sim 2\left[\frac{\sin x}{1} - \frac{\sin 2x}{2} + \frac{\sin 3x}{3} - \cdots\right].$$

The graphs of the partial sums $S_1 = b_1 \sin x$ and $S_3 = \sum_{n=1}^{3} b_n \sin nx$ are shown in Figure 1602.

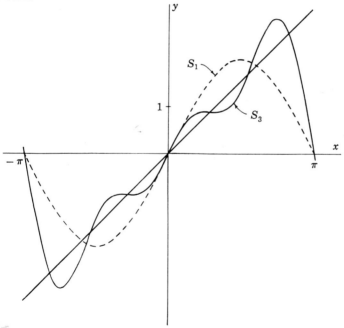

FIG. 1602

Example 2. Find the Fourier series of the function $f \in R_{2\pi}$, defined for the interval $[-\pi, \pi]$:
$$f(x) = |x|, \quad -\pi \leq x \leq \pi.$$

Solution. Since $f(x)$ is even, $b_n = 0$, $n = 1, 2, \cdots$,

$$a_0 = \frac{1}{\pi} \int_{-\pi}^{\pi} |x| \, dx = \frac{2}{\pi} \int_0^{\pi} x \, dx = \pi,$$

$$a_n = \frac{2}{\pi} \int_0^{\pi} x \, d\left(\frac{\sin nx}{n}\right) = -\frac{2 \cos nx}{n^2 \pi}\bigg]_0^{\pi} = \begin{cases} -4/n^2\pi, & n = 1, 3, \cdots, \\ 0, & n = 2, 4, 6, \cdots. \end{cases}$$

Therefore
$$f(x) \sim \frac{\pi}{2} - \frac{4}{\pi}\left[\frac{\cos x}{1^2} + \frac{\cos 3x}{3^2} + \frac{\cos 5x}{5^2} + \cdots\right].$$

The graphs of the partial sums $S_1 = \frac{1}{2}a_0 + a_1 \cos x$ and $S_3 = \frac{1}{2}a_0 + a_1 \cos x + a_3 \cos 3x$ are shown in Figure 1603.

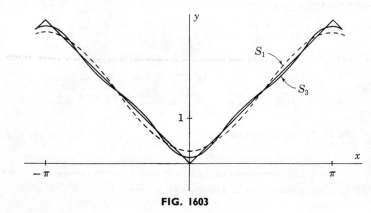

FIG. 1603

1607. EXERCISES

In Exercises 1–6, find the Fourier series of the given function $f(x)$. Draw graphs of the function $f(x)$ and of the first three distinct partial sums of its Fourier series.

1. $f(x) = \begin{cases} -1, & -\pi < x < 0, \\ 1, & 0 < x < \pi; \end{cases}$ $\quad f(-\pi) = f(0) = f(\pi) = 0.$

2. $f(x) = \begin{cases} 0, & -\pi < x < 0, \\ 1, & 0 < x < \pi; \end{cases}$ $\quad f(-\pi) = f(0) = f(\pi) = \frac{1}{2}.$

3. $f(x) = \begin{cases} 0, & -\pi < x < 0, \\ 1, & 0 < x < \frac{1}{2}\pi, \\ 0, & \frac{1}{2}\pi < x < \pi; \end{cases}$ $\quad f(0) = f(\frac{1}{2}\pi) = \frac{1}{2}.$

4. $f(x) = \begin{cases} 0, & -\pi < x < \frac{1}{2}\pi, \\ 1, & \frac{1}{2}\pi < x < \pi; \end{cases}$ $\quad f(-\pi) = f(\frac{1}{2}\pi) = f(\pi) = \frac{1}{2}.$

5. $f(x) = \begin{cases} 0, & -\pi < x < 0, \\ x, & 0 < x < \pi; \end{cases}$ $\quad f(-\pi) = f(\pi) = \frac{1}{2}\pi.$

6. $f(x) = \begin{cases} 0, & -\pi < x < 0, \\ x, & 0 < x < \frac{1}{2}\pi, \\ 0, & \frac{1}{2}\pi < x < \pi; \end{cases}$ $\quad f(\frac{1}{2}\pi) = \frac{1}{4}\pi.$

7. Verify that each of the Examples, § 1602, is a linear space, and prove the statement of Note 1, § 1602.

8. Show that the set of all real-valued functions defined on a fixed interval and possessing there both maximum and minimum values is not a linear space.

9. Prove the equivalence of the two definitions of the space $R_{2\pi}$, given in § 1603, and that this space is a linear space.

10. Prove that the definition of $f(x)$ provided by (2) of the Note, § 1603, is meaningful. That is, prove that the integer n exists and that, although n may not be uniquely determined, the value of $f(x)$ is.

11. Prove Theorem I, § 1604.

12. Prove Theorem II, § 1604. (Cf. Ex. 21, § 402, Ex. 15, below.)

13. Prove Theorem III, § 1604. *Hint:* For (iv), write
$$d(f, h) = \|(f - g) + (g - h)\|.$$

*14. Prove that equality, in the Schwarz inequality of Note 4, § 1604, holds if and only if one of the two functions is at a zero distance from some scalar multiple of the other. Hence prove that for the space $R_{2\pi}$ the triangle inequality becomes an equality if and only if one of the two functions is at a zero distance from some nonnegative scalar multiple of the other. *Hint for "only if":* Assume without loss of generality that $\|f\| = \|g\| = (f,g) = 1$. Then $\|f - g\| = 0$.

*15. Prove that whenever a linear space has an operation (f, g) on pairs of its members subject to the two properties of Theorem I, § 1604, and the third property that $(f, f) \geq 0$ for all members f of the space, then a "norm" defined $\|f\| = (f,f)^{\frac{1}{2}}$ has the four properties listed in Theorem II, § 1604. Furthermore, prove that the Schwarz inequality $|f, g| \leq \|f\| \cdot \|g\|$ is a further consequence. *Hint:* Prove the Schwarz inequality first. Cf. Exercise 20, § 402.

*16. Establish the following *law of cosines* (cf. Example 1, § 1604):
$$\|f - g\|^2 = \|f\|^2 + \|g\|^2 - 2\|f\| \cdot \|g\|(f/\|f\|, g/\|g\|)$$
for any functions f and g of $R_{2\pi}$ for which $\|f\| \cdot \|g\| \neq 0$. Also prove the *parallelogram law* for arbitrary $f, g \in R_{2\pi}$:
$$\|f + g\|^2 + \|f - g\|^2 = 2\|f\|^2 + 2\|g\|^2,$$
and that f and g are orthogonal if and only if
$$\|f + g\|^2 = \|f\|^2 + \|g\|^2.$$

*17. Justify the statements of Note 2, § 1605.

*18. Justify writing the Fourier series of a function f of $R_{2\pi}$ in the form

(1) $$\sum_{-\infty}^{+\infty} c_n e^{inx}$$

where, for every n, c_n and c_{-n} are complex conjugates, and

(2) $$c_n = \frac{1}{2\pi} \int_{-\pi}^{\pi} f(x) e^{inx} \, dx.$$

Prove the analogue of Note 1, § 1606, in case the series (1) converges *uniformly* on $[-\pi, \pi]$. (Cf. Chapter 15.)

1608. A CONVERGENCE THEOREM. THE SPACE $S_{2\pi}$

The question of *convergence* of the Fourier series of a given function f of $R_{2\pi}$ has not yet been raised. (Cf. § 1621.) One reason is that our approach to the topic of Fourier series has been based on approximation in the sense of least squares (§ 1605) rather than in the sense of the convergence of a series of functions. That is, we have considered, so far, approximation based only on integration over an entire interval, and not on functional values at particular values of the independent variable. Another reason for deferring consideration of this question of convergence is that it is an extremely difficult one to answer. Even if the given function is everywhere continuous the full story is not understood, although this much is known: *it is possible for the Fourier series of a continuous function to be divergent at some point.*† The twentieth century Russian mathematician Kolmogoroff has given an example of a function (integrable in the Lebesgue sense) whose Fourier series is *everywhere* divergent.‡

There are really two parts to the question of convergence: (*i*) Does the series converge at all? (*ii*) If the series converges does it represent the given function; that is, does it converge to $f(x)$? A trivial example shows that the Fourier series of a given function may converge everywhere, but fail to represent the function everywhere. Such an example is provided by any function that vanishes except on a nonempty finite set of points.

Although continuity of a function is not *sufficient* for convergence of its Fourier series to the function, neither is it *necessary*. That is, it is possible for a discontinuous function to be represented (even *everywhere*) by its Fourier series, provided its discontinuities are sufficiently mild, and provided it is sufficiently docile between its points of discontinuity.§ It turns out that a guarantee of good behavior is provided by the property of *sectional smoothness* (§ 408). For such functions we state a convergence theorem now and give a proof in § 1616. (It is shown in *RV*, § 1520, that a generalization of this theorem is obtained by replacing the words *sectionally smooth* by the words *of bounded variation*.)

Theorem. *Let $f(x)$ be periodic with period 2π, and sectionally smooth (§ 408) on the interval $[-\pi, \pi]$. Then the Fourier series of $f(x)$ converges for every real number x to the limit*

(1) $$\frac{f(x+) + f(x-)}{2}.$$

In particular, the series converges to the value of the given function at every point of continuity of the function, and if at every point of discontinuity the value of the function is defined as the average of its two one-sided limits there:

(2) $$f(x) = \frac{f(x+) + f(x-)}{2},$$

then its Fourier series represents the function everywhere. (Cf. Fig. 1604.)

---

† For a discussion of such an example, due to L. Fejér, see E. C. Titchmarsh, *The Theory of Functions* (Oxford Press, 1932), page 416.

‡ This example is given in A. Zygmund, *Trigonometrical Series* (New York, Dover Publications, 1955), page 175.

§ For a discussion of the unusual behavior of the Fourier series of a function near a point of discontinuity, known as the *Gibbs phenomenon*, see *RV*, § 1527.

§ 1608] A CONVERGENCE THEOREM. THE SPACE $S_{2\pi}$

For notational convenience and conceptual simplification, we introduce a new linear space (cf. Ex. 6, § 409):

Definition. *The space $S_{2\pi}$ consists of all members f of $R_{2\pi}$ that are sectionally smooth on the closed interval $[-\pi, \pi]$, and that possess there the averaging property* (2), *and its special case*:

$$(3) \qquad f(-\pi) = f(\pi) = \frac{f(-\pi+) + f(\pi-)}{2}.$$

We have as an immediate Corollary to the preceding theorem:

Corollary. *The Fourier series of any member of $S_{2\pi}$ converges and represents that function everywhere.*

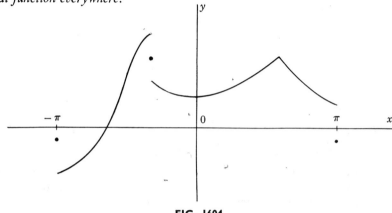

FIG. 1604

Example 1. Since the function $f(x)$ of Example 1, § 1606, satisfies the conditions of the preceding Theorem, it is represented everywhere by its Fourier series. In particular, for the interval $(-\pi, \pi)$,

$$(4) \qquad x = 2\left[\frac{\sin x}{1} - \frac{\sin 2x}{2} + \frac{\sin 3x}{3} - \cdots\right].$$

Substitution of $x = \frac{1}{2}\pi$ gives a familiar evaluation (cf. Ex. 31, § 1211):

$$(5) \qquad 1 - \frac{1}{3} + \frac{1}{5} - \frac{1}{7} + \cdots = \frac{\pi}{4}.$$

Example 2. As in the preceding Example, we have from Example 2, § 1606, for the interval $[-\pi, \pi]$:

$$(6) \qquad |x| = \frac{\pi}{2} - \frac{4}{\pi}\left[\frac{\cos x}{1^2} + \frac{\cos 3x}{3^2} + \frac{\cos 5x}{5^2} + \cdots\right].$$

Substitution of $x = 0$ gives the evaluation (cf. Ex. 11, § 1121):

$$(7) \qquad 1 + \frac{1}{3^2} + \frac{1}{5^2} + \frac{1}{7^2} + \cdots = \frac{\pi^2}{8}.$$

From this we can derive an important evaluation of a value of the Zeta-function (cf. the footnote, p. 388, and the Note, § 1120),

$$(8) \qquad \zeta(2) = 1 + \frac{1}{2^2} + \frac{1}{3^2} + \frac{1}{4^2} + \cdots = \frac{\pi^2}{6}:$$

Since the sum of the even-numbered terms of (8) is

$$\frac{1}{2^2} + \frac{1}{4^2} + \cdots = \frac{1}{4}\left[\frac{1}{1^2} + \frac{1}{2^2} + \cdots\right] = \frac{1}{4}\zeta(2),$$

we have, by subtraction, the sum of the odd-numbered terms as given in (7): $\frac{3}{4}\zeta(2) = \frac{1}{8}\pi^2$; (8) follows. (Cf. Example 1, § 1609.)

1609. BESSEL'S INEQUALITY. PARSEVAL'S EQUATION

Two important relations between the Fourier coefficients of a function and its norm are stated below. Proof of the first is given in this section, but that of the second is deferred to § 1621.

Theorem I. Bessel's Inequality. *If* $\Phi = (\phi_1, \phi_2, \cdots)$ *is an orthonormal family in* $R_{2\pi}$, *and if* $\alpha_1, \alpha_2, \cdots$ *are the Fourier coefficients of a given function* f *with respect to* Φ *(that is, if* $\alpha_n = (f, \phi_n), n = 1, 2, \cdots$*), then the series* $\Sigma \alpha_n^2$ *converges, and*

(1) $$\sum_{n=1}^{+\infty} \alpha_n^2 \leq \|f\|^2.$$

Proof. For any finite subfamily $\Psi = (\psi_1, \cdots, \psi_n)$ of Φ, the relation (2), § 1605, together with the inequality $[d(f, \psi)]^2 \geq 0$ for the particular Fourier coefficients $\beta_i = (f, \psi_i), i = 1, \cdots, n$, gives directly an inequality of the form (1) for an arbitrary subset of the coefficients α_n. In particular, since all partial sums of the series in (1) satisfy the inequality $\Sigma \alpha_n^2 \leq \|f\|^2$, so does the complete sum.

In case the orthonormal family Φ of Bessel's inequality is the special family of trigonometric functions Φ of § 1606, Bessel's inequality becomes an equality (cf. Ex. 18, § 1612). In the notation of § 1606:

Theorem II. Parseval's Equation. *If* $f \in R_{2\pi}$, *and if* $\frac{1}{2}a_0 + a_1 \cos x + b_1 \sin x + \cdots$ *is the Fourier series of* f, *then*

(2) $$\frac{1}{2}a_0^2 + a_1^2 + b_1^2 + a_2^2 + b_2^2 + \cdots = \frac{1}{\pi}\int_{-\pi}^{\pi} f[(x)]^2\, dx.$$

NOTE 1. In the Euclidean space E_3 (cf. Note 1, § 1605), Bessel's inequality takes such forms as

$$v_1^2 + v_2^2 + v_3^2 \leq |\vec{v}|^2,$$
$$v_1^2 + v_2^2 \leq |\vec{v}|^2,$$
$$v_1^2 \leq |\vec{v}|^2,$$

depending on whether the orthonormal family Φ consists of the three vectors $\vec{i}, \vec{j}$, and $\vec{k}$, the two vectors $\vec{i}$ and $\vec{j}$, or the single vector $\vec{i}$. In the first case Bessel's inequality becomes an equality:

$$v_1^2 + v_2^2 + v_3^2 = |\vec{v}|^2.$$

(Cf. Ex. 18, § 1612.)

§ 1609] BESSEL'S INEQUALITY. PARSEVAL'S EQUATION

As a corollary to Bessel's inequality we have:

Theorem III. Riemann's Lemma. *For any infinite orthonormal family $\Phi = (\phi_1, \phi_2, \cdots)$ in $R_{2\pi}$, and any function f of $R_{2\pi}$, the limit of the nth Fourier coefficient is zero:*

(3) $$\lim_{n \to +\infty} \alpha_n = \lim_{n \to +\infty} (f, \phi_n) = 0.$$

NOTE 2. In the notation of § 1606, Riemann's Lemma becomes:

(4) $$\lim_{n \to +\infty} a_n = \lim_{n \to +\infty} b_n = 0.$$

Example 1. From Example 1, § 1606,

$$4\left[\frac{1}{1^2} + \frac{1}{2^2} + \frac{1}{3^2} + \cdots\right] = \frac{1}{\pi}\int_{-\pi}^{\pi} x^2 \, dx = \frac{2\pi^2}{3}.$$

This is equivalent (cf. Example 2, § 1608): to

(5) $$\zeta(2) = \frac{1}{1^2} + \frac{1}{2^2} + \frac{1}{3^2} + \cdots + \frac{1}{n^2} + \cdots = \frac{\pi^2}{6}.$$

Example 2. Use Parseval's equation and Example 2, § 1606, to obtain the evaluation (cf. the footnote, p. 388):

(6) $$\zeta(4) = \frac{1}{1^4} + \frac{1}{2^4} + \frac{1}{3^4} + \cdots + \frac{1}{n^4} + \cdots = \frac{\pi^4}{90}.$$

Solution. Parseval's equation, for this example, is

(7) $$\frac{\pi^2}{2} + \frac{16}{\pi^2}\left[\frac{1}{1^4} + \frac{1}{3^4} + \frac{1}{5^4} + \cdots\right] = \frac{2\pi^2}{3},$$

whence

(8) $$\frac{1}{1^4} + \frac{1}{3^4} + \frac{1}{5^4} + \cdots = \frac{\pi^4}{96}.$$

By the method used in Example 2, § 1608, to evaluate $\zeta(2)$, we infer (6) from (8).

Example 3. The trigonometric series

$$\sin x + \sin 2x + \sin 3x + \cdots$$

is not a Fourier series because (4) fails. The series

(9) $$\sin x + \frac{\sin 2x}{\sqrt{2}} + \frac{\sin 3x}{\sqrt{3}} + \cdots,$$

although (4) holds, is not a Fourier series (of a Riemann integrable function) since the series Σb_n^2 is divergent (cf. (2)). (Cf. Note 3, below.)

*NOTE 3. It is of interest that although the trigonometric series (9) is not a Fourier series, it does converge for every value of x. This can be seen (assuming that x is not a multiple of π, in which case the result is trivial) by invoking the Abel test of Exercise 27, § 1117, with $a_n = \sin nx$ and $b_n = 1/\sqrt{n}$. In this case

$$\sum_{i=1}^{n} a_i = \frac{2 \sin x \sin \tfrac{1}{2}x + 2 \sin 2x \sin \tfrac{1}{2}x + \cdots + 2 \sin nx \sin \tfrac{1}{2}x}{2 \sin \tfrac{1}{2}x}.$$

With the aid of the identity $2 \sin \alpha \sin \beta = \cos(\alpha - \beta) - \cos(\alpha + \beta)$, this can be written

$$\frac{1}{2 \sin \tfrac{1}{2}x}\{\cos \tfrac{1}{2}x - \cos \tfrac{3}{2}x + \cos \tfrac{3}{2}x - \cos \tfrac{5}{2}x + \cdots + \cos(n - \tfrac{1}{2})x - \cos(n + \tfrac{1}{2})x\},$$

so that
$$\left|\sum_{i=1}^{n} a_i\right| = \tfrac{1}{2}|\csc \tfrac{1}{2}x| \cdot |\cos \tfrac{1}{2}x - \cos(n+\tfrac{1}{2})x| \leq |\csc \tfrac{1}{2}x|,$$
and the partial sums of the series $\Sigma \sin nx$ are bounded.

1610. COSINE SERIES. SINE SERIES

If $f(x) \in R_{2\pi}$ and if $f(x)$ is *even* (that is, if $f(-x) = f(x)$ for all x), then every coefficient b_n of its Fourier series, (3), § 1606, vanishes (why?), and the series becomes a pure **cosine series**:

(1) $$f(x) \sim \tfrac{1}{2}a_0 + a_1 \cos x + a_2 \cos 2x + \cdots,$$

where the coefficients can be written (Ex. 15, § 1612):

(2) $$a_n = \frac{2}{\pi} \int_0^\pi f(x) \cos nx \, dx, \quad n = 0, 1, 2, \cdots.$$

If $f(x)$ is a given function defined only on $[0, \pi]$, and integrable there, its domain of definition can be extended immediately to the full interval $[-\pi, \pi]$ by requiring that it be even there, and then to $(-\infty, +\infty)$ by periodicity (with period 2π). The series (1), subject to the formula (2), is called the cosine series of the given function defined on $[0, \pi]$.

In a similar fashion, if $f(x) \in R_{2\pi}$, and if $f(x)$ is *odd* (that is, if $f(-x) = -f(x)$ for all x), then the Fourier series of f becomes a pure **sine series**:

(3) $$f(x) \sim b_1 \sin x + b_2 \sin 2x + \cdots,$$

where the coefficients can be written (Ex. 15, § 1612):

(4) $$b_n = \frac{2}{\pi} \int_0^\pi f(x) \sin nx \, dx, \quad n = 1, 2, \cdots.$$

If $f(x)$ is a given function defined only on $[0, \pi]$, and integrable there, and if it is redefined (if necessary) to have the value 0 at $x = 0$ and $x = \pi$ ($f(0) = f(\pi) = 0$), then its domain of definition can be extended immediately to the interval $[-\pi, \pi]$ by requiring that it be odd there, and then to $(-\infty, +\infty)$ by periodicity. The series (3), subject to (4), is called the sine series of the given function defined on $[0, \pi]$.

Example 1. Examples 1 and 2, § 1606, give the sine and cosine series respectively, for the function x, defined on $[0, \pi]$.

Example 2. Find the cosine and sine series for the function x^2 on $[0, \pi]$. Then use Parseval's equation to verify the evaluation of $\zeta(4)$ (Example 2, § 1609) and to obtain an evaluation of $\zeta(6)$.

Solution. Integration by parts gives, for (2) and (4):

$$a_0 = \frac{2\pi^2}{3}, \quad a_n = (-1)^n \frac{4}{n^2}, \quad n = 1, 2, \cdots,$$

$$b_n = \frac{(-1)^{n+1} 2\pi}{n} - \frac{4[1 + (-1)^{n+1}]}{n^3 \pi}, \quad n = 1, 2, \cdots.$$

Parseval's equation for the resulting cosine series for x^2 on the interval $[-\pi, \pi]$ is

$$\frac{2\pi^4}{9} + 16\left(\frac{1}{1^4} + \frac{1}{2^4} + \cdots\right) = \frac{2}{\pi}\int_0^\pi x^4\,dx = \frac{2\pi^4}{5},$$

and the desired evaluation $\zeta(4) = \pi^4/90$ follows immediately.

For the sine series, Parseval's equation gives

$$4\pi^2\zeta(2) - 32\cdot\frac{15}{16}\zeta(4) + \frac{64}{\pi^2}\cdot\frac{63}{64}\zeta(6) = \frac{2\pi^4}{5},$$

whence

(5) $$\zeta(6) = \frac{1}{1^6} + \frac{1}{2^6} + \frac{1}{3^6} + \frac{1}{4^6} + \cdots = \frac{\pi^6}{945}.$$

NOTE. Since any function $f(x)$ can be written as the sum of an even function and an odd function:

(6) $$f(x) = \frac{f(x) + f(-x)}{2} + \frac{f(x) - f(-x)}{2},$$

any trigonometric Fourier series can be obtained by "adding" a cosine series and a sine series.

Example 3. Find the Fourier series of $f(x)$, defined:

$$f(x) = \begin{cases} 0, & -\pi < x < 0, \\ x, & 0 < x < \pi. \end{cases}$$

Solution. Using (6), we have

$$f(x) = \frac{|x|}{2} + \frac{x}{2},$$

and therefore, from Examples 1 and 2, § 1606, the Fourier series is

$$\frac{\pi}{4} + \sum_{n=1}^{+\infty}\left\{-\frac{1}{n^2\pi}[1 + (-1)^{n+1}]\cos nx + \frac{(-1)^{n+1}}{n}\sin nx\right\}.$$

1611. OTHER INTERVALS

Let $f(x)$ be a periodic function with domain $(-\infty, +\infty)$ and of period $2p$ (notationally, $2p$ is more convenient here than p), integrable on the interval $[-p, p]$ (or any interval of length $2p$). It is merely a matter of a change of scale associated with the positive factor π/p to adapt the expansion formulas of § 1606, and their derivation, to the Fourier series of $f(x)$ for the interval $[-p, p]$ (Ex. 16, § 1612):

(1) $$f(x) \sim \frac{1}{2}a_0 + a_1\cos\frac{\pi x}{p} + b_1\sin\frac{\pi x}{p} + a_2\cos\frac{2\pi x}{p} + b_2\sin\frac{2\pi x}{p} + \cdots,$$

where

(2) $$\begin{cases} a_n = \dfrac{1}{p}\displaystyle\int_{-p}^p f(x)\cos\frac{n\pi x}{p}\,dx, & n = 0, 1, 2, \cdots, \\[2mm] b_n = \dfrac{1}{p}\displaystyle\int_{-p}^p f(x)\sin\frac{n\pi x}{p}\,dx, & n = 1, 2, \cdots. \end{cases}$$

NOTE 1. A function $f(x)$ defined and integrable on $[0, p]$ can be expanded in either a cosine series or a sine series. (Cf. Ex. 17, § 1612.)

Example. Find the Fourier series for the function $f(x) = |x|$ on the interval $[-10, 10]$, as a periodic function of period 20.

First Solution. Substitution in (2), with $p = 10$, gives
$$a_0 = 10, \quad a_n = -\frac{20}{n^2\pi^2}[1 + (-1)^{n+1}], \quad b_n = 0, \quad n = 1, 2, 3, \cdots,$$
and therefore, for $|x| \leq 10$,

(3) $\quad |x| = 5 - \dfrac{40}{\pi^2}\left[\dfrac{1}{1^2}\cos\dfrac{\pi x}{10} + \dfrac{1}{3^2}\cos\dfrac{3\pi x}{10} + \dfrac{1}{5^2}\cos\dfrac{5\pi x}{10} + \cdots\right].$

Second Solution. Substitution of $x = \pi t/10$ in the Fourier series obtained in Example 2, § 1606, gives
$$\frac{\pi}{10}|t| = \frac{\pi}{2} - \frac{4}{\pi}\left[\frac{1}{1^2}\cos\frac{\pi t}{10} + \frac{1}{3^2}\cos\frac{3\pi t}{10} + \frac{1}{5^2}\cos\frac{5\pi t}{10} + \cdots\right],$$
which reduces immediately to (3) except for notation.

NOTE 2. For simplicity, most of the results in the remaining portions of this chapter are restricted to periodic functions with period 2π (instead of period $2p$). There is no essential loss of generality in this restriction, which is made for the sake of manipulative simplification. A simple scale factor converts statements in terms of period 2π to apply to the appropriate period.

1612. EXERCISES

1. By substituting $x = \frac{1}{2}\pi$ in each of the Fourier series of Exercises 1–4, § 1607, obtain equation (5), § 1608.

2. By substituting $x = \pi$ and $x = \frac{1}{2}\pi$ in the Fourier series of Exercises 5 and 6, respectively, § 1607, obtain equation (7), § 1608.

3. Show that Parseval's equation for each of Exercises 1 and 2, § 1607, reduces to (7), § 1608.

In Exercises 4–6, find the Fourier series for the given function of the space $R_{2\pi}$, with the averaging condition (2), § 1608. Then use Parseval's equation to obtain the indicated evaluation of the Zeta-function. (Cf. the footnote, p. 388; also Ex. 20, below; Ex. 5, § 1624.)

4. $f(x) = x^3$; $\zeta(6) = \pi^6/945$.
5. $f(x) = x^4$; $\zeta(8) = \pi^8/9450$.
6. $f(x) = x^5$; $\zeta(10) = \pi^{10}/93555$.

In Exercises 7–10, find the cosine series and the sine series for the given function of the space $R_{2\pi}$, with the averaging condition (2), § 1608. Draw graphs of the function and of the first three distinct partial sums of the requested series.

7. $f(x) = \begin{cases} 0, & 0 < x < \frac{1}{2}\pi, \\ 1, & \frac{1}{2}\pi < x < \pi. \end{cases}$
8. $f(x) = \begin{cases} x, & 0 < x < \frac{1}{2}\pi, \\ \pi - x, & \frac{1}{2}\pi < x < \pi. \end{cases}$
9. $f(x) = \cos x.$
10. $f(x) = \sin x.$

In Exercises 11–14, find the Fourier series of the periodic function of period $2p$ for the given interval $[-p, p]$, with the averaging condition (2), § 1608. Draw graphs of the function and of the first three distinct partial sums of the requested series.

11. $f(x) = \begin{cases} -1, & -1 < x < 0, \\ 1, & 0 < x < 1, \end{cases}$ where $p = 1$.

12. $f(x) = x, \ -2 < x < 2,$ where $p = 2$.

13. $f(x) = \begin{cases} 0, & -4 < x < 0, \\ x, & 0 < x < 4, \end{cases}$ where $p = 4$.

14. $f(x) = \begin{cases} 0, & -6 < x < 0, \\ x, & 0 < x < 3, \\ 0, & 3 < x < 6, \end{cases}$ where $p = 6$.

15. Establish formulas (2) and (4), § 1610.

16. Establish formulas (1) and (2), § 1611.

17. Establish formulas for cosine series and sine series coefficients, for functions of period $2p$.

★18. Let $\Phi = (\phi_1, \phi_1, \cdots, \phi_n)$ be a finite orthonormal family in $R_{2\pi}$, and let $f \in R_{2\pi}$. Prove that Bessel's inequality $\sum_{k=1}^{n} \alpha_k^2 \leq \|f\|^2$ is an equality if and only if f is at a zero distance from some linear combination of $\phi_1, \cdots, \phi_n$. (Cf. Note 2, § 1605.)

★19. The **Bernoulli polynomials**, $B_0(x), B_1(x), B_2(x), \cdots$, are defined by the following expansion, where the numbers $B_0, B_1, B_2, \cdots$ are the Bernoulli numbers (cf. Ex. 38, § 1211):

(1) $\quad e^{xt} \cdot \dfrac{t}{e^t - 1} = \sum_{n=0}^{+\infty} \dfrac{(xt)^n}{n!} \cdot \sum_{n=0}^{+\infty} \dfrac{B_n}{n!} t^n = \sum_{n=0}^{+\infty} \dfrac{B_n(x)}{n!} t^n.$

Prove that $B_n(x)$ is a polynomial of degree n given by the formula

(2) $\quad B_n(x) = \binom{n}{0} B_0 x^n + \binom{n}{1} B_1 x^{n-1} + \cdots + \binom{n}{n-1} B_{n-1} x + \binom{n}{n} B_n.$

Show that $B_n(0) = B_n, n \geq 0$, and by (2), Exercise 38, § 1211, and (2) above, that $B_n(1) = B_n, n \geq 2$. Furthermore, show that for $n \geq 1$:

(3) $\quad \begin{cases} B_{n+1}(x) = B_{n+1} + (n+1) \int_0^x B_n(t)\, dt, \\ B_{n+1}'(x) = (n+1) B_n(x), \\ \int_0^1 B_n(x)\, dx = 0. \end{cases}$

Obtain the first five Bernoulli polynomials.

★20. Obtain the following Fourier cosine and sine series for Bernoulli polynomials on the closed interval $0 \leq x \leq 1$ (cf. Ex. 19):

(4) $\quad B_{2n}(x) = (-1)^{n+1} \dfrac{2(2n)!}{(2\pi)^{2n}} \sum_{k=1}^{+\infty} \dfrac{\cos 2k\pi x}{k^{2n}}, \quad n \geq 1,$

(5) $\quad B_{2n+1}(x) = (-1)^{n+1} \dfrac{2(2n+1)!}{(2\pi)^{2n+1}} \sum_{k=1}^{+\infty} \dfrac{\sin 2k\pi x}{k^{2n+1}}, \quad n \geq 1.$

From (4) derive the formula for values of the Riemann Zeta-function (cf. § 1108) for positive even integers:

(6) $\quad \zeta(2n) = \sum_{k=1}^{+\infty} \dfrac{1}{k^{2n}} = (-1)^{n+1} \dfrac{(2\pi)^{2n} B_{2n}}{2(2n)!}.$

Infer that the Bernoulli numbers alternate in sign. *Hint:* Show that (5) holds for $n = 0$ in the open interval $0 < x < 1$. Then use mathematical induction, with the aid of (3) and integration by parts.

*1613. PARTIAL SUMS OF FOURIER SERIES

Since, by Riemann's Lemma (§ 1609), the general term of the Fourier series of any function f of $R_{2\pi}$ tends toward 0, it is sufficient in any considerations of convergence to limit our investigations to partial sums of an odd number of terms. We introduce the notation:

$$(1) \qquad S_n(x) \equiv \frac{a_0}{2} + \sum_{k=1}^{n}(a_k \cos kx + b_k \sin kx).$$

In terms of the given function f of $R_{2\pi}$, $S_n(x)$ can be written:

$$(2) \quad S_n(x) = \frac{1}{2\pi}\int_{-\pi}^{\pi} f(t)\,dt + \frac{1}{\pi}\sum_{k=1}^{n}\left\{\cos kx \int_{-\pi}^{\pi} f(t)\cos kt\,dt + \sin kx \int_{-\pi}^{\pi} f(t)\sin kt\,dt\right\}$$

$$= \frac{1}{\pi}\int_{-\pi}^{\pi}\left\{\frac{1}{2} + \sum_{k=1}^{n}\cos k(t-x)\right\}f(t)\,dt.$$

We now use the identity (Ex. 1, § 1624)

$$(3) \qquad \frac{1}{2} + \sum_{k=1}^{n}\cos 2k\alpha = \frac{\sin(2n+1)\alpha}{2\sin\alpha}.$$

After substituting first $\alpha = \frac{1}{2}(t - x)$ and then $t - x = u$, we can write the right-hand member of (2) as follows:

$$(4) \quad \frac{1}{2\pi}\int_{-\pi}^{\pi}\frac{\sin(n+\frac{1}{2})(t-x)}{\sin\frac{1}{2}(t-x)}f(t)\,dt = \frac{1}{2\pi}\int_{-\pi-x}^{\pi-x}\frac{\sin(n+\frac{1}{2})u}{\sin\frac{1}{2}u}f(x+u)\,du.$$

By periodicity, we have **Dirichlet's Integral**

$$(5) \qquad S_n(x) = \frac{1}{2\pi}\int_{-\pi}^{\pi}\frac{\sin(n+\frac{1}{2})u}{\sin\frac{1}{2}u}f(x+u)\,du.$$

By writing this integral as the sum of two, and finally using the substitution $v = -u$ to convert the second integral, below, we obtain for $S_n(x)$:

$$\frac{1}{2\pi}\int_{0}^{\pi}\frac{\sin(n+\frac{1}{2})u}{\sin\frac{1}{2}u}f(x+u)\,du + \frac{1}{2\pi}\int_{-\pi}^{0}\frac{\sin(n+\frac{1}{2})v}{\sin\frac{1}{2}v}f(x+v)\,dv,$$

or

$$(6) \qquad S_n(x) = \frac{1}{2\pi}\int_{0}^{\pi}\frac{\sin(n+\frac{1}{2})u}{\sin\frac{1}{2}u}[f(x+u) + f(x-u)]\,du.$$

Any question of convergence of a Fourier series can thus be formulated in terms of the limit of the partial sums $S_n(x)$ in equation (6).

*1614. FUNCTIONS WITH ONE-SIDED LIMITS

Under the assumption that the function $f(x)$, whose Fourier series is under consideration, has one-sided limits at a particular point x, we are now in a position to establish a criterion for the partial sums of the Fourier series to converge at x to the average of these two limits:

(1) $$S_n(x) \to \frac{f(x+) + f(x-)}{2}.$$

We start by writing equation (6), § 1613, for the function $f(x)$ that is identically 1 (cf. Ex. 2, § 1624):

(2) $$1 = \frac{1}{\pi} \int_0^\pi \frac{\sin(n + \tfrac{1}{2})u}{\sin \tfrac{1}{2}u}\, du.$$

Multiplying both members of (2) by the constant $\tfrac{1}{2}[f(x+) + f(x-)]$, and subtracting the resulting expressions from the two members of (6), § 1613, we have

(3) $$S_n(x) - \frac{f(x+) + f(x-)}{2} = \frac{1}{2\pi} \int_0^\pi \frac{\sin(n + \tfrac{1}{2})u}{\sin \tfrac{1}{2}u}[f(x+u) - f(x+)]\, du$$
$$+ \frac{1}{2\pi} \int_0^\pi \frac{\sin(n + \tfrac{1}{2})u}{\sin \tfrac{1}{2}u}[f(x-u) - f(x-)]\, du.$$

Therefore, a necessary and sufficient condition for the limit relation (1) to hold is that the right-hand member of (3) converge to 0 as $n \to +\infty$. We are thus led to a study of the two integrals on the right of (3).

*1615. THE RIEMANN-LEBESGUE THEOREM

The following corollary of Riemann's Lemma (Theorem III, § 1609) will be needed. It is a special case of a result known as the **Riemann-Lebesgue Theorem**:

Theorem. *If $g(x)$ is integrable on $[0, \pi]$, then*

$$\lim_{n \to +\infty} \int_0^\pi [\sin(n + \tfrac{1}{2})x] \cdot g(x)\, dx = 0.$$

Proof. By a standard trigonometric identity the integral whose limit is sought can be written as the sum of the following two:

$$\int_0^\pi \sin nx \left[\cos \frac{x}{2} \cdot g(x)\right] dx + \int_0^\pi \cos nx \left[\sin \frac{x}{2} \cdot g(x)\right] dx.$$

Each of these tends toward zero by Riemann's Lemma applied in turn to the function identically 0 for $-\pi < x < 0$ and equal to $\cos \dfrac{x}{2} \cdot g(x)$ for $0 < x < \pi$, and to the function identically 0 for $-\pi < x < 0$ and equal to $\sin \dfrac{x}{2} \cdot g(x)$ for $0 < x < \pi$.

*1616. PROOF OF THE CONVERGENCE THEOREM

The stage is now set for a simple proof of the convergence theorem for Fourier series of a sectionally smooth function, as given in § 1608. By (3), § 1614, we have only to show that

$$(1) \quad \lim_{n \to +\infty} \int_0^\pi [\sin(n + \tfrac{1}{2})u] \cdot g(u)\, du = 0,$$

where $g(u)$ is equal to either of the following:

$$(2) \quad \frac{u}{\sin \tfrac{1}{2}u} \cdot \frac{f(x+u) - f(x+)}{u}, \quad \frac{u}{\sin \tfrac{1}{2}u} \cdot \frac{f(x-u) - f(x-)}{u}.$$

Owing to the existence of both a right-hand and a left-hand derivative for $f(x)$ at every point x, each of the functions (2) has a right-hand limit at $u = 0$ (cf. § 303, Ex. 45, § 308) and is therefore integrable on $[0, \pi]$ (cf. Theorem III, § 403). By the Theorem, § 1615, (1) follows immediately, and the proof is complete.

*1617. FEJÉR'S SUMMABILITY THEOREM

If the condition of sectional smoothness of the function $f(x)$ is completely dropped from the hypotheses of the convergence theorem of § 1608, convergence of the Fourier series cannot be inferred. However, as long as the existence of the one-sided limits at the point in question is retained, a useful and striking result can be obtained if, instead of considering *convergence*, we turn our attention to *summability* $(C, 1)$ (that is, summability by Cesàro means of order 1; cf. Ex. 20, § 1117).

Theorem. Fejér's Summability Theorem. *If $f \in R_{2\pi}$, if $\tfrac{1}{2}a_0 + a_1 \cos x + b_1 \sin x + \cdots$ is its Fourier series, if*

$$(1) \quad S_n(x) \equiv \tfrac{1}{2}a_0 + a_1 \cos x + b_1 \sin x + \cdots + a_n \cos nx + b_n \sin nx,$$

and if

$$(2) \quad \sigma_n(x) \equiv \frac{S_0(x) + S_1(x) + \cdots + S_{n-1}(x)}{n},$$

then

$$(3) \quad \sigma_n(x) \to \frac{f(x+) + f(x-)}{2}$$

at any point x at which the function f has two one-sided limits. In particular, the Fourier series is summable to the value $f(x)$ at every point x at which $f(x) = \tfrac{1}{2}[f(x+) + f(x-)]$, and therefore at every point of continuity of the function f.

Proof. Defining

$$(4) \quad \phi(x, u) \equiv [f(x+u) - f(x+)] + [f(x-u) - f(x-)],$$

we can write (3), § 1614, in the form

(5) $$S_n(x) - \frac{f(x+) + f(x-)}{2} = \frac{1}{2\pi} \int_0^\pi \frac{\sin(n + \tfrac{1}{2})u}{\sin \tfrac{1}{2}u} \phi(x, u) \, du.$$

Consequently,

(6) $$\sigma_n(x) - \frac{f(x+) + f(x-)}{2} = \sum_{k=0}^{n-1} \frac{1}{n} \left[S_k(x) - \frac{f(x+) + f(x-)}{2} \right]$$

$$= \frac{1}{2n\pi} \int_0^\pi \frac{1}{\sin \tfrac{1}{2}u} \left\{ \sum_{k=0}^{n-1} \sin(k + \tfrac{1}{2})u \right\} \phi(x, u) \, du.$$

By means of the identity $\sum_{k=0}^{n-1} \sin(k + \tfrac{1}{2})u = \sin^2 \tfrac{1}{2}nu / \sin \tfrac{1}{2}u$ (Ex. 3, § 1624), equation (6) can be written

(7) $$\sigma_n(x) - \frac{f(x+) + f(x-)}{2} = \frac{1}{2n\pi} \int_0^\pi \frac{\sin^2 \tfrac{1}{2}nu}{\sin^2 \tfrac{1}{2}u} \phi(x, u) \, du,$$

and the proof of the theorem reduces to showing

(8) $$\frac{1}{n} \int_0^\pi \frac{\sin^2 \tfrac{1}{2}nu}{\sin^2 \tfrac{1}{2}u} \phi(x, u) \, du \to 0.$$

The next step is to show that (8) can be replaced by

(9) $$\frac{1}{n} \int_0^\pi \frac{\sin^2 \tfrac{1}{2}nu}{u^2} \phi(x, u) \, du \to 0.$$

The equivalence of (8) and (9) follows from the following inequality (where it is assumed that for all x, $|f(x)| < K$):

$$\left| \int_0^\pi \sin^2 \tfrac{1}{2}nu \left(\frac{1}{\sin^2 \tfrac{1}{2}u} - \frac{1}{(\tfrac{1}{2}u)^2} \right) \phi(x, u) \, du \right|$$

$$\leq \int_0^\pi \left| \csc^2 \tfrac{1}{2} u - \frac{4}{u^2} \right| \cdot 4K \, du < +\infty.$$

(Give the details.)

To prove (9), let x be a point at which $f(x+)$ and $f(x-)$ exist. Then, since

(10) $$\lim_{u \to 0+} \phi(x, u) = 0,$$

corresponding to $\epsilon > 0$ there exists a $\delta > 0$ such that $|\phi(x, \delta)| < 2\epsilon/\pi$. Then (with $v = \tfrac{1}{2}nu$)

(11) $$\left| \frac{1}{n} \int_0^\delta \frac{\sin^2 \tfrac{1}{2}nu}{u^2} \phi(x, u) \, du \right| < \frac{2\epsilon}{n\pi} \int_0^\delta \frac{\sin^2 \tfrac{1}{2}nu}{u^2} \, du$$

$$= \frac{\epsilon}{\pi} \int_0^{\tfrac{1}{2}n\delta} \frac{\sin^2 v}{v^2} \, dv < \frac{\epsilon}{\pi} \int_0^{+\infty} \frac{\sin^2 v}{v^2} \, dv = \frac{\epsilon}{2}.$$

(For the evaluation of the improper integral, cf. Ex. 14, § 1409.)

For the remaining part of the proof we use the following inequality (with $|f(x)| < K$):

(12) $$\left| \frac{1}{n} \int_\delta^\pi \frac{\sin^2 \tfrac{1}{2}nu}{u^2} \phi(x, u) \, du \right| \leq \frac{1}{n} \int_\delta^\pi \frac{1}{\delta^2} \cdot 4K \, du.$$

This quantity is less than $\tfrac{1}{2}\epsilon$ if $n > 8\pi K/\delta^2 \epsilon$.

Corollary. *If a Fourier series converges at a point x where $f(x+)$ and $f(x-)$ exist, it converges to $\frac{1}{2}[f(x+) + f(x-)]$.*

Proof. Any convergent series is summable to the sum of the series (cf. Ex. 20, § 1117).

*1618. UNIFORM SUMMABILITY

Inasmuch as each $\sigma_n(x)$ defined in (2), § 1617, is everywhere continuous, it follows that whenever the Fourier series of a function $f(x)$ possessing one-sided limits is uniformly summable over some interval (in the sense that the limit (3), § 1617, is a uniform limit over that interval) the function $f(x)$ must be continuous over that interval. A converse of this fact is the following:

Theorem. *If $f(x)$ is continuous on an open interval J, and if I is any closed interval contained in J, then the convergence of the Cesàro means to the function $f(x)$ is uniform throughout I:*

$$(1) \qquad \sigma_n(x) \xrightarrow{\rightarrow} f(x), \quad x \in I.$$

If $f(x)$ is everywhere continuous, the limit (1) is uniform for all x.

Proof. We may assume without loss of generality that the length of the interval J does not exceed π. Let I_1 be a closed interval within J and containing I in its interior, and let $\eta > 0$ be so small that if $x \in I$ and $0 \leq u < \eta$, then $x \pm u \in I_1$.

Turning our attention to the details of the proof in § 1617, we see that our task is to show that the limit (9), § 1617, is uniform for $x \in I$, since the estimate relating the quantities in (8) and (9), § 1617, is independent of x. Since the inequality (12), § 1617, is also independent of x, the only remaining detail is to show that the number δ, as used in (11), § 1617, can be chosen independently of x. But this follows from the *uniform* continuity of $f(x)$ on the closed interval I_1: corresponding to $\epsilon > 0$ there exists $\delta > 0$ such that $x_1, x_2 \in I_1$, $|x_1 - x_2| < \delta$ imply $|f(x_1) - f(x_2)| < \epsilon/\pi$; we require that $\delta < \eta$ and let $x \in I$; then, since $0 \leq u \leq \delta$, $x \pm u \in I_1$.

*1619. WEIERSTRASS'S THEOREM

Weierstrass's famous theorem on the uniform approximation of a continuous function by means of polynomials is a simple consequence of the uniform summability theorem of § 1618. We give two forms of Weierstrass's theorem, the first for *trigonometric polynomials*, in distinction to the *algebraic polynomials* defined in Exercise 19, § 208.

Theorem I. *If $f(x)$ is continuous on a closed interval I of length less than $2p$, and if ϵ is an arbitrary positive number, there exists a **trigonometric polynomial***

$$(1) \quad T(x) = \frac{1}{2} a_0 + a_1 \cos \frac{\pi x}{p} + b_1 \sin \frac{\pi x}{p} + \cdots + a_n \cos \frac{n\pi x}{p} + b_n \sin \frac{n\pi x}{p}$$

such that for all $x \in I$,

$$(2) \qquad |f(x) - T(x)| < \epsilon.$$

Proof. If the interval I is $[a, b]$, extend the domain of $f(x)$ to the closed interval $[a, a + 2p]$ by linear interpolation in $(b, a + 2p)$, the value of f at $a + 2p$ being the same as that at a: $f(a + 2p) \equiv f(a)$. Then extend the domain to $(-\infty, +\infty)$ by periodicity with period $2p$. Then $f(x)$ is everywhere continuous, so that the Cesàro means $\sigma_n(x)$ of the Theorem, § 1618 (with the appropriate adjustment from 2π to $2p$ as indicated in Note 2, § 1611), converge uniformly to $f(x)$ everywhere, and in particular on I. Since every $\sigma_n(x)$ is a trigonometric polynomial of the type required, the proof is complete.

Theorem II. *If $f(x)$ is continuous on a closed interval I and if ϵ is an arbitrary positive number, there exists a polynomial $P(x)$ such that for all $x \in I$,*

(3) $$|f(x) - P(x)| < \epsilon.$$

Proof. We first find a trigonometric polynomial $T(x)$ according to Theorem I, such that for all $x \in I$,
$$|f(x) - T(x)| < \tfrac{1}{2}\epsilon.$$
Then, since $T(x)$ is a finite sum of sines and cosines it is represented by its Maclaurin series, which converges uniformly on any bounded set and in particular on I. We can therefore find a partial sum $P(x)$ of this Maclaurin series such that on I,
$$|T(x) - P(x)| < \tfrac{1}{2}\epsilon.$$
In combination, these two inequalities give (3).

NOTE. It should be observed that the Theorem, § 1618, implies immediately that if $f(x)$ is everywhere continuous and periodic with period $2p$, and if $\epsilon > 0$, there exists a trigonometric polynomial (1) such that the inequality (2) holds for *all* real x.

*1620. DENSITY OF TRIGONOMETRIC POLYNOMIALS

In any distance space S a subset A is said to be **dense** if and only if corresponding to an arbitrary member s of S and an arbitrary positive number ϵ there exists a member a of A such that $d(a, s) < \epsilon$. For the space of real numbers this definition is equivalent to that given in Theorem VI, § 113, as the reader may easily verify.

We shall use Weierstrass's uniform approximation theorem (Theorem I, § 1619) to establish an important density theorem. (Cf. *RV*, § 1524.)

Theorem. *The trigonometric polynomials*

(1) $\quad T(x) = \tfrac{1}{2}a_0 + a_1 \cos x + b_1 \sin x + \cdots + a_n \cos nx + b_n \sin nx$

are dense in the space $R_{2\pi}$. That is, if $f(x)$ is any periodic function with period 2π integrable on $[-\pi, \pi]$ and if $\epsilon > 0$, there exists a trigonometric polynomial $T(x)$ of the form (1) such that

(2) $$d(f, T) = \|f - T\| = \left\{ \int_{-\pi}^{\pi} [f(x) - T(x)]^2 \, dx \right\}^{\frac{1}{2}} < \epsilon.$$

We start the proof by means of two lemmas.

Lemma I. *The space $C_{2\pi}$ of continuous periodic functions with period 2π is dense in the space $R_{2\pi}$.*

Proof. If $f \in R_{2\pi}$ and $\epsilon > 0$, we first find a step-function σ, for the interval $[-\pi, \pi]$, such that $d(\sigma, f) < \frac{1}{2}\epsilon$, as follows: If $|f(x)| < K$ on $[-\pi, \pi]$ we find step-functions σ and τ such that $-K \leq \sigma(x) \leq f(x) \leq \tau(x) \leq K$ on $[-\pi, \pi]$ and $\int_{-\pi}^{\pi} [\tau(x) - \sigma(x)]\, dx < \epsilon^2/8K$. Then:

$$[d(\sigma, f)]^2 = \int_{-\pi}^{\pi} [f(x) - \sigma(x)]^2\, dx \leq 2K \int_{-\pi}^{\pi} [f(x) - \sigma(x)]\, dx$$

$$\leq 2K \int_{-\pi}^{\pi} [\tau(x) - \sigma(x)]\, dx < \epsilon^2/4,$$

whence $d(\sigma, f) < \frac{1}{2}\epsilon$, as desired.

We next find a continuous function $g \in C_{2\pi}$ such that $d(g, \sigma) < \frac{1}{2}\epsilon$, by constructing its graph from straight line segments—horizontal pieces from the graph of σ and slanting pieces that can be thought of as ladders or ramps for scaling the jumps of σ. It is not hard to see that when this is done, $-K \leq \sigma(x) \leq g(x) \leq K$, and that since σ has only finitely many jumps, if these "ladders" are made nearly vertical we can ensure the inequality $\int_{-\pi}^{\pi} [g(x) - \sigma(x)]\, dx < \epsilon^2/8K$. (The student should draw a figure and provide the details; cf. Ex. 4, § 1624.) As in the preceding paragraph, then,

$$[d(g, \sigma)]^2 = \int_{-\pi}^{\pi} [g(x) - \sigma(x)]^2\, dx$$

$$\leq 2K \int_{-\pi}^{\pi} [g(x) - \sigma(x)]\, dx < \epsilon^2/4,$$

and $d(g, \sigma) < \frac{1}{2}\epsilon$.

In conjunction, the two inequalities obtained above give, by the triangle inequality:

$$d(g, f) \leq d(g, \sigma) + d(\sigma, f) < \tfrac{1}{2}\epsilon + \tfrac{1}{2}\epsilon = \epsilon.$$

Lemma II. *The trigonometric polynomials* (1) *are dense in the space $C_{2\pi}$ of Lemma I.*

Proof. For a given function $g \in C_{2\pi}$ and $\epsilon > 0$, by Weierstrass's theorem (in the form of the Note, § 1619) there exists a trigonometric polynomial $T(x)$ of the form (1) such that for all x:

$$|T(x) - g(x)| < \min(\epsilon^2/2\pi, 1),$$

and hence

$$[d(T, g)]^2 = \int_{-\pi}^{\pi} [T(x) - g(x)]^2\, dx$$

$$\leq \int_{-\pi}^{\pi} |T(x) - g(x)|\, dx < \epsilon^2,$$

and $d(T, g) < \epsilon$.

The details of the proof of the Theorem are now completed with the aid of the triangle inequality. If $f \in R_{2\pi}$ and $\epsilon > 0$, we first find $g \in C_{2\pi}$ such that $d(f, g) < \tfrac{1}{2}\epsilon$, and then T of the form (1) such that $d(g, T) < \tfrac{1}{2}\epsilon$. Then $d(f, T) \leq d(f, g) + d(g, T) < \tfrac{1}{2}\epsilon + \tfrac{1}{2}\epsilon = \epsilon$.

*1621. SOME CONSEQUENCES OF DENSITY

In this section we shall establish five important theorems all of which are corollaries of the Theorem, § 1620, concerning the density in $R_{2\pi}$ of the trigonometric polynomials.

We denote by $\Phi = (\phi_1, \phi_2, \cdots)$ the orthonormal family of Example 2, § 1604:

(1) $$\frac{1}{\sqrt{2\pi}}, \frac{\cos x}{\sqrt{\pi}}, \frac{\sin x}{\sqrt{\pi}}, \frac{\cos 2x}{\sqrt{\pi}}, \frac{\sin 2x}{\sqrt{\pi}}, \cdots,$$

and, for any $f \in R_{2\pi}$, by α_n its nth Fourier coefficient:

(2) $$\alpha_n \equiv (f, \phi_n) = \int_{-\pi}^{\pi} f(x)\phi_n(x)\,dx, \quad n = 1, 2, \cdots.$$

We record for current convenience three facts emerging directly from least squares principles and formula (2), § 1605, where $\beta_1, \beta_2, \cdots$ are arbitrary real numbers:

(3) $$\left\| f - \sum_{i=1}^{n} \alpha_i \phi_i \right\| \leq \left\| f - \sum_{i=1}^{n} \beta_i \phi_i \right\|,$$

(4) $$\left\| f - \sum_{i=1}^{n} \alpha_i \phi_i \right\|^2 = \|f\|^2 - \sum_{i=1}^{n} \alpha_i^2.$$

A third fact is implied by (4):

(5) $$\left\| f - \sum_{i=1}^{n} \alpha_i \phi_i \right\| \downarrow \quad \text{as } n \uparrow.$$

Theorem I. Parseval's Equation. *If $f \in R_{2\pi}$,*

(6) $$\sum_{i=1}^{+\infty} \alpha_i^2 = \|f\|^2.$$

Proof. The density of the trigonometric polynomials means that the right-hand member of (3) can be made arbitrarily small, and consequently so can the left-hand member. By (4) and (5) the two members of (4) must both have zero limits, and (6) follows.

The preceding proof also establishes:

Theorem II. *If $f \in R_{2\pi}$,*

(7) $$\lim_{n \to +\infty} \left\| f - \sum_{i=1}^{n} \alpha_i \phi_i \right\| = 0.$$

NOTE 1. Formula (7) defines what is meant by the statement that the series $\sum_{i=1}^{+\infty} \alpha_i \phi_i$ **converges in the mean** to the function f, or that f is the **limit in the mean** of the series $\Sigma \alpha_i \phi_i$. Theorem II, then, states that *any member of $R_{2\pi}$ is the limit in the mean of its Fourier series.* This is written:

$$(8) \qquad f = \underset{n \to +\infty}{\text{l.i.m.}} \sum_{i=1}^{n} \alpha_i \phi_i.$$

The property of the orthonormal sequence $\{\phi_n\}$ stated in Theorem II is sometimes expressed in either of the following two forms: (i) the sequence $\{\phi_n\}$ is **closed** in $R_{2\pi}$; (ii) the sequence $\{\phi_n\}$ is a **basis** in $R_{2\pi}$. (Cf. *RV*, § 1532.)

Theorem III. *A member f of $R_{2\pi}$ is orthogonal to every member of Φ if and only if it has zero norm:* $\|f\| = 0$.

Proof. If $f \perp \phi_n$ for every $n = 1, 2, \cdots$, then $\alpha_n = (f, \phi_n) = 0$, $n = 1, 2, \cdots$. Therefore both members of Parseval's equation (6) vanish. Conversely, by the Schwarz inequality, $|(f, \phi_n)| \leq \|f\| \cdot 1$ and the vanishing of $\|f\|$ implies that of α_n, $n = 1, 2, \cdots$.

Theorem IV. *Two members, f and g, of $R_{2\pi}$ have identical Fourier coefficients if and only if they are a zero distance apart:* $\|f - g\| = 0$.

Proof. The two functions f and g have identical Fourier coefficients if and only if their difference $f - g$ has all of its Fourier coefficients equal to 0; that is, if and only if $f - g$ is orthogonal to every member of Φ. By Theorem III, that is true if and only if $\|f - g\| = 0$.

Theorem V. *The orthonormal family Φ is maximal in the sense that the only orthonormal family Ψ containing Φ is Φ itself.*

Proof. Suppose Ψ is an orthonormal family containing Φ and at least one additional member g. Then, by the definition of an orthonormal family, applied to Ψ, g is orthogonal to every remaining member of Ψ, and hence to every member of Φ. By Theorem III, $\|g\| = 0$, in contradiction to the property of an orthonormal family that all of its members have norm equal to unity.

NOTE 2. The property of the orthonormal sequence $\{\phi_n\}$ stated in Theorem V is sometimes expressed by the statement that the sequence $\{\phi_n\}$ is **complete**. (Cf. *RV*, § 1532.)

*1622. FURTHER REMARKS

We conclude our discussion of the more theoretical aspects of Fourier series by stating two theorems. (For proofs, cf. *RV*, §§ 1525, 1526.)

Theorem I. *Let $f \in S_{2\pi}$ and assume that f is continuous on an open interval J. If I is any closed interval contained in J, then the convergence of the Fourier series of f is uniform throughout I:*

$$(1) \qquad S_n(x) \rightrightarrows f(x), \quad x \in I.$$

If f is everywhere continuous (as well as being periodic with period 2π) and sectionally smooth on $[-\pi, \pi]$, then the limit (1) is uniform for all x.

NOTE. Let f be assumed merely to be everywhere continuous, and periodic with period 2π. Then, by the Note, § 1619, there is a trigonometric polynomial T approximating f uniformly for all x. By the techniques of § 1620 it is seen that T also approximates f in the sense that the norm $\|T - f\|$ is small. Finally, by the least squares principle as used in § 1621, the partial sums S_n of the Fourier series of f make the norm $\|S_n - f\|$ even smaller, for large n. One is tempted to infer from all this that since S_n is a *better* approximation to f, at least in the sense of the norm, and since T is a good approximation to f in *two* senses, S_n should also be a good approximation to f in the *uniform* sense as well. That this conclusion is completely unjustified is shown by the fact that a continuous function *may* have a divergent Fourier series (cf. § 1608). In other words, *some* condition such as sectional smoothness is essential in the statement of Theorem I. (A weaker alternative condition is bounded variation—cf. *RV*, § 1525.)

Theorem II. *Any Fourier series can be integrated term by term. That is, if $f \in R_{2\pi}$, with Fourier series*

(2) $\quad \tfrac{1}{2}a_0 + a_1 \cos x + b_1 \sin x + a_2 \cos 2x + b_2 \sin 2x + \cdots,$

and if a and b are any two real numbers, then

(3) $\quad \int_a^b f(x)\,dx = \int_a^b \tfrac{1}{2}a_0\,dx + \int_a^b a_1 \cos x\,dx + \int_a^b b_1 \sin x\,dx + \cdots.$

The series on the right of (3) *always converges to the left-hand member of* (3), *whether the series* (2) *converges or not.*

*1623. OTHER ORTHONORMAL SYSTEMS

Fourier series are only one of many important expansions of functions in terms of orthonormal systems. One of these, called **Legendre series**, is based on *algebraic polynomials* (instead of trigonometric polynomials) on the interval $[-1, 1]$. Other series extend the principle of orthogonality to include an additional factor in the defining integrand called a "weight function." Some series use improper integrals over the interval $(0, +\infty)$ or the interval $(-\infty, +\infty)$. (For a brief introduction, and further references, cf. *RV*, §§ 1535–1538.)

*1624. EXERCISES

*1. Prove the identity (3), § 1613, with the aid of the identity
$$2 \cos A \sin B = \sin (A + B) - \sin (A - B).$$

*2. Prove (2), § 1614, directly by means of (3), § 1613.

*3. Prove the identity given between (6) and (7), § 1617.

*4. Provide the figure and details requested in the proof of Lemma I, § 1620.

*5. By integrating the Fourier sine series for the function $\pi/4$ (cf. Ex. 1, § 1607) an odd number of times, evaluate $\zeta(2n)$, for a few values of n. (Cf. Exs. 4–6, 20, § 1612.)

1625. APPLICATIONS OF FOURIER SERIES. THE VIBRATING STRING

We discuss in this section one of the more basic applied problems for which the techniques of Fourier analysis are especially appropriate. This is the problem of the *vibrating string*, of considerable significance in the history of Fourier series. (For a more complete discussion, see H. S. Carslaw, *Fourier's Series and Integrals*, 3rd ed. (New York, Dover Publications, 1930) and W. W. Rogosinski, *Fourier Series* (New York, Chelsea Publishing Co., 1950).)

Assume that a stretched string of uniform composition is vibrating with fixed endpoints and in a fixed plane. As a matter of convenience, we shall take units and a coordinate system such that the string is vibrating in the xy-plane, with its endpoints fixed at the two points $(0, 0)$ and $(\pi, 0)$ on the x-axis. The motion of the string is determined when the displacement y of the point of the string whose abscissa is x is known for any instant of time t:

(1) $\qquad y = u(x, t)$.

The general problem of the vibrating string is to determine a function (1) that satisfies certain prescribed conditions or restrictions.

Our first assumption is that the function (1) satisfies a certain partial differential equation, whose derivation is based on physical considerations (and will not be given here),† known as the **wave equation**:

(2) $\qquad \dfrac{\partial^2 u}{\partial t^2} = a^2 \dfrac{\partial^2 u}{\partial x^2},$

where a is a positive constant depending on the string.

We shall now give detailed consideration to a particularization of the vibrating-string problem, known as the *plucked-string* problem. For this we assume that the string is initially deformed to fit a prescribed graph:

(3) $\qquad y = f(x)$,

and then instantaneously released. Taking $t = 0$ to be the instant that the string is released from rest, and expressing the fact that the endpoints remain fixed, we have the following four conditions:

(4) $\quad u(0, t) = 0,$
(5) $\quad u(\pi, t) = 0,$ } **Boundary conditions**

(6) $\quad u(x, 0) = f(x),$
(7) $\quad u_2(x, 0) = 0,$ } **Initial conditions**

for $0 \leq x \leq \pi$ and $t \geq 0$, and where $u_2(x, t)$ means $\partial u/\partial t$. We now seek a solution of the differential equation (2), subject to the boundary conditions (4) and (5) and the initial conditions (6) and (7).

† For a derivation, see D. V. Widder, *Advanced Calculus* (Englewood Cliffs, N.J., Prentice-Hall, 1961), p. 414.

A useful technique, often employed in the solving of partial differential equations, is to find solutions of the differential equation that have a particularly simple form, namely that of the product of factors each of which is a function of precisely one of the independent variables. In the present circumstance we are looking for a function of the form

(8) $$u = X(x)T(t),$$

where $X(x)$ is a function of x alone, and $T(t)$ is a function of t alone, that satisfies (2). Substitution gives the equation

(9) $$T''X = a^2 T X''.$$

Proceeding formally, and not concerning ourselves with minor obstacles like dividing by 0 (cf. Ex. 3, § 1627), we have the identity

(10) $$\frac{X''}{X} = \frac{T''}{a^2 T}.$$

Since the left-hand member of (10) is independent of t, and the right-hand member is independent of x, they must both be equal to the same constant, which might conceivably be positive, zero, or negative. The student is asked in Exercise 4, § 1627, to show that *positive* values for this constant have no significance for this particular problem.

Let us see now why the value *zero* must be rejected. The main idea ahead of us, after we have found solutions of (2) of the factored form XT, is to combine such solutions in a fashion that gives a new solution that satisfies the boundary and initial conditions as well. The first step in this direction is to *impose the boundary conditions* (4) *and* (5) *and the initial condition* (7) *on the solutions in factored form*. Now, if the left-hand member of (10) were identically equal to zero, the function X would have the form $X = ax + b$. But then, the boundary conditions (4) and (5) would require a and b to be zero, and only the trivial function identically zero would result.

We therefore set the two members of (10) equal to a *negative* constant, and obtain the two equations:

(11) $$X''(x) + \lambda^2 X(x) = 0, \quad T''(t) + \lambda^2 a^2 T(t) = 0,$$

where $\lambda > 0$.

Solving (11) (by elementary differential equations theory; cf. Example 2, § 1513), and forming the product, we have the following *solutions of* (2) *in factored form*:

(12) $$u = (A \cos \lambda x + B \sin \lambda x)(C \cos \lambda a t + D \sin \lambda a t).$$

If these solutions are to be nontrivial, the boundary conditions (4) and (5) require that $A = 0$ and $\lambda = 1, 2, 3, \cdots$. The initial condition (7) requires that $D = 0$. Our conclusion is that the nontrivial solutions of (2) in factored form that satisfy (4), (5), and (7) are multiples of

(13) $$\sin nx \cos nat,$$

where n is a positive integer.

We are now ready to put things together. Because of the form of the differential equation (2) (it is "linear") and that of the conditions (4), (5), and (7) it follows readily (check the details) that any finite linear combination of functions of the form (13) is again a solution of (2), (4), (5), and (7). It is not too much to hope that the same may be said for a suitable *limit* of such finite linear combinations, in the form of an *infinite series*:

$$(14) \qquad u(x, t) = \sum_{n=1}^{+\infty} b_n \sin nx \cos nat.$$

Assuming that this is possible, we are led to the following form for the remaining initial condition (6):

$$(15) \qquad u(x, 0) = f(x) = \sum_{n=1}^{+\infty} b_n \sin nx.$$

In other words, a solution (14) is obtained, at least formally, by expanding the prescribed function $f(x)$ in a sine series. If the function $f(x)$ is sufficiently well-behaved, the legitimacy of such operations as term by term differentiation is not difficult to supply. For uniqueness of the solution, see Exercises 5, 6, § 1627.

Examples for detailed study are given in Exercises 1, 2, § 1627. The student is asked in Exercise 7, § 1627, to work out similar details for the problem of the *struck string*.

NOTE. The values of λ, $\lambda = \lambda_n = n = 1, 2, 3, \cdots$, found in the solution of the differential equation $X'' + \lambda^2 X = 0$, subject to the boundary conditions of the problem, are called **characteristic values** or **eigenvalues**. The corresponding functions $\sin nx$, $n = 1, 2, 3, \cdots$, are called **characteristic functions** or **eigenfunctions**. The set of all characteristic values is called the **spectrum**. The characteristic values of λ give the *resonant frequencies*: if the string produces musical tones, the value $\lambda = 1$ corresponds to the *fundamental*, $\lambda = 2$ to the *first octave*, $\lambda = 3$ the *fifth* above the octave, and higher values of λ to higher *harmonics* or *overtones*.

1626. A HEAT CONDUCTION PROBLEM

Consider a thin rod located on the x-axis for $0 \leq x \leq \pi$, with an initial temperature distribution

$$(1) \qquad \text{initial temperature} = f(x)$$

along the rod. Furthermore, assume that the rod is insulated except at the endpoints, which are maintained at a fixed temperature which can be assumed to be 0 (for a suitable temperature scale). If the temperature at the point x at the time t is denoted

$$(2) \qquad \text{temperature} = u(x, t),$$

the physical theory of heat conduction demands that this function satisfy the **heat equation**:

$$(3) \qquad \frac{\partial u}{\partial t} = a^2 \frac{\partial^2 u}{\partial x^2},$$

where $a > 0$.

The problem as stated calls for a solution of the partial differential equation (3) subject to the boundary and initial conditions:

(4) $u(0, t) = 0,$ } **Boundary conditions**
(5) $u(\pi, t) = 0,$
(6) $u(x, 0) = f(x),$ **Initial condition**

for $0 \leq x \leq \pi$ and $t \geq 0$.

Adopting the method of the preceding section, with $u = X(x)T(t)$, we find

(7) $$\frac{X''}{X} = \frac{T'}{a^2 T} = -\lambda^2,$$

where $\lambda > 0$, with the result that any nontrivial solutions of (3) in factored form that satisfy (4) and (5) are multiples of

(8) $$\sin nx e^{-n^2 a^2 t},$$

where n is a positive integer.

Finally, if $f(x)$ is a sufficiently well-behaved function, a solution of the system (3), (4), (5), (6) exists having the form

(9) $$u(x, t) = \sum_{n=1}^{+\infty} b_n \sin nx e^{-n^2 a^2 t},$$

where the coefficients b_n are those of the sine series expansion of $f(x)$:

(10) $$u(x, 0) = f(x) = \sum_{n=1}^{+\infty} b_n \sin nx.$$

For uniqueness of the solution, and examples for detailed study, see Exercises 9–11, § 1627.

NOTE. The technique of solutions in factored form used in this and the preceding section is not applicable to all applied vibrational and heat-flow problems, even fairly simple ones. Other methods, such as integral equations, will be found in many references in which applied problems are discussed. See in particular R. V. Churchill, *Fourier Series and Boundary Value Problems* (New York, McGraw-Hill Book Co., 1941) and *Modern Operational Mathematics in Engineering* (New York, McGraw-Hill Book Co., 1944); R. Courant and D. Hilbert, *Methods of Mathematical Physics* (New York, Interscience Publishers, 1953); F. B. Hildebrand, *Advanced Calculus for Engineers* (Englewood Cliffs, N.J., Prentice-Hall, 1949) and *Methods of Applied Mathematics* (Englewood Cliffs, N.J., Prentice-Hall, 1952); and D. Jackson, *Fourier Series and Orthogonal Polynomials* (Carus Mathematical Monographs, 1941).

1627. EXERCISES

1. Solve the problem of the plucked string if the initial condition (6), § 1625, is given by the function
$$f(x) = \begin{cases} 2px/\pi, & 0 \leq x \leq \tfrac{1}{2}\pi, \\ 2p(\pi - x)/\pi, & \tfrac{1}{2}\pi \leq x \leq \pi. \end{cases}$$

2. Discuss the vibrations of a plucked string if the initial shape is that of a single arch of a sine curve: $f(x) = p \sin x$. Show that the string retains the same

general shape. Complete the discussion by considering functions of the form $f(x) = p \sin kx$, and their linear combinations.

3. Prove that any solution of (2), § 1625, that has the form (8), § 1625, and is not identically zero, must be such that $X(x)$ and $T(t)$ satisfy equations of the form $X'' = kX$ and $T'' = a^2kT$ simultaneously. *Hint:* If $X(x)$ is not identically zero, let x_1 and x_2 be such that $X''(x_1) = k_1 X(x_1)$ and $X''(x_2) = k_2 X(x_2)$, where $k_1 \neq k_2$ and $X(x_1)X(x_2) \neq 0$. Then $T(t)$ must vanish identically.

4. Prove that the constant of (10), § 1625, cannot be positive if the boundary conditions of the problem are satisfied.

*5. Prove uniqueness of the solution of the plucked-string problem of § 1625, assuming continuous existence of the second-order partial derivatives involved, as follows: First assume without loss of generality that $f(x) = 0$, identically. For each t expand the solution in a sine series: $u(x, t) = \Sigma \phi_n(t) \sin nx$, where

(1) $$\phi_n(t) = \frac{2}{\pi} \int_0^\pi u(x, t) \sin nx \, dx.$$

Differentiate (1) twice using Leibnitz's rule (§ 930), then with the aid of (2), (4), (5), § 1625, obtain the relation $\phi_n''(t) = -n^2 a^2 \phi_n(t)$, whence $\phi_n(t) = A_n \cos nat + B_n \sin nat$. Finally, from (1), $\phi_n(0) = \phi_n'(0) = 0$, and hence $A_n = B_n = 0$.

*6. Show that the change of variables $r = x + at$, $s = x - at$ transforms the wave equation (2), § 1625, to $\partial^2 u/\partial r \partial s = 0$. Conclude that any solution must have the form (cf. Ex. 35, § 915):

(2) $$u(x, t) = \phi(x + at) + \psi(x - at).$$

Show that the initial conditions (6) and (7), § 1625, with $f(x) = 0$, become $\phi(x) + \psi(x) = \phi'(x) - \psi'(x) = 0$. Use this to construct a new proof of the uniqueness theorem of Exercise 5.

7. The problem of the **struck string** is that of solving the differential equation (2), § 1625, subject to the boundary conditions (4) and (5), § 1625, and the two initial conditions

(3) $\quad\quad\quad\quad\quad\quad\quad u(x, 0) = 0,$
(4) $\quad\quad\quad\quad\quad\quad\quad u_2(x, 0) = f(x),$

for $0 \leq x \leq \pi$ and $t \geq 0$. Assuming such legitimacy as term by term differentiation, obtain the solution

$$u = \sum_{n=1}^{+\infty} c_n \sin nx \sin nat, \quad \text{where} \quad c_n = \frac{2}{na\pi} \int_0^\pi f(x) \sin nx \, dx.$$

Show that this solution is unique (cf. Exs. 5, 6).

8. Solve the problem of the struck string (Ex. 7) for the function $f(x)$ that is equal to the positive constant p when x is in the interval $[\frac{1}{2}\pi - h, \frac{1}{2}\pi + h]$ and identically zero otherwise.

9. Solve the heat conduction problem subject to (3), (4), (5), (6), § 1626, for the function $f(x)$ of Exercise 1.

10. Solve the heat conduction problem subject to (3), (4), (5), (6), § 1626, for the function

$$f(x) = A \sin 2x + B \sin 7x.$$

*11. State and prove a uniqueness theorem for the solution of the heat conduction problem as described in § 1626. (Cf. Ex. 5.)

★12. The problem of heat conduction in a *rod with insulated ends* is that of solving the differential equation (3), § 1626, subject to the initial condition (6), § 1626, and the two boundary conditions

(5) $$u_1(0, t) = 0,$$
(6) $$u_1(\pi, t) = 0,$$

for $t \geq 0$. Assuming legitimacy of the operations involved, find the general solution, and show that it is unique (cf. Ex. 11).

★13. Solve the heat conduction problem of Exercise 12 for the case of the function $f(x)$ given in Exercise 1.

★14. Discuss the solution of the heat conduction problem of a rod whose endpoints are maintained at constant, but distinct, temperatures as follows: Show that a solution of the differential equation (3), § 1626, subject to the conditions

(7) $$u(0, t) = \alpha, \quad u(\pi, t) = \beta, \quad u(x, 0) = f(x)$$

can be obtained by adding a solution $\sigma(x, t)$ of the **steady-state problem** (3), § 1626, with conditions

(8) $$u(0, t) = \alpha, \quad u(\pi, t) = \beta, \quad u(x, 0) = L(x) \equiv \alpha + (\beta - \alpha)x/\pi$$

and a solution $\tau(x, t)$ of the **transient problem**, (3), § 1626, with conditions

(9) $$u(0, t) = 0, \quad u(\pi, t) = 0, \quad u(x, 0) = f(x) - L(x).$$

Show that these solutions are unique, and that $\sigma(x, t) = L(x)$. (Cf. Ex. 11.)

★15. Use Exercise 14 to solve the problem of a rod whose temperature distribution is initially that of the steady state with end temperatures of 10° and 20°, if these end temperatures are suddenly changed to 50° and 80°, respectively, and then maintained at these temperatures.

17

Vector Analysis

1701. INTRODUCTION

In Chapter 7 (Solid Analytic Geometry and Vectors) we introduced some of the most basic concepts and operations of vectors in E_3. These had principally to do with linear combinations of vectors, and the scalar or dot product of two vectors, and were adequate for the purposes of Chapter 7, Chapter 8 (Arcs and Curves), Chapter 9 (Partial Differentiation), and Chapter 16 (Fourier Series). However, for the objectives of the final two chapters, Chapter 18 (Line and Surface Integrals) and Chapter 19 (Differential Geometry), we shall find it fruitful to extend the subject matter of vectors. It will be the specific aim of this chapter to introduce and study the algebraic operation of the vector or cross product of two vectors, and to define and discuss the differential calculus of vectors and vector operators.

1702. THE VECTOR OR OUTER OR CROSS PRODUCT

The *scalar product* of two vectors was defined in § 707 in terms of the components of these vectors in a given rectangular coordinate system, and shown in Note 6, § 707, to be independent of the rectangular coordinate system. The *vector product* of two vectors will be defined in this section in terms of components in a given *right-handed* rectangular coordinate system, and proved in the following section (Note 1, § 1703) to be independent of such a coordinate system.

In Theorem II, § 708, it was shown that if $\vec{u} = (u_1, u_2, u_3)$ and $\vec{v} = (v_1, v_2, v_3)$ are any two vectors, the vector

(1) $$\left(\begin{vmatrix} u_2 & u_3 \\ v_2 & v_3 \end{vmatrix}, \begin{vmatrix} u_3 & u_1 \\ v_3 & v_1 \end{vmatrix}, \begin{vmatrix} u_1 & u_2 \\ v_1 & v_2 \end{vmatrix} \right)$$

is orthogonal to both $\vec{u}$ and $\vec{v}$. We now give this vector a title, according to the following definition:

Definition. *The **vector product** or **outer product** or **cross product** of the two vectors $\vec{u} = (u_1, u_2, u_3)$ and $\vec{v} = (v_1, v_2, v_3)$, denoted $\vec{u} \times \vec{v}$, is the vector*

§ 1702]   THE VECTOR OR OUTER OR CROSS PRODUCT   555

defined by (1).   *In terms of the basic triad of unit vectors $\vec{i}, \vec{j}$, and $\vec{k}$ ((5), § 704):*

$$(2) \quad \vec{u} \times \vec{v} \equiv \begin{vmatrix} u_2 & u_3 \\ v_2 & v_3 \end{vmatrix} \vec{i} + \begin{vmatrix} u_3 & u_1 \\ v_3 & v_1 \end{vmatrix} \vec{j} + \begin{vmatrix} u_1 & u_2 \\ v_1 & v_2 \end{vmatrix} \vec{k}$$

or, in a more compact determinant form:

$$(3) \quad \vec{u} \times \vec{v} \equiv \begin{vmatrix} \vec{i} & \vec{j} & \vec{k} \\ u_1 & u_2 & u_3 \\ v_1 & v_2 & v_3 \end{vmatrix}.$$

NOTE 1. The vector product has some properties that correspond precisely to similar properties of the scalar product (Note 2, § 707). This is true, in particular, of the homogeneous and distributive laws, given below. On the other hand, the contrast is sharp in the case of the commutative law, which breaks down completely with the vector product. The proofs of the following three laws are simple and routine, involving no more than elementary properties of second-order determinants, and are left to the student in Exercise 1, § 1705.

(i) *The following* **anticommutative** *or* **skew-commutative** *law holds for all vectors $\vec{u}$ and $\vec{v}$:*

$$\vec{u} \times \vec{v} = -\vec{v} \times \vec{u}.$$

(ii) *The following* **homogeneous** *law holds for all scalars x and y and vectors $\vec{u}$ and $\vec{v}$:*

$$(x\vec{u}) \times (y\vec{v}) = (xy)(\vec{u} \times \vec{v}).$$

(iii) *The following* **distributive** *laws hold for all vectors $\vec{u}, \vec{v}$, and $\vec{w}$:*

$$\vec{u} \times (\vec{v} + \vec{w}) = \vec{u} \times \vec{v} + \vec{u} \times \vec{w},$$
$$(\vec{u} + \vec{v}) \times \vec{w} = \vec{u} \times \vec{w} + \vec{v} \times \vec{w}.$$

If $\vec{u}$ and $\vec{v}$ are noncollinear vectors (and hence nonzero), they can be written:

$$(4) \quad \begin{aligned} \vec{u} &= |\vec{u}| \cdot (\lambda, \mu, \nu), \\ \vec{v} &= |\vec{v}| \cdot (\lambda', \mu', \nu'), \end{aligned}$$

where (λ, μ, ν) and (λ', μ', ν') are unit vectors in the direction of $\vec{u}$ and $\vec{v}$, respectively. By property (ii) of Note 1,

$$(5) \quad \vec{u} \times \vec{v} = |\vec{u}| \cdot |\vec{v}| \cdot \left(\begin{vmatrix} \mu & \nu \\ \mu' & \nu' \end{vmatrix}, \begin{vmatrix} \nu & \lambda \\ \nu' & \lambda' \end{vmatrix}, \begin{vmatrix} \lambda & \mu \\ \lambda' & \mu' \end{vmatrix} \right).$$

By Theorem I, § 708, the vector in parentheses on the right of (5) has magnitude $\sin \theta$, where θ is the angle between $\vec{u}$ and $\vec{v}$ ($0 < \theta < \pi$). We therefore have the formula for the *magnitude of the cross product*:

$$(6) \quad |\vec{u} \times \vec{v}| = |\vec{u}| \cdot |\vec{v}| \cdot \sin \theta.$$

The precise *direction* of $\vec{u} \times \vec{v}$ is discussed in the following section, although it is already established that this vector is orthogonal to both $\vec{u}$ and $\vec{v}$.

NOTE 2. Formula (6) can be extended to include all vectors $\vec{u}$ and $\vec{v}$, whether they are collinear or not. In the first place, if $\vec{u}$ and $\vec{v}$ are collinear nonzero vectors, then $\theta = 0$ or $\theta = \pi$, and since either vector is a scalar multiple of the other, both members of (6) vanish. In the second place, if either $\vec{u}$ or $\vec{v}$ is the zero vector, both members of (6) vanish even though θ (and therefore $\sin \theta$) is indeterminate. We conclude from these considerations that <u>two vectors $\vec{u}$ and $\vec{v}$ are collinear if and only if their vector product is the zero vector</u>:

(7) $$\vec{u} \times \vec{v} = \vec{0}.$$

In particular, the vector product of any vector and itself is zero; that is, the equation

(8) $$\vec{u} \times \vec{u} = \vec{0}$$

is an identity, true of all vectors $\vec{u}$.

NOTE 3. The magnitude of the vector product of two vectors is equal to the area of the parallelogram determined by them. (Give a proof in Ex. 2, § 1705.)

NOTE 4. The associative law fails for the vector product. This is discussed in Note 2, § 1704.

NOTE 5. The following multiplication table holds for vector products of the basic unit vectors $\vec{i}, \vec{j}$, and $\vec{k}$ (in a right-handed coordinate system), where the first factor is listed at the left and the second factor is listed at the top:

(9)

$\times$	$\vec{i}$	$\vec{j}$	$\vec{k}$
$\vec{i}$	$\vec{0}$	$\vec{k}$	$-\vec{j}$
$\vec{j}$	$-\vec{k}$	$\vec{0}$	$\vec{i}$
$\vec{k}$	$\vec{j}$	$-\vec{i}$	$\vec{0}$

(Verify (9) in Ex. 3, § 1705.) A similar multiplication table holds for *any* system of mutually orthogonal unit vectors.

Example 1.

$$(1, -3, 2) \times (5, 0, -4) = \left(\begin{vmatrix} -3 & 2 \\ 0 & -4 \end{vmatrix}, \begin{vmatrix} 2 & 1 \\ -4 & 5 \end{vmatrix}, \begin{vmatrix} 1 & -3 \\ 5 & 0 \end{vmatrix} \right) = (12, 14, 15);$$

$$(3\vec{i} + \vec{k}) \times (\vec{j} - 7\vec{k}) = \begin{vmatrix} \vec{i} & \vec{j} & \vec{k} \\ 3 & 0 & 1 \\ 0 & 1 & -7 \end{vmatrix} = -\vec{i} + 21\vec{j} + 3\vec{k}.$$

Example 2. Prove that $\vec{u} \times \vec{v} = \vec{v} \times \vec{u}$ if and only if $\vec{u}$ and $\vec{v}$ are collinear.

Solution. By the anticommutative law (*i*) of Note 1, $\vec{u} \times \vec{v} = \vec{v} \times \vec{u}$ if and only if $\vec{u} \times \vec{v} = -\vec{u} \times \vec{v}$, or $\vec{u} \times \vec{v} = \vec{0}$. The conclusion follows from (7) of Note 2.

1703. THE TRIPLE SCALAR PRODUCT. ORIENTATION IN SPACE

Let $\vec{u} = (u_1, u_2, u_3)$, $\vec{v} = (v_1, v_2, v_3)$, and $\vec{w} = (w_1, w_2, w_3)$ be any three noncoplanar vectors, with a common initial point. Then $\vec{u}$ and $\vec{v}$ are noncollinear vectors, and their vector product $\vec{u} \times \vec{v}$ is a nonzero vector which, by Theorem II, § 708, is orthogonal to both $\vec{u}$ and $\vec{v}$. Since $\vec{w}$ is a nonzero vector that does not lie in the plane of $\vec{u}$ and $\vec{v}$, $\vec{w}$ cannot be orthogonal to $\vec{u} \times \vec{v}$, which is orthogonal to the plane of $\vec{u}$ and $\vec{v}$. Consequently, the scalar product of $\vec{u} \times \vec{v}$ and $\vec{w}$ is nonzero. Since this scalar product has the form

(1) $\quad \begin{vmatrix} u_2 & u_3 \\ v_2 & v_3 \end{vmatrix} w_1 + \begin{vmatrix} u_3 & u_1 \\ v_3 & v_1 \end{vmatrix} w_2 + \begin{vmatrix} u_1 & u_2 \\ v_1 & v_2 \end{vmatrix} w_3,$

we have the conclusion:

(2) $\quad (\vec{u} \times \vec{v}) \cdot \vec{w} = \begin{vmatrix} u_1 & u_2 & u_3 \\ v_1 & v_2 & v_3 \\ w_1 & w_2 & w_3 \end{vmatrix}$

is different from zero.

The expression (2), involving an ordered triad of vectors and having a scalar value, is called the **triple scalar product** of the vectors $\vec{u}$, $\vec{v}$, and $\vec{w}$, whether these vectors are coplanar or not. Expanding the determinant in (2) according to the elements in the first row, we see that this same triple scalar product can also be written in the form $\vec{u} \cdot (\vec{v} \times \vec{w})$. Another notation is $[\vec{u}, \vec{v}, \vec{w}]$, so that:

(3) $\quad [\vec{u}, \vec{v}, \vec{w}] = (\vec{u} \times \vec{v}) \cdot \vec{w} = \vec{u} \cdot (\vec{v} \times \vec{w}) = \begin{vmatrix} u_1 & u_2 & u_3 \\ v_1 & v_2 & v_3 \\ w_1 & w_2 & w_3 \end{vmatrix}.$

By elementary properties of determinants (give a proof in Ex. 4, § 1705):

(4) $\quad [\vec{u}, \vec{v}, \vec{w}] = [\vec{v}, \vec{w}, \vec{u}] = [\vec{w}, \vec{u}, \vec{v}] = -[\vec{u}, \vec{w}, \vec{v}] = -[\vec{w}, \vec{v}, \vec{u}] = -[\vec{v}, \vec{u}, \vec{w}].$

These formulas show that *the triple scalar product is unchanged by any cyclic permutation of the three vectors, and is changed in sign by the interchange of any two.*

Since, for noncoplanar vectors, the expression (2) is either positive or negative, the following definition is meaningful:

Definition. *In a right-handed coordinate system an ordered triad of noncoplanar vectors $\vec{u}$, $\vec{v}$, and $\vec{w}$ is **right-handed** or **left-handed** (alternatively, has **positive** or **negative orientation**) according as their triple scalar product (2) is positive or negative.*

A convenient test for determining whether an ordered triad of three noncoplanar vectors has positive or negative orientation is given in the following theorem, framed in terms of a right-handed screw (cf. § 701):

Theorem I. *Let $\vec{u}$, $\vec{v}$, and $\vec{w}$ be noncoplanar vectors with a common initial point, and let θ be the angle between $\vec{u}$ and $\vec{v}$ $(0 < \theta < \pi)$. Then the ordered triad $(\vec{u}, \vec{v}, \vec{w})$ is right-handed or left-handed according as a right-handed screw whose axis is orthogonal to $\vec{u}$ and $\vec{v}$ at their point of intersection advances in a direction that makes an acute or obtuse angle with $\vec{w}$ when turned through the angle θ that rotates the vector $\vec{u}$ into a new vector having the same direction as $\vec{v}$.*

Proof. Assume first that the direction in which the right-handed screw advances makes an acute angle with $\vec{w}$. Place the three vectors $\vec{u}$, $\vec{v}$, and $\vec{w}$ with initial points at the origin, and then make a rigid rotation that carries the vector $\vec{u}$ into coincidence with a portion of the positive half of the x-axis, with $\vec{v}$ lying in either the first or second quadrant of the xy-plane. Then, since the coordinate system is right-handed, the vector $\vec{w}$ must make an acute angle with the z-axis. Now deform continuously the vector $\vec{u}$ into the vector $\vec{i}$ (keeping $\vec{u}$ on the positive half of the x-axis), the vector $\vec{v}$ into the vector $\vec{j}$ (keeping $\vec{v}$ in the first and second quadrants of the xy-plane), and the vector $\vec{w}$ into the vector $\vec{k}$ (maintaining an acute angle between $\vec{w}$ and $\vec{k}$). During these continuous deformations the value of the determinant in (2) must vary continuously (as a function of time, say, as a parameter), with a final value of

$$\begin{vmatrix} 1 & 0 & 0 \\ 0 & 1 & 0 \\ 0 & 0 & 1 \end{vmatrix} = 1 > 0.$$

By the intermediate value property of continuous functions (Theorems III, IV, § 213), the value of the determinant in (2) must have been positive initially, since its value varies continuously and can never vanish.

In case the angle in question is obtuse, a similar deformation continuously transforms the vectors $\vec{u}, \vec{v},$ and $\vec{w}$ into the vectors $\vec{i}, \vec{j},$ and $-\vec{k}$, respectively, with a final value for the determinant of

$$\begin{vmatrix} 1 & 0 & 0 \\ 0 & 1 & 0 \\ 0 & 0 & -1 \end{vmatrix} = -1 < 0.$$

The original value of the determinant must then also have been negative. This completes the proof.

As a simple corollary to this theorem, we have:

Theorem II. *If $\vec{u}$ and $\vec{v}$ are noncollinear vectors, then the ordered triad $\vec{u}, \vec{v}, \vec{u} \times \vec{v}$ has positive orientation. The vector product $\vec{u} \times \vec{v}$ can therefore be described as the vector with magnitude $|\vec{u}| \cdot |\vec{v}| \cdot \sin \theta$, orthogonal to both $\vec{u}$ and $\vec{v}$, and directed so that $\vec{u}, \vec{v},$ and $\vec{u} \times \vec{v}$ form a right-handed triad.*

Proof. We must test the triple scalar product $[\vec{u}, \vec{v}, \vec{u} \times \vec{v}]$ for sign. By (2), this is equal to $(\vec{u} \times \vec{v}) \cdot (\vec{u} \times \vec{v}) = |\vec{u} \times \vec{v}|^2 > 0$.

NOTE 1. As a corollary to Theorem II, we have the fact that *the vector product of any two vectors is independent of the right-handed rectangular coordinate system.* In contrast to the scalar product (cf. Note 6, § 707), the vector product is not preserved if the orientation of the coordinate system is reversed. It is basically for this reason that we shall consistently adhere to right-handed systems. It would be equally logical to have insisted on left-handed systems throughout. (Cf. Example 2, § 1708.)

NOTE 2. Let $(\vec{u}, \vec{v}, \vec{w})$ be an ordered triad of noncoplanar vectors with positive orientation, and a common initial point, and consider the parallelepiped having these vectors as three adjacent edges. Recalling from (7), § 707, the formula:

$$\vec{u} \cdot \vec{v} = (\text{comp}_u \vec{v})|\vec{u}|,$$

we can write the triple scalar product

(5) $\qquad (\vec{u} \times \vec{v}) \cdot \vec{w} = (\text{comp}_{u \times v} \vec{w})|\vec{u} \times \vec{v}|,$

where $\operatorname{comp}_{u \times v} \vec{w}$ is the component of $\vec{w}$ perpendicular to the plane of $\vec{u}$ and $\vec{v}$, and where $|\vec{u} \times \vec{v}|$ is the area of the parallelogram defined by $\vec{u}$ and $\vec{v}$. Interpreting this parallelogram as the base of the parallelepiped, we have in (5) the product of this base area and the altitude of the parallelepiped. In other words, *the triple scalar product of any right-handed triad of noncoplanar vectors with a common initial point is the volume of the parallelepiped determined by them.* Similarly, if the triad is left-handed, their triple scalar product is the negative of the volume of the corresponding parallelepiped. (Cf. Ex. 16, § 1012.)

NOTE 3. Three vectors are coplanar if and only if their triple scalar product vanishes. (Cf. Ex. 5, § 1705.)

Example. Find a condition that the four points (x, y, z), (x_1, y_1, z_1), (x_2, y_2, z_2), and (x_3, y_3, z_3) be coplanar.

Solution. This condition is equivalent to the coplanarity of the three vectors $(x - x_3, y - y_3, z - z_3)$, $(x_1 - x_3, y_1 - y_3, z_1 - z_3)$ and $(x_2 - x_3, y_2 - y_3, z_2 - z_3)$, or the vanishing of

$$\begin{vmatrix} x - x_3 & y - y_3 & z - z_3 \\ x_1 - x_3 & y_1 - y_3 & z_1 - z_3 \\ x_2 - x_3 & y_2 - y_3 & z_2 - z_3 \end{vmatrix} = \begin{vmatrix} x & y & z & 1 \\ x_1 & y_1 & z_1 & 1 \\ x_2 & y_2 & z_2 & 1 \\ x_3 & y_3 & z_3 & 1 \end{vmatrix}.$$

(Cf. Ex. 12, § 712.)

NOTE 4. It is important to recognize that the concepts of *right-handedness* and *left-handedness*, both for a rectangular coordinate system and for an ordered triad of vectors, are based on observations of the material world around us. (Without reference to the physical universe it would be impossible to explain to an extraterrestrial being what we mean by clockwise motion, a right-handed screw, or a southpaw pitcher.) In a true sense, then, these two concepts are *nonmathematical* and *nonintrinsic*. What *is* mathematical about right-handedness and left-handedness (whether applied to coordinate systems or to triads of vectors) is whether two systems are *similar* or *opposite*. The sign of the determinant D in (2) determines how an ordered triad of vectors is oriented *relative* to the coordinate system (if $D > 0$ they are similar; if $D < 0$ they are opposite). Since it is *convenient* to speak of right-handedness and left-handedness as if they were intrinsic properties, and since proofs depend only on relative and not absolute orientation, we shall continue to use this language.

1704. THE TRIPLE VECTOR PRODUCT

In the algebra of vectors one frequently encounters vector products where one of the factors is itself a vector product. Two useful expansion formulas for such **triple vector products** are:

(1) $\quad (\vec{u} \times \vec{v}) \times \vec{w} = (\vec{u} \cdot \vec{w})\vec{v} - (\vec{v} \cdot \vec{w})\vec{u},$
(2) $\quad \vec{u} \times (\vec{v} \times \vec{w}) = (\vec{u} \cdot \vec{w})\vec{v} - (\vec{u} \cdot \vec{v})\vec{w}.$

Since the anticommutative law for vector products (Note 1, § 1702) implies the equivalence of the two formulas (1) and (2) (show this in Ex. 6, § 1705), it will be sufficient to establish (1). Before doing so, however, let us observe that (1) is trivially satisfied if $\vec{u}$ and $\vec{v}$ are collinear since in this case one is a scalar multiple of the other and both members of (1) reduce to the zero vector. Assuming now that $\vec{u}$ and $\vec{v}$ are noncollinear vectors with a common initial point, so that $\vec{u} \times \vec{v}$ is a nonzero vector orthogonal to both $\vec{u}$ and $\vec{v}$, we can see immediately that the triple vector product on the left of (1), since it is orthogonal to $\vec{u} \times \vec{v}$, must lie in the plane determined by $\vec{u}$ and $\vec{v}$ and must therefore be some linear combination of $\vec{u}$ and $\vec{v}$. It is the *particular* linear combination prescribed by (1) that we wish now to establish.

In order to prove (1), construct a right-handed triad of unit vectors $(\vec{e}, \vec{f}, \vec{g})$ as follows: Let $\vec{e}$ be a unit vector in the direction of $\vec{u}$, $\vec{f}$ a unit vector orthogonal to $\vec{e}$ and lying in the plane of $\vec{u}$ and $\vec{v}$, and let $\vec{g} \equiv \vec{e} \times \vec{f}$. In terms of these unit vectors we can represent $\vec{u}$, $\vec{v}$, and $\vec{w}$ in the form:

$$\vec{u} = k\vec{e},$$
$$\vec{v} = a\vec{e} + b\vec{f},$$
$$\vec{w} = x\vec{e} + y\vec{f} + z\vec{g}.$$

Substitution in the left-hand member of (1) gives

$$[(k\vec{e}) \times (a\vec{e} + b\vec{f})] \times [x\vec{e} + y\vec{f} + z\vec{g}]$$

which, by the algebraic laws of § 1702 (Note 5, § 1702, in particular), is equal to

$$(kb\vec{g}) \times (x\vec{e} + y\vec{f} + z\vec{g}) = kbx\vec{f} - kby\vec{e}.$$

The right-hand member of (1), similarly, is equal to

$$(kx)(a\vec{e} + b\vec{f}) - (ax + by)(k\vec{e}) = kbx\vec{f} - kby\vec{e},$$

and the derivation of (1) is complete.

NOTE 1. A mnemonic device for formulas (1) and (2) is that the outer factor is always dotted first with the remote inner factor and then with the adjacent inner factor. Another is that the term containing a scalar multiple of the middle factor always bears the plus sign.

NOTE 2. Show that *the associative law for vector products*,

(3) $\qquad\qquad\qquad \vec{u} \times (\vec{v} \times \vec{w}) = (\vec{u} \times \vec{v}) \times \vec{w},$

fails. Find a necessary and sufficient set of conditions on $\vec{u}$, $\vec{v}$, and $\vec{w}$ for the equation (3) to hold.

Solution. Let $\vec{u} = \vec{i}$, $\vec{v} = \vec{j}$, $\vec{w} = \vec{j}$. Then the two members of (3) are $\vec{0}$ and $-\vec{i}$, respectively. By (1) and (2), equation (3) holds if and only if

(4) $\qquad\qquad\qquad (\vec{v} \cdot \vec{w})\vec{u} = (\vec{u} \cdot \vec{v})\vec{w}.$

It follows almost immediately that (3) holds if and only if *either $\vec{u}$ and $\vec{w}$ are collinear or $\vec{v}$ is orthogonal to both $\vec{u}$ and $\vec{w}$*.

1705. EXERCISES

1. Prove the three laws of Note 1, § 1702.
2. Prove the statement of Note 3, § 1702.
3. Verify the multiplication table (9) of Note 5, § 1702.
4. Prove formulas (4), § 1703.

5. Prove the statement of Note 3, § 1703.

6. Use the anticommutative law of vector products to establish the equivalence of formulas (1) and (2), § 1704.

7. If $\vec{u} = 3\vec{i} - \vec{j} - 5\vec{k}$ and $\vec{v} = \vec{i} + 2\vec{j} - \vec{k}$, find $\vec{u} \times \vec{v}$, $\vec{v} \times \vec{u}$, and $(\vec{u} + \vec{v}) \times (\vec{u} + 2\vec{v})$.

8. Verify the distributive laws of (iii), Note 1, § 1702, for the vectors $\vec{u} = 4\vec{i} - \vec{j} - 3\vec{k}$, $\vec{v} = 2\vec{i} + 5\vec{j}$, and $\vec{w} = \vec{i} + 3\vec{j} - 2\vec{k}$.

9. Verify formulas (2) and (3), § 1703, and evaluate, for the vectors of Exercise 8. Is the triad $(\vec{u}, \vec{v}, \vec{w})$ right-handed or left-handed?

10. Verify formulas (1) and (2), § 1704, and evaluate, for the vectors of Exercise 8.

11. Find the area of the triangle with vertices $(3, 5, 2)$, $(1, -1, 6)$, and $(-2, 1, 4)$, by evaluating the magnitude of a vector product.

12. Find the volume of the parallelepiped that has as three of its adjacent edges the radius vectors of the three points of Exercise 11.

13. Suppose a clock is lying face up in a horizontal plane. Let the vectors $\vec{u}$ and $\vec{v}$ be represented by the hour hand and the minute hand, respectively, and let $\vec{w}$ be a nonzero vertical vector directed upward. Find the times between noon and 2:00 P.M. when the ordered triad $(\vec{u}, \vec{v}, \vec{w})$ is (a) right-handed; (b) left-handed; (c) coplanar.

14. Show that *the cancellation law for vector products, in the form*

(1) $\qquad \vec{u} \times \vec{v} = \vec{u} \times \vec{w} \quad \text{implies} \quad \vec{v} = \vec{w}$

fails, but that it is valid if (1) holds for *all* vectors $\vec{u}$. (Cf. Example 2, § 707.)

In Exercises 15–19, establish the given formula.

15. $|\vec{u} \times \vec{v}|^2 + (\vec{u} \cdot \vec{v})^2 = |\vec{u}|^2 |\vec{v}|^2$.

16. $(\vec{s} \times \vec{t}) \cdot (\vec{u} \times \vec{v}) = \begin{vmatrix} \vec{s} \cdot \vec{u} & \vec{s} \cdot \vec{v} \\ \vec{t} \cdot \vec{u} & \vec{t} \cdot \vec{v} \end{vmatrix}$.

Hint: Write the first member as $\vec{s} \cdot [\vec{t} \times (\vec{u} \times \vec{v})]$.

17. $[\vec{u} \times \vec{v}, \vec{v} \times \vec{w}, \vec{w} \times \vec{u}] = [\vec{u}, \vec{v}, \vec{w}]^2$.

18. $\vec{u} \times (\vec{v} + \vec{w}) + \vec{v} \times (\vec{w} + \vec{u}) + \vec{w} \times (\vec{u} + \vec{v}) = \vec{0}$.

19. $\vec{u} \times (\vec{v} \times \vec{w}) + \vec{v} \times (\vec{w} \times \vec{u}) + \vec{w} \times (\vec{u} \times \vec{v}) = \vec{0}$.

20. If $\vec{u}$, $\vec{v}$, and $\vec{w}$ are the sides of a triangle, directed cyclically so that $\vec{u} + \vec{v} + \vec{w} = \vec{0}$, prove that

$$\vec{u} \times (\vec{v} + \vec{w}) - \vec{v} \times (\vec{w} + \vec{u}) = \vec{w} \times (\vec{u} + \vec{v}) = \vec{0}.$$

From this, and the distributive law, deduce the law of sines of trigonometry.

21. Show that the area A of a parallelogram with adjacent sides $\vec{u}$ and $\vec{v}$ is

$$A = \begin{vmatrix} \vec{u} \cdot \vec{u} & \vec{u} \cdot \vec{v} \\ \vec{v} \cdot \vec{u} & \vec{v} \cdot \vec{v} \end{vmatrix}^{\frac{1}{2}}.$$

(Cf. Ex. 16.)

22. Assume that $\vec{u}$, $\vec{v}$, and $\vec{w}$ are the radius vectors of three noncollinear points lying in a plane Π. Prove that the vector

$$\vec{u} \times \vec{v} + \vec{v} \times \vec{w} + \vec{w} \times \vec{u}$$

is orthogonal to Π.

*23. If $\vec{v}_1$, $\vec{v}_2$, and $\vec{v}_3$ is an arbitrary set of three linearly independent vectors, its **reciprocal set** is defined:

$$\vec{V}_1 \equiv \frac{\vec{v}_2 \times \vec{v}_3}{[\vec{v}_1, \vec{v}_2, \vec{v}_3]}, \quad \vec{V}_2 \equiv \frac{\vec{v}_3 \times \vec{v}_1}{[\vec{v}_1, \vec{v}_2, \vec{v}_3]}, \quad \vec{V}_3 \equiv \frac{\vec{v}_1 \times \vec{v}_2}{[\vec{v}_1, \vec{v}_2, \vec{v}_3]}.$$

Prove:

(i) $\vec{v}_m \cdot \vec{V}_n = \begin{cases} 0 \text{ if } m \neq n, \\ 1 \text{ if } m = n; \end{cases}$

(ii) $[\vec{v}_1, \vec{v}_2, \vec{v}_3][\vec{V}_1, \vec{V}_2, \vec{V}_3] = 1$;

(iii) the reciprocal set of $\vec{V}_1, \vec{V}_2, \vec{V}_3$ is $\vec{v}_1, \vec{v}_2, \vec{v}_3$. Show that a set of vectors is **self-reciprocal** (that is, $\vec{V}_m = \vec{v}_m$, $m = 1, 2, 3$) if and only if it is an orthogonal set of unit vectors.

*24. If $\vec{v}_1, \vec{v}_2$, and $\vec{v}_3$ are linearly independent, show that any vector $\vec{u}$ can be represented as a linear combination of these three by means of the formula:

(2) $\qquad \vec{u} = (\vec{u} \cdot \vec{V}_1)\vec{v}_1 + (\vec{u} \cdot \vec{V}_2)\vec{v}_2 + (\vec{u} \cdot \vec{V}_3)\vec{v}_3,$

where $\vec{V}_1, \vec{V}_2, \vec{V}_3$ is the reciprocal set of $\vec{v}_1, \vec{v}_2, \vec{v}_3$ (Ex. 23).

*25. Establish the formula

$$[\vec{u}_1, \vec{u}_2, \vec{u}_3][\vec{v}_1, \vec{v}_2, \vec{v}_3] = \begin{vmatrix} \vec{u}_1 \cdot \vec{v}_1 & \vec{u}_1 \cdot \vec{v}_2 & \vec{u}_1 \cdot \vec{v}_3 \\ \vec{u}_2 \cdot \vec{v}_1 & \vec{u}_2 \cdot \vec{v}_2 & \vec{u}_2 \cdot \vec{v}_3 \\ \vec{u}_3 \cdot \vec{v}_1 & \vec{u}_3 \cdot \vec{v}_2 & \vec{u}_3 \cdot \vec{v}_3 \end{vmatrix}.$$

Hint: If $[\vec{v}_1, \vec{v}_2, \vec{v}_3] = 0$, express one $\vec{v}_m$ as a linear combination of the other two. If $[\vec{v}_1, \vec{v}_2, \vec{v}_3] \neq 0$, express $\vec{u}_1, \vec{u}_2$, and $\vec{u}_3$ in terms of the reciprocal set of $\vec{v}_1, \vec{v}_2, \vec{v}_3$, and compute $[\vec{u}_1, \vec{u}_2, \vec{u}_3]$. (Cf. Exs. 23–24.)

*26. Use the result of Exercise 24, and the geometric interpretation of $[\vec{u}, \vec{v}, \vec{w}]$ to show that the volume V of any parallelepiped determined by $\vec{u}, \vec{v}$, and $\vec{w}$ is

$$V = \begin{vmatrix} \vec{u} \cdot \vec{u} & \vec{u} \cdot \vec{v} & \vec{u} \cdot \vec{w} \\ \vec{v} \cdot \vec{u} & \vec{v} \cdot \vec{v} & \vec{v} \cdot \vec{w} \\ \vec{w} \cdot \vec{u} & \vec{w} \cdot \vec{v} & \vec{w} \cdot \vec{w} \end{vmatrix}^{\frac{1}{2}}.$$

(Cf. Ex. 21.)

*27. Use Exercises 16 and 25 to establish the multiplication theorem for determinants quoted in the first paragraph of § 922, for the special cases of second- and third-order determinants.

1706. COORDINATE TRANSFORMATIONS

The student is already familiar with transformations of coordinates in plane analytic geometry, which are introduced primarily for the purpose of simplifying equations in two variables. Such simplification is accomplished by means of two fundamental types of transformations, *translations* and *rotations*, the rotations being taken about the origin.

In a similar manner and for similar reasons we shall be interested in simplifying equations in three variables by means of two basic transformations, translations and rotations.

Consider two rectangular coordinate systems, as indicated in Figure 1701, and assume for simplicity that both are right-handed. Any point in space has two sets of coordinates, one for each set of axes. The basic problem is to find a relationship between these two sets of coordinates that will give the coordinates of any point in either system in terms of its coordinates in the other system. We shall obtain this relationship by resolving the general problem into two simpler parts.

As indicated in Figure 1701, let a third system be introduced in such a way that its origin coincides with the origin of one coordinate system and its axes are respectively parallel to those of the other. In this way the problem is

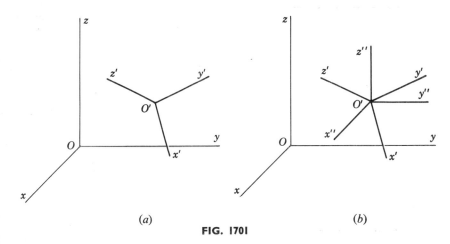

FIG. 1701

reduced to two simpler ones associated with the coordinate systems illustrated in Figure 1702 and appropriately labeled.

Once the relationship between the primed and unprimed coordinates has been obtained for each of these two special types of transformations, the more

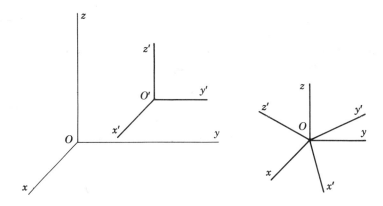

(a) Translation. Axes parallel (b) Rotation. Origins identical
FIG. 1702

general problem is resolved. For example (cf. Fig. 1701b), if x', y', and z' are expressed in terms of x'', y'', and z'', and the latter are expressed in terms of x, y, and z, the primed coordinates are immediately expressible in terms of the unprimed coordinates.

1707. TRANSLATIONS

Inspection of Figure 1702a indicates that if the coordinates of O' and a point p in the xyz system are (x_0, y_0, z_0) and (x, y, z), respectively, and the coordinates of p in the $x'y'z'$ system are (x', y', z'), then the two sets of coordinates of p are related by the equations

* $\begin{cases} (1) \qquad x = x' + x_0, \quad y = y' + y_0, \quad z = z' + z_0, \\ (2) \qquad x' = x - x_0, \quad y' = y - y_0, \quad z' = z - z_0. \end{cases}$

In terms of vectors, these equations are:

(3) $\qquad \overrightarrow{Op} = \overrightarrow{O'p} + \overrightarrow{OO'}, \quad \overrightarrow{O'p} = \overrightarrow{Op} - \overrightarrow{OO'}.$

Since, by definition, a vector is unchanged by translation:

(4) $\qquad \overrightarrow{p'q'} = \overrightarrow{pq},$

there is relatively little interest to us at present in this particular type of transformation. We turn our attention to rotations.

1708. ROTATIONS

Consider, as indicated in Figure 1702b, two right-handed rectangular coordinate systems with a common origin, let the unit vectors in the directions of the x-, y-, and z-axis be designated as usual by $\vec{i}, \vec{j}$, and $\vec{k}$, respectively, and let the unit vectors in the system with primed coordinates be designated correspondingly by $\vec{i}', \vec{j}'$, and $\vec{k}'$. Furthermore, let the direction cosines in the xyz system of the x'-axis, the y'-axis, and the z'-axis be $(\lambda_{11}, \lambda_{21}, \lambda_{31})$, $(\lambda_{12}, \lambda_{22}, \lambda_{32})$, and $(\lambda_{13}, \lambda_{23}, \lambda_{33})$, respectively. It follows, then, that the direction cosines in the $x'y'z'$ system of the x-axis, the y-axis, and the z-axis are $(\lambda_{11}, \lambda_{12}, \lambda_{13})$, $(\lambda_{21}, \lambda_{22}, \lambda_{23})$, and $(\lambda_{31}, \lambda_{32}, \lambda_{33})$, respectively, as indicated by the diagram:

(1)

	x'	y'	z'
x	λ_{11}	λ_{12}	λ_{13}
y	λ_{21}	λ_{22}	λ_{23}
z	λ_{31}	λ_{32}	λ_{33}

	$\vec{i}'$	$\vec{j}'$	$\vec{k}'$
$\vec{i}$	λ_{11}	λ_{12}	λ_{13}
$\vec{j}$	λ_{21}	λ_{22}	λ_{23}
$\vec{k}$	λ_{31}	λ_{32}	λ_{33}

If p is a general point, with coordinates (x, y, z) in the xyz system and coordinates (x', y', z') in the $x'y'z'$ system, then its radius vector can be written in two ways:

(2) $\qquad \begin{cases} \overrightarrow{Op} = x\vec{i} + y\vec{j} + z\vec{k}, \\ \overrightarrow{Op} = x'\vec{i}' + y'\vec{j}' + z'\vec{k}'. \end{cases}$

The coordinates x, y, and z can be solved from (2) in terms of the coordinates x', y', and z', and conversely, by the process of equating the two right-hand members of (2) and forming appropriate dot products. For example,

$$\vec{i} \cdot (x\vec{i} + y\vec{j} + z\vec{k}) = \vec{i} \cdot (x'\vec{i}' + y'\vec{j}' + z'\vec{k}')$$

gives the equation $x = \lambda_{11}x' + \lambda_{12}y' + \lambda_{13}z'$. The complete sets of equations follow:

(3) $\begin{cases} x = \lambda_{11}x' + \lambda_{12}y' + \lambda_{13}z', & x' = \lambda_{11}x + \lambda_{21}y + \lambda_{31}z, \\ y = \lambda_{21}x' + \lambda_{22}y' + \lambda_{23}z', & y' = \lambda_{12}x + \lambda_{22}y + \lambda_{32}z, \\ z = \lambda_{31}x' + \lambda_{32}y' + \lambda_{33}z', & z' = \lambda_{13}x + \lambda_{23}y + \lambda_{33}z. \end{cases}$

(Complete the derivation of (3)). These equations can be remembered and constructed easily with the aid of (1).

Either proceeding directly or by substituting in equations (3) $(x', y', z') = (1, 0, 0), \cdots, (x, y, z) = (1, 0, 0), \cdots$, we see that the basic unit vectors are related by systems of equations similar to (3):

(4) $\begin{cases} \vec{i} = \lambda_{11}\vec{i}' + \lambda_{12}\vec{j}' + \lambda_{13}\vec{k}', & \vec{i}' = \lambda_{11}\vec{i} + \lambda_{21}\vec{j} + \lambda_{31}\vec{k}, \\ \vec{j} = \lambda_{21}\vec{i}' + \lambda_{22}\vec{j}' + \lambda_{23}\vec{k}', & \vec{j}' = \lambda_{12}\vec{i} + \lambda_{22}\vec{j} + \lambda_{32}\vec{k}, \\ \vec{k} = \lambda_{31}\vec{i}' + \lambda_{32}\vec{j}' + \lambda_{33}\vec{k}'; & \vec{k}' = \lambda_{13}\vec{i} + \lambda_{23}\vec{j} + \lambda_{33}\vec{k}. \end{cases}$

The transformation of coordinates prescribed by equations (3) is called a **rotation**, with **rotation matrix**

(5) $$A = \begin{pmatrix} \lambda_{11} & \lambda_{12} & \lambda_{13} \\ \lambda_{21} & \lambda_{22} & \lambda_{23} \\ \lambda_{31} & \lambda_{32} & \lambda_{33} \end{pmatrix}.$$

The second set of equations (3) shows that the rotation matrix for the inverse transformation, called the **inverse** of the matrix A and denoted A^{-1}, is obtained from A by interchanging rows and columns:

(6) $$A^{-1} = \begin{pmatrix} \lambda_{11} & \lambda_{21} & \lambda_{31} \\ \lambda_{12} & \lambda_{22} & \lambda_{32} \\ \lambda_{13} & \lambda_{23} & \lambda_{33} \end{pmatrix}.$$

Such a matrix is called **orthogonal**. Thus, every rotation matrix is orthogonal. For further discussion of rotation and other orthogonal matrices, see the author's *Solid Analytic Geometry* (New York, Appleton-Century-Crofts, Inc., 1947).

The elements of the rotation matrix A admit many interrelationships, according to the theorem (a *cofactor* of any element of A being defined as the cofactor of that same element in the determinant of A):

Theorem. *The three row vectors of A are a right-handed orthogonal set of unit vectors; the three column vectors of A are a right-handed orthogonal set of unit vectors; each element of A is equal to its own cofactor in A; the determinant of A is equal to 1.*

Proof. The first two statements follow immediately from the representation of the fundamental triad of unit vectors in each system in terms of those of the other. The third statement is a consequence of the fact that each row is the cross product of the other two in cyclic sequence. The last statement follows from the preceding ones if the determinant of A is expanded with respect to the elements of, say, the first row.

NOTE. Equations (3) may be interpreted as defining a *mapping* or *point transformation* (cf. § 612), where there is only *one* coordinate system and the point $p:(x, y, z)$ is transformed into a *new* point $p':(x', y', z')$. Under these circumstances, where there is no change in the assumptions made regarding the elements of the matrix A, the transformation is again called a **rotation**. In case the rows (or columns) of A form a left-handed orthogonal set of unit vectors, the point-transformation is called a **skew-rotation**.

Example 1. Prove algebraically that the scalar product of two vectors is independent of the rectangular coordinate system.

Solution. Let the fundamental triads of unit vectors in two coordinate systems be $(\vec{i}, \vec{j}, \vec{k})$ and $(\vec{i}', \vec{j}', \vec{k}')$, whether these two coordinate systems are related by a translation, rotation, or a combination of the two (or even in case one system is right-handed and the other left-handed). Let $\vec{u}$ and $\vec{v}$ be any two vectors, represented as follows in the two systems:

(7) $$\begin{cases} \vec{u} = u_1\vec{i} + u_2\vec{j} + u_3\vec{k} = u_1'\vec{i}' + u_2'\vec{j}' + u_3'\vec{k}', \\ \vec{v} = v_1\vec{i} + v_2\vec{j} + v_3\vec{k} = v_1'\vec{i}' + v_2'\vec{j}' + v_3'\vec{k}'. \end{cases}$$

The problem is to establish the equality:

(8) $$u_1 v_1 + u_2 v_2 + u_3 v_3 = u_1' v_1' + u_2' v_2' + u_3' v_3'.$$

By expanding with the aid of the distributive law and the nature of the triad $(\vec{i}', \vec{j}', \vec{k}')$, we have for $\vec{u} \cdot \vec{v}$, which is equal to the left-hand side of (8) by definition:

$$\vec{u} \cdot \vec{v} = (u_1'\vec{i}' + u_2'\vec{j}' + u_3'\vec{k}') \cdot (v_1'\vec{i}' + v_2'\vec{j}' + v_3'\vec{k}')$$
$$= u_1'v_1'\vec{i}' \cdot \vec{i}' + u_1'v_2'\vec{i}' \cdot \vec{j}' + u_1'v_3'\vec{i}' \cdot \vec{k}' + \cdots$$
$$= u_1'v_1' + u_2'v_2' + u_3'v_3',$$

and with this verification of (8) the proof is complete.

Example 2. Prove algebraically that the vector product of two vectors is independent of the right-handed rectangular coordinate system.

Solution. With the notation (7) of Example 1, and the additional assumption that both triads $(\vec{i}, \vec{j}, \vec{k})$ and $(\vec{i}', \vec{j}', \vec{k}')$ are right-handed, we wish to establish the equality:

(9) $$\begin{vmatrix} \vec{i} & \vec{j} & \vec{k} \\ u_1 & u_2 & u_3 \\ v_1 & v_2 & v_3 \end{vmatrix} = \begin{vmatrix} \vec{i}' & \vec{j}' & \vec{k}' \\ u_1' & u_2' & u_3' \\ v_1' & v_2' & v_3' \end{vmatrix}.$$

The left-hand side of (9) is by definition equal to

$$\vec{u} \times \vec{v} = (u_1'\vec{i}' + u_2'\vec{j}' + u_3'\vec{k}') \times (v_1'\vec{i}' + v_2'\vec{j}' + v_3'\vec{k}')$$

which, by the distributive law and right-handed orientation, is equal to

$$u_1'v_1'\vec{i}' \times \vec{i}' + u_1'v_2'\vec{i}' \times \vec{j}' + u_1'v_3'\vec{i}' + \vec{k}' + \cdots$$
$$= \vec{0} + u_1'v_2'\vec{k}' - u_1'v_3'\vec{j}' + \cdots$$
$$= (u_2'v_3' - u_3'v_2')\vec{i}' + (u_3'v_1' - u_1'v_3')\vec{j}' + (u_1'v_2' - u_2'v_1')\vec{k}'.$$

Since this last expression is by definition equal to the right-hand member of (9), the proof is complete.

1709. EXERCISES

In Exercises 1–4, supply the missing elements of the rotation matrix.

1. $\begin{pmatrix} \frac{3}{7} & -\frac{6}{7} & \frac{2}{7} \\ \frac{6}{7} & \frac{2}{7} & -\frac{3}{7} \\ & & \end{pmatrix}$

2. $\begin{pmatrix} \frac{6}{19} & & \frac{15}{19} \\ \frac{10}{19} & & \frac{6}{19} \\ \frac{15}{19} & & \end{pmatrix}$

3. $\begin{pmatrix} \frac{1}{3} & & \frac{2}{3} \\ & \frac{1}{3} & \frac{2}{3} \\ & & \end{pmatrix}$

4. $\begin{pmatrix} & \frac{1}{\sqrt{2}} & -\frac{1}{\sqrt{2}} \\ & \frac{1}{\sqrt{3}} & \\ & & \end{pmatrix}$

In Exercises 5 and 6, write the rotation equations for a transformation of coordinates to a new set of axes, two of which have the given directions in the original coordinate system.

5. $x': \left(\frac{1}{\sqrt{3}}, \frac{1}{\sqrt{3}}, \frac{1}{\sqrt{3}}\right)$; $z': \left(\frac{1}{\sqrt{2}}, 0, -\frac{1}{\sqrt{2}}\right)$.

6. $y': \left(\frac{1}{\sqrt{14}}, \frac{2}{\sqrt{14}}, \frac{3}{\sqrt{14}}\right)$; $z': \left(\frac{2}{\sqrt{5}}, -\frac{1}{\sqrt{5}}, 0\right)$.

7. A coordinate system, with $x'y'z'$ axes, is obtained from an original xyz system by the 45° rotation of axes about the z-axis that places the x'-axis in the first quadrant of the xy-plane. Write the rotation matrix for the corresponding coordinate transformation.

8. Find the axis of the point rotation with matrix

$$\begin{pmatrix} \frac{2}{7} & \frac{6}{7} & -\frac{3}{7} \\ \frac{3}{7} & \frac{2}{7} & \frac{6}{7} \\ \frac{6}{7} & -\frac{3}{7} & -\frac{2}{7} \end{pmatrix}.$$

Hint: Find a point other than the origin that is invariant, or unchanged by the transformation.

9. Find the angle of the rotation of Exercise 8. *Hint:* Use a point whose radius vector is orthogonal to the axis of the rotation.

10. Prove algebraically that the magnitude of a vector $\vec{v} = (v_1, v_2, v_3)$, defined to be $\sqrt{v_1^2 + v_2^2 + v_3^2}$, is independent of the rectangular coordinate system.

11. Prove algebraically that if the angle θ between two vectors $\vec{u} = (u_1, u_2, u_3)$ and $\vec{v} = (v_1, v_2, v_3)$ is *defined* by the equation $\cos\theta = (\vec{u} \cdot \vec{v})/|\vec{u}| |\vec{v}|$, then θ is independent of the rectangular coordinate system.

12. Suppose somebody attempts to define a "circle product" of two vectors $\vec{u} = (u_1, u_2, u_3)$ and $\vec{v} = (v_1, v_2, v_3)$ to be the scalar:

$$\vec{u} \circ \vec{v} \equiv u_1 v_2 + u_2 v_3 + u_3 v_1.$$

Prove that this circle product depends on the right-handed orthogonal coordinate system, and is therefore *not* a legitimate vector operation. (Cf. Ex. 13.)

13. Suppose somebody attempts to define a "square product" of two vectors $\vec{u} = (u_1, u_2, u_3)$ and $\vec{v} = (v_1, v_2, v_3)$ to be the vector:

$$\vec{u} \square \vec{v} \equiv (u_1 v_1, u_2 v_2, u_3 v_3).$$

Prove that this square product depends on the right-handed orthogonal coordinate system, and is therefore *not* a legitimate vector operation. (Cf. Ex. 12.)

14. Let two rotations be given by equations:

(1) $\begin{cases} x = \lambda_{11} x' + \lambda_{12} y' + \lambda_{13} z', & x' = \mu_{11} x'' + \mu_{12} y'' + \mu_{13} z'', \\ y = \lambda_{21} x' + \lambda_{22} y' + \lambda_{23} z', & y' = \mu_{21} x'' + \mu_{22} y'' + \mu_{23} z'', \\ z = \lambda_{31} x' + \lambda_{32} y' + \lambda_{33} z'; & z' = \mu_{31} x'' + \mu_{32} y'' + \mu_{33} z''. \end{cases}$

Show that the resultant rotation, obtained by composition of the given two rotations, is given by the system of equations:

(2) $\begin{cases} x = \nu_{11} x'' + \nu_{12} y'' + \nu_{13} z'', \\ y = \nu_{21} x'' + \nu_{22} y'' + \nu_{23} z'', \quad \nu_{mn} = \sum_{k=1}^{3} \lambda_{mk} \mu_{kn}, \; m, n = 1, 2, 3. \\ z = \nu_{31} x'' + \nu_{32} y'' + \nu_{33} z''; \end{cases}$

The rotation matrix of (2) is called the **product** of the two rotation matrices of (1).

1710. SCALAR AND VECTOR FIELDS. VECTOR FUNCTIONS

In many of the applications of vector analysis, it is common to speak of a real-valued function $f(x, y, z)$, defined over a region R in E_3, as a **scalar function of position**, or as a **scalar field**. (The use of the word *field* in this connection has no relation to its use in Chapter 1 with the axioms of the real number system.) Similarly, a vector-valued function $\vec{v}(x, y, z)$, defined over a region R in E_3 is called a **vector function of position**, or a **vector field**.

Of more immediate interest is a vector-valued function of a single scalar, which can be pictured as a variable vector whose initial point is the origin and whose terminal point traces out a space curve as the independent scalar variable (thought of as time) varies over an interval. Continuity of this moving *radius vector* is equivalent to continuity of the moving point. The vector formulation of this follows:

Definition. *If*

(1) $$\vec{r}(t) = x(t)\vec{i} + y(t)\vec{j} + z(t)\vec{k}$$

*is a vector-valued function of a scalar variable t, then $\vec{r}(t)$ is **continuous** at $t = t_0$ if and only if it is defined there and corresponding to an arbitrary positive number ϵ there exists a positive number $\delta = \delta(\epsilon)$ such that $|t - t_0| < \delta$ implies $|\vec{r}(t) - \vec{r}(t_0)| < \epsilon$ for values of t for which $\vec{r}(t)$ is defined.*

By Theorem I, § 612, the vector function (1) is continuous if and only if all of its components, $x(t)$, $y(t)$, and $z(t)$ are continuous.

For *limits* of vector valued functions of a scalar, formulations and remarks similar to those just made apply.

1711. ORDINARY DERIVATIVES OF VECTOR FUNCTIONS

Consider the radius vector $\vec{r}$ of a point p defined parametrically by (1), § 1710. By the remarks of the next-to-the-last paragraph of § 1710, if $x(t)$, $y(t)$, and $z(t)$ are assumed to be continuous, $\vec{r}(t)$ is also continuous. Furthermore, since

(1) $$\frac{\Delta \vec{r}}{\Delta t} \equiv \frac{\vec{r}(t + \Delta t) - \vec{r}(t)}{\Delta t} = \frac{\Delta x}{\Delta t}\vec{i} + \frac{\Delta y}{\Delta t}\vec{j} + \frac{\Delta z}{\Delta t}\vec{k},$$

the difference quotient $\Delta \vec{r}/\Delta t$ has a (vector) limit if and only if the three components of $\vec{r}(t)$ are differentiable. This limit is defined to be the **derivative** of $\vec{r}(t)$, and we have:

(2) $$\vec{r}'(t) = \frac{d\vec{r}}{dt} \equiv \lim_{\Delta t \to 0} \frac{\Delta \vec{r}}{\Delta t} = \frac{dx}{dt}\vec{i} + \frac{dy}{dt}\vec{j} + \frac{dz}{dt}\vec{k}.$$

A curve with vector formula or parametrization $\vec{r}(t)$ thus has a nonzero tangent vector at any point p where $\vec{r}'(t)$ exists and is nonzero, and any tangent vector at p is a multiple of $\vec{r}'(t)$. (Cf. § 805.)

From (2) we infer that a radius vector function $\vec{r}(t)$ is (continuously) differentiable if and only if its components are. A **smooth curve** (§ 805) can now be redefined as a curve whose vector parametrization $\vec{r}(t)$ has a continuous nonvanishing derivative $\vec{r}'(t)$.

If the parameter t in (1), § 1710, is *arclength* s the vector (2) is the **unit tangent vector** in the direction of increasing s (§ 805), sometimes called the **unit forward tangent**. If the parameter t in (1), § 1710, is *time*, the vector (2) is (by definition) the **velocity vector** $\vec{V}$ of the moving point p. Its magnitude $V = |\vec{V}|$ is the **speed** of p. Note that (according to these definitions) *velocity is a vector and speed is a nonnegative scalar*.

Higher-order derivatives are defined in an obvious fashion. For example, **acceleration** is a vector-valued function of time, defined:

$$\vec{A}(t) \equiv \frac{d}{dt}(\vec{V}(t)) = \frac{d^2}{dt^2}(\vec{r}(t)).$$

The direction of the acceleration vector $\vec{A}$ in relation to the velocity vector $\vec{V}$ and the curve C traced out by the radius vector $\vec{r}$ is discussed in § 1903, on kinematics. (Cf. the Example below.)

The following laws are a few of those governing differentiation of vector-valued functions of a scalar parameter. The scalar c and the vector $\vec{c}$ are assumed to be constant. The variable s is arc length. The remaining letters represent differentiable functions of a parameter t.

Primes indicate differentiation with respect to t. (Give proofs in Ex. 3, § 1715.)

(i) $\vec{c}' = \vec{0}$.
(ii) $(\vec{u} + \vec{v})' = \vec{u}' + \vec{v}'$.
(iii) $(f\vec{v})' = f\vec{v}' + f'\vec{v}$.
(iiia) $(c\vec{v})' = c\vec{v}'$.
(iiib) $(f\vec{c})' = f'\vec{c}$.
(iv) $(\vec{u} \cdot \vec{v})' = \vec{u} \cdot \vec{v}' + \vec{u}' \cdot \vec{v}$.
(v) $(\vec{u} \times \vec{v})' = \vec{u} \times \vec{v}' + \vec{u}' \times \vec{v}$.
(vi) $[\vec{u}, \vec{v}, \vec{w}]' = [\vec{u}', \vec{v}, \vec{w}] + [\vec{u}, \vec{v}', \vec{w}] + [\vec{u}, \vec{v}, \vec{w}']$.

Example. Let $\vec{u}$ be a differentiable function of a parameter t, and assume that the magnitude of $\vec{u}$ is constant. Show that $\vec{u}$ and its derivative $\vec{u}'$ are orthogonal. In particular show that for any particle moving with constant speed the velocity and acceleration vectors are orthogonal, and since the velocity vector must be tangent to the trajectory, the acceleration vector must be perpendicular to it.

Solution. Assuming $|\vec{u}|^2 = \vec{u} \cdot \vec{u}$ to be constant, we differentiate this dot product, using formula (iv):
$$\vec{u} \cdot \vec{u}' + \vec{u}' \cdot \vec{u} = 2\vec{u} \cdot \vec{u}' = 0.$$

This establishes the orthogonality of $\vec{u}$ and $\vec{u}'$. The remaining parts of the solution follow from the definition of speed as the magnitude of the velocity vector, and acceleration as the derivative with respect to time of the velocity vector.

1712. THE GRADIENT OF A SCALAR FIELD

In § 910, the gradient of a real-valued function $f(x, y, z)$ of three real variables was defined to be a vector whose components are the three first partial derivatives. In vector notation this vector can be written $\frac{\partial f}{\partial x}\vec{i} + \frac{\partial f}{\partial y}\vec{j} + \frac{\partial f}{\partial z}\vec{k}$. It is convenient at this time to introduce a *symbolic operator*, called **del**, denoted

(1) $$\vec{\nabla} = \frac{\partial}{\partial x}\vec{i} + \frac{\partial}{\partial y}\vec{j} + \frac{\partial}{\partial z}\vec{k},$$

and defined by its effect when operating on a scalar function f of a point in E_3 to produce the gradient of f:

(2) $$\operatorname{grad} f = \vec{\nabla} f = \frac{\partial f}{\partial x}\vec{i} + \frac{\partial f}{\partial y}\vec{j} + \frac{\partial f}{\partial z}\vec{k}.$$

We restate in vector form some of the results of § 910:

1. The directional derivative of a scalar function $f(x, y, z)$ in the direction of the unit vector $\vec{u}$ is

(3) $$\frac{df}{ds} = (\operatorname{grad} f) \cdot \vec{u}.$$

2. The normal derivative of f is the magnitude of its gradient:

(4) $$\frac{df}{dn} = |\operatorname{grad} f|.$$

THE GRADIENT OF A SCALAR FIELD

3. The equations of the normal plane and tangent line to a curve C with parametric representation $\vec{r}(t)$, at the point p_0 where $\vec{r}_0 = \vec{r}(t_0)$, can be expressed ($\vec{r}_0'$ being the value of $\vec{r}'$ at p_0):

(5) Normal plane: $\vec{r}_0' \cdot (\vec{r} - \vec{r}_0) = 0$,
(6) Tangent line: $\vec{r} - \vec{r}_0 = \lambda \vec{r}_0'$, λ a scalar parameter.

4. The equations of the tangent plane and normal line to the surface $f(x, y, z) = 0$, at the point p_0 where $\vec{r}_0 = x_0\vec{i} + y_0\vec{j} + z_0\vec{k}$ (grad f_0 being the value of grad f at p_0):

(7) Tangent plane: $(\text{grad } f_0) \cdot (\vec{r} - \vec{r}_0) = 0$,
(8) Normal line: $\vec{r} - \vec{r}_0 = \lambda \text{ grad } f_0$, λ a scalar parameter.

NOTE 1. Since the directional derivative of a scalar-valued function f of a point p in E_3 is independent of the coordinate system used to specify the point p, it follows that the vector grad f (in spite of the fact that it is defined in terms of a given rectangular coordinate system) is also independent of the coordinate system. This question is discussed further in § 1716.

NOTE 2. A vector field $\vec{v}$ that has the form

$$\vec{v} = \text{grad } f,$$

for some scalar field f, is called a **conservative field**. The scalar field f is called a **scalar potential** for $\vec{v}$.

The following laws are a few of those governing the gradient operator, c representing a constant scalar and all other letters representing differentiable scalar functions of a point in E_3. (Give proofs in Ex. 4, § 1715.)

  (i)   grad $c = \vec{0}$.
  (ii) grad $(f + g) = \text{grad } f + \text{grad } g$.
  (iii) grad $(fg) = f \text{ grad } g + g \text{ grad } f$.
  (iiia) grad $(cf) = c \text{ grad } f$.
  (iiib) grad $(f^2) = 2f \text{ grad } f$.

For the gradient of a scalar product of two vector fields, see Exercise 31, § 1715.

Example. If $\vec{r} = x\vec{i} + y\vec{j} + z\vec{k}$, if $r = |\vec{r}|$, and if ϕ is a differentiable function of r, establish the formula

$$\text{grad } (\phi(r)) = \frac{\phi'(r)}{r} \vec{r}.$$

Solution. By definition, and the formula $r = \sqrt{x^2 + y^2 + z^2}$,

$$\text{grad } (\phi(r)) = \frac{\partial}{\partial x}(\phi(r))\vec{i} + \frac{\partial}{\partial y}(\phi(r))\vec{j} + \frac{\partial}{\partial z}(\phi(r))\vec{k}$$

$$= \phi'(r)\frac{x}{r}\vec{i} + \phi'(r)\frac{y}{r}\vec{j} + \phi'(r)\frac{z}{r}\vec{k}$$

$$= \frac{\phi'(r)}{r}(x\vec{i} + y\vec{i} + z\vec{k}) = \frac{\phi'(r)}{r}\vec{r}.$$

1713. THE DIVERGENCE AND CURL OF A VECTOR FIELD

Let
$$\vec{v} = \vec{v}(p) = v_1\vec{i} + v_2\vec{j} + v_3\vec{k}$$
denote a vector field with components v_1, v_2, and v_3,[†] each of which is thus a scalar-valued function of the point p in E_3. We use the symbolic operator del $= \vec{\nabla}$ to define the following two functions of $\vec{v}$:

1. The **divergence** of $\vec{v}$ is the scalar-valued function denoted and defined:

(1) $$\operatorname{div} \vec{v} = \vec{\nabla} \cdot \vec{v} = \left(\frac{\partial}{\partial x}\vec{i} + \frac{\partial}{\partial y}\vec{j} + \frac{\partial}{\partial z}\vec{k}\right) \cdot (v_1\vec{i} + v_2\vec{j} + v_3\vec{k})$$
$$\equiv \frac{\partial v_1}{\partial x} + \frac{\partial v_2}{\partial y} + \frac{\partial v_3}{\partial z}.$$

2. The **curl** of $\vec{v}$ is the vector-valued function denoted and defined:

(2) $$\operatorname{curl} \vec{v} = \vec{\nabla} \times \vec{v} = \begin{vmatrix} \vec{i} & \vec{j} & \vec{k} \\ \frac{\partial}{\partial x} & \frac{\partial}{\partial y} & \frac{\partial}{\partial z} \\ v_1 & v_2 & v_3 \end{vmatrix}$$
$$\equiv \left(\frac{\partial v_3}{\partial y} - \frac{\partial v_2}{\partial z}\right)\vec{i} + \left(\frac{\partial v_1}{\partial z} - \frac{\partial v_3}{\partial x}\right)\vec{j} + \left(\frac{\partial v_2}{\partial x} - \frac{\partial v_1}{\partial y}\right)\vec{k}.$$

NOTE 1. It should be appreciated that the determinant in (2) is merely a convenient symbolic means of writing the vector function expanded in the second line of (2).

NOTE 2. As with the gradient of a scalar field, the divergence and curl of a vector field are independent of the (right-handed) coordinate system (cf. § 1716; also cf. Note 1, § 1819, § 1824).

Example. Find the divergence and curl of the vector field $x^2y\vec{i} + xz\vec{j} + yz\vec{k}$ at the point $(4, 3, 2)$.

Solution. The divergence and curl are respectively
$$\frac{\partial}{\partial x}(x^2y) + \frac{\partial}{\partial y}(xz) + \frac{\partial}{\partial z}(yz) = 2xy + y,$$
$$\begin{vmatrix} \vec{i} & \vec{j} & \vec{k} \\ \frac{\partial}{\partial x} & \frac{\partial}{\partial y} & \frac{\partial}{\partial z} \\ x^2y & xz & yz \end{vmatrix} = (z-x)\vec{i} + (0-0)\vec{j} + (z - x^2)\vec{k},$$
which have the particular values at $(4, 3, 2)$: 27 and $-2\vec{i} - 14\vec{k}$.

The following laws are a few of those governing the divergence and curl operators, $\vec{c}$ representing a constant vector and all other letters representing

[†] The subscripts indicate components, and *not* partial derivatives.

differentiable scalar and vector functions of a point in E_3. (Give proofs in Exs. 5-6, § 1715.)

(i) div $\vec{c} = 0$.
(ii) div $(\vec{u} + \vec{v}) = $ div $\vec{u} + $ div $\vec{v}$.
(iii) curl $\vec{c} = \vec{0}$.
(iv) curl $(\vec{u} + \vec{v}) = $ curl $\vec{u} + $ curl $\vec{v}$.
(v) div $(f\vec{v}) = ($grad $f) \cdot \vec{v} + f$ div $\vec{v}$.
(vi) div $(\vec{u} \times \vec{v}) = \vec{v} \cdot $ curl $\vec{u} - \vec{u} \cdot $ curl $\vec{v}$.
(vii) curl $(f\vec{v}) = f$ curl $\vec{v} + ($grad $f) \times \vec{v}$.

For the curl of the cross product of two vector fields, see Exercise 32, § 1715.

NOTE 3. Standard differentiation formulas of elementary calculus (cf. formulas (ii) and (iiia), § 1711) show that the operations d/dt, grad, div, and curl are all *linear*; that is, if a and b are constants, f and g scalar fields, and $\vec{u}$ and $\vec{v}$ vector fields:

$$\frac{d}{dt}(af + bg) = a\frac{df}{dt} + b\frac{dg}{dt}, \quad \text{grad } (af + bg) = a \text{ grad } f + b \text{ grad } g,$$

div $(a\vec{u} + b\vec{v}) = a$ div $\vec{u} + b$ div $\vec{v}$, curl $(a\vec{u} + b\vec{v}) = a$ curl $\vec{u} + b$ curl $\vec{v}$.

(Give proofs in Ex. 7, § 1715.)

1714. RELATIONS AMONG VECTOR OPERATIONS

In the preceding two sections we have defined three operations involving vectors, and it might appear at first glance that they could be combined in pairs in $3 \cdot 3 = 9$ ways. However, four of these are meaningless (for example, the curl of a divergence is impossible since the divergence is scalar valued, and the curl operates only on vectors). Of the five that have meaning, the two involving the curl just once vanish identically:

(i) curl grad $f = \vec{0}$,
(ii) div curl $\vec{v} = 0$,

as can be seen by explicit expansion (Ex. 8, § 1715).

One relation is already familiar (Exs. 26, 27, § 924); if f is a scalar field, denote by Δf or $\nabla^2 f$ the **Laplacian** of f, defined:

(iii) $\Delta f = \nabla^2 f \equiv \vec{\nabla} \cdot (\vec{\nabla} f) = $ div grad f

$$= \text{Lapl } f = \frac{\partial^2 f}{\partial x^2} + \frac{\partial^2 f}{\partial y^2} + \frac{\partial^2 f}{\partial z^2}.$$

Finally, if we define the **Laplacian** of a vector field $\vec{v} = v_1 \vec{i} + v_2 \vec{j} + v_3 \vec{k}$ to be the vector $\nabla^2 \vec{v} \equiv \nabla^2 v_1 \vec{i} + \nabla^2 v_2 \vec{j} + \nabla^2 v_3 \vec{k}$, or

(1) $\qquad$ Lapl $\vec{v} \equiv $ Lapl $v_1 \vec{i} + $ Lapl $v_2 \vec{j} + $ Lapl $v_3 \vec{k}$,

we have the two equivalent relations (Ex. 9, § 1715):

(iv) curl curl $\vec{v} = $ grad div $\vec{v} - $ Lapl $\vec{v}$,
(v) grad div $\vec{v} = $ curl curl $\vec{v} + $ Lapl $\vec{v}$.

Example 1. A vector field $\vec{v}$ is **irrotational** if and only if its curl vanishes identically:
$$\operatorname{curl} \vec{v} = \vec{0}.$$
By (i), then, the gradient of any scalar field is irrotational; equivalently, any conservative vector field (Note 2, § 1712) is irrotational. (The converse is false; cf. Example 2, § 1825.) Verify directly that the vector field
$$\operatorname{grad} xy^2z^3 = y^2z^3\vec{i} + 2xyz^3\vec{j} + 3xy^2z^2\vec{k}$$
is irrotational.

Solution. By direct evaluation,
$$\begin{vmatrix} \vec{i} & \vec{j} & \vec{k} \\ \dfrac{\partial}{\partial x} & \dfrac{\partial}{\partial y} & \dfrac{\partial}{\partial z} \\ y^2z^3 & 2xyz^3 & 3xy^2z^2 \end{vmatrix} = 0\vec{i} + 0\vec{j} + 0\vec{k} = \vec{0}.$$

(Cf. Note 2, § 1825, for further discussion of irrotational vector fields.)

Example 2. A vector field $\vec{v}$ is **solenoidal** if and only if its divergence vanishes identically:
$$\operatorname{div} \vec{v} = 0.$$
By (ii), then, the curl of any vector field is solenoidal. Show that the cross product of any two gradients is solenoidal:
$$\operatorname{div} (\operatorname{grad} f \times \operatorname{grad} g) = 0.$$
Solution. By formula (vi), § 1713, and (i), above
$$\operatorname{div} (\operatorname{grad} f \times \operatorname{grad} g) = \operatorname{grad} g \cdot \vec{0} - \operatorname{grad} f \cdot \vec{0} = 0.$$
(Cf. §§ 1819, 1826, for further discussion of solenoidal vector fields.)

1715. EXERCISES

In the Exercises of this section f, g, and h are scalar fields, $\vec{u}$, $\vec{v}$, and $\vec{w}$ are vector fields, $\vec{c}$ is a constant vector, $\vec{r}$ is the radius vector of the point (x, y, z): $\vec{r} = x\vec{i} + y\vec{j} + z\vec{k}$, $r = |\vec{r}|$, and ϕ is a real-valued function of a real variable. All derivatives indicated in the formulas are assumed to exist.

1. Find grad div $\vec{v}$ if $\vec{v} = x^2y^2z^3\vec{i} + xy^3z^3\vec{j} + xy^2z^4\vec{k}$.
2. Find curl curl $\vec{v}$ if $\vec{v}$ is the vector field of Exercise 1.
3. Prove laws (i)–(vi), § 1711.
4. Prove laws (i)–(iiib), § 1712.
5. Prove laws (i)–(iv), § 1713.
6. Prove laws (v)–(vii), § 1713.
7. Prove the statements of Note 3, § 1713.
8. Prove laws (i) and (ii), § 1714.
*9. Prove laws (iv) and (v), § 1714.
*10. Find the Laplacian of the vector $\vec{v}$ of Exercise 1. Then, with the aid of Exercises 1 and 2, verify formulas (iv) and (v), § 1714, for this example.
11. Show that the triple scalar product of the gradients of three scalar fields is equal to their Jacobian with respect to the variables x, y, and z:
$$[\operatorname{grad} f, \operatorname{grad} g, \operatorname{grad} h] = \frac{\partial(f, g, h)}{\partial(x, y, z)}.$$

In Exercises 12–14, establish the formula involving the two scalar fields f and g.

12. $\operatorname{grad} \dfrac{f}{g} = \dfrac{g \operatorname{grad} f - f \operatorname{grad} g}{g^2}$.

13. $\operatorname{div}(f \operatorname{grad} g - g \operatorname{grad} f) = f \operatorname{Lapl} g - g \operatorname{Lapl} f$.

14. $\operatorname{Lapl}(fg) = f \operatorname{Lapl} g + 2 \operatorname{grad} f \cdot \operatorname{grad} g + g \operatorname{Lapl} f$.

15. For a vector function $\vec{v}$ of a scalar parameter t, prove:
$$\vec{v} \cdot \frac{d}{dt}(\vec{v}) = |\vec{v}| \frac{d}{dt}(|\vec{v}|).$$

In Exercises 16–23, establish the formula involving the radius vector $\vec{r}$.

16. $\operatorname{grad}(\vec{c} \cdot \vec{r}) = \vec{c}$.

17. $\operatorname{div}(\vec{c} \times \vec{r}) = 0$.

18. $\operatorname{curl}(\vec{c} \times \vec{r}) = 2\vec{c}$.

19. $\operatorname{grad} \phi(r) = \dfrac{\phi'(r)}{r} \vec{r}$; $\operatorname{grad} r^n = nr^{n-2}\vec{r}$.

20. $\operatorname{div}(\phi(r)\vec{r}) = 3\phi(r) + r\phi'(r)$; $\operatorname{div}(r^n \vec{r}) = (n+3)r^n$.

21. $\operatorname{curl}(\phi(r)\vec{r}) = \vec{0}$: $\phi(r)\vec{r}$ is irrotational.

22. $\operatorname{Lapl}(\phi(r)) = \phi''(r) + \dfrac{2}{r}\phi'(r)$; $\operatorname{Lapl} r^n = n(n+1)r^{n-2}$.

23. $\operatorname{grad} \operatorname{div}(\phi(r)\vec{r}) = \left[\phi''(r) + \dfrac{4}{r}\phi'(r)\right]\vec{r}$; $\operatorname{grad} \operatorname{div}(r^n \vec{r}) = n(n+3)r^{n-2}\vec{r}$.

24. If arc length s is an increasing function of time for a particle moving along a trajectory C with velocity and acceleration vectors $\vec{V}$ and $\vec{A}$, respectively, obtain the formulas
$$\vec{V} = |\vec{V}| \frac{d\vec{r}}{ds}; \quad \vec{A} = |\vec{V}| \frac{d\vec{V}}{ds}.$$

25. Prove that $f \operatorname{grad} f$ is irrotational: $\operatorname{curl}(f \operatorname{grad} f) = \vec{0}$.

26. Prove that the cross product of two irrotational vector fields is solenoidal: $\operatorname{div}(\vec{u} \times \vec{v}) = 0$. In particular, $\operatorname{grad} f \times \operatorname{grad} g$ is solenoidal.

27. Prove that the gradient of any harmonic function f ($\operatorname{Lapl} f = 0$) is *both* irrotational and solenoidal.

*28. Establish the differentiation formulas for triple cross products:
$$\{(\vec{u} \times \vec{v}) \times \vec{w}\}' = (\vec{u}' \times \vec{v}) \times \vec{w} + (\vec{u} \times \vec{v}') \times \vec{w} + (\vec{u} \times \vec{v}) \times \vec{w}',$$
$$\{\vec{u} \times (\vec{v} \times \vec{w})\}' = \vec{u}' \times (\vec{v} \times \vec{w}) + \vec{u} \times (\vec{v}' \times \vec{w}) + \vec{u} \times (\vec{v} \times \vec{w}').$$

*29. Show by the example of the helix

(1) $\qquad\qquad\qquad \vec{v} = \cos t\, \vec{i} + \sin t\, \vec{j} + t\, \vec{k}$

that the law of the mean:

(2) $\qquad\qquad\qquad \vec{v}(t_2) - \vec{v}(t_1) = \vec{v}'(t_3)(t_2 - t_1),$

where $t_1 < t_3 < t_2$ fails for differentiable vector functions. *Hint:* It is obvious that (2) fails if t_1 and t_2 differ by a multiple of 2π. To show that (2) can *never* hold, with $t_1 \neq t_2$, equate the squares of the magnitudes of its two members. (Cf. the Note, § 1513.)

*30. In spite of the failure of the law of the mean for vector-valued functions (cf. Ex. 29), show that if $\vec{v}' = \vec{0}$ identically in the parameter t, then $\vec{v}$ is a constant vector. (Cf. § 306.)

★31. Interpret and establish the formula
$$\text{grad}\,(\vec{u}\cdot\vec{v}) = (\vec{v}\cdot\vec{\nabla})\vec{u} + (\vec{u}\cdot\vec{\nabla})\vec{v} + \vec{v}\times\text{curl}\,\vec{u} + \vec{u}\times\text{curl}\,\vec{v}.$$

★32. Interpret and establish the formula
$$\text{curl}\,(\vec{u}\times\vec{v}) = (\vec{v}\cdot\vec{\nabla})\vec{u} - (\text{div}\,\vec{u})\vec{v} - (\vec{u}\cdot\vec{\nabla})\vec{v} + (\text{div}\,\vec{v})\vec{u}.$$

★1716. INDEPENDENCE OF THE COORDINATE SYSTEM

It was pointed out in the Note, § 1712, that although the gradient of a scalar function is defined in terms of a given rectangular coordinate system, it is nevertheless independent of the rectangular coordinate system used. This conclusion followed from the particular way in which the gradient vector is related to the concept of directional derivative. In this section we shall discuss the question of independence of coordinate system, with special objective to establish this independence for the gradient, divergence, and curl operators. All coordinate systems considered are assumed to be right-handed rectangular.

For present purposes we denote a vector field $\vec{v}$:

(1) $$\vec{v} = v_1\vec{i}_1 + v_2\vec{i}_2 + v_3\vec{i}_3,$$

where $\vec{i}_1, \vec{i}_2, \vec{i}_3$ are the fundamental triad of a right-handed rectangular coordinate system, with coordinates x', y', z', respectively corresponding to $\vec{i}_1, \vec{i}_2, \vec{i}_3$.

When the radius vector $\vec{r} = x\vec{i} + y\vec{j} + z\vec{k}$ is considered as a function of arc length s along any smooth curve C, the unit tangent vector in the direction of increasing s is given by the derivative with respect to s (cf. § 1711):

$$\frac{d\vec{r}}{ds} = \frac{dx}{ds}\vec{i} + \frac{dy}{ds}\vec{j} + \frac{dz}{ds}\vec{k}.$$

As special cases of this formula, we have for the three axes in the $x'y'z'$ coordinate system (the coordinates x, y, z being considered as functions of the coordinates x', y', z'):

(2) $$\begin{cases} \vec{i}_1 = \dfrac{\partial x}{\partial x'}\vec{i} + \dfrac{\partial y}{\partial x'}\vec{j} + \dfrac{\partial z}{\partial x'}\vec{k}, \\ \vec{i}_2 = \dfrac{\partial x}{\partial y'}\vec{i} + \dfrac{\partial y}{\partial y'}\vec{j} + \dfrac{\partial z}{\partial y'}\vec{k}, \\ \vec{i}_3 = \dfrac{\partial x}{\partial z'}\vec{i} + \dfrac{\partial y}{\partial z'}\vec{j} + \dfrac{\partial z}{\partial z'}\vec{k}. \end{cases}$$

I. To prove that the gradient of a scalar field f is independent of the coordinate system we consider f as a function of the three variables x', y', z' and write:

(3) $$\frac{\partial f}{\partial x'} = \frac{\partial f}{\partial x}\frac{\partial x}{\partial x'} + \frac{\partial f}{\partial y}\frac{\partial y}{\partial x'} + \frac{\partial f}{\partial z}\frac{\partial z}{\partial x'},$$

§1716] INDEPENDENCE OF THE COORDINATE SYSTEM

with similar expressions for $\partial f/\partial y'$ and $\partial f/\partial z'$. By (3), the definition of grad f, and (2):

(4) $\quad \dfrac{\partial f}{\partial x'} = (\operatorname{grad} f) \cdot \vec{i}_1, \quad \dfrac{\partial f}{\partial y'} = (\operatorname{grad} f) \cdot \vec{i}_2, \quad \dfrac{\partial f}{\partial z'} = (\operatorname{grad} f) \cdot \vec{i}_3.$

Since equations (4) are equivalent to the single vector equation

(5) $\quad \operatorname{grad} f = \dfrac{\partial f}{\partial x'}\vec{i}_1 + \dfrac{\partial f}{\partial y'}\vec{i}_2 + \dfrac{\partial f}{\partial z'}\vec{i}_3,$

the proof of the invariance of the gradient is complete.

II. We now establish independence of the coordinate system for the divergence of a vector field $\vec{v}$, with the aid of equation (5) and the following special case of (v), §1713, where $\vec{c}$ is a constant vector:

(6) $\quad\quad\quad\quad \operatorname{div}(f\vec{c}) = (\operatorname{grad} f) \cdot \vec{c}:$

(7) $\quad \operatorname{div} \vec{v} = \operatorname{div}(v_1\vec{i}_1) + \operatorname{div}(v_2\vec{i}_2) + \operatorname{div}(v_3\vec{i}_3)$
$= (\operatorname{grad} v_1) \cdot \vec{i}_1 + (\operatorname{grad} v_2) \cdot \vec{i}_2 + (\operatorname{grad} v_3) \cdot \vec{i}_3$
$= \left(\dfrac{\partial v_1}{\partial x'}\vec{i}_1 + \dfrac{\partial v_1}{\partial y'}\vec{i}_2 + \dfrac{\partial v_1}{\partial z'}\vec{i}_3\right) \cdot \vec{i}_1$
$+ \left(\dfrac{\partial v_2}{\partial x'}\vec{i}_1 + \dfrac{\partial v_2}{\partial y'}\vec{i}_2 + \dfrac{\partial v_2}{\partial z'}\vec{i}_3\right) \cdot \vec{i}_2$
$+ \left(\dfrac{\partial v_3}{\partial x'}\vec{i}_1 + \dfrac{\partial v_3}{\partial y'}\vec{i}_2 + \dfrac{\partial v_3}{\partial z'}\vec{i}_3\right) \cdot \vec{i}_3$
$= \dfrac{\partial v_1}{\partial x'} + \dfrac{\partial v_2}{\partial y'} + \dfrac{\partial v_3}{\partial z'},$

as desired.

III. For the curl of a vector field we proceed as in II, for the divergence, using the special case of (vii), §1713, where $\vec{c}$ is a constant vector:

(8) $\quad\quad\quad\quad \operatorname{curl}(f\vec{c}) = (\operatorname{grad} f) \times \vec{c}:$

(9) $\quad \operatorname{curl} \vec{v} = \operatorname{curl}(v_1\vec{i}_1) + \operatorname{curl}(v_2\vec{i}_2) + \operatorname{curl}(v_3\vec{i}_3)$
$= (\operatorname{grad} v_1) \times \vec{i}_1 + (\operatorname{grad} v_2) \times \vec{i}_2 + (\operatorname{grad} v_3) \times \vec{i}_3$
$= \left(\dfrac{\partial v_1}{\partial x'}\vec{i}_1 + \dfrac{\partial v_1}{\partial y'}\vec{i}_2 + \dfrac{\partial v_1}{\partial z'}\vec{i}_3\right) \times \vec{i}_1$
$+ \left(\dfrac{\partial v_2}{\partial x'}\vec{i}_1 + \dfrac{\partial v_2}{\partial y'}\vec{i}_2 + \dfrac{\partial v_2}{\partial z'}\vec{i}_3\right) \times \vec{i}_2$
$+ \left(\dfrac{\partial v_3}{\partial x'}\vec{i}_1 + \dfrac{\partial v_3}{\partial y'}\vec{i}_2 + \dfrac{\partial v_3}{\partial z'}\vec{i}_3\right) \times \vec{i}_3.$

We now expand this expression for curl $\vec{v}$ by the distributive law, and use the fact that the vectors $\vec{i}_1, \vec{i}_2, \vec{i}_3$ form a right-handed orthogonal set of unit vectors (so that $\vec{i}_2 \times \vec{i}_1 = -\vec{i}_3, \vec{i}_3 \times \vec{i}_1 = \vec{i}_2, \cdots$):

(10) $\quad \operatorname{curl} \vec{v} = -\dfrac{\partial v_1}{\partial y'}\vec{i}_3 + \dfrac{\partial v_1}{\partial z'}\vec{i}_2 + \cdots$
$= \left(\dfrac{\partial v_3}{\partial y'} - \dfrac{\partial v_2}{\partial z'}\right)\vec{i}_1 + \left(\dfrac{\partial v_1}{\partial z'} - \dfrac{\partial v_3}{\partial x'}\right)\vec{i}_2 + \left(\dfrac{\partial v_2}{\partial x'} - \dfrac{\partial v_1}{\partial y'}\right)\vec{i}_3,$

and the proof is complete.

NOTE 1. In §§ 1819 and 1824 there will be developed geometric representations of the divergence and curl that are by their very nature independent of the (right-handed rectangular) coordinate system. For an additional demonstration of invariance, see Note 1, § 1718.

NOTE 2. If the vectors $\vec{i}_1, \vec{i}_2, \vec{i}_3$ form a *left-handed* system, then formulas (5) and (7) are unchanged, and (10) is changed by a factor of -1. (Cf. Ex. 5, § 1719.)

Example. Prove that the **trivergence** of a vector field $\vec{v} = v_1\vec{i} + v_2\vec{j} + v_3\vec{k}$ defined by the equation

(11) $$\text{triv } \vec{v} = \vec{\Box} \cdot \vec{v} \equiv v_1 + v_2 + v_3,$$

depends on the right-handed rectangular coordinate system.

Solution. Let $\vec{v}$ be a given fixed constant unit vector. If a coordinate system is chosen so that $\vec{v} = \vec{i}, \vec{j}$, or $\vec{k}$, then triv $\vec{v} = 1$. If the direction is reversed, then triv $\vec{v} = -1$. If axes are chosen so that $\vec{v}$ lies in the xy-plane, starting at the origin and bisecting the first quadrant, then triv $\vec{v} = \sqrt{2}$. If $\vec{v}$ makes equal acute angles with all axes, then triv $\vec{v} = \sqrt{3}$ (cf. Ex. 9, § 709).

*1717. CURVILINEAR COORDINATES. ORTHOGONAL COORDINATES

In the preceding discussion of vectors and vector operations it has been assumed that the underlying coordinate system is right-handed rectangular Cartesian. The student is already familiar with some of the advantages of other coordinate systems such as cylindrical and spherical coordinates (cf. §§ 808, 1019, 1020). In many of the applications of mathematics it is convenient to consider more general coordinate systems, and to formulate in terms of such systems the basic vector operations of gradient, divergence, and curl.

We recall first (§§ 805, 1711) that if

(1) $$\vec{r}(t) = x(t)\vec{i} + y(t)\vec{j} + z(t)\vec{k}$$

is a smooth curve C, then

(2) $$\vec{r}'(t) = x'(t)\vec{i} + y'(t)\vec{j} + z'(t)\vec{k}$$

is a nonzero vector tangent to C.

Let us assume now that throughout a certain region Ω of space a change of coordinates is given by continuously differentiable functions

(3) $$\begin{cases} x = x(u, v, w), \\ y = y(u, v, w), \\ z = z(u, v, w), \end{cases} \quad \begin{array}{l} u = u(x, y, z), \\ v = v(x, y, z), \\ w = w(x, y, z), \end{array}$$

in a one-to-one manner, and that throughout this region the Jacobian of the transformation is positive:†

(4) $$J = J(u, v, w) = \frac{\partial(x, y, z)}{\partial(u, v, w)} > 0.$$

---

† Since J is continuous over the region Ω, the assumption that J never vanishes implies that J is of one sign (cf. Theorem III, § 610). If J were negative, an interchange of two variables would make J positive. As will be seen, the sign of J is related to the question of orientation.

§ 1717] CURVILINEAR COORDINATES

At any fixed point of the region Ω let us consider the three curves obtained by letting the variables u, v, and w, each acting in the role of the parameter t in (1), vary one at a time. Associated with these three **coordinate curves**, then, are the following three tangent vectors (partial differentiation notation being used in an obvious fashion):

(5)
$$\begin{cases} \dfrac{\partial \vec{r}}{\partial u} = \dfrac{\partial x}{\partial u} \vec{i} + \dfrac{\partial y}{\partial u} \vec{j} + \dfrac{\partial z}{\partial u} \vec{k}, \\ \dfrac{\partial \vec{r}}{\partial v} = \dfrac{\partial x}{\partial v} \vec{i} + \dfrac{\partial y}{\partial v} \vec{j} + \dfrac{\partial z}{\partial v} \vec{k}, \\ \dfrac{\partial \vec{r}}{\partial w} = \dfrac{\partial x}{\partial w} \vec{i} + \dfrac{\partial y}{\partial w} \vec{j} + \dfrac{\partial z}{\partial w} \vec{k}. \end{cases}$$

Since the triple scalar product of these three vectors is the Jacobian $J(u, v, w)$, which is positive, each of the vectors is nonzero (and each of the three coordinate curves is therefore smooth), and the three vectors form a *positive triad*. If we denote their magnitudes by h_1, h_2, and h_3, respectively:

(6)
$$h_1 \equiv \left| \dfrac{\partial \vec{r}}{\partial u} \right|, \quad h_2 \equiv \left| \dfrac{\partial \vec{r}}{\partial v} \right|, \quad h_3 \equiv \left| \dfrac{\partial \vec{r}}{\partial w} \right|,$$

and their unit forward tangent vectors by $\vec{i}_1$, $\vec{i}_2$, and $\vec{i}_3$, respectively, we have the two (equivalent) sets of formulas:

(7)
$$\begin{cases} \dfrac{\partial \vec{r}}{\partial u} = h_1 \vec{i}_1, \quad \dfrac{\partial \vec{r}}{\partial v} = h_2 \vec{i}_2, \quad \dfrac{\partial \vec{r}}{\partial w} = h_3 \vec{i}_3, \\ \vec{i}_1 = \dfrac{1}{h_1} \dfrac{\partial \vec{r}}{\partial u}, \quad \vec{i}_2 = \dfrac{1}{h_2} \dfrac{\partial \vec{r}}{\partial v}, \quad \vec{i}_3 = \dfrac{1}{h_3} \dfrac{\partial \vec{r}}{\partial w}. \end{cases}$$

The (u, v, w) coordinate system is said to be **orthogonal**, and u, v, and w are called **orthogonal coordinates** if and only if the three coordinate curves intersect orthogonally throughout the given region Ω; that is, if and only if the three vectors $\vec{i}_1$, $\vec{i}_2$, and $\vec{i}_3$ are mutually orthogonal. In this case, from the positive orientation of these three vectors,

(8)
$$\vec{i}_2 \times \vec{i}_3 = \vec{i}_1, \quad \vec{i}_3 \times \vec{i}_1 = \vec{i}_2, \quad \vec{i}_1 \times \vec{i}_2 = \vec{i}_3.$$

Furthermore, by substitution of $h_1 \vec{i}_1$, $h_2 \vec{i}_2$, and $h_3 \vec{i}_3$, from (7), in the triple scalar product of the three vectors of (5), we have (since the triple scalar product of $\vec{i}_1$, $\vec{i}_2$, and $\vec{i}_3$ is unity):

(9)
$$J = J(u, v, w) = \dfrac{\partial(x, y, z)}{\partial(u, v, w)} = h_1 h_2 h_3.$$

Example 1. For cylindrical coordinates, with $u = \rho$, $v = \theta$, $w = z$, $x = \rho \cos \theta$, $y = \rho \sin \theta$, $z = z$, the vectors (5) are

(10)
$$\begin{cases} \cos \theta \, \vec{i} + \sin \theta \, \vec{j}, \\ -\rho \sin \theta \, \vec{i} + \rho \cos \theta \, \vec{j}, \\ \vec{k}. \end{cases}$$

These are mutually orthogonal, with magnitudes $h_1 = 1$, $h_2 = \rho$, $h_3 = 1$. Equation (9) becomes $J = h_1 h_2 h_3 = \rho$. (Cf. (10), § 922.)

Example 2. For spherical coordinates, with $u = r$, $v = \phi$, $w = \theta$, $x = r \sin \phi \cos \theta$, $y = r \sin \phi \sin \theta$, $z = r \cos \phi$, the vectors (5) are

(11) $\quad \begin{cases} \sin \phi \cos \theta \, \vec{i} + \sin \phi \sin \theta \, \vec{j} + \cos \phi \, \vec{k}, \\ r \cos \phi \cos \theta \, \vec{i} + r \cos \phi \sin \theta \, \vec{j} - r \sin \phi \, \vec{k}, \\ -r \sin \phi \sin \theta \, \vec{i} + r \sin \phi \cos \theta \, \vec{j}. \end{cases}$

These are mutually orthogonal, with magnitudes $h_1 = 1$, $h_2 = r$, $h_3 = r \sin \phi$. Equation (9) becomes $J = h_1 h_2 h_3 = r^2 \sin \phi$. (Cf. (11), § 922.)

*1718. VECTOR OPERATIONS IN ORTHOGONAL COORDINATES

Our objective in this section will be the derivation of formulas (2)–(4), below, for the gradient, divergence, and curl in general orthogonal coordinates with positive Jacobian J. The notation of the preceding section is used. The letters f and $\vec{F}$ denote scalar and vector fields, respectively, the latter having components F_1, F_2, F_3 with respect to the unit vectors $\vec{i}_1, \vec{i}_2, \vec{i}_3$:

(1) $\quad \vec{F} = F_1 \vec{i}_1 + F_2 \vec{i}_2 + F_3 \vec{i}_3.$

In the uvw-system the following formulas hold:

(2) $\quad \operatorname{grad} f = \dfrac{1}{h_1} \dfrac{\partial f}{\partial u} \vec{i}_1 + \dfrac{1}{h_2} \dfrac{\partial f}{\partial v} \vec{i}_2 + \dfrac{1}{h_3} \dfrac{\partial f}{\partial w} \vec{i}_3,$

(3) $\quad \operatorname{div} \vec{F} = \dfrac{1}{h_1 h_2 h_3} \left[\dfrac{\partial}{\partial u}(h_2 h_3 F_1) + \dfrac{\partial}{\partial v}(h_3 h_1 F_2) + \dfrac{\partial}{\partial w}(h_1 h_2 F_3) \right],$

(4) $\quad \operatorname{curl} \vec{F} = \dfrac{1}{h_2 h_3} \left[\dfrac{\partial (h_3 F_3)}{\partial v} - \dfrac{\partial (h_2 F_2)}{\partial w} \right] \vec{i}_1$

$\quad + \dfrac{1}{h_3 h_1} \left[\dfrac{\partial (h_1 F_1)}{\partial w} - \dfrac{\partial (h_3 F_3)}{\partial u} \right] \vec{i}_2 + \dfrac{1}{h_1 h_2} \left[\dfrac{\partial (h_2 F_2)}{\partial u} - \dfrac{\partial (h_1 F_1)}{\partial v} \right] \vec{i}_3$

$\quad = \dfrac{1}{h_1 h_2 h_3} \begin{vmatrix} h_1 \vec{i}_1 & h_2 \vec{i}_2 & h_3 \vec{i}_3 \\ \dfrac{\partial}{\partial u} & \dfrac{\partial}{\partial v} & \dfrac{\partial}{\partial w} \\ h_1 F_1 & h_2 F_2 & h_3 F_3 \end{vmatrix}.$

Before embarking on the derivations of the preceding formulas we record the formula for the Laplacian $\nabla^2 f$ of a scalar field f, obtained by combining (2) and (3) (the Laplacian of f being the divergence of its gradient):

(5) $\quad \operatorname{Lapl} f = \dfrac{1}{h_1 h_2 h_3} \left[\dfrac{\partial}{\partial u} \left(\dfrac{h_2 h_3}{h_1} \dfrac{\partial f}{\partial u} \right) + \dfrac{\partial}{\partial v} \left(\dfrac{h_3 h_1}{h_2} \dfrac{\partial f}{\partial v} \right) + \dfrac{\partial}{\partial w} \left(\dfrac{h_1 h_2}{h_3} \dfrac{\partial f}{\partial w} \right) \right].$

In order to prove (2) we observe that

$$\dfrac{\partial f}{\partial u} = \dfrac{\partial f}{\partial x} \dfrac{\partial x}{\partial u} + \dfrac{\partial f}{\partial y} \dfrac{\partial y}{\partial u} + \dfrac{\partial f}{\partial z} \dfrac{\partial z}{\partial u} = (\operatorname{grad} f) \cdot \dfrac{\partial \vec{r}}{\partial u},$$

by (5), § 1717, with similar expressions for $\partial f/\partial v$ and $\partial f/\partial w$. By (7), § 1717, we have the three formulas

$$(\text{grad } f) \cdot \vec{i}_1 = \frac{1}{h_1}\frac{\partial f}{\partial u}, \quad (\text{grad } f) \cdot \vec{i}_2 = \frac{1}{h_2}\frac{\partial f}{\partial v}, \quad (\text{grad } f) \cdot \vec{i}_3 = \frac{1}{h_3}\frac{\partial f}{\partial w},$$

which are equivalent to (2).

As special cases of (2) we have

(6) $$\text{grad } u = \frac{\vec{i}_1}{h_1}, \quad \text{grad } v = \frac{\vec{i}_2}{h_2}, \quad \text{grad } w = \frac{\vec{i}_3}{h_3},$$

and hence

(7) $$\text{grad } f = \frac{\partial f}{\partial u}\text{grad } u + \frac{\partial f}{\partial v}\text{grad } v + \frac{\partial f}{\partial w}\text{grad } w.$$

To prove (3), we use (8), § 1717, and (6), above, to write (1) in the form

(8) $\vec{F} = (h_2 h_3 F_1)(\vec{\nabla} v \times \vec{\nabla} w) + (h_3 h_1 F_2)(\vec{\nabla} w \times \vec{\nabla} u) + (h_1 h_2 F_3)(\vec{\nabla} u \times \vec{\nabla} v).$

The reason behind the use of this rather artificial form for (1) is that the divergence of the cross product of two gradients vanishes, by (vi), § 1713, and (i), § 1714. (Cf. Ex. 26, § 1715.) Therefore, by (v), § 1713:

(9) $\text{div } \vec{F} = \text{grad }(h_2 h_3 F_1) \cdot (\vec{\nabla} v \times \vec{\nabla} w) + \text{grad }(h_3 h_1 F_2) \cdot (\vec{\nabla} w \times \vec{\nabla} u)$
$+ \text{grad }(h_1 h_2 F_3) \cdot (\vec{\nabla} u \times \vec{\nabla} v).$

By (7), above, the first term on the right of (9) is

$$\left[\frac{\partial}{\partial u}(h_2 h_3 F_1)\vec{\nabla} u + \frac{\partial}{\partial v}(h_2 h_3 F_1)\vec{\nabla} v + \frac{\partial}{\partial w}(h_2 h_3 F_1)\vec{\nabla} w\right] \cdot (\vec{\nabla} v \times \vec{\nabla} w)$$

$$= \frac{\partial}{\partial u}(h_2 h_3 F_1)[\vec{\nabla} u, \vec{\nabla} v, \vec{\nabla} w],$$

since any triple scalar product of three vectors two of which are identical vanishes. By (6), $[\vec{\nabla} u, \vec{\nabla} v, \vec{\nabla} w] = 1/h_1 h_2 h_3$, and the first term of (3) is the result. The other two terms are found in a similar fashion.

The derivation of (4) is similar. We use (6) to write (1) in the form

(10) $$\vec{F} = h_1 F_1 \text{grad } u + h_2 F_2 \text{grad } v + h_3 F_3 \text{grad } w.$$

By (vii), § 1713, and (i), § 1714,

$$\text{curl } \vec{F} = \vec{\nabla}(h_1 F_1) \times \vec{\nabla} u + \vec{\nabla}(h_2 F_2) \times \vec{\nabla} v + \vec{\nabla}(h_3 F_3) \times \vec{\nabla} w.$$

Application of (7) three times, together with (6) and the fact that $\vec{i}_1 \times \vec{i}_1 = \vec{i}_2 \times \vec{i}_2 = \vec{i}_3 \times \vec{i}_3 = \vec{0}$, gives

$$\left[\frac{\partial(h_1 F_1)}{\partial v}\frac{\vec{i}_2}{h_2} + \frac{\partial(h_1 F_1)}{\partial w}\frac{\vec{i}_3}{h_3}\right] \times \frac{\vec{i}_1}{h_1} + \left[\frac{\partial(h_2 F_2)}{\partial u}\frac{\vec{i}_1}{h_1} + \frac{\partial(h_2 F_2)}{\partial w}\frac{\vec{i}_3}{h_3}\right] \times \frac{\vec{i}_2}{h_2}$$
$$+ \left[\frac{\partial(h_3 F_3)}{\partial u}\frac{\vec{i}_1}{h_1} + \frac{\partial(h_3 F_3)}{\partial v}\frac{\vec{i}_2}{h_2}\right] \times \frac{\vec{i}_3}{h_3}.$$

The desired formulas (4) now result from (8), § 1717.

NOTE 1. The formulas established in this section provide an alternative proof of the independence of the right-handed rectangular coordinate system, for the gradient, divergence, and curl. For a translation or rotation of axes, $h_1 = h_2 = h_3 = 1$, and equations (2)–(4) reduce to the familiar ones used to define these operations.

NOTE 2. If the Jacobian $J = \partial(x, y, z)/\partial(u, v, w)$ is negative, then $J = -h_1 h_2 h_3$, formulas (2), (3), and (5) are unchanged, and (4) is changed by a factor of -1. (Cf. Ex. 6, § 1719.)

NOTE 3. Formulas (2)–(5) are given explicit formulation for cylindrical and spherical coordinates in the Examples that follow. For corresponding treatment of several other curvilinear coordinate systems, see M. R. Spiegel, *Vector Analysis* (New York, Schaum Publishing Co., 1959), pp. 137–141.

Example 1. In cylindrical coordinates (cf. Example 1, § 1717, Ex. 26, § 924), formulas (2)–(5) become:

$$\operatorname{grad} f = \frac{\partial f}{\partial \rho} \vec{i}_1 + \frac{1}{\rho} \frac{\partial f}{\partial \theta} \vec{i}_2 + \frac{\partial f}{\partial z} \vec{i}_3,$$

$$\operatorname{div} \vec{F} = \frac{1}{\rho} \left[\frac{\partial(\rho F_1)}{\partial \rho} + \frac{\partial F_2}{\partial \theta} + \frac{\partial(\rho F_3)}{\partial z} \right],$$

$$\operatorname{curl} \vec{F} = \frac{1}{\rho} \left[\frac{\partial F_3}{\partial \theta} - \frac{\partial(\rho F_2)}{\partial z} \right] \vec{i}_1 + \left[\frac{\partial F_1}{\partial z} - \frac{\partial F_3}{\partial \rho} \right] \vec{i}_2$$
$$+ \frac{1}{\rho} \left[\frac{\partial(\rho F_2)}{\partial \rho} - \frac{\partial F_1}{\partial \theta} \right] \vec{i}_3,$$

$$\operatorname{Lapl} f = \frac{1}{\rho} \left[\frac{\partial}{\partial \rho} \left(\rho \frac{\partial f}{\partial \rho} \right) + \frac{\partial}{\partial \theta} \left(\frac{1}{\rho} \frac{\partial f}{\partial \theta} \right) + \frac{\partial}{\partial z} \left(\rho \frac{\partial f}{\partial z} \right) \right].$$

Example 2. In spherical coordinates (cf. Example 2, § 1717, Ex. 27, § 924), formulas (2)–(5) become:

$$\operatorname{grad} f = \frac{\partial f}{\partial r} \vec{i}_1 + \frac{1}{r} \frac{\partial f}{\partial \theta} \vec{i}_2 + \frac{1}{r \sin \phi} \frac{\partial f}{\partial \phi} \vec{i}_3,$$

$$\operatorname{div} \vec{F} = \frac{1}{r^2 \sin \phi} \left[\frac{\partial(r^2 \sin \phi \, F_1)}{\partial r} + \frac{\partial(r \sin \phi \, F_2)}{\partial \theta} + \frac{\partial(r F_3)}{\partial \phi} \right],$$

$$\operatorname{curl} \vec{F} = \frac{1}{r^2 \sin \phi} \left[\frac{\partial(r \sin \phi \, F_3)}{\partial \theta} - \frac{\partial(r F_2)}{\partial \phi} \right] \vec{i}_1$$
$$+ \frac{1}{r \sin \phi} \left[\frac{\partial F_1}{\partial \phi} - \frac{\partial(r \sin \phi \, F_3)}{\partial r} \right] \vec{i}_2 + \frac{1}{r} \left[\frac{\partial(r F_2)}{\partial r} - \frac{\partial F_1}{\partial \theta} \right] \vec{i}_3,$$

$$\operatorname{Lapl} f = \frac{1}{r^2 \sin \phi} \left[\frac{\partial}{\partial r} \left(r^2 \sin \phi \frac{\partial f}{\partial r} \right) + \frac{\partial}{\partial \phi} \left(\sin \phi \frac{\partial f}{\partial \phi} \right) + \frac{\partial}{\partial \theta} \left(\frac{1}{\sin \phi} \frac{\partial f}{\partial \theta} \right) \right].$$

*1719. EXERCISES

*1. Show that the triad of vectors (10), § 1717, is right-handed.

*2. Show that the triad of vectors (11), § 1717, is right-handed.

*3. Find the vectors $\vec{i}_1, \vec{i}_2, \vec{i}_3$ of (7), § 1717, for cylindrical coordinates. Then solve for $\vec{i}, \vec{j}, \vec{k}$ in terms of $\vec{i}_1, \vec{i}_2, \vec{i}_3$.

*4. Same as Exercise 3, for spherical coordinates.

*5. Prove the statements of Note 2, § 1716.

*6. Prove the statements of Note 2, § 1718.

*7. Prove that the **radiant** of a scalar field f, defined by the equation
$$\operatorname{rad} f = \vec{\Box} f \equiv \frac{\partial^2 f}{\partial y\, \partial z}\vec{i} + \frac{\partial^2 f}{\partial z\, \partial x}\vec{j} + \frac{\partial^2 f}{\partial x\, \partial y}\vec{k}$$
depends on the right-handed rectangular coordinate system.

*8. Prove that in a system of general curvilinear coordinates, whether orthogonal or not but subject to the condition $\partial(x, y, z)/\partial(u, v, w) \neq 0$, the following two ordered triads of vectors are reciprocal systems (Ex. 23, § 1705):
$$\left(\frac{\partial \vec{r}}{\partial u}, \frac{\partial \vec{r}}{\partial v}, \frac{\partial \vec{r}}{\partial w}\right), \ (\operatorname{grad} u, \operatorname{grad} v, \operatorname{grad} w).$$

18
Line and Surface Integrals

1801. INTRODUCTION

In earlier chapters single and multiple integrals were introduced, and their properties investigated. In § 417 a second kind of single integral, the Riemann-Stieltjes integral, was studied. In the following section (§ 1802) we propose for consideration a third kind of single integral, a *line integral*, which is defined by means of the standard Riemann or definite integral (but which is fundamentally more closely related to the Riemann-Stieltjes integral). By means of Green's Theorem in the plane (§ 1805), these line integrals have a basic and interesting relation to the double integral of Chapter 13. After the notion of *surface area* is explored, the concept of *surface integral*, a new kind of double integral, becomes possible (§ 1816). The divergence theorem (§ 1819) relates these surface integrals to the triple integrals of Chapter 13. With the aid of Stokes's Theorem (§ 1824), line integrals in space are related to both surface integrals and the existence of a "scalar potential" (§ 1825). The concept of "vector potential" is treated briefly in § 1826. Finally, the Exercises of § 1828 present some of the more elementary properties of "exterior differential forms," by means of which an economical synthesis and unification of many of the results of this chapter is possible.

1802. LINE INTEGRALS IN THE PLANE

Definition I. *If C is a smooth arc† in E_2 (self-intersecting or not), with parametrization $x = f(t)$, $y = g(t)$, $a \leq t \leq b$, (cf. § 805), if $H = H(x, y)$ is defined and continuous on C, and if $\gamma(t)$ is a continuously differentiable function on $[a, b]$, then the **line integral** $\int_C H \, d\gamma$ is defined to be the Riemann integral:*

(1) $$\int_C H \, d\gamma \equiv \int_a^b H(f(t), g(t))\gamma'(t) \, dt.$$

† For convenience we shall permit the *endpoints* of the arc C to be either regular or singular points.

Important special cases of (1) are provided by replacing $\gamma(t)$ by the parametrization functions $f(t)$ and $g(t)$, and by the arc length $s(t)$ (cf. Note 1, below). We are thus led to line integrals of the four types:

$$\text{(2)} \quad \int_C P\,dx, \quad \int_C Q\,dy, \quad \int_C P\,dx + Q\,dy, \quad \int_C H\,ds,$$

where P and Q are continuous on C and the third integral of (2) is defined as the sum of the first two.

NOTE 1. In the representation (2), § 805, arc length $s(t)$ has the continuous positive derivative $s'(t) = [(f'(t))^2 + (g'(t))^2]^{\frac{1}{2}}$. *By convention, $s(t)$ is always chosen to be a strictly increasing function of the parameter.* Reversal of the orientation on C (that is, the sense in which C is traversed), then, changes the signs of the first three integrals of (2), but has no effect on the fourth. (Prove this last statement in Ex. 1, § 1804.)

In order to permit consideration of line integrals for arcs with "corners," we introduce the definitions:

Definition II. *A sectionally smooth arc is an arc that is composed of a finite number of smooth pieces attached sequentially end-to-end.*

Definition III. *For a sectionally smooth arc C any line integral (1) or (2) is defined to be the sum of the corresponding line integrals over the pieces that constitute C.*

NOTE 2. If C is a horizontal line segment, $\int_C Q\,dy = 0$, and if C is a vertical line segment, $\int_C P\,dx = 0$. (Give a proof in Ex. 2, § 1804.)

NOTE 3. The integral (1) is equal to the limit

$$\text{(3)} \quad \lim_{|\mathfrak{N}|\to 0} \sum_{i=1}^n H(f(t_i), g(t_i))(\gamma(a_i) - \gamma(a_{i-1})),$$

where $\mathfrak{N}$ is a net: $a = a_0 < a_1 < \cdots < a_n = b$, and $a_{i-1} \leq t_i \leq a_i$, $i = 1, 2, \cdots, n$, since, by Bliss's theorem (Theorem XV, § 401), the integral (1) is the limit of the expression provided from the preceding sum by the law of the mean:

$$\lim_{|\mathfrak{N}|\to 0} \sum_{i=1}^n H(f(t_i), g(t_i))\gamma'(t_i')\,\Delta t_i.$$

A generalization of the definition (1), applicable to rectifiable arcs in general, is given by the Riemann-Stieltjes integral (§ 417), the line integral on the left of (1) being *defined* in this case by a limit of the form (3). In this more general situation the integral on the right of (1) may not exist. (Cf. *RV*, Ex. 1, § 1331.)

NOTE 4. If the arc C is *simple*, the values of the integrals in (2) are independent of the parametrization, provided in the first three integrals the orientation is preserved. (Cf. *RV*, § 1107.) Furthermore, for a sectionally smooth simple arc the line integral (1) or (2) is independent of the decomposition involved in Definition III. (Cf. Ex. 34, § 1804.)

Example 1. Let C be the quarter circle given in polar coordinates by $\rho = 1$, $0 \leq \theta \leq \tfrac{1}{2}\pi$, oriented in the direction of increasing θ. Evaluate

$$I = \int_C xy\, dx + (x^2 + y^2)\, dy$$

in three ways, with parameter θ, x, and y.

Solution. With parameters θ, x, and y, respectively,

$$I = \int_0^{\frac{\pi}{2}} [-\sin^2 \theta \cos \theta + \cos \theta]\, d\theta = \left[-\frac{\sin^3 \theta}{3} + \sin \theta \right]_0^{\frac{\pi}{2}} = \frac{2}{3},$$

$$I = \int_1^0 \left[x\sqrt{1 - x^2} - \frac{x}{\sqrt{1 - x^2}} \right] dx = \frac{2}{3},$$

$$I = \int_0^1 \left[y\sqrt{1 - y^2}\, \frac{-y}{\sqrt{1 - y^2}} + 1 \right] dy = \frac{2}{3}.$$

Example 2. Evaluate $\int_C x\, ds$, where C is the arc of Example 1.

Solution. With parameter θ,

$$\int_C x\, ds = \int_0^{\frac{\pi}{2}} \cos \theta\, d\theta = 1.$$

Example 3. Let C be the perimeter of a square, with sides C_1 from $(1, 0)$ to $(1, 1)$, C_2 from $(1, 1)$ to $(0, 1)$, C_3 from $(0, 1)$ to $(0, 0)$, and C_4 from $(0, 0)$ to $(1, 0)$. Evaluate

$$I = \int_C xy\, dx + (x^2 + y^2)\, dy.$$

Solution. With the aid of Note 4, we have

$$I = \int_0^1 (1 + y^2)\, dy + \int_1^0 x\, dx + \int_1^0 y^2\, dy + \int_0^1 0\, dx = \tfrac{4}{3} - \tfrac{1}{2} - \tfrac{1}{3} + 0 = \tfrac{1}{2}.$$

A basic relation between the last two integrals of (2) for a smooth arc C with unit tangent vector $\vec{\tau} = \dfrac{dx}{ds}\vec{i} + \dfrac{dy}{ds}\vec{j} = \cos \theta\, \vec{i} + \sin \theta\, \vec{j}$ (cf. § 805) is

(4) $\displaystyle \int_C P\, dx + Q\, dy = \int_C \left[P\frac{dx}{ds} + Q\frac{dy}{ds} \right] ds$

$$= \int_C [P \cos \theta + Q \sin \theta]\, ds = \int_C \vec{F} \cdot \vec{\tau}\, ds,$$

where $\vec{F} = P\vec{i} + Q\vec{j}$. If we denote by $\overrightarrow{ds}$ the symbolic vector differential expression:

(5) $\overrightarrow{ds} \equiv \vec{\tau}\, ds = dx\, \vec{i} + dy\, \vec{j},$

the last integral above can be compressed to $\displaystyle \int_C \vec{F} \cdot \overrightarrow{ds}$.

1803. INDEPENDENCE OF PATH AND EXACT DIFFERENTIALS

Let P and Q be continuous in a region R of E_2. For such functions we introduce two concepts, and establish an important relation between them.

Definition I. *The statement* $\int_C P\,dx + Q\,dy$ *is **independent of the path** in R means that for any two points of R and any two sectionally smooth paths (arcs) C_1 and C_2 joining these two points and lying in R the values of the two line integrals are equal:*

$$(1) \qquad \int_{C_1} P\,dx + Q\,dy = \int_{C_2} P\,dx + Q\,dy.$$

The following theorem is an almost immediate consequence of Definition I:

Theorem I. *The integral* $\int_C P\,dx + Q\,dy$ *is independent of the path in R if and only if for every sectionally smooth closed curve C lying in R:*

$$(2) \qquad \int_C P\,dx + Q\,dy = 0.$$

Proof. We first assume (2), and let C_1 and C_2 be any two sectionally smooth arcs with common endpoints and lying in R. Letting C be the closed curve constructed from C_1 and $-C_2$ (with the orientation of C_2 reversed) we have

$$\int_C P\,dx + Q\,dy = \int_{C_1} P\,dx + Q\,dy - \int_{C_2} P\,dx + Q\,dy = 0,$$

so that $\int_{C_1} P\,dx + Q\,dy = \int_{C_2} P\,dx + Q\,dy$.

On the other hand, if $\int_C P\,dx + Q\,dy$ is independent of the path in R and if C is any sectionally smooth closed curve lying in R, we have only to choose an arbitrary pair of points of C to decompose C into two parts that provide two sectionally smooth arcs by means of which the argument of the preceding paragraph can be reversed.

NOTE 1. The equivalence between independence of path and the vanishing of the line integral (2) remains valid (with a proof identical to that of Theorem I) in case all arcs and curves under consideration are polygonal in nature (broken line segments and closed polygons).

Definition II. *The expression*

$$P\,dx + Q\,dy$$

*is called an **exact differential** in R if and only if there exists a differentiable function $\phi = \phi(x, y)$ defined in R such that*

$$(3) \qquad \frac{\partial \phi}{\partial x} = P, \quad \frac{\partial \phi}{\partial y} = Q$$

throughout R. A scalar function ϕ satisfying the two preceding equations (3) is called a **scalar potential** for the vector function $P\vec{i} + Q\vec{j}$ that is equal to the **gradient of** ϕ:

(4) $$P\vec{i} + Q\vec{j} = \operatorname{grad} \phi \equiv \frac{\partial \phi}{\partial x} \vec{i} + \frac{\partial \phi}{\partial y} \vec{j},$$

and also for the exact differential $d\phi = P\,dx + Q\,dy$.

NOTE 2. Theorem II, § 912, implies that whenever a scalar potential exists for a region, it is unique except for an additive constant.

For exact differentials $P\,dx + Q\,dy$, line integrals are easily evaluated by means of the theorem:

Theorem II. *If $P\,dx + Q\,dy$ is an exact differential, with scalar potential ϕ, in a region R, and if C is a sectionally smooth arc in R joining the point (x_0, y_0) to the point (x_1, y_1), then*

(5) $$\int_C d\phi = \int_C P\,dx + Q\,dy = \phi(x_1, y_1) - \phi(x_0, y_0).$$

Proof. We need prove this theorem only for the case when C is smooth, since the more general case of a sectionally smooth arc can be obtained from the special case by adding finitely many expressions of the form (5). Accordingly, we evaluate with the help of the chain rule (cf. § 907):

$$\int_a^b P(f(t), g(t))f'(t)\,dt + \int_a^b Q(f(t), g(t))g'(t)\,dt$$
$$= \int_a^b \left[\frac{\partial \phi}{\partial x}(f(t), g(t))f'(t) + \frac{\partial \phi}{\partial y}(f(t), g(t))g'(t) \right] dt$$
$$= \int_a^b \frac{d}{dt} \phi(f(t), g(t))\,dt = \phi(f(t), g(t)) \Big]_a^b = \phi(x_1, y_1) - \phi(x_0, y_0).$$

Example 1. Evaluate

$$I = \int_C 2xy\,dx + (x^2 + y^2)\,dy,$$

where C is the first-quadrant portion of the ellipse $9x^2 + 4y^2 = 36$, extending from $(2, 0)$ to $(0, 3)$.

Solution. The expression $P\,dx + Q\,dy$ is exact, being the differential of the function $\phi = x^2 y + \tfrac{1}{3} y^3$. Therefore, by (5), $I = \phi(0, 3) - \phi(2, 0) = 9$.

We are now ready to establish the following relation between the two new concepts introduced in this section.

Theorem III. *Under the assumption that P and Q are continuous in the region R, the line integral $\int_C P\,dx + Q\,dy$ is independent of the path in R if and only if $P\,dx + Q\,dy$ is an exact differential in R.*

§ 1803] INDEPENDENCE OF PATH

Proof. If $P\,dx + Q\,dy$ is an exact differential, the form of (5) guarantees independence of path.

We now assume that the line integral $\int_C P\,dx + Q\,dy$ is independent of the path, and construct a scalar potential ϕ, as follows: Letting (x_0, y_0) be any point of R, to be held fixed, we define ϕ by the formula

(6) $$\phi(x, y) \equiv \int_C P\,dx + Q\,dy,$$

where C is a sectionally smooth arc in R from (x_0, y_0) to (x, y). The independence of path ensures that ϕ is a single-valued function. By the Theorem of § 905, since P and Q are assumed to be continuous it is sufficient to establish the two formulas

(7) $$\frac{\partial \phi}{\partial x} = P, \quad \frac{\partial \phi}{\partial y} = Q,$$

for the function (6). We give the details for the first of these two formulas and leave the second to the student in Exercise 3, § 1804.

If (x_1, y_1) is an arbitrary point of R, and if Δx is numerically less than the radius of a circular neighborhood of (x_1, y_1) lying in R, we wish to prove:

(8) $$\lim_{\Delta x \to 0} \frac{\phi(x_1 + \Delta x, y_1) - \phi(x_1, y_1)}{\Delta x} = P(x_1, y_1)$$

The numerator on the left is the difference between two line integrals the second of which is evaluated over an arc C in R from (x_0, y_0) to (x_1, y_1). Since the first of these two line integrals, extending from (x_0, y_0) to $(x_1 + \Delta x, y_1)$ is independent of the path, by assumption, this path can be chosen to be made up of two parts, C and a horizontal line segment Γ from (x_1, y_1) to $(x_1 + \Delta x, y_1)$. When this is done, the numerator in question simplifies, since the line integrals over C cancel out:

(9) $$\phi(x_1 + \Delta x, y_1) - \phi(x_1, y_1) = \int_\Gamma P\,dx + Q\,dy.$$

By Note 2, § 1802, $\int_\Gamma Q\,dy = 0$, and the line integral on the right of (9) simplifies to:

(10) $$\int_{x_1}^{x_1 + \Delta x} P(x, y_1)\,dx,$$

where we have chosen the parameter to be x itself, for greatest simplicity. By the First Mean Value Theorem for Integrals (Theorem XI, § 401), (10) is equal to $P(\xi, y_1)\Delta x$, where ξ is between x_1 and $x_1 + \Delta x$. Division by Δx gives

(11) $$\frac{\phi(x_1 + \Delta x, y_1) - \phi(x_1, y_1)}{\Delta x} = P(\xi, y_1),$$

and by continuity of P, (7) follows, and the proof is complete.

NOTE 3. The equivalence between independence of path of $\int_C P\,dx + Q\,dy$ and exactness of $P\,dx + Q\,dy$ remains valid (with a proof identical to that of Theorem III) in case all arcs under consideration are polygonal in nature (broken line segments).

We conclude with a necessary condition for $P\,dx + Q\,dy$ to be exact, under an additional assumption regarding the derivatives of P and Q:

Theorem IV. *If P and Q and their partial derivatives $\dfrac{\partial P}{\partial y}$ and $\dfrac{\partial Q}{\partial x}$ are continuous in a region R and if the differential expression $P\,dx + Q\,dy$ is exact in R, then*

(12) $$\frac{\partial P}{\partial y} = \frac{\partial Q}{\partial x}.$$

Proof. If ϕ is a scalar potential for $P\,dx + Q\,dy$, then since $\partial \phi/\partial x = P$ and $\partial \phi/\partial y = Q$, equation (12) is equivalent to the equality of the two mixed partial derivatives of ϕ:

$$\frac{\partial^2 \phi}{\partial y\, \partial x} = \frac{\partial^2 \phi}{\partial x\, \partial y},$$

which is true by § 903.

NOTE 4. The vector notation of (4) and (5), § 1802, permits compression of some of the statements and formulas of this section. We adapt from three dimensions (cf. Example 1, § 1714) the concepts of a **conservative field** $\vec{F}$ (the gradient of a scalar field: $\vec{F} = \operatorname{grad} \phi$) and an **irrotational field** $\vec{F}$ (one whose curl vanishes: if $\vec{F} = P\vec{i} + Q\vec{j}$, where P and Q are functions of x and y alone, $\operatorname{curl} \vec{F} = \left(\dfrac{\partial Q}{\partial x} - \dfrac{\partial P}{\partial y}\right)\vec{k} = \vec{0}$). We have, then, for portions of Theorems II–IV:

Theorem II: $\int_C \operatorname{grad} \phi \cdot \vec{ds} = \phi(x_1, y_1) - \phi(x_0, y_0).$

Theorem III: $\int_C \vec{F} \cdot \vec{ds}$ *is independent of the path in R if and only if $\vec{F}$ is conservative in R. In particular, if C is closed,* $\int_C \operatorname{grad} \phi \cdot \vec{ds} = 0.$

Theorem IV: *A conservative vector field is irrotational.*

Example 2. Show that the differential expression
$$x^2\,dx + 2xy\,dy$$
is not exact in any region R of E_2.

Solution. In this case $\dfrac{\partial P}{\partial y} = 0$ and $\dfrac{\partial Q}{\partial x} = 2y$, and these two partial derivatives are equal only on the x-axis.

Example 3. Show that the differential expression
$$y^2\,dx + 2xy\,dy$$
is exact in E_2.

Solution. Since $\dfrac{\partial P}{\partial y} = \dfrac{\partial Q}{\partial x} = 2y$, it is at least *possible* for $y^2\,dx + 2xy\,dy$ to be exact. To show that it *is* exact, we seek $\phi(x, y)$ by integrating $P = y^2$ with respect to x to obtain as a first tentative form for ϕ:

(12) $$\phi(x, y) = xy^2 + \chi(y),$$

where the function $\chi(y)$, to be determined, corresponds to a constant of integration. We could also integrate Q with respect to y and compare the result with (12). Another method is to differentiate (12) with respect to y and equate to Q: $2xy + \chi'(y) = 2xy$. We therefore have a scalar potential $\phi = xy^2$, by setting $\chi(y) = 0$ (or more generally, $\phi = xy^2 + k$, where $\chi(y) = k$, a constant).

Example 4. Show that the differential expression

(13) $$-\frac{y}{x^2 + y^2}\,dx + \frac{x}{x^2 + y^2}\,dy,$$

in spite of the fact that equation (11) is satisfied is *not* exact in the "punctured plane" R, defined to be the plane E_2 with the origin deleted.

Solution. We first verify by differentiation:

$$\frac{\partial P}{\partial y} = \frac{\partial Q}{\partial x} = \frac{y^2 - x^2}{(x^2 + y^2)^2}.$$

On the other hand, if C is the circumference of the circle given parametrically by

$$x = a\cos\theta;\quad y = a\sin\theta,\quad 0 \le \theta \le 2\pi,\quad a > 0,$$

then, with θ as the parameter:

$$\int_C P\,dx + Q\,dy = \frac{a^2}{a^2}\int_0^{2\pi}[(-\sin\theta)(-\sin\theta) + \cos^2\theta]\,d\theta = 2\pi \ne 0.$$

By Theorems I and III, $\displaystyle\int_C P\,dx + Q\,dy$ is not independent of the path in R, and therefore $P\,dx + Q\,dy$ is not exact in R. No scalar potential for $P\,dx + Q\,dy$ can exist for R.

1804. EXERCISES

1. Prove the statement of the last sentence of Note 1, § 1802.
2. Prove Note 2, § 1802.
3. Show that the function ϕ defined by (5), § 1803, has the property $\partial\phi/\partial y = Q$ of (6), § 1803.

In Exercises 4–11, find the value of the given line integral.

4. $\displaystyle\int_C (x + 2y)\,dx$ along $y = x^2$ from $(0, 0)$ to $(1, 1)$.

5. $\displaystyle\int_C (x + 2y)\,dy$ along $y = x^2$ from $(0, 0)$ to $(1, 1)$.

6. $\displaystyle\int_C (x^3 + y)\,dx$ along $y = 4 - x^2$ from $(2, 0)$ to $(0, 4)$.

7. $\displaystyle\int_C (x^2 + y)\,dy$ along $x = t^3$, $y = t^2$ from $(0, 0)$ to $(-8, 4)$.

8. $\int_C (e^y + y) \, dx$ along $2x + 3y = 6$ from $(0, 2)$ to $(3, 0)$.

9. $\int_C x \, dy$ along $x = \dfrac{6t}{t^3 + 1}$, $y = \dfrac{6t^2}{t^3 + 1}$ from $(0, 0)$ to $(3, 3)$.

10. $\int_C x \, ds$ along $y = x^3$, from $(0, 0)$ to $(2, 8)$.

11. $\int_C y \, ds$ along $x = a \cos^3 t$, $y = a \sin^3 t$ from $(a, 0)$ to $(0, a)$.

In Exercises 12–19, evaluate the line integral for the simple closed curve C in the counterclockwise sense, where C is the frontier of the prescribed region R in E_2.

12. $\int_C (2x + 3y) \, dy$; $x > 0$, $y > 0$, $2x + 3y < 6$.

13. $\int_C (x^2 + y) \, dx$; $0 < y < 4 - x^2$.

14. $\int_C x^2 y \, dx + (2x + 1) y^2 \, dy$; $|x| < 1$, $|y| < 1$.

15. $\int_C xy \, dx + x^2 \, dy$; $0 < y < x^{\frac{3}{2}}$, $0 < x < 1$.

16. $\int_C y^2 \, dx + x \, dy$; R is the inside of the hypocycloid $x = a \cos^3 t$, $y = a \sin^3 t$, $a > 0$.

17. $\int_C y^n \, dx + x^n \, dy$, $n = 0, 1, 2, 3, \cdots$; $x^2 + y^2 < a^2$, $a > 0$.

18. $\int_C (x^2 - y^2) \, dx + 2xy \, dy$; $x > 0$, $y > 0$, $x^2 + y^2 < a^2$, $a > 0$.

19. $\int_C y|y| \, dx - x|x| \, dy$; $|x| + |y| < 1$.

In Exercises 20–23, show that the given differential expression is not exact in any region of E_2.

20. $x \, dx + x \, dy$.
21. $(x^2 - y^2) \, dx + 2xy \, dy$.
22. $(x^3 - 3xy^2) \, dx + (3x^2 y - y^3) \, dy$.
23. $e^x \cos y \, dx + e^x \sin y \, dy$.

In Exercises 24–27, show that the given differential expression is exact in E_2 by finding a scalar potential there.

24. $y \, dx + x \, dy$.
25. $e^x \sin y \, dx + e^x \cos y \, dy$.
26. $(3x^2 y - y^3) \, dx + (x^3 - 3xy^2 + y) \, dy$.
27. $(\cos x \cosh y - \sin x \sinh y) \, dx + (\sin x \sinh y + \cos x \cosh y) \, dy$.
28. If R is E_2 with $(0, 0)$ deleted, show that the differential expression

$$\frac{2xy}{(x^2 + y^2)^2} \, dx + \frac{y^2 - x^2}{(x^2 + y^2)^2} \, dy$$

is exact in R, by finding a scalar potential there.

In Exercises 29–33, evaluate the line integral $\int_C P\,dx + Q\,dy$, where $P\,dx + Q\,dy$ is the differential of the indicated exercise, and C is a sectionally smooth arc extending from the point $(1, 0)$ to the point $(0, 1)$, and lying in the region under consideration.

29. Ex. 24. **30.** Ex. 25. **31.** Ex. 26. **32.** Ex. 27. **33.** Ex. 28.
★34. Prove the statements of Note 4, § 1802.

1805. GREEN'S THEOREM IN THE PLANE

The subject for discussion in this section is the relationship between a certain line integral and a certain double integral, where the line integral is formed over closed curves that constitute the complete frontier of the closed

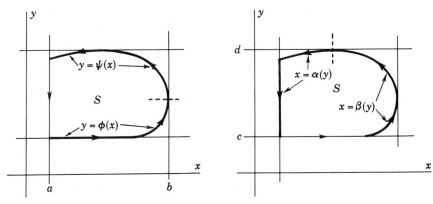

FIG. 1801

region of double integration. For this purpose it will be necessary to restrict the curves and regions to be considered, and to explain carefully their interrelationships. For a discussion of Green's theorem under much more general hypotheses, and a further reference, see T. M. Apostol, *Mathematical Analysis* (Reading, Mass., Addison-Wesley Publishing Company, 1957), pages 289–292.

We start by defining an **elementary compact region** to be a compact region S described simultaneously by inequalities $\phi(x) \leq y \leq \psi(x)$, $a \leq x \leq b$ and inequalities $\alpha(y) \leq x \leq \beta(y)$, $c \leq y \leq d$, where $\phi(x)$ and $\psi(x)$ are continuous and sectionally smooth for $a \leq x \leq b$ and $\alpha(y)$ and $\beta(y)$ are continuous and sectionally smooth for $c \leq y \leq d$ (cf. Fig. 1801). By the nature of S the functions ϕ, ψ, α, and β are made up of monotonic parts, and the entire frontier of S is a sectionally smooth simple closed curve C. This curve C will now be considered to be **oriented** as shown in Figure 1801: to the right on the lower curve $y = \phi$, to the left on the upper curve $y = \psi$, upward on the right-hand curve $x = \beta$, and downward on the left-hand curve $x = \alpha$. (Show in Ex. 1, § 1809, that this definition is self-consistent and free from self-contradictions.) The frontier curve C with the orientation just assigned is called the **boundary** of the compact region S. Intuitively it is convenient to think of the orientation of C as being such that when we traverse C in the direction specified, *we keep the compact region S to the left.* In this direction C is being traversed in the **positive** sense; in the reverse direction C is being traversed in the **negative** sense. If no statement to the contrary is made, the positive sense or orientation will be assumed.

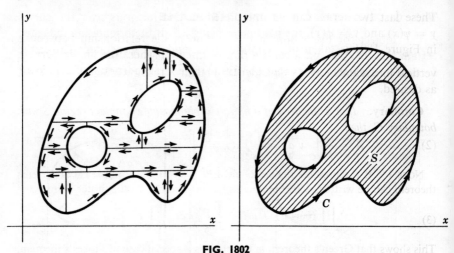

FIG. 1802

Finally, we define a **finitely decomposable** compact region S to be a compact region that is a finite union of elementary compact regions any two of which have at most frontier points in common and such that whenever two of these elementary compact regions have an arc of their boundaries in common, this arc has opposite orientations for the two elementary compact regions. The **boundary** of S consists of those oriented arcs of the boundaries of the elementary compact regions that belong to exactly *one* of these elementary compact regions (except possibly for endpoints). (Cf. Fig. 1802.) Intuitively, we think of all internal lines of subdivision as "cancelling out," and the boundary C (possibly consisting of several parts) being traced out with S always to the left.

Theorem. Green's Theorem in the Plane. *Let S be a finitely decomposable compact region in E_2, with boundary C, and let P, Q, $\partial P/\partial y$, and $\partial Q/\partial x$ be defined and continuous in an open set containing S.* Then

$$(1) \qquad \int_C P\,dx + Q\,dy = \iint_S \left(\frac{\partial Q}{\partial x} - \frac{\partial P}{\partial y}\right) dA.$$

Proof. Because of the additivity properties of line and double integrals, it can be assumed that S is itself an elementary compact region. (Justify this assumption in Ex. 2, § 1809.) We give the details of the proof of (1) for the terms involving P (those for Q being entirely similar—cf. Ex. 3, § 1809). We use the notation of Figure 1801, and use iterated integrals (§ 1008):

$$\iint_S \frac{\partial P}{\partial y} dA = \int_a^b \int_{\phi(x)}^{\psi(x)} \frac{\partial P}{\partial y} dy\, dx = \int_a^b \left[P \right]_{y=\phi(x)}^{y=\psi(x)} dx$$

$$= \int_a^b P(x, \psi(x))\, dx - \int_a^b P(x, \phi(x))\, dx.$$

These last two terms can be interpreted as line integrals over the curves $y = \psi(x)$ and $y = \phi(x)$, the parameter being x. If the orientation indicated in Figure 1801 is taken into consideration, and the zero integral over the vertical line segment at $x = a$ added, the result is the line integral $-\int_C P\, dx$, as desired.

Corollary. *If A is the area of a finitely decomposable compact region S with boundary C, then*

(2) $\qquad A = -\int_C y\, dx = \int_C x\, dy = \frac{1}{2}\int_C (-y\, dx + x\, dy).$

NOTE 1. If $\vec{F}$ represents the vector field $P\vec{i} + Q\vec{j} + 0\vec{k}$, formula (1) of Green's theorem can be written, with the notation of (4) and (5), § 1802, in the form

(3) $\qquad \iint_S (\operatorname{curl} \vec{F}) \cdot \vec{k}\, dA = \int_C \vec{F} \cdot \vec{\tau}\, ds = \int_C \vec{F} \cdot d\vec{s}.$

This shows that Green's theorem in the plane is a special case of Stokes's theorem, presented in § 1824.

NOTE 2. Let $\vec{n}$ be the **unit outer normal vector** at any point of the directed boundary C of a finitely decomposable compact region S, defined as the unit vector whose amplitude α is $\frac{1}{2}\pi$ less than the amplitude θ of the positively directed unit tangent $\vec{\tau}$: $\alpha = \theta - \frac{1}{2}\pi$. If $\vec{F} = f\vec{i} + g\vec{j} + 0\vec{k}$ represents the vector field $Q\vec{i} - P\vec{j}$, then the left-hand member of (1) becomes (cf. (4), § 1802):

$$\int_C P\, dx + Q\, dy = \int_C (f \sin\theta - g \cos\theta)\, ds = \int_C (f\cos\alpha + g\sin\alpha)\, ds = \int_C \vec{F} \cdot \vec{n}\, ds.$$

Therefore, since the right-hand member of (1) is $\iint_S \left(\frac{\partial f}{\partial x} + \frac{\partial g}{\partial y} \right) dA$, the formula (1) of Green's theorem can be written in the form

(4) $\qquad \iint_S \operatorname{div} \vec{F}\, dA = \int_C \vec{F} \cdot \vec{n}\, ds.$

This shows that Green's theorem in the plane is a two-dimensional analogue of the divergence theorem in space, presented in § 1819.

If ϕ is a scalar field that is continuously differentiable in an open set containing a finitely decomposable compact set S with (oriented) boundary C, then the **normal derivative** of ϕ with respect to the curve C, at any point of C, is defined:

(5) $\qquad \dfrac{d\phi}{dn} \equiv (\operatorname{grad} \phi) \cdot \vec{n}.$

Note 3. The concepts of *elementary compact region* and *finitely decomposable compact region* are not independent of the rectangular coordinate system. Indeed, a simple rotation through 45° may convert an elementary compact region into a compact region that is not finitely decomposable. (For an example, cf. Ex. 39, § 1809.)

Example 1. Verify Green's theorem for the line integral and compact region of Example 3, § 1802:

$$\int_C xy\, dx + (x^2 + y^2)\, dy; \quad 0 \leq x \leq 1, \quad 0 \leq y \leq 1.$$

Solution. The right-hand member of equation (1) is

$$\int_0^1 \int_0^1 (2x - x)\, dy\, dx = \frac{1}{2},$$

in agreement with the evaluation of Example 3, § 1802.

Example 2. Compute the area of the compact region $0 \leq y \leq 4 - x^2$ by means of the Corollary to Green's theorem.

Solution. The three quantities of (2) are respectively:

$$\int_C -y\, dx = -\int_2^{-2} (4 - x^2)\, dx = \frac{32}{3},$$

$$\int_C x\, dy = \int_0^4 \sqrt{4 - y}\, dy - \int_4^0 \sqrt{4 - y}\, dy = \frac{32}{3},$$

$$\frac{1}{2}\int_C (-y\, dx + x\, dy) = \frac{1}{2}\left(\frac{32}{3} + \frac{32}{3}\right) = \frac{32}{3}.$$

1806. LOCAL EXACTNESS

It was established in Theorem IV, § 1803, that the partial differential equation $\partial P/\partial y = \partial Q/\partial x$ must hold throughout any region in which the differential expression $P\, dx + Q\, dy$ is exact and the partial derivatives involved are continuous. We can now use Green's theorem to give a necessary and sufficient condition for the relation $\partial P/\partial y = \partial Q/\partial x$ to hold. To this end we introduce a definition:

Definition. *The expression $P\, dx + Q\, dy$ is **locally exact** in a region R of E_2 if and only if it is exact in some neighborhood of every point of R.*

Theorem. *If P and Q and their partial derivatives $\dfrac{\partial P}{\partial y}$ and $\dfrac{\partial Q}{\partial x}$ are continuous in a region R, then $P\, dx + Q\, dy$ is locally exact in R if and only if at every point of R*

$$(1) \qquad \frac{\partial P}{\partial y} = \frac{\partial Q}{\partial x}.$$

Proof. If $P\, dx + Q\, dy$ is exact in some neighborhood of some point of R then, by Theorem IV, § 1803, the equation (1) must hold at that point.

If $P\,dx + Q\,dy$ is exact in *some* neighborhood of *every* point of R, (1) must hold throughout R.

Assuming that equation (1) holds throughout R and that (x_0, y_0) is an arbitrary point of R, we let $N = N_{(x_0, y_0)}$ be a neighborhood of (x_0, y_0) that lies in R. If (x, y) is any point of N, where $x \neq x_0$ and $y \neq y_0$, let C_1 and C_2 be the two simple paths from (x_0, y_0) to (x, y) each of which is made up of two sides of the closed rectangle S that has sides parallel to the coordinate axes and (x_0, y_0) and (x, y) as opposite vertices. If C_3 denotes the boundary of S, then by Green's theorem:

$$\int_C P\,dx + Q\,dy = \iint_S \left(\frac{\partial Q}{\partial x} - \frac{\partial P}{\partial y}\right) dA = 0,$$

and $\int_{C_1} P\,dx + Q\,dy = \int_{C_2} P\,dx + Q\,dy$. Therefore, if $\phi(x, y)$ is defined in N:

$$\phi(x, y) \equiv \int_C P\,dx + Q\,dy,$$

where C is made up of *at most one* directed line segment parallel to each coordinate axis, the function ϕ is uniquely defined. The details of the proof that the relations

(2) $$\frac{\partial \phi}{\partial x} = P, \quad \frac{\partial \phi}{\partial y} = Q$$

must hold at an arbitrary point (x_1, y_1) of N are almost identical with those of the last part of the proof of Theorem III, § 1803. In particular, for the equation $\partial \phi/\partial x = P$, the result is obtained by insisting that any integration along a vertical line segment precede any integration along a horizontal line segment. (Give the remaining details of the proof of both equations (2) in Ex. 4, § 1809.)

NOTE 1. In the language of Note 4, § 1803, the theorem of this section can be stated simply: *A vector field is locally conservative if and only if it is irrotational.*

NOTE 2. A property like *exactness*, which is defined for an entire region at once, is known as a property **in the large**. A property like local exactness, which is defined in terms of individual points and their neighborhoods, is known as a **local** property, or a property **in the small**. Independence of path is another property in the large. Any property defined by a differential equation (like (1)) is a local property.

NOTE 3. Let C_1 and C_2 be two sectionally smooth simple closed curves such that $-C_1$ (the orientation of C_1 being reversed) and C_2 constitute the total boundary of a finitely decomposable compact region S (cf. Fig. 1803). Assume, furthermore, that P, Q, $\dfrac{\partial P}{\partial y}$, and $\dfrac{\partial Q}{\partial x}$ are continuous in an open set containing S and that equation (1) holds throughout S. Then, by Green's theorem,

$$\int_{C_2} P\,dx + Q\,dy - \int_{C_1} P\,dx + Q\,dy = \iint \left(\frac{\partial Q}{\partial x} - \frac{\partial P}{\partial y}\right) dA = 0,$$

and

(3) $$\int_{C_1} P\,dx + Q\,dy = \int_{C_2} P\,dx + Q\,dy.$$

Equation (3) is useful in the evaluation of line integrals of locally exact differentials.

Example. If C is the ellipse $17x^2 + 31y^2 = 41$, directed counterclockwise, then
$$\int_C -\frac{y}{x^2+y^2}\,dx + \frac{x}{x^2+y^2}\,dy = 2\pi,$$
since this line integral must, by Note 3, be equal to a corresponding one for a small circle about the origin (cf. Example 4, § 1803).

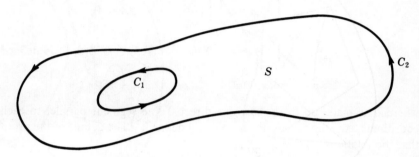

FIG. 1803

1807. SIMPLY- AND MULTIPLY-CONNECTED REGIONS

The region R of Figure 1802 contains two "holes" or "islands." Precisely what is meant by a "hole" or an "island" we shall not attempt to say, but we *shall* explain exactly what is meant by the *absence* of these objects. The technical term for this absence of holes or islands is *simple-connectedness*. The simplest definition of a simply-connected region is one that contains the inside of every Jordan curve (Definition, § 1027) that lies in the region.† We have chosen a more complicated definition for two reasons: (*i*) the definition given below leads directly to the desired results of the following section; (*ii*) the definition given below extends in a simple and natural way to higher-dimensional spaces. This definition puts into precise terms what one would mean by "shrinking a curve to a point without leaving the region."

Definition I. *A region R of E_2 is **simply-connected** if and only if for every closed polygon C: $[p_1p_2\cdots p_np_1]$ lying in R there exists a point p in R such that for every vertex p_i of C, $i = 1, 2, \cdots, n$, there exists a simple smooth arc C_i from p to p_i lying in R and having the property that for every $i = 1, 2, \cdots, n$ the three arcs C_i, $\overline{p_ip_{i+1}}$, and $-C_{i+1}$ (where $C_{n+1} \equiv C_1$, $p_{n+1} \equiv p_1$, and the negative sign indicates a reversal of orientation) form the boundary Γ_i (with*

---

† For discussion of this definition, see T. M. Apostol, *Mathematical Analysis* (Reading, Mass., Addison-Wesley Publishing Company, 1957) and M. H. A. Newman, *Elements of the Topology of Plane Sets of Points* (New York, Cambridge University Press, 1954).

§ 1807] SIMPLY- AND MULTIPLY-CONNECTED REGIONS 599

*positive or negative orientation) of a finitely decomposable compact region S_i lying in R. A region is **multiply-connected** if and only if it fails to be simply-connected.* (Cf. Fig. 1804.)

Intuitively, one can usually think of "shrinking" the polygonal curve C to the point p by permitting each vertex p_i to approach p along the arc C_i, at an appropriately adjusted rate, so that no part of the "variable curve C" leaves

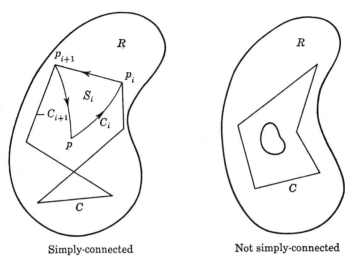

Simply-connected   Not simply-connected

FIG. 1804

the simply-connected region R. In the presence of an "island" it is possible to surround such an object by a polygonal curve C that cannot be shrunk to a point without passing across the island, and hence outside R. If the shrinking process indicated above can always be done linearly, we have the concept of a *star-shaped* region (a special kind of simply-connected region), and its simplification, a *convex region* (cf. Ex. 33, § 618):

Definition II. *A region R in E_2 is **star-shaped** if and only if there exists a point p in R such that for every q in R the segment $\overline{pq}$ lies in R.* (Cf. Fig. 1805.) *A region R in E_2 is **convex** if and only if for any two points p and q of R the segment $\overline{pq}$ lies in R.*

NOTE 1. Any convex region is star-shaped, and any star-shaped region is simply-connected (cf. Ex. 5, § 1809).

Examples. Some examples of convex regions in E_2 are (*i*) the unit disk $x^2 + y^2 < 1$, (*ii*) the inside of any square, rectangle, or more generally, any "convex" polygon, (*iii*) the open first quadrant $x > 0$, $y > 0$, (*iv*) an open half-plane, such as $x > 0$. The union of the two half-planes $x > 0$ and $y > 0$ is star-shaped but not convex. So is the plane with either the nonnegative or the nonpositive part of the x-axis (or y-axis) deleted. If in this last example the closed unit circle $x^2 + y^2 \leq 1$ is also deleted the remaining region is simply-connected but not star-shaped. The punctured plane R obtained by deleting from E_2 the single point $(0, 0)$ is multiply-connected; although this is intuitively *plausible* it is *not obvious*, and a *proof* is provided in the Example, § 1808. Similarly, the region between two concentric circles is multiply connected (cf. Ex. 6, § 1809).

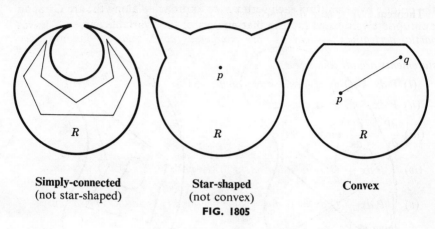

Simply-connected **Star-shaped** **Convex**
(not star-shaped) (not convex)
FIG. 1805

NOTE 2. Any region obtained by "moderately distorting" a simply-connected region is again simply-connected. For example, a region shaped like a starfish can be distorted to one shaped like an octopus (with fewer tentacles than normal, of course), which is then also simply-connected.

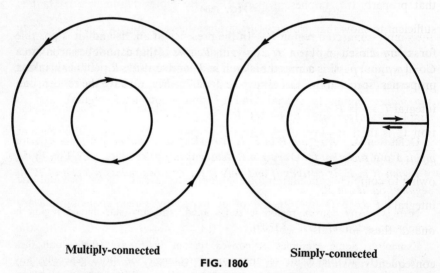

Multiply-connected **Simply-connected**
FIG. 1806

NOTE 3. It is often possible and convenient to replace a multiply-connected region by a simply-connected region by means of deleting the points of arcs joining disjoint simple closed curves of the boundary (such arcs are called cuts, or cross-cuts). This is indicated in Figure 1806 for the region between two concentric circles.

1808. EQUIVALENCES IN SIMPLY-CONNECTED REGIONS

The principal concepts introduced in the preceding sections of this chapter all coalesce in the presence of simple-connectedness, according to the theorem:

Theorem. *If R is a simply-connected region of E_2, and if P, Q, and their partial derivatives $\dfrac{\partial P}{\partial y}$ and $\dfrac{\partial Q}{\partial x}$ are continuous in R, then the following five statements are all equivalent:*

(i) $P\,dx + Q\,dy$ *is locally exact in* R;

(ii) $P\,dx + Q\,dy$ *is exact in* R;

(iii) $\dfrac{\partial P}{\partial y} = \dfrac{\partial Q}{\partial x}$ *throughout* R;

(iv) $\displaystyle\int_C P\,dx + Q\,dy$ *is independent of the path in* R;

(v) $\displaystyle\int_C P\,dx + Q\,dy = 0$ *for every sectionally smooth closed curve C lying in* R.

Proof. By Theorems I and III, § 1803, properties *(ii)*, *(iv)*, and *(v)* are equivalent; by the Theorem of § 1806, properties *(i)* and *(iii)* are equivalent; and by definition, property *(ii)* implies property *(i)*. It remains only to show that property *(iii)* implies property *(ii)*. By Notes 1 and 3, § 1803, it is sufficient to show that property *(iii)* implies the vanishing of $\displaystyle\int_C P\,dx + Q\,dy$ for every closed polygon C lying in R. Accordingly, we assume that $C = [p_1 p_2 \cdots p_n p_1]$ is a closed polygon in R, and that p is a point having the properties specified in Definition I, § 1807. We now evaluate the line integral $I_i \equiv \displaystyle\int_{\Gamma_i} P\,dx + Q\,dy$, $i = 1, 2, \cdots, n$, where the curve of integration Γ_i is the boundary of S_i with the orientation that directs the segment $\overline{p_i p_{i+1}}$ from p_i to p_{i+1}. On the one hand, the sum of these integrals, $\displaystyle\sum_{i=1}^{n} I_i$, owing to complete cancellation on the duplicated arcs C_i, is the original line integral $\displaystyle\int_C P\,dx + Q\,dy$. On the other hand, by Green's theorem, every one of these integrals is equal to $\pm \displaystyle\iint_{S_i} \left(\dfrac{\partial Q}{\partial x} - \dfrac{\partial P}{\partial y} \right) dA = 0$. With the consequent vanishing of $\displaystyle\int_C P\,dx + Q\,dy$, as desired, the proof is complete.

Example. It was shown in Example 4, § 1803, that

(1) $$-\frac{y}{x^2 + y^2}\,dx + \frac{x}{x^2 + y^2}\,dy$$

is *not* exact in the punctured plane R obtained by deleting the origin from E_2, although the equation $\partial P/\partial y = \partial Q/\partial x$ holds throughout R. This shows (by the preceding Theorem) that R is multiply-connected. In fact, *whenever* a point is deleted from any region, the resulting region is multiply-connected (cf. Ex. 7, § 1809). Since the region R_1 obtained from the plane by deleting the nonpositive part of the x-axis is star-shaped and therefore simply-connected (cf. the Examples, § 1807), the expression (1) is exact in R_1. A scalar potential for (1) that is valid throughout R_1 is compositely defined (cf. § 301):

$$(2) \quad \phi(x, y) \equiv \begin{cases} \operatorname{Arctan} \dfrac{y}{x}, & x > 0; \\ \operatorname{Arccot} \dfrac{x}{y}, & y > 0; \\ \operatorname{Arccot} \dfrac{x}{y} - \pi, & y < 0. \end{cases}$$

(The student should verify this statement in Ex. 8, § 1809, showing in particular that the definition of ϕ is unambiguous.)

1809. EXERCISES

1. Show that the definition given in § 1805 of the orientation of the boundary C of an elementary compact region S is self-consistent. For example, show that whether the orientation of a portion of C is determined by its being a part of the upper curve of S or by its being as a part of the right-hand curve of S, this orientation is unique.

2. Show that in the proof of Green's theorem in the plane, § 1805, it may be assumed without loss of generality (cf. § 321) that S is an elementary compact region.

3. Complete the proof of Green's theorem in the plane, § 1805, by showing that

$$\iint_S \frac{\partial Q}{\partial x} \, dA = \int_C Q \, dy.$$

4. Complete the details in the proof of the Theorem, § 1806.

5. Prove the statements of Note 1, § 1807.

6. Prove that the region between two concentric circles is multiply-connected.

7. Prove that whenever a point is deleted from a region, the resulting region is multiply-connected.

8. Verify that the function ϕ of (2), § 1808, is a scalar potential for the differential expression (1), § 1808.

In Exercises 9–16, use Green's theorem to evaluate the line integral of the indicated Exercise of § 1804.

9. Ex. 12.   **10.** Ex. 13.   **11.** Ex. 14.   **12.** Ex. 15.
13. Ex. 16.   **14.** Ex. 17.   **15.** Ex. 18.   **16.** Ex. 19.

In Exercises 17–20, use at least one of formulas (2), § 1805, to evaluate the indicated area.

17. Inside the ellipse $x = a \cos t$, $y = b \sin t$, $a, b > 0$.

18. Inside the hypocycloid $x = a \cos^3 t$, $y = a \sin^3 t$, $a > 0$.

19. Inside the loop of the folium of Descartes (cf. § 319): $x = 3at(t^3 + 1)^{-1}$, $y = 3at^2(t^3 + 1)^{-1}$, $a > 0$.

20. Inside the loop of the strophoid $x = a(1 - t^2)(1 + t^2)^{-1}$, $y = xt$, $a > 0$.

★21. Show that the first moments of a plane region with constant density δ equal to unity (cf. § 1017) can be expressed:

$$M_y = -\int_C xy\, dx = \frac{1}{2}\int_C x^2\, dy = \frac{1}{4}\int_C -2xy\, dx + x^2\, dy,$$

$$M_x = -\frac{1}{2}\int_C y^2\, dx = \int_C xy\, dy = \frac{1}{4}\int_C -y^2\, dx + 2xy\, dy.$$

Obtain similar formulas for second moments.

★22. Derive the following form for Green's theorem in the plane, in polar coordinates:

$$\int_C P\, d\rho + Q\, d\theta = \iint_S \frac{1}{\rho}\left[\frac{\partial Q}{\partial \rho} - \frac{\partial P}{\partial \theta}\right] dA,$$

with a careful statement of conditions. Deduce the following formulas for the area A of S:

$$A = -\int_C \rho\theta\, d\rho = \frac{1}{2}\int_C \rho^2\, d\theta.$$

Check by finding the area of the circle $\rho = 2a \cos \theta$.

23. Show that Green's theorem in the form of (3), § 1805, where $\vec{F}$ has the form $\phi \, \text{grad}\, \psi$, where ϕ and ψ are scalar fields, reduces to

$$\iint_S \frac{\partial(\phi, \psi)}{\partial(x, y)} dA = \int_C \phi \, \text{grad}\, \psi \cdot \vec{ds}.$$

24. The Laplacian and directional derivative of a scalar field ϕ in two dimensions are defined and denoted in a manner completely analogous to those employed in three dimensions (cf. Ex. 9, § 911). Prove that Green's theorem in the form of (4), § 1805, when the vector field $\vec{F}$ is *conservative* ($\vec{F} = \text{grad}\, \phi$; cf. Note 4, § 1803), becomes:

$$\iint_S \text{Lapl}\, \phi\, dA = \int_C \frac{d\phi}{dn}\, ds.$$

In particular, if ϕ is *harmonic* (Exs. 19-22, § 904) in an open set containing S, show that

$$\int_C \frac{d\phi}{dn}\, ds = 0.$$

25. Replacing $\vec{F}$ by $\phi\vec{F}$ in (4), § 1805, obtain the formula

$$\iint_S [\phi\, \text{div}\, \vec{F} + (\text{grad}\, \phi) \cdot \vec{F}]\, dA = \int_C \phi\vec{F} \cdot \vec{n}\, ds.$$

26. With the aid of Exercise 25, derive the two **Green's identities** (cf. § 1820):

(1) $$\iint_S [\phi\, \text{Lapl}\, \psi + \text{grad}\, \phi \cdot \text{grad}\, \psi]\, dA = \int_C \phi \frac{d\psi}{dn}\, ds,$$

(2) $$\iint_S [\phi\, \text{Lapl}\, \psi - \psi\, \text{Lapl}\, \phi]\, dA = \int_C \left[\phi \frac{d\psi}{dn} - \psi \frac{d\phi}{dn}\right] ds.$$

In particular, show that if ϕ and ψ are harmonic in an open set containing S:

$$\int_C \phi \frac{d\psi}{dn}\, ds = \int_C \psi \frac{d\phi}{dn}\, ds.$$

27. Prove that if a scalar field ϕ is harmonic throughout a finitely decomposable compact region S, then its values in S are completely determined by its values on the boundary C. That is, show that if ϕ and ψ are harmonic in an open set containing S and equal on C, then $\phi = \psi$ in S. (Cf. Ex. 22, § 1810.) *Hints:* First prove the statement for the case $\psi = 0$ by using Green's identity (1) with ϕ playing the role of both scalar fields. Then consider the difference $\phi - \psi$.

★28. If R is the punctured plane E_2 with the origin deleted, and if $P\,dx + Q\,dy$ is locally exact in R, show that it is exact in R if and only if $\int_C P\,dx + Q\,dy = 0$ when C is the unit circle $x^2 + y^2 = 1$. *Hint:* Assuming $\int_C P\,dx + Q\,dy = 0$ for $x^2 + y^2 = 1$, use Note 3, § 1806, to show that $\int_C P\,dx + Q\,dy = 0$ when C is the boundary of *any* disk or rectangle with sides parallel to the coordinate axes and situated so that C lies in R. Then use the ideas of the proof of the Theorem, § 1806, to construct a scalar potential.

In Exercises 29–35, the region R is the punctured plane E_2 with the origin deleted. Show that the given differential expression $P\,dx + Q\,dy$ is locally exact in R. Evaluate $\int_C P\,dx + Q\,dy$, where C is the boundary of the unit disk $x^2 + y^2 \le 1$, and determine whether $P\,dx + Q\,dy$ is exact in R or not. If $P\,dx + Q\,dy$ is exact in R, find a scalar potential. (Cf. Ex. 28.)

★29. $x(x^2 + y^2)^{-1}\,dx + y(x^2 + y^2)^{-1}\,dy$.
★30. $x(x^2 + y^2)^{-2}\,dx + y(x^2 + y^2)^{-2}\,dy$.
★31. $-x^2 y(x^2 + y^2)^{-2}\,dx + x^3(x^2 + y^2)^{-2}\,dy$.
★32. $-xy^2(x^2 + y^2)^{-2}\,dx + x^2 y(x^2 + y^2)^{-2}\,dy$.
★33. $-y^5(x^2 + y^2)^{-3}\,dx + xy^4(x^2 + y^2)^{-3}\,dy$.
★34. $-x^3 y^2(x^2 + y^2)^{-3}\,dx + x^4 y(x^2 + y^2)^{-3}\,dy$.
★35. $-x^2 y^3(x^2 + y^2)^{-3}\,dx + x^3 y^2(x^2 + y^2)^{-3}\,dy$.

★36. Find a scalar potential for the differential expression of Exercise 31 for the region got by deleting from E_2 the nonpositive x-axis. (Cf. the Example, § 1808.)

★37. If $f(x, y) \equiv -y/(x^2 + y^2)$, and if R and R_1 are the regions defined in the Example, § 1808, show that if Q is a differentiable function such that $\partial Q/\partial x = f$ throughout R_1, then Q must have the form $Q(x, y) = \phi(x, y) + \psi(y)$, where $\phi(x, y)$ is defined in (2), § 1808. Thus show that there is no function Q differentiable throughout R and satisfying there the equation $\partial Q/\partial x = f$.

★38. Let S be a finitely decomposable compact region in the uv-plane, with positively oriented boundary C, and assume that the transformation given by twice continuously differentiable functions $x = x(u, v)$, $y = y(u, v)$ maps S onto a finitely decomposable compact region R in the xy-plane in a one-to-one bicontinuous manner, the boundary C being mapped onto the positively oriented boundary Γ of R. Furthermore, assume that $f(x, y)$ is a continuous function defined over R and equal there to a function of the form $\partial Q/\partial x$, where Q is differentiable over R (cf. Ex. 37). Obtain a proof of the Transformation Theorem for double integrals (§ 1026), under these assumptions, by applying Green's theorem as follows (with $P \equiv 0$, and $g(u, v) \equiv f(x(u, v), y(u, v)))$:

$$\iint_R \frac{\partial Q}{\partial x} \, dA = \int_\Gamma Q \, dy = \int_C Q(x(u,v), y(u,v)) \left[\frac{\partial y}{\partial u} du + \frac{\partial y}{\partial v} dv \right]$$

$$= \iint_S \left\{ \frac{\partial}{\partial u} \left[Q(x(u,v), y(u,v)) \frac{\partial y}{\partial v} \right] - \frac{\partial}{\partial v} \left[Q(x(u,v), y(u,v)) \frac{\partial y}{\partial u} \right] \right\} dA$$

$$= \iint_S \frac{\partial Q}{\partial x} \left[\frac{\partial x}{\partial u} \frac{\partial y}{\partial v} - \frac{\partial x}{\partial v} \frac{\partial y}{\partial u} \right] dA = \iint_S g \frac{\partial(x, y)}{\partial(u, v)} \, dA.$$

*39. Show by the following example that without stipulations of the type given in Exercise 38 the method outlined there is inadequate for proving the Transformation Theorem of § 1026, even when S is a simple square and $f(x, y)$ is identically equal to 1: Let S be the square $0 \leq u \leq 1, 0 \leq v \leq 1$, let $\phi(t) \equiv \frac{1}{10} t^6 \sin\frac{1}{t}$ for $0 < t \leq 1$, $\phi(0) = 0$, and let $x(u, v) \equiv u + \phi(v), y(u, v) \equiv \phi(u) + v$. Show that $x(u, v), y(u, v)$, and their Jacobian $J(u, v) \equiv \partial(x, y)/\partial(u, v)$, together with their partial derivatives of order 1 and 2, are continuous, with $J(u, v) > 0$, over S. Show that the mapping of S onto its image R in the xy-plane is one-to-one and bicontinuous, but that R fails to be finitely decomposable. Show that if R is rotated through 45°, the result is not only finitely decomposable, but is an elementary compact region.

*1810. ANALYTIC FUNCTIONS OF A COMPLEX VARIABLE. EXERCISES

*1. Let $w = f(z)$ be a complex-valued function of the complex variable $z = x + iy$, defined over a region R of E_2. At any point $z = x + iy$ of R, f is said to be **differentiable** if and only if the limit

(1) $$\lim_{z \to 0} \frac{f(z + \Delta z) - f(z)}{\Delta z}$$

exists and is finite. In case (1) exists and is finite its value is called the **derivative** of f at the point z, and denoted $dw/dz = f'(z)$. A function whose derivative exists and is continuous in a neighborhood of a point is called **analytic** there.† Prove that sums, differences, products, and quotients of analytic functions are analytic, provided division by zero is not involved, and establish laws similar to familiar laws for real variables (e.g., $(fg)' = fg' + f'g$). Conclude that a rational function is analytic wherever defined.

*2. Let $w = u(x, y) + iv(x, y) = f(z) = f(x + iy)$ be an analytic function with real and imaginary parts $u(x, y)$ and $v(x, y)$, respectively. Derive the formula:

$$\frac{dw}{dz} = \frac{\partial u}{\partial x} + i\frac{\partial v}{\partial x} = \frac{\partial v}{\partial y} - i\frac{\partial u}{\partial y},$$

† In any complete textbook on analytic functions of a complex variable it is proved that the mere existence of a derivative throughout a region implies its continuity there. In the very brief introduction encompassed within these exercises we shall find it expedient to assume continuity of the derivative as part of the definition of analyticity.

and thereby the Cauchy-Riemann differential equations (cf. Ex. 23, § 904):

(2) $$\frac{\partial u}{\partial x} = \frac{\partial v}{\partial y}, \quad \frac{\partial u}{\partial y} = -\frac{\partial v}{\partial x}.$$

Assuming continuous differentiability, conclude that the real and imaginary parts of any analytic function are harmonic (cf. Exs. 19–22, § 904, and Ex. 13 below). *Hint:* First let $\Delta z = \Delta x$ and then let $\Delta z = i \Delta y$, in (1).

*3. Prove that if u and v are continuously differentiable in a region R, and if the Cauchy-Riemann differential equations (2) hold throughout R, then the function $f(z) = f(x + iy) \equiv u(x, y) + iv(x, y)$ is analytic there. *Hint:* Use the fundamental increment formula of § 905 to write

$$\Delta f = \Delta u + i \Delta v = \frac{\partial u}{\partial x} \Delta x + \frac{\partial u}{\partial y} \Delta y + \varepsilon_1 \Delta x + \varepsilon_2 \Delta y$$
$$+ i \left[\frac{\partial v}{\partial x} \Delta x + \frac{\partial v}{\partial y} \Delta y + \varepsilon_3 \Delta x + \varepsilon_4 \Delta y \right]$$
$$= \left(\frac{\partial u}{\partial x} + i \frac{\partial v}{\partial x} \right)(\Delta x + i\Delta y) + (\varepsilon_1 + i\varepsilon_3) \Delta x + (\varepsilon_2 + i\varepsilon_4) \Delta y.$$

*4. Prove that a function with identically vanishing derivative throughout a region is constant there. *Hint:* Cf. Theorem II, § 912.

*5. Prove that an analytic function whose modulus is constant throughout a region R is itself constant throughout R. *Hint:* If $u^2 + v^2 > 0$, obtain the system $u(\partial u/\partial x) + v(\partial v/\partial x) = 0$, $v(\partial u/\partial x) - u(\partial v/\partial x) = 0$.

*6. Prove that e^z, $\cos z$, and $\sin z$ (Exs. 26, 29, § 1509) are analytic throughout E_2, with derivatives:

$$\frac{d}{dz}(e^z) = e^z, \quad \frac{d}{dz}(\cos z) = -\sin z, \quad \frac{d}{dz}(\sin z) = \cos z.$$

*7. If $f(z) = u(x, y) + iv(x, y)$ is continuous on a sectionally smooth arc C, the **line integral** $\int_C f(z) \, dz$ is defined:

(3) $$\int_C f(z) \, dz \equiv \int_C (u \, dx - v \, dy) + i \int_C v \, dx + u \, dy.$$

Prove the following two forms of the **Cauchy Integral Theorem**: (*i*) *If the function $f(z)$ is the (continuous) derivative of an analytic function $F(z)$ throughout a region R, and if C is a sectionally smooth closed curve in R, then $\int_C f(z) \, dz = 0$.* (*ii*) *If $f(z)$ is analytic in a simply-connected region R, and if C is a sectionally smooth closed curve in R, then $\int_C f(z) \, dz = 0$.* *Hints* (*i*) If $F = U(x, y) + iV(x, y)$ show that U and V are scalar potentials for $u \, dx - v \, dy$ and $v \, dx + u \, dy$, respectively. (*ii*) Use the Cauchy-Riemann differential equations (2).

*8. If C is a circle with center at the origin, oriented positively, and if n is an arbitrary integer (positive or negative), prove that

$$\int_C z^n \, dz = \begin{cases} 0 & \text{if } n \neq -1, \\ 2\pi i & \text{if } n = -1. \end{cases}$$

(Cf. Example 4, § 1803.)

★9. Assuming $f(z)$ continuous on a sectionally smooth arc C of length L, with $|f(z)| \leq M$ there, prove that

(4) $$\left| \int_C f(z)\, dz \right| \leq ML.$$

Hint: By Note 3, § 1802,

$$\int_C f(z)\, dz = \lim_{|\mathfrak{N}| \to 0} \sum_{i=1}^{n} [u(x(a_i), y(a_i))\, \Delta x_i - v(x(a_i), y(a_i))\, \Delta y_i$$
$$+ iv(x(a_i), y(a_i))\, \Delta x_i - iu(x(a_i), y(a_i))\, \Delta y_i]$$
$$= \lim_{|\mathfrak{N}| \to 0} \sum_{i=1}^{n} f(z(a_i))\, \Delta z_i,$$

whence

$$\left| \int_C f(z)\, dz \right| \leq M \lim_{|\mathfrak{N}| \to 0} \sum_{i=1}^{n} |\Delta z_i|.$$

★10. If $g(z)$ is continuous on a sectionally smooth arc C, if R is a region having no points in common with C, and if the function f is defined at an arbitrary point a of R:

(5) $$f(a) \equiv \int_C \frac{g(z)}{z - a}\, dz,$$

prove that f is analytic in R, and that

(6) $$f'(a) = \int_C \frac{g(z)}{(z - a)^2}\, dz.$$

Hint: Write

$$\frac{f(a + \Delta z) - f(a)}{\Delta z} - \int_C \frac{g(z)}{(z - a)^2}\, dz$$
$$= \int_C g(z) \left[\frac{1}{(z - a - \Delta z)(z - a)} - \frac{1}{(z - a)^2} \right] dz$$
$$= \int_C g(z) \frac{\Delta z}{(z - a - \Delta z)(z - a)^2}\, dz.$$

★11. Prove that the function f defined by (5) can be differentiated any number of times, its nth derivative being given by

(7) $$f^{(n)}(a) = n! \int_C \frac{g(z)}{(z - a)^{n+1}}\, dz.$$

Hint: First prove (7) for $n = 2$ and $n = 3$; then use mathematical induction.

★12. If $f(z)$ is analytic at every point of a compact elementary region S, with boundary C (positively oriented), and if a is a point interior to S, prove the **Cauchy Integral Formula**:

(8) $$f(a) = \frac{1}{2\pi i} \int_C \frac{f(z)}{z - a}\, dz.$$

Hints: Choose about a a small circle Γ of radius r, lying in the interior of S, and partition the finitely decomposable compact region remaining after the inside of Γ has been removed from S, by means of vertical and horizontal lines through a. By these means show that

$$\frac{1}{2\pi i} \int_C \frac{f(z)}{z - a}\, dz = \frac{1}{2\pi i} \int_\Gamma \frac{f(z)}{z - a}\, dz.$$

Then (cf. Ex. 8) show that $f(a) = \dfrac{1}{2\pi i}\int_\Gamma \dfrac{f(a)}{z-a}\,dz$, and obtain the inequality:

$$\left|\frac{1}{2\pi i}\int_\Gamma \frac{f(z)}{z-a}\,dz - \frac{1}{2\pi i}\int_\Gamma \frac{f(a)}{z-a}\,dz\right| = \frac{1}{2\pi}\left|\int_\Gamma \frac{f(z)-f(a)}{z-a}\,dz\right|$$

$$\leq \max_{z\in\Gamma}\left|f(z)-f(a)\right|\frac{2\pi r}{2\pi r}.$$

Finally consider the limit of this final expression, as $r \to 0$.

*13. Infer from Exercises 11 and 12 that if $f(z)$ is analytic in a neighborhood of the point a, then it has (analytic) derivatives of all orders, with formula:

(9) $$f^{(n)}(a) = \frac{n!}{2\pi i}\int_C \frac{f(z)}{(z-a)^{n+1}}\,dz.$$

*14. Prove **Morera's Theorem**: *If $f(z)$ is continuous in a region R and if $\int_C f(z)\,dz = 0$ for every sectionally smooth closed curve C in R, then $f(z)$ is analytic in R.* Hint: Since both $u\,dx - v\,dy$ and $v\,dx + u\,dy$ are exact in R (cf. Theorem III, § 1803), there exist real-valued functions U and V such that $dU = u\,dx - v\,dy$ and $dV = v\,dx + u\,dy$. Then $F \equiv U + iV$ is analytic with f as its derivative. Therefore (Ex. 13), f is also analytic.

*15. Prove **Liouville's Theorem**: *If $f(z)$ is analytic throughout the plane E_2, and bounded there, then $f(z)$ is constant.* Hint: Let C be a large circle of radius R, with center a and surrounding b. Consider the relations:

$$|f(a) - f(b)| = \left|\frac{1}{2\pi i}\int_C f(z)\frac{a-b}{(z-a)(z-b)}\,dz\right|$$

$$\leq \frac{|a-b|}{2\pi}\cdot M\cdot \frac{2\pi R}{R(R-|a-b|)} = \frac{|a-b|M}{R-|a-b|}.$$

*16. Prove the **Fundamental Theorem of Algebra** (cf. § 1515): *Any nonconstant polynomial has at least one zero.* Hint: Assume $P(z)$ is never 0, define $f(z) \equiv [P(z)]^{-1}$. Then (Ex. 1), $f(z)$ is analytic in E_2, $|f(z)| \leq 1$ for $|z| >$ some R, and $f(z)$ is bounded for $|z| \leq R$. (Cf. Ex. 15.)

*17. Assume that $\{f_n\}$ is a sequence of functions analytic in a region R and converging uniformly there to a function $f(z)$. Prove that $f(z)$ is analytic in R. Conclude that any function represented by a power series is analytic within the circle of convergence (cf. Theorem IX, § 1511). *Hint:* By the extension to complex variables of the Theorem, § 1305, $f(z)$ is continuous. Let C be an arbitrary sectionally smooth closed curve in R, of length L. Then

$$\left|\int_C (f_n(z) - f(z))\,dz\right| \leq \max_{z\in C}|f_n(z) - f(z)|\cdot L.$$

(Cf. Ex. 14.)

*18. Using $\dfrac{1}{1-r} = 1 + r + r^2 + \cdots + r^{n-1} + \dfrac{r^n}{1-r}$, with $r = (z-a)/(\zeta - a)$, $\zeta \neq z$, $\zeta \neq a$, obtain the formula

(10) $$\frac{1}{\zeta - z} = \frac{1}{\zeta - a} + \frac{z-a}{(\zeta-a)^2} + \cdots + \frac{(z-a)^{n-1}}{(\zeta-a)^n} + \frac{(z-a)^n}{(\zeta-a)^n(\zeta-z)}.$$

Let a be the center of a circle C of radius ρ, and let z be inside C and ζ a variable on C itself. Multiply (10) by $f(\zeta)/2\pi i$ and integrate with respect to ζ around C to obtain (cf. Ex. 13) **Taylor's Formula with a Remainder**:

(11) $$f(z) = f(a) + \frac{f'(a)}{1!}(z-a) + \frac{f''(a)}{2!}(z-a)^2 + \cdots$$
$$+ \frac{f^{(n-1)}(a)}{(n-1)!}(z-a)^{n-1} + R_n(z),$$

where

(12) $$R_n(z) = \frac{1}{2\pi i}\int_C \frac{(z-a)^n f(\zeta)}{(\zeta-a)^n(\zeta-z)}\,d\zeta.$$

If $|f(\zeta)| \leq M$, show that
$$|R_n(z)| \leq \frac{M\rho}{\rho - |z-a|}\left(\frac{|z-a|}{\rho}\right)^n,$$
and hence (Theorem XIII, § 204) show that any function f analytic in a circle about a is represented there by its **Taylor series**:

(13) $$f(z) = f(a) + \frac{f'(a)}{1!}(z-a) + \frac{f''(a)}{2!}(z-a)^2 + \cdots$$
$$+ \frac{f^{(n)}(a)}{n!}(z-a)^n + \cdots.$$

*19. Show that the analytic functions e^z, $\cos z$, and $\sin z$ (cf. Ex. 6) are represented throughout E_2 by their Taylor series:

(14) $$e^z = 1 + z + \frac{z^2}{2!} + \cdots + \frac{z^n}{n!} + \cdots,$$

(15) $$\cos z = 1 - \frac{z^2}{2!} + \frac{z^4}{4!} - \cdots + (-1)^n \frac{z^{2n}}{(2n)!} + \cdots,$$

(16) $$\sin z = z - \frac{z^3}{3!} + \frac{z^5}{5!} - \cdots + (-1)^{n-1}\frac{z^{2n-1}}{(2n-1)!} + \cdots.$$

*20. Assume that $f(z)$ is analytic in a region R and that $f(z) = 0$ on a set of points S having a limit point a in R. Prove that $f(z) = 0$ throughout R. Conclude that whenever two functions analytic in a region R are identical on a subset having a limit point in R, these functions are identical throughout R. *Hints*: 1. By continuity $f(a) = 0$. Let C be the boundary of a closed disk with center at a and lying in R, assume there exists an $n \geq 1$ such that $f^{(n)}(a) \neq 0$, and let k be the smallest such n. Then $f(z) = (z-a)^k g(z)$, where $g(z)$, being represented by a power series, is analytic inside C and vanishing on S. Since $g(a) = 0, f^{(k)}(a) = 0$. (!) Therefore $f(z) = 0$ everywhere inside C. 2. Let b be an arbitrary point of R and connect a to b by a broken line segment Γ. Show that $f(z)$ must vanish everywhere along Γ by iteration of the ideas in 1, preceding.

NOTE. Exercise 20 can be used to validate formulas like those of Exercise 19, as follows: The two members of (14), for example, are analytic functions in E_2, and since they are identical on the x-axis they must also be identical throughout E_2.

*21. Prove the **Maximum Modulus Theorem**: *If $f(z)$ is analytic and nonconstant in a bounded region R and continuous on its closure $\bar{R}$, then the maximum value of $|f(z)|$ on $\bar{R}$ cannot be attained within R, and must therefore be attained at a frontier point of R.* *Hint*: Assume $\max_{z \in \bar{R}} |f(z)| = M$ is attained at a point a of R. If

$|f(z)|$ is identically equal to M in any neighborhood of a, then $f(z)$ must be constant throughout R (Exs. 5, 20). Assume $|f(z)| < M$ at some point of the boundary C of a closed disk of radius r with center at a and lying in R, and consider an arc C_1 of C subtending an angle $\alpha > 0$ such that $|f(z)| \leq M - \epsilon < M$ on C_1. Then

$$|f(a)| \leq \frac{1}{2\pi} \left| \int_{C_1} \frac{f(z)}{z-a} dz \right| + \frac{1}{2\pi} \left| \int_{C_2} \frac{f(z)}{z-a} dz \right|$$

$$\leq \frac{1}{2\pi r}(M - \epsilon) \cdot \alpha r + \frac{1}{2\pi r} M(2\pi - \alpha)r = M - \frac{\epsilon \alpha}{2\pi} < M. \quad \text{(Contradiction.)}$$

*22. Prove that if $f(z)$ is a function analytic throughout a finitely decomposable compact region S then its values in S are completely determined by its values on the boundary C. That is, show that if f and g are analytic throughout S and equal on C, then $f = g$ in S. (Cf. Ex. 27, § 1809.) *Hints:* Two methods are available: (*i*) the maximum modulus theorem of Ex. 21 applied to $f - g$; (*ii*) Ex. 27, § 1809, with Ex. 2 above.

*23. If a function $f(z)$ is analytic in a region R and vanishes at a point a of R but not identically throughout R, prove that there exists a unique positive integer k such that $f(z) = (z - a)^k g(z)$, where $g(z)$ is analytic in R and $g(a) \neq 0$. The number k is called the **order of the zero** a. *Hint:* Use Exs. 18, 20.

*24. Assume that $f(z)$ is analytic in a deleted neighborhood N_a of a point a, and that it is bounded in N_a. Prove that the point a is a **removable singularity** in the sense that there exists a function $g(z)$ identical with $f(z)$ in N_a and analytic at a as well. *Hint:* If C is an *arbitrary* circle about a and lying in N_a, define for any z inside C:

$$g(z) \equiv \frac{1}{2\pi i} \int_C \frac{f(\zeta)\, d\zeta}{\zeta - z},$$

and make use of a very small supplementary circle about a to show that $g(z) = f(z)$ for $z \neq a$.

*25. If a function $f(z)$ is analytic in a deleted neighborhood N_a of a point a, and if $\lim_{z \to a} f(z) = \infty$, $f(z)$ is said to have a **pole** at a. Prove that if $f(z)$ has a pole at a, then there exists a unique positive integer k such that $f(z) = g(z)/(z-a)^k$, where $g(z)$ is analytic in N_a and analytic and nonzero at a. Hence obtain the following infinite series representation for $f(z)$ in N_a:

(17) $$f(z) = \frac{a_{-k}}{(z-a)^k} + \cdots + \frac{a_{-1}}{z-a} + a_0 + a_1(z-a) + \cdots + a_n(z-a)^n + \cdots,$$

where $a_{-k} \neq 0$. If C is a positively oriented circle about a, lying in N_a, prove that

(18) $$\frac{1}{2\pi i} \int_C f(z)\, dz = a_{-1}.$$

The positive integer k is called the **order of the pole** at a, and a_{-1} is called the **residue** of $f(z)$ at a.

1811. SURFACE ELEMENTS

An *arc* was defined in § 803 to be a continuous image of a closed interval. It is possible to proceed similarly and define a *surface* to be a continuous image of a closed rectangle. The problems pertaining to surfaces, however, are much deeper and more complex than those pertaining to curves—indeed, many fundamental questions still remain unanswered—and we shall therefore give a much more restrictive definition. All surfaces considered here are in the space E_3.

We start by defining a basic "building block" for surfaces:

Definition I. *A **surface element** is a one-to-one image S of a closed rectangle R: $a \leq u \leq b$, $c \leq v \leq d$ under a mapping*

(1) $$x = x(u, v), \quad y = y(u, v), \quad z = z(u, v),$$

where each of the mapping functions (1) is continuously differentiable,† and the three Jacobians

(2) $$j_1 \equiv \frac{\partial(y, z)}{\partial(u, v)}, \quad j_2 \equiv \frac{\partial(z, x)}{\partial(u, v)}, \quad j_3 \equiv \frac{\partial(x, y)}{\partial(u, v)}$$

never vanish simultaneously. **Inner** *and* **boundary points** *of S are points that correspond (under the mapping (1)) to interior and boundary points, respectively, of R.* **Vertices** *and* **edges** *of S are the images under (1) of vertices and (closed) edges, respectively, of R. The variables u and v are called* **parameters** *for the surface element S.*

NOTE 1. For surfaces in E_3 we shall distinguish between the concepts of *boundary points* and *frontier points*. For example, any point of a surface element S, whether inner or boundary, is a limit point of the complement of S and therefore a frontier point of S.

NOTE 2. Under the mapping (1) the lines $u =$ constant and $v =$ constant are carried into **coordinate curves** lying in S, with tangent vectors

(3) $$\begin{cases} \dfrac{\partial \vec{r}}{\partial u} = \dfrac{\partial x}{\partial u} \vec{i} + \dfrac{\partial y}{\partial u} \vec{j} + \dfrac{\partial z}{\partial u} \vec{k}, \\ \dfrac{\partial \vec{r}}{\partial v} = \dfrac{\partial x}{\partial v} \vec{i} + \dfrac{\partial y}{\partial v} \vec{j} + \dfrac{\partial z}{\partial v} \vec{k}, \end{cases}$$

whose cross product, by (2), is equal to

(4) $$\frac{\partial \vec{r}}{\partial u} \times \frac{\partial \vec{r}}{\partial v} = j_1 \vec{i} + j_2 \vec{j} + j_3 \vec{k}.$$

The condition that the Jacobians (2) never vanish simultaneously is equivalent to the nonvanishing of the vector (4), and therefore to the noncollinearity of the vectors (3). It follows in particular that the coordinate curves defined above are smooth (§ 805). (Cf. Fig. 1807.) The vector (4), being orthogonal to both vectors (3), is said to be **normal** to the surface element S; that is, the line normal to S at any point is *by definition* collinear with (4). The student is asked in Exercise 16, § 1815, to show analytically that the preceding definition of the line normal to S at any point is independent of the parametrization, and that this definition is consistent with the one given in § 910.

Example 1. Obtain each of the plane regions of Figure 1808 explicitly as a surface element. The quadrilateral in part (*a*) is assumed to be strictly convex;

† For points (u, v) on the edges of R, one-sided derivatives are used.

that is, the interior angle at each vertex is less than π radians. The region in part (b) is bounded by the circle $\rho = a > 0$, the rays $\theta = \pm\tfrac{1}{4}\pi$, and the line $x = \tfrac{1}{4}\sqrt{2}a$.

Solution. (a) Let $l_i = a_i x + b_i y + c_i$ be a linear expression such that the graph of $l_i = 0$ is the ith line, and such that for any point (x, y) in the interior of

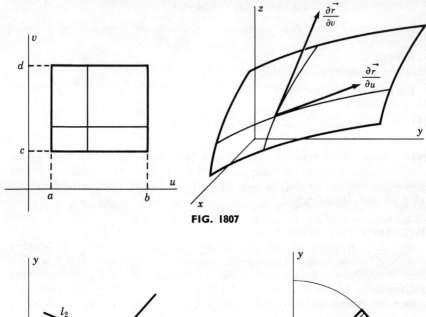

FIG. 1807

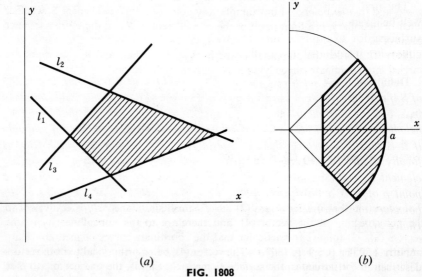

(a) (b)
FIG. 1808

the quadrilateral the value of l_i is positive, $i = 1, 2, 3, 4$. Then a mapping of the required type is given by the equations

(5) $$u = \frac{l_1}{l_1 + l_2}, \quad v = \frac{l_3}{l_3 + l_4},$$

$0 \leq u \leq 1$, $0 \leq v \leq 1$. Solving (5) gives equations of the form $x = x(u, v)$, $y = y(u, v)$, $z = 0$. (Check the details.)

(b) A preliminary transformation

(6) $$\rho = (2 - \tfrac{1}{2}\sqrt{2} \sec \theta)u + (-1 + \tfrac{1}{2}\sqrt{2} \sec \theta)a, \quad \theta = v$$

transforms the rectangle $\tfrac{1}{2}a \leq u \leq a$, $-\tfrac{1}{4}\pi \leq v \leq \tfrac{1}{4}\pi$ into the set in the $\rho\theta$-plane bounded above and below by the lines $\theta = \pm \tfrac{1}{4}\pi$, on the right by the line $\rho = a$, and on the left by the curve $\rho = \tfrac{1}{4}\sqrt{2}a \sec \theta$. A second transformation

(7) $$x = \rho \cos \theta, \quad y = \rho \sin \theta$$

transforms the new set into the desired closed region of Figure 1808.

Example 2. Any convex plane quadrilateral in E_3 can be obtained from Example 1(a), and suitable translations and rotations in space.

Example 3. The mappings

(8) $$\begin{cases} x = \rho \cos \theta, & y = \rho \sin \theta, \quad z = c\rho, \\ a \leq \rho \leq b, & \alpha \leq \theta \leq \beta < \alpha + 2\pi, \end{cases}$$

(9) $$\begin{cases} x = c \cos \theta, & y = c \sin \theta, \quad z = z, \\ a \leq z \leq b, & \alpha \leq \theta \leq \beta < \alpha + 2\pi, \end{cases}$$

give strips wrapped partway around the cone $z^2 = c^2(x^2 + y^2)$ and the cylinder $x^2 + y^2 = c^2$, respectively.

1812. SMOOTH SURFACES

The first requirement to be placed on a surface is that it be suitably well-behaved locally. The following definition eliminates self-intersecting surfaces, for example. Surfaces with edges and corners—like that of a cube—are introduced in § 1818.

Definition I. *A set S in E_3 is **locally a surface at a point** p: (x_0, y_0, z_0) of S if and only if there exist (i) a surface element S_1 containing p and contained in S, and (ii) a neighborhood N_p of p such that every point of S that lies in N_p belongs to S_1: $N_p \cap S \subset S_1$. A set S in E_3 is **locally a surface** if and only if it is locally a surface at every one of its points. A point p of S (that is locally a surface) is called an **inner point** of S if and only if the surface element S_1 can be chosen so that p is an inner point of S_1: otherwise a point p of S is a **boundary point** of S. A **vertex** of S is a point of the boundary of S that is a vertex of the surface element S_1 for every possible choice of S_1.*

*NOTE 1. The property that a set in E_3 may have of being locally a surface was described in § 910 by means of a definition which, at least in outward appearance, seems distinct from the one just given: *A (smooth) surface is the set of points (x, y, z) whose coordinates satisfy an equation of the form $f(x, y, z) = 0$, where f has a nonzero continuous gradient.* That these two definitions are equivalent, for local properties, can be seen as follows: (i) If a set is given locally by equations of the form (1), § 1811, where j_3 (say) does not vanish, then by the Implicit Function Theorem (§ 932), u and v can be solved locally in terms of x and y, the results being substituted to give z as a continuously differentiable function $\phi(x, y)$ of x and y. We then define $f(x, y, z) \equiv \phi(x, y) - z$. (ii) If $f(x, y, z)$ has a continuous nonzero gradient, assume for definiteness that $\partial f/\partial z \neq 0$ (locally). Then z can be solved as a continuously differentiable function $\phi(x, y)$ of x and y, locally, and we have the representation sought: $x = u, y = v, z = \phi(u, v)$.

We are now ready to give our restrictive definition of a surface:

Definition II. *A set S in E_3 is a **smooth surface** if and only if (i) S is locally a surface and (ii) S is the union of a finite number of surface elements every pair of which have in common at most boundary points of these two*

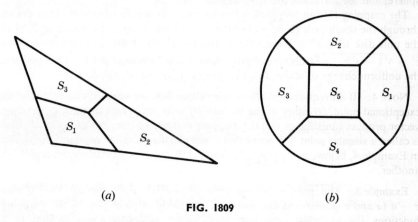

(a)   (b)

FIG. 1809

*surface elements. A **closed smooth surface** is a smooth surface with no boundary points. A **connected smooth surface** is a smooth surface S having the property that if S_i and S_j are any two of its surface elements, there exists a chain of surface elements of S, $S_i = S_{k_1}, S_{k_2}, \cdots, S_{k_m} = S_j$, such that every consecutive pair has a nonempty intersection: $S_{k_1} \cap S_{k_2} \neq \emptyset, S_{k_2} \cap S_{k_3} \neq \emptyset, \cdots, S_{k_{m-1}} \cap S_{k_m} \neq \emptyset$.*

NOTE 2. According to Definition II every smooth surface is bounded. The cylinder $x^2 + y^2 = c^2$ (Example 3, § 1811) is therefore not a smooth surface, but the portion restricted by $a \leq z \leq b$ is.

Example 1. A plane triangle and a plane circle are obtained as smooth surfaces by fitting together elements of the type given in Example 1, § 1811, as indicated in Figure 1809.

NOTE 3. With the help of the smooth surfaces shown in Figure 1809, by the application of composition of transformations (§ 922), we may think of a smooth surface as being made up of suitable combinations of images of squares, triangles, and circles.

Example 2. Show that the sphere $x^2 + y^2 + z^2 = a^2, a > 0$, is a closed smooth surface.

Solution. To show that the sphere is a smooth surface it is sufficient to obtain the hemisphere $z = \sqrt{a^2 - x^2 - y^2}$ as the image of the circle $u^2 + v^2 \leq 1$ by means of continuously differentiable mapping functions (cf. Example 1). Such a set of functions is

(1) $$x = \frac{2u}{1 + u^2 + v^2} a, \quad y = \frac{2v}{1 + u^2 + v^2} a, \quad z = \frac{1 - (u^2 + v^2)}{1 + u^2 + v^2} a.$$

The condition that the vector $(\partial \vec{r}/\partial u) \times (\partial \vec{r}/\partial v)$ be nonzero is satisfied, since its magnitude is $4a^2(1 + u^2 + v^2)^{-2} \geq a^2$.

Since the sphere is of a homogeneous character, and the north pole, $(0, 0, a)$, is an inner point for the mapping given above, there are no boundary points on the sphere, and the surface is therefore closed.

The mapping (1) can be visualized as follows: By means of straight lines drawn through the south pole $(0, 0, -1)$ of the unit sphere, project the points $(u, v, 0)$ of the unit disk $u^2 + v^2 \leq 1$ onto the points (x', y', z') of the northern hemisphere $z' = \sqrt{1 - x'^2 - y'^2}$ (this type of projection is called **stereographic**). Then perform the uniform change of scale: $(x', y', z') \to (x, y, z) = (ax', ay', az')$.

NOTE 4. If all requirements for a smooth surface are satisfied except that at an exceptional point, (i) either of the vectors $\partial \vec{r}/\partial u$ or $\partial \vec{r}/\partial v$ fails to exist, or (ii) their vector product vanishes, or (iii) the mapping is many-to-one, this exceptional point is called a **singular point**. A point that is not singular is called **regular**. As shown in Example 4, below, a point may be singular in one representation and regular in another.

Example 3. The portion of the cone $z^2 = c^2(x^2 + y^2)$ between $z = 0$ and $z = a$ (a and c positive) is the image of the circle $u^2 + v^2 \leq 1$ by the mapping functions

$$x = (a/c)u, \quad y = (a/c)v, \quad z = a\sqrt{u^2 + v^2}.$$

The origin is a singular point because $\partial z/\partial u$ and $\partial z/\partial v$ fail to exist there.

This same portion of the cone is the union of the images of the two rectangles $0 \leq \rho \leq a/c, 0 \leq \theta \leq \pi$ and $\pi \leq \theta \leq 2\pi$, suitably patched, under the mapping

$$x = \rho \cos \theta, \quad y = \rho \sin \theta, \quad z = c\rho.$$

In this case, not only does the vector $\partial \vec{r}/\partial \theta$ vanish when $\rho = 0$, but the mapping is many-to-one when $\rho = 0$.

Example 4. The hemisphere $z = \sqrt{a^2 - x^2 - y^2}$, $a > 0$, is the union of the images of the two rectangles $0 \leq \phi \leq \frac{1}{2}\pi, 0 \leq \theta \leq \pi$ and $\pi \leq \theta \leq 2\pi$ under the mapping

$$x = a \sin \phi \cos \theta, \quad y = a \sin \phi \sin \theta, \quad z = a \cos \phi.$$

For this particular mapping the north pole $(0, 0, a)$ is a singular point.

This same hemisphere is the union of the images of the two rectangles $0 \leq \rho \leq a$, $0 \leq \theta \leq \pi$, and $\pi \leq \theta \leq 2\pi$ under the mapping

$$x = \rho \cos \theta, \quad y = \rho \sin \theta, \quad z = \sqrt{a^2 - \rho^2}.$$

The north pole is again a singular point, and for this particular mapping so is every point on the equator, $z = 0, \rho = a$, since $\partial z/\partial \rho$ is infinite there.

Example 5. The region in E_3 prescribed by the inequalities $2 < x^2 + y^2 + z^2 < 5$ has a boundary surface consisting of the two connected spheres $x^2 + y^2 + z^2 = 2$ and $x^2 + y^2 + z^2 = 5$.

1813. SCHWARZ'S EXAMPLE

Since surfaces bear a slight resemblance to curves, it might at first be thought that surface area could be defined as the least upper bound of the areas of all "inscribed polyhedra" in much the same way that arc length is defined as the least upper bound of the lengths of all inscribed polygons (§ 804). An ingenious example due to H. A. Schwarz† shows that even with as simple a surface as the lateral surface of a right circular cylinder (if the base radius is a and the altitude h, the familiar formula for the area is $2\pi ah$), the process suggested above may lead to an infinite "area." We proceed to show this startling result.

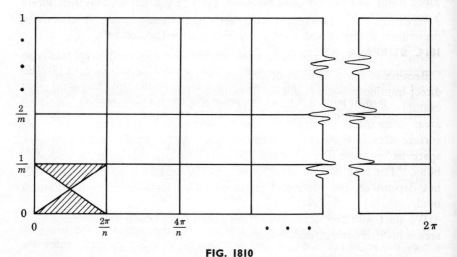

FIG. 1810

Let a right circular cylinder of base radius 1 and altitude 1 be cut along a generating line and rolled flat into a rectangle of sides 1 and 2π. Then subdivide this rectangle into smaller congruent rectangles (by means of $m - 1$ equally spaced lines parallel to the sides of length 2π and $n - 1$ equally spaced lines parallel to the sides of length 1), and each of these smaller rectangles into four triangles (by means of the diagonals of the rectangle), as shown in Figure 1810. If the big rectangle is again wrapped into the original cylinder, the vertices of the $4mn$ triangles of Figure 1810 become vertices of $4mn$ plane triangles in space that constitute a polyhedron inscribed in the cylinder. The idea now is to show that if m is allowed to increase enough more rapidly than n, the total area of this polyhedron becomes unbounded. The triangles responsible for this phenomenon are the $2mn$ spatial triangles corresponding to those of the type that are shaded in Figure 1810. Since each of these spatial triangles has base equal to

† *Gesammelte Mathematische Abhandlungen*, Vol. 2, p. 309. (Berlin, Julius Springer, 1890.)

2 sin (π/n) and altitude greater than $1 - \cos(\pi/n) = 2\sin^2(\pi/2n)$, the sum of the areas of these $2mn$ spatial triangles is greater than

$$S_{mn} \equiv \left[4n \sin \frac{\pi}{n}\right]\left[m \sin^2 \frac{\pi}{2n}\right].$$

The limit, as $n \to +\infty$, of the first bracketed quantity is 4π (cf. §§ 315–317) while if $m = n^3$ (for example) the second bracketed quantity has an infinite limit, as $n \to +\infty$. This completes the desired demonstration.

For further discussion of the history and general problem of surfaces and surface area the reader is referred to two articles in the *American Mathematical Monthly*: "What Is the Area of a Surface?" (vol. 50, pp. 139–141, 1943) by Tibor Radó, and "Curves and Surfaces" (vol. 51, pp. 1–11, 1944) by J. W. T. Youngs.

1814. SURFACE AREA

Because of the difficulties of attempting a definition of surface area by a direct limiting process (some of these difficulties are apparent in Schwarz's example, § 1813), and because the most useful method of handling surface areas is by means of double integrals, we take a direct course and *define* surface area in terms of double integrals. Although this procedure does spare us many vexing problems, it does not free us from all responsibility. For example, we must show that the area of a given surface does not depend on the particular parametrization (functional representation) used.

We start with the definition of the area of a surface element, and then seek a justification for that definition:

Definition. *If S is a surface element defined parametrically: $x = x(u, v)$, $y = y(u, v)$, $z = z(u, v)$, where (u, v) belongs to the rectangle R: $a \leq u \leq b$, $c \leq v \leq d$, then the **area** $A(S)$ of S is:*

(1) $\quad A(S) \equiv \iint_R \left\{\left[\frac{\partial(y, z)}{\partial(u, v)}\right]^2 + \left[\frac{\partial(z, x)}{\partial(u, v)}\right]^2 + \left[\frac{\partial(x, y)}{\partial(u, v)}\right]^2\right\}^{\frac{1}{2}} du\, dv$

$\quad = \iint_R [j_1^2 + j_2^2 + j_3^2]^{\frac{1}{2}} du\, dv = \iint_R \left|\frac{\partial \vec{r}}{\partial u} \times \frac{\partial \vec{r}}{\partial v}\right| du\, dv$

(cf. (2), (4), § 1811).

Our justification for this definition consists of three parts:

I. *Invariance.* If the same surface element S is given in a second way: $x = f(\tilde{u}, \tilde{v})$, $y = g(\tilde{u}, \tilde{v})$, $z = h(\tilde{u}, \tilde{v})$, where $(\tilde{u}, \tilde{v})$ belongs to a rectangle $\tilde{R}$, then a one-to-one correspondence between R and $\tilde{R}$ results: $\tilde{u} = \tilde{u}(u, v)$, $\tilde{v} = \tilde{v}(u, v)$. Furthermore, these functions are continuously differentiable, as can be seen at any point by considering $\tilde{u}$, $\tilde{v}$, and z (say) as continuously differentiable functions of x and y, and obtaining $\tilde{u}(u, v)$ and $\tilde{v}(u, v)$ by composition (§ 922). Finally, since

$$\frac{\partial(y, z)}{\partial(u, v)} = \frac{\partial(y, z)}{\partial(\tilde{u}, \tilde{v})} \frac{\partial(\tilde{u}, \tilde{v})}{\partial(u, v)}, \frac{\partial(z, x)}{\partial(u, v)} = \cdots,$$

we have, for $A(S)$,

$$\iint_R \left\{ \left[\frac{\partial(y, z)}{\partial(\tilde{u}, \tilde{v})}\right]^2 + \left[\frac{\partial(z, x)}{\partial(\tilde{u}, \tilde{v})}\right]^2 + \left[\frac{\partial(x, y)}{\partial(\tilde{u}, \tilde{v})}\right]^2 \right\}^{\frac{1}{2}} \left|\frac{\partial(\tilde{u}, \tilde{v})}{\partial(u, v)}\right| du\, dv,$$

which reduces to the desired formula (1), $\tilde{u}$ and $\tilde{v}$ replacing u and v, respectively, by the Fundamental Transformation Theorem for double integrals (§ 1026).

II. *Intuitive appeal.* If we think of a "differential rectangle" dR, very "small," with sides du and dv, then the parallelogram $d\Pi$ in space with sides $(\partial \vec{r}/\partial u)\, du$ and $(\partial \vec{r}/\partial v)\, dv$ (evaluated at a corner of dR) closely approximates the set that is the image of dR (cf. the discussion of § 1027). The area of $d\Pi$, which is the absolute value of the vector product of these two vectors, should closely approximate the area of this image of dR. From the point of view of differentials, then, we can think of the preceding area as a "differential of area" for the differential surface element:

$$(2) \qquad dS = \left|\left(\frac{\partial \vec{r}}{\partial u} du\right) \times \left(\frac{\partial \vec{r}}{\partial v} dv\right)\right| = \left|\frac{\partial \vec{r}}{\partial u} \times \frac{\partial \vec{r}}{\partial v}\right| du\, dv.$$

This leads to formula (1), by integration. (Cf. § 1027.)

III. *Variables x, y, z.* Suppose the surface element is located in space in such a way that one of the variables x, y, z can be solved for in terms of the other two. To be specific, let us assume that S is given by an equation of the form

$$(3) \qquad z = f(x, y),$$

for $(x, y) \in R$. Then equation (1) becomes

$$(4) \qquad A(S) = \iint_R \sqrt{\left(\frac{\partial z}{\partial x}\right)^2 + \left(\frac{\partial z}{\partial y}\right)^2 + 1}\, dx\, dy = \iint_R \sec \gamma\, dx\, dy,$$

where γ is the acute angle between the normal to the surface and the z-axis. (To see that the radical in (4) is equal to $\sec \gamma$, write the equation of S in the form $F(x, y, z) \equiv f(x, y) - z = 0$; by § 910, a set of direction numbers for the normal are $f_x(x, y), f_y(x, y), -1$; to obtain direction cosines we divide by the square root of the sum of the squares.) This is the result that would be expected by considering a differential of area dS, tangent to the surface S and lying directly over a differential of area dA in the xy-plane, with $dA = dS \cos \gamma$, or $dS = \sec \gamma\, dA$.

The area of a more general smooth surface is now defined to be the sum of the areas of its constituent elements. It is not difficult to see that this concept is again independent of the particular representation, or decomposition, used (cf. Ex. 17, § 1815), and that it can be extended to surfaces with edges and corners (cf. § 1818). Furthermore, by much the same principle, the rectangle R can be replaced by a much less restrictive compact region of integration.

NOTE 1. In practice, for much the same reason as indicated in the Note, § 1013, formula (1) can be used even in the presence of singular points and in the event that a convergent improper integral arises. (Cf. Example 1, below.)

NOTE 2. A useful alternative form for equation (1) can be expressed in terms of the squares of the magnitudes of the vectors $\partial \vec{r}/\partial u$ and $\partial \vec{r}/\partial v$ and their inner product:

(5)
$$\begin{cases} E \equiv \left|\dfrac{\partial \vec{r}}{\partial u}\right|^2 = \left(\dfrac{\partial x}{\partial u}\right)^2 + \left(\dfrac{\partial y}{\partial u}\right)^2 + \left(\dfrac{\partial z}{\partial u}\right)^2, \\ F \equiv \dfrac{\partial \vec{r}}{\partial u} \cdot \dfrac{\partial \vec{r}}{\partial v} = \dfrac{\partial x}{\partial u}\dfrac{\partial x}{\partial v} + \dfrac{\partial y}{\partial u}\dfrac{\partial y}{\partial v} + \dfrac{\partial z}{\partial u}\dfrac{\partial z}{\partial v}, \\ G \equiv \left|\dfrac{\partial \vec{r}}{\partial v}\right|^2 = \left(\dfrac{\partial x}{\partial v}\right)^2 + \left(\dfrac{\partial y}{\partial v}\right)^2 + \left(\dfrac{\partial z}{\partial v}\right)^2. \end{cases}$$

The square of the magnitude of the cross product $\dfrac{\partial \vec{r}}{\partial u} \times \dfrac{\partial \vec{r}}{\partial v}$ is:

$$\left|\dfrac{\partial \vec{r}}{\partial u} \times \dfrac{\partial \vec{r}}{\partial v}\right|^2 = \left|\dfrac{\partial \vec{r}}{\partial u}\right|^2 \left|\dfrac{\partial \vec{r}}{\partial v}\right|^2 \sin^2 \theta$$
$$= \left|\dfrac{\partial \vec{r}}{\partial u}\right|^2 \left|\dfrac{\partial \vec{r}}{\partial v}\right|^2 - \left[\left|\dfrac{\partial \vec{r}}{\partial u}\right|\left|\dfrac{\partial \vec{r}}{\partial v}\right|\cos \theta\right]^2 = EG - F^2.$$

Consequently, the formula (1) for area can be written:

(6)
$$A(S) = \iint_R \sqrt{EG - F^2}\, du\, dv.$$

NOTE 3. If the independent variables in the representation of a surface are the polar coordinates ρ, θ, with $x = \rho \cos \theta$, $y = \rho \sin \theta$, $z = z(\rho, \theta)$, formula (1) becomes

(7)
$$A(S) = \iint_R \left[\rho^2 \left(\dfrac{\partial z}{\partial \rho}\right)^2 + \left(\dfrac{\partial z}{\partial \theta}\right)^2 + \rho^2\right]^{\frac{1}{2}} d\rho\, d\theta.$$

Two special cases are of interest:

(i) For a *surface of revolution* about the z-axis, $\partial z/\partial \theta = 0$, and (7) becomes

(8)
$$A(S) = 2\pi \int_{\rho_1}^{\rho_2} \rho \sqrt{1 + \left(\dfrac{\partial z}{\partial \rho}\right)^2}\, d\rho = 2\pi \int_{s_1}^{s_2} \rho\, ds,$$

where ds is the differential of arc length for the curve $z = z(\rho)$ in a $z\rho$-plane. When expressed in terms of the variables x and y this becomes a familiar formula, from elementary calculus, for the area of the surface obtained by revolving an arc about the x-axis (y-axis):

(9)
$$A(S) = 2\pi \int_{s_1}^{s_2} y\, ds \; \left(2\pi \int_{s_1}^{s_2} x\, ds\right).$$

(ii) For a plane area in the $z = 0$ plane, formula (7) reduces to the familiar formula for plane areas in polar coordinates: $\iint_R \rho\, d\rho\, d\theta$.

Example 1. Compute the surface area of a sphere of radius a, using each of the three representations of Examples 2 and 4, § 1812.

Solution. From Example 2, § 1812, if R represents the unit circle $u^2 + v^2 \leq 1$, we have (by symmetry):

$$A(S) = 2\iint_R \frac{4a^2}{(1 + u^2 + v^2)^2} \, dA = 4\pi a^2.$$

The integrals for the representation of Example 4, § 1812, are

$$A(S) = 2\int_0^{2\pi}\int_0^{\frac{\pi}{2}} a^2 \sin\phi \, d\phi \, d\theta = 4\pi a^2,$$

$$A(S) = 2\int_0^{2\pi}\int_0^a \frac{a\rho}{\sqrt{a^2 - \rho^2}} \, d\rho \, d\theta = 4\pi a^2.$$

In the last case, a convergent improper integral is involved.

NOTE 4. From formula (9), with the help of formulas from § 1023 for the centroid of an arc of length L (assuming constant density equal to unity):

$$(\bar{x}, \bar{y}) = \left(\frac{1}{L}\int_{s_1}^{s_2} x \, ds, \frac{1}{L}\int_{s_1}^{s_2} y \, ds\right),$$

we have immediately a derivation of the **Theorem of Pappus** (cf. Ex. 15, § 1022): *The area of the surface S obtained by revolving about the y-axis (x-axis) an arc of length L and centroid $(\bar{x}, \bar{y})$ is given by the formula $A(S) = 2\pi\bar{x}L$ ($A(S) = 2\pi\bar{y}L$).*

Example 2. Let S be the surface of the torus generated by revolving the circle $(x - b)^2 + y^2 = a^2 (0 < a < b)$ about the y-axis. By the Theorem of Pappus (Note 4), $A(S) = 4\pi^2 ab$. The Theorem of Pappus also provides a simple means of computing the centroid of a homogeneous semicircular thin rod (cf. the Example, § 1023). If this rod is described by the arc $C: \rho = a > 0$, $0 \leq \theta \leq \pi$, $\bar{x} = 0$, and $\bar{y}$ can be computed from the formula for the area of the sphere generated by revolving C about the x-axis: $4\pi a^2 = 2\pi\bar{y} \cdot \pi a$. Therefore the centroid is $\left(0, \dfrac{2a}{\pi}\right)$. (Cf. Ex. 16, § 1022.)

1815. EXERCISES

1. Find the area of the surface generated by revolving the arch of the curve $y = \sin x$, $0 \leq x \leq \pi$, about (*a*) the x-axis; (*b*) the y-axis.

2. Find the surface area of each spheroid ($a > b > 0$):

(*a*) $\dfrac{x^2}{a^2} + \dfrac{y^2}{b^2} + \dfrac{z^2}{b^2} = 1;$ (*b*) $\dfrac{x^2}{a^2} + \dfrac{y^2}{a^2} + \dfrac{z^2}{b^2} = 1.$

3. Find the area of the surface generated by revolving the cycloid $x = a(t - \sin t)$, $y = a(1 - \cos t)$, $0 \leq t \leq 2\pi$, about (*a*) the x-axis; (*b*) the y-axis.

4. Find the area of the surface generated by revolving the hypocycloid $x = a\cos^3 t$, $y = a\sin^3 t$ about either coordinate axis.

In Exercises 5–12, find the area of the given surface subject to the specified restrictions.

5. Sphere of radius a; between two parallel planes that cut it and are separated by a distance $b > 0$.

6. Plane $ax + by + cz + d = 0$, $c \neq 0$; inside the cylinder $x^2 + y^2 = r^2$, $r > 0$.
7. Cylinder $y^2 + z^2 = a^2$; in the first octant bounded by the planes $x = 0$, $y = 0$, $z = kx$, $a > 0$, $k > 0$.
8. Sphere $x^2 + y^2 + z^2 = a^2$; inside the cylinder $x^2 + y^2 = ax$, $a > 0$.
9. Cylinder $x^2 + y^2 = ax$; inside the sphere $x^2 + y^2 + z^2 = a^2$, $a > 0$.
10. Hyperbolic paraboloid $z = xy$; inside the cylinder $x^2 + y^2 = a^2$, $a > 0$.
11. Half-cone $z = \rho$; inside the cylinder $\rho = 2a \cos \theta$, $a > 0$.
12. Cylinder $x^2 + z^2 = a^2$; inside the cylinder $y^2 + z^2 = a^2$, $a > 0$.
13. Find the total surface area of the volume common to the two cylinders of Exercise 12.
14. Show that the area of that portion of the half-cone $z = k\sqrt{x^2 + y^2}$, $k > 0$, directly above a region in the xy-plane of area A is $A\sqrt{k^2 + 1}$.
15. Find the area of the torus of Example 2, § 1814, by integration with parameters ϕ and θ: $x = (b + a \cos \phi) \cos \theta$, $y = (b + a \cos \phi) \sin \theta$, $z = a \sin \phi$, $-\pi \leq \phi \leq \pi$, $0 \leq \theta \leq 2\pi$.
16. Give an analytic proof of the two statements given in the last sentence of Note 2, § 1811. *Hint:* Use the transformation formulas of I, § 1814.
*17. Prove that the area of a smooth surface is independent of the particular representation, or decomposition, used. *Hint:* If S is simultaneously the union of pieces $S_1, S_2, \cdots, S_m$ and $T_1, T_2, \cdots, T_n$, consider S_i as the union of $S_i \cap T_1$, $S_i \cap T_2, \cdots, S_i \cap T_n$, $i = 1, 2, \cdots, m$, and show that $A(S_i) = \sum_{j=1}^{n} A(S_i \cap T_j)$ by considering the decompositions of the corresponding rectangles R_i, $i = 1, 2, \cdots, m$, associated with the intersections $S_i \cap T_j$. (Cf. §§ 1006, 1026.)

1816. SURFACE INTEGRALS

If $f(x, y, z)$ is a function defined and continuous for all points (x, y, z) of a smooth surface S, the **surface integral**

(1) $$\iint_S f(x, y, z) \, dS$$

can be defined in much the same way that a line integral $\int_C f(x, y) \, ds$ was in § 1802, in terms of its parametric representation. For a *surface element* the integral takes the form

(2) $$\iint_S f(x, y, z) \, dS \equiv \iint_R \phi(u, v)\sqrt{j_1^2 + j_2^2 + j_3^2} \, du \, dv$$
$$= \iint_R \phi(u, v)\sqrt{EG - F^2} \, du \, dv,$$

where $\phi(u, v) \equiv f(x(u, v), y(u, v), z(u, v))$, j_1, j_2, and j_3 are the Jacobians (2), § 1811, E, F, and G are defined in Note 2, § 1814, and R is the rectangle of integration, $a \leq u \leq b$, $c \leq v \leq d$. For a more general smooth surface the integral (1) is defined as the sum of the integrals (2), for the constituent surface elements. The value of (2) is independent of the decomposition and the parametrization. (Cf. I, § 1814, Ex. 16, § 1815; also cf. Note 4, § 1802.)

One particular form of the integral (2) should be sufficient to indicate the manner in which the formulas of § 1814 can be adapted to surface integrals in general. If $z = z(x, y)$ is the equation of a surface element S, then

$$(3) \quad \iint_S f(x, y, z) \, dS = \iint_R f(x, y, z(x, y)) \sqrt{\left(\frac{\partial z}{\partial x}\right)^2 + \left(\frac{\partial z}{\partial y}\right)^2 + 1} \, dx \, dy$$

$$= \iint_R f(x, y, z(x, y)) \sec \gamma \, dx \, dy,$$

from (4), § 1814. This relation can also be expressed in the form (with $g = f \sec \gamma$):

$$(4) \quad \iint_S g(x, y, z) \cos \gamma \, dS = \iint_R g(x, y, z(x, y)) \, dx \, dy.$$

Note. If the integrand in (1) is identically 1, the value of the surface integral is the area of S.

Example. By means of formulas analogous to those of §§ 1017 and 1021 for first moments and centroids, find the centroid of a homogeneous hemispherical surface, using the three parametrizations of Examples 2 and 4, § 1812.

Solution. With a density function $\delta(x, y, z)$ defined on a surface S, **first moments** relative to the coordinate planes are defined:

$$M_{yz} \equiv \iint_S \delta(x, y, z) x \, dS, \quad M_{zx} \equiv \iint_S \delta(x, y, z) y \, dS, \quad M_{xy} \equiv \iint_S \delta(x, y, z) z \, dS,$$

and the **mass** and **centroid** are:

$$M \equiv \iint_S \delta(x, y, z) \, dS, \quad (\bar{x}, \bar{y}, \bar{z}) \equiv \left(\frac{M_{yz}}{M}, \frac{M_{zx}}{M}, \frac{M_{xy}}{M}\right).$$

Let us consider the hemisphere $z = \sqrt{a^2 - x^2 - y^2}$, $a > 0$, with density δ identically equal to 1. Then M becomes the area $2\pi a^2$, and by symmetry $\bar{x} = \bar{y} = 0$. It remains only to evaluate M_{xy} and compute $\bar{z}$. With the parametrizations of Examples 2 and 4, § 1812 (cf. Example 1, § 1814), M_{xy} has the evaluations:

$$\iint_S z \, dS = 4a^3 \iint_{u^2+v^2 \leq 1} \frac{1 - (u^2 + v^2)}{(1 + u^2 + v^2)^3} \, du \, dv$$

$$= 4a^3 \int_0^{2\pi} \int_0^1 \frac{1 - \rho^2}{(1 + \rho^2)^3} \rho \, d\rho \, d\theta = \pi a^3,$$

$$\iint_S z \, dS = a^3 \int_0^{2\pi} \int_0^{\frac{\pi}{2}} \cos \phi \sin \phi \, d\phi \, d\theta = \pi a^3,$$

$$\iint_S z \, dS = a \int_0^{2\pi} \int_0^a \sqrt{a^2 - \rho^2} \frac{\rho}{\sqrt{a^2 - \rho^2}} \, d\rho \, d\theta = \pi a^3.$$

Finally, $\bar{z} = \pi a^3 / 2\pi a^2 = \frac{1}{2} a$.

1817. ORIENTABLE SMOOTH SURFACES

The requirement that a smooth surface S be locally a surface (Definition II, § 1812) implies that through every point of S passes a unique normal line collinear with the vector $(\partial \vec{r}/u) \times (\partial \vec{r}/v)$. At each point of S, therefore, there are exactly two normal directions (normal unit vectors). Whether it is possible to pick out *one* of these directions to serve as *the* normal direction, and to do this in a consistent continuous manner for the entire surface S, depends on whether or not S enjoys the following property "in the large":

Definition. *A smooth surface S is **orientable** if and only if there exists a vector-valued function*

$$\vec{n} = \vec{n}(p) \tag{1}$$

*defined, continuous, normal to S, and of unit length at every point p of S. Otherwise S is **nonorientable**.*

Example 1. The cylindrical surface $x^2 + y^2 = a^2, 0 \leq z \leq h, a > 0$, is orientable. One suitable family (1) is $(x/a)\vec{i} + (y/a)\vec{j}$. Another is the negative of this.

Example 2. The sphere $x^2 + y^2 + z^2 = a^2, a > 0$, is orientable. Two families of normal directions are

$$\vec{n} = \frac{x}{a}\vec{i} + \frac{y}{a}\vec{j} + \frac{z}{a}\vec{k} \tag{2}$$

and

$$\vec{n} = -\frac{x}{a}\vec{i} - \frac{y}{a}\vec{j} - \frac{z}{a}\vec{k}. \tag{3}$$

Example 3. The **Möbius band** is a simple example of a nonorientable smooth surface. A model can be constructed by pasting together the two ends of a long strip of paper after a half-twist has been given to one of them (cf. Fig. 1811).† If

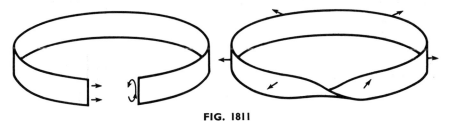

FIG. 1811

one starts attaching thumbtacks to represent unit vectors, in one portion of this surface, and then proceeds continuously around the entire strip, one eventually finds thumbtacks being placed "back-to-back," in contradiction to the definition of orientability. For this reason, the Möbius band is also called a "one-sided"

† The following parametrization for the Möbius band is given in R. Courant, *Differential and Integral Calculus* (New York, Nordemann Publishing Company, Inc., 1937), vol. 2, p. 379: $x = 2\cos u + v \sin \frac{1}{2}u \cos u$, $y = 2 \sin u + v \sin \frac{1}{2}u \sin u$, $z = v \cos \frac{1}{2}u$, $0 \leq u \leq 2\pi$, $-1 \leq v \leq 1$.

surface. For further discussion of one-sided surfaces, see D. Hilbert and S. Cohn-Vossen, *Geometry and the Imagination* (New York, Chelsea Publishing Company, 1952), pp. 302–312.

If a region Ω of E_3 has a closed smooth surface S as its frontier, then S is of necessity orientable. This can be seen as follows: At an arbitrary point p of S let one variable of x, y, z be determined as a continuously differentiable function of the other two, say $z = f(x, y)$. Then for points (x, y) near p the region Ω either contains points "immediately above" (x, y) or "immediately below" (x, y) and not both, since otherwise the part of S near p would not be a part of the frontier of Ω. For the same reason, one of the two normal directions at p points "in the direction of Ω" and the other points "away from Ω." These two directions are called the **inner normal** and the **outer normal** respectively. Either one can be chosen to fit the preceding Definition, but for future purposes we shall choose as *the* normal direction the *outer* normal direction. The continuity of the resulting function $\vec{n}(p)$ is a consequence of the continuity, as a function of x and y for the point p discussed above, of the vector

$$\left[\left(\frac{\partial z}{\partial x}\right)^2 + \left(\frac{\partial z}{\partial y}\right)^2 + 1\right]^{-\frac{1}{2}} \left[\frac{\partial z}{\partial x}\vec{i} + \frac{\partial z}{\partial y}\vec{j} - \vec{k}\right].$$

It follows as a simple corollary of the preceding remarks that any portion of a smooth surface that is the boundary of a region is orientable. Equivalently, any smooth surface that can be *extended* to surround a region is orientable. The Möbius band of Example 3, therefore, cannot be a part of the smooth boundary of any region.

If ϕ is a scalar field that is continuously differentiable in an open set containing an orientable smooth surface S, then the **normal derivative** of ϕ with respect to the surface S, at any point of S, is defined:

(4) $$\frac{d\phi}{dn} \equiv \operatorname{grad} \phi \cdot \vec{n}$$

In the particular case when S is the surface $\phi(x, y, z) = \phi(x_0, y_0, z_0)$, oriented by the direction of grad ϕ, the normal derivative of ϕ with respect to S at the point (x_0, y_0, z_0) reduces to the normal derivative of ϕ, as defined in § 910.

1818. SURFACES WITH EDGES AND CORNERS

In order to describe the kind of surface that is most simply exemplified by a cube and a tetrahedron, that is, a surface with edges and corners, we are forced to drop the requirement that the set under consideration be locally a surface at every point. One effect of this loss is that it is essential to be much more careful than was the case for Definition II, § 1812, in prescribing the exact manner in which the surface elements shall be permitted to combine. We must avoid, for example, having two surface elements touching at a common vertex without having "adjacent" elements to "fill the gap" on one side or the other.

Definition I. *A **sectionally smooth surface** is a set S in E_3 that can be represented as the union of a finite number of surface elements $S_1, S_2, \cdots, S_n$ having the following three properties: (i) the intersection of any two of these surface elements is either empty, a common vertex of the two, or a common edge of the two; (ii) if two surface elements S_i and S_j of S have exactly one vertex p in common, there exists a chain of surface elements of S, $S_i = S_{k_1}$, $S_{k_2}, \cdots, S_{k_m} = S_j$, every one of which has p as a vertex and such that every consecutive pair $\{S_{k_1}, S_{k_2}\}, \cdots, \{S_{k_{m-1}}, S_{k_m}\}$ has a common edge; (iii) no three surface elements have more than one point in common.* A **connected sectionally smooth surface** *is a sectionally smooth surface S having the property that if S_i and S_j are any two of its surface elements, there exists a chain of surface elements of S, $S_i = S_{k_1}, S_{k_2}, \cdots, S_{k_m} = S_j$ such that every consecutive pair has a nonempty intersection:*

$$S_{k_1} \cap S_{k_2} \neq \emptyset, S_{k_2} \cap S_{k_3} \neq \emptyset, \cdots, S_{k_{m-1}} \cap S_{k_m} \neq \emptyset.$$

As with smooth surfaces, so with sectionally smooth surfaces in general, *surface area* is defined to be the sum of the areas of the individual surface elements. A similar remark applies to surface integrals in general. It is the concept of *orientability* that becomes appreciably more difficult, since any family of normal unit vectors is necessarily discontinuous at an edge or corner. One way out of the difficulty is considered in this section. Another is treated in § 1822. Henceforth, unless explicit statement to the contrary is made, *the single unmodified word* **surface** *should be interpreted to mean sectionally smooth surface*.

Let S be a (sectionally smooth) surface (connected or not) that is the frontier of a region Ω in E_3. For such a set an orientation can be assigned by letting $\bar{n}(p)$ be the *outer unit normal* (cf. § 1817) whenever p is a regular point, and by permitting $\bar{n}(p)$ to be either undefined or multiple-valued at every point p belonging to an edge or vertex. By the **boundary** of a region in E_3 whose frontier is a surface S, we shall mean the surface S *together with the outer normal orientation*.

Example. If Ω is the region between two concentric spheres, its boundary consists of the two spherical surfaces. On the outer sphere the outer normal points away from the common center of the two spheres, while on the inner sphere the outer normal is directed toward their common center.

1819. THE DIVERGENCE THEOREM

In a manner similar to that employed for Green's theorem in the plane, we define an **elementary compact region** in E_3 to be a compact region described simultaneously by three sets of inequalities: (i) $z_1(x, y) \leq z \leq z_2(x, y)$ for an elementary compact region R_{xy} in the xy-plane having the property that the set of all (x, y, z) such that $(x, y) \in R_{xy}$ and $z = z_1(x, y)$ is a surface and the set of all (x, y, z) such that $(x, y) \in R_{xy}$ and $z = z_2(x, y)$ is a surface, and two other sets of inequalities of the form (ii) $x_1(y, z) \leq x \leq x_2(y, z)$ and (iii) $y_1(z, x) \leq y \leq y_2(z, x)$, similarly restricted (Fig. 1812). A **finitely decomposable compact region** W is a compact region that is a finite union of elementary compact regions any two of which have at most frontier points in common. It follows that whenever two of these elementary compact

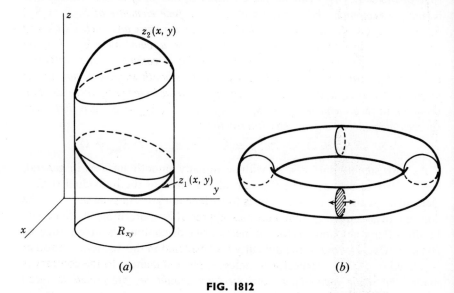

(a) (b)

FIG. 1812

regions have a surface of their boundaries in common, this surface has opposite orientations for the two elementary compact regions. The **boundary** of W consists of the total oriented surface made up of those oriented boundary surfaces of the elementary compact regions constituting W that belong to exactly *one* of these elementary compact regions (except possibly for boundary arcs). The torus in Figure 1812(b), when cut as shown into four elementary regions has four inner circular surfaces that "cancel out" because of opposite orientations.

We now come to a theorem of great importance in applied fields, particularly physics, and of basic theoretical significance in relation to the divergence function. Although it is usually known as the *divergence theorem*, it is also referred to as *Gauss's theorem* and *Green's theorem*.

Theorem. Divergence Theorem. *Let W be a finitely decomposable compact region in E_3, and let S be its boundary, with outer normal orientation, the outer normal unit vector being*

(1) $$\vec{n}(p) = \cos \alpha \, \vec{i} + \cos \beta \, \vec{j} + \cos \gamma \, \vec{k}.$$

Let P, Q, R be continuously differentiable throughout an open set containing W, and let

(2) $$\vec{F}(p) \equiv P(x, y, z)\vec{i} + Q(x, y, z)\vec{j} + R(x, y, z)\vec{k}.$$

Then

(3) $$\iiint_W \left(\frac{\partial P}{\partial x} + \frac{\partial Q}{\partial y} + \frac{\partial R}{\partial z}\right) dV = \iint_S (P \cos \alpha + Q \cos \beta + R \cos \gamma) \, dS$$

or, in vector notation,

(4) $$\iiint_W \operatorname{div} \vec{F} \, dV = \iint_S \vec{F} \cdot \vec{n} \, dS = \iint_S \vec{F} \cdot d\vec{S}.$$

That is, *the volume integral, over W, of the divergence of the vector field $\vec{F}$ is equal to the surface integral, over S, of its outer normal component.*

Proof. As in § 1805 we reduce the proof to the case where W is an elementary compact region and, furthermore, to the simple case of the equality between the two third terms of (3):

(5) $$\iiint_W \frac{\partial R}{\partial z} \, dV = \iint_S R \cos \gamma \, dS.$$

We start by evaluating the left-hand member of (5) as an iterated integral (cf. (xii), § 1011, § 1805, Fig. 1812(a)):

(6) $$\iint_{R_{xy}} \left\{ \int_{z_1(x,y)}^{z_2(x,y)} \frac{\partial R}{\partial z} \, dz \right\} dA$$
$$= \iint_{R_{xy}} R(x, y, z_2(x, y)) \, dA - \iint_{R_{xy}} R(x, y, z_1(x, y)) \, dA.$$

We now look at the right-hand member of (5). Since $\cos \gamma = 0$ on S except for the upper and lower surfaces $z = z_2(x, y)$ and $z = z_1(x, y)$, the surface integral of (5) consists of the corresponding two parts. For the first of these, since γ is acute, this angle is the same as the angle γ of formula (3), § 1816, and the part of $\iint_S R \cos \gamma \, dS$ corresponding to the upper surface of W becomes $\iint_{R_{xy}} R(x, y, z_2(x, y)) \cos \gamma \sec \gamma \, dA$, the first term on the right of (6). For the lower surface of W, $z = z_1(x, y)$, the angle γ of (5), above, is the supplement of the angle γ of (3), § 1816. The change in sign resulting from this fact accounts for the second term on the right of (6), and the proof is complete.

One of the more significant consequences of the divergence theorem is a representation of the divergence of a vector field at a fixed point p in terms of the limiting behavior of the field normal to a small spherical surface S surrounding $p : (x_0, y_0, z_0)$:

(7) $$\operatorname{div} \vec{F} = \lim_{r \to 0} \frac{1}{V} \iint_S \vec{F} \cdot \vec{n} \, dS,$$

where V is the volume of the sphere W of radius r about p, and S is the surface of this sphere. To prove formula (7) we use the fact that $\phi(x, y, z) \equiv \operatorname{div} \vec{F}$ is a continuous function, and hence, by the Mean Value Theorem for Multiple Integrals (cf. Theorem XIII, § 1006),

$$\frac{1}{V}\iint_S \vec{F} \cdot \vec{n}\, dS = \frac{1}{V}\iiint_W \phi(x, y, z)\, dV = \phi(\xi, \eta, \zeta),$$

where $(\xi, \eta, \zeta) \in \Omega$. Therefore $\phi(\xi, \eta, \zeta) \to \phi(x_0, y_0, z_0)$, and (7) follows.

NOTE 1. Formula (7) not only demonstrates that the divergence of a vector field, as previously defined, is independent of the coordinate system, but displays the appropriateness of the term *divergence*. Suppose, for definiteness, that the vector $\vec{F}(p)$ is the velocity of a homogeneous fluid at a point p. Then the quantity $\vec{F} \cdot \vec{n}$ is the normal component of the velocity, and the integral $\iint_S \vec{F} \cdot \vec{n}\, dS$ measures the resultant rate at which the fluid is passing out, or *diverging*, through the surface S. For an *incompressible* fluid the net change of mass through a surface S that is the boundary of a fixed region Ω is zero, and div $\vec{F}$ must vanish. In general, a solenoidal vector field (one with identically vanishing divergence—cf. Example 2, § 1714) can be thought of as the velocity vector field of a fluid without "sources" or "sinks." (Solenoidal vector fields are discussed further in § 1826.)

NOTE 2. Formula (3) of the divergence theorem is sometimes written

(8) $$\iiint_W \left(\frac{\partial P}{\partial x} + \frac{\partial Q}{\partial y} + \frac{\partial R}{\partial z}\right) dV = \iint_S P\, dy\, dz + Q\, dz\, dx + R\, dx\, dy$$

(cf. (4), § 1816). If this form of the right-hand member is used it should be understood that each double differential (e.g., $dx\, dy$) represents a "differential area" in a coordinate plane and not on the surface S itself.

Example. Let S be a surface that is met by an arbitrary ray from the origin O in at most one point, and let the normal to S be taken so that the angle θ between this normal and the radius vector at any point of S satisfies the inequalities $0 \leq \theta \leq \frac{1}{2}\pi$. The **solid angle** ω subtended at the origin by S consists of all points belonging to rays from the origin that meet S. If Σ is the portion of the surface of the unit sphere $x^2 + y^2 + z^2 = 1$ that belongs to a solid angle ω, the area of Σ is by definition the **measure** of ω, written $\mu(\omega)$. Assuming that the set of points Φ between the origin and S, including the origin and S, is a finitely decomposable compact region, show that the measure of ω is given by the formula (where $r^2 = x^2 + y^2 + z^2$ and dr/dn is a normal derivative, defined in (4), § 1817):

(9) $$\mu(\omega) = \iint_S \frac{\cos\theta}{r^2}\, dS = \iint_S \frac{1}{r^2}\frac{dr}{dn}\, dS.$$

Solution. Let $\epsilon > 0$ be so small that the portion σ of the surface of the sphere $r = \epsilon$ that belongs to ω is between the origin and S, and that the compact region W consisting of all points of Φ whose distance from the origin is at least ϵ is finitely decomposable. If we apply the divergence theorem to the vector field $|\vec{r}|^{-3}\vec{r} = r^{-3}\vec{r}$, for the compact region W, we have

(10) $$\iiint_W \text{div}\left(\frac{\vec{r}}{r^3}\right) dV = \iint_S \frac{\vec{r} \cdot \vec{n}}{r^3} dS + \iint_\sigma \frac{\vec{r} \cdot \vec{n}}{r^3} dS$$
$$= \iint_S \frac{\cos\theta}{r^2} dS - \iint_\sigma \frac{1}{\epsilon^2} dS.$$

A direct computation shows that

$$\text{div}\left(\frac{\vec{r}}{r^3}\right) = \frac{\partial}{\partial x}[x(x^2+y^2+z^2)^{-\frac{3}{2}}] + \frac{\partial}{\partial y}[y(x^2+y^2+z^2)^{-\frac{3}{2}}] + \cdots$$
$$= (x^2+y^2+z^2)^{-\frac{5}{2}}[(-2x^2+y^2+z^2)+\cdots] = 0.$$

Therefore the two surface integrals in the last expression of (10) are equal. An entirely similar development shows that if Σ is permitted to play the role of S in (10):

(11) $$\iint_\Sigma 1 \, dS = \iint_\sigma \frac{1}{\epsilon^2} dS.$$

Equating equals to equals, we have the first part of the conclusion. The last expression in (9) is equal to its predecessor since for $r = (x^2+y^2+z^2)^{\frac{1}{2}}$, $\text{grad } r = r^{-1}\vec{r}$ (a unit vector), and $dr/dn = \cos\theta$.

A particular case is of interest. If W is a finitely decomposable compact region with surface S, and if the origin O is a point in the interior of W such that every ray from O meets S in exactly one point, then $\iint_S \frac{\cos\theta}{r^2} dS = 4\pi$. Of course, if the origin is not a point of W (including S), then the divergence theorem guarantees that $\iint_S \frac{\cos\theta}{r^2} dS = 0$.

1820. GREEN'S IDENTITIES

The divergence theorem permits simple proofs of the following two relations (1) and (2), where the compact region W and the surface S are assumed to be subject to the restrictions described in § 1819, and the functions $f = f(x, y, z)$ and $g = g(x, y, z)$ are continuously differentiable. These are known as **Green's first and second identities**, respectively (cf. Ex. 26, § 1809);

(1) $$\iiint_W [f \text{ Lapl } g + \text{grad } f \cdot \text{grad } g] \, dV = \iint_S f \frac{dg}{dn} dS,$$

(2) $$\iiint_W [f \text{ Lapl } g - g \text{ Lapl } f] \, dV = \iint_S \left[f \frac{dg}{dn} - g \frac{df}{dn}\right] dS.$$

The symbols $\frac{df}{dn}$ and $\frac{dg}{dn}$ represent normal derivatives (cf. (4), § 1817):

$$\frac{df}{dn} \equiv \text{grad } f \cdot \vec{n}, \quad \frac{dg}{dn} \equiv \text{grad } g \cdot \vec{n}.$$

To prove (1) we first use the divergence theorem with $\vec{F}$ replaced by $f\vec{F}$ in (4), § 1819, together with property (v), § 1713 ($\operatorname{div}(f\vec{v}) = (\operatorname{grad} f) \cdot \vec{v} + f \operatorname{div} \vec{v}$):

(3) $$\iiint_W [f \operatorname{div} \vec{F} + (\operatorname{grad} f) \cdot \vec{F}] \, dV = \iint_S f\vec{F} \cdot \vec{n} \, dS.$$

This becomes (1) if $\vec{F} = \operatorname{grad} g$. Equation (2) follows immediately from (1).

Two special cases of (1) should be noted:

(4) $$\iiint_W [f \operatorname{Lapl} f + |\operatorname{grad} f|^2] \, dV = \iint_S f \frac{df}{dn} \, dS,$$

(5) $$\iiint_W \operatorname{Lapl} f \, dV = \iint_S \frac{df}{dn} \, dS.$$

1821. HARMONIC FUNCTIONS

A function $f(x, y, z)$ is said to be **harmonic** in a region Ω if and only if it has continuous second-order partial derivatives and satisfies Laplace's differential equation there (cf. § 1714):

(1) $$\operatorname{Lapl} f = \operatorname{div} \operatorname{grad} f = \frac{\partial^2 f}{\partial x^2} + \frac{\partial^2 f}{\partial y^2} + \frac{\partial^2 f}{\partial z^2} = 0.$$

Harmonic functions are important in certain applied fields such as fluid dynamics, electromagnetism, heat conduction, and elasticity. For a brief introduction, see W. Kaplan, *Advanced Calculus* (Reading, Mass., Addison-Wesley Press, Inc., 1952). For a more extended treatment, see Sir Harold and Lady Jeffreys, *Methods of Mathematical Physics* (New York, Cambridge University Press, 1956).

One of the important and interesting properties of a harmonic function is that its values throughout any bounded region are completely determined by its values on the surface of that region. More precisely:

Theorem. *If W is a finitely decomposable compact region, and if $f(x, y, z)$ and $g(x, y, z)$ are harmonic in an open set containing W, and identical on the surface S of W, then f and g are identical throughout W.*

Proof. The theorem is an immediate consequence of the following lemma (let $h \equiv f - g$):

Lemma. *If $h(x, y, z)$ is harmonic in an open set containing W, in the preceding Theorem, and if h vanishes identically on the surface S, then h vanishes identically throughout W.*

Proof of Lemma. Under these assumptions, by (4), § 1820, since $\operatorname{Lapl} h$ vanishes identically throughout W and h vanishes identically on S,

(1) $$\iiint_W |\operatorname{grad} h|^2 \, dV = 0.$$

Since the integrand in (1) is nonnegative and continuous, and the integral $= 0$, this integrand must vanish identically throughout W (cf. Theorem XII, § 1006, (*vii*), § 1011). Therefore all three first partial derivatives of h vanish identically in W, and thus h must be a constant there (Theorem II, § 912). Since h vanishes on S its constant value must be 0.

1822. EXERCISES

In the following exercises the symbol W represents a finitely decomposable compact region, and S its surface. The unit outer normal to the surface S is denoted $\vec{n}$. All functions appearing in formulas are assumed to possess continuous derivatives of as high an order as is desired.

1. Compute the surface area of the sphere $r = 2a \cos \phi$, $a > 0$, using the spherical coordinates ϕ and θ as parameters.

2. Evaluate $\iint_S (x + z^2) \, dS$, where S is the portion of the half-cone $z = \rho$ ($\rho \geq 0$) within the cylinder $\rho = 2 \cos \theta$.

3. Find the centroid of the surface of Exercise 8, § 1815.

4. Find the moment of inertia of a homogeneous spherical surface of radius a with respect to a diameter.

5. Find the moment of inertia with respect to the z-axis of the surface of the torus of Example 2, § 1814, and Exercise 15, § 1815, assuming a homogeneous density distribution.

6. Find the attractive force between the homogeneous hemispherical surface $r = 2a \cos \phi$, $z \geq a > 0$, and a unit mass at the origin.

7. Show that the volume V of W can be expressed by each of the following formulas

$$V = \iint_S x \, dy \, dz = \iint_S y \, dz \, dx = \iint_S z \, dx \, dy$$

$$= \frac{1}{3} \iint_S (x \cos \alpha + y \cos \beta + z \cos \gamma) \, dS = \frac{1}{3} \iint_S \vec{r} \cdot \overrightarrow{dS}.$$

8. Show that the x-coordinate of the centroid of W can be expressed by each of the following formulas:

$$\bar{x} = \frac{1}{2V} \iint_S x^2 \, dy \, dz = \frac{1}{V} \iint_S xy \, dz \, dx = \frac{1}{V} \iint_S xz \, dx \, dy.$$

9. Show that the moment of inertia of W with respect to the z-axis, assuming constant density 1, can be expressed by each of the following formulas:

$$I_z = \frac{1}{3} \iint_S (x^3 + 3xy^2) \, dy \, dz = \frac{1}{3} \iint_S (3x^2y + y^3) \, dz \, dx = \iint_S (x^2z + y^2z) \, dx \, dy.$$

10. Evaluate $\iint_S (x^3 \cos \alpha + y^3 \cos \beta + z^3 \cos \gamma) \, dS$, where S is the surface of the sphere $x^2 + y^2 + z^2 = a^2$, $a > 0$. (Cf. Ex. 9.)

11. Verify the divergence theorem for the vector field $x^2\vec{i} + y^2\vec{j} + z^2\vec{k}$ and the cube $0 \le x \le 1, 0 \le y \le 1, 0 \le z \le 1$.

12. Verify the divergence theorem for the vector field $xz\vec{i} + 2yz\vec{j} + 3xy\vec{k}$ and the cylinder $x^2 + y^2 \le 4, 0 \le z \le 3$.

13. Use the divergence theorem to evaluate
$$\iint_S x^2\,dy\,dz + y^2\,dz\,dx + z^2\,dx\,dy$$
if S is the surface of the sphere $(x - x_0)^2 + (y - y_0)^2 + (z - z_0)^2 = a^2$.

14. Derive the following formula for the Laplacian of a scalar field f:
$$\mathrm{Lapl}\,f = \lim_{r \to 0} \frac{1}{V} \iint_S \frac{df}{dn}\,dS.$$
(Cf. (7), § 1819.) Discuss independence of the coordinate system.

15. Establish the formula
$$\iiint_W r\,dV = \frac{1}{4} \iint_S r\vec{r} \cdot \vec{dS},$$
and verify it for the sphere $r \le a$.

16. Show that the volume of W is
$$V = \frac{1}{3} \iint_S r\frac{dr}{dn}\,dS.$$
Is this formula valid if the origin is a point of W?

17. Prove that a vector field $\vec{F}$ is solenoidal in a region Ω if and only if whenever S is the boundary of a finitely decomposable compact region W lying in Ω,
$$\iint_S \vec{F} \cdot \vec{dS} = 0.$$

18. Prove that a scalar field f is harmonic in a region Ω if and only if whenever S is the boundary of a finitely decomposable compact region W lying in Ω,
$$\iint_S \frac{df}{dn}\,dS = 0.$$

19. Discuss the analogue for three dimensions of Note 3, § 1806, and establish the formula
$$\iint_{S_1} P\,dy\,dz + Q\,dz\,dx + R\,dx\,dy = \iint_{S_2} P\,dy\,dz + Q\,dz\,dx + R\,dx\,dy.$$

20. Interpret and establish the formulas:
$$\iiint_W \mathrm{curl}\,\vec{F}\,dV = \iint_S \vec{n} \times \vec{F}\,dS,$$
$$\iiint_W \mathrm{grad}\,\phi\,dV = \iint_S \vec{n}\,\phi\,dS.$$

1823. ORIENTABLE SECTIONALLY SMOOTH SURFACES

For the divergence theorem of § 1819 we defined the orientation of the boundary S, consisting of one or more connected pieces of a closed surface,

§ 1823] ORIENTABLE SECTIONALLY SMOOTH SURFACES

of a region Ω of space in terms of this region by means of the concept of *outer normal*. For the theorem of the following section, a single connected surface will be under consideration, but this surface will not in general bound a region. On the contrary, it may itself have a boundary, consisting of one or more curves. The problem now will be to define orientability for such a surface, in case edges and corners exist, and to relate the orientation of the surface to the orientation of its boundary. We start with a surface element.

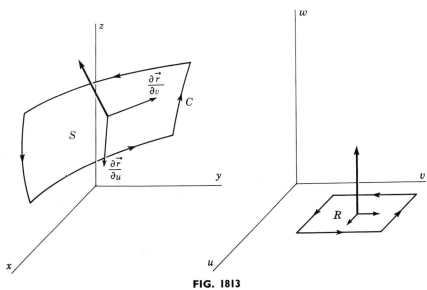

FIG. 1813

Definition I. *Let S be a surface element $x(u, v)$, $y(u, v)$, $z(u, v)$ for (u, v) belonging to the rectangle R: $a \leq u \leq b$, $c \leq v \leq d$, and let R be located in the uv-plane of a right-handed uvw coordinate system. Then the **orientation** of S relative to a given orientation of its boundary C is in the direction of*

$$\frac{\partial \vec{r}}{\partial u} \times \frac{\partial \vec{r}}{\partial v} \quad \text{or} \quad -\frac{\partial \vec{r}}{\partial u} \times \frac{\partial \vec{r}}{\partial v}$$

according as the orientation of C corresponds to a counterclockwise or clockwise orientation of the boundary of R, as viewed from a point on the positive w-axis. (Cf. Fig. 1813.)

From an intuitive point of view, this means that the normal to S is directed relative to the boundary C according to the right-handed convention as follows: If the surface element S were to be set into rotation about an axis normal to S, at some point, the direction of rotation being determined by the sense of C, then a screw with right-handed threading would advance in the positive normal direction. Alternatively, if one thinks of traveling around on one side of S, along the directed boundary C, in such a way as to keep the inside of S to one's left, then one is on the positively oriented side of S.

In order to verify the statements of the preceding paragraph, as well as to show that the orientation defined above is independent of the parametrization, we represent the cross product vector of Definition I in terms of two of the coordinates x, y, z, say for definiteness x and y (cf. (4), § 1811, (7), § 922):

$$\frac{\partial \vec{r}}{\partial u} \times \frac{\partial \vec{r}}{\partial v} = \left[\frac{\partial(y, z)}{\partial(x, y)} \vec{i} + \frac{\partial(z, x)}{\partial(x, y)} \vec{j} + \frac{\partial(x, y)}{\partial(x, y)} \vec{k} \right] \frac{\partial(x, y)}{\partial(u, v)}$$

$$= \left[-\frac{\partial z}{\partial x} \vec{i} - \frac{\partial z}{\partial y} \vec{j} + \vec{k} \right] \frac{\partial(x, y)}{\partial(u, v)}.$$

The direction of this vector, then, is upward or downward according as the Jacobian $\partial(x, y)/\partial(u, v)$ is positive or negative; in other words, according as

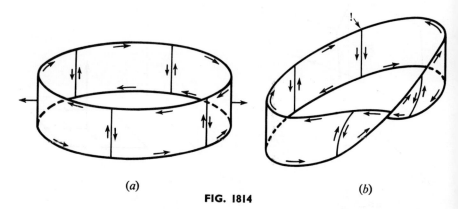

(a) (b)

FIG. 1814

the mapping of R to the projection of S into the xy-plane, $x = x(u, v)$ $y = y(u, v)$ preserves or reverses orientation. (Cf. § 1027.)

We are now in a position to define orientability for a surface with edges and corners when the concept of "outer normal" may be meaningless:

Definition II. *A (sectionally smooth) surface S is **orientable** if and only if the boundaries C_i of the surface elements constituting S can be oriented in such a way that whenever an arc is a common edge of two of these surface elements the two orientations given this arc by the two surface elements are opposite. The **orientation** of S at the inner points of the surface elements is that determined by the orientations of C_i. A surface that is not orientable is **nonorientable**.*

Hereafter, when the terms *orientable surface* and *boundary* are used, it will be assumed that both the surface and its boundary are provided with orientations in a consistent manner, as described above.

Example 1. The curved surface of a right circular cylinder, and the Möbius band (cf. Examples 1, 3, § 1817) are again orientable and nonorientable, respectively, as shown in Figure 1814.

NOTE. As with smooth surfaces, discussed in the first paragraph of § 1817, any surface that is part of a surface bounding a region in space is orientable.

Example 2. The surface consisting of the three triangles lying in the coordinate planes with vertices $(0, 0, 0)$, $(1, 0, 0)$, $(0, 1, 0)$, $(0, 0, 1)$ is orientable (cf. Fig. 1815). It is part of the boundary of a tetrahedron.

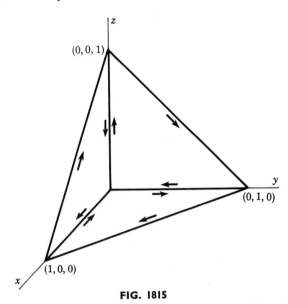

FIG. 1815

1824. STOKES'S THEOREM

The concept of line integral, defined in § 1802, extends readily to higher-dimensional spaces, in particular to E_3. For this case we shall be interested in line integrals of the forms

(1) $$\int_C P\,dx + Q\,dy + R\,dz,$$

(2) $$\int_C f\,ds,$$

where C consists of a finite number of sectionally smooth curves, and P, Q, R, and f are defined and continuous, as functions of the point (x, y, z), on C. As with plane curves, if the orientation on C is reversed, the sign of (1) is changed, and that of (2) is unchanged. If $\vec{F} = P\vec{i} + Q\vec{j} + R\vec{k}$, and if $\vec{\tau}$ represents the forward unit tangent vector to C, then (1) can be written:

(3) $$\int_C P\,dx + Q\,dy + R\,dz = \int_C \vec{F} \cdot \vec{\tau}\,ds = \int_C \vec{F} \cdot \vec{ds}.$$

(Cf. (4), (5), § 1802.)

With these concepts, and those of § 1823, in mind we can extend Green's theorem in the plane to E_3:

Theorem. Stokes's Theorem. *Let S be an orientable surface with boundary C, consisting of one or more simple closed curves. Let the unit normal vector to S, $\vec{n} = \cos\alpha\,\vec{i} + \cos\beta\,\vec{j} + \cos\gamma\,\vec{k}$ and the unit tangent vector $\vec{\tau}$ to C be related according to the right-handed convention of Definition II, § 1823. If P, Q, and R are continuously differentiable in an open set containing S, then*

$$(4)\quad \iint_S \left[\left(\frac{\partial R}{\partial y} - \frac{\partial Q}{\partial z}\right)\cos\alpha + \left(\frac{\partial P}{\partial z} - \frac{\partial R}{\partial x}\right)\cos\beta + \left(\frac{\partial Q}{\partial x} - \frac{\partial P}{\partial y}\right)\cos\gamma\right] dS$$

$$= \int_C P\,dx + Q\,dy + R\,dz$$

or, in vector notation with $\vec{F} \equiv P\vec{i} + Q\vec{j} + R\vec{k}$,

$$(5)\quad \iint_S (\operatorname{curl} \vec{F}) \cdot \vec{n}\,dS = \iint_S (\operatorname{curl} \vec{F}) \cdot \vec{dS} = \int_C \vec{F} \cdot \vec{\tau}\,ds = \int_C \vec{F} \cdot \vec{ds}.$$

That is, the surface integral of the normal component of the curl is equal to the line integral of the tangential component of the vector.

Proof. Because of additive properties of line and surface integrals, it is sufficient to prove the theorem for a single surface element S: $x(u, v)$, $y(u, v)$, $z(u, v)$, for (u, v) in the rectangle R: $a \leq u \leq b$, $c \leq v \leq d$, with counterclockwise orientation in the uv-plane for the boundary Γ of R. Since the direction cosines of $\vec{n}$ are j_1/σ, j_2/σ, j_3/σ, where $\sigma = \sqrt{j_1^2 + j_2^2 + j_3^2}$, the left-hand member of (4) becomes

$$(6)\quad \iint_R \left[\left(\frac{\partial R}{\partial y} - \frac{\partial Q}{\partial z}\right)j_1 + \left(\frac{\partial P}{\partial z} - \frac{\partial R}{\partial x}\right)j_2 + \left(\frac{\partial Q}{\partial x} - \frac{\partial P}{\partial y}\right)j_3\right] dA,$$

as a consequence of the definition of a surface integral, § 1816. The right-hand member of (4) can be written

$$(7)\quad \int_\Gamma \left[P\left(\frac{\partial x}{\partial u}du + \frac{\partial x}{\partial v}dv\right) + Q\left(\frac{\partial y}{\partial u}du + \frac{\partial y}{\partial v}dv\right) + R\left(\frac{\partial z}{\partial u}du + \frac{\partial z}{\partial v}dv\right)\right]$$

$$= \int_\Gamma \left[P\frac{\partial x}{\partial u} + Q\frac{\partial y}{\partial u} + R\frac{\partial z}{\partial u}\right]du + \left[P\frac{\partial x}{\partial v} + Q\frac{\partial y}{\partial v} + R\frac{\partial z}{\partial v}\right]dv.$$

A simple application of Green's theorem converts (7) to

$$(8)\quad \iint_R \left\{\frac{\partial}{\partial u}\left[P\frac{\partial x}{\partial v} + Q\frac{\partial y}{\partial v} + R\frac{\partial z}{\partial v}\right] - \frac{\partial}{\partial v}\left[P\frac{\partial x}{\partial u} + Q\frac{\partial y}{\partial u} + R\frac{\partial z}{\partial u}\right]\right\} dA.$$

If (8) is expanded, by means of formulas typified by

$$\frac{\partial P}{\partial u} = \frac{\partial P}{\partial x}\frac{\partial x}{\partial u} + \frac{\partial P}{\partial y}\frac{\partial y}{\partial u} + \frac{\partial P}{\partial z}\frac{\partial z}{\partial u},$$

24 terms are obtained of which 12 cancel immediately. The remaining 12 are identical with the expansion of (6), and the proof is complete.

NOTE. A special case of Stokes's theorem is of interest: Under the assumptions of Stokes's theorem, and the further assumption that S is a closed surface (having no boundary points), the surface integrals on the left of (4) and (5) vanish:

$$\iint_S (\operatorname{curl} \vec{F}) \cdot \vec{n} \, dS = \iint_S (\operatorname{curl} \vec{F}) \cdot \vec{dS} = 0.$$

Much as the divergence theorem gives a direct meaning to the divergence function, Stokes's theorem gives a direct meaning to the curl function. Let $p = (x_0, y_0, z_0)$ be any point in the region of definition of the vector field curl $\vec{F}$, and let curl $\vec{F}_0$ denote the value of curl $\vec{F}$ at the point p. Let S be a small oriented circular disk with center p, radius r, area $A = \pi r^2$, and boundary C, let θ be the angle between curl $\vec{F}$ and the normal unit vector $\vec{n}$ to S, and let θ_0 be the value of θ for the particular vector curl $\vec{F}_0$. Then the component of curl $\vec{F}$ normal to S at any point of S is $(\operatorname{curl} \vec{F}) \cdot \vec{n} = |\operatorname{curl} \vec{F}| \cos \theta$. By the Mean Value Theorem for Multiple Integrals (Theorem XIII, § 1006), owing to the continuity of all functions involved and the planar nature of S, the integral of $(\operatorname{curl} \vec{F}) \cdot \vec{n}$ over S can be expressed

(9) $$\iint_S (\operatorname{curl} \vec{F}) \cdot \vec{n} \, dS = A |\operatorname{curl} \vec{F}| \cos \theta,$$

where $|\operatorname{curl} \vec{F}|$ and $\cos \theta$ are evaluated at some point of S. If we divide both members of (9) by the area $A = \pi r^2$, and let $r \to 0$ while holding $\vec{n}$ fixed, we have in the limit, by Stokes's theorem:

(10) $$|\operatorname{curl} \vec{F}_0| \cos \theta_0 = \lim_{r \to 0} \frac{1}{A} \int_C \vec{F} \cdot \vec{\tau} \, ds.$$

From this formula we can describe the curl of a vector field at any point p as a vector whose direction is normal to the circular disk about p for which the limit in (10) is maximum, and whose magnitude is that maximum value. If $\vec{F}$ represents the velocity vector of a fluid, formula (10) demonstrates the appropriateness of the term *curl*. Any point p where the curl $\vec{F} \neq \vec{0}$ is called a **vortex**. In general an irrotational vector field (one with identically vanishing curl—cf. Example 3, § 1714), can be thought of as a velocity vector field without vortices.

Example 1. Verify Stokes's theorem for the vector field $z\vec{i} + x^2\vec{j} + y^3\vec{k}$ and the surface of Example 2, § 1823.

Solution. The surface integral in (4) is $\iint_S (3y^2 \cos \alpha + \cos \beta + 2x \cos \gamma) \, dS$, which decomposes into the three surface integrals in the yz-, zx-, and xy-planes, respectively:

$$-\int_0^1 \int_0^{1-z} 3y^2 \, dy \, dz - \int_0^1 \int_0^{1-x} dz \, dx - \int_0^1 \int_0^{1-y} 2x \, dx \, dy = -\frac{13}{12}.$$

The line integral in (4), when expressed in terms of segments in the yz-, zx-, and xy-planes, respectively, becomes

$$\int_{(0,0,1)}^{(0,1,0)} y^3 \, dz + \int_{(1,0,0)}^{(0,0,1)} z \, dx + \int_{(0,1,0)}^{(1,0,0)} x^2 \, dy$$

$$= \int_1^0 (1-z)^3 \, dz + \int_1^0 (1-x) \, dx + \int_1^0 (1-y)^2 \, dy = -\frac{13}{12}.$$

Example 2. Verify formula (10) for the vector field $\vec{F} = -y\vec{i} + x\vec{j}$, and the disk S in the xy-plane $x^2 + y^2 \leq r^2$, oriented positively.

Solution. By direct computation, curl $\vec{F} = 2\vec{k}$, $\vec{n} = \vec{k}$, $\cos\theta = 1$, and the left-hand member of (10) is equal to 2. On the other hand, with the polar coordinate θ as parameter,

$$\vec{F} \cdot \vec{\tau} = (-r\sin\theta\,\vec{i} + r\cos\theta\,\vec{j}) \cdot (-\sin\theta\,\vec{i} + \cos\theta\,\vec{j}) = r,$$

$$\int_C \vec{F} \cdot \vec{\tau}\, ds = \int_0^{2\pi} r \cdot r\, d\theta = 2\pi r^2,$$

and the right-hand member of (10) is $\lim\limits_{r \to 0} \dfrac{2\pi r^2}{\pi r^2} = 2$.

1825. INDEPENDENCE OF PATH. SCALAR POTENTIAL

Most of the concepts and results of §§ 1806–1808, for two dimensions, extend to three. In this section we treat some of these matters but minimize the details that are routine extensions of those of §§ 1806–1808. (Cf. Exs. 1–4, § 1827.)

If the functions P, Q, and R are continuous throughout a region Ω, the expression $P\, dx + Q\, dy + R\, dz$ is called an **exact differential** in Ω if and only if there exists a differentiable function ϕ, defined throughout Ω, such that

(1) $$\frac{\partial \phi}{\partial x} = P, \quad \frac{\partial \phi}{\partial y} = Q, \quad \frac{\partial \phi}{\partial z} = R,$$

and hence

(2) $$P\, dx + Q\, dy + R\, dz = \frac{\partial \phi}{\partial x} dx + \frac{\partial \phi}{\partial y} dy + \frac{\partial \phi}{\partial z} dz = d\phi.$$

In this case ϕ is called a **scalar potential** for the differential expression $P\, dx + Q\, dy + R\, dz$ and for the vector field $\vec{F} = P\vec{i} + Q\vec{j} + R\vec{k} = \text{grad } \phi$, in Ω. A differential expression, or a vector field, is **locally exact** in a region if and only if it is exact in some neighborhood of every point of the region.

Theorem I. *If $P\, dx + Q\, dy + R\, dz$ is an exact differential, with scalar potential ϕ, in a region Ω, and if C is a sectionally smooth arc in Ω joining the point (x_0, y_0, z_0) to the point (x_1, y_1, z_1), then*

(3) $$\int_C d\phi = \int_C P\, dx + Q\, dy + R\, dz = \phi(x_1, y_1, z_1) - \phi(x_0, y_0, z_0).$$

Theorem II. *If P, Q, and R are continuous throughout the region Ω, then $P\, dx + Q\, dy + R\, dz$ is an exact differential in Ω if and only if the line integral*

§ 1825] INDEPENDENCE OF PATH. SCALAR POTENTIAL   639

$\int_C P\,dx + Q\,dy + R\,dz$ is independent of the path or, equivalently, if and only if $\int_C P\,dx + Q\,dy + R\,dz$ vanishes for every sectionally smooth closed curve in Ω.

Theorem III. *Let* $P = P(x, y, z)$, $Q = Q(x, y, z)$, *and* $R = R(x, y, z)$ *be continuously differentiable throughout a region* Ω. *Then* $P\,dx + Q\,dy + R\,dz$ *is locally exact in* Ω *if and only if*

(4) $$\frac{\partial R}{\partial y} = \frac{\partial Q}{\partial z}, \quad \frac{\partial P}{\partial z} = \frac{\partial R}{\partial x}, \quad \frac{\partial Q}{\partial x} = \frac{\partial P}{\partial y}$$

throughout Ω; *that is, if and only if*

(5) $$\operatorname{curl} \vec{F} = \operatorname{curl}(P\vec{i} + Q\vec{j} + R\vec{k})$$

vanishes identically in Ω.

Definition I. *A region* Ω *of* E_3 *is* **simply-connected** *if and only if for every closed polygon* $C: [p_1 p_2 \cdots p_n p_1]$ *lying in* Ω *there exists a point* p *in* Ω *such that for every vertex* p_i *of* C, $i = 1, 2, \cdots, n$, *there exists a smooth arc* C_i *from* p *to* p_i *lying in* Ω *and having the property that for every* $i = 1, 2, \cdots, n$ *the three arcs* C_i, $\overline{p_i p_{i+1}}$, *and* $-C_{i+1}$ (*where* $C_{n+1} \equiv C_1$, $p_{n+1} \equiv p_1$, *and the negative sign indicates a reversal of orientation*) *form the boundary of an orientable surface* S_i *lying in* Ω. *A region is* **multiply-connected** *if and only if it fails to be simply-connected.*

Definition II. *A region* Ω *in* E_3 *is* **star-shaped** *if and only if there exists a point* p *in* Ω *such that for every* q *in* Ω *the segment* $\overline{pq}$ *lies in* Ω. *A region* Ω *is* **convex** *if and only if the line segment* $\overline{pq}$ *lies in* Ω *whenever* p *and* q *do.*

NOTE 1. As in E_2, every convex region is star-shaped, and every star-shaped region is simply-connected.

Theorem IV. *If* Ω *is a simply-connected region of* E_3, *and if* P, Q, *and* R *are continuously differentiable in* Ω, *then* (*under the assumption of sectional smoothness for curves* C *lying in* Ω), *the following five statements are all equivalent:*

(i) $P\,dx + Q\,dy + R\,dz$ *is locally exact in* Ω;
(ii) $P\,dx + Q\,dy + R\,dz$ *is exact in* Ω;
(iii) $\operatorname{curl}(P\vec{i} + Q\vec{j} + R\vec{k})$ *vanishes identically in* Ω;
(iv) $\int_C P\,dx + Q\,dy + R\,dz$ *is independent of the path;*
(v) $\int_C P\,dx + Q\,dy + R\,dz = 0$ *for closed curves* C.

NOTE 2. Recall that a conservative vector field is one that is the gradient of some scalar field, that an irrotational vector field is one whose curl vanishes identically, and that a conservative vector field is always irrotational (cf. Example 1, § 1714). Theorem III states, in part, that *in any simply-connected region an irrotational vector field is always conservative*.

Examples 1. Some examples of convex regions in E_3 are (*i*) the unit sphere $x^2 + y^2 + z^2 < 1$, (*ii*) the inside of a cube or tetrahedron or any regular polyhedron, (*iii*) the open first octant $x > 0$, $y > 0$, $z > 0$, (*iv*) any open half-space, such as $x > 0$. The union of the two half-spaces $x > 0$ and $y > 0$ is star-shaped but not convex. So is the space E_3 with the nonnegative portion of the x-axis deleted, or with the half-plane $z = 0$, $x \leq 0$ deleted. If, in either example of the preceding sentence, the closed unit sphere $x^2 + y^2 + z^2 \leq 1$ is also deleted, the remaining region is simply-connected but not star-shaped. The region obtained by deleting from E_3 either the single point $(0, 0, 0)$ or the closed unit sphere is simply connected.

Examples 2. Show that each of the following three regions is multiply-connected: (*i*) E_3 with the entire z-axis deleted; (*ii*) E_3 with the cylinder $x^2 + y^2 \leq 1$ deleted; (*iii*) the inside of the torus obtained by revolving about the z-axis the circle $y = 0$, $(x - 2)^2 + z^2 < 1$.

Solution. For each of these regions the vector field

$$(6) \qquad \vec{F} = -\frac{y}{x^2 + y^2}\vec{i} + \frac{x}{x^2 + y^2}\vec{j}$$

is continuously differentiable and locally exact, but for the closed curve $C: x = 2\cos\theta$, $y = 2\sin\theta$, $z = 0$ $(\theta\uparrow)$, $\int_C \vec{F} \cdot \vec{\tau}\, ds = 2\pi \neq 0$. By Theorem IV, this would be impossible if any of the regions were simply-connected. For each of the prescribed regions the vector field (6) is irrotational but fails to be conservative (cf. Note 2).

⋆1826. VECTOR POTENTIAL

In the preceding section it was shown that a conservative vector field is always irrotational, and that in a simply-connected region an irrotational vector field is conservative. In a multiply-connected region a vector field whose curl vanishes identically may fail to possess a scalar potential.

In the present section we shall consider a problem somewhat similar to the one just described. If a vector field $\vec{F}$ is the curl of a vector field $\vec{G}$, the vector field $\vec{G}$ is called a **vector potential** for $\vec{F}$. By (*ii*), § 1714, since div curl $\vec{G}$ vanishes identically, any vector field that possesses a vector potential must be solenoidal (cf. Example 2, § 1714). As shown in the following Example, if no restrictions are placed on the region under consideration, a solenoidal vector field may possess no vector potential. In contrast to the problem of existence of a scalar potential for an irrotational vector field, the property of simple-connectedness is irrelevant and is replaced by a somewhat similar property indicated in the Example below.

We shall content ourselves here with proving the existence of a vector potential for a given solenoidal vector field in the particularly simple case where the region is the interior of a rectangular parallelepiped with edges parallel to the coordinate axes, $\Omega: a_1 < x < a_2, b_1 < y < b_2, c_1 < z < c_2$.

Let the given solenoidal vector field be $\vec{F} = P\vec{i} + Q\vec{j} + R\vec{k}$, where P, Q, and R are assumed to be continuously differentiable in Ω, and where

(1) $$\operatorname{div} \vec{F} = \frac{\partial P}{\partial x} + \frac{\partial Q}{\partial y} + \frac{\partial R}{\partial z} = 0$$

identically in Ω. We seek continuously differentiable functions u, v, and w such that

(2) $$\frac{\partial w}{\partial y} - \frac{\partial v}{\partial z} = P, \frac{\partial u}{\partial z} - \frac{\partial w}{\partial x} = Q, \frac{\partial v}{\partial x} - \frac{\partial u}{\partial y} = R.$$

It is a matter of routine differentiation (cf. Theorem I, § 405, Theorem II, § 930), with the aid of (1), to verify that if (x_0, y_0, z_0) is any fixed point of Ω, then the following functions satisfy (2) (the details are left as an exercise for the reader):

(3) $$u = \int_{z_0}^{z} Q(x, y, \zeta) \, d\zeta - \int_{y_0}^{y} R(x, \eta, z_0) \, d\eta, \ v = -\int_{z_0}^{z} P(x, y, \zeta) \, d\zeta, \ w = 0.$$

It is clear from the asymmetry of (3) that a vector potential is by no means unique.

Example. Show that if $\vec{r} = x\vec{i} + y\vec{j} + z\vec{k}$, and $r = |\vec{r}|$, and if Ω is the region $x^2 + y^2 + z^2 > 0$, then the vector field

$$\vec{F} = \frac{\vec{r}}{r^3}$$

is solenoidal in Ω, but has no vector potential there.

Solution. Showing that $\vec{F}$ is solenoidal is straightforward:

$$\frac{\partial}{\partial x}\{(x^2 + y^2 + z^2)^{-\frac{3}{2}} x\} + \frac{\partial}{\partial y}\{(x^2 + y^2 + z^2)^{-\frac{3}{2}} y\} + \cdots$$
$$= (x^2 + y^2 + z^2)^{-\frac{5}{2}} [(-2x^2 + y^2 + z^2) + (x^2 - 2y^2 + z^2) + \cdots] = 0.$$

If S denotes the surface of the sphere $x^2 + y^2 + z^2 = a^2$, $a > 0$, then $\vec{n} = \vec{r}/r$, and $\vec{F} \cdot \vec{n} = 1/r^2$, with $r = a$. Therefore $\iint_S \vec{F} \cdot \vec{n} \, dS = \frac{1}{a^2} \cdot 4\pi a^2 = 4\pi$ (cf. the Example, § 1819). On the other hand, if $\vec{F}$ were the curl of a vector potential, since S is a closed surface the integral $\iint_S \vec{F} \cdot \vec{n} \, dS$ would vanish by the Note, § 1824.

Although the region Ω of this example is simply-connected, it suffers from a pathology somewhat similar to that of multiple-connectedness in the plane. In this case it is closed surfaces of a general "spherical character" (rather than curves of a general "circular character") that cannot be shrunk to a point in the region.

1827. EXERCISES

In Exercises 1–4, prove the indicated Theorem of § 1825.
1. Theorem I. **2.** Theorem II. **3.** Theorem III. **4.** Theorem IV.
5. Evaluate the line integral

$$\int_C 2xy \, dx + (6y^2 - xz) \, dy + 10z \, dz,$$

where C is the path extending from $(0, 0, 0)$ to $(1, 1, 1)$ (a) along the twisted cubic $x = t$, $y = t^2$, $z = t^3$; (b) along the polygonal arc joining the points $(0, 0, 0)$, $(0, 0, 1)$, $(0, 1, 1)$, $(1, 1, 1)$ linearly, in that order; (c) along the straight line segment from $(0, 0, 0)$ to $(1, 1, 1)$.

In Exercises 6–7, evaluate the line integral

$$\int_C z\, dx + x\, dy + y\, dz$$

for the prescribed arc C.

6. The helix $x = a \cos t$, $y = a \sin t$, $z = \mu t$, $a\mu \neq 0$, t increasing from 0 to $\tfrac{1}{2}\pi$.

7. The ellipse $x^2 + y^2 = 8$, $z = x + y$, z increasing, from the point $(2, -2, 0)$ to the point $(2, 2, 4)$.

In Exercises 8–11, evaluate the line integral by finding a scalar potential.

8. $\displaystyle\int_{(1,1,1)}^{(2,2,2)} (x^2 + y^2z^2)\, dx + (y + 2xyz^3)\, dy + (z^3 + 3xy^2z^2)\, dz.$

9. $\displaystyle\int_{(0,0,0)}^{(1,5,\tfrac{1}{2}\pi)} e^x \sin yz\, dx + e^x z \cos yz\, dy + (e^x y \cos yz - \sin z)\, dz.$

10. $\displaystyle\int_{(1,1,1)}^{(e,e,e)} \left(\ln yz + \frac{y+z}{x}\right) dx + \left(\ln zx + \frac{z+x}{y}\right) dy + \left(\ln xy + \frac{x+y}{z}\right) dz.$

11. $\displaystyle\int_{(0,0,0)}^{(1,1,1)} e^{-x^2y^2z^2}(yz\, dx + zx\, dy + xy\, dz).$

12. Evaluate $\displaystyle\int_C x^2y\, ds$, where C is the portion of the helix $x = a \cos \lambda t$, $y = a \sin \lambda t$, $z = \mu t$, $a > 0$, $\lambda\mu \neq 0$, for $t_1 \leq t \leq t_2$.

13. Evaluate $\displaystyle\int_C z^2\, ds$, where C is the circle $x^2 + y^2 + z^2 = a^2$, $x = y$, $a > 0$.

In Exercises 14–17, verify Stokes's theorem for the given vector field and surface.

14. $(x + 2y)\vec{i} + (y + 2z)\vec{j} + (z + 2x)\vec{k}$; the triangle with vertices $(1, 0, 0)$, $(0, 1, 0)$, $(0, 0, 1)$.

15. $y\vec{i} + 2x\vec{j} + z\vec{k}$; the hemisphere $z = \sqrt{1 - x^2 - y^2}$.

16. $z\vec{i} + x\vec{j} + y\vec{k}$; $z = x^2 - y^2$ for $x^2 + y^2 \leq a^2$, $a > 0$.

17. $-y\vec{i} + x\vec{j} + xy\vec{k}$; $z = a^2 - x^2 - y^2$ for $z \geq 0$.

*18. Let $\vec{r} = x\vec{i} + y\vec{j} + z\vec{k}$, $r = |\vec{r}|$, and let Ω be any region of E_3 not containing $(0, 0, 0)$. If $\psi = \psi(x, y, z)$ is a continuous scalar field in Ω, prove that $\psi\vec{r}$ is a conservative vector field in Ω. Infer that $\displaystyle\int_C \psi\vec{r} \cdot \vec{ds}$ is independent of the path C in Ω.

*19. Let C be the (positively oriented) boundary of a smooth plane surface S in E_3, and let $\vec{n} = \cos \alpha\, \vec{i} + \cos \beta\, \vec{j} + \cos \gamma\, \vec{k}$ be the (positive) unit normal of S. Prove that the area of S can be expressed in the form

$$A(S) = \frac{1}{2}\int_C \begin{vmatrix} dx & dy & dz \\ \cos \alpha & \cos \beta & \cos \gamma \\ x & y & z \end{vmatrix}.$$

In Exercises 20–21, find a vector potential for the given vector field in E_3.

*20. $y^2z\vec{i} + xz^2\vec{j} + x^2y\vec{k}.$

*21. $2xyz\vec{i} - x^2y\vec{j} + (x^2z - yz^2)\vec{k}.$

*22. Assume that C is a sectionally smooth closed curve that is the boundary of a sectionally smooth orientable surface S, with unit normal $\cos \alpha \vec{i} + \cos \beta \vec{j} + \cos \gamma \vec{k}$, that lies in an open parallelepiped Ω with sides parallel to the coordinate axes. Prove that if $P\vec{i} + Q\vec{j} + R\vec{k}$ is a solenoidal vector field in Ω, then the value of

$$\iint_S (P \cos \alpha + Q \cos \beta + R \cos \gamma)\, dS$$

is independent of the surface S, for a given C.

*23. If S is any closed orientable sectionally smooth surface with unit normal $\vec{n}$, and if $\vec{c}$ is any constant vector, show that

$$\iint_S \vec{c} \cdot \vec{n}\, dS = 0.$$

Interpret the integral $\iint_S \vec{n}\, dS$ and show that $\iint_S \vec{n}\, dS = \vec{0}$. Conclude that for any closed polyhedron, if $\vec{S}_i$ is an outer normal vector for the ith face, with magnitude equal to the area of that face, then $\sum_{i=1}^{n} \vec{S}_i = \vec{0}$. *Hint:* Find a vector potential for $\vec{c}$, and use the Note, § 1824.

*1828. EXTERIOR DIFFERENTIAL FORMS.† EXERCISES

In the following exercises it is assumed that every real-valued function concerned has continuous derivatives of whatever order may be indicated, in a region Ω of E_3. Every curve C is assumed to be sectionally smooth, oriented, and to lie in Ω. Every surface S is assumed to be sectionally smooth, orientable with normal vector $\vec{n}$, with positively oriented boundary C, and to lie in Ω. Every compact region W is assumed to be finitely decomposable, with outer-oriented boundary S, and to lie in Ω. Reversal of orientation on a curve or surface is indicated by a minus sign, thus: $-C$, $-S$. The notation $[\vec{u}, \vec{v}, \vec{w}]$ is that of the triple scalar product of three vectors (cf. § 1703).

*1. A **0-form** is a scalar field, that is, a real-valued function with domain Ω, denoted

(1) $$\omega = F = F(x, y, z).$$

A **1-form** is a real-valued function (also called **functional**) whose domain of definition is the set of sectionally smooth arcs C lying in Ω, denoted

(2) $$\omega = P\, dx + Q\, dy + R\, dz,$$

where P, Q, and R are scalar fields, with value on C:

(3) $$\omega(C) = \int_C \omega = \int_C P\, dx + Q\, dy + R\, dz.$$

A **2-form** is a real-valued function (also called **functional**) whose domain of definition is the set of sectionally smooth surfaces S lying in Ω, denoted

(4) $$\omega = A\, dy \wedge dz + B\, dz \wedge dx + C\, dx \wedge dy,\ddagger$$

---

† For a more extensive treatment of exterior differential forms, see R. C. Buck, *Advanced Calculus* (New York, McGraw-Hill Book Co., 1956) and H. K. Nickerson, D. C. Spencer, and N. E. Steenrod, *Advanced Calculus* (Princeton, D. Van Nostrand Co., 1959).

‡ The symbol $\wedge$ is informally called *wedge*.

where A, B, and C are scalar fields, with value on S:

(5) $$\omega(S) = \iint_S \omega \equiv \iint_S \{A[\vec{n},\vec{j},\vec{k}] + B[\vec{n},\vec{k},\vec{i}] + C[\vec{n},\vec{i},\vec{j}]\}\, dS$$
$$= \iint_S (A\vec{n}\cdot\vec{i} + B\vec{n}\cdot\vec{j} + C\vec{n}\cdot\vec{k})\, dS = \iint_S (A\cos\alpha + B\cos\beta + C\cos\gamma)\, dS.$$

A **3-form** is a real-valued function (also called **functional**) whose domain of definition is the set of finitely decomposable compact regions W lying in Ω, denoted

(6) $$\omega = G\, dx \wedge dy \wedge dz,$$

where G is a scalar field, with value on W:

(7) $$\omega(W) = \iiint_W \omega \equiv \iiint_W G[\vec{i},\vec{j},\vec{k}]\, dV = \iiint_W G\, dV.$$

Prove that any **exterior differential form**, (1), (2), (4), or (6), is identically 0 over its domain of definition if and only if all scalar functions involved in its definition are identically 0. State and prove a corresponding relation for *equality* of two differential forms. Define an operation of **addition** for two differential forms of the same class (e.g., $(P\, dx + Q\, dy + R\, dz) + (U\, dx + V\, dy + W\, dz) \equiv (P+U)x\, d + (Q+V)\, dy + (R+W)\, dz$) so that all forms of the same class constitute an additive commutative group (cf. Ex. 26, § 103). Extend the definitions of 2-forms and 3-forms, in a natural way, to include such forms as $dx \wedge dz$ and $dx \wedge dx \wedge dz$, and thus justify such relations as:

(8) $$dz \wedge dy = -dy \wedge dz, \quad dx \wedge dz = -dz \wedge dx, \quad \cdots,$$
(9) $$dx \wedge dx = dy \wedge dy = dz \wedge dz = 0,$$
(10) $dx \wedge dy \wedge dz = dy \wedge dz \wedge dx = dz \wedge dx \wedge dy$
$$= -dx \wedge dz \wedge dy = -dz \wedge dy \wedge dx = -dy \wedge dx \wedge dz,$$
(11) $$dx \wedge dx \wedge dy = dy \wedge dz \wedge dy = \cdots = 0.$$

Establish relations of the type exemplified by
$$\iint_S \omega_1 + \omega_2 = \iint_S \omega_1 + \iint_S \omega_2, \quad \iint_{-S} \omega = \iint_S (-\omega) = -\iint_S \omega.$$

★2. Multiplication of forms is defined in terms of expansions typified:

(12) $F \wedge (A\, dy \wedge dz + B\, dz \wedge dx + C\, dx \wedge dy)$
$$= FA\, dy \wedge dz + FB\, dz \wedge dx + FC\, dx \wedge dy,$$
(13) $(P\, dx + Q\, dy + R\, dz) \wedge (U\, dx + V\, dy + W\, dz)$
$$= (QW - RV)\, dy \wedge dz + (RU - PW)\, dz \wedge dx + (PV - QU)\, dx \wedge dy$$
(14) $(A\, dy \wedge dz + B\, dz \wedge dx + C\, dx \wedge dy) \wedge (P\, dx + Q\, dy + R\, dz)$
$$= (AP + BQ + CR)\, dx \wedge dy \wedge dz,$$
(15) (1-form) $\wedge$ (3-form) = (2-form) $\wedge$ (2-form) = 0, (2-form) $\wedge$ (3-form) = 0.

Show how formal expansions of the left-hand members give formulas (12)–(15). Prove that multiplication of forms is associative, and that it is commutative except when both forms are 1-forms, in which case multiplication is anticommutative (cf. (1), § 1702). Show how multiplication by a 0-form corresponds to multiplication of a vector by a scalar, how the product of two 1-forms corresponds to a vector product, how the product of a 1-form and a 2-form corresponds to a scalar product, and how the product of three 1-forms corresponds to a triple scalar product. Discuss such relations as the following: (i) $dx \wedge dy$ is the same, whether considered as a basic 2-form or as a product of two 1-forms; (ii) $dx \wedge dy \wedge dz$ is the same, whether considered as a basic 3-form, or as a product of a 1-form and a 2-form, or as a product of three 1-forms.

★3. Differentiation of forms is defined:

(16) d(0-form) = 1-form:
$$dF \equiv \frac{\partial F}{\partial x} dx + \frac{\partial F}{\partial y} dy + \frac{\partial F}{\partial z} dz.$$

(17) d(1-form) = 2-form:
$$d(P\,dx + Q\,dy + R\,dz) \equiv dP \wedge dx + dQ \wedge dy + dR \wedge dz.$$

(18) d(2-form) = 3-form:
$$d(A\,dy \wedge dz + B\,dz \wedge dx + C\,dx \wedge dy)$$
$$= dA \wedge dy \wedge dz + dB \wedge dz \wedge dx + dC \wedge dx \wedge dy.$$

(19) d(3-form) = 0.

Derive the following formulas:

(20) $d(P\,dx + Q\,dy + R\,dz)$
$$= \left(\frac{\partial R}{\partial y} - \frac{\partial Q}{\partial z}\right) dy \wedge dz + \left(\frac{\partial P}{\partial z} - \frac{\partial R}{\partial x}\right) dz \wedge dx + \left(\frac{\partial Q}{\partial x} - \frac{\partial P}{\partial y}\right) dx \wedge dy,$$

(21) $d(A\,dy \wedge dz + B\,dz \wedge dx + C\,dx \wedge dy) = \left(\frac{\partial A}{\partial x} + \frac{\partial B}{\partial y} + \frac{\partial C}{\partial z}\right) dx \wedge dy \wedge dz.$

Obtain a relation between the operation of gradient, curl, and divergence, and the differential of a 0-form, a 1-form, and a 2-form, respectively.

★4. Derive the formula

(22) $$du \wedge dv \wedge dw = \frac{\partial(u, v, w)}{\partial(x, y, z)} dx \wedge dy \wedge dz.$$

★5. Show that the divergence theorem (§ 1819) can be expressed:

(23) $$\iiint_\Omega d\omega = \iint_S \omega.$$

★6. Show that Stokes's theorem (§ 1824) can be expressed:

(24) $$\iint_S d\omega = \int_C \omega.$$

★7. Prove that if ω is any form, then

(25) $$d\,d\omega = 0.$$

Show that if ω is a 0-form this corresponds to the vector formula curl grad $\vec{v} = \vec{0}$, and that if ω is a 1-form this corresponds to the vector formula div curl $\vec{v} = 0$.

★8. Show that if $d\omega = 0$, then ω has the form $d\omega_1$, locally. Show that this corresponds, if ω is a 1-form, to the local existence of a scalar potential for an irrotational vector field and, if ω is a 2-form, to the local existence of a vector potential for a solenoidal vector field.

19

Differential Geometry

1901. INTRODUCTION

In Chapters 8 and 18 the fundamental definitions introducing space curves and surfaces were given and modestly explored. In this chapter we shall investigate more thoroughly the geometry of curves and surfaces in E_3, and obtain certain applications to the kinematics of a particle moving along a space curve. In this presentation we shall limit ourselves almost exclusively to local properties (that is, properties associated with points in a neighborhood of a given point). For a more complete treatment, historical commentary, and further references, see D. J. Struik, *Differential Geometry* (Reading, Mass., Addison-Wesley Publishing Company, Inc., 1950).

All curves and surfaces considered in this chapter are assumed to be smooth. When derivatives of order higher than the first are involved, it will tacitly be assumed that these derivatives exist and are continuous.

1902. CURVATURE. OSCULATING PLANE

Let $\vec{r} = \vec{r}(t)$ be the variable radius vector of a point moving along a smooth space curve C, the parameter being t (which can often be thought of informally as time), and let s represent arc length along C measured from a fixed point and increasing in the direction of increasing t. The assumption of smoothness implies that

(1) $$\left|\frac{d\vec{r}}{dt}\right|^2 = \left(\frac{dx}{dt}\right)^2 + \left(\frac{dy}{dt}\right)^2 + \left(\frac{dz}{dt}\right)^2 = \left(\frac{ds}{dt}\right)^2 > 0$$

(cf. §§ 805, 809), so that t and hence $\vec{r}$ become differentiable functions of s. For certain purposes we shall on occasion wish to consider $\vec{r}$ as a function of s as a parameter.

Our first formula recalls (§§ 809, 1711) the notation for $d\vec{r}/ds$, the *unit tangent vector* in the direction of increasing s:

(2) $$\vec{\tau} = \frac{d\vec{r}}{ds}.$$

As the point p moves along the curve C the vector $\vec{\tau}$, though fixed in magnitude, will in general vary in direction. Its rate of change is important.

§ 1902] CURVATURE. OSCULATING PLANE

In the first place, differentiating the identity $\vec{\tau} \cdot \vec{\tau} = 1$, with respect to s, we obtain, from (4), § 1711,

(3) $$\vec{\tau} \cdot \frac{d\vec{\tau}}{ds} + \frac{d\vec{\tau}}{ds} \cdot \vec{\tau} = 2 \frac{d\vec{\tau}}{ds} \cdot \vec{\tau} = 0,$$

and hence the fact that *the vector*

(4) $$\frac{d\vec{\tau}}{ds} = \frac{d^2\vec{r}}{ds^2} = \frac{d^2x}{ds^2}\vec{i} + \frac{d^2y}{ds^2}\vec{j} + \frac{d^2z}{ds^2}\vec{k}$$

is normal to C. Its magnitude is defined to be the **curvature** and the reciprocal of its magnitude the **radius of curvature** of C at the point in question, denoted, respectively,

(5) $$K \equiv \left|\frac{d\vec{\tau}}{ds}\right|, \quad R \equiv \frac{1}{K}.$$

The direction of (4) is called the **principal normal**, with the notation

(6) $$\vec{\nu} \equiv R \frac{d\vec{\tau}}{ds} = R \frac{d^2\vec{r}}{ds^2},$$

$$\frac{d\vec{\tau}}{ds} = \frac{d^2\vec{r}}{ds^2} = K\vec{\nu}.$$

The plane through a point p of C and containing both $\vec{\tau}$ and $\vec{\nu}$ is the **osculating plane** of C at p. The vector (4), with initial point p has as its terminal point the **center of curvature** of C at p. The **circle of curvature**, or **osculating circle**, is the circle lying in the osculating plane whose center is the center of curvature and whose radius is the radius of curvature.

If $\vec{r}$ is the radius vector of a general point p: (x, y, z), and $\vec{r}_0$ that of a fixed point p_0: (x_0, y_0, z_0) on a smooth curve C whose curvature K at p_0 is not zero, then a necessary and sufficient condition for p to lie in the osculating plane to C at p_0 is that the three vectors $\vec{r} - \vec{r}_0$, $\vec{\tau}$, and $\vec{\nu}$ be collinear; that is, that the triple scalar product $[\vec{r} - \vec{r}_0, \vec{\tau}, \vec{\nu}]$ equal 0. By (6), § 710, this gives the equation of the osculating plane in rectangular coordinates:

(7) $$\begin{vmatrix} x - x_0 & y - y_0 & z - z_0 \\ x_0' & y_0' & z_0' \\ x_0'' & y_0'' & z_0'' \end{vmatrix} = \begin{vmatrix} x & y & z & 1 \\ x_0 & y_0 & z_0 & 1 \\ x_0' & y_0' & z_0' & 0 \\ x_0'' & y_0'' & z_0'' & 0 \end{vmatrix} = 0,$$

where primes indicate differentiation with respect to arc length s. In terms of a more general parameter t, we have the following relations, where primes now indicate differentiation with respect to t:

(8) $$\vec{r}' = \frac{d\vec{r}}{ds} s' = s'\vec{\tau},$$

(9) $$\vec{r}'' = s''\vec{\tau} + s'^2 \frac{d\vec{\tau}}{ds} = s''\vec{\tau} + Ks'^2\vec{\nu}.$$

Therefore $[\vec{r} - \vec{r}_0, \vec{r}', \vec{r}''] = Ks'^3 [\vec{r} - \vec{r}_0, \vec{\tau}, \vec{\nu}]$, and *equation* (7) *remains valid for the osculating plane in case the primes indicate differentiation with respect to a general parameter.*

Equations (8) and (9) furnish a formula for the curvature. Taking a vector product, we have

(10) $\quad \vec{r}' \times \vec{r}'' = s'\vec{\tau} \times (s''\vec{\tau} + Ks'^2\vec{v}) = Ks'^3\vec{\tau} \times \vec{v},$

and since $\vec{\tau}$ and $\vec{v}$ are orthogonal unit vectors, their cross product is a unit vector, so that on taking magnitudes and using (1), we have:

(11) $$K = \frac{|\vec{r}' \times \vec{r}''|}{|\vec{r}'|^3}.$$

In the special case that the curve C lies in the xy-plane, (11) reduces to $K = |x'y'' - y'x''|[x'^2 + y'^2]^{-\frac{3}{2}}$, in agreement with (2), § 811.

The osculating plane of a curve C at a point p_0 can be described as the plane through p_0 that "hugs the curve C" more closely than any other. To see this, assume that the coordinates x, y, and z, as functions of s, have bounded third derivatives in a neighborhood of p_0: (x_0, y_0, z_0). Then, by Taylor's formula with a remainder (§ 1204), with $\Delta s \equiv s - s_0$ where $x_0 = x(s_0), \cdots,$

$$x = x_0 + x_0' \Delta s + \tfrac{1}{2}x_0'' \Delta s^2 + O(\Delta s^3),$$

the primes indicating differentiating with respect to s, with similar expansions for y and z. Hence

$$\vec{r} = \vec{r}_0 + \vec{r}_0' \Delta s + \tfrac{1}{2}\vec{r}_0'' \Delta s^2 + \vec{O}(\Delta s^3),$$

where $\vec{O}(\Delta s^3)$ represents a vector whose magnitude is $O(\Delta s^3)$ as $\Delta s \to 0$. Using equations (2) and (6), we can write this "vector expansion" in the form.

(12) $\quad \vec{r} - \vec{r}_0 = (\Delta s)\vec{\tau} + (\tfrac{1}{2}K\Delta s^2)\vec{v} + \vec{O}(\Delta s^3).$

Since the first term on the right-hand side of (12) is in the direction of the tangent line to C at p_0, and the sum of the first two terms is a vector in the osculating plane, we can conclude that the vector from p_0 to p differs from a vector in a general tangent plane (that is, one containing the tangent line) by a vector whose magnitude is $O(\Delta s^2)$, whereas it differs from a vector in the osculating plane by a vector whose magnitude is $O(\Delta s^3)$.

Example. Find the vectors $\vec{\tau}$ and $\vec{v}$, the curvature and the radius of curvature and the equation of the osculating plane at a general point of the helix

(13) $\quad x = a \cos t, \quad y = a \sin t, \quad z = bt, \quad ab \neq 0,$

which winds around the cylinder $x^2 + y^2 = a^2$. (This helix is called **right-handed** or **left-handed** according as b is *positive* or *negative*; cf. the Note, § 1904.)

First Solution. Setting $s \equiv \int_0^t \left[\left(\frac{dx}{d\theta}\right)^2 + \left(\frac{dy}{d\theta}\right)^2 + \left(\frac{dz}{d\theta}\right)^2 \right]^{\frac{1}{2}} d\theta = \sqrt{a^2 + b^2}\, t,$

we have in terms of the parameter s, with $c \equiv \sqrt{a^2 + b^2}$:

$$\vec{r} = a \cos \frac{s}{c} \vec{i} + a \sin \frac{s}{c} \vec{j} + b \frac{s}{c} \vec{k},$$

(14) $\quad \vec{\tau} = \dfrac{d\vec{r}}{ds} = -\dfrac{a}{c} \sin \dfrac{s}{c} \vec{i} + \dfrac{a}{c} \cos \dfrac{s}{c} \vec{j} + \dfrac{b}{c} \vec{k},$

$\quad K\vec{v} = \dfrac{d^2\vec{r}}{ds^2} = -\dfrac{a}{c^2} \cos \dfrac{s}{c} \vec{i} - \dfrac{a}{c^2} \sin \dfrac{s}{c} \vec{j}.$

These equations provide the vectors $\vec{\tau}$ and $\vec{\nu}$ $\left(\vec{\nu} = -\dfrac{a}{|a|}\cos\dfrac{s}{c}\vec{i} - \dfrac{a}{|a|}\sin\dfrac{s}{c}\vec{j}\right)$, and the constant curvature $K = |a|/c^2 = |a|/(a^2 + b^2)$. The radius of curvature is therefore $R = (a^2 + b^2)/|a|$. From (7) and (14) the equation of the osculating plane, at the point given by a fixed value of t, reduces to $bx \sin t - by \cos t + az = abt$.

Second Solution. In terms of the parameter t,

(15)
$$\vec{r} = a \cos t\, \vec{i} + a \sin t\, \vec{j} + bt\vec{k},$$
$$\vec{r}' = -a \sin t\, \vec{i} + a \cos t\, \vec{j} + b\vec{k},$$
$$\vec{r}'' = -a \cos t\, \vec{i} - a \sin t\, \vec{j}.$$

From (1), $s' = ds/dt = \sqrt{a^2 + b^2}$, and hence $s'' = 0$. From (11), $K = |a|/(a^2 + b^2)$, so that by (8) and (9), with $c \equiv \sqrt{a^2 + b^2}$, $\vec{\tau} = \vec{r}'/c$ and $\vec{\nu} = \vec{r}''/|a|$, in agreement with the results of the first solution.

1903. APPLICATIONS TO KINEMATICS

Formulas related to a curve C, when the coordinates are expressed in terms of time t as a parameter, have special interpretation. For example, as shown in § 1711, the velocity and acceleration vectors are

(1) $\quad$ velocity $= \vec{V} = \dfrac{d\vec{r}}{dt}$; $\quad$ acceleration $= \vec{A} = \dfrac{d\vec{v}}{dt} = \dfrac{d^2\vec{r}}{dt^2}.$

Since $\left(\dfrac{ds}{dt}\right)^2 = \left(\dfrac{dx}{dt}\right)^2 + \left(\dfrac{dy}{dt}\right)^2 + \left(\dfrac{dz}{dt}\right)^2$, and $\dfrac{d\vec{r}}{dt} = \dfrac{dx}{dt}\vec{i} + \dfrac{dy}{dt}\vec{j} + \dfrac{dz}{dt}\vec{k}$, we have for the speed,

(2) $\quad$ speed $= V = |\vec{V}| = \left|\dfrac{d\vec{r}}{dt}\right| = \dfrac{ds}{dt},$

where it is assumed that the distance s increases with increasing t.

From (9), § 1902, we have immediately the fact that the acceleration vector always lies in the osculating plane of the trajectory of the particle under consideration, with the following resolution into components, one in the direction of the tangent and the other in the direction of the principal normal:

(3) $$\vec{A} = \dfrac{dV}{dt}\vec{\tau} + \dfrac{V^2}{R}\vec{\nu}.$$

The tangential component is the time rate of change of the speed, and the principal normal component is equal to the square of the speed divided by the radius of curvature. Two special cases of (3) are of interest: (*i*) for linear motion the acceleration vector is in the direction of motion and equal to the rate of change of the speed; (*ii*) for uniform motion (constant speed) the acceleration vector is normal to the curve and directed toward the center of curvature, of magnitude V^2/R.

Example. For motion of a particle in the Euclidean plane, resolve the velocity and acceleration vectors into radial and transverse components; that is, into components in the direction of and normal to the radius vector of the moving point.

Solution. At the point p with polar coordinates ρ, θ let $\vec{\xi}$ and $\vec{\eta}$ be unit vectors in the direction of increasing ρ and θ, respectively (cf. Fig. 1901). Then we have the relations

(4) $$\vec{\xi} = \vec{i} \cos \theta + \vec{j} \sin \theta, \quad \vec{\eta} = -\vec{i} \sin \theta + \vec{j} \cos \theta,$$

and, by differentiation, and using dots to mean d/dt,

(5) $$\frac{d\vec{\xi}}{dt} = (-\vec{i} \sin \theta + \vec{j} \cos \theta)\dot{\theta} = \vec{\eta}\dot{\theta}, \quad \frac{d\vec{\eta}}{dt} = -\vec{\xi}\dot{\theta}.$$

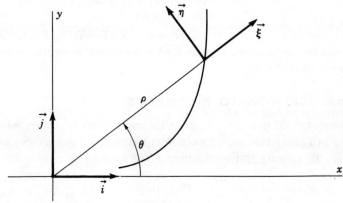

FIG. 1901

Therefore, differentiating $\vec{r} = \rho\vec{\xi}$ and substituting from (5):

(6) $$\vec{V} = \frac{d\vec{r}}{dt} = \dot{\rho}\vec{\xi} + (\rho\dot{\theta})\vec{\eta},$$

(7) $$\vec{A} = \frac{d^2\vec{r}}{dt^2} = (\ddot{\rho} - \rho\dot{\theta}^2)\vec{\xi} + (2\dot{\rho}\dot{\theta} + \rho\ddot{\theta})\vec{\eta}.$$

The components can be read from equations (6) and (7). For example, the transverse component of the acceleration is $2\dot{\rho}\dot{\theta} + \rho\ddot{\theta}$.

1904. TORSION. THE FRENET FORMULAS

As a point p moves along a smooth curve C the osculating plane to C at p will in general turn or twist in space. It is of interest to study the rate of this turning.

We start by defining, in terms of the orthogonal unit vectors $\vec{\tau}$ and $\vec{\nu}$ a third unit vector orthogonal to both:

(1) $$\vec{\beta} \equiv \vec{\tau} \times \vec{\nu}.$$

This vector $\vec{\beta}$ is called the **binormal** direction to the curve C at p. The three vectors $\vec{\tau}$, $\vec{\nu}$, and $\vec{\beta}$ constitute a right-handed *fundamental orthonormal triad* moving with the point p. In pairs these vectors determine three mutually perpendicular planes through p: the **normal plane** is the plane of $\vec{\nu}$ and $\vec{\beta}$; the **osculating plane** is the plane of $\vec{\tau}$ and $\vec{\nu}$; the **rectifying plane** is (by definition) the plane of $\vec{\tau}$ and $\vec{\beta}$.

TORSION. THE FRENET FORMULAS

The derivative $d\vec{\beta}/ds$ measures the rate at which the osculating plane turns. We shall now determine the nature of this vector.

By differentiating the two equations (cf. § 1711)
$$\vec{\beta} \cdot \vec{\beta} = 1, \quad \vec{\beta} \cdot \vec{\tau} = 0,$$
we have, with the help of (6), § 1902 (and $\vec{\beta} \cdot \vec{\nu} = 0$):
$$2\frac{d\vec{\beta}}{ds} \cdot \vec{\beta} = 0, \quad \vec{\beta} \cdot \frac{d\vec{\tau}}{ds} + \frac{d\vec{\beta}}{ds} \cdot \vec{\tau} = \frac{d\vec{\beta}}{ds} \cdot \vec{\tau} = 0.$$
Therefore $\dfrac{d\vec{\beta}}{ds}$ is orthogonal to both $\vec{\beta}$ and $\vec{\tau}$, and hence must be collinear with $\vec{\nu}$. We define the quantity T, called the **torsion** of C at p, by the equation

(2) $$\frac{d\vec{\beta}}{ds} = -T\vec{\nu}.$$

The reciprocal of the absolute value of the torsion is called the **radius of torsion**: $|T|^{-1}$.

Before discussing the sign of T which, in contrast to the nonnegative sign of K may be either positive or negative, we investigate one more rate-of-change formula. With (6), § 1902, and (2), above, we have the rates of change of two of the fundamental triad vectors expressed in terms of the triad system. It remains to find $d\vec{\nu}/ds$. To do this we write, from (1), $\vec{\nu} = \vec{\beta} \times \vec{\tau}$ and differentiate, using (v), § 1711:
$$\frac{d\vec{\nu}}{ds} = \vec{\beta} \times \frac{d\vec{\tau}}{ds} + \frac{d\vec{\beta}}{ds} \times \vec{\tau} = K\vec{\beta} \times \vec{\nu} - T\vec{\nu} \times \vec{\tau}$$
$$= -K\vec{\tau} + T\vec{\beta}.$$

This and the two previous formulas alluded to were obtained in 1847 by the French mathematician F. Frenet, and are known as the **Frenet formulas** (or as the Frenet-Serret formulas, after Frenet and the nineteenth century French mathematician J. A. Serret who also made use of them):

(3) $$\begin{cases} \dfrac{d\vec{\tau}}{ds} = & K\vec{\nu}, \\ \dfrac{d\vec{\nu}}{ds} = -K\vec{\tau} & + T\vec{\beta}, \\ \dfrac{d\vec{\beta}}{ds} = & -T\vec{\nu} \end{cases}.$$

A formula for the torsion is given by formulas (8) and (9), § 1902, together with the following expansion, obtained by differentiating (9), § 1902, with respect to the parameter t:
$$\vec{r}''' = s'''\vec{\tau} + s's''\frac{d\vec{\tau}}{ds} + \frac{d}{dt}(Ks'^2)\vec{\nu} + Ks'^3\frac{d\vec{\nu}}{ds},$$
which, by (3), is a vector of the form

(4) $$\vec{r}''' = A\vec{\tau} + B\vec{\nu} + KTs'^3\vec{\beta}.$$

(cf. Ex. 17, § 1906). The triple scalar product of $\vec{r}'$, $\vec{r}''$, and $\vec{r}'''$ can therefore be written:

$$[\vec{r}', \vec{r}'', \vec{r}'''] = [s'\vec{\tau}, s''\vec{\tau} + Ks'^2\vec{\nu}, A\vec{\tau} + B\vec{\nu} + KTs'^3\vec{\beta}]$$
$$= [s'\vec{\tau}, Ks'^2\vec{\nu}, KTs'^3\vec{\beta}] = K^2Ts'^6.$$

By (10), § 1902, $K^2s'^6 = |\vec{r}' \times \vec{r}''|^2$, and therefore

(5) $$T = \frac{[\vec{r}', \vec{r}'', \vec{r}''']}{|\vec{r}' \times \vec{r}''|^2},$$

where primes indicate differentiation with respect to the parameter t.

Finally, the fundamental triad of vectors can be easily expressed in terms of the vector $\vec{r}$ and its derivatives. For the general parameter t, under the assumption that s increases as t increases, formula (10), § 1902, gives $\vec{r}' \times \vec{r}'' = |\vec{r}' \times \vec{r}''|\vec{\beta}$. Using this, formula (8), § 1902, and the fact that $\vec{\nu} = \vec{\beta} \times \vec{\tau}$, we have:

(6) $$\vec{\tau} = \frac{\vec{r}'}{|\vec{r}'|}, \quad \vec{\nu} = \frac{(\vec{r}' \times \vec{r}'') \times \vec{r}'}{|\vec{r}' \times \vec{r}''| \cdot |\vec{r}'|}, \quad \vec{\beta} = \frac{\vec{r}' \times \vec{r}''}{|\vec{r}' \times \vec{r}''|}.$$

If the parameter is arc length s, equations (6) become:

(7) $$\vec{\tau} = \vec{r}', \quad \vec{\nu} = \frac{\vec{r}''}{|\vec{r}''|} = \frac{\vec{r}''}{K}, \quad \vec{\beta} = \frac{\vec{r}' \times \vec{r}''}{|\vec{r}' \times \vec{r}''|} = \frac{\vec{r}' \times \vec{r}''}{K}.$$

Example 1. Find the torsion and discuss the fundamental orthonormal triad for the helix $x = a \cos t$, $y = a \sin t$, $z = bt$, $ab \neq 0$, of the Example, § 1902.

Solution. With the notation of § 1902 ($c \equiv \sqrt{a^2 + b^2}$):

(8) $$\begin{cases} \vec{r} = a \cos t\, \vec{i} + a \sin t\, \vec{j} + bt\vec{k}, \\ \vec{r}' = -a \sin t\, \vec{i} + a \cos t\, \vec{j} + b\vec{k}, \\ \vec{r}'' = -a \cos t\, \vec{i} - a \sin t\, \vec{j}, \\ \vec{r}''' = a \sin t\, \vec{i} - a \cos t\, \vec{j}, \end{cases}$$

from which $|\vec{r}' \times \vec{r}''| = |ac|$ and $[\vec{r}', \vec{r}'', \vec{r}'''] = a^2b$, so that the torsion T is equal to the constant b/c^2. From (6),

$$\vec{\tau} = \frac{\vec{r}'}{c} = -\frac{a}{c} \sin t\, \vec{i} + \frac{a}{c} \cos t\, \vec{j} + \frac{b}{c} \vec{k},$$

$$\vec{\beta} = \frac{\vec{r}' \times \vec{r}''}{|\vec{r}' \times \vec{r}''|} = \frac{a}{|a|c}(b \sin t\, \vec{i} - b \cos t\, \vec{j} + a\vec{k}),$$

so that (as with the Example, § 1902);

$$\vec{\nu} = \vec{\beta} \times \vec{\tau} = \frac{a}{|a|}(-\cos t\, \vec{i} - \sin t\, \vec{j}).$$

From these equations it is easily seen that if a point p is moving upward along the helix (that is, if b is positive and t increasing) the direction of the tangent is forward and upward, making a constant acute angle with the z-axis, the principal normal is directed horizontally inward, pointing toward the z-axis, and the binormal points backward and upward, making a constant acute angle with the z-axis.

NOTE. The torsion of the preceding helix has the same sign as b, and is therefore positive or negative according as the helix is right-handed or left-handed. It is for this reason that the negative sign is introduced in equation (2). In general, a curve is called **right-handed** or **left-handed** at a point according as the torsion is positive or negative there.

Example 2. Find the curvature, torsion, the fundamental orthonormal triad, and the equations of the normal, rectifying, and osculating planes for the twisted cubic:
$$x = t, \quad y = t^2, \quad z = t^3.$$
(Cf. Fig. 1902 for a graph.)

Solution. With primes indicating differentiation with respect to t, we have

(9)
$$\begin{cases} \vec{r} = t\vec{i} + t^2\vec{j} + t^3\vec{k} \\ \vec{r}' = \vec{i} + 2t\vec{j} + 3t^2\vec{k} = s'\vec{\tau} \\ \vec{r}'' = \quad\quad 2\vec{j} + 6t\vec{k} = s''\vec{\tau} + Ks'^2\vec{v} \\ \vec{r}''' = \quad\quad\quad\quad 6\vec{k} \end{cases}$$

and hence
$$s' = |\vec{r}'| = (1 + 4t^2 + 9t^4)^{\frac{1}{2}},$$
$$\vec{r}' \times \vec{r}'' = 6t^2\vec{i} - 6t\vec{j} + 2\vec{k},$$
$$|\vec{r}' \times \vec{r}''| = 2(1 + 9t^2 + 9t^4)^{\frac{1}{2}},$$
$$[\vec{r}', \vec{r}'', \vec{r}'''] = 12.$$

From equations (11), §1902, and (5), §1904, we have:
$$K = 2(1 + 9t^2 + 9t^4)^{\frac{1}{2}}(1 + 4t^2 + 9t^4)^{-\frac{3}{2}},$$
$$T = 3(1 + 9t^2 + 9t^4)^{-1}.$$

Since $T > 0$, the curve is right-handed.

From (6):
$$\vec{\tau} = (\vec{i} + 2t\vec{j} + 3t^2\vec{k})(1 + 4t^2 + 9t^4)^{-\frac{1}{2}},$$
$$\vec{\beta} = (3t^2\vec{i} - 3t\vec{j} + \vec{k})(1 + 9t^2 + 9t^4)^{-\frac{1}{2}},$$
whence
$$\vec{v} = \vec{\beta} \times \vec{\tau} = [(-9t^3 - 2t)\vec{i} + (1 - 9t^4)\vec{j}$$
$$+ (6t^3 + 3t)\vec{k}] \cdot (1 + 4t^2 + 9t^4)^{-\frac{1}{2}}(1 + 9t^2 + 9t^4)^{-\frac{1}{2}}.$$

The equations of the planes requested (for any fixed value of t) are therefore:
normal plane: $(x - t) + 2t(y - t^2) + 3t^2(z - t^3) = 0,$
rectifying plane: $(-9t^3 - 2t)(x - t) + (1 - 9t^4)(y - t^2) + (6t^3 + 3t)(z - t^3) = 0,$
osculating plane: $3t^2(x - t) - 3t(y - t^2) + (z - t^3) = 0.$

1905. LOCAL BEHAVIOR

In order to study the behavior of a curve C in the neighborhood of a point p at which the curvature and torsion do not vanish, we choose a convenient coordinate system as follows: Let the origin be at p, and the axes be such that $\vec{i} = \vec{\tau}$, $\vec{j} = \vec{v}$, $\vec{k} = \vec{\beta}$ at this point. Furthermore, assume that C is sufficiently "regular" near p to permit the following Taylor expansions of the coordinates in terms of arc length s measured from p:

so that
$$x = x_0's + \tfrac{1}{2}x_0''s^2 + \tfrac{1}{6}x_0'''s^3 + O(s^4),$$
$$y = y_0's + \tfrac{1}{2}y_0''s^2 + \tfrac{1}{6}y_0'''s^3 + O(s^4),$$
$$z = z_0's + \tfrac{1}{2}z_0''s^2 + \tfrac{1}{6}z_0'''s^3 + O(s^4),$$
(1)
$$\vec{r} = \vec{r}_0's + \tfrac{1}{2}\vec{r}_0''s^2 + \tfrac{1}{6}\vec{r}_0'''s^3 + \vec{O}(s^4),$$

where $\vec{O}(s^4)$ is a vector whose magnitude is $O(s^4)$, as $s \to O$, and primes indicate differentiation with respect to s.

We observe first that $s' = 1$ and $s'' = 0$. Therefore, by the Frenet formulas, equations (8) and (9), § 1902, and (4), § 1904, take the particularly simple forms
$$\vec{r}' = \vec{\tau},$$
$$\vec{r}'' = K\vec{\nu},$$
$$\vec{r}''' = A\vec{\tau} + B\vec{\nu} + KT\vec{\beta}.$$

Substitution in (1), with $\vec{i} = \vec{\tau}, \vec{j} = \vec{\nu}, \vec{k} = \vec{\beta}$, gives
$$\vec{r} = \left(s + \frac{A}{6}s^3\right)\vec{i} + \left(\frac{K}{2}s^2 + \frac{B}{6}s^3\right)\vec{j} + \frac{KT}{6}s^3\vec{k} + \vec{O}(s^4).$$

If, in each coefficient, we neglect terms of higher degree, and if we also neglect the term $\vec{O}(s^4)$, we have the following approximation for $\vec{r}$, for small s:

(2) $\qquad\qquad\qquad \vec{r}: s\vec{i} + \tfrac{1}{2}Ks^2\vec{j} + \tfrac{1}{6}KTs^3\vec{k}.$

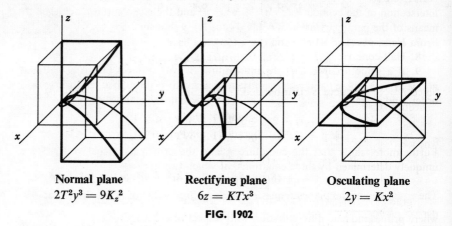

Normal plane	Rectifying plane	Osculating plane
$2T^2y^3 = 9K_z^2$	$6z = KTx^3$	$2y = Kx^2$

FIG. 1902

In other words, the curve C behaves near p in the same way as the twisted cubic

(3) $\qquad\qquad x = s, \quad y = \tfrac{1}{2}Ks^2, \quad z = \tfrac{1}{6}KTs^3,$

near the origin.

The projections of (3) on the normal, rectifying, and osculating plane reveal this behavior explicitly:

(4) $\begin{cases} \text{projection on the normal plane:} & 2T^2y^3 = 9Kz^2, \\ \text{projection on the rectifying plane:} & 6z = KTx^3, \\ \text{projection on the osculating plane:} & 2y = Kx^2. \end{cases}$

Figure 1902 shows these projections for the twisted cubic: $x = t$, $y = t^2$, $z = t^3$ for $p = (0, 0, 0)$. At this point $K = 2$ and $T = 3$ (cf. Example 2, § 1904), and equations (3) reduce to $x = s$, $y = s^2$, $z = s^3$. The projection equations (4) simplify to $y^3 = z^2$, $z = x^3$, and $y = x^2$.

1906. EXERCISES

In Exercises 1–2, find the curvature, torsion, and the fundamental triad of unit vectors, and write the equations of the normal, osculating, and rectifying planes, at a general point of the given curve.

1. Intersection of the plane $x = y$ and the cylinder $z = x^2$.
2. $x = t$, $y = t^2$, $z = \frac{2}{3}t^3$.

In Exercises 3–4, find the curvature, torsion, and the fundamental triad of unit vectors at a general point of the given curve.

3. $x = a(3t - t^3)$, $y = 3at^2$, $z = a(3t + t^3)$, $a > 0$.
4. $x = \dfrac{1 - t^2}{t}$, $y = \dfrac{2 + t^2}{t}$, $z = \dfrac{3t - 3}{t}$, $t > 0$.

In Exercises 5–6, find the curvature and torsion at a general point of the given curve.

5. $x = t \cos t$, $y = t \sin t$, $z = t$.
6. $x = t - \sin t$, $y = 1 - \cos t$, $z = ct$.

In Exercises 7–8, find the curvature and torsion at a general point of the curve of intersection of the cylinder $(x - a)^2 + y^2 = a^2$ and the given surface, achieved by means of the parametrization $x = a(1 + \cos 2t)$, $y = a \sin 2t$.

7. Hemisphere $x^2 + y^2 + z^2 = a^2$, $z \geq 0$.
8. Half-cone $x^2 + y^2 = z^2$, $z \geq 0$.

9. If a smooth curve C is the intersection of two surfaces $f(x, y, z) = 0$ and $g(x, y, z) = 0$, where f and g have continuous derivatives through at least the second order, and if $\operatorname{grad} f \times \operatorname{grad} g \neq \vec{0}$, show that the unit tangent vector is

$$\vec{\tau} = \pm \frac{\operatorname{grad} f \times \operatorname{grad} g}{|\operatorname{grad} f \times \operatorname{grad} g|}.$$

Furthermore, show that the curvature K and the principal normal vector $\vec{v}$ are uniquely determined by three equations, of which two are:

$$\vec{r}' \cdot \vec{r}'' = 0,$$

$$\operatorname{grad} f \cdot \vec{r}'' + f_{11}x'^2 + f_{22}y'^2 + f_{33}z'^2 + 2f_{12}x'y' + 2f_{13}x'z' + 2f_{23}y'z' = 0,$$

where primes indicate differentiation with respect to arc length.

10. Prove that if a particle is moving in such a way that both its velocity vector and its acceleration vector have constant magnitudes, then its trajectory has constant curvature.

11. Prove that if a particle is moving in a plane in such a way that its acceleration vector and its radius vector are collinear, then (in the notation of § 1903) $\rho^2 \dot{\theta}$ is constant. This leads immediately to Kepler's law of areas. Explain.

12. Show that for plane curves the Frenet formulas reduce to

$$\frac{d\vec{\tau}}{ds} = K\vec{v}, \quad \frac{d\vec{v}}{ds} = -K\vec{\tau}.$$

13. Prove that if

$$\vec{d} = T\vec{\tau} + K\vec{\beta}$$

the Frenet formulas can be written

$$\frac{d\vec{\tau}}{ds} = \vec{d} \times \vec{\tau}, \quad \frac{d\vec{v}}{ds} = \vec{d} \times \vec{v}, \quad \frac{d\vec{\beta}}{ds} = \vec{d} \times \vec{\beta}.$$

The vector $\vec{d}$, as an axis of instantaneous rotation, was studied by the French mathematicians G. Darboux and E. Cartan. For further comments, see D. J. Struik, *Differential Geometry* (Reading, Mass., Addison-Wesley Press, Inc., 1950), page 22.

14. Prove that a smooth curve is a straight line if and only if its curvature vanishes identically.

15. Prove that a smooth curve whose parametrization functions have continuous second-order derivatives is a plane curve if and only if its torsion vanishes identically. Take into consideration extreme cases such as the identical vanishing of the curvature (cf. Ex. 14).

16. Show that if primes indicate differentiation with respect to arc length:
$$K' = \frac{\vec{r}'' \cdot \vec{r}'''}{|\vec{r}''|}.$$

17. Evaluate the coefficients A and B of formula (4), § 1904, for $\vec{r}'''$.

18. Prove that the curvature of a space curve at any point is the same as the curvature of its projection on the osculating plane at that point. Assume continuity and differentiability conditions as desired.

19. Show that the center of curvature of a curve C at a point whose radius vector is $\vec{r} = \vec{r}(t)$ is represented by the radius vector
$$\vec{r} + R\vec{v}.$$
Use this result to derive formula (4), § 813, for the center of curvature for a plane curve:
$$\left(x - \frac{y'(x'^2 + y'^2)}{x'y'' - y'x''}, y + \frac{x'(x'^2 + y'^2)}{x'y'' - y'x''} \right).$$

20. Show that the locus of the centers of curvature of the helix
$$x = a \cos t, \quad y = a \sin t, \quad z = bt, \quad a > 0,$$
is another helix, lying in a cylinder coaxial with that of the first helix. Show that the original helix is the locus of the centers of curvature of this second helix.

21. Show that if C is a given curve and Γ the locus of its centers of curvature, then the tangent vectors to C and Γ at corresponding points are orthogonal. Assume continuity and differentiability conditions as desired.

22. If $\vec{r} = \vec{r}(t)$ is the radius vector of a general point on a curve C, if $\vec{r}, \vec{r}', \cdots, \vec{r}^{(n)}$ are continuous in a neighborhood of a number $t = t_0$, and if $\vec{r}' = \vec{r}'' = \cdots = \vec{r}^{(n-1)} = \vec{0}$ at $t = t_0$, but $\vec{r}^{(n)} \neq \vec{0}$ there, show that $\vec{r}^{(n)}$, at the point in question, is tangent to C. *Hint:* Use the Extended Law of the Mean (§ 307).

23. Show that the curve C:
$$\vec{r}(t) = e^{at}(\cos bt\, \vec{i} + \sin bt\, \vec{j} + c\vec{k}), \quad abc \neq 0,$$
lies on a cone and intersects all rulings of this cone (that is, all straight lines lying in the cone) at the same angle. Find the curvature and torsion of C at a general point.

24. A general **helix** is a curve for which the tangent vector $\vec{\tau}$ makes a constant angle with a fixed line in space. Prove that a curve is a helix if and only if its curvature and torsion are in a constant ratio:
$$\frac{T}{K} = \text{a constant.}$$
Prove that a curve is a helix if and only if the vector $T\vec{\tau} + K\vec{\beta}$ is of constant direction. Assume continuity and differentiability conditions as desired. *Hints:* If $\vec{c}$ is a constant vector such that $\vec{c} \cdot \vec{\tau} = $ a constant, show that $\vec{c} \cdot \vec{v} = 0$ so that $\vec{c}$ must

have the form $\vec{c} = a\vec{\tau} + b\vec{\beta}$. Differentiate with respect to s, showing that a and b must be constants. On the other hand, if $d(a\vec{\tau} + b\vec{\beta})/ds = \vec{0}$ for some a and b (not both zero), then $\vec{\tau} \cdot (a\vec{\tau} + b\vec{\beta})$ is a constant.

25. Show that in Exercises 2, 3, 20, and 23 the curve is a helix by the criterion of Exercise 24. In each case find a constant vector $\vec{c}$ such that $\vec{c} \cdot \vec{\tau}$ is constant, and verify that $T\vec{\tau} + K\vec{\beta}$ is a multiple of $\vec{c}$.

1907. CURVES ON A SURFACE. FIRST FUNDAMENTAL FORM

In this and the following sections we shall be interested in certain *local* properties of a surface. We therefore restrict our attention to a single surface element S:

(1) $$\vec{r}(u, v) = x(u, v)\vec{i} + y(u, v)\vec{j} + z(u, v)\vec{k},$$

where

(2) $$\frac{\partial \vec{r}}{\partial u} \times \frac{\partial \vec{r}}{\partial v} \neq \vec{0}$$

for $a \le u \le b$, $c \le v \le d$.

Let $u(t)$, $v(t)$ be the equations of a smooth curve in the uv-rectangle $a \le u \le b$, $c \le v \le d$; that is, $u'(t)$ and $v'(t)$ are continuous and do not vanish simultaneously. By means of the mapping functions (1), this curve is carried into a smooth curve C:

(3) $$\vec{r}(t) = x(t)\vec{i} + y(t)\vec{j} + z(t)\vec{k},$$

lying in the surface element S. The fact that C is smooth follows directly from the expansion

(4) $$\frac{d\vec{r}}{dt} = \frac{\partial \vec{r}}{\partial u}\frac{du}{dt} + \frac{\partial \vec{r}}{\partial v}\frac{dv}{dt}.$$

The fact that $d\vec{r}/dt$ is nonzero can be seen as follows: since the vector product (2) of the two vectors $\partial \vec{r}/\partial u$ and $\partial \vec{r}/\partial v$ is nonzero, these two vectors are noncollinear (Note 2, §1702) and thus linearly independent (Note, §704); therefore their linear combination (4) could vanish only if both coefficients $u'(t)$ and $v'(t)$ were to vanish, and this is impossible by assumption.

Arclength along C is obtained from (4):

(5) $$\left(\frac{ds}{dt}\right)^2 = \frac{d\vec{r}}{dt} \cdot \frac{d\vec{r}}{dt} = E\left(\frac{du}{dt}\right)^2 + 2F\frac{du}{dt}\frac{dv}{dt} + G\left(\frac{dv}{dt}\right)^2,$$

where (cf. §1814):

(6) $$\begin{cases} E \equiv \left|\frac{\partial \vec{r}}{\partial u}\right|^2 = \left(\frac{\partial x}{\partial u}\right)^2 + \left(\frac{\partial y}{\partial u}\right)^2 + \left(\frac{\partial z}{\partial u}\right)^2, \\ F \equiv \frac{\partial \vec{r}}{\partial u} \cdot \frac{\partial \vec{r}}{\partial v} = \frac{\partial x}{\partial u}\frac{\partial x}{\partial v} + \frac{\partial y}{\partial u}\frac{\partial y}{\partial v} + \frac{\partial z}{\partial u}\frac{\partial z}{\partial v}, \\ G \equiv \left|\frac{\partial \vec{r}}{\partial v}\right|^2 = \left(\frac{\partial x}{\partial v}\right)^2 + \left(\frac{\partial y}{\partial v}\right)^2 + \left(\frac{\partial z}{\partial v}\right)^2. \end{cases}$$

Relation (5) can be written in differential form:

(7) $$ds^2 = d\vec{r} \cdot d\vec{r} = E\,du^2 + 2F\,du\,dv + G\,dv^2,$$

or in integral form:

(8) $$s = \int_{t_0}^{t_1} \left[E\left(\frac{du}{dt}\right)^2 + 2F\frac{du}{dt}\frac{dv}{dt} + G\left(\frac{dv}{dt}\right)^2 \right]^{\frac{1}{2}} dt.$$

The *quadratic form* (7), in the variables du and dv, is called the **first fundamental form** of the surface S. Since the vector $d\vec{r}/dt$ can never vanish the quadratic form (7) is always positive, unless *both* quantities du and dv are zero. A quadratic form having this property is called **positive definite** (cf. § 1911). A necessary and sufficient condition that a quadratic form $Ex^2 + 2Fxy + Gy^2$ be positive definite is that $EG - F^2$ be positive (cf. Ex. 4, § 1912). The positiveness of $EG - F^2$ in the present context can be seen directly from the expression (cf. § 1814):

(9) $$\left| \frac{\partial \vec{r}}{\partial u} \times \frac{\partial \vec{r}}{\partial v} \right|^2 = \left| \frac{\partial \vec{r}}{\partial u} \right|^2 \left| \frac{\partial \vec{r}}{\partial v} \right|^2 - \left(\frac{\partial \vec{r}}{\partial u} \cdot \frac{\partial \vec{r}}{\partial v} \right)^2.$$

Suppose that, in addition to (3), a second regular curve Γ, which is the image under (1) of

(10) $$u = \phi(\bar{t}), \quad v = \psi(\bar{t}),$$

is under consideration, and assume that these two curves intersect at a point p with an angle η between their forward tangents. By (4) a forward tangent vector to C is

(11) $$d\vec{r}_C = \frac{\partial \vec{r}}{\partial u}\,du + \frac{\partial \vec{r}}{\partial v}\,dv.$$

Similarly, a forward tangent vector to Γ is

(12) $$d\vec{r}_\Gamma = \frac{\partial \vec{r}}{\partial u}\,d\phi + \frac{\partial \vec{r}}{\partial v}\,d\psi.$$

Therefore,

(13) $$\cos \eta = \frac{d\vec{r}_C \cdot d\vec{r}_\Gamma}{|d\vec{r}_C|\,|d\vec{r}_\Gamma|} = \frac{\dfrac{\partial \vec{r}}{\partial u} \cdot \dfrac{\partial \vec{r}}{\partial u}\,du\,d\phi + \cdots}{\sqrt{d\vec{r}_C \cdot d\vec{r}_C}\sqrt{d\vec{r}_\Gamma \cdot d\vec{r}_\Gamma}}$$
$$= \frac{E\,du\,d\phi + F(du\,d\psi + dv\,d\phi) + G\,dv\,d\psi}{\sqrt{E\,du^2 + 2F\,du\,dv + G\,dv^2}\sqrt{E\,d\phi^2 + 2F\,d\phi\,d\psi + G\,d\psi^2}}.$$

From (13) we have the following *necessary and sufficient condition that C and Γ be orthogonal*:

(14) $$E\,du\,d\phi + F(du\,d\psi + d\phi\,dv) + G\,dv\,d\psi = 0.$$

Of special interest are the **parametric curves**:

(15) $u =$ parameter, $v =$ constant; $u =$ constant, $v =$ parameter.

For such curves, formulas previously obtained simplify as follows:

(3): $\begin{cases} \vec{r}(u) = x(u, v_0)\vec{i} + y(u, v_0)\vec{j} + z(u, v_0)\vec{k}, \\ \vec{r}(v) = x(u_0, v)\vec{i} + y(u_0, v)\vec{j} + z(u_0, v)\vec{k}; \end{cases}$

(4): $\dfrac{d\vec{r}}{du} = \dfrac{\partial \vec{r}}{\partial u}, \quad \dfrac{d\vec{r}}{dv} = \dfrac{\partial \vec{r}}{\partial v};$

(7): $ds^2 = E\, du^2, \quad ds^2 = G\, dv^2;$

(8): $s = \int E\, du, \quad s = \int G\, dv;$

(11) and (12): $d\vec{r}(u) = \dfrac{\partial \vec{r}}{\partial u}\, du, \quad d\vec{r}(v) = \dfrac{\partial \vec{r}}{\partial v}\, dv;$

(13): $\cos \eta = \dfrac{F\, du\, d\psi}{\sqrt{E\, du^2}\sqrt{G\, d\psi^2}} = \dfrac{F}{\sqrt{EG}},$

where η is the angle between the parametric curves;

(14): $F = 0,$

this being the condition that the parametric curves be orthogonal.

Example. Show that for the hemisphere $z = \sqrt{a^2 - x^2 - y^2}$ the parametric curves are orthogonal if the parameters are either the spherical coordinates ϕ and θ or the cylindrical coordinates ρ and θ, and evaluate $\cos \eta$, in (13), if the parameters are x and y.

Solution. In spherical and cylindrical coordinates,

$$\dfrac{\partial \vec{r}}{\partial \phi} \cdot \dfrac{\partial \vec{r}}{\partial \theta} = a^2(\cos\phi\cos\theta\,\vec{i} + \cos\phi\sin\theta\,\vec{j} - \sin\phi\,\vec{k}) \cdot (-\sin\phi\sin\theta\,\vec{i} + \sin\phi\cos\theta\,\vec{j}) = 0,$$

$$\dfrac{\partial \vec{r}}{\partial \rho} \cdot \dfrac{\partial \vec{r}}{\partial \theta} = \left(\cos\theta\,\vec{i} + \sin\theta\,\vec{j} - \dfrac{\rho}{\sqrt{a^2 - \rho^2}}\,\vec{k}\right) \cdot (-\rho\sin\theta\,\vec{i} + \rho\cos\theta\,\vec{j}) = 0.$$

In rectangular coordinates,

$$\dfrac{\partial \vec{r}}{\partial x} = \vec{i} - x(a^2 - x^2 - y^2)^{-\frac{1}{2}}\vec{k}, \quad \dfrac{\partial \vec{r}}{\partial y} = \vec{j} - y(a^2 - x^2 - y^2)^{-\frac{1}{2}}\vec{k}.$$

Therefore,

$$E = (a^2 - y^2)(a^2 - x^2 - y^2)^{-1},$$
$$F = xy(a^2 - x^2 - y^2)^{-1},$$
$$G = (a^2 - x^2)(a^2 - x^2 - y^2)^{-1},$$

and
$$\cos \eta = xy[(a^2 - x^2)(a^2 - y^2)]^{-\frac{1}{2}}.$$

1908. INTERSECTIONS OF SMOOTH SURFACES

The following theorem specifying a local property near a point common to two surfaces will be important in the sequel:

Theorem. *If two surfaces both of which are smooth at a common point p have distinct tangent planes there, their intersection is a smooth curve in a neighborhood of p.*

Proof. For convenience we choose a coordinate system such that the point p is the origin $(0, 0, 0)$, and for one surface S_1 the tangent plane at p is $z = 0$. This surface, then, can be specified in a neighborhood of the origin by an equation where z is a differentiable function of the parameters x and y, subject to the following condition:

(1) $\quad\quad\quad S_1: z = f(x, y), \quad f_x(0, 0) = f_y(0, 0) = 0.$

In a neighborhood of p the other surface S_2 can be specified by an equation of the form $F(x, y, z) = 0$ (cf. Note 1, § 1812), where F is continuously differentiable, and since the normal at p is not the z-axis not both F_x and F_y can vanish there:

(2) $\quad\quad\quad S_2: F(x, y, z) = 0, \quad [F_x(0, 0, 0)]^2 + [F_y(0, 0, 0)]^2 \neq 0.$

We now wish to solve the equations in (1) and (2) simultaneously in order to write the coordinates x, y, z of the intersection C as continuously differentiable functions of one parameter, either x or y, in a neighborhood of the origin. In order to do this we substitute from (1) into (2) and seek to solve in the resulting equation:

(3) $\quad\quad\quad \phi(x, y) \equiv F(x, y, f(x, y)) = 0,$

for x or y in terms of the other. Since, at the origin,

$$\phi_x(0, 0) = F_x(0, 0, 0) + F_z(0, 0, 0)f_x(0, 0) = F_x(0, 0, 0),$$
$$\phi_y(0, 0) = F_y(0, 0, 0) + F_z(0, 0, 0)f_y(0, 0) = F_y(0, 0, 0),$$

not both ϕ_x and ϕ_y can vanish, and it follows from the implicit function theorem (§ 932) that (3) can be solved, and a smooth curve C of intersection is obtained. The regularity of C follows trivially, since if either x or y is the parameter, $x'^2 + y'^2 + z'^2 \geq 1$.

*Note. With the notation of the preceding proof, if $f(x, y)$ and $F(x, y, z)$ have continuous second-order partial derivatives near the origin, d^2y/dx^2 (or d^2x/dy^2) exists and is continuous near the origin. This follows, for the case where $F_y(0, 0, 0) \neq 0$ (to be specific), from differentiating

$$\frac{dy}{dx} = -\frac{\phi_x}{\phi_y} = -\frac{F_x(x, y, f(x, y)) + F_z(x, y, f(x, y))f_x(x, y)}{F_y(x, y, f(x, y)) + F_z(x, y, f(x, y))f_y(x, y)}.$$

1909. PLANE SECTIONS. MEUSNIER'S THEOREM

In this section we address ourselves to the following situation and question: A fixed straight line L is tangent to a smooth surface S (that is, perpendicular to the normal to S) at a point p, while a variable plane $\Pi = \Pi(\omega)$ containing L, and making a variable acute angle ω with the normal line to S at p, rotates about L, cutting the surface S in curves $C = C(\omega)$, called **sections**. How do the curvatures $K = K(\omega)$ of these curves C at the point p vary with ω?

§ 1909]   PLANE SECTIONS. MEUSNIER'S THEOREM

Example. Let S be a sphere of radius R, and let L be tangent to S at the "north pole" (cf. Fig. 1903*a*). The curves C, in this case, are circles whose radii are given by the formula $R(\omega) = R \cos \omega$.

The following theorem, due to the French mathematician J. B. Meusnier (1754–1793), shows that the relationships of the preceding example are more generally representative than one might at first believe:

Theorem. Meusnier's Theorem. *Under the assumptions and with the notation of the first paragraph above, the curvature $K(\omega)$ of a general section is*

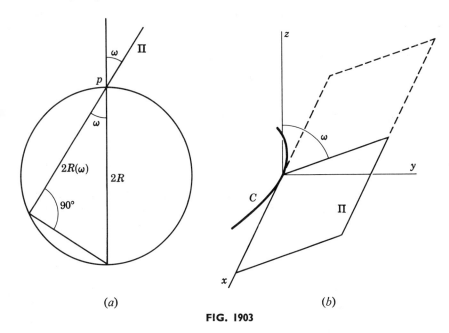

FIG. 1903

related to the **normal curvature** $K_n \equiv K(0)$ *of the* **normal section** $C(0)$, *at the point p, by the formula:*

(1) $$K = K(\omega) = K_n \sec \omega.$$

If the **normal radius of curvature** $R_n \equiv 1/K_n$ *is nonzero:*

(2) $$R = R(\omega) = R_n \cos \omega.$$

The osculating circles of C at p all lie on the sphere for which the osculating circle of the normal section at p is a great circle.

Proof. Let us observe in the first place (cf. § 1908) that all curves $C(\omega)$, for $0 \leq \omega < \tfrac{1}{2}\pi$, are smooth in a neighborhood of p, and therefore that all curvatures $K(\omega)$ exist—although it is possible they all vanish. In the second place, consideration of Figure 1903 readily shows that the only part of the proof that is nontrivial is formula (1), which we now proceed to establish.

Choose a coordinate system such that p is the origin, L is the x-axis, and S has the equation $z = f(x, y)$, where $f_x(0, 0) = f_y(0, 0) = 0$. The plane Π, then, has the equation $y = az$, where $a = \pm \tan \omega$. If the curve $C = C(\omega)$ is given in terms of x as parameter by the equations $x = x$, $y = a\phi(x)$, $z = \phi(x)$, then near (and at) the origin:

(3)
$$\begin{cases} \vec{r}(x) = x\vec{i} + a\phi(x)\vec{j} + \phi(x)\vec{k}; & \vec{r}(0) = \vec{0}; \\ \vec{r}'(x) \quad \vec{i} + a\phi'(x)\vec{j} + \phi'(x)\vec{k}; & \vec{r}'(0) = \vec{i}; \\ \vec{r}''(x) = \quad a\phi''(x)\vec{j} + \phi''(x)\vec{k}; & \vec{r}''(0) = a\phi''(0)\vec{j} + \phi''(0)\vec{k}. \end{cases}$$

At the origin, $|\vec{r}'| = 1$, and $|\vec{r}' \times \vec{r}''| = |-\phi''(0)\vec{j} + a\phi''(0)\vec{k}|$. Therefore the curvature at the origin is, by (11), § 1902,

$$K = K(\omega) = \sqrt{a^2 + 1} \, |\phi''(0)| = \sec \omega \, |\phi''(0)|.$$

By substitution, $K_n = K(0) = |\phi''(0)|$, and the proof is complete.

1910. NORMAL SECTIONS. MEAN AND TOTAL CURVATURE

From Meusnier's theorem, § 1909, we know the curvature of any plane section of a surface as soon as we know that of the corresponding normal section. For a given point p of a surface, each *tangent direction* has its own normal curvature. Our problem in this section, and the following, is to study how the normal curvature varies with the tangent direction; that is, how the normal curvature changes as the tangent line L to S at p rotates about p in the tangent plane.

As in the proof of Meusnier's theorem, we shall choose a coordinate system so that the point p is the origin, and the surface S is given near p by an equation

(1) $$z = f(x, y),$$

where $f(0, 0) = f_x(0, 0) = f_y(0, 0) = 0$. We shall also assume that the second-order partial derivatives of f are continuous near $(0, 0)$. A further requirement for the coordinate system is convenient, and is made possible by the lemma:

Lemma. *If $f(x, y)$ has continuous second-order partial derivatives near $(0, 0)$, a rotation of axes*

(2) $$\begin{cases} x = x' \cos \theta - y' \sin \theta, \\ y = x' \sin \theta + y' \cos \theta \end{cases}$$

exists such that if

(3) $$g(x', y') \equiv f(x' \cos \theta - y' \sin \theta, x' \sin \theta + y' \cos \theta),$$

then

(4) $$g_{x'y'}(0, 0) = 0.$$

Proof. If $f_{xy}(0,0) = 0$, there is nothing to be done; the zero rotation with $\theta = 0$ suffices. We now assume $f_{xy}(0,0) \neq 0$, and compute by the chain rule:

$$g_{x'y'} = \frac{\partial}{\partial x}\left(\frac{\partial f}{\partial x}\frac{\partial x}{\partial y'} + \frac{\partial f}{\partial y}\frac{\partial y}{\partial y'}\right)\frac{\partial x}{\partial x'} + \frac{\partial}{\partial y}\left(\frac{\partial f}{\partial x}\frac{\partial x}{\partial y'} + \frac{\partial f}{\partial y}\frac{\partial y}{\partial y'}\right)\frac{\partial y}{\partial x'}$$

$$= \frac{\partial}{\partial x}(-f_x \sin\theta + f_y \cos\theta)\cos\theta + \frac{\partial}{\partial y}(-f_x \sin\theta + f_y \cos\theta)\sin\theta$$

$$= f_{xy}(\cos^2\theta - \sin^2\theta) - (f_{xx} - f_{yy})\sin\theta\cos\theta.$$

Evaluated at the origin, with the notation

(5) $\qquad A \equiv f_{xx}(0,0), \quad B \equiv f_{xy}(0,0), \quad C \equiv f_{yy}(0,0),$

the last relation above can be written

$$g_{x'y'}(0,0) = B\cos 2\theta - \tfrac{1}{2}(A - C)\sin 2\theta.$$

Therefore, (4) is produced by any rotation such that $\cot 2\theta = (A - C)/2B$.

Throughout the remainder of this section we shall assume axes chosen so that p is the point $(0, 0, 0)$ and the surface S is given near p by (1), where

(6) $\qquad f_x(0,0) = f_y(0,0) = f_{xy}(0,0) = 0,$

By introducing either a rotation of π about the x-axis, or a rotation of $\tfrac{1}{2}\pi$ about the z-axis, or both, we can, and shall, assume in addition the following inequality:

(7) $\qquad f_{xx}(0,0) \geq |f_{yy}(0,0)|.$

(Justify (7) in Ex. 1, § 1912.)

We now consider the family of tangent directions

(8) $\qquad\qquad \cos\theta \vec{i} + \sin\theta \vec{j}$

to S at p, and for each such direction the normal section C_θ of S by the vertical plane through p:

(9) $\qquad\qquad x\sin\theta - y\cos\theta = 0.$

As parameter for C_θ we shall use the polar coordinate ρ, where $x = \rho\cos\theta$, $y = \rho\sin\theta$, *permitting ρ to be positive, negative, or zero.*

In preceding sections curvature of a space curve was defined as a nonnegative quantity. In the remaining discussion, where we shall be considering certain curves (normal sections) in relation to a *surface*, it will be important to define the concept of *normal curvature as a signed quantity*. The reason for this is that we shall thereby be able to distinguish between surfaces like ellipsoids that bend around "in the same sense," and saddle-shaped surfaces like hyperbolic paraboloids that bend "in different senses for different directions."

In order to define normal curvature we choose one of the two available normal directions to the surface S at the point p in a quite arbitrary manner. This arbitrariness is unavoidable, since the orientation of a surface element can always be reversed by an interchange of the parameters. Even if the surface is oriented according to the assumptions (6) and (7) no unique normal in any *absolute* sense can be defined, as shown by the hyperbolic paraboloid $z = x^2 - y^2$ since a rotation of π about the x-axis followed by a rotation of $\frac{1}{2}\pi$ about the z-axis would induce a reversal of orientation. It will turn out that this manner of orientation is unimportant since the significant question is whether two "principal curvatures" have the same sign or opposite signs.

Relative to the surface (1), with the restrictions (6) and (7), we define the normal $\vec{n}$ at the origin:

(10) $$\vec{n} = \vec{k},$$

and the normal curvature of the section C_θ corresponding to the direction $\vec{\tau} = \cos\theta\,\vec{i} + \sin\theta\,\vec{j}$:

(11) $$K = K_\theta \equiv \frac{d\vec{\tau}}{ds} \cdot \vec{n} = \frac{d^2\vec{r}}{ds^2} \cdot \vec{k}.$$

In other words, the normal curvature satisfies the inequalities $K \geq 0$ or $K \leq 0$ according as the normal section is concave upward or downward, and its magnitude is the curvature as defined earlier.

We now compute K using the parameter ρ and the equations:

(12) $$\begin{cases} \vec{r}(\rho) = \rho\cos\theta\,\vec{i} + \rho\sin\theta\,\vec{j} + f(\rho\cos\theta, \rho\sin\theta)\vec{k}, \\ \vec{r}'(\rho) = \cos\theta\,\vec{i} + \sin\theta\,\vec{j} \\ \phantom{\vec{r}'(\rho) =} + [f_x(\rho\cos\theta, \rho\sin\theta)\cos\theta + f_y(\rho\cos\theta, \rho\sin\theta)\sin\theta]\vec{k}, \\ \vec{r}''(\rho) = [f_{xx}\cos^2\theta + 2f_{xy}\cos\theta\sin\theta + f_{yy}\sin^2\theta]\vec{k}, \end{cases}$$

(13) $$\begin{cases} \dfrac{d\vec{r}}{ds} = \dfrac{d\vec{r}}{d\rho}\dfrac{d\rho}{ds}, \\ \dfrac{d^2\vec{r}}{ds^2} = \dfrac{d^2\vec{r}}{d\rho^2}\left(\dfrac{d\rho}{ds}\right)^2 + \dfrac{d\vec{r}}{d\rho}\dfrac{d^2\rho}{ds^2}. \end{cases}$$

Therefore, at the origin ($\rho = 0$), from (5), (6), and (11):

$$\begin{aligned} K &= \left[\frac{d^2\vec{r}}{d\rho^2}\left(\frac{d\rho}{ds}\right)^2 + \frac{d\vec{r}}{d\rho}\frac{d^2\rho}{ds^2}\right] \cdot \vec{k} \\ &= \left[(A\cos^2\theta + C\sin^2\theta)\vec{k}(1) + (\cos\theta\,\vec{i} + \sin\theta\,\vec{j})\frac{d^2\rho}{ds^2}\right] \cdot \vec{k} \\ &= A\cos^2\theta + C\sin^2\theta. \end{aligned}$$

We see that with $\theta = 0$, $K = A =$ normal curvature of the section by the xz-plane, denoted:

(14) $$K_1 \equiv \text{normal curvature, for the plane } y = 0,$$

and similarly, for $\theta = \frac{1}{2}\pi$, $K = C =$ normal curvature of the section by the yz-plane, denoted:

(15) $$K_2 \equiv \text{normal curvature, for the plane } x = 0.$$

Furthermore, it is clear from the identities

$$A \cos^2 \theta + C \sin^2 \theta = A - (A - C) \sin^2 \theta = C + (A - C) \cos^2 \theta,$$

and the assumption (6) that $A \geq |C|$, that K_1 is the *maximum normal curvature* and K_2 is the *minimum normal curvature*. The conclusion just obtained, written in the form

(16) $$K = K_\theta = K_1 \cos^2 \theta + K_2 \sin^2 \theta,$$

is known as **Euler's theorem.** Together with Meusnier's theorem (§ 1909) it tells the complete story of the curvature of every plane section of a surface.

The **mean curvature** and **total** (or **Gaussian**) **curvature** of a surface are defined in terms of the **principal curvatures** K_1 and K_2:

(17) $$\begin{cases} \text{mean curvature} \equiv \tfrac{1}{2}(K_1 + K_2) \\ \text{total curvature} \equiv K_1 K_2. \end{cases}$$

NOTE 1. By virtue of the conventions (7) and (10), the mean curvature turns out to be nonnegative. However, if the orientation is reversed, the sign of the mean curvature is changed. In other words, *the mean curvature depends in sign on the orientation of the surface*. In contrast to this, since a change of sign in each of the two factors K_1 and K_2 has no effect on their product, *the total curvature is independent of the orientation of the surface*. It is possible to define total curvature exclusively in terms of the surface itself independently of the manner in which the surface is embedded in E_3 or any other space. In this sense it is an *intrinsic property*. For a discussion of this fact, by means of tensor calculus, see L. P. Eisenhart, *Differential Geometry* (Princeton University Press, 1940), pages 146–152.

A point of a surface is called an **elliptic, hyperbolic,** or **parabolic** point according as the total curvature of the surface there is positive, negative, or zero. The reason for this is that if the terms of Taylor's formula (§ 1204), through those of the second degree are used as an approximation for z in (1), then z is approximately equal, for small values of x and y, to

$$f(x_0, y_0) + f_x(x_0, y_0)(x - x_0) + f_y(x_0, y_0)(y - y_0)$$
$$+ \tfrac{1}{2}[f_{xx}(x - x_0)^2 + 2f_{xy}(x - x_0)(y - y_0) + f_{yy}(y - y_0)^2],$$

or, with the simplifying assumption of this section, to

(18) $$\tfrac{1}{2}(K_1 x^2 + K_2 y^2).$$

From this it follows that plane sections parallel to and near the tangent plane at p are approximately ellipses, hyperbolas, or parallel lines (assuming that not both K_1 and K_2 vanish) according as p is an elliptic, hyperbolic, or parabolic point. (A pair of parallel lines constitute a degenerate parabola.)

The reader may wish to verify the following properties of a surface near an elliptic or hyperbolic point (cf. Ex. 2, § 1912):

(*i*) The tangent plane at an elliptic point is on one side only of the surface near that point.

(*ii*) The tangent plane at a hyperbolic point crosses and lies on both sides of the surface near that point. There are exactly two directions at a hyperbolic point for which the normal curvature vanishes (these directions are the **asymptotic directions** there).

NOTE 2. The **Dupin indicatrix** at a given point of a surface, where not both K_1 and K_2 vanish, is defined by the equation $K_1 x^2 + K_2 y^2 = \pm 1$ (cf. (16) and (18)) as follows: at an elliptic or parabolic point the $\pm$ sign is chosen to give a real graph, either an ellipse or a pair of parallel lines; at a hyperbolic point both signs are used, giving a pair of conjugate hyperbolas (whose common asymptotes are in the asymptotic directions). If $K_1 = K_2 = 0$ the Dupin indicatrix is the empty set.

Example. Find the principal, mean, and total curvatures, and the asymptotic directions, of the hyperbolic paraboloid $z = -x^2 + 2y^2$ at the origin.

Solution. A rotation through $\frac{1}{2}\pi$ converts the equation to the form $z = f(x, y) = 2x^2 - y^2$, so that at the origin $f_{xx} = 4 > 0 > f_{yy} = -2$. Hence the principal curvatures are

$$K_1 = 4, \quad K_2 = -2.$$

Therefore the mean curvature is $\frac{1}{2}(4 - 2) = 1$, and the total curvature is -8. The asymptotic directions are those for which

$$K_1 \cos^2 \theta + K_2 \sin^2 \theta = 4 \cos^2 \theta - 2 \sin^2 \theta = 0 \quad \text{or} \quad \tan \theta = \pm \sqrt{2}.$$

1911. SECOND FUNDAMENTAL FORM

For a surface element S:

$$x = x(u, v), \quad y = y(u, v), \quad z = z(u, v),$$

where x, y, and z are assumed to have continuous second-order partial derivatives, we shall define, in terms of this particular parametrization, the unit normal vector (cf. § 1811):

$$(1) \qquad \vec{n} \equiv \frac{\dfrac{\partial \vec{r}}{\partial u} \times \dfrac{\partial \vec{r}}{\partial v}}{\left| \dfrac{\partial \vec{r}}{\partial u} \times \dfrac{\partial \vec{r}}{\partial v} \right|} = \frac{\dfrac{\partial \vec{r}}{\partial u} \times \dfrac{\partial \vec{r}}{\partial v}}{\sqrt{EG - F^2}}.$$

If C is a normal section of S, with unit tangent vector $\vec{\tau}$, then since $\vec{\tau} \cdot \vec{n} = 0$,

$$(2) \qquad \frac{d\vec{\tau}}{ds} \cdot \vec{n} + \vec{\tau} \cdot \frac{d\vec{n}}{ds} = 0,$$

and consequently, from (11), § 1910, the normal curvature is

$$(3) \qquad K = \frac{d\vec{\tau}}{ds} \cdot \vec{n} = -\vec{\tau} \cdot \frac{d\vec{n}}{ds} = -\frac{d\vec{r}}{ds} \cdot \frac{d\vec{n}}{ds}$$
$$= -\frac{(d\vec{r}/dt) \cdot (d\vec{n}/dt)}{(ds/dt)^2} = -\frac{\vec{r}\,' \cdot \vec{n}\,'}{\vec{r}\,' \cdot \vec{r}\,'},$$

where primes indicate differentiation with respect to an arbitrary parameter. In differential form, (3) is

$$(4) \qquad K = -\frac{d\vec{r} \cdot d\vec{n}}{d\vec{r} \cdot d\vec{r}}.$$

SECOND FUNDAMENTAL FORM

Our object now will be to express (4) in terms of the parameters u and v. To this end we write two consequences of (1):

(5) $$\frac{\partial \vec{r}}{\partial u} \cdot \vec{n} = 0, \quad \frac{\partial \vec{r}}{\partial v} \cdot \vec{n} = 0,$$

and hence, by differentiation,

(6) $$\begin{cases} \dfrac{\partial^2 \vec{r}}{\partial u^2} \cdot \vec{n} + \dfrac{\partial \vec{r}}{\partial u} \cdot \dfrac{\partial \vec{n}}{\partial u} = 0, & \dfrac{\partial^2 \vec{r}}{\partial u \partial v} \cdot \vec{n} + \dfrac{\partial \vec{r}}{\partial u} \cdot \dfrac{\partial \vec{n}}{\partial v} = 0, \\[6pt] \dfrac{\partial^2 \vec{r}}{\partial v^2} \cdot \vec{n} + \dfrac{\partial \vec{r}}{\partial v} \cdot \dfrac{\partial \vec{n}}{\partial v} = 0, & \dfrac{\partial^2 \vec{r}}{\partial u \partial v} \cdot \vec{n} + \dfrac{\partial \vec{r}}{\partial v} \cdot \dfrac{\partial \vec{n}}{\partial u} = 0. \end{cases}$$

Noting in passing that these equations imply in particular that

(7) $$\frac{\partial \vec{r}}{\partial u} \cdot \frac{\partial \vec{n}}{\partial v} = \frac{\partial \vec{r}}{\partial v} \cdot \frac{\partial \vec{n}}{\partial u} = -\frac{\partial^2 \vec{r}}{\partial u \partial v} \cdot \vec{n},$$

we proceed to evaluate (4) by writing:

(8) $$d\vec{r} = \frac{\partial \vec{r}}{\partial u} du + \frac{\partial \vec{r}}{\partial v} dv, \quad d\vec{n} = \frac{\partial \vec{n}}{\partial u} du + \frac{\partial \vec{n}}{\partial v} dv,$$

from which, by (6) and (7), we obtain

(9) $$K = -\frac{\dfrac{\partial \vec{r}}{\partial u} \cdot \dfrac{\partial \vec{n}}{\partial u} du^2 + 2 \dfrac{\partial \vec{r}}{\partial u} \cdot \dfrac{\partial \vec{n}}{\partial v} du\, dv + \dfrac{\partial \vec{r}}{\partial v} \cdot \dfrac{\partial \vec{n}}{\partial v} dv^2}{\dfrac{\partial \vec{r}}{\partial u} \cdot \dfrac{\partial \vec{r}}{\partial u} du^2 + 2 \dfrac{\partial \vec{r}}{\partial u} \cdot \dfrac{\partial \vec{r}}{\partial v} du\, dv + \dfrac{\partial \vec{r}}{\partial v} \cdot \dfrac{\partial \vec{r}}{\partial v} dv^2}$$

$$= \frac{\dfrac{\partial^2 \vec{r}}{\partial u^2} \cdot \vec{n}\, du^2 + 2 \dfrac{\partial^2 \vec{r}}{\partial u \partial v} \cdot \vec{n}\, du\, dv + \dfrac{\partial^2 \vec{r}}{\partial v^2} \cdot \vec{n}\, dv^2}{E\, du^2 + 2F\, du\, dv + G\, dv^2}.$$

(Cf. (6), § 1907.) This expression for K can now be written:

(10) $$K = \frac{e\, du^2 + 2f\, du\, dv + g\, dv^2}{E\, du^2 + 2F\, du\, dv + G\, dv^2},$$

where, from (1),

(11) $$\begin{cases} e \equiv \dfrac{\partial^2 \vec{r}}{\partial u^2} \cdot \vec{n} = \dfrac{\left[\dfrac{\partial \vec{r}}{\partial u}, \dfrac{\partial \vec{r}}{\partial v}, \dfrac{\partial^2 \vec{r}}{\partial u^2}\right]}{\sqrt{EG - F^2}} = \dfrac{1}{\sqrt{EG - F^2}} \begin{vmatrix} x_u & y_u & z_u \\ x_v & y_v & z_v \\ x_{uu} & y_{uu} & z_{uu} \end{vmatrix}, \\[18pt] f \equiv \dfrac{\partial^2 \vec{r}}{\partial u \partial v} \cdot \vec{n} = \dfrac{\left[\dfrac{\partial \vec{r}}{\partial u}, \dfrac{\partial \vec{r}}{\partial v}, \dfrac{\partial^2 \vec{r}}{\partial u \partial v}\right]}{\sqrt{EG - F^2}} = \dfrac{1}{\sqrt{EG - F^2}} \begin{vmatrix} x_u & y_u & z_u \\ x_v & y_v & z_v \\ x_{uv} & y_{uv} & z_{uv} \end{vmatrix}, \\[18pt] g \equiv \dfrac{\partial^2 \vec{r}}{\partial v^2} \cdot \vec{n} = \dfrac{\left[\dfrac{\partial \vec{r}}{\partial u}, \dfrac{\partial \vec{r}}{\partial v}, \dfrac{\partial^2 \vec{r}}{\partial v^2}\right]}{\sqrt{EG - F^2}} = \dfrac{1}{\sqrt{EG - F^2}} \begin{vmatrix} x_u & y_u & z_u \\ x_v & y_v & z_v \\ x_{vv} & y_{vv} & z_{vv} \end{vmatrix}. \end{cases}$$

The denominator of (10) is the *first fundamental form* discussed in § 1907. Its coefficients involve only the first-order partial derivatives of x, y, and z. The numerator of (10) is known as the **second fundamental form** of the surface S. Its coefficients involve both the first- and second-order partial derivatives of x, y, and z. Equation (10), then, expresses the normal curvature K for the direction corresponding to differentials du and dv as the quotient of the second fundamental form divided by the first.

The next problem is to find the maximum and minimum values, K_1 and K_2, of K. To do this we introduce an angle α defined by the equations

(12) $$\cos \alpha = du/\sqrt{du^2 + dv^2}, \quad \sin \alpha = dv/\sqrt{du^2 + dv^2},$$

and seek the maximum and minimum values of

(13) $$K(\alpha) \equiv \frac{e \cos^2 \alpha + 2f \cos \alpha \sin \alpha + g \sin^2 \alpha}{E \cos^2 \alpha + 2F \cos \alpha \sin \alpha + G \sin^2 \alpha}.$$

We already know from § 1907 that the denominator in (13) is always positive, so that $K(\alpha)$ is finite for all α, and from § 1910 that $K(\alpha)$ has exactly one maximum value and one minimum value. We proceed to examine the equation $K'(\alpha) = 0$.

It is a routine matter to verify that the equation $K'(\alpha) = 0$ is equivalent to

(14) $$(Ef - Fe) \cos^2 \alpha + (Eg - Ge) \cos \alpha \sin \alpha + (Fg - Gf) \sin^2 \alpha = 0,$$

and, as a consequence, to the following two equations involving K, as given in (13):

(15) $$\begin{cases} (EK - e) \cos \alpha + (FK - f) \sin \alpha = 0, \\ (FK - f) \cos \alpha + (GK - g) \sin \alpha = 0. \end{cases}$$

(Do this verification in Ex. 3, § 1912.) From (15) we find that the extreme values of K must satisfy the equation

(16) $$\begin{vmatrix} EK - e & FK - f \\ FK - f & GK - g \end{vmatrix}$$
$$= (EG - F^2)K^2 - (Eg - 2Ff + Ge)K + (eg - f^2) = 0,$$

and consequently ((17), § 1910):

(17) $$\begin{cases} \text{mean curvature} = \dfrac{K_1 + K_2}{2} = \dfrac{Eg - 2Ff + Ge}{2(EG - F^2)}, \\ \\ \text{total curvature} = K_1 K_2 = \dfrac{eg - f^2}{EG - F^2}. \end{cases}$$

One simple consequence of (17) is worth mentioning: *A point is an elliptic, hyperbolic, or parabolic point according as $eg - f^2$ is positive, negative, or zero.* (This follows from basic definitions and the fact that $EG - F^2$ is always positive.) A quadratic form $Ax^2 + 2Bxy + Cy^2$ is called **definite** in case its values are always of one sign and are zero only for $x = y = 0$; it is called **semidefinite** in case its values are always of one sign or zero; it is called **indefinite** in case its values include both positive and negative numbers (cf. § 1907). The student is asked in Exercise 4, § 1912, to show that the form $Ax^2 + 2Bxy + Cy^2$, where $A^2 + B^2 + C^2 > 0$, is definite, indefinite, or semidefinite but not definite according as $AC - B^2$ is positive, negative, or zero. It follows that *a point is an elliptic, hyperbolic, or parabolic point according as the second fundamental form is definite, indefinite, or semidefinite but not definite*. An **asymptotic direction** can now be defined as one for which the second fundamental form vanishes. An **asymptotic curve** is a curve lying in the surface the direction of which at every point is an asymptotic direction.

Example 1. Write the first and second fundamental forms for each point of the hyperbolic paraboloid $z = -x^2 + 2y^2$ of the Example, § 1910. Also determine its mean and total curvature at each point.

Solution. With $x = u$ and $y = v$, we have

$$\vec{r} = x\vec{i} + y\vec{j} + (-x^2 + 2y^2)\vec{k},$$

$$\frac{\partial \vec{r}}{\partial x} = \vec{i} \qquad - 2x\vec{k},$$

$$\frac{\partial \vec{r}}{\partial y} = \qquad \vec{j} \quad + 4y\vec{k},$$

$$\frac{\partial^2 \vec{r}}{\partial x^2} = -2\vec{k}, \quad \frac{\partial^2 \vec{r}}{\partial x \, \partial y} = \vec{0}, \quad \frac{\partial^2 \vec{r}}{\partial y^2} = 4\vec{k},$$

so that the first and second fundamental forms are:

I: $E \, du^2 + 2F \, du \, dv + G \, dv^2 = (1 + 4x^2) \, dx^2 - 16xy \, dx \, dy + (1 + 16y^2) \, dy^2$,
II: $\quad e \, du^2 + 2f \, du \, dv + g \, dv^2 = (-2 \, dx^2 + 4 \, dy^2)(1 + 4x^2 + 16y^2)^{-\frac{1}{2}}$.

Therefore the mean and total curvatures are:

$$\text{mean curvature} = \frac{K_1 + K_2}{2} = \frac{4 + 16x^2 - 2 - 32y^2}{2(1 + 4x^2 + 16y^2)^{\frac{3}{2}}} = \frac{1 + 8x^2 - 16y^2}{[1 + 4x^2 + 16y^2]^{\frac{3}{2}}},$$

$$\text{total curvature} = K_1 K_2 = \frac{-8}{(1 + 4x^2 + 16y^2)^2}.$$

Since the total curvature is negative, every point is hyperbolic. The asymptotic directions are given by the vanishing of the second fundamental form: $-2 \, dx^2 + 4 \, dy^2 = 0$, and provide the "rulings" (straight lines in the surface):

$$\vec{r} = (x \pm \sqrt{2}t)\vec{i} + (y - t)\vec{j} + [(2y^2 - x^2) - (4y \pm 2\sqrt{2}x)t]\vec{k},$$

where $(x, y, 2y^2 - x^2)$ is a fixed point in the surface and t is a parameter for the rulings. (For a discussion of rulings of quadric surfaces see the author's *Solid Analytic Geometry* (New York, Appleton-Century-Crofts, Inc., 1947), pp. 150–158.)

Example 2. Same as Example 1 for the hemisphere $z = \sqrt{a^2 - x^2 - y^2}$, where $a > 0$.

Solution. Instead of the parameters x and y, we shall use the spherical coordinates ϕ and θ:

$$\vec{r} = a \sin \phi \cos \theta \, \vec{i} + a \sin \phi \sin \theta \, \vec{j} + a \cos \phi \, \vec{k},$$

$$\frac{\partial \vec{r}}{\partial \phi} = a \cos \phi \cos \theta \, \vec{i} + a \cos \phi \sin \theta \, \vec{j} - a \sin \phi \, \vec{k},$$

$$\frac{\partial \vec{r}}{\partial \theta} = -a \sin \phi \sin \theta \, \vec{i} + a \sin \phi \cos \theta \, \vec{j},$$

$$\frac{\partial^2 \vec{r}}{\partial \phi^2} = -a \sin \phi \cos \theta \, \vec{i} - a \sin \phi \sin \theta \, \vec{j} - a \cos \phi \, \vec{k},$$

$$\frac{\partial^2 \vec{r}}{\partial \phi \, \partial \theta} = -a \cos \phi \sin \theta \, \vec{i} + a \cos \phi \cos \theta \, \vec{j},$$

$$\frac{\partial^2 \vec{r}}{\partial \theta^2} = -a \sin \phi \cos \theta \, \vec{i} - a \sin \phi \sin \theta \, \vec{j}.$$

The first and second fundamental forms are

I: $a^2 \, d\phi^2 + a^2 \sin^2 \phi \, d\theta^2$,

II: $[-a^3 \sin \phi \, d\phi^2 - a^3 \sin^3 \phi \, d\theta^2]/a^2 \sin \phi = -a \, d\phi^2 - a \sin^2 \phi \, d\theta^2$.

Since the quotient of these forms is

$$K = \frac{-a \, d\phi^2 - a \sin^2 \phi \, d\theta^2}{a^2 \, d\phi^2 + a^2 \sin^2 \phi \, d\theta^2} = -\frac{1}{a},$$

the sphere is a surface of constant normal curvature. The negative sign indicates an inward direction, since the order of the parameters ϕ, θ is associated with the outer normal. The mean curvature is $-1/a$, and the total curvature is $1/a^2$. Every point is elliptic.

1912. EXERCISES

1. Show that there is no loss of generality in assuming the inequality (7), § 1910.

2. Prove the statements (*i*) and (*ii*), near the end of § 1910, concerning the tangent plane at an elliptic or hyperbolic point.

3. Verify that equations (14) and (15), § 1911, are equivalent to the equation $K'(\alpha) = 0$.

4. Assume that $\psi(x, y) \equiv Ax^2 + 2Bxy + Cy^2$ is a quadratic form where $A^2 + B^2 + C^2 > 0$. Prove that ψ is definite, indefinite, or semidefinite but not definite according as $AC - B^2$ is positive, negative, or zero. *Hint:* If $A \neq 0$, write $\psi = A^{-1}[(Ax + By)^2 + (AC - B^2)y^2]$.

5. Prove that every point of the surface (*a*) $z = x^2 + 3y^2$ is an elliptic point, (*b*) $z = xy$ is a hyperbolic point, (*c*) $z = x^2$ is a parabolic point. (Cf. Ex. 11.)

In Exercises 6–8, find the mean and total curvature at an arbitrary point.

6. $z = \dfrac{a^3}{xy}$.

7. $3z = ax^3 + by^3$.

8. $x = a(u + v)$, $y = b(u - v)$, $z = uv$.

9. Find the principal, mean, and total curvatures of the paraboloid $z = 1 - \rho^2 = 1 - x^2 - y^2$.

10. Find the principal curvatures of the surface $z = ae^{-\rho^2} = ae^{-x^2-y^2}$, and determine which points are elliptic, which are hyperbolic, and which are parabolic.

11. Prove that the surface $z = f(x, y)$ is elliptic, hyperbolic, or parabolic at a point according as $f_{11}f_{22} - f_{12}^2$ is positive, negative, or zero there. Show that at a point where the gradient of f vanishes f has an extremum if the point is an elliptic point of the graph, and f fails to have an extremum if the point is a hyperbolic point of the graph.

12. Prove that at a point where $f = F = 0$ the principal curvatures are e/E and g/G.

13. Prove that if the principal curvatures are unequal they are given by orthogonal directions, and that if they are equal they are given by arbitrary tangential directions.

14. Prove that the average of the normal curvatures for an arbitrary pair of orthogonal tangential directions is constant, at a given point.

15. Let C be a plane curve lying on one side of a line L in the same plane, and let S be the surface obtained by revolving C about L. Prove that the magnitudes of the principal curvatures of S at a point p of C are: (*i*) the curvature of C at p and (*ii*) the reciprocal of the distance between p and the point of intersection of L and the normal to C at p. Show that if C is the curve $y = f(x) > 0$ and L is the x-axis the principal curvatures are given by the formulas

(1) $$\frac{f''}{[1 + f'^2]^{\frac{3}{2}}} \quad \text{and} \quad -\frac{1}{f[1 + f'^2]^{\frac{1}{2}}},$$

and that if C is the curve $y = f(x)$, $x > 0$, and L is the y-axis, the principal curvatures are given by the formulas

(2) $$\frac{f''}{[1 + f'^2]^{\frac{3}{2}}} \quad \text{and} \quad \frac{f'}{x[1 + f'^2]^{\frac{1}{2}}}.$$

Hint: Let L be the z-axis and use polar coordinates for both C and S: $z = f(\rho)$.

16. The surface obtained by revolving about the x-axis the tractrix (1), Exercise 26, § 814, is called the **pseudosphere**. Prove that the pseudosphere is a surface of *constant negative* (total) *curvature*. (Cf. Ex. 15.)

17. A **minimal surface** can be defined as one whose mean curvature is identically zero. Prove that a surface is a minimal surface if and only if its Dupin indicatrix (cf. Note 2, § 1910) is a rectangular hyperbola, or the empty set.

18. Show that the surface obtained by revolving about the x-axis the catenary $y = a \cosh(x/a)$ (cf. § 507, Ex. 35, § 320) is a minimal surface (cf. Ex. 17).

19. Show that the surface $z = a \operatorname{Arctan} \dfrac{y}{x}$ is a minimal surface (cf. Ex. 17), and find its total curvature at an arbitrary point.

20. A general **helicoid** is defined as a surface generated by a moving ray (half-line) that issues orthogonally from a fixed line. If the fixed line is the z-axis, such a surface can be represented (at least locally) by the radius vector $\vec{r} = \rho \cos\theta \vec{i} + \rho \sin\theta \vec{j} + f(\theta)\vec{k}$, as a function of the polar coordinates ρ and θ. Show that a general helicoid is a minimal surface (cf. Ex. 17) and find its total curvature at an arbitrary point. Particularize this to the **right helicoid**, $f(\theta) = c\theta$, $c > 0$.

21. Prove that a surface is a plane if and only if $e = f = g = 0$, identically. (Cf. Ex. 35, § 915.)

★22. Prove that any cone or cylinder (whether quadric or not) has total curvature identically zero (at all regular points).

★23. Prove that the osculating plane of any asymptotic curve is a tangent plane to the surface. *Hint:* Use (3), § 1911, $K = 0$, and a Frenet formula.

⋆24. Prove that ellipsoids, hyperboloids of two sheets, and elliptic paraboloids have positive total curvature at all points, that hyperboloids of one sheet and hyperbolic paraboloids have negative total curvature at all points, and that quadric cones and cylinders have zero total curvature at all (regular) points. (Cf. Ex. 22.)

⋆25. An **umbilic** is a point of a surface at which the principal curvatures are equal and nonzero. Prove that at an umbilic the Dupin indicatrix (cf. Note 2, § 1910) is a circle and all normal curvatures are equal. (Cf. Ex. 13.)

⋆26. Prove that any umbilic of a quadric surface lies on a plane of symmetry. (Cf. Ex. 25.) *Hints:* Show that axes can be chosen so that the quadric has an equation of the form

(3) $$ax^2 + by^2 + cz^2 + 2fyz + 2gxz - 2z = 0,$$

with an umbilic at the origin. Show that equality of normal curvatures implies $a = b$ and thus the reducibility of (3), by rotation about the z-axis, to the form

(4) $$a(x^2 + y^2) + cz^2 + 2gxz - 2z = 0.$$

⋆27. Prove that the ellipsoid $\frac{x^2}{a^2} + \frac{y^2}{b^2} + \frac{z^2}{c^2} = 1$, where $a > b > c > 0$, has exactly four umbilics (cf. Ex. 25), given by:

$$\frac{x^2}{a^2} = \frac{a^2 - b^2}{a^2 - c^2}, \quad \frac{y^2}{b^2} = 0, \quad \frac{z^2}{c^2} = \frac{b^2 - c^2}{a^2 - c^2}.$$

Hint: Removing provisionally the order-relation assumptions for a, b, and c, let $y = 0$ be a plane of symmetry containing an umbilic (cf. Ex. 26), and let x and y be parameters. Show that $F = f = 0$, and solve for e/E and g/G (cf. Ex. 12), obtaining the fact that b must lie between a and c.

⋆28. Prove that the hyperboloid of two sheets $\frac{x^2}{a^2} + \frac{y^2}{b^2} - \frac{z^2}{c^2} = -1$, where $a > b > 0$, has exactly four umbilics (cf. Ex. 25), given by:

$$\frac{x^2}{a^2} = 0, \quad \frac{y^2}{b^2} = \frac{a^2 - b^2}{b^2 + c^2}, \quad \frac{z^2}{c^2} = \frac{a^2 + c^2}{b^2 + c^2}.$$

(Cf. Ex. 27.)

⋆29. Prove that the elliptic paraboloid $\frac{x^2}{a^2} + \frac{y^2}{b^2} + 2z = 0$, where $a > b > 0$, has exactly two umbilics (cf. Ex. 25), given by:

$$\frac{x^2}{a^2} = 0, \quad \frac{y^2}{b^2} = \frac{a^2 - b^2}{a^2}, \quad 2z = \frac{b^2 - a^2}{a^2}.$$

(Cf. Exs. 28–29.)

NOTE. Another approach to the subject of umbilics of quadric surfaces is given in the author's *Solid Analytic Geometry* (New York, Appleton-Century-Crofts, Inc., 1947) Exs. 74–76, § 156.

Answers

§ 111, page 16

12. $x > a + |b|$.
13. $-1 < x < 5$.
14. $x \leq -5$ or $x \geq -1$.
15. $x > 2$.
16. $x < 3$.
17. No values.
18. All values.
19. $-\sqrt{3} \leq x \leq -1$ or $1 \leq x \leq \sqrt{3}$.
20. $-3 < x < 5$.
21. No values.
22. $x < -\frac{5}{3}$ or $x > 5$.
23. $4 < x < 6$.
24. $2 < x < 4$ or $6 < x < 12$.
25. $|x| > |a|$.
26. $|x| < |a|$ or $x = a (a \neq 0)$.
27. $|x - 1| > 2$.
28. If $a = b$, no values; if $a < b$: $-|b| < x < 0$ or $x > |b|$; if $a > b$: $0 < x < |b|$ or $x < -|b|$.

§ 114, page 20

7. $ad = bc$.

§ 203, page 28

11. $\dfrac{2n}{2n - 1}$.
12. $\dfrac{(-1)^{n+1}}{n^2 + 2}$.
13. $\dfrac{(-1)^{n-1}}{(n - 1)!}$.
14. $(2n - 2)!$.
15. $1 \cdot 3 \cdot 5 \cdots (2n + 1) = \dfrac{(2n + 1)!}{2^n n!}$.
16. $a_{2n} = 2$, $a_{4n-1} = 3$, $a_{4n-3} = 1$.
17. $a_{3n} = a_{3n-1} = n$, $a_{3n-2} = -n$.
18. n even, $2^n n!$; n odd, $\dfrac{(2n)!}{2^n n!}$.
19. 2.
20. 1.
21. $\frac{1}{2}$.
22. $+\infty$.
23. ∞.
24. $-\infty$.
25. $N(\epsilon) \equiv 1$.
26. $N(\epsilon) \equiv 1/\epsilon$.
27. $N(\epsilon) \equiv 1/\epsilon$.
28. $N(B) \equiv B$.
29. $N(B) \equiv \frac{1}{5}(7B + 2)$.
30. $N(B) \equiv |B| + 10$.

§ 208, page 39

21. 4.
22. $\frac{7}{13}$.
23. $\frac{11}{9}$.
24. $-\frac{1}{4}$.
25. $3a^2$.
26. ma^{m-1}.
33. (a) $-\infty$; (b) $+\infty$; (c) ∞; (d) $+\infty$; (e) $-\infty$; (f) ∞.
34. (c) 0; (e) 1; (f) 0; (a), (b), (d): there is no limit.
37. $\frac{5}{6}$.
38. $+\infty$.
39. $-\infty$.
40. $+\infty$.
41. 0.
42. $-\infty$.
59. 6; $\delta(\epsilon) \equiv \dfrac{\epsilon}{3}$.
60. 9; $\delta(\epsilon) \equiv \min\left(1, \dfrac{\epsilon}{7}\right)$.
61. 28; $\delta(\epsilon) \equiv \min\left(1, \dfrac{\epsilon}{22}\right)$.
62. $-\frac{1}{5}$; $\delta(\epsilon) \equiv \min(1, 20\epsilon)$.

63. $\frac{3}{7}$; $\delta(\epsilon) \equiv \min\left(1, \frac{\epsilon}{5}\right)$.
64. 6; $\delta(\epsilon) \equiv \min\left(\frac{1}{8}, \frac{\epsilon}{42}\right)$.
65. 0; $N(\epsilon) \equiv \frac{5}{\epsilon}$.
66. 0; $N(\epsilon) \equiv -\max\left(1, \frac{1}{\epsilon}\right)$.
67. 3; $N(\epsilon) \equiv \frac{17}{\epsilon}$.
68. $\frac{5}{3}$; $N(\epsilon) \equiv \max\left(1, \frac{1}{3\epsilon}\right)$.
69. $+\infty$; if $B > 0$, $\delta(B) \equiv \min\left(1, \frac{1}{B}\right)$; otherwise $\delta(B) \equiv 392$.
70. $-\infty$; if $B < 0$, $\delta(B) \equiv \min\left(1, \frac{1}{-2B}\right)$; otherwise $\delta(B) \equiv 1$.

§ 216, page 50

26. $\delta(\epsilon) \equiv \min(3, \epsilon)$.
27. $\delta(\epsilon) \equiv \min\left(1, \frac{\epsilon^2}{5}\right)$.
28. $\delta(\epsilon) \equiv \min\left(1, \frac{\epsilon}{5}\right)$.
29. $\delta(\epsilon) \equiv \min(5, \epsilon)$.
30. $\delta(\epsilon) \equiv \min(5, 6\epsilon)$.
31. $\delta(\epsilon) \equiv \min\left(\frac{7}{8}, \frac{\epsilon}{2}\right)$.

§ 220, page 57

14. 1; 0.
15. $+\infty$; $-\infty$.
16. 1; 0.
17. 1; 0.

§ 223, page 61

15. $+\infty$; $-\infty$.
16. 1; -1.
17. 1; -1.
18. 1; -1.
19. 1; -1.
20. 1; -1.

§ 225, page 65

3. $\delta(\epsilon) \equiv \frac{\epsilon}{2}$.
4. $\delta(\epsilon) \equiv \frac{\epsilon}{4}$.
5. $\delta(\epsilon) \equiv 2\epsilon$.
6. $\delta(\epsilon) \equiv \epsilon^2$.
7. $\delta(\epsilon) \equiv \epsilon$.
8. $\delta(\epsilon) \equiv \frac{\epsilon^2}{4}$.
13. $\delta(\epsilon) \equiv \min\left(\frac{x_0}{2}, \frac{x_0^2 \epsilon}{2}\right)$.
14. $\delta(\epsilon) \equiv \min\left(1, \frac{\epsilon}{1 + 2x_0}\right)$.

§ 304, page 73

1. $2x - 4$.
2. $3x^2$.
3. $-\frac{2}{x^3}$.
4. $-\frac{22}{(5x-4)^2}$.
5. $\frac{1}{2\sqrt{x}}$.
6. $\frac{1}{3\sqrt[3]{x^2}}$.
15. Yes.
16. Yes.
17. No.
18. No.
19. No.
20. Yes
21. $n > 1$; $n > 1$.
22. $f'(x) = nx^{n-1} \sin\frac{1}{x} - x^{n-2} \cos\frac{1}{x}$, $x > 0$; $f'(0) = 0$, $n > 1$. $n > 1$. $n > 2$.

§ 308, page 82

1. π, 2π, or 3π.
2. $2 - \sqrt{2}$.
3. $e - 1$.
4. $\frac{1}{2}(a + b)$.

ANSWERS   675

5. $\dfrac{1}{e-1}$.

6. $\frac{1}{2}$.

7. $\dfrac{a+b}{2}$.

8. $\frac{13}{6}$.

9. $1 - {}^{n+1}\!\sqrt{1-b}$.

10. $2/\sqrt[3]{\ln 27}$.

§ 310, page 88

1. Relative maximum = 9; relative minimum = 5; increasing on $(-\infty, 1]$ and $[3, +\infty)$; decreasing on $[1, 3]$.
2. Absolute maximum = 1; absolute minimum = -1; increasing on $[-1, 1]$; decreasing on $(-\infty, -1]$ and $[1, +\infty)$.
3. Relative maximum = $3\sqrt[3]{20}/25$; relative minimum = 0; increasing on $[0, \frac{2}{5}]$; decreasing on $(-\infty, 0]$ and $[\frac{2}{5}, +\infty)$.
4. Relative maximum = $2\sqrt{3}/9$; absolute minimum = 0; increasing on $[0, \frac{1}{3}]$ and $[1, +\infty)$; decreasing on $[\frac{1}{3}, 1]$.
5. Maximum = $\frac{1}{2}$; minimum = -1.
6. Maximum = -4; minimum = $-109/16$.
7. Maximum = 2; minimum = $-9/8$.
8. No maximum; minimum = $-1/e$.
15. (a) $x = 20$; (b) $x = 25$; (c) $x = 21$; (d) no profit possible.
16. (a) $x = 20$; (b) $x = 37.4$; (c) full speed, $x = 60$.
19. If $t \leq s$, $x = b$; if $t > s$, $x = \min(b, as/\sqrt{t^2 - s^2})$.
20. $r \leq -1$: no minimum; $-1 < r \leq -\frac{1}{2}$: minimum $= -a^2/4(r+1)$; $-\frac{1}{2} \leq r \leq 0$: minimum $= ra^2$; $r \geq 0$: minimum $= 0$.

§ 313, page 94

11. $\epsilon(\Delta x) = (6x^2 - 5)\Delta x + 4x\,\Delta x^2 + \Delta x^3$.
12. $\epsilon(\Delta x) = \Delta x/x^2(x + \Delta x)$.
13. $\epsilon(\Delta x) = -\sqrt{x}\,\Delta x/2(x + \sqrt{x^2 + x\,\Delta x})^2$.
15. 10.48810.
16. h.
17. -0.06.
18. x.
19. x.
20. $\dfrac{1}{2} - \dfrac{\sqrt{3}}{2}\left(x - \dfrac{\pi}{3}\right)$.
21. x.
22. $1 + \dfrac{h}{n}$.
23. 0.
24. h.
25. $1 + x$.
26. 0.

§ 318, page 101

1. $\frac{13}{3}$.
2. $\frac{2}{3}$.
3. $\frac{1}{3}$.
4. $-\dfrac{\pi}{2}$.
5. $\frac{1}{24}$.
6. 1.
7. $\frac{1}{16}$.
8. 1.
9. $\dfrac{\ln a}{\ln b}$.
10. 2.
11. $\frac{8}{3}$.
12. 1.
13. 1.
14. Meaningless.
15. $\frac{1}{2}$.
16. 0.
17. 0.
18. $+\infty$.
19. 0.
20. $-\dfrac{2a}{\pi}$.
21. -1.
22. $\frac{1}{2}$.
23. 1.
24. e^2.
25. e.
26. e^3.
27. 0.
28. 1.
29. e^{-2}.
30. 0.

§ 402, page 117

19. $\dfrac{\pi}{4}$.

§ 407, page 129

3. 0.
4. $\sin x^2$.
5. $-\sin x^2$.
6. $3x^2 \sin x^6$.
7. $4x^3 \sin x^8 - 3x^2 \sin x^6$.
8. $2x \cos x^2 \sin(\sin^2 x^2) - \sin 2x \sin(\sin^4 x)$.

§ 409, page 133

3. No. No.

§ 411, page 135

14. $-\frac{1}{6}\sin^5 x \cos x - \frac{5}{24}\sin^3 x \cos x - \frac{5}{16}\sin x \cos x + \frac{5}{16}x + C$.
15. $\frac{1}{5}\sin x \cos^4 x + \frac{4}{15}\sin x \cos^2 x + \frac{8}{15}\sin x + C$.
16. $-\frac{5}{4}\cot^4 \frac{x}{5} + \frac{5}{2}\cot^2 \frac{x}{5} + 5 \ln \sin \frac{x}{5} + C$.
17. $\frac{1}{6}\sec^5 x \tan x + \frac{5}{24}\sec^3 x \tan x + \frac{5}{16}\sec x \tan x + \frac{5}{16}\ln|\sec x + \tan x| + C$.
18. $\frac{1}{8}[(4x^3 - 6x)\sin 2x + (-2x^4 + 6x^2 - 3)\cos 2x] + C$.
19. $\frac{1}{3}(x^2 + 4x)^{\frac{3}{2}} - (x+2)\sqrt{x^2 + 4x} + 4\ln|x + 2 + \sqrt{x^2 + 4x}| + C$.
20. $\frac{1}{24}(8x^2 - 10x - 1)\sqrt{x^2 + x + 1} + \frac{7}{16}\ln|x + \frac{1}{2} + \sqrt{x^2 + x + 1}| + C$.
21. $-\frac{1}{20}(4x^2 + 21x + 105)(6x - x^2)^{\frac{3}{2}} + \frac{189}{8}(x - 3)\sqrt{6x - x^2} + \frac{1701}{8}\text{Arcsin}\frac{x-3}{3} + C$.

§ 416, page 143

1. $\frac{\pi}{2}$.
2. 0.
3. Divergent.
4. 2.
5. 2.
6. Divergent.
7. $\frac{\pi}{2}$.
8. $\frac{1}{(k-1)(\ln 2)^{k-1}}$, $k > 1$; divergent, $k \leq 1$.
9. 2.
10. $\frac{\pi}{2}$.
15. Convergent.
16. Divergent.
17. Convergent.
18. Convergent.
19. Divergent.
20. Convergent.
21. Convergent.
22. Convergent.
23. Divergent.
24. Convergent.

§ 418, page 151

1. $\frac{2}{3}$.
2. 3.
3. $\frac{\pi}{4}$.
4. $-(1 + e + e^2)$.
5. $e + e^{-1} - 2$.
6. $\pi - 2$.

§ 506, page 163

1. $\frac{1}{5}\ln|\sec 5x| + C$.
2. $\frac{1}{4}\ln|\sec 4x + \tan 4x| + C$.
3. $\text{Arcsin}(x/\sqrt{2}) + C$, $x^2 < 2$.
4. $\frac{1}{2}(x+2)\sqrt{x^2 + 4x} - 2\ln|x + 2 + \sqrt{x^2 + 4x}| + C$, $x \geq 0$ or $x \leq -4$.
5. $\frac{1}{2}\ln|2x - 1 + \sqrt{4x^2 - 4x + 5}| + C$.
6. $\frac{6x-1}{12}\sqrt{x - 3x^2} + \frac{\sqrt{3}}{72}\text{Arcsin}(6x - 1) + C$, $0 < x < \frac{1}{3}$.
7. $\frac{1}{\sqrt{5}}\ln|(\sqrt{5 - 2x^2} - \sqrt{5})/x| + C$, $|x| < \sqrt{5/2}$.
8. $\sec\sqrt{x}\tan\sqrt{x} + \ln|\sec\sqrt{x} + \tan\sqrt{x}| + C$, $x > 0$.
9. $\frac{1}{\sqrt{109}}\ln\left|\frac{6x + 5 - \sqrt{109}}{6x + 5 + \sqrt{109}}\right| + C$.
10. $\frac{2}{\sqrt{59}}\text{Arctan}\frac{6x + 5}{\sqrt{59}} + C$.

ANSWERS

§ 509, page 167

1. $3 \sinh 3x$.
2. $\sinh 2x$.
3. $-\text{sech}^2 (2 - x)$.
4. $\coth x^2 - 2x^2 \text{csch}^2 x^2$.
5. $2 \coth 2x$.
6. $e^{ax}(a \cosh bx + b \sinh bx)$.
7. $\dfrac{4}{\sqrt{1 + 16x^2}}$.
8. $\dfrac{e^x}{\sqrt{e^{2x} - 1}}, x > 0$.
9. $\dfrac{2x}{1 - x^4}, |x| < 1$.
10. $-\csc x, x \neq n\pi$.
11. $\tfrac{1}{6} \ln \cosh 6x + C$.
12. $\ln (\sinh e^x) + C$.
13. $\sinh x + \tfrac{1}{3} \sinh^3 x + C$.
14. $x - \tfrac{1}{10} \tanh 10x + C$.
15. $\tfrac{1}{4} \sinh 2x - \tfrac{1}{2}x + C$.
16. $\tfrac{1}{4}[(x^4 - 1) \tanh^{-1} x + \tfrac{1}{3}x^3 + x] + C$.
17. $\cosh^{-1} \dfrac{x}{\sqrt{2}} + C$.
18. $\dfrac{1}{2} \sinh^{-1} \dfrac{2x - 1}{2} + C$.
19. $-\dfrac{2}{\sqrt{109}} \coth^{-1} \dfrac{6x + 5}{\sqrt{109}} + C$.
20. $\dfrac{2}{\sqrt{109}} \tanh^{-1} \dfrac{6x + 5}{\sqrt{109}} + C$.

§ 605, page 179

1. (a) and (b) no, (c) yes, (d)–(g) no, (h) and (i) $a \leq x \leq b, c \leq y \leq d$, (j) $a < x < b, c < y < d$, (k) the four edges of the rectangle, including the corners.
2. (a) yes, (b)–(g) no, (h) and (i) the entire plane, (j) the set itself, (k) the two coordinate axes.
3. (a) no, (b)–(d) yes, (e)–(g) no, (h) and (i) the set itself, (j) $0 < \rho^2 < \cos 2\theta$, (k) $\rho^2 = \cos 2\theta$.
4. (a)–(g) no, (h), (i), and (k) the set together with all (x, y) such that $x = 0$ and $-1 \leq y \leq 1$, (j) empty.
5. (a)–(g) no, (h) and (i) all (x, y) such that $y \leq [x]$, (j) all points of the set except for x integral and $y \geq x - 1$, (k) all (x, y) such that $y = [x]$ together with all (x, y) such that x is integral and $x - 1 \leq y < x$.
6. (a)–(g) no, (h) all (x, y) of the form (m, n) or $\left(m + \dfrac{1}{p}, n\right)$ or $\left(m, n + \dfrac{1}{q}\right)$, (i) and (k) the set together with (h), (j) empty.

§ 611, page 189

1. No.
2. Yes. 0.
3. Yes. 0.
4. No.
5. No.
6. No.
7. Yes. 0.
8. Yes. 0.
9. Yes. 0.
10. Yes. 0.

§ 618, page 199

14. $\delta(\epsilon) = \epsilon/2$.
15. $\delta(\epsilon) = \epsilon/2e^a$.

§ 705, page 210

11. Third minus the second.
12. One-third the second + one-third the third.
13. 3 times 2nd − 5 times 3rd + 2 times 4th.
14. $(k + 2)$ times 2nd − $(k + 3)$ times 3rd − k times 4th.
15. Coplanar, not collinear.
16. Collinear.
17. Noncoplanar.
18. Coplanar, not collinear.

§ 709, page 217

3. (b), (d), and (f).
4. 45° or 135°.
5. 1, 0, 0.
6. 0, 1, 0.
7. 0, 0, 1.
8. $-\dfrac{1}{\sqrt{2}}, -\dfrac{1}{\sqrt{2}}, 0$.
9. $\dfrac{1}{\sqrt{3}}, \dfrac{1}{\sqrt{3}}, \dfrac{1}{\sqrt{3}}$.
10. $-\dfrac{2}{11}, -\dfrac{6}{11}, \dfrac{9}{11}$.

11. $\frac{2}{7}, \frac{3}{7}, -\frac{6}{7}$.
12. $-\frac{4}{21}, -\frac{8}{21}, \frac{19}{21}$.
13. $-\frac{2}{15}, \frac{2}{3}, \frac{11}{15}$.
14. $(24, 12, 8)$.
15. $\frac{3}{13}\sqrt{2}$.
16. $\frac{2}{3}$.
17. $\frac{1}{6}\sqrt{2}$.
18. $\frac{1}{2}$.
19. Yes.
20. Yes.
21. No.
22. Yes.
23. $-57, -\frac{19}{66}\sqrt{6}, -\frac{57}{121}(-6, 2, 9), -\frac{19}{6}(2, 1, -1)$.
24. $3, 1/\sqrt{14}, \frac{1}{3}(2, 2, -1), \frac{3}{14}(-2, 3, -1)$.
25. $1, 1, 1$.
26. $1, 0, 0$.
27. $2, 3, 0$.
28. $1, 2, -3$.
29. $3, 5, 1$.
30. $3, -2, 4$.

§ 712, page 222

3. $4x + y - 3z + 10 = 0$.
4. $x + 2y - 4z - 21 = 0$.
5. $2x + 5y - z + 9 = 0$.
6. $2x - 2y + 7z + 4 = 0$.
7. $4x + 3y - 7 = 0$.
8. $3x - y + 2z = 0$.
9. $2x - 3y - 6z + 6 = 0$.
10. $3x + 5y - z - 2 = 0$.
11. $2x - 7y + 3z - 6 = 0$.
17. $11x + 14y - 3z = 0$.
18. $12x + 13y - z + 5 = 0$.
19. $5x + 5y + 3 = 0$.
20. $4x + y + 3z + 9 = 0$.
21. $5x + 5y + 14 = 0$.
22. $10x + 5y + 5z + 17 = 0$.

26. $\frac{x-5}{2} = \frac{y-1}{3} = \frac{z-8}{-3}$; $x = 5 + 2t, y = 1 + 3t, z = 8 - 3t$.

27. $\frac{x+3}{4} = \frac{y-2}{-1} = \frac{z}{1}$; $x = -3 + 4t, y = 2 - t, z = t$.

28. $\frac{x-3}{0} = \frac{y-4}{1} = \frac{z-1}{2}$; $x = 3, y = 4 + t, z = 1 + 2t$.

29. $\frac{x-7}{1} = \frac{y+9}{0} = \frac{z+2}{0}$; $x = 7 + t, y = -9, z = -2$.

30. $\frac{x-3}{5} = \frac{y-5}{8} = \frac{z-6}{-1}$; $x = 3 + 5t, y = 5 + 8t, z = 6 - t$.

31. $\frac{x+2}{1} = \frac{y-1}{-3} = \frac{z}{-2}$; $x = -2 + t, y = 1 - 3t, z = -2t$.

32. $\frac{x}{3} = \frac{y-5}{1} = \frac{z-1}{0}$; $x = 3t, y = 5 + t, z = 1$.

33. $\frac{x-1}{0} = \frac{y+3}{1} = \frac{z+4}{0}$; $x = 1, y = -3 + t, z = -4$.

34. $3x - 2y - 6z - 17 = 0$.
35. $4x - 5z - 20 = 0$.
36. $5x - 3y + 6z - 28 = 0$.
37. $6x + 5y - 8 = 0$; $x + 5z - 3 = 0$; $y - 6z + 2 = 0$.
38. $(3, 5, -2)$.
39. $(-2, -3, 1)$.
40. $(1, -1, 0)$.
41. $(-2, -4, 1)$.
42. $(1, 0, 1)$.
43. $(2, 3, -1)$.
44. $x - y - z - 2 = 0$.
45. $3x - 4y + z + 8 = 0$.
46. $2x - y - 3z = 0$.
47. $2x - y - 3 = 0$.

§ 717, page 229

1. $x^2 + y^2 - 4 = 0$; circle.
2. $xz = 1$; hyperbola.
3. $(x + 2y - 1)^2 = 0$; two coincident straight lines.
4. Vacuous.
5. $3, -1$.
6. 2 (once).
7. $(x - 2)^2 + (y + 5)^2 + (z - 1)^2 = 36$.
8. $(x - 3)^2 + y^2 + (z + 4)^2 = 4$.
9. $(x - 4)^2 + (y - 1)^2 + (z - 3)^2 = 9$.
10. $(x + 2)^2 + (y + 1)^2 + (z - 5)^2 = 140$.

ANSWERS 679

11. $(3, -1, -2)$; real; 5.
12. $(-1, -7, 5)$; imaginary.
13. $(-\frac{3}{2}, 2, \frac{1}{2})$; point sphere.
14. $(\frac{1}{3}, -\frac{5}{3}, 1)$; real; $\frac{2}{3}$.
15. $x^2 + y^2 + z^2 - x - y - z = 0$.
16. $x^2 + y^2 + z^2 - 3x + 27y - 13z + 62 = 0$.
18. $2x - y + 4z = 34$.
19. $2x - y + 4z = 13$.
20. Circle, center $(-3, -3, 2)$, radius $\sqrt{19}$, direction numbers of normal 7, 4, -4.
21. Circle, center $(0, 2, 5)$, radius 5, direction numbers of normal 1, 3, 1.
22. Point $\left(-\frac{8}{11}, \frac{3}{11}, \frac{42}{11}\right)$.
23. $x^2 + y^2 + z^2 - 4x - 3z - 10 = 0$.
24. $3x^2 + 3y^2 + 3z^2 - 2x + 8y + 23z + 3 = 0$.
25. z-axis.
26. y-axis.
27. x-axis.
28. z-axis.
29. y-axis.
30. x-axis.
31. $y^2 + z^2 = 9$.
32. $x^2 + y^2 = 25$.
33. $x^2 + z^2 = 4z$.
34. $(x - 3)^2 + (y - 4)^2 = 25$.
43. (a) $x^2 + 2y^2 + 2z^2 + 8$; (b) $x^2 + 2y^2 + z^2 = 8$.
44. (a) $4x^2 - 9y^2 - 9z^2 = 5$; (b) $4x^2 + 4y^2 - 9z^2 = 5$.
45. (a) and (b) $6x^2 + 6y^2 + 6z^2 = 7$.
46. (a) $4x^2 - 9y^2 - 9z^2 - 24x + 36 = 0$; (b) $4x^2 - 9y^2 + 4z^2 + 36y - 36 = 0$.
47. (a) $y = 2$; (b) $x^2 + y^2 = 4$.
48. (a) $x^{\frac{2}{3}} + (y^2 + z^2)^{\frac{1}{3}} = 1$; (b) $(x^2 + y^2)^{\frac{1}{3}} + z^{\frac{2}{3}} = 1$.
49. z-axis; $x^2 + z = 2, y = 0$.
50. x-axis; $x^2 - 4y^2 = 8, z = 0$.
51. x-axis; $x^2 - 4z^2 = 0$ (or $x - 2z = 0$ or $x + 2z = 0$), $y = 0$.
52. y-axis; $z = 2y^2 - y + 1, x = 0$.
53. $4a^2(x^2 + z^2) = (x^2 + y^2 + z^2 + a^2 - b^2)^2$.

§719, page 235

3. 3, 6, 2.
4. $\sqrt{6}, \sqrt{3}, \sqrt{2}$.
5. Oblate; y-axis.
6. Prolate; z-axis.
7. Oblate; x-axis.
8. Prolate; x-axis.
14. 1 sheet; transverse: 3, 6; conjugate: 2.
15. 2 sheets; transverse: 3; conjugate: 6, 2.
16. 1 sheet; all $= 1$.
17. 2 sheets; all $= 1$.
26. Imaginary cone.
27. Real elliptic cylinder.
28. Hyperbolic cylinder.
29. Imaginary parallel planes.
30. 1-sheeted hyp. of rev.; y-axis.
31. Elliptic paraboloid.
32. Real parallel planes.
33. Right. circ. cone; x-axis.
34. Imag. elliptic cylinder.
35. Ell. parab. of rev.; y-axis.
36. Prolate spheroid; z-axis.
37. Hyperbolic paraboloid.
38. Imaginary ellipsoid.
39. Coincident planes.
40. Imag. intersecting planes.
41. Sphere.
42. 2-sheeted hyp. of rev.; x-axis.
43. Parabolic cylinder.
44. Real intersecting planes.
45. Hyperbolic paraboloid.
46. Right circular cylinder.
47. Sphere.
48. Real parallel planes.
49. Parabolic cylinder.
50. Paraboloid of revolution; $x^2 + y^2 + 4z = 0$.
51. Right circular cone; $x^2 + y^2 - z^2 = 0$.
52. Right circular cone; $x^2 + y^2 - 3z^2 = 0$.
53. Oblate spheroid; $x^2 + y^2 + 2z^2 = 4$.
54. Prolate spheroid.
55. Sphere.

§ 807, page 247

1. $\int_0^2 \sqrt{1 + 9x^4}\, dx$.
2. $\sqrt{17} + \frac{1}{4} \ln (4 + \sqrt{17})$.
3. πa.
4. $\frac{k}{2} [\alpha\sqrt{1 + \alpha^2} + \ln (\alpha + \sqrt{1 + \alpha^2})]$.
5. $10n\pi$.
6. $6a$.

7. $6a$.
8. $a \sinh \dfrac{b}{a}$.
9. $\sqrt{2} + \ln(\sqrt{2} + 1)$.
10. $\sqrt{2}(e^{\pi/2} - 1)$.
11. $x = r\cos(s/r),\ y = r\sin(s/r)$.
12. $x = 2 \operatorname{Arccos} \dfrac{4-s}{4} - \dfrac{(4-s)\sqrt{8s-s^2}}{8},\ y = \dfrac{8s - s^2}{8},\ 0 \leq s \leq 8$.
13. $x = [(3 - 2s)/3]^{\frac{3}{2}},\ y = [2s/3]^{\frac{3}{2}},\ 0 \leq s \leq \frac{3}{2}$.
14. $\theta = \ln[(s + \sqrt{2})/\sqrt{2}],\ \rho = (s + \sqrt{2})/\sqrt{2}$.

§ 810, page 251

1. $(3, 0, 2),\ (3, 3\sqrt{3}, -1)$.
2. $(4, \pi, 1),\ (4\sqrt{3}, \pi/6, 0)$.
3. $(-7, 0, 0),\ (-3, 3\sqrt{3}, -6\sqrt{3})$.
4. $(5, \pi/2, 3\pi/2),\ (\sqrt{3}, \operatorname{Arccos} 3^{-\frac{1}{2}}, \pi/4)$.
5. Cylinder, half-plane, cylinder.
6. Sphere, half-cone, surface of revolution.
7. $x^2 + y^2 - z^2 = 0$.
8. $\rho = \cos \theta$.
9. $x^2 + y^2 + z^2 = 9$.
10. $\phi = \pi/3,\ 2\pi/3$.
13. $\sqrt{a^2\lambda^2 + \mu^2} \cdot |t_2 - t_1|$.
14. $\sqrt{a^2\lambda^2 + \mu^2} \cdot |t_2 - t_1|$.
15. $\dfrac{t\sqrt{a^2 + \mu^2 + a^2\lambda^2 t^2}}{2} + \dfrac{a^2 + \mu^2}{2a\lambda} \ln \dfrac{a\lambda t + \sqrt{a^2 + \mu^2 + a^2\lambda^2 t^2}}{\sqrt{a^2 + \mu^2}}$.
16. $\dfrac{vt\sqrt{1 + \lambda^2(\sin^2\alpha)t^2}}{2} + \dfrac{v}{2\lambda \sin \alpha} \ln[\lambda(\sin\alpha)t + \sqrt{1 + \lambda^2(\sin^2\alpha)t^2}]$.

§ 814, page 257

1. $R = \frac{1}{2};\ (0, \frac{1}{2})$.
2. $R = \frac{5}{2}\sqrt{5};\ (-4, \frac{7}{2})$.
3. $R = \frac{13}{6}\sqrt{13};\ (\frac{16}{3}, -\frac{11}{2})$.
4. $R = \frac{13}{12}\sqrt{5};\ (\frac{7}{22}, \frac{40}{11})$.
5. $R = 6;\ (3, -4)$.
6. $R = 4;\ (\pi, -2)$.
7. $R = \frac{4}{3};\ (\frac{2}{3}, \pi)$.
8. $R = \frac{1}{3};\ (\pm\frac{2}{3}, 0)$.
14. $3\sqrt{3}/2$.
19. $\alpha = (3y^2 + 2p^2)/2p,\ \beta = -y^3/p^2$.
20. $\alpha = 2x^3,\ \beta = -2y^3$.
21. $\alpha = -t^2 - \frac{9}{2}t^4,\ \beta = \frac{4}{3}t + 4t^3$.
22. $\alpha = \frac{3}{2}(1 + 2t^2 - t^4),\ \beta = -4t^3$.
23. $\alpha = x + 2\sqrt{y}(x + y),\ \beta = y + 2\sqrt{x}(x + y)$.
24. $2\alpha = 3\cos\theta - \cos 3\theta,\ 2\beta = 3\sin\theta + \sin 3\theta$.
30. $x = \cos s + s\sin s,\ y = \sin s - s\cos s$.

§ 904, page 264

1. $u_x = \sin(x+y) + (x-y)\cos(x+y)$,
$u_y = -\sin(x+y) + (x-y)\cos(x+y)$.
2. $u_x = e^y + ye^x,\ u_y = xe^y + e^x$.
3. $f_x = x/\sqrt{x^2+y^2+z^2},\ f_y = y/\sqrt{x^2+y^2+z^2},\ f_z = z/\sqrt{x^2+y^2+z^2}$.
4. $x_r = \sin\phi\cos\theta,\ x_\phi = r\cos\phi\cos\theta,\ x_\theta = -r\sin\phi\sin\theta$.
5. $f_{xx} = 2A,\ f_{xy} = 2B,\ f_{yy} = 2C$.
6. $z_{xx} = -\dfrac{2y}{(x+y)^3},\ z_{xy} = \dfrac{x-y}{(x+y)^3},\ z_{yy} = \dfrac{2x}{(x+y)^3}$.
7. $u_{xx} = 0,\ u_{yy} = xe^y,\ u_{zz} = -y\sin z,\ u_{xy} = e^y,\ u_{xz} = 0,\ u_{yz} = \cos z$.
8. $u_{xx} = \dfrac{2xyz}{(x^2+y^2)^2},\ u_{yy} = \dfrac{-2xyz}{(x^2+y^2)^2},\ u_{zz} = 0$,
$u_{xy} = \dfrac{(y^2-x^2)z}{(x^2+y^2)^2},\ u_{xz} = \dfrac{-y}{x^2+y^2},\ u_{yz} = \dfrac{x}{x^2+y^2}$.

§ 909, page 272

1. $0;\ 6t$.
2. $3r^2e^{3s},\ 3r^3e^{3s};\ 4r,\ 0$.
9. $f'(g(x) + h(x))(g'(x) + h'(x))$.
10. $f'(g(x)h(x))(g(x)h'(x) + g'(x)h(x))$.
13. $x^x(\ln x + 1);\ x^{x^x}x^{x-1}[x\ln x(\ln x + 1) + 1]$.

ANSWERS

§ 911, page 278

1. $x = -ar$, $y = b$, $z = \frac{1}{2}\pi c + cr$ (r a parameter); $2ax - 2cz + \pi c^2 = 0$.
2. $x = -\frac{1}{2}\pi ar$, $y = \frac{1}{2}\pi b + br$, $z = \frac{1}{2}\pi c + cr$ (r a parameter);
   $\pi ax - 2by - 2cz + \pi b^2 + \pi c^2 = 0$.
3. $x = 0$, $y = t$, $z = -t$; $y = z$.
4. $\dfrac{x - x_1}{Ax_1 + Fy_1 + Ez_1} = \dfrac{y - y_1}{Fx_1 + By_1 + Dz_1} = \dfrac{z - z_1}{Ex_1 + Dy_1 + Cz_1}$;
   $Ax_1x + By_1y + Cz_1z + D(y_1z + z_1y) + E(x_1z + z_1x) + F(x_1y + y_1x) = G$.
7. $2\sqrt{61}$; $\frac{8}{3}$. 8. $\sqrt{41}$; $-7/\sqrt{3}$. 10. $2\sqrt{5}$, $-2 - \sqrt{3}$.

§ 915, page 286

1. $\theta = \frac{1}{2}$.
2. $\dfrac{e}{2} = e^\theta \left[\sin\dfrac{\theta\pi}{6} + \dfrac{\pi}{6}\cos\dfrac{\theta\pi}{6}\right]$.
3. $\theta = \frac{1}{3}$.
4. $1 = 0 + \ln(1 + \theta e - \theta) + \dfrac{2(\theta e - \theta)}{1 + \theta e - \theta} - \dfrac{1}{2}\left(\dfrac{\theta e - \theta}{1 + \theta e - \theta}\right)^2$.
5. $2\frac{1}{12}$ sq. ft.; $\frac{1}{2}$ sq. in. 6. 1.387 in.; 0.003 in.
7. $1 + \frac{1}{2}(x - y)$; y.
8. $a > 0$, minimum value $-a^3$; $a = 0$, no extremum; $a < 0$, maximum value $-a^3$.
9. Maximum value b/e; minimum value min $(a, 0)/e$.
10. (a) and (c) Minimum of 0 at origin; (b) saddle point at origin.
11. (a) and (b) Minimum of 0 at origin; (c) saddle point at origin.
12. (a) Minimum of 0 at origin; (b) and (c) saddle point at origin.
13. (a) Minimum of 6 for $-1 \leq x + 2y \leq 5$;
    (b) Minimum of 2 for $0 \leq x \leq 2$, $y = 0$;
    (c) Minimum of 2 for $-1 \leq x \leq 1$, $y = 0$.
17. $p = 0$. If $\Sigma x_i^2 = \Sigma y_i^2$, $\theta = \frac{1}{4}\pi$ or $\frac{3}{4}\pi$ (arbitrary if $\Sigma x_i y_i = 0$). If $\Sigma x_i^2 \neq \Sigma y_i^2$, $\tan 2\theta = 2(\Sigma x_i y_i)/(\Sigma x_i^2 - \Sigma y_i^2)$.
19. Maximum $\frac{1}{4}$ at $(1, \frac{1}{2})$; minimum $-\frac{1}{4}$ at $(\frac{1}{2}, 1)$.
20. Maximum $\frac{1}{9}\sqrt{3}$ at $x = \frac{1}{6}(\sqrt{15} + \sqrt{3})$, $y = \frac{1}{6}(\sqrt{15} - \sqrt{3})$;
    Minimum $-\frac{1}{9}\sqrt{3}$ at $x = \frac{1}{6}(\sqrt{15} - \sqrt{3})$, $y = \frac{1}{6}(\sqrt{15} + \sqrt{3})$.
21. If $a > b$, minimum $= \sqrt{a^2 - b^2}$; if $a = b$, no minimum; if $a < b$, minimum $= 0$.
22. (a) no extrema; (b) maximum e, minimum $-e$.
23. If $a \leq 0$, no extrema; if $a > 0$, no maximum; minimum $= -\dfrac{1}{a}e^{-(a^2+1)}$ at $x = -a - \dfrac{1}{a}$, $y = a$.
25. Maximum 25 at $(0, 3)$; minimum 2 at $(1, -1)$.
29. $6 = 0 + 6\theta^2 + 6\theta^2 + 6\theta^2$.   34. $f(x, y) = ax + by + c$.
35. (a) $f(x, y) = g(x) + h(y)$; (b) $f(x, y) = axy + bx + cy + d$; (c) It does not exist.

§ 918, page 292

1. $-\dfrac{2x + y}{x + 6y}$. 2. $-\left(\dfrac{y}{x}\right)^{\frac{1}{3}}$.
3. $-\dfrac{\sin y + y \cos x}{\sin x + x \cos y}$. 4. $(a - cx + acy)^{-1}$.
5. $-\dfrac{Cy + Bz}{Bx + Ay}$, $-\dfrac{Cx + Az}{Bx + Ay}$. 6. $\dfrac{x^2 - ayz}{axy - z^2}$, $\dfrac{y^2 - axz}{axy - z^2}$.
7. $\dfrac{z}{x(1 + yz)}$, $\dfrac{-z^2}{1 + yz}$. 8. $\dfrac{1}{x(\cot z - 1)}$, $\dfrac{1}{y(1 - \cot z)}$.
11. $\dfrac{4z - 6x}{15u^4}$, $\dfrac{8yz - 9y^2}{15u^4}$. 12. $\dfrac{3x^2 - 6x^2z + 6xz^2}{(1 - 2z)(1 - 3u^2)}$, $\dfrac{3y^2 - 6y^2z + 6yz^2}{(1 - 2z)(1 - 3u^2)}$.
13. $\dfrac{2 - \cos z}{(1 - \cos z)(1 - e^u)}$, $\dfrac{2 - \cos z}{(1 - \cos z)(1 - e^u)}$.

14. $\dfrac{f_3g_1 - f_1g_3}{f_4g_3}, \dfrac{f_3g_2 - f_2g_3}{f_4g_3}$.   16. $-\dfrac{f_2}{f_1}, \dfrac{1-f_3}{f_2}, \dfrac{f_1}{1-f_3}$.

17. $\dfrac{g_1 - f_2}{f_1}, \dfrac{g_2}{f_2 - g_1}, \dfrac{f_1}{g_2}$.   18. $\dfrac{f_2 - 2y}{2x - f_1}, \dfrac{f_3}{2y - f_2}, \dfrac{2x - f_1}{f_3}$.

19. $-\dfrac{f_1 + xzf_2}{f_1 + yzf_2}, \dfrac{1 - f_1 - xyf_2}{f_1 + xzf_2}, \dfrac{f_1 + yzf_2}{1 - f_1 - xyf_2}$.   20. $\dfrac{g' - f_1}{f_2 + f_3g'}$.

§ 920, page 295

1. The circle $x^2 + y^2 = 9$.   2. The parabola $x^2 + 4y = 0$.
3. There is no envelope.
4. The parabola $y^2 = x + \tfrac{1}{4}$ if $t \geq \tfrac{1}{4}$; no envelope if $t < \tfrac{1}{4}$.
5. The lines $y = \pm 2x$.   6. The semicubical parabola $27x^2 = 4y^3$.
8. $27y^2 = 32$.
9. The parabola $2y = x^2$.
10. The line $y = 0$ and the circle $2x^2 + 2y^2 = y$.
11. The hypocycloid $x^{\frac{2}{3}} + y^{\frac{2}{3}} = a^{\frac{2}{3}}$.
12. The parabola $(x - y)^2 - 2c(x + y) + c^2 = 0$.
13. Two conjugate rectangular hyperbolas, $xy = \pm k$.
14. The hypocycloid $x^{\frac{2}{3}} + y^{\frac{2}{3}} = c^{\frac{2}{3}}$.
15. The square $|x| + |y| = c$.
16. The lemniscate $(x^2 + y^2)^2 = 4(x^2 - y^2)$.

§ 924, page 306

1. $\dfrac{du}{dx} = \dfrac{x^2u - xv^2}{u^3 + v^3}, \dfrac{dv}{dx} = \dfrac{-xu^2 - x^2v}{u^3 + v^3}$.

2. $\dfrac{\partial u}{\partial x} = \dfrac{e^v - 1}{e^{u+v} - 1}, \dfrac{\partial u}{\partial y} = \dfrac{-e^v - 1}{e^{u+v} - 1}, \dfrac{\partial v}{\partial x} = \dfrac{e^u - 1}{e^{u+v} - 1}, \dfrac{\partial v}{\partial y} = \dfrac{e^u + 1}{e^{u+v} - 1}$.

3. $\dfrac{du}{dx} = \dfrac{u(u - 2x)}{(u - v)(u - w)}, \dfrac{dv}{dx} = \dfrac{v(v - 2x)}{(v - u)(v - w)}, \dfrac{dw}{dx} = \dfrac{w(w - 2x)}{(w - u)(w - v)}$.

4. If $D \equiv 1 + \cos u \cos v \cos w$, $\dfrac{\partial u}{\partial x} = 1/D$, $\dfrac{\partial u}{\partial y} = -\cos v/D$, $\dfrac{\partial v}{\partial x} = \cos u \cos v/D$, $\dfrac{\partial v}{\partial y} = 1/D$, $\dfrac{\partial w}{\partial x} = -\cos u/D$, $\dfrac{\partial w}{\partial y} = \cos u \cos v/D$.

5. $\left(\dfrac{\partial u}{\partial x}\right)_y = \dfrac{z - x}{u - z}, \left(\dfrac{\partial u}{\partial x}\right)_z = \dfrac{y - x}{u - y}, \left(\dfrac{\partial x}{\partial u}\right)_y = \dfrac{u - z}{z - x}, \left(\dfrac{\partial x}{\partial u}\right)_z = \dfrac{u - y}{y - x}$.

6. $\left(\dfrac{\partial u}{\partial x}\right)_{yz} = \dfrac{v - x}{u - v}, \left(\dfrac{\partial u}{\partial x}\right)_{yv} = \dfrac{z - x}{u - z}, \left(\dfrac{\partial u}{\partial x}\right)_{zv} = \dfrac{y - x}{u - y}$.

10. $-(x^2 + y^2)^{-2}$.

§ 929, page 320

1. $4\sqrt{a^2 + b^2}$.   2. $(\sqrt{2}, 1)$.   3. $24\sqrt{3}\,abc$.
4. 48.   6. (a) $\tfrac{1}{2}$; (b) 3.   8. $(0, \pm\sqrt{10}, 2\sqrt{5})$.
9. $\sqrt{10}$ and $3\sqrt{2}$.
11. (a) 6912, for $x = 6, y = 4, z = 2$; (b) $\tfrac{25}{2}$, for $x = y = 5, z = 2$; (c) $\tfrac{5}{2}$, for $x = 8, y = z = 2$; (d) $14 - 4\sqrt{6}$, for $x = 12 - 2\sqrt{6}, y = 2\sqrt{6} - 2, z = 2$.
20. If $a_k = \max(a_1, a_2, \cdots, a_n)$, let $x_k = 1, x_i = 0, i \neq k$.
21. $x': 2x + y = 0$; $y': x - 2y = 0, 4x'^2 + 9y'^2 = 36$.
22. $\sqrt{6}, 2, \sqrt{2}$.

§ 931, page 324

1. $\dfrac{2 \sin x^2}{x}$.   2. $\dfrac{4}{x} \sinh x^2$.

ANSWERS

§ 935, page 333

8. $u^2 + v^2 = 1$.
9. $u^2 - v^2 = 4$.
10. $u^2 + v^2 - w = 0$.
11. $u + 2v - w^2 = 0$.

§ 1010, page 350

1. $\int_0^1 \int_{x^3}^{x^{\frac{1}{3}}} f(x,y)\,dy\,dx$, $\int_0^1 \int_{y^3}^{y^{\frac{1}{3}}} f(x,y)\,dx\,dy$, $\tfrac{3}{20}$.

2. $\int_0^a \int_{x^2/a}^{\sqrt{2ax-x^2}} f(x,y)\,dy\,dx$, $\int_0^a \int_{a-\sqrt{a^2-y^2}}^{\sqrt{ay}} f(x,y)\,dx\,dy$, $\dfrac{a^2}{12}(3\pi - 4)$.

3. $\int_{-5}^{-4} \int_{-\sqrt{25-x^2}}^{\sqrt{25-x^2}} f(x,y)\,dy\,dx + \int_{-4}^{4} \int_{3x/4}^{\sqrt{25-x^2}} f(x,y)\,dy\,dx$,

$\int_{-3}^{3} \int_{-\sqrt{25-y^2}}^{4y/3} f(x,y)\,dx\,dy + \int_{3}^{5} \int_{-\sqrt{25-y^2}}^{\sqrt{25-y^2}} f(x,y)\,dx\,dy$, $25\pi/2$.

4. $\int_{-3}^{-2} \int_{x^2}^{\frac{1}{2}(15+\sqrt{129+40x})} f(x,y)\,dy\,dx + \int_{-2}^{1} \int_{\frac{1}{2}(15-\sqrt{129+40x})}^{\frac{1}{2}(15+\sqrt{129+40x})} f(x,y)\,dy\,dx$

$+ \int_{1}^{4} \int_{x^2}^{\frac{1}{2}(15+\sqrt{129+40x})} f(x,y)\,dy\,dx$, $\int_{1}^{4} \int_{\frac{1}{10}(y^2-15y+24)}^{\sqrt{y}} f(x,y)\,dx\,dy$

$+ \int_{4}^{9} \int_{-\sqrt{y}}^{\sqrt{y}} f(x,y)\,dx\,dy + \int_{9}^{16} \int_{\frac{1}{10}(y^2-15y+24)}^{\sqrt{y}} f(x,y)\,dx\,dy$, $58\tfrac{5}{6}$.

9. $\int_{-2}^{1} \int_{-y}^{\sqrt{2-y}} f(x,y)\,dx\,dy + \int_{1}^{2} \int_{-\sqrt{2-y}}^{\sqrt{2-y}} f(x,y)\,dx\,dy$.

10. $\int_0^8 \int_{x^2/32}^{\sqrt[3]{x}} f(x,y)\,dy\,dx$.

11. $\int_{\frac{1}{2}}^{1} \int_{1}^{2x} f(x,y)\,dy\,dx + \int_{1}^{2} \int_{x}^{2x} f(x,y)\,dy\,dx + \int_{2}^{4} \int_{x}^{4} f(x,y)\,dy\,dx$.

12. $\int_0^1 \int_0^1 f(x,y)\,dx\,dy + \int_1^{1+e} \int_{\phi(y)}^{1} f(x,y)\,dx\,dy$, where $x = \phi(y)$ is the inverse of $y = x + e^x$.

13. $\tfrac{7}{12}$.   14. 45.   15. $\dfrac{a^4}{8}$.

16. $\tfrac{1}{9}(19 - 18a - 18a^2 - 10a^3 - 3a^4)$, where a is the (positive) real root of $e^{-x} = x$.

§ 1012, page 355

1. $\int_{-4}^{4} \int_{-\sqrt{16-x^2}}^{\sqrt{16-x^2}} \int_{7}^{23-x^2-y^2} f(x,y,z)\,dz\,dy\,dx$, 128π.

2. $\int_{-1}^{1} \int_{-1+|x|}^{1-|x|} \int_{-1+|x|+|y|}^{1-|x|-|y|} f(x,y,z)\,dz\,dy\,dx$, $\tfrac{4}{3}$.

5. $\int_0^2 \int_{-\sqrt{2x-x^2}}^{\sqrt{2x-x^2}} \int_0^{\sqrt{2x-x^2-y^2}} f(x,y,z)\,dz\,dy\,dx$,

$\int_{-1}^{1} \int_{1-\sqrt{1-y^2}}^{1+\sqrt{1-y^2}} \int_0^{\sqrt{2x-x^2-y^2}} f(x,y,z)\,dz\,dx\,dy$,

$$\int_0^2 \int_0^{\sqrt{2x-x^2}} \int_{-\sqrt{2x-x^2-z^2}}^{\sqrt{2x-x^2-z^2}} f(x,y,z)\, dy\, dz\, dx,$$

$$\int_0^1 \int_{1-\sqrt{1-z^2}}^{1+\sqrt{1-z^2}} \int_{-\sqrt{2x-x^2-z^2}}^{\sqrt{2x-x^2-z^2}} f(x,y,z)\, dy\, dx\, dz,$$

$$\int_{-1}^1 \int_0^{\sqrt{1-y^2}} \int_{1-\sqrt{1-y^2-z^2}}^{1+\sqrt{1-y^2-z^2}} f(x,y,z)\, dx\, dz\, dy,$$

$$\int_0^1 \int_{-\sqrt{1-z^2}}^{\sqrt{1-z^2}} \int_{1-\sqrt{1-y^2-z^2}}^{1+\sqrt{1-y^2-z^2}} f(x,y,z)\, dx\, dy\, dz.$$

6. $\frac{1}{540}$.

§ 1015, page 360

1. $\pi a^4/16$.   **2.** $5\pi a^4/128$.

3. $\dfrac{1}{5}\int_0^{\pi/4} \sec^5\theta\, d\theta = [7\sqrt{2} + 3\ln(\sqrt{2}+1)]/40$.

4. $16\pi/3$.   **5.** $3\pi/2$.   **6.** $\pi a^2/8$.
7. $(\frac{1}{3}\pi + \frac{1}{2}\sqrt{3})a^2$.   **8.** $\pi - \frac{3}{2}\sqrt{3}$.   **9.** $\frac{32}{9}a^3$.
10. $\frac{2}{9}(3\pi - 4)a^3$.   **11.** $8\pi(10\sqrt{5} - 19)/3$.   **12.** $\pi/2$.

13. $\displaystyle\int_0^a \int_0^{\pi/4} F(\rho,\theta)\rho\, d\theta\, d\rho + \int_a^{\sqrt{2}a} \int_{\text{Arccos}(a/\rho)}^{\pi/4} F(\rho,\theta)\rho\, d\theta\, d\rho$.

§ 1018, page 363

1. $\bar{x} = \frac{3}{5}b,\ \bar{y} = \frac{3}{8}\sqrt{2bp}$.   **2.** $\bar{x} = \bar{y} = \dfrac{256a}{315\pi}$.

3. $\bar{x} = 0,\ \bar{y} = \dfrac{8\pi + 3\sqrt{3}}{4\pi + 6\sqrt{3}}\, a$.   **4.** $\bar{x} = \bar{y} = \dfrac{128a}{105\pi}$.

5. $\bar{x} = \frac{2}{3},\ \bar{y} = \frac{1}{2}$.   **6.** $\bar{x} = \bar{y} = \dfrac{8a}{5\pi}$.

7. $\bar{x} = \dfrac{6a}{5},\ \bar{y} = 0$.   **8.** $\bar{x} = \bar{y} = \dfrac{16a}{35}$.

9. $\frac{1}{4}\pi ab^3 = \frac{1}{4}b^2 A$.   **10.** $\dfrac{(3\pi+8)a^4}{48} = \dfrac{3\pi+8}{48}a^2 A$.

11. $\dfrac{bh^4k}{5} = \dfrac{3}{5}h^2 M$.   **12.** $\dfrac{63\pi k}{20}a^5 = 1.89\, a^2 M$.

§ 1022, page 368

1. $\frac{2}{3}\pi(16\sqrt{2} - 17)$.   **2.** $V = \frac{2}{3}\pi a^3(\sqrt{2}-1),\ \bar{z} = \frac{3}{8}a(\sqrt{2}+1)$.
3. $\frac{2}{3}\pi(13\sqrt{13} - 41)$.   **4.** $\frac{1}{2}\pi a^4 hk = \frac{1}{2}a^2 M$.
5. $\bar{x} = \dfrac{3}{8}a,\ \bar{y} = \dfrac{3\pi}{16}a,\ \bar{z} = \dfrac{3\pi}{32}ab$.   **6.** $\frac{1}{10}\pi a^4 hk = \frac{3}{10}a^2 M$.

7. Centroid on axis of symmetry, distance $\frac{3}{8}a$ from center (a = radius), $I = \frac{4}{15}\pi a^5 k = \frac{2}{5}a^2 M$.

8. $M = \frac{2}{3}\pi a^3(1-\cos\alpha)k,\ \bar{z} = \frac{3}{8}a(1+\cos\alpha)$.

9. $M = \dfrac{\pi^2 a^4 k}{8},\ \bar{z} = \dfrac{16a}{15\pi},\ I_z = \dfrac{\pi^2 a^6 k}{16} = \dfrac{a^2}{2}M$.

ANSWERS

10. $M = \pi a^4 k$; $I = \dfrac{4a^2}{9} M$.

11. $M = abck$, $I_{yz} = \dfrac{a^2 M}{3}$, $I_x = \dfrac{(b^2 + c^2)M}{3}$, $I_0 = \dfrac{(a^2 + b^2 + c^2)M}{3}$.

12. $(\bar{x}, \bar{y}, \bar{z}) = \left(\dfrac{a}{4}, \dfrac{b}{4}, \dfrac{c}{4}\right)$, $I_z = \dfrac{a^2 + b^2}{10} M$.

13. $\bar{z} = \dfrac{3(b^4 - a^4)}{8(b^3 - a^3)}$.

14. $M = \tfrac{1}{3}[\sqrt{2} + \ln(1 + \sqrt{2})]a^4 k$, $I = \tfrac{1}{20}[7\sqrt{2} + 3\ln(1 + \sqrt{2})]a^6 k$.

16. Centroid: distance $4a/3\pi$ from center; volume: $2\pi^2 a^2 b$.

§ 1025, page 372

1. $\dfrac{GMh}{(a^2 + h^2)^{3/2}}$.

2. $\dfrac{2GM}{a^2}\left(1 - \dfrac{h}{\sqrt{a^2 + h^2}}\right)$.

3. $\dfrac{GM}{2ab} \ln \left[\dfrac{c+a}{c-a} \cdot \dfrac{b + \sqrt{(c-a)^2 + b^2}}{b + \sqrt{(c+a)^2 + b^2}}\right]$.

4. $\dfrac{2GM}{a^2 h}[h + \sqrt{a^2 + (c-h)^2} - \sqrt{a^2 + c^2}]$.

5. $\dfrac{GM}{a^2}(\sqrt{2} - 1)$.

6. $\dfrac{6GM}{a^2}\left(1 - \dfrac{h}{\sqrt{a^2 + h^2}}\right)$.

§ 1028, page 379

1. $\tfrac{8}{15}$.

2. $\dfrac{\sqrt{3}\pi}{18}$.

3. $\tfrac{1}{8}(b-a)(d-c)$.

4. $\tfrac{1}{6}(e^3 - e)$.

5. $\tfrac{1}{6}[\sqrt{2} + \ln(1 + \sqrt{2})]$.

6. n odd, 0; n even, $\dfrac{1 \cdot 3 \cdot 5 \cdots (n-1)}{2 \cdot 4 \cdot 6 \cdots (n+2)} \cdot 2\pi(a^2 + b^2)^{n/2}$.

7. $-\tfrac{11}{36}$.

8. $\bar{x} = \bar{y} = \bar{z} = \tfrac{1}{4}$.

§ 1107, page 385

1. $\dfrac{n}{n+1}$; 1.

2. $\dfrac{n}{2n+1}$; $\tfrac{1}{2}$.

3. $1 - (\tfrac{2}{3})^n$; 1.

4. $\dfrac{n(n+1)}{2}$; $+\infty$.

5. $2 + \tfrac{1}{2} + \tfrac{1}{6} + \tfrac{1}{12}$; $a_1 = 2$, $a_n = \dfrac{1}{n(n-1)}$, $n > 1$.

6. $1 + 2 - 2 + 2$; $a_1 = 1$, $a_n = 2(-1)^n$, $n > 1$.

7. $1.3 - 0.39 + 0.117 - 0.0351$; $1.3(-0.3)^{n-1}$.

8. $2 - 0.7 - 0.21 - 0.063$; $a_1 = 2$, $a_n = -0.7(0.3)^{n-2}$, $n > 1$.

9. $\tfrac{36}{5}$. **10.** Divergent. **11.** 0.3.

12. $\dfrac{10{,}201}{201}$. **13.** $\tfrac{5}{9}$. **14.** $\tfrac{2957}{900}$.

15. $\tfrac{679}{110}$. **16.** $\tfrac{3}{7}$. **17.** Divergent.
18. Divergent. **19.** Convergent. **20.** Convergent.
21. Convergent. **22.** Divergent. **23.** Divergent.
24. Convergent. **25.** Convergent. **26.** Convergent.

§ 1111, page 393

1. Divergent. **2.** Divergent. **3.** Convergent.
4. Convergent. **5.** Convergent. **6.** Divergent.

7. Divergent.   8. Convergent.   9. Convergent.
10. Convergent.   11. Convergent.   12. Divergent.
13. Convergent.   14. Convergent.
15. Convergent for $\alpha > \frac{1}{2}$; divergent for $\alpha \leq \frac{1}{2}$.
16. Convergent.   17. Divergent.
18. Convergent for $\alpha > 1$; divergent for $\alpha \leq 1$.
19. Convergent.
20. Convergent for $0 < r < 1$; divergent for $r \geq 1$ unless $\alpha = k\pi$.

§ 1113, page 397

1. Divergent.   2. Convergent.   3. Divergent.
4. Convergent for $p > 2$; divergent for $p \leq 2$.

§ 1117, page 401

1. Conditionally convergent.   2. Divergent.
3. Conditionally convergent.   4. Absolutely convergent.
5. Absolutely convergent.   6. Absolutely convergent.
7. Absolutely convergent for $x > 0$; divergent for $x \leq 0$.
8. Absolutely convergent for $|r| < 1$; divergent for $|r| \geq 1$.
9. Absolutely convergent for $p > 1$; conditionally convergent for $p \leq 1$.
10. Absolutely convergent for $p > 2$; conditionally convergent for $0 < p \leq 2$; divergent for $p \leq 0$.

§ 1121, page 410

13. 2.718.   14. 0.6931.
15. 1.202.   16. 0.1775.
17. 0.7854.   18. 0.4055.

§ 1202, page 415

1. Absolutely convergent for $-\infty < x < +\infty$.
2. Absolutely convergent for $|x| < 1$; conditionally convergent for $x = \pm 1$.
3. Absolutely convergent for $-\infty < x < +\infty$.
4. Absolutely convergent for $x = 0$.
5. Absolutely convergent for $-2 \leq x \leq 0$.
6. Absolutely convergent for $-\infty < x < +\infty$.
7. Absolutely convergent for $4 < x < 6$; conditionally convergent for $x = 4$.
8. Absolutely convergent for $0 < x < 2$.
9. Absolutely convergent for $|x| < 1$; conditionally convergent for $x = 1$.
10. Absolutely convergent for $|x| \leq 1$.
11. Absolutely convergent for $x < 2$ or $x > 4$; conditionally convergent for $x = 2$.
12. Absolutely convergent if $|x| \neq \dfrac{2n+1}{2}\pi$; conditionally convergent if $|x| = \dfrac{2n+1}{2}\pi$.

§ 1211, page 433

1. $1 + \dfrac{x^2}{2!} + \dfrac{x^4}{4!} + \dfrac{x^6}{6!} + \cdots$.
2. $1 - x^2 + x^4 - x^6 + \cdots$.
3. $x - \dfrac{x^3}{3} + \dfrac{x^5}{5} - \dfrac{x^7}{7} + \cdots$.
4. $1 + \dfrac{x}{1! \cdot 2} + \dfrac{x^2}{2! \cdot 4} + \dfrac{x^3}{3! \cdot 8} + \cdots$.
5. $1 - \dfrac{x^4}{2!} + \dfrac{x^8}{4!} - \dfrac{x^{12}}{6!} + \cdots$.
6. $\ln 2 + \dfrac{3x}{1 \cdot 2} - \dfrac{9x^2}{2 \cdot 4} + \dfrac{27x^3}{3 \cdot 8} - \cdots$.
7. $1 + \dfrac{1}{2}x^4 + \dfrac{1 \cdot 3}{2 \cdot 4}x^8 + \dfrac{1 \cdot 3 \cdot 5}{2 \cdot 4 \cdot 6}x^{12} + \cdots$.
8. $2 + \dfrac{2}{8}x - \dfrac{2 \cdot 1}{8 \cdot 16}x^2 + \dfrac{2 \cdot 1 \cdot 3}{8 \cdot 16 \cdot 24}x^3 - \dfrac{2 \cdot 1 \cdot 3 \cdot 5}{8 \cdot 16 \cdot 24 \cdot 32}x^4 + \cdots$.

ANSWERS

9. $2\left[x + \dfrac{x^3}{3} + \dfrac{x^5}{5} + \dfrac{x^7}{7} + \cdots\right].$

10. $x - \dfrac{x^3}{3\cdot 3!} + \dfrac{x^5}{5\cdot 5!} - \dfrac{x^7}{7\cdot 7!} + \cdots.$

11. $x - \dfrac{x^3}{3} + \dfrac{x^5}{2!\cdot 5} - \dfrac{x^7}{3!\cdot 7} + \cdots.$

12. $\dfrac{x^3}{3} - \dfrac{x^7}{3!\cdot 7} + \dfrac{x^{11}}{5!\cdot 11} - \dfrac{x^{15}}{7!\cdot 15} + \cdots.$

13. $\dfrac{1}{2} - \dfrac{x}{4} + \dfrac{x^3}{48} - \dfrac{x^5}{480}.$

14. $1 + x - \dfrac{x^3}{3} - \dfrac{x^4}{6} + \dfrac{x^5}{30}.$

15. $x - \dfrac{1}{3}x^3 + \dfrac{2}{15}x^5 - \dfrac{17}{315}x^7.$

16. $1 + x + \dfrac{x^2}{2} + \dfrac{x^3}{2} + \dfrac{3x^4}{8} + \dfrac{37x^5}{120}.$

17. $-\dfrac{x^2}{6} - \dfrac{x^4}{180} - \dfrac{x^6}{2835}.$

18. $-\dfrac{x^2}{2} - \dfrac{x^4}{12} - \dfrac{x^6}{45} - \dfrac{17x^8}{2520}.$

19. $\cos a\left[1 - \dfrac{(x-a)^2}{2!} + \dfrac{(x-a)^4}{4!} - \cdots\right]$
$- \sin a\left[(x-a) - \dfrac{(x-a)^3}{3!} + \dfrac{(x-a)^5}{5!} - \cdots\right].$

20. $\dfrac{\sqrt{2}}{2}\left[1 - \left(x - \dfrac{\pi}{4}\right) - \dfrac{\left(x - \dfrac{\pi}{4}\right)^2}{2!} + \dfrac{\left(x - \dfrac{\pi}{4}\right)^3}{3!} + - - \cdots\right].$

21. $1 + \dfrac{x-e}{e} - \dfrac{(x-e)^2}{2e^2} + \dfrac{(x-e)^3}{3e^3} - \cdots.$

22. $1 + m(x-1) + \dfrac{m(m-1)}{2!}(x-1)^2 + \cdots.$

23. $(1-x)^{-2}.$
24. $\dfrac{1}{2}[(x^2+1)\operatorname{Arctan} x - x].$
25. $(1+x^2)^{-2}.$
26. $3\ln\dfrac{3}{2} - 1.$
38. $B_0 = 1,\ B_1 = -\dfrac{1}{2},\ B_2 = \dfrac{1}{6},\ B_4 = -\dfrac{1}{30},\ B_6 = \dfrac{1}{42},\ B_8 = -\dfrac{1}{30},\ B_{10} = \dfrac{5}{66}.$

§ 1214, page 438

1. $\dfrac{1}{2}.$
2. $-\dfrac{1}{2}.$
3. $\dfrac{3}{8}.$
4. $\dfrac{1}{4}.$
5. $-\dfrac{1}{7!}.$
6. $1.$
7. $\dfrac{4}{3}.$
8. $-2.$
9. $2e^2.$
10. $0.$
11. $7.38906.$
12. $1.64872.$
13. $1.0986.$
14. $0.8776.$
15. $2.15443.$
16. $0.10033.$
17. $0.4931.$
18. $0.0976.$
19. $0.4920.$
20. $0.7468.$

§ 1216, page 440

1. $1 + x + y + \dfrac{x^2}{2} + xy + \dfrac{y^2}{2} + \dfrac{x^3}{6} + \dfrac{x^2y}{2} + \dfrac{xy^2}{2} + \dfrac{y^3}{6}.$

2. $(x-1) + (y-1) - \dfrac{(x-1)^2}{2} - \dfrac{(y-1)^2}{2}.$

§ 1304, page 446

37. $\dfrac{\ln \epsilon}{\ln \sin(\frac{1}{2}\pi - \eta)}.$

38. $\dfrac{b}{\epsilon}.$

39. $\dfrac{1}{\epsilon\eta}.$

40. $\dfrac{2b}{\epsilon^2}.$

41. $\max\left(\dfrac{2}{\eta}, \dfrac{6}{\epsilon\eta}\right).$

42. $\dfrac{\ln \epsilon}{\ln(1-\eta)}.$

§ 1311, page 457

7. $\ln 4 - 1$.
8. $\frac{1}{4}\pi - \frac{1}{2}\ln 2$.

§ 1315, page 461

12. $\delta(\epsilon) = \epsilon\eta$.
13. $\delta(\epsilon) = \epsilon/b$.
14. $\delta(\epsilon) = \epsilon/\max(|a|, |b|)$.
15. $N(\epsilon) = K/\epsilon$.

§ 1403, page 466

1. Absolutely convergent.
2. Absolutely convergent.
3. Divergent.
4. Conditionally convergent.
5. Conditionally convergent, $0 < p < 1$.
6. Absolutely convergent, $0 < p < 2$, conditionally convergent, $-1 < p \leq 0$.
7. Absolutely convergent, $-1 < p < 1$, conditionally convergent, $1 \leq p < 2$.
8. Absolutely convergent, $1 < p < 3$.

§ 1412, page 484

9. $\frac{3}{8}\sqrt{\pi}$.
10. $\sqrt{\pi/5}$.
11. $\sqrt{2\pi}$.
12. $\pi/8$.
18. $s^{-1}, s > 0$.
19. $\frac{1}{2}\sqrt{\pi}s^{-\frac{3}{2}}, s > 0$.
20. $(s-a)^{-1}, s > a$.
21. $\Gamma(p)/(s-a)^p, s > a$.
22. $b[(s-a)^2 + b^2]^{-1}, s > a$.
23. $b[(s-a)^2 - b^2]^{-1}, s > a + |b|$.

§ 1420, page 495

1. π.
2. -4π.
3. Divergent.
4. 2π.
5. $\frac{4}{3}\pi$.
6. -8π.
7. 8π.
8. Divergent.

§ 1509, page 503

6. $5 + 85i$.
7. $\frac{22}{89} + \frac{53}{89}i$.
8. $32i$.
9. $10^{-18}[\cos 18 \operatorname{Arcsin} \frac{4}{5} + \sin 18 \operatorname{Arcsin} \frac{4}{5}]$.
10. i.
11. $\frac{1}{2}(-1 + i\sqrt{3})$.
12. $-1, \frac{1}{2}(1 \pm i\sqrt{3})$.
13. $2(\cos 36° + i \sin 36°)$, $2(\cos 108° + i \sin 108°)$, $2(\cos 180° + i \sin 180°) = -2$, $2(\cos 252° + i \sin 252°)$, $2(\cos 324° + i \sin 324°)$.
14. $\sqrt[4]{5}(\cos 67\frac{1}{2}° + i \sin 67\frac{1}{2}°)$; other amplitudes are $157\frac{1}{2}°, 247\frac{1}{2}°, 337\frac{1}{2}°$.
15. $\sqrt[5]{2}(\cos 30° + i \sin 30°) = \frac{1}{2}\sqrt[5]{2}(\sqrt{3} + i)$; other amplitudes are $102°, 174°, 246°, 318°$.
27. $\frac{1}{2}(4n + 1)\pi i$.
29. $\sin i = \frac{1}{2}i(e - e^{-1})$, $\cos i = \frac{1}{2}(e^{-1} - e)$, $\arccos 4 = 2n\pi \pm i \ln(4 + \sqrt{15})$, $\arcsin 2 = \frac{1}{2}(4n + 1)\pi \pm i \ln(2 + \sqrt{3})$.

§ 1514, page 513

18. $\dfrac{i}{\sqrt{2\pi}\,s}[e^{-i\alpha s} - 1]$.
19. $(2\pi)^{-\frac{1}{2}}(\alpha + is)^{-1}$.
20. $\dfrac{\sqrt{2}}{2}e^{-\frac{1}{2}s^2}$.

§ 1607, page 528

1. $\dfrac{4}{\pi}\left[\dfrac{\sin x}{1} + \dfrac{\sin 3x}{3} + \dfrac{\sin 5x}{5} + \cdots\right]$.
2. $\dfrac{1}{2} + \dfrac{2}{\pi}\left[\dfrac{\sin x}{1} + \dfrac{\sin 3x}{3} + \dfrac{\sin 5x}{5} + \cdots\right]$.
3. $\dfrac{1}{4} + \dfrac{1}{\pi}\sum_{n=1}^{+\infty}\left[\dfrac{1}{n}\sin \tfrac{1}{2}n\pi \cos nx + \dfrac{1}{n}(1 - \cos \tfrac{1}{2}n\pi)\sin nx\right]$.

ANSWERS

4. $\dfrac{1}{4} + \dfrac{1}{\pi} \sum\limits_{n=1}^{+\infty} \left[-\dfrac{1}{n} \sin \tfrac{1}{2}n\pi \cos nx + \dfrac{1}{n}(\cos \tfrac{1}{2}n\pi - \cos n\pi) \sin nx \right].$

5. $\dfrac{\pi}{4} + \sum\limits_{n=1}^{+\infty} \left[-\dfrac{1+(-1)^{n+1}}{n^2\pi} \cos nx + \dfrac{(-1)^{n+1}}{n} \sin nx \right].$

6. $\dfrac{\pi}{16} + \sum\limits_{n=1}^{+\infty} \left\{ \left[\dfrac{1}{2n} \sin \tfrac{1}{2}n\pi - \dfrac{1}{n^2\pi}(1 - \cos \tfrac{1}{2}n\pi) \right] \cos nx \right.$
$\left. + \left[-\dfrac{1}{2n} \cos \tfrac{1}{2}n\pi + \dfrac{1}{n^2\pi} \sin \tfrac{1}{2}n\pi \right] \sin nx \right\}.$

§ 1612, page 536

4. $b_n = (-1)^{n+1} \dfrac{2\pi^2}{n} + (-1)^n \dfrac{12}{n^3}.$

5. $a_0 = \dfrac{2\pi^4}{5},\ a_n = (-1)^n \dfrac{8\pi^2}{n^2} + (-1)^{n+1} \dfrac{48}{n^4},\ n > 0.$

6. $b_n = (-1)^{n+1} \dfrac{2\pi^4}{n} + (-1)^n \dfrac{40\pi^2}{n^3} + (-1)^{n+1} \dfrac{240}{n^5}.$

7. $\dfrac{1}{2} + \dfrac{2}{\pi} \sum\limits_{n=1}^{+\infty} (-1)^n \cos(2n-1)x/(2n-1);\ \dfrac{2}{\pi} \sum\limits_{n=1}^{+\infty} \dfrac{1}{n}(\cos \tfrac{1}{2}n\pi - \cos n\pi) \sin nx.$

8. $\dfrac{\pi}{4} - \dfrac{8}{\pi}\left[\dfrac{\cos 2x}{2^2} + \dfrac{\cos 6x}{6^2} + \dfrac{\cos 10x}{10^2} + \cdots \right];\ \dfrac{4}{\pi}\left[\dfrac{\sin x}{1^2} - \dfrac{\sin 3x}{3^2} + \dfrac{\sin 5x}{5^2} - \cdots \right].$

9. $\cos x;\ \dfrac{4}{\pi}\left[\dfrac{2 \sin 2x}{2^2 - 1} + \dfrac{4 \sin 4x}{4^2 - 1} + \dfrac{6 \sin 6x}{6^2 - 1} + \cdots \right].$

10. $\dfrac{2}{\pi} - \dfrac{4}{\pi}\left[\dfrac{\cos 2x}{2^2 - 1} + \dfrac{\cos 4x}{4^2 - 1} + \dfrac{\cos 6x}{6^2 - 1} + \cdots \right];\ \sin x.$

11. $\dfrac{4}{\pi}\left[\dfrac{\sin \pi x}{1} + \dfrac{\sin 3\pi x}{3} + \dfrac{\sin 5\pi x}{5} + \cdots \right].$

12. $\dfrac{4}{\pi}\left[\dfrac{\sin \tfrac{1}{2}\pi x}{1} - \dfrac{\sin \tfrac{2}{2}\pi x}{2} + \dfrac{\sin \tfrac{3}{2}\pi x}{3} - \cdots \right].$

13. $1 + \dfrac{4}{\pi} \sum\limits_{n=1}^{+\infty} \left\{ -\dfrac{1+(-1)^{n+1}}{n^2\pi} \cos \dfrac{n\pi x}{4} + \dfrac{(-1)^{n+1}}{n} \sin \dfrac{n\pi x}{4} \right\}.$

14. $\dfrac{3}{8} + \dfrac{6}{\pi} \sum\limits_{n=1}^{+\infty} \left\{ \left[\dfrac{1}{2n} \sin \tfrac{1}{2}n\pi - \dfrac{1}{n^2\pi}(1 - \cos \tfrac{1}{2}n\pi) \right] \cos \dfrac{n\pi x}{6} \right.$
$\left. + \left[-\dfrac{1}{2n} \cos \tfrac{1}{2}n\pi + \dfrac{1}{n^2\pi} \sin \tfrac{1}{2}n\pi \right] \sin \dfrac{n\pi x}{6} \right\}.$

17. $a_n = \dfrac{2}{p}\int_0^p f(x) \cos \dfrac{n\pi x}{p}\,dx,\ b_n = \dfrac{2}{p}\int_0^p f(x) \sin \dfrac{n\pi x}{p}\,dx.$

19. $B_0(x) = 1,\ B_1(x) = x - \tfrac{1}{2},\ B_2(x) = x^2 - x + \tfrac{1}{6},\ B_3(x) = x^3 - \tfrac{3}{2}x^2 + \tfrac{1}{2}x,$
$B_4(x) = x^4 - 2x^3 + x^2 - \tfrac{1}{30}.$

§ 1627, page 551

1. $u = \dfrac{8p}{\pi^2} \sum\limits_{n=1}^{+\infty} (-1)^{n+1} \dfrac{\sin(2n-1)x \cos(2n-1)at}{(2n-1)^2}.$

8. $u(x,t) = \dfrac{4p}{\pi a} \sum\limits_{n=1}^{+\infty} \left[\dfrac{(-1)^{n+1} \sin(2n-1)h}{(2n-1)^2} \sin(2n-1)x \sin(2n-1)at \right].$

9. $u(x,t) = \dfrac{8p}{\pi^2} \sum\limits_{n=1}^{+\infty} (-1)^{n+1} \dfrac{\sin(2n-1)x e^{-(2n-1)^2 a^2 t}}{(2n-1)^2}.$

10. $u(x, t) = A \sin 2x e^{-4a^2 t} + B \sin 7x e^{-49a^2 t}$.

12. $u(x, t) = \frac{1}{2}a_0 + \sum_{n=1}^{+\infty} a_n \cos nx e^{-n^2 a^2 t}$, $a_n = \frac{2}{\pi} \int_0^\pi f(x) \cos nx\, dx$, $n = 0, 1, 2, \cdots$.

13. $u(x, t) = \frac{p}{2} - \frac{16p}{\pi^2} \sum_{n=1}^{+\infty} \frac{\cos(4n-2)x e^{-(4n-2)^2 a^2 t}}{(4n-2)^2}$.

15. $50 + \frac{30x}{\pi} - \frac{200}{\pi} \sum_{n=1}^{+\infty} \frac{\sin(2n-1)x e^{-(2n-1)^2 a^2 t}}{(2n-1)} + \frac{40}{\pi} \sum_{n=1}^{+\infty} \frac{\sin 2nx e^{-4n^2 a^2 t}}{2n}$.

§ 1705, page 560

7. $11\vec{i} - 2\vec{j} + 7\vec{k}$, $-11\vec{i} + 2\vec{j} - 7\vec{k}$, $11\vec{i} - 2\vec{j} + 7\vec{k}$.
9. -47, left-handed.
10. $-54\vec{i} + 52\vec{j} + 51\vec{k}$, $11\vec{i} + 26\vec{j} + 6\vec{k}$.
11. $3\sqrt{21}$. 12. 112.
13. (a) $12:32\tfrac{8}{11} < t < 1:05\tfrac{5}{11}$, $1:38\tfrac{2}{11} < t \leq 2:00$;
    (b) $12:00 < t < 12:32\tfrac{8}{11}$, $1:05\tfrac{5}{11} < t < 1:38\tfrac{2}{11}$;
    (c) $12:00$, $12:32\tfrac{8}{11}$, $1:05\tfrac{5}{11}$.

§ 1709, page 567

1. $\tfrac{2}{7}, \tfrac{3}{7}, \tfrac{6}{7}$. 2. $\tfrac{10}{19}, -\tfrac{15}{19}, \tfrac{6}{19}, -\tfrac{10}{19}$.
3. $-\tfrac{2}{3}, -\tfrac{2}{3}, -\tfrac{2}{3}, -\tfrac{2}{3}, -\tfrac{1}{3}$. 4. $0, \pm\tfrac{1}{\sqrt{3}}, \tfrac{1}{\sqrt{3}}, \tfrac{2}{\sqrt{6}}, \mp\tfrac{1}{\sqrt{6}}, \mp\tfrac{1}{\sqrt{6}}$.
5. One set: $x' = \frac{1}{\sqrt{3}}(x+y+z)$; $y' = \frac{1}{\sqrt{6}}(x-2y+z)$; $z' = \frac{1}{\sqrt{2}}(x-z)$.
6. One set: $x' = \frac{1}{\sqrt{70}}(3x + 6y - 5z)$; $y' = \frac{1}{\sqrt{14}}(x+2y+3z)$; $z' = \frac{1}{\sqrt{5}}(2x-y)$.
7. $\begin{pmatrix} \tfrac{1}{2}\sqrt{2} & -\tfrac{1}{2}\sqrt{2} & 0 \\ \tfrac{1}{2}\sqrt{2} & \tfrac{1}{2}\sqrt{2} & 0 \\ 0 & 0 & 1 \end{pmatrix}$. 8. $\frac{x}{3} = \frac{y}{3} = \frac{z}{1}$.
9. $\operatorname{Arccos}(-\tfrac{5}{14})$.

§ 1715, page 574

1. $9y^2 z^3 \vec{i} + 18xyz^3 \vec{j} + 27xy^2 z^2 \vec{k}$.
2. $(7y^2 z^3 - 2x^2 z^3 - 6x^2 y^2 z)\vec{i} + (12xyz^3 - 6xy^3 z)\vec{j} + (15xy^2 z^2 - 2xz^4)\vec{k}$.
10. $(2y^2 z^3 + 2x^2 z^3 + 6x^2 y^2 z)\vec{i} + (6xyz^3 + 6xy^3 z)\vec{j} + (2xz^4 + 12xy^2 z^2)\vec{k}$.

§ 1719, page 582

3. $\vec{i} = \cos\theta \vec{i}_1 - \sin\theta \vec{i}_2$, $\vec{j} = \sin\theta \vec{i}_1 + \cos\theta \vec{i}_2$, $\vec{k} = \vec{i}_3$.
4. $\vec{i} = \sin\phi\cos\theta \vec{i}_1 + \cos\phi\cos\theta \vec{i}_2 - \sin\theta \vec{i}_3$,
   $\vec{j} = \sin\phi\sin\theta \vec{i}_1 + \cos\phi\sin\theta \vec{i}_2 + \cos\theta \vec{i}_3$,
   $\vec{k} = \cos\phi \vec{i}_1 - \sin\phi \vec{i}_2$.

§ 1804, page 591

4. $\tfrac{7}{6}$. 5. $\tfrac{5}{3}$. 6. $-\tfrac{28}{3}$. 7. 72.
8. $\tfrac{3}{2}(e^2 + 1)$. 9. $\tfrac{15}{2}$. 10. $\sqrt{145} + \tfrac{1}{12}\ln(12 + \sqrt{145})$.
11. $\tfrac{3}{8}a^2$. 12. 6. 13. $-\tfrac{32}{3}$. 14. $\tfrac{4}{3}$.
15. $\tfrac{3}{8}$. 16. $\tfrac{3}{8}\pi a^2$. 17. 0. 18. $\tfrac{4}{3}a^3$.
19. $-\tfrac{8}{3}$. 24. xy. 25. $e^x \sin y$. 26. $x^3 y - xy^3 + \tfrac{1}{2}y^2$.
27. $\sin x \cosh y + \cos x \sinh y$. 28. $-\dfrac{y}{x^2 + y^2}$.
29. 0. 30. $\sin 1$. 31. $\tfrac{1}{2}$.
32. 0. 33. -1.

ANSWERS

§ 1809, page 602

17. πab.
18. $\frac{3}{8}\pi a^2$.
19. $\frac{3}{2}a^2$.
20. $\frac{1}{2}a^2(4 - \pi)$.
29. 0, exact, $\frac{1}{2}\ln(x^2 + y^2)$.
30. 0, exact, $-\frac{1}{2}(x^2 + y^2)^{-1}$.
31. π, not exact.
32. 0, exact, $\frac{1}{2}y^2(x^2 + y^2)^{-1}$.
33. $\frac{3}{4}\pi$, not exact.
34. 0, exact, $-\frac{1}{4}x^4(x^2 + y^2)^{-2}$.
35. $\frac{1}{4}\pi$, not exact.
36. $\dfrac{xy}{2(x^2 + y^2)} + \frac{1}{2}\phi(x, y)$, where ϕ is the function of the Example, § 1808.

§ 1815, page 620

1. (a) $2\pi(\sqrt{2} + \ln(1 + \sqrt{2}))$; (b) $2\pi\int_0^{\pi} x\sqrt{1 + \cos^2 x}\, dx$.

2. (a) $2\pi b^2 + \dfrac{2\pi a^2 b}{\sqrt{a^2 - b^2}}\,\text{Arcsin}\,\dfrac{\sqrt{a^2 - b^2}}{a}$; (b) $2\pi a^2 + \dfrac{2\pi ab^2}{\sqrt{a^2 - b^2}}\ln\dfrac{a + \sqrt{a^2 - b^2}}{b}$.

3. (a) $\frac{64}{3}\pi a^2$; (b) $16\pi^2 a^2$.
4. $\frac{12}{5}\pi a^2$.
5. $2\pi ab$.
6. $\dfrac{\pi r^2}{c}\sqrt{a^2 + b^2 + c^2}$.
7. a^2/k.
8. $2a^2(\pi - 2)$.
9. $4a^2$.
10. $\frac{2}{3}\pi[(1 + a^2)^{\frac{3}{2}} - 1]$.
11. $\pi a^2\sqrt{2}$.
12. $16a^2$.
13. $16a^2$.

§ 1822, page 631

1. $\displaystyle\int_0^{2\pi}\int_0^{\frac{\pi}{2}} 4a^2 \cos\phi \sin\phi\, d\phi\, d\theta$.
2. $\frac{5}{2}\sqrt{2}\pi$.
3. $(2a/3(\pi - 2), 0, 0)$.
4. $\frac{2}{3}a^2 M$.
5. $\frac{1}{2}M(3a^2 + 2b^2)$.
6. $\dfrac{GM}{2a^2}(2 - \sqrt{2})$.
10. $\frac{12}{5}\pi a^5$.
11. 3.
12. 54π.
13. $\frac{8}{3}\pi a^3(x_0 + y_0 + z_0)$.
15. πa^4.
16. Yes.

§ 1827, page 641

5. (a) 43/6; (b) 8; (c) 22/3.
6. $\frac{1}{4}\pi a^2$.
7. $12 - 2\pi$.
8. $70\frac{7}{12}$.
9. $e - 1$.
10. $6e$.
11. $\displaystyle\int_0^1 e^{-t^2}\, dt$.
12. $\dfrac{a^3\sqrt{a^2\lambda^2 + \mu^2}}{3\lambda}(\cos^3 \lambda t_1 - \cos^3 \lambda t_2)$.
13. πa^3.
14. -3.
15. π.
16. πa^2.
17. $2\pi a^2$.
20. $u = \dfrac{xz^3}{3} - \dfrac{x^2 y^2}{2}$, $v = -\dfrac{y^2 z^2}{2}$, $w = 0$.
21. $u = -x^2 yz$, $v = -xyz^2$, $w = 0$.

§ 1906, page 655

1. At $(x_0, y_0, z_0) = (t, t, t^2)$, $K = (1 + 2t^2)^{-\frac{3}{2}}$, $T = 0$; $\vec{\tau} = (2 + 4t^2)^{-\frac{1}{2}}(\vec{i} + \vec{j} + 2t\vec{k})$, $\vec{v} = (1 + 2t^2)^{-\frac{1}{2}}(-t\vec{i} - t\vec{j} + \vec{k})$, $\vec{\beta} = 2^{-\frac{1}{2}}(\vec{i} - \vec{j})$; normal plane: $x + y + 2tz = 2t + 2t^3$, osculating plane: $x = y$, rectifying plane: $tx + ty - z = t^2$.

2. $K = T = 2(1 + 2t^2)^{-2}$; $\vec{\tau} = (1 + 2t^2)^{-1}(\vec{i} + 2t\vec{j} + 2t^2\vec{k})$, $\vec{v} = (1 + 2t^2)^{-1}(-2t\vec{i} + (1 - 2t^2)\vec{j} + 2t\vec{k})$, $\vec{\beta} = (1 + 2t^2)^{-1}(2t^2\vec{i} - 2t\vec{j} + \vec{k})$; normal plane: $(x - t) + 2t(y - t^2) + 2t^2(z - \frac{2}{3}t^3) = 0$, osculating plane: $2t^2 x - 2ty + z = \frac{2}{3}t^3$, rectifying plane: $2t(x - t) + (-1 + 2t^2)(y - t^2) - 2t(z - \frac{2}{3}t^3) = 0$.

ANSWERS

3. $K = T = [3a(1 + t^2)^2]^{-1}$;
$\vec{\tau} = (\sqrt{2}(1 + t^2))^{-1}((1 - t^2)\vec{i} + 2t\vec{j} + (1 + t^2)\vec{k})$,
$\vec{\nu} = (1 + t^2)^{-1}(-2t\vec{i} + (1 - t^2)\vec{j})$,
$\vec{\beta} = (\sqrt{2}(1 + t^2))^{-1}((-1 + t^2)\vec{i} - 2t\vec{j} + (1 + t^2)\vec{k})$.

4. $K = 6\sqrt{3}t^3(14 - 2t^2 + 2t^4)^{-\frac{3}{2}}$, $T = 0$;
$\vec{\tau} = (14 - 2t^2 + 2t^4)^{-\frac{1}{2}}((-1 - t^2)\vec{i} + (-2 + t^2)\vec{j} + 3\vec{k})$,
$\vec{\nu} = (42 - 6t^2 + 6t^4)^{-\frac{1}{2}}((5 - t^2)\vec{i} + (-4 - t^2)\vec{j} + (-1 + 2t^2)\vec{k})$,
$\vec{\beta} = 3^{-\frac{1}{2}}(\vec{i} + \vec{j} + \vec{k})$.

5. $K = (8 + t^2 + t^4)^{\frac{1}{2}}(2 + t^2)^{-\frac{3}{2}}$, $T = (6 + t^2)(8 + t^2 + t^4)^{-1}$.

6. $K = (c^2 + 4\sin^4 \frac{1}{2}t)^{\frac{1}{2}}(c^2 + 4\sin^4 \frac{1}{2}t)^{-\frac{3}{2}}$, $T = -c(c^2 + 4\sin^4 \frac{1}{2}t)^{-1}$.

7. $K = 2a(32a^2 - 3z^2)^{\frac{1}{2}}(8a^2 - z^2)^{-\frac{3}{2}}$,
$T = 12ay/z(32a^2 - 3z^2)$, $z \neq 0$,
$T = 3/8a$ at $(2a, 0, 0)$, $T = -3/8a$ at $(0, 0, 0)$.

8. $K = 2a(32a^2 - 3z^2)^{\frac{1}{2}}(8a^2 - z^2)^{-\frac{3}{2}}$,
$T = -12ay/z(32a^2 - 3z^2)$, $z \neq 0$,
T indeterminate at $(0, 0, 0)$.

17. $A = s''' - s'^3K^2$, $B = 3Ks's'' + s'^2K'$.

23. $K = e^{-at}b(a^2 + b^2)^{\frac{1}{2}}(a^2 + b^2 + a^2c^2)^{-1}$, $T = e^{-at}abc(a^2 + b^2 + a^2c^2)^{-1}$.

25. Ex. 2: $\vec{c} = \vec{i} + \vec{k}$; Exs. 3, 20, 23: $\vec{c} = \vec{k}$.

§ 1912, page 670

6. $\dfrac{a^3(x^2 + y^2 + z^2)}{(y^2z^2 + z^2x^2 + x^2y^2)^{\frac{3}{2}}}$, $\dfrac{3a^6}{(y^2z^2 + z^2x^2 + x^2y^2)^2}$.

7. $\dfrac{ax + by + 3abxyz}{(1 + a^2x^4 + b^2y^4)^{\frac{3}{2}}}$, $\dfrac{4abxy}{(1 + a^2x^4 + b^2y^4)^2}$.

8. $\dfrac{2ab(a^2 - b^2 + uv)}{[4a^2b^2 + a^2(u - v)^2 + b^2(u + v)^2]^{\frac{3}{2}}}$, $\dfrac{4a^2b^2}{[4a^2b^2 + a^2(u - v)^2 + b^2(u + v)^2]^2}$.

9. $-\dfrac{2}{(1 + 4\rho^2)^{\frac{3}{2}}}$, $-\dfrac{2}{(1 + 4\rho^2)^{\frac{1}{2}}}$, $-\dfrac{2 + 4\rho^2}{(1 + 4\rho^2)^{\frac{3}{2}}}$, $\dfrac{4}{(1 + 4\rho^2)^2}$.

10. $2a(2\rho^2 - 1)e^{-\rho^2}[1 + 4a^2\rho^2e^{-2\rho^2}]^{-\frac{3}{2}}$, $-2ae^{-\rho^2}[1 + 4a^2\rho^2e^{-2\rho^2}]^{-\frac{1}{2}}$; elliptic: $\rho < \frac{1}{2}\sqrt{2}$, hyperbolic: $\rho > \frac{1}{2}\sqrt{2}$, parabolic: $\rho = \frac{1}{2}\sqrt{2}$.

19. $-\dfrac{a^2}{(x^2 + y^2 + a^2)^2}$.

20. $-\dfrac{f'^2}{(\rho^2 + f'^2)^2}$, $-\dfrac{c^2}{(\rho^2 + c^2)^2}$.

INDEX

(The numbers refer to pages)

Abel's test for improper integrals, 465, 467 (Exs. 9-11)
  uniform convergence, 472
Abel's test for series, 403 (Exs. 27, 28)
  uniform convergence, 448 (Exs. 47, 48)
Abel's theorem, 455-457
  on product series, 406, 458 (Ex. 20)
Absolute convergence, of improper integrals, 464
  of infinite products, 493 (Ex. 1)
  of series, 399-401
Absolute maximum (minimum), 85
Absolute value, 14-16, 205, 498-499
Acceleration, 569
Addition and subtraction, 2, 496
  of series, 405-407, 425
  of vectors, 206
Algebra, fundamental theorem, 515-516, 608 (Ex. 16)
Algebraic function, 168
Algebraic number, 168
Almost everywhere, 112, 342
Alternating harmonic series, 399
Alternating integrals, 464-466
Alternating p-series, 401
Alternating series, 398, 448 (Ex. 49)
Amplitude, 499
Analytic function, 421, 605-610
Angle, solid, 628
Anticommutative law, 555
Antiderivative, 128
Applications of Fourier series, 548-553
Approximations by differentials, 92-94, 281-282
Arc, 239
  Peano, 241
  rectifiable, 241
  sectionally smooth, 585
  simple, 239
  smooth, 244
  space-filling, 241
Arc length, 241-247, 251

Archimedean property, 19
Arcs and curves, 238-260, 584-602, 646-670
Arc-wise connected set, 198
Area, 337-342, 349
  bounded sets without, 342
  criterion for, 341
  differential element of, 377
  inner (outer), 339-342
  invariance of, 337, 617
  of a parallelogram, 351 (Ex. 27), 561 (Ex. 21)
  of a surface, 617-620
Argument, 499
Arithmetic mean, 7 (Ex. 21), 321 (Ex. 19)
Associative law, 2, 9, 207, 497, 560
Asymptotes, 103, 104
Asymptotic curve, 669
Asymptotic direction, 665
Attraction, 370-372
Axes, coordinate, 203
Axiom, of completeness, 17-18
  of induction, 9
Axioms, categorical, 20
  of a field, 1
  of addition and multiplication, 2
  of an ordered field, 5
  of Peano, 1

Basic triad of unit vectors, 208
Basis, 208, 546
Bernoulli numbers, 434 (Exs. 38, 39)
Bernoulli polynomials, 537 (Exs. 19-20)
Bessel's inequality, 532
Beta function, 143, 482-484
Between, 7 (Ex. 21)
Bicontinuous mapping, 192
Big O notation, 141, 388
Binary number, 21 (Ex. 4)
Binomial coefficient, 12 (Exs. 24, 25), 424

694　INDEX

Binomial function, 423
Binomial series, 423
Binomial theorem, 12 (Ex. 25), 423
Binormal, 650
Bliss's theorem, 116, 126 (Ex. 7), 238, 352 (Ex. 31)
Bolzano-Weierstrass theorem, 201 (Ex. 29)
Bonnet form of the second mean value theorem, 153 (Ex. 26), 154 (Ex. 28)
Bound, upper, 17
　lower, 18
Boundary, 177, 594, 611, 613, 625, 626
Boundary conditions, 548-553
Bounded function, 48
Bounded sequence, 29, 194
Bounded sequences, fundamental theorem, 53-54, 194
Bounded set, 18
Bounded variation, 151
Bracket function, 24

Cancellation law, 3, 4 (Ex. 7), 215, 561 (Ex. 14)
Cartesian form, 500
Categorical axioms, 20
Cauchy criterion, for functions, 60-61, 462 (Ex. 20)
　for improper integrals, 463, 471
　for sequences, 56-57, 195, 448 (Ex. 45), 508
　for series, 403 (Ex. 22), 448 (Ex. 46)
　for uniform convergence, 448 (Exs. 45, 46), 462 (Ex. 20), 471
Cauchy form of the remainder, 419
Cauchy inequality, 13 (Ex. 32), 120 (Ex. 20), 402 (Ex. 14), 522, 529 (Ex. 14)
Cauchy integral formula, 607 (Ex. 12)
Cauchy integral theorem, 606 (Ex. 7)
Cauchy principal value, 145 (Ex. 30)
Cauchy sequence, 56, 195, 508
Cauchy-Riemann differential equations, 265 (Ex. 23), 308 (Ex. 25), 605 (Exs. 2, 3)
Center of curvature, 254, 647
Centroid, 211 (Ex. 27), 362, 367, 369, 622
Cesàro summability, 403 (Ex. 20), 447 (Ex. 44), 540, 542
Chain rule, 69, 269-272, 302, 307 (Ex. 22)
Change of variables, 69, 269-272, 302, 307 (Ex. 22)
　in multiple integrals, 373-379

Characteristic equation, 318, 321 (Ex. 22)
Characteristic function, 550
Characteristic value, 550
Circle of convergence, 508
Circle of curvature, 254, 647
Circle, osculating, 647
Circular neighborhood, 173
Classification of numbers and functions, 167-170
Closed curve, 239
Closed disk, 174
Closed interval, 15
Closed orthonormal sequence, 546
Closed region, 177
Closed set, 175
Closed smooth surface, 614
Closed unit disk, 174
Closure of a set, 177
Collinear vectors, 205
Commutative group, 5 (Ex. 26)
Commutative law, 2, 10, 207, 214, 555
Compact region, 177
Compact set, 176, 196
Compactness and continuity, 196-197
Comparison tests, for improper integrals, 140-143, 463, 469
　for series, 387-389, 445
Complement of a set, 174
Complete ordered field, 1
Complete orthonormal sequence, 546
Completeness, axiom of, 17-18
Complex number, 496
Complex variables, 496-516, 529 (Ex. 18), 605-610
Complex-valued functions of a real variable, 509-513
Components of a vector, 204, 214
Composite function, 69, 269-272, 302, 307 (Ex. 22)
　continuity and limit of, 44, 46, 182-183, 187, 190 (Exs. 22, 25-29)
Composite number, 12 (Ex. 22), 504 (Ex. 20)
Composition of ordinates, 102
Composition of rotations, 568 (Ex. 14)
Composition of series, 427-430
Composition of transformations, 302
Computation of series, 407-409
Computations with series, 435-438
Conditional convergence, for improper integrals, 464
　for series, 399-401
Cone, 233
Conicoid, 231-235
Conjugate, 501
Connected sectionally smooth surface, 625

INDEX

Connected set, 198
  arc-wise, 198
Connected smooth surface, 614
Conservative field, 571, 590, 640, 645 (Ex. 8)
Constant function, 23
Constant of integration, 128
Constraints, extrema with, 308-319
Content, Jordan, 337
Continuity, 42-48, 186-189, 191-193, 506-507, 568
  and compactness, 196-197
  and integrability, 115, 345, 353
  and uniform convergence, 448, 460, 473
  for complex variables, 506-507
  from the right or left, 43-44
  negation of, 47 (Ex. 14)
  of a composite function, 44, 46, 182-183, 187, 190 (Exs. 22, 25-29)
  of inverse, 50, 192, 330
  one-sided, 43-44
  sectional, 131
  sequential criterion, 59
  uniform, 63-65, 198-199
Continuity theorems, 46-48, 50, 115, 187-189, 196-199, 345, 353
Continuous image, 192
Continuous nondifferentiable function, 68, 454 (Ex. 30)
Continuously differential function, 67, 267
Convergence, absolute (conditional) 399-401, 493 (Ex. 1), 464
  circle of, 508
  in the mean, 545-546
  interval of, 412-415
  of improper integrals, 136, 139, 468, 494
  of infinite products, 486
  of sequences, 26, 194, 442, 507
  of series, 381, 444, 508, 529-531, 540, 546
  radius of, 413, 508
  uniform, 442, 444, 459, 468
Convex region, 599, 639
Convex set, 202 (Ex. 33)
Coordinate axes, 203
Coordinate curves, 579, 611
Coordinate system, right-handed (left-handed), 204
Coordinate transformation, 300-304, 562-566
Coordinates, curvilinear, 578-582
  cylindrical, 249, 303, 308 (Ex. 26), 364, 582
  of a vector, 204
  orthogonal, 578-582
  spherical, 250, 304, 308 (Ex. 27), 365, 582
Coplanar points, 559
Coplanar vectors, 206
Cosine series, 534
Cosine transform, 514 (Ex. 17)
Cosines, direction, 211-213
  law of, 529 (Ex. 16)
Covering, open, 202 (Ex. 31)
Cracked china definition of double integral, 338-339
Cramer's rule, 296
Criterion, Cauchy (see Cáuchy criterion)
Criterion, for area, 341
  for integrability, 121, 345, 353
  for volume, 352
Critical point, 85, 284
Critical value, 85, 284
Cross product, 554-555
Cross-cut, 600
Cubical neighborhood, 178
Curl of a vector field, 572, 637, 645 (Ex. 3)
  in cylindrical coordinates, 582
  in orthogonal coordinates, 580
  in spherical coordinates, 582
Curvature, 252-257, 647
  center of, 254, 647
  circle of, 254, 647
  Gaussian, 665, 668
  mean, 665, 668
  normal, 661
  principal, 665
  radius of, 254, 647
  total, 665, 668
Curve, 240
  asymptotic, 669
  closed, 239
  coordinate, 579
  Jordan, 378
  oriented, 593-594
  parametric, 658
  simple closed, 239, 378
  smooth, 244, 569
Curve tracing, 102-106
Curves and arcs, 238-260, 584-602, 646-670
Curves on a surface, 657-660
Curvilinear coordinates, 578-582
Cut, 600
Cylinder, 226
Cylindrical coordinates, 249, 303, 308 (Ex. 26), 364, 582
Cylindrical shells, 356 (Ex. 18)

Decimal, 21 (Ex. 5)
Dedekind's theorem, 21 (Ex. 15)

INDEX

Definite integral, 110-133, 512
Definite quadratic form, 668
Del, 570
Deleted neighborhood, 35, 174
De Moivre's theorem, 501
Density, of irrational numbers, 20 (Ex. 2)
  of rational numbers, 20
  of trigonometric polynomials, 543-545
Density function, 361, 366, 369, 622
Dependence, functional, 304-306, 331-333
  linear, 207
Dependent variable, 23
Derivative, directional, 276, 278 (Ex. 9)
  from the right (left), 71
  infinite, 68
  normal, 277, 595, 624
  one-sided, 70-73
  ordinary, 67, 510, 569, 605 (Ex. 1)
  partial, 261-263
  right-hand (left-hand), 71
Derived series, 431
Determinant, derivative of, 273 (Ex. 15)
  functional, 296-306
  maximum value of, 316-317
Diameter of a set, 200 (Ex. 18), 338
Difference, between numbers, 2
  between sequences, 30
  between series, 405, 425
  between sets, 173
Differentiable function, 67, 267, 510, 569, 605 (Ex. 1)
  continuously, 67, 267, 605 (Ex. 1)
Differential, 90-94, 268-269
  approximation by, 92-94, 281-282
  exact, 587, 638
  locally exact, 596, 638
  total, 268
Differential element of area, 377
Differential equation, second-order, 512-513
Differential forms, exterior, 643-645
Differential geometry, 646-672
Differentiation, 67-109, 261-334
  and uniform convergence, 451, 461, 474
  implicit, 288, 296, 325
  of power series, 431
  under the integral sign, 321-324, 474
Dini's theorem, 454 (Ex. 29)
Direct function, 50
Directed distance from a plane to a point, 223 (Ex. 13)
Direction, 205
  normal, 277
Direction angles, cosines and numbers, 211-213

Directional derivative, 276, 278 (Ex. 9)
Directrix of a cylinder, 226
Dirichlet's integral, 538
Dirichlet's theorem, 404
Discontinuity, infinite, 45
  jump, 45
  removable, 44, 187, 610 (Ex. 24)
Disjoint sets, 173
Disk, 174
Distance, between a point and a set, 200 (Ex. 16)
  between two sets, 200 (Ex. 17)
  from a plane to a point, 223 (Ex. 14)
  in a function space, 522
  in Euclidean spaces, 178
Distance space, 522
Distributive law, 2, 10, 207, 214, 555
Divergence of a vector field, 572, 627, 645 (Ex. 3)
  in cylindrical coordinates, 582
  in orthogonal coordinates, 580
  in spherical coordinates, 582
Divergence theorem, 625-629, 645 (Ex. 5)
Division, 2
Domain of definition, 22
Dominance, for improper integrals, 140-143, 469
  for series, 387-389, 445
Dominant terms, 103
Dot product, 213
Double integral, 335-339
  in polar coordinates, 356-360
  theorems on, 342-346, 374
  transformation of, 373-379
Duhamel's principle, 238
Dummy variable, 12 (Ex. 26), 127
Dupin indicatrix, 666

e, 52 (Ex. 33), 156 (Ex. 5), 157 (Ex. 9), 395 (Ex. 33), 434 (Ex. 34), 458 (Ex. 1), 493 (Ex. 3)
Edges and corners, 611, 624-625
Eigenfunction, 550
Eigenvalue, 550
Element, of area, 377
  surface, 610-613
Elementary compact region, 593, 625
Elementary function, 169-170
Ellipsoid, 231
Elliptic integral, 169, 248 (Ex. 24), 458 (Ex. 21)
Elliptic paraboloid, 234
Elliptic point, 665
Empty set, 172
End-point, 15
Envelope, 293-295

INDEX

Epsilon-neighborhood, 16
Equality of mixed partial derivatives, 263
Equivalence of neighborhoods, 173-174
Error, relative, 281
Euclidean spaces, 23, 173-178, 203-237
Euler formula, 500
Euler's constant, 489
Euler's theorem, 272, 274 (Ex. 31), 665
Evaluation of improper integrals, 475-477
Even function, 118 (Exs. 7, 9-11), 429
Evolute, 255-257
Exact differential, 587, 638
  locally, 596
Existence theorem, for implicit functions, 325-330
  for inverse transformations, 330-331
Expansion of functions, 419-421, 526
Exponential function, 155-158, 421, 458 (Ex. 1), 505 (Exs. 26, 28), 509 (Ex. 32), 511, 514 (Ex. 10), 609 (Ex. 19)
Exponents, laws of, 12 (Exs. 17-21), 14 (Ex. 7)
Extended law of the mean, 80-82, 280
Exterior differential forms, 643-645
Extrema, 85-88, 282-285
  with constraints, 308-319

Factor, 12 (Ex. 22)
Factorial, 12 (Ex. 24), 480
Fejér's summability theorem, 540
Field, 1, 497
  complete ordered, 1, 18
  conservative, 571, 590, 640, 645 (Ex. 8)
  finite, 5 (Ex. 27)
  irrotational, 574, 590, 640, 645 (Ex. 8)
  ordered, 5
  scalar, 568
  solenoidal, 574, 640, 645 (Ex. 8)
  vector, 568
Finite field, 5 (Ex. 27)
Finitely decomposable compact region, 594, 626
First derivative test, 86, 90 (Exs. 22, 23)
First fundamental form, 658
First mean value theorem for integrals, 116, 117 (Exs. 5, 6), 152 (Ex. 22)
First octant, 204
Fourier coefficients, 524-525
Fourier series, 517-553
  convergence theorem, 530
  cosine series, 534
  generalized, 526
  integration of, 547
  sine series, 534
  uniform convergence, 546
  with complex coefficients, 529 (Ex. 18)
Fourier transform, 514 (Exs. 16-20)
Frenet formulas, 651, 655 (Exs. 12, 13)
Fresnel integrals, 169
Frontier, 177
Fubini's theorem, 349-350, 353
Function, 22, 181, 190
  algebraic, 168
  analytic, 421, 605-610
  beta, 143, 482-484
  binomial, 423
  bounded, 48
  bracket, 24
  composite, 69, 269-272, 302, 307 (Ex. 22)
  constant, 23
  continuous, 42-48, 186-189, 191-193, 506-507, 568
  continuously differentiable, 67, 267
  decreasing, 49
  defined by power series, 458-459
  density, 361, 366, 369, 622
  differentiable, 67, 267, 510, 569, 605 (Ex. 1)
  direct, 50
  elementary, 169-170
  even, 118 (Exs. 7, 9-11), 429
  exponential, 155-158, 421, 458 (Ex. 1), 505 (Exs. 26, 28), 509 (Ex. 32), 511, 514 (Ex. 10), 609 (Ex. 19)
  factorial, 12 (Ex. 24), 480
  gamma, 143, 480-482
  greatest integer, 24
  harmonic, 265 (Exs. 19-22), 604 (Ex. 27), 605 (Ex. 2), 630-631
  homogeneous, 272
  hyperbolic, 163-167
  implicit, 288-292, 296-300, 325-330
  increasing, 49
  inverse, 50, 80, 192, 300, 303, 330
  inverse hyperbolic, 165-167
  linear, 520, 573
  locally homogeneous, 274 (Ex. 31)
  logarithmic, 155-158, 422, 458 (Ex. 1), 505 (Ex. 27)
  monotonic, 49, 52 (Exs. 23-25), 84 (Exs. 37, 38)
  multiple-valued, 23-25, 111, 190 (Ex. 32), 505 (Exs. 27, 28)
  odd, 118 (Exs. 8-11), 429
  of several variables, 172-202
  periodic, 519-520
  polynomial, 40 (Exs. 17-19), 168, 509 (Exs. 23-29)
  power, 505 (Ex. 28)
  rational, 40 (Ex. 20), 168
  real-valued, 23, 181

Function—(*Continued*)
  sectionally smooth, 132, 529-531
  signum, 36, 45
  single-valued, 22
  transcendental, 168
  trigonometric, 158-160, 247, 422, 459 (Ex. 2), 505 (Ex. 29), 509 (Ex. 32), 514 (Ex. 8), 609 (Ex. 19)
Function space, 518-519
Functional, 643 (Ex. 1)
Functional dependence, 304-306, 331-333
Functional determinant, 296-306
Functions, orthogonal, 521
  orthonormal, 523, 547
Fundamental form, first, 658
  second, 668
Fundamental increment formula, 266-268
Fundamental theorem, for double integrals, 347-350
  for triple integrals, 353
  of algebra, 515-516, 608 (Ex. 16)
  of integral calculus, 127, 131 (Exs. 19, 21)
  of mathematical induction, 9
  on bounded sequences, 53-54, 194, 508
Fundamental triad, 650

Gamma function, 143, 480-482
Gauss's test, 397 (Ex. 7)
Gaussian curvature, 665, 668
Gaussian integer, 504 (Ex. 20)
Generalized Fourier series, 526
Generalized iterated limit, 190 (Ex. 32)
Generator of a cylinder, 226
Generatrix of a surface of revolution, 228
Geometric mean, 321 (Ex. 19)
Geometric series, 383
Gradient of a scalar field, 276, 570, 588
  in cylindrical coordinates, 582
  in orthogonal coordinates, 580
  in spherical coordinates, 582
Graph, 23, 201 (Ex. 23)
Graphing, 102-106
Greatest integer function, 24
Greatest lower bound, 18
Green's identities, 603 (Ex. 26), 629-630
Green's theorem, 593-596, 603 (Exs. 22-24), 626
Group, 4 (Ex. 25)
  commutative, 5 (Ex. 26)
Grouping of terms, 404

Hadamard inequality, 317
Harmonic function, 265 (Exs. 19-22), 604 (Ex. 27), 605 (Ex. 2), 630-631
Harmonic series, 388

Heat conduction, 550-553
Heat equation, 550
Heine-Borel theorem, 202 (Exs. 31, 32)
Helicoid, 671 (Ex. 20)
Helix, 652, 656 (Exs. 20, 24)
Higher-dimensional spaces, 178
Hölder summable, 403 (Ex. 21), 447 (Ex. 44)
Homeomorphism, 192
Homogeneous function, 272
  locally, 274 (Ex. 31)
Homogeneous law, 214, 555
l'Hospital's rule, 95-101
Hyperbolic functions, 163-167
  inverse, 165-167
Hyperbolic paraboloid, 234
Hyperbolic point, 665
Hyperboloid, of one sheet, 232
  of two sheets, 233
Hypergeometric series, 397 (Ex. 6)

i, 498
Identity transformation, 303
Image, 192
  continuous, 192
  inverse, 192
Imaginary axis, 498
Imaginary number, 497
Imaginary part, 498
Implicit function, 288-292, 296-300, 325-330
Implicit function theorem, 325-330
Improper integral, 136-146, 463-495
  alternating, 464-466
  Cauchy principal value, 145 (Ex. 30)
  evaluation of, 475-477
  multiple, 493-495
  uniform convergence, 467-475
In the large, 597
In the small, 597
Increment formula, fundamental, 266-268
Indefinite integral, 128
Indefinite quadratic form, 669
Independence, functional, 304-306, 331-333
  linear, 207
  of coordinate system, 215, 351 (Ex. 26), 566, 567 (Exs. 10-13), 576-578
  of parametrization, 241
  of path, 587-591, 638-640
Independent variable, 23
Indeterminate forms, 95-101, 435
Indicatrix, Dupin, 666
Induction, axiom of, 9
  mathematical, 7-11
Inductive set, 8

INDEX

Inequality, 6
  Bessel's, 532
  Hadamard, 317
  Minkowski, 13 (Ex. 33), 120 (Ex. 21), 402 (Ex. 14), 522
  Schwarz (Cauchy), 13 (Ex. 32), 120 (Ex. 20), 402 (Ex. 14), 522, 529 (Ex. 14)
  triangle, 15, 180 (Ex. 22), 207, 499, 521, 522
Infimum, 18
Infinite derivative, 68
Infinite discontinuity, 45
Infinite interval, 15
Infinite limit, 26, 36, 182
Infinite product, 486-488
Infinite sequence, 25, 193-196, 507-508
Infinite series (see Series), 381-459, 507-508, 517-553
Infinitesimal, 69, 92
Initial conditions, 548-553
Initial point, 205
Inner area, 339-342
Inner normal, 624
Inner point, 611, 613
Inner product, 213, 520
Inside of a curve, 378
Integer, 13
  Gaussian, 504 (Ex. 20)
  negative, 13
  positive, 7-11
Integrability criterion, 121, 345, 353
Integral, definite (Riemann), 110-133, 512
  double, 335-350, 356-360
  elliptic, 169, 248 (Ex. 24), 458 (Ex. 21)
  improper, 136-146, 463-495
  indefinite, 128
  iterated, 347-350, 353, 358, 364-366
  line, 584-602, 606 (Ex. 7), 635
  multiple, 335-380
  Riemann, 110-133, 512
  Riemann-Stieltjes, 146-151
  surface, 621-622
  triple, 352-354, 364-366
Integral form for arc length, 242-247
Integral form of the remainder, 419
Integral points, 176
Integral test, 384
Integrating factor, 513
Integration, 110-154, 335-380, 512, 584-602, 606 (Ex. 7), 635
  and uniform convergence, 449, 461, 473
  by parts, 130 (Exs. 10-13), 148
  by substitution, 128, 152 (Ex. 18)
  formulas, 161-162

  of Fourier series, 547
  of power series, 431
Intercept, 225
Intercept form for equation of plane, 219
Interchange of order, 475
Interior, 176
Interior point, 15, 176
Intermediate-value property, 48, 52 (Ex. 22), 84 (Ex. 40), 189
Intersection, of sets, 173
  of smooth surfaces, 659
Interval, 15
  of convergence, 412-415
Intrinsic orientation, 559
Invariance of area, 337, 617
Inverse function (mapping, transformation), 50, 80, 192, 300, 303, 330
  continuity of, 50, 192, 330
  derivative of, 70, 303, 330
Inverse hyperbolic functions, 165-167
Inverse image, 192
Involute, 255-257
Irrational number, 14
Irrationality, of e, 168, 434 (Ex. 34)
  of π, 168
Irrotational vector field, 574, 590, 640, 645 (Ex. 8)
Isolated point, 43, 186
Isomorphism, 20, 497
Iterated integral, 347-350, 353, 358, 364-366
Iterated limit, 184-186
  generalized, 190 (Ex. 32)

Jacobian, 296-306
  and multiple integral, 373-379
Jacobian matrix, 313
Jordan content, 337
Jordan curve, 378
Jump discontinuity, 45

Kepler's equation, 170
Kinematics, 649-650
Kummer's test, 395

Lagrange form of the remainder, 418
Lagrange multipliers, 317-319
Lamina, 361
Laplace transform, 485 (Exs. 17-23)
Laplace's equation, 265 (Exs. 19-22), 304, 308 (Exs. 26, 27), 605 (Ex. 2), 630
Laplacian, 573, 580, 630, 632 (Ex. 14)
  in cylindrical coordinates, 308 (Ex. 26), 582

Laplacian—(*Continued*)
  in orthogonal coordinates, 580
  in spherical coordinates, 308 (Ex. 27), 582
Law, anticommutative, 555
  associative, 2, 9, 207, 497, 560
  cancellation, 3, 4 (Ex. 7), 215, 561 (Ex. 14)
  commutative 2, 10, 207, 214, 555
  distributive, 2, 10, 207, 214, 555
  homogeneous, 214, 555
  of cosines, 529 (Ex. 16)
  of exponents, 12 (Exs. 17-21), 14 (Ex. 7)
  of sines, 561 (Ex. 20)
  of trichotomy, 6
  skew-commutative, 555
  transitive, 6
Law of the mean, 75-82, 279-280, 511, 575 (Ex. 29)
  extended, 80-82, 280
  generalized, 77
Least squares, 287 (Exs. 16-18), 524-525
Least upper bound, 17
Lebesgue measure zero, 342
Left-hand derivative, 71
Left-handed coordinate system, 204
Left-handed orientation, 557
Legendre duplication formula, 485 (Ex. 13)
Legendre series, 547
Leibnitz's rule, 74 (Ex. 23), 321-324
Length of arc, 241-247, 251
l'Hospital's rule, 95-101
Limit, along a path, 183
  for complex variables, 506-507
  in the mean, 545-546
  infinite, 26, 36, 182
  iterated, 184-186, 190 (Ex. 32)
  of a composite function, 44, 46, 182-183, 187, 190 (Exs. 22, 25-29)
  of a function, 34-39, 181-185, 506-507
  of a sequence, 26, 381, 507
  of a variable, 38
  one-sided, 34-35
  uniform, 442, 444, 459, 468
Limit inferior, of a function, 62 (Ex. 10), 201 (Exs. 26-28)
  of a sequence, 58 (Ex. 11), 201 (Ex. 25)
Limit point, of a sequence, 58 (Ex. 7), 201 (Ex. 25)
  of a set, 175
Limit superior, of a function, 62 (Ex. 10), 201 (Exs. 26-28)
  of a sequence, 58 (Ex. 11), 201 (Ex. 25)

Limit theorems, for functions, 38, 187
  for sequences, 29, 193-196
Line, 220-222
  of intersection of two planes, 219
  of regression, 287 (Ex. 18)
Line integral, 584-602, 606 (Ex. 7), 635
Line segment, 180 (Ex. 21)
Linear combination, 207, 518
Linear dependence and independence, 207
Linear function, 520, 573
Linear space, 207, 518-519
  normed, 521
Liouville's theorem, 608 (Ex. 15)
Little o notation, 141, 388
Local exactness, 596-598, 638
Local property, 597
Locally a surface, 613
Locally homogeneous function, 274 (Ex. 31)
Locus problems, 237 (Exs. 50-55)
Logarithmic function, 155-158, 422, 458 (Ex. 1), 505 (Ex. 27)
Lower bound, 18
Lower semicontinuity, 63 (Ex. 22)
Lowest terms, 12 (Ex. 23)

Maclaurin series, 417, 421-425, 509 (Ex. 32), 609 (Ex. 19)
  for $\cos x$, 422, 509 (Ex. 32), 609 (Ex. 19)
  for e^x, 421, 509 (Ex. 32), 609 (Ex. 19)
  for $\ln(1+x)$, 422
  for $\sin x$, 422, 509 (Ex. 32), 609 (Ex. 19)
  for $(1+x)^m$, 423
Magnitude of a vector, 205-207
Mapping, 190-193, 300-304, 330-331
  bicontinuous, 192
  continuous, 192
  inverse, 192, 300, 303, 330
  open, 331
  topological, 192
Mass, 361, 366, 369, 622
Mathematical induction, 7-11
Matrix, Jacobian, 313
  orthogonal, 565
  rotation, 565
Maxima and minima, 85-88, 282-285
  with constraints, 308-319
Maximum modulus theorem, 609 (Ex. 21)
Maximum value, 48, 75, 85, 188, 282
  of a determinant, 316-317
Mean, arithmetic, 7 (Ex. 21), 321 (Ex. 19)
  geometric, 321 (Ex. 19)

Mean, law of the, 75-82, 279-280, 511, 575 (Ex. 29)
  extended, 80-82, 280
  generalized, 77
Mean center, 287 (Ex. 16)
Mean convergence, 545-546
Mean curvature, 665, 668
Mean value of a function, 120 (Ex. 22)
Mean value theorem, for derivatives, 75-82, 279-280, 511, 575 (Ex. 29)
  for double integrals, 346
  for Riemann integrals, 116, 117 (Exs. 5, 6), 130 (Ex. 14), 153 (Exs. 25, 26), 154 (Ex. 28)
  for Riemann-Stieltjes integrals, 152 (Ex. 22), 153 (Ex. 24)
Measure zero, 342
Medians of a triangle, 210 (Ex. 20)
Membership notation, 172
Mertens' theorem, 406
Metric space, 522
Meusnier's theorem, 661
Midpoint of a segment, 210 (Ex. 23)
Minimal surface, 671 (Ex. 17)
Minimum value, 48, 75, 85, 188, 282
Minkowski inequality, 13 (Ex. 33), 120 (Ex. 21), 402 (Ex. 14), 522
Mixed partial derivatives, 262-263
  equality of, 263
Möbius band, 623, 634
Modulus, 498-499
Moments, 362, 367, 369, 622
Monkey saddle, 283
Monotonic function, 49, 52 (Exs. 23-25), 84 (Exs. 37, 38)
Monotonic sequence, of functions, 454 (Ex. 29)
  of numbers, 32
  of sets, 201 (Ex. 30)
Morera's theorem, 608 (Ex. 14)
Multiple integrals, 335-380
  and Jacobians, 373-379
  improper, 493-495
  transformation of, 373-379
Multiple-valued function, 23-25, 111, 190 (Ex. 32), 505 (Exs. 27, 28)
Multiplication, of determinants, 300, 562 (Ex. 27)
  of matrices, 568 (Ex. 14)
  of numbers, 2, 496
  of series, 405-407, 410 (Exs. 4, 5, 8-10), 425
  of vectors, 213, 554, 557, 559
Multiplication table, 5 (Ex. 27), 214, 556
Multipliers, Lagrange, 317-319
Multiply-connected region, 599, 639

Natural numbers, 7-9
Negation, of continuity, 47 (Ex. 14)
  of uniform continuity, 65, 199 (Ex. 7)
  of uniform convergence, 443, 459, 469
Negative, of a number, 2
  of a vector, 205
Negative integer, 13
Negative number, 6
Negative orientation, 557
Neighborhood, 16, 173, 178
  circular, 173
  cubical, 178
  deleted, 35, 174
  of infinity, 27
  spherical, 178
  square, 173
Nested intervals theorem, 201 (Ex. 30)
Net, 110, 335, 352, 356
Nondifferentiable continuous function, 68, 454 (Ex. 30)
Nonnegative series, 383
Norm, in a linear space, 521
  of a net, 110, 335
  of a partition, 338
Normal, inner, 624
  outer, 624
  principal, 647
  to a surface, 277, 307 (Ex. 18), 611
Normal curvature, 661
Normal derivative, 277, 595, 624
Normal direction, 277, 307 (Ex. 18), 611
Normal equations, 287 (Ex. 18)
Normal plane, 277, 650
Normal radius of curvature, 661
Normal section, 661-666
Normalization, of a function, 523
  of a vector, 205
Normed linear space, 521
Notational pitfalls, 290, 299
Number, algebraic, 168
  binary, 21 (Ex. 4)
  complex, 497
  composite, 12 (Ex. 22)
  imaginary, 497
  integral, 13, 504 (Ex. 20)
  irrational, 14
  natural, 7-9
  negative, 6
  positive, 5
  prime, 12 (Ex. 22)
  pure imaginary, 498
  rational, 14
  real, 1-21
  transcendental, 168
Numbers, direction, 211-213

Octant, first, 204
Odd function, 118 (Exs. 8-11), 429

One, 2
One-sided derivative, 70-73
One-sided limit, 34-35
Open covering, 202 (Ex. 31)
Open disk, 174
Open interval, 15
Open mapping, 331
Open set, 175
Open unit disk, 174
Order of magnitude, 141, 388
Order of pole, 610 (Ex. 25)
Order of zero, 610 (Ex. 23)
Ordered field, 5
Ordered pair, 22
Orientable surface, 623-624, 632-635
Orientation, and sign of Jacobian, 377
  in space, 556-559
  of a surface, 623-624, 632-635
Oriented curve, 593-594, 633-634
Origin, 15
Orthogonal coordinates, 578-582
Orthogonal functions, 521
Orthogonal matrix, 565
Orthogonal projection, 214, 525
Orthogonal vectors, 214, 216
Orthonormal family, 523, 547
Orthonormal sequence, closed, 546
  complete, 546
Osculating circle, 259 (Ex. 28), 647
Osculating plane, 647
Outer area, 339-342
Outer normal, 595, 624
Outer product, 554-555
Outside of a curve, 378

Pappus, theorem of, 368 (Ex. 15), 620
Parabolic point, 665
Paraboloid, 234
Parallelepiped, volume of, 356 (Ex. 16), 559, 562 (Ex. 26)
Parallelogram, area of, 351 (Ex. 27), 561 (Ex. 21)
Parallelogram law, 529 (Ex. 16)
Parameter, 77, 106, 239, 468, 611
Parametric curves, 658
Parametric equations, 77, 84 (Ex. 39), 106, 220, 239
Parentheses, introduction of, 404
Parseval's equation, 532, 545
Partial differentiation, 261-334
Partition, 338, 353
Parts, integration by, 130 (Exs. 10-13), 148
Pascal's triangle, 12 (Ex. 24)
Path, independence of, 587-591, 638-640
  limit along, 183
Peano arc, 241

Peano axioms, 1
Periodic function, 519-520
π, 158 (Ex. 2), 437
Plane, Euclidean, 23
  normal, 277, 650
  osculating, 647
  perpendicular to two planes, 219
  rectifying, 650
  tangent, 277
  through three points, 219, 222 (Ex. 12)
Plane section, 660
Planes, 218-220
  line of intersection of, 219
  through a line, 223 (Ex. 16)
Plucked string, 548-552
Point, boundary, 177, 594, 611, 613, 625, 626
  elliptic, 665
  frontier, 176
  hyperbolic, 665
  inner, 611, 613
  interior, 15, 176
  limit, 58 (Ex. 7), 175, 201 (Ex. 25)
  parabolic, 665
  regular (singular), 244, 276, 615
Polar coordinates, 356-360, 619
Polar form, 499
Polar net, 356
Pole, 610 (Ex. 25)
Polygon, inscribed, 241
Polynominal, 40 (Exs. 17-19), 168, 509 (Exs. 23-29)
  Bernoulli, 537 (Exs. 19, 20)
  trigonometric, 542-545
Positive definite quadratic form, 284, 658
Positive integer, 7-11
Positive orientation, 557
Positive series, 383
Positiveness, 5
  of unity, 6 (Ex. 6)
Potential, scalar, 571, 588, 638-640, 645 (Ex. 8)
  vector, 640-641, 645 (Ex. 8)
Power function, 505 (Ex. 28)
Power series, 412-440, 455-459, 508, 609 (Ex. 19)
  differentiation of, 431
  for complex variables, 508
  functions defined by, 458-459
  integration of, 431
  substitution of, 427-430
Preimage, 192
Prime numbers, 12 (Ex. 22), 504 (Ex. 20)
  relatively, 12 (Ex. 23)
Primitive, 128
Principal curvature, 665

Principal normal, 647
Principal value (Cauchy), 145 (Ex. 30)
Probability integral, 169, 359
Product, cross, 554-555
  dot, 213
  infinite, 486-488
  inner, 213, 520
  of matrices, 568 (Ex. 14)
  of sequences, 30
  of series, 405-407, 410 (Exs. 4, 5, 8-10), 425
  outer, 554-555
  scalar, 213
  triple scalar, 557
  triple vector, 559
  vector, 554-555
Projection, orthogonal, 214, 525
  stereographic, 615
p-series, 385, 388
  alternating, 401
Pseudo-metric space, 522
Pseudosphere, 671 (Ex. 16)
Pure imaginary number, 498
Pythagorean theorem, 523

Quadratic form, 318, 321 (Ex. 22), 658, 668
  definite, 668
  indefinite, 669
  positive definite, 284
  semidefinite, 669
Quadratic formula, 504 (Ex. 17)
Quadric cone, 233
Quadric surfaces, 231-235
Quotient, of numbers, 2
  of polynomials, 41 (Exs. 35, 36)
  of sequences, 30

Raabe's test, 396
Radiant of a scalar field, 583 (Ex. 7)
Radius of convergence, 413, 414, 508
Radius of curvature, 254, 647
  normal, 661
Radius vector, 205, 568
Range of values, 22
Rank of a matrix, 314
Ratio test, 390, 394 (Ex. 28), 401, 447 (Ex. 36)
Rational function, 40 (Ex. 20), 168
Rational number, 14
Rational points, 176
Real axis, 498
Real number, 1-21
Real part, 498
Real variable, 23
Rearrangement of terms, 404

Reciprocal of a number, 2
Reciprocal set of vectors, 561 (Ex. 23), 583 (Ex. 8)
Rectangular form, 500
Rectifiable arc, 241, 341
Rectifying plane, 650
Reduction formulas, 133
Refined tests for series, 395
Region, 177
  closed, 177
  compact, 177
  convex, 599, 639
  multiply-connected, 599, 639
  simply-connected, 598-602, 639
  star-shaped, 599, 639
Regression, line of, 287 (Ex. 18)
Regular point, 244, 276, 615
Relations among vector operations, 573-574
Relative error, 281
Relative maximum (minimum), 85, 282
Relative orientation, 559
Relatively prime, 12 (Ex. 23)
Removable discontinuity, 44, 187
Removable singularity, 610 (Ex. 24)
Representation by series, 419
Residue, 610 (Ex. 25)
Resultant of vectors, 206
Revolution, surface of, 227, 619
Riemann integral, 110-133, 512
Riemann-Lebesgue theorem, 539
Riemann-Stieltjes integral, 146-151
Riemann Zeta-function, 388, 409, 531, 533, 534, 536 (Exs. 4-6), 537 (Ex. 20), 547 (Ex. 5)
Riemann's lemma, 533
Right-hand derivative, 71
Right-handed coordinate system, 204
Right-handed orientation, 557
Rolle's theorem, 75
Root, square, 7 (Ex. 22)
Root test, 392, 394 (Ex. 29)
Roots, of complex numbers, 502-503
  of unity, 502
Rotation matrix, 565
Rotations, 564-566
Ruling of a cylinder, 226

Saddle point, 283
Scalar, 205, 519
Scalar field, 568
Scalar multiple, 205
Scalar potential, 571, 588, 638-640, 645 (Ex. 8)
Scalar product, 213
  triple, 557

704 INDEX

Schwarz (Cauchy) inequality, 13 (Ex. 32), 120 (Ex. 20), 402 (Ex. 14), 522, 529 (Ex. 14)
Schwarz's example, 616-617
Second derivative test, 86
Second fundamental form, 668
Second mean value theorem for integrals, 130 (Ex. 14), 153 (Exs. 24-26), 154 (Ex. 28)
Section, 225, 660
  normal, 661-666
Sectional continuity and smoothness, 131, 529-531
Sectionally smooth arc, 585
Sectionally smooth surface, 625
  orientable, 634
Segment, 180 (Ex. 21)
Semiaxes, 318, 321 (Exs. 22, 23)
Semicontinuity, 63 (Ex. 22)
Semidefinite quadratic form, 669
Separation of variables, 549, 551
Sequence, 25
  bounded, 29, 194
  Cauchy, 56, 195, 508
  convergent, 26
  of complex numbers, 507-508
  of points, 193-196
Sequential criterion for continuity, 59
Sequential criterion for limits, 60
Series, 381-459, 508, 517-553
  addition of, 405-407, 425
  alternating, 398
  alternating harmonic, 399
  computation of, 407-409
  computations with, 435-438
  cosine, 534
  derived, 431
  Fourier, 517-553
  geometric, 383
  harmonic, 388
  hypergeometric, 397 (Ex. 6)
  Maclaurin, 417, 421-425, 509 (Ex. 32), 609 (Ex. 19)
  multiplication of, 405-407, 410 (Exs. 4, 5, 8-10), 425
  nonnegative, 383
  of complex numbers, 507-508
  p-series, 385, 388, 401
  positive, 383
  power, 412-440, 455-459, 508, 609 (Ex. 19)
  product, 405-407, 410 (Exs. 4, 5, 8-10), 425
  representation by, 419
  sine, 534
  subtraction of, 405-407, 425
  summable, 403 (Exs. 20, 21)

Taylor, 416-417, 438-440, 609 (Ex. 18)
  trigonometric, 526
  uniform convergence of, 444
Set, arc-wise connected, 198
  bounded, 18
  closed, 175
  compact, 176, 196
  convex, 202 (Ex. 33)
  inductive, 8
  open, 175
Set notation, 172-173
Sigma notation, 12 (Ex. 26)
Signum function, 36, 45
Simple arc, 239
Simple closed curve, 378
Simply-connected region, 598-602, 639
Simpson's rule, 119 (Ex. 17)
Sine of angle between two vectors, 216
Sine series, 534
Sine transform, 514 (Ex. 17)
Singular point, 244, 276, 615
Singularity, removable, 610 (Ex. 24)
Skew-commutative law, 555
Skew-rotation, 566
Smooth curve, 244, 569
Smooth surface, 276, 613-615
  orientable, 623
Smoothness, sectional, 131
Solenoidal vector field, 574, 640, 645 (Ex. 8)
Solid analytic geometry, 203-237
Solid angle, 628
Space, distance, 522
  function, 518-519
  linear, 207, 518-519
  metric, 522
  pseudo-metric, 522
  vector, 207, 519
Space-filling arc, 241
Spectrum, 550
Speed, 569
Sphere, 225
Spherical coordinates, 250, 304, 308 (Ex. 27), 365, 582
Spherical neighborhood, 178
Spheroid, 232
Square neighborhood, 173
Square root, 7 (Ex. 22)
Steady state, 553 (Ex. 14)
Star-shaped region, 599, 639
Step-function, 120, 344
Stereographic projection, 615
Stirling's formula, 490
Stokes's theorem, 595, 635-638, 645 (Ex. 6)
String, plucked, 548-552
  struck, 552 (Exs. 7, 8)

INDEX

unwinding, 260 (Ex. 29)
vibrating, 548-552
Struck string, 552 (Exs. 7, 8)
Subsequence, 27
Subset, 172
Substitution, integration by, 128, 152 (Ex. 18)
  of power series, 427-430
Subtraction, 2, 496
  of series, 405-407, 425
  of sets, 173
  of vectors, 206
Sum, of numbers, 2, 496
  of sequences, 30
  of series, 405-407, 425
  of vectors, 206
Summability, Cesàro, 403 (Ex. 20), 447 (Ex. 44), 540, 542
  Hölder, 403 (Ex. 21), 447 (Ex. 44)
  of Fourier series, 540
Summation notation, 12 (Ex. 26)
Supremum, 17
Surface, 225, 276, 613-615, 625
  minimal, 671 (Ex. 17)
  of revolution, 227, 619
  orientable, 623, 634
  sectionally smooth, 625
  smooth, 276, 613-615
  with edges and corners, 624-625
Surface area, 617-620
Surface element, 610-613
Surface integral, 621-622
Symmetric form for equations of line, 220

Tangent line, 277, 307 (Ex. 17)
Tangent plane, 277
Tangent vector, 275
  unit, 569
Taylor series, 416-417, 438-440, 609 (Ex. 18)
Taylor's formula with a remainder, 80-82, 280, 418-419, 609 (Ex. 18)
  Cauchy form, 419
  integral form, 419
  Lagrange form, 418
Terminal point, 205
Topological mapping, 192
Topologically equivalent sets, 192
Torsion, 651
  radius of, 651
Torus, 231 (Ex. 53), 369 (Ex. 16, 17)
Total curvature, 665, 668
Total differential, 268
Trace, 225
Tractrix, 258 (Ex. 26)
Transcendental function, 168

Transcendental number, 168
Transform, Fourier, 514 (Exs. 16-20)
  Laplace, 485 (Exs. 17-23)
Transformation, 190-193, 300-304, 330-331
  continuous, 192
  coordinate, 300-304, 562-566
  identity, 303
  inverse, 300, 303
  of multiple integrals, 373-379
Transformation theorem for double integrals, 374, 604 (Ex. 38)
Transient, 553 (Ex. 14)
Transitive law, 6
Translations, 564
Trapezoidal rule, 118 (Ex. 16)
Triangle inequality, 15, 180 (Ex. 22), 207, 499, 521, 522
Trichotomy, law of, 6
Trigonometric form, 499
Trigonometric functions, 158-160, 247, 422, 459 (Ex. 2), 505 (Ex. 29), 509 (Ex. 32), 514 (Ex. 8), 609 (Ex. 19)
Trigonometric polynomials, 542
  density of, 543-545
Trigonometric series, 526
Triple integrals, 352-354
  cylindrical coordinates, 364
  spherical coordinates, 365
Triple scalar product, 557
Triple vector product, 559
Trivergence of a vector field, 578
Twisted cubic, 653, 654

Umbilic, 672 (Exs. 25-29)
Uniform approximation theorem of Weierstrass, 542
Uniform continuity, 63-65, 84 (Ex. 43), 198-199
  sequential criterion for, 59
Uniform convergence, 441-462, 467-475
  and continuity, 448-449, 460, 473
  and differentiation, 451-453, 461, 474
  and integration, 449-451, 461, 473
  of Fourier series, 546
  of improper integrals, 467-475
Uniform limits, 441-462, 467-475
Uniform summability, 447 (Ex. 44)
  of Fourier series, 542
Union of sets, 172-173
Unique factorization, 504 (Ex. 21)
Uniqueness, of limit, 29, 38
  of negative, 4 (Ex. 1)
  of reciprocal, 4 (Ex. 8)
  of square root, 7 (Ex. 22)
  of unity, 4 (Ex. 6)
  of zero, 3

Uniqueness theorem for power series, 417
Unit disk, 174
Unit vector, 205
   outer normal, 595
   tangent, 275, 569
Unity, 2
   positiveness of, 6 (Ex. 6)
   roots of, 502
Unwinding string, 260 (Ex. 29)
Upper bound, 17
Upper semicontinuity, 63 (Ex. 22)

Variable, dependent, 23
   independent, 23
   real, 23
Variation, bounded, 151
Vector, 204
Vector analysis, 204-217, 554-583
Vector field, 568
   conservative, 571, 590, 640, 645 (Ex. 8)
   irrotational, 574, 590, 640, 645 (Ex. 8)
   solenoidal, 574, 640, 645 (Ex. 8)
Vector potential, 640-641, 645 (Ex. 8)
Vector product, 554-555
   triple, 559
Vector space, 207, 519
Velocity vector, 569
Vectors, collinear, 205
   coplanar, 206
   linearly dependent (independent), 207
   orthogonal, 214, 216
Vertex, 234, 611, 613
Very nearly everywhere, 342, 353
Vibrating string, 548-552
   plucked string, 548-552
   struck string, 552 (Exs. 7, 8)
Volume, 337, 352, 353, 359, 364-366
   by cylindrical shells, 356 (Ex. 18)
   of a parallelepiped, 356 (Ex. 16), 559, 562 (Ex. 26)
Vortex, 637

Wallis's formulas, 146 (Ex. 36)
Wallis's infinite product for π, 488
Wave equation, 548
Weierstrass continuous nondifferentiable function, 68
Weierstrass infinite product for $1/\Gamma(\alpha)$, 491
Weierstrass M-test, 445, 470
Weierstrass uniform approximation theorem, 542
Well-ordering principle, 9
Without loss of generality, 108

Zero, 2
Zero vector, 205
Zeta-function of Riemann, 388, 409, 531, 533, 534, 536 (Exs. 4-6), 537 (Ex. 20), 547 (Ex. 5)